T0298429

# The Biology of Sea Turtles

## Volume III

# CRC
# MARINE BIOLOGY
## SERIES

The late Peter L. Lutz, Founding Editor
David H. Evans, Series Editor

**PUBLISHED TITLES**

*Biology of Marine Birds*
    E.A. Schreiber and Joanna Burger

*Biology of the Spotted Seatrout*
    Stephen A. Bortone

*The Biology of Sea Turtles, Volume II*
    Peter L. Lutz, John A. Musick, and Jeanette Wyneken

*Early Stages of Atlantic Fishes: An Identification Guide for the Western Central North Atlantic*
    William J. Richards

*The Physiology of Fishes, Third Edition*
    David H. Evans and James B. Claiborne

*Biology of the Southern Ocean, Second Edition*
    George A. Knox

*Biology of the Three-Spined Stickleback*
    Sara Östlund-Nilsson, Ian Mayer, and Felicity Anne Huntingford

*Biology and Management of the World Tarpon and Bonefish Fisheries*
    Jerald S. Ault

*Methods in Reproductive Aquaculture: Marine and Freshwater Species*
    Elsa Cabrita, Vanesa Robles, and Paz Herráez

*Sharks and Their Relatives II: Biodiversity, Adaptive Physiology, and Conservation*
    Jeffrety C. Carrier, John A. Musick, and Michael R. Heithaus

*Artificial Reefs in Fisheries Management*
    Stephen A. Bortone, Frederico Pereira Brandini, Gianna Fabi, and Shinya Otake

*Biology of Sharks and Their Relatives, Second Edition*
    Jeffrey C. Carrier, John A. Musick, and Michael R. Heithaus

*The Biology of Sea Turtles, Volume III*
    Jeanette Wyneken, Kenneth J. Lohmann, and John A. Musick

# The
# Biology of
# Sea Turtles

## Volume III

*Edited by*
Jeanette Wyneken
Kenneth J. Lohmann
John A. Musick

CRC Press
Taylor & Francis Group
Boca Raton  London  New York

CRC Press is an imprint of the
Taylor & Francis Group, an **informa** business

CRC Press
Taylor & Francis Group
6000 Broken Sound Parkway NW, Suite 300
Boca Raton, FL 33487-2742

© 2013 by Taylor & Francis Group, LLC
CRC Press is an imprint of Taylor & Francis Group, an Informa business

No claim to original U.S. Government works

ISBN-13: 978-1-4398-7307-6 (hbk)

---

### Library of Congress Cataloging-in-Publication Data

---

The biology of sea turtles / edited by Peter L. Lutz and John A. Musick.
    p. cm.--(CRC marine science series)
    Includes bibliographical references (p. ) and index.
  ISBN 978-1-4398-7307-6
  1. Sea turtles. I. Lutz, Peter L. II. Musick, John A. III. Series: Marine science series.

QL666.C536B56 1996
597.92--dc20                                           96-36432

---

**Visit the Taylor & Francis Web site at**
**http://www.taylorandfrancis.com**

**and the CRC Press Web site at**
**http://www.crcpress.com**

*This book is dedicated to the memory of founding editor, Peter L. Lutz, who valued good science and reminded us to "Fight the Good Fight!"*

# Contents

# Preface

Since the first volume of *The Biology of Sea Turtles* was published in 1997, the field has grown and matured in ways few of the authors would have predicted. Volume III provides an updated view of several fields covered in that original volume and brings together the best in the field to develop a comprehensive go-to resource. There have been significant advances in physiology, foraging, genetics, and health. Life history is now partitioned into three chapters, covering age determination, predator–prey interactions, and mortality from a major source, bycatch. Several new areas have emerged and grown since the original volume was conceived. These include in vivo imaging of structure, spatial distributions of marine turtles at sea, epibiosis, and climatic effects. Two chapters, imprinting and parasitology, bring forward areas that we identified as a need. This volume, like its predecessors, grew from the many collegial discussions with the participants of the annual International Sea Turtle Society Symposium (formerly the Workshops on Sea Turtle Biology and Conservation). Sadly, as discussions for this third volume were taking place, our lead editor, Peter L. Lutz, passed away. We introduce Kenneth Lohmann as our third editor who helped to guide this volume.

**Jeanette Wyneken**
**Kenneth J. Lohmann**
**John A. (Jack) Musick**
*Boca Raton, Florida*

# Preface

# Acknowledgments

The book was made possible because of the hard work and wealth of knowledge offered by the contributors. We thank the selfless external reviewers for careful reading and constructive criticisms and suggestions. Their perspectives have made this book far better than we could have otherwise expected. We thank our editors at CRC Press, John Sulzycki and David Fausel, for their guidance throughout the process. The cover photos for this book were taken by Jim Abernethy (www.jimabernethyimagery.com). His view of sea turtles merges art with science and reminds us how important it is to learn about the lives of these animals where they live—at sea.

# Editors

**Jeanette Wyneken**, PhD, is an associate professor of biological sciences at Florida Atlantic University in Boca Raton. She received her BA from Illinois Wesleyan University in 1978 and her PhD in biology from the University of Illinois in 1988. She was a research associate at the University of Illinois before serving as a research faculty member at Florida Atlantic University for several years. In 2000, she joined the faculty at Florida Atlantic University. Dr. Wyneken is a comparative and functional morphologist and a marine conservation biologist. Her studies and approaches are diverse; she integrates aspects of morphology and development into her studies of growth, energetics, migratory behavior, feeding and environmental sex determination, and medical imaging of reptiles. In addition to teaching vertebrate anatomy and development, she developed and taught university courses on the biology of sea turtles at Harbor Branch Oceanographic Institution in Florida and worked with colleagues to develop a similar conservation-based sea turtle biology course at Duke University Marine Laboratory in Beaufort, North Carolina. Dr. Wyneken is a former president of the Annual Sea Turtle Symposium (now The International Sea Turtle Society). She served as convener of the 7th International Congress for Vertebrate Morphology and has organized symposia on various aspects of the biology of turtles and vertebrate morphology. Dr. Wyneken is a member of the Association of Ichthyologists and Herpetologists, the Society for the Study of Amphibians and Reptiles, the Herpetologists' League, the Association of Reptilian and Amphibian Veterinarians, the International Sea Turtle Society, Sigma Xi, the Society of Integrative and Comparative Biology, and the IUCN Marine Turtle Specialist Group. She has authored more than 50 peer-reviewed papers, 9 book chapters, and 1 book, *The Anatomy of Sea Turtles*, and has coedited *The Biology of Sea Turtles*, Volume II, and *The Biology of Turtles*.

**Kenneth J. Lohmann**, PhD, is the Charles P. Postelle, Jr. Distinguished Professor of Biology at the University of North Carolina at Chapel Hill. He received his BS from Duke University, his MS from the University of Florida, and his PhD from the University of Washington, after which he carried out postdoctoral work at the University of Illinois, the Marine Biological Laboratory, and the University of Washington Friday Harbor Laboratories. His research interests focus on the behavior and neurobiology of marine animals, with a particular emphasis on unusual sensory systems and how animals use the Earth's magnetic field to guide long-distance migrations. He has published research on diverse invertebrate and vertebrate animals, including more than 50 peer-reviewed studies on sea turtles. He has also taught university courses in marine biology, invertebrate zoology, neurobiology, and scientific writing and has conducted courses in Sweden and Ecuador. He serves on the advisory board of the Galapagos Science Center and, in recognition of his contributions to the field of animal navigation, has been named a fellow of the Royal Institute of Navigation.

**John A. (Jack) Musick**, PhD, is the Marshall Acuff Professor Emeritus in Marine Science at the Virginia Institute of Marine Science (VIMS), College of William and Mary, where he has served on the faculty since 1967. He received his BA in biology from Rutgers University in 1962 and his MA and PhD in biology from Harvard University in 1964 and 1969, respectively. While at VIMS, he successfully mentored 37 master's and 49 PhD students. Dr. Musick has been awarded the Thomas Ashley Graves Award for Sustained Excellence in Teaching from the College of William and Mary, the Outstanding Faculty Award from the State Council on Higher Education in Virginia, and the Excellence in Fisheries Education Award by the American Fisheries Society. In 2008, he was awarded The Lifetime Achievement Award in Science by the State of Virginia. He has published more than 150 scientific papers and coauthored or edited 21 books focused on the ecology and conservation of sharks, marine fisheries management, and sea turtle ecology. In 1985, he was elected a fellow by the American Association for the Advancement of Science. He has received distinguished service awards from both the American Fisheries Society and the American Elasmobranch Society (AES), for which he has served as president. In 2009, the AES recognized him as a distinguished fellow. Dr. Musick has also served as president of the Annual Sea Turtle Symposium (now the International Sea Turtle Society) and as a member of the World Conservation Union (IUCN) Marine Turtle Specialist Group, for which he currently serves on the Red List Authority. Dr. Musick served as cochair of the IUCN Shark Specialist Group for nine years and is currently the vice chair for science. Since 1979, Dr. Musick has served on numerous stock assessment and scientific and statistics committees for the Atlantic States Marine Fisheries Commission (ASMFC), the Mid-Atlantic Fisheries Management Council, the National Marine Fisheries Service, and the Chesapeake Bay Stock Assessment Program. He has chaired the ASMFC Shark Management Technical Committee and ASMFC Summer Flounder Scientific and Statistics Committee. His consultancies have included analyses of sea turtle/long-line interactions in the Canadian and U.S. Atlantic swordfish and tuna long-line fisheries. Many of Dr. Musick's research papers over the last decade have been devoted to problems focused on fisheries bycatch of long-lived marine animals such as sharks and sea turtles.

# Contributors

**Larisa Avens**
National Marine Fisheries Service
National Oceanic and Atmospheric
    Administration
Beaufort, North Carolina

**Natalie C. Ban**
Australian Research Council Centre of
    Excellence for Coral Reef Studies
James Cook University
Townsville, Queensland, Australia

**J. Roger Brothers**
Department of Biology
The University of North Carolina
    at Chapel Hill
Chapel Hill, North Carolina

**Peter H. Dutton**
Protected Resources Division
Southwest Fisheries Science Center
La Jolla, California

**Nancy N. FitzSimmons**
School of Environment
Griffith University
Brisbane, Queensland, Australia

**Mark Flint**
School of Veterinary Science
The University of Queensland
Gatton, Queensland, Australia

and

College of Veterinary Medicine
University of Florida
Gainesville, Florida

**Michael G. Frick**
Department of Biology
University of Florida
Gainesville, Florida

**Kerstin A. Fritsches**
School of Biomedical Sciences
University of Queensland
Brisbane, Queensland, Australia

**Mariana M.P.B. Fuentes**
Australian Research Council Centre of
    Excellence for Coral Reef Studies
James Cook University
Townsville, Queensland, Australia

**Ellis C. Greiner**
College of Veterinary Medicine
University of Florida
Gainesville, Florida

**Mark Hamann**
School of Earth and Environmental Sciences
James Cook University
Townsville, Queensland, Australia

**Elliott L. Hazen**
Southwest Fisheries Science Center
National Oceanic and Atmospheric
    Administration
Pacific Grove, California

**Michael R. Heithaus**
Department of Biological Sciences
Florida International University
Miami, Florida

**Michael P. Jensen**
National Marine Fisheries Service
National Oceanic and Atmospheric
    Administration
La Jolla, California


**T. Todd Jones**
National Marine Fisheries Service
National Oceanic and Atmospheric
    Administration
Honolulu, Hawaii


**Jennifer M. Keller**
Hollings Marine Laboratory
National Institute of Standards and Technology
Charleston, South Carolina


**Rebecca Lewison**
Department of Biology
San Diego State University
San Diego, California


**Catherine M.F. Lohmann**
Department of Biology
The University of North Carolina
    at Chapel Hill
Chapel Hill, North Carolina


**Kenneth J. Lohmann**
Department of Biology
The University of North Carolina
    at Chapel Hill
Chapel Hill, North Carolina


**Jeffrey C. Mangel**
Pro Delphinus
Lima, Peru

and

College of Life and Environmental Sciences
University of Exeter
Exeter, United Kingdom


**Katherine L. Mansfield**
Cooperative Institute for Marine and
    Atmospheric Studies
Florida International University
and
National Marine Fisheries Service
National Oceanic and Atmospheric
    Administration
Miami, Florida


**Sara M. Maxwell**
Hopkins Marine Station
Stanford University
Pacific Grove, California

and

Marine Conservation Institute
Glen Ellen, California


**Véronique J.L. Mocellin**
School of Earth and Environmental Sciences
James Cook University
Townsville, Queensland, Australia


**Joseph B. Pfaller**
Department of Biology
University of Florida
Gainesville, Florida

and

Caretta Research Project
Savannah, Georgia


**Nathan F. Putman**
Department of Fisheries and Wildlife
Oregon State University
Corvallis, Oregon


**Vincent S. Saba**
National Marine Fisheries Service
National Oceanic and Atmospheric
    Administration
and
Geophysical Fluid Dynamics Laboratory
Princeton University
Princeton, New Jersey

**Jeffrey A. Seminoff**
National Marine Fisheries Service
National Oceanic and Atmospheric
    Administration
La Jolla, California


**Joanna Alfaro-Shigueto**
ProDelphinus
Lima, Peru

and

College of Life and Environmental Sciences
University of Exeter
Exeter, United Kingdom


**Melissa L. Snover**
Forest and Rangeland Ecosystem Science
    Center
United States Geological Survey
Corvallis, Oregon

**Bryan Wallace**
The Oceanic Society
Washington, District of Columbia

and

Duke University Marine Laboratory
Nicholas School of the Environment
Beaufort, North Carolina


**Eric J. Warrant**
Department of Biology
University of Lund
Lund, Sweden


**Amanda Southwood Williard**
Department of Biology & Marine Biology
University of North Carolina at Wilmington
Wilmington, North Carolina


**Jeanette Wyneken**
Department of Biological Sciences
Florida Atlantic University
Boca Raton, Florida

# 1 Physiology as Integrated Systems

*Amanda Southwood Williard*

## CONTENTS

## 1.1 INTEGRATIVE APPROACHES TO STUDYING PHYSIOLOGY

The discipline of integrative biology explores the structure and function of biological systems at multiple levels of organization. Within this field, physiologists investigate how living organisms function by studying processes at the molecular, cellular, and organismal levels. Data collected through controlled laboratory experiments and carefully designed field studies contribute to our understanding of physiological responses to variable and challenging environmental conditions, and provide insight into the limits of physiological performance. Modern, integrative studies of animal physiology have roots in the field of physiological ecology, a discipline that incorporates a comparative approach to investigate physiological and biochemical mechanisms and their adaptive significance.

Increasingly, we recognize the need for integrating data on physiological attributes of individual organisms with population-level effects, particularly for species with threatened or endangered status. Abiotic factors (e.g., temperature, humidity, salinity, and oxygen ($O_2$) availability) and biotic factors (e.g., social interactions, prey availability, and predation risk) have profound effects

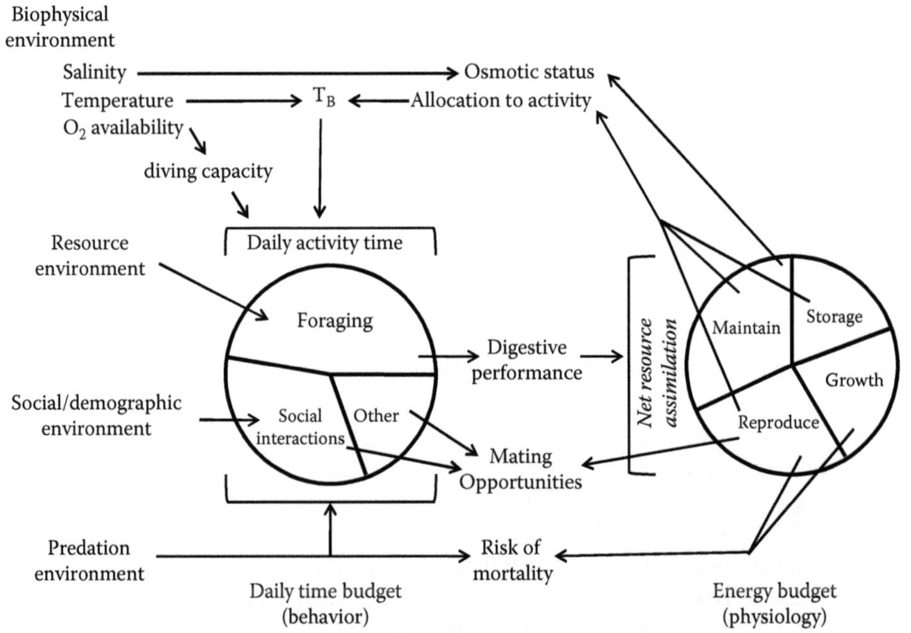


FIGURE 1.1 A summary of the interplay between environmental factors, animal behavior, animal physiology, and population effects. The effects of multiple biotic and abiotic factors on animal energetics and physiological processes may influence population demography through effects on birth rate, mortality rate, immigration, and emigration. For threatened and endangered species of sea turtles, this information is critical for crafting effective management and conservation strategies. (Adapted from Dunham, A.E. et al., *Physiol. Zool.*, 62, 335, 1989.)

on physiological processes, which may in turn affect an individual's ability to acquire resources from the environment and allocate energy to activities critical to survival and fitness (Figure 1.1). An understanding of physiological functioning of individual organisms is, therefore, quite important for assessing the potential impacts of natural or anthropogenic environmental perturbations on a given population (Dunham et al., 1989; Carey, 2005; Tracy et al., 2006; Wikelski and Cooke, 2006; Chown and Gaston, 2008).

Sea turtles are highly migratory reptiles that inhabit the marine realm and spend the majority of their time submerged beneath the ocean surface. The metabolic adaptations and physiological mechanisms underlying their capacity for long-distance movements, prolonged dive durations, and maintenance of the proper balance of water and salts in their body fluids have been the subject of intense interest for many years. Early investigations of sea turtle physiology were conducted in laboratory settings, and provided detailed information about physiological responses to submergence, salt-loading, and various temperature treatments under tightly controlled conditions. Over the past several decades, the development of sophisticated remote-monitoring technologies has permitted investigations of the physiology of sea turtles freely swimming at sea. Integration of behavioral and physiological data collected from both laboratory and field studies has yielded a clearer picture of how sea turtles are adapted for a marine existence and how they respond to alterations in their environment.

This chapter will highlight aspects of sea turtle physiology central to their ability to exploit the marine environment. Our exploration of sea turtle physiology will begin with an overview of metabolism and energetics, as all physiological tasks performed by an animal depend on the acquisition, processing, and allocation of chemical energy. Within the broad framework of metabolic physiology, we will consider the effects of temperature on biochemical and physiological processes, discuss mechanisms to manage $O_2$ stores while diving, and explore the means by which sea turtles regulate body fluid composition. Aspects of physiological function at multiple levels of biological organization will be presented for each topic covered. The chapter will conclude with illustrations of the relevance of physiological studies to sea turtle conservation.

## 1.2 METABOLIC PHYSIOLOGY

Energetics is the study of energy transformations that occur within living organisms, in particular with regard to the use of chemical energy to perform physiological work. Food ingested by an organism is digested, assimilated, and ultimately used as substrate for biochemical pathways that produce adenosine triphosphate (ATP) within the cells. Adenosine triphosphate, in turn, serves as the main source of chemical energy to power cellular processes and physiological work. Animals primarily rely on the $O_2$-dependent (i.e., aerobic) metabolic pathways of cellular respiration to produce the ATP required by cells. Cellular respiration takes place within the mitochondria of cells and, in addition to generating ATP, results in production of carbon dioxide ($CO_2$), water, and heat. Rates of ATP production by cellular respiration are typically well matched with rates of ATP utilization within the cell (Hochachka and Somero, 2002). Consequently, the metabolic rate of an organism (i.e., the rate at which the organism utilizes chemical energy) may be determined either directly by measuring heat production or indirectly by measuring rates of $O_2$ consumption ($\dot{V}O_2$) or $CO_2$ production ($\dot{V}CO_2$) due to cellular respiration. Metabolic rate provides an index of the "cost of living" for an organism and valuable information about the impacts of varying environmental, behavioral, and physiological conditions on energetic requirements of an organism.

The most common means to assess aerobic metabolic rates of sea turtles is the use of respirometry to determine $\dot{V}O_2$ (mL $O_2$ min$^{-1}$ kg$^{-1}$). There are many variations in respirometry techniques that may be used with sea turtles (Wallace and Jones, 2008), but all involve the use of a metabolic chamber to monitor the partial pressure of $O_2$ ($PO_2$) in the air the turtle breathes. The metabolic chamber may be a mask placed over the turtle's nostrils and mouth, an air-filled dome into which a diving turtle is trained to surface and breathe, or a dry box into which the turtle is placed. In an open-flow system, there is a constant flow of air through the metabolic chamber. The $PO_2$ of the air flowing into and out of the metabolic chamber is monitored, and the difference between these two measurements represents the amount of $O_2$ removed from the chamber (e.g., consumed) by the turtle. In a closed system, the turtle is placed in a metabolic chamber of known volume and the decline in $PO_2$ within the sealed chamber provides a measure of the amount of $O_2$ consumed by the turtle over a given period of time.

Respirometry has been used to document $\dot{V}O_2$ of various age classes of loggerhead (*Caretta caretta*), green (*Chelonia mydas*), olive ridley (*Lepidochelys olivacea*), and leatherback turtles (*Dermochelys coriacea*) (Table 1.1). Researchers have used this technique to assess metabolic rates associated with specific physiological states, and to assess the effects of body size, temperature, and activity level on metabolic rate. The following discussion of metabolic physiology of sea turtles will summarize the findings of these studies, and integrate the information on whole animal metabolism provided by respirometry with current knowledge of metabolic functioning at lower levels of biological organization. We will also consider how anaerobic means of ATP production may augment aerobic metabolism during periods of strenuous activity or prolonged submergence. Finally, we will discuss alternative techniques to assess metabolic rates of free-ranging sea turtles under natural conditions.

### 1.2.1 DO SEA TURTLES HAVE "REPTILIAN" METABOLIC RATES?

It is fitting to open our discussion of sea turtle metabolism by considering how their metabolic rates compare with other taxa. Migratory sea turtles have a relatively active lifestyle, and their large body size as adults permits effective heat retention and regional elevation of body temperatures during activity (Standora et al., 1982). The massive leatherback turtle, in particular, has been shown to maintain high, stable internal body temperatures over a broad range of $T_W$

## TABLE 1.1

## Range of Values Reported in the Literature for Mean $\dot{V}O_2$ of Green, Loggerhead, Olive Ridley, and Leatherback Turtles at Temperatures between 20°C and 30°C

| Species | Average Mass Range (kg) | $\bar{X}\,\dot{V}O_2$ (ml min$^{-1}$ kg$^{-1}$) |
|---|---|---|
| Green turtle (*Chelonia mydas*) | | |
| Hatchling[a,b] | 0.025–0.031 | 1.61–20.80 |
| Immature[c,d] | 1.14–24.10 | 0.46–5.60 |
| Adult[e,f] | 128–142 | 0.40–3.43 |
| Loggerhead turtle (*Caretta caretta*) | | |
| Hatchling[b,g] | 0.018–0.022 | 3.64–17.37 |
| Immature[h,i] | 9.5–25.8 | 0.11–1.40 |
| Olive ridley turtle (*Lepidochelys olivacea*) | | |
| Hatchling[j] | 0.013–0.019 | 2.10–27.78 |
| Leatherback turtle (*Dermochelys coriacea*) | | |
| Hatchling[b,j] | 0.044–0.048 | 2.50–10.00 |
| Adult[k,l] | 300–366 | 0.25–5.04 |

Data reported encompass both resting and active $\dot{V}O_2$. The reader is referred to Wallace and Jones (2008) for a detailed accounting of metabolic rate for specific size classes at specific temperatures and activity levels.

[a] Prange and Ackerman (1974).
[b] Wyneken (1997).
[c] Southwood et al. (2003).
[d] Butler et al. (1984).
[e] Enstipp et al. (2011).
[f] Prange and Jackson (1976).
[g] Lutcavage and Lutz (1986).
[h] Hochscheid et al. (2004).
[i] Lutcavage et al. (1987).
[j] Jones et al. (2007).
[k] Lutcavage et al. (1990).
[l] Paladino et al. (1990).

(James and Mrosovsky, 2004; Southwood et al., 2005; Casey et al., 2010). These aspects of sea turtle biology have led investigators to question whether the metabolic rates of sea turtles might be elevated in relation to other reptiles.

One way to address this issue is through analyses of the allometric relationships between resting metabolic rate (RMR) and mass (M) for sea turtles and reptiles in general: $RMR = aM^b$ (Figure 1.2). In this equation, the proportionality coefficient a represents the intercept and b represents the slope of the power regression relating RMR to M (Schmidt-Nielsen, 1984). The exponent b reflects the nature of the relationship between metabolic rate and mass, or how metabolic rate changes with mass. The proportionality coefficient a in allometric equations provides information about the metabolic intensity of a given taxa, and can be used for comparisons between taxa as long as the slopes of power regressions for the groups are the same. Wallace and Jones (2008) used available data from the literature to generate power regressions relating RMR and M for green turtles ($RMR_{green} = 0.494\,M^{-0.207}$) and leatherback turtles ($RMR_{leatherback} = 0.768\,M^{-0.169}$). Proportionality coefficients and slopes of the power regressions for these two sea turtles species were compared with values reported for reptiles

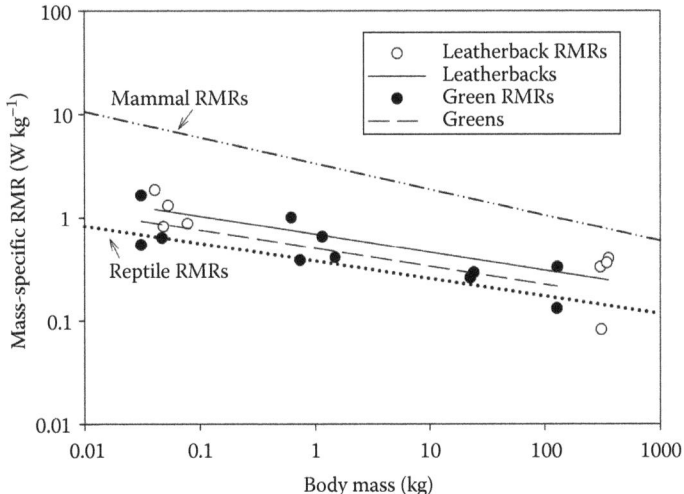

**FIGURE 1.2** Power regressions for mass-specific resting metabolic rates (RMRs) versus body mass for leatherback turtles (open circles, solid line; $RMR_{leatherback} = 0.735 \ M^{-0.177}$) and green turtles (filled circles, dashed line; $RMR_{green} = 0.554 \ M^{-0.171}$) compared to allometric equations for reptiles (dotted line: $RMR_{reptile} = 0.378 \ M^{-0.17}$) and mammals (dash-dot-dash line: $RMR_{mammal} = 3.35 \ M^{-0.25}$). There was no significant difference in the proportionality coefficient for leatherback turtles, green turtles, or reptiles. (From Wallace, B.P. and Jones, T.T., *J. Exp. Mar. Biol. Ecol.*, 356, 8, 2008.)

($RMR_{reptile} = 0.378 \ M^{-0.17}$; Bennett, 1982) and mammals ($RMR_{mammal} = 3.35 \ M^{-0.25}$; Bennett, 1982) (Figure 1.2). There was no significant difference in the exponent b between the sea turtle equations and the equations for reptiles or mammals, so comparisons of metabolic intensity of sea turtles relative to these groups based on proportionality coefficients are valid. The proportionality coefficients for green (0.494) and leatherback (0.768) turtles were not significantly different from each other or from the proportionality coefficient for reptiles (0.378); however, they were significantly lower than mammalian proportionality coefficient (3.35). This analysis indicated that despite differences in ecology and thermal biology, the metabolic rates of sea turtles are within the range typical of reptiles and significantly lower than that of mammalian endotherms.

## 1.2.2 METABOLIC CAPACITY

### 1.2.2.1 Aerobic Metabolism

Fundamental differences exist between the aerobic metabolic capacity of ectotherms, which rely primarily on environmental sources of heat to warm tissues, and endotherms, which use endogenous heat generated by metabolism to warm tissues. When temperature and activity status are accounted for, metabolic rates of endotherms (e.g., birds and mammals) are typically 5–10 times higher than metabolic rates of comparably sized ectothermic vertebrates (e.g., fish, amphibians, reptiles; Bennett and Dawson, 1976). The disparity in metabolic rate between these groups is reflected by differences in metabolic machinery within cells, as well as differences in the organ systems involved in $O_2$ delivery to tissues. Mitochondrial density, mitochondrial surface area, activity of enzymes involved in oxidative metabolism, and ion transport metabolism due to activity of the $Na^+$–$K^+$–ATPase pump are significantly lower in reptiles compared with mammals (Bennett, 1972; Else and Hulbert, 1981; Hulbert and Else, 1981). In addition to the limitations on oxidative metabolism at the cellular level, structural features of the cardiorespiratory system in reptiles place limitations on delivery of $O_2$ to the metabolizing tissues. There is a great deal of variation in complexity of reptile lungs; however, the surface area available for pulmonary gas exchange in reptiles is typically much less than that of birds and mammals (Perry, 1983). And while the three-chambered heart of non-crocodilian reptiles permits mixing of oxygenated and de-oxygenated blood (i.e., cardiac shunting), which may be advantageous under certain physiological and ecological conditions (Hicks and Wang, 1996), the use of a single ventricle to pump blood into both the pulmonary and systemic circuits places a ceiling on blood pressures, which is determined by the pressures that the lung can withstand (Burggren et al., 1997). As reptiles, sea turtles are bound by their phylogeny and must function within these constraints.

Aerobic scope, measured as the difference between the maximal sustainable rate of $O_2$ consumption ($\dot{V}O_{2max}$) and resting rate of $O_2$ consumption ($\dot{V}O_{2rest}$) (Fry, 1947), or factorial aerobic scope ($\dot{V}O_{2max}/\dot{V}O_{2rest}$) are often used as measures of an animal's capacity for aerobic metabolism and sustained activity (Bennett, 1982). Factorial aerobic scopes of 3–12 (i.e., $\dot{V}O_{2max}$ 3–12 times higher than $\dot{V}O_{2rest}$) are typical for reptiles (Bennett and Dawson, 1976), whereas mammals and birds adapted for endurance activity may have factorial aerobic scopes within the range of 30–65 (Bishop, 1999). Care must be taken to account for the effects of temperature and mass on factorial aerobic scope when making comparisons between taxa, but in general species with high factorial aerobic scopes are those with highest capacity for sustained activity. Reptiles typically have limited scopes for aerobic activity when compared with birds and mammals, and sea turtles appear to follow this general trend for reptiles.

Whole animal $\dot{V}O_2$ at rest and during various types of activity (crawling, swimming, nesting) has been investigated for several species and age classes of sea turtles (Table 1.1). Factorial aerobic scopes for green, loggerhead, olive ridley, and leatherback turtle hatchlings following emergence and during the frenzy swimming and post-frenzy stages fall within the range of 1.4–4.2 (Wyneken, 1997; Jones et al., 2007). A two- to threefold difference between $\dot{V}O_{2rest}$ and $\dot{V}O_2$ during routine activity has been reported for immature loggerhead (Lutz et al., 1989) and green turtles (Davenport et al., 1982), although it is difficult to say whether a true representation of factorial aerobic scope is reflected by the data since $\dot{V}O_{2max}$ may not have been obtained. Slightly higher factorial aerobic scopes (3–4) were reported for immature green turtles swimming at sustained speeds up to 0.35 m s$^{-1}$ in a swim flume at 25°C (Prange, 1976). In a similarly designed experiment, Butler et al. (1984) found that immature green turtles swam steadily at speeds of 0.4–0.6 m s$^{-1}$ in $T_W$ of 27°C–30°C. The $\dot{V}O_2$ at the highest swim speed was 2.83 times higher than resting $\dot{V}O_2$, but the authors speculate that the design of their swim flume prevented turtles from swimming at their maximal speed and that $\dot{V}O_{2max}$, and thus the factorial aerobic scope, could actually be higher than what they recorded. Increases in swim speed were matched by increases in $\dot{V}O_2$ and respiratory frequency, and there was a positive linear relationship between heart rate and $\dot{V}O_2$. Blood lactate levels measured after 10 min of swimming at speeds of 0.5 m s$^{-1}$ were no different than resting levels. Taken together, these observations suggest that steady-state swimming in immature green turtles over the range of speeds tested was powered primarily by aerobic metabolism (Butler et al., 1984).

The factorial aerobic scope for adult green turtles is considerably higher than that observed for juveniles of this species. Jackson and Prange (1979) reported that $\dot{V}O_2$ of active adult female green turtles on a nesting beach in Costa Rica was 10 times higher than $\dot{V}O_{2rest}$, whereas factorial aerobic scopes for immature green turtles is between 2 and 4 (Prange, 1976; Butler et al., 1984). Differences in the scaling exponents for resting and maximal $\dot{V}O_2$ of green turtles may account for the higher factorial aerobic scope observed in larger turtles. Prange and Jackson (1976) calculated a slope of −0.17 for the regression relating mass-specific $\dot{V}O_2$ to body mass in resting green turtles, but the relationship is almost directly proportional (slope = −0.06) for active turtles. This pattern suggests that as green turtles get larger, their capacity for sustained aerobic activity increases, which may have important implications for the ontogenetic timing and energetic costs of migration at different life stages. This trend may apply to other species of sea turtles as well. For example, the fourfold difference between resting and maximal metabolic rates in adult female leatherback turtles on a nesting beach in Costa Rica (Paladino et al., 1990) is approximately two times higher than factorial aerobic scopes reported for hatchling leatherback turtles (1.39–2.50; Wyneken, 1997; Jones et al., 2007).

Factorial aerobic scopes for sea turtles fall well within the range reported for other species of reptiles (3–12; Bennett and Dawson, 1976). If sea turtles are not exceptional with regard to their aerobic capacity, then why are they the only reptiles to undertake long-distance migrations on the order of hundreds to thousands of kilometers? The answer lies partly in the low cost of transport for animals that use swimming as their primary mode of locomotion and the relationship between cost of transport and body size (Tucker, 1970; Schmidt-Nielsen, 1972). The cost of transport is a reflection of the metabolic cost to move a given mass a given distance. Fully aquatic reptiles have low mass-specific net cost of transport (NCT, measured as the slope of the relationship between aerobic metabolic rate and speed; Schmidt-Nielsen, 1972) compared with terrestrial forms. In other words, the metabolic cost for a sea turtle to swim a given distance is much lower than the cost for a similarly sized terrestrial reptile to walk the same distance. Additionally, mass-specific NCT decreases as animal size increases, so the large size attained by adult sea turtles contributes to locomotor economy. This point is underscored by the modest increase in $\dot{V}O_2$ (1.3–1.9 $\dot{V}O_{2rest}$) observed by Enstipp et al. (2011) for adult green turtles during routine swimming in a large swim channel. Sea turtles, with their large body size, streamlined body form, reduced shell, highly modified limbs, and powerstroke gait are well adapted for efficient locomotion in the marine environment (Wyneken, 1997), and their economical means of movement permits them to travel great distances despite the fact that they have aerobic metabolic capacities typical of reptiles and lower than that of many endothermic vertebrates.

Characteristics of the pulmonary system in sea turtles appear to be well suited for meeting elevated levels of $O_2$ demand during sustained activity. The complex, multichambered lungs of sea turtles allow for rapid and efficient exchange of respiratory gases, with high tidal volumes and high expiratory flow rates (Tenney et al., 1974; Perry, 1983; Gatz et al., 1987; Lutcavage et al., 1989). These features are advantageous for promoting $O_2$ uptake to support elevated metabolic rates during sustained exercise, and also facilitate rapid gas exchange during surface intervals between dives. In green turtles, increases in activity are well matched by increases in breathing frequency, pulmonary blood flow, and heart rate to meet increased $O_2$ demands within the tissues and metabolizing cells (West et al., 1982; Butler et al., 1984; Southwood et al., 2003).

The majority of data on the metabolic physiology of sea turtles are derived from investigations of whole animal metabolic rates. Penick et al. (1998) provide the only published reports of tissue metabolic rates for sea turtles. The mean in vitro rate of $O_2$ consumption in skeletal muscle tissue is 119.5 $\mu$L $O_2$ g$^{-1}$ h$^{-1}$ at 35°C for immature green turtles and 60.5 $\mu$L $O_2$ g$^{-1}$ h$^{-1}$ at 35°C for much larger adult leatherback turtles (Penick et al., 1996, 1998). The large disparity in body mass and differences in methodology between studies make comparisons of mass-specific rates of tissue metabolism for these two species and between these species and other reptiles difficult. It would be informative to gather data on tissue metabolic rates for sea turtles and other reptiles using specimens of similar size.

A comprehensive assessment of correlates of aerobic capacity at the cellular level has not yet been conducted for sea turtles, but there are some data available on mitochondrial enzyme activity in muscle tissue. Comparisons of enzyme activities between sea turtles and other species of reptile are complicated by the numerous factors that may differ between studies, including muscle fiber type composition at the biopsy site, environmental temperatures experienced by the animal, assay temperatures at which enzyme activity is measured, and body size. Nevertheless, a general discussion of the limited data available provides a comparative framework to discuss the metabolic capacity of sea turtles and guidance for areas of future research. Activity of citrate synthase (CS), an enzyme that participates in the citric acid cycle, in the iliotibialis hip abductor muscle of immature green turtles ($\bar{X}$ mass = 15.3 kg) at Heron Island, Australia, during the summer was between 1 and 2 $\mu$mol min$^{-1}$ gram wet tissue$^{-1}$ when measured at assay temperatures between 15°C and 30°C (Southwood, 2002; Southwood et al., 2006). These values are comparable to values reported for CS activity in tail muscle of immature and adult alligators (mass range 0.9–54.5 kg; 1–5 $\mu$mol min$^{-1}$ gram wet tissue$^{-1}$) captured during the summer at a wildlife refuge in Louisiana, the United States

(Seebacher et al., 2003). Citrate synthase activity in the flexor tibialis muscle of captive immature green turtles (mass range = 18.9–43.1 kg; 10–30 µmol min$^{-1}$ gram wet tissue$^{-1}$) maintained at 26°C was considerably higher than values reported for iliotibialis muscle of similarly sized wild green turtles at comparable temperatures (Southwood 2002; Southwood et al., 2003a). The difference in CS activity could be due to differences in mass and nutritional factors between captive and wild turtles, but may also reflect variation in fiber type composition between muscles. Based on the limited data available, CS activity in sea turtle muscle tissue falls within the range observed for aquatic reptiles (Seebacher et al., 2003; Guderley and Seebacher, 2011; Southwood and Harden, 2011), and the capacity for flux through aerobic metabolic pathways is likely to be similar as well.

It is difficult to draw solid conclusions regarding aerobic metabolic capacity of sea turtles based on trends in one mitochondrial enzyme from one tissue type. Additional data on mitochondrial enzyme activity from skeletal muscle and visceral organs (i.e., heart, brain, and liver) and information on mitochondrial density and surface area would provide a more complete picture of the capacity for flux through aerobic ATP-producing pathways within the cells of sea turtles.

### 1.2.2.2 Anaerobic Metabolism

It is widely assumed that sea turtles rely chiefly on oxidative metabolism to supply the ATP necessary to power routine activity and long-distance migration, since aerobic pathways are much more efficient at transferring energy from substrate molecules to ATP than are anaerobic pathways (Hochachka and Somero, 2002). For example, oxidative metabolism yields approximately 38 molecules of ATP for every molecule of glucose substrate whereas the primary biochemical pathway for anaerobic metabolism in reptiles (i.e., anaerobic glycolysis coupled with fermentation) yields only two molecules of ATP for every molecule of glucose substrate. In addition to being an inefficient means of producing ATP, anaerobic metabolism also generates lactic acid which readily dissociates into a lactate anion and a proton. Accumulation of lactic acid may therefore result in disruptions to acid–base balance of body fluids and subsequent disturbances in molecular and cellular functioning. Anaerobic glycolysis is insufficient as a means to provide energy to support sustained activity due to the self-limiting nature of this pathway. It is, however, a critically important means of ATP production during short duration bouts of high-intensity activity when there is an $O_2$ supply–demand mismatch, and potentially during long-duration dives as $O_2$ stores become depleted (see Section 1.3.2).

Anaerobic metabolism may be assessed by measuring the accumulation of lactate during a bout of activity or a dive. Lactate may be distributed differentially within the body so, ideally, whole body lactate levels should be used to assess anaerobic energetics. This terminal approach has rarely been used with sea turtles, given their protected status and large body size attained as adults, so blood lactate levels are generally used as a proxy for anaerobic metabolism during activity. Baldwin et al. (1989) found high levels of blood lactate during nest emergence and beach crawling stages of hatchling dispersal in loggerhead (3.2–5.5 mM L$^{-1}$) and green turtles (8.9–9.2 mM L$^{-1}$). The high-intensity activity associated with initial stages of dispersal are referred to as the "hatchling frenzy" (Carr and Ogren, 1960), and this study demonstrated that aerobic metabolism is supplemented by anaerobic metabolism during this critical period when hatchlings are very vulnerable to predation (Baldwin et al., 1989). Elevated blood lactate levels (6.5 mM L$^{-1}$) have also been documented for adult female green turtles engaged in the arduous task of nesting (Jackson and Prange, 1979).

A comparison of activity of enzymes involved in anaerobic glycolysis (pyruvate kinase; PK) and fermentation (lactate dehydrogenase; LDH) with mitochondrial enzyme activity (CS) in sea turtle muscle tissue provides useful perspective on the relative importance of anaerobic means of ATP production for locomotory activity. Activities of PK (400–800 µmol min$^{-1}$ gram wet tissue$^{-1}$) and LDH (600–1200 µmol min$^{-1}$ gram wet tissue$^{-1}$) in the iliotibialis hip abductor muscle of immature green turtles during the summer at Heron Island, Australia, are between 2 and 3 orders of magnitude higher than CS activity (1–2 µmol min$^{-1}$ gram wet tissue$^{-1}$) measured over the same range of assay temperatures (Southwood 2002; Southwood et al., 2006). Equally large differences in activity of anaerobic and aerobic enzymes are observed in skeletal muscle of alligators (Seebacher et al., 2003; Guderley and Seebacher, 2011), diamondback terrapins (Southwood and Harden, 2011), and the Australian longneck turtle (*Chelodina longicollis*; Seebacher et al., 2004). While this pattern would not be expected to apply to the same degree in highly aerobic visceral organs such as the heart, brain, and lungs (Guderley and Seebacher, 2011), it is reflective of high capacity for anaerobic ATP production in locomotory muscles and indicative of the importance of these pathways in the activity metabolism of sea turtles and other reptiles. The high efficiency of aquatic locomotion in sea turtles permits routine, sustained activity to be powered by aerobic metabolism, but sea turtles likely resort to anaerobic metabolism to supplement ATP production during episodes of intense activity, particularly during activities on land.

### 1.2.3 TEMPERATURE EFFECTS ON METABOLISM

Alterations in body temperature ($T_B$) affect structure and function of metabolic enzymes, rates of flux through biochemical pathways, and thus rates of cellular and organismal metabolism. With the notable exception of the leatherback turtle (see Section 1.2.3.3), $T_B$ of sea turtles varies predictably with water temperatures ($T_W$) and is generally no more than a few degrees higher than $T_W$ (Standora et al., 1982; Sakamoto et al., 1990; Sato et al., 1994, 1995, 1998). Spatial and temporal variation in thermal environment may, therefore, have pronounced effects on metabolism, physiology, and behavior of sea turtles. The nature of the metabolic response to changes in $T_B$ depends on the magnitude of the temperature change and the duration of exposure to a new thermal regime. Studies of temperature effects on metabolism at various levels of biological organization have been conducted with green, loggerhead, and leatherback turtles, and results of these studies are described below in the context of acute and prolonged exposure to a new thermal regime.

#### 1.2.3.1 Acute Effects

A common way to assess the direct effect of temperature on physiological or biochemical processes is to calculate the thermal coefficient ($Q_{10}$), which provides an index of thermal sensitivity. Typical $Q_{10}$ values for metabolic rate in reptiles fall within the range of 2–3 (Bennett, 1976, 1982), indicating that metabolic rate increases two- to threefold with a 10°C increase in temperature. Kraus and Jackson (1980) measured $\dot{V}O_2$ of fasted, immature green turtles (0.48–1.24 kg) acclimated to 25°C and then acutely exposed to 15°C, 25°C, or 35°C; turtles exhibited $Q_{10}$ values of 2.1–2.6 over the range of temperatures tested. Lutz et al. (1989) measured $\dot{V}O_2$ of loggerhead turtles (4.3–22.7 kg) fasted for 2 days at temperatures of 10°C, 15°C, 20°C, and 30°C, and found that the $Q_{10}$ for $\dot{V}O_2$ of resting and active turtles was 2.4.

The acute effects of temperature on metabolic processes at lower levels of biological organization have been investigated in a limited number of studies. Southwood et al. (2003a) measured activity of the enzymes CS, PK, and LDH from muscle tissue of captive and wild immature green turtles at assay temperatures ranging from 15°C to 30°C and calculated $Q_{10}$ values as an indicator of thermal sensitivity of enzyme function. The $Q_{10}$ values for CS, PK, and LDH activity in green turtle muscle (range 1.20–1.69) were low, but fell well within the range of values reported for these enzymes in other species of aquatic turtles and alligators (Seebacher et al., 2003, 2004; Guderley and Seebacher, 2011; Southwood Williard and Harden, 2011). Additionally, the $Q_{10}$ values for activities of the glycolytic enzyme PK (1.66–1.69) and mitochondrial enzyme CS (1.20–1.44) in green turtle muscle correspond well with $Q_{10}$ values reported for green turtle muscle tissue metabolism over the range of 12.5°C–27.5°C (1.31–1.56; Penick et al., 1996).

The available data demonstrate that the metabolic responses of sea turtles to acute changes in temperature are similar to responses observed for other reptiles. One remarkable exception is a report by Penick et al. (1998) of thermal independence ($Q_{10} = 1$) of muscle tissue metabolism in leatherback turtles. The acute effects of temperature on whole animal $\dot{V}O_2$ have not yet been investigated in leatherback turtles. Future studies to investigate thermal dependence of metabolism at the molecular and organismal level in this species are warranted in light of its unique thermal biology and migratory patterns (see Section 1.2.3.3).

### 1.2.3.2 Seasonal Changes in Temperature

Long-term exposure to a new thermal regime, such as occurs on a seasonal basis, may elicit physiological responses that are not detectable in acute laboratory studies. Furthermore, changes in photoperiod, food availability, and risk of predation may contribute to seasonal adjustments in physiology and behavior of sea turtles in their natural environment (Huey, 1982; Tsuji, 1988). Maintenance of activity year-round is advantageous if the energetic benefits outweigh the costs (Tsuji, 1988; Guderley and St-Pierre, 2002; Seebacher, 2005). On the other hand, if resources are limited during the colder winter months, then hypometabolism, decreased activity, and dormancy would be favored. Sea turtles exhibit both of these strategies.

Year-round activity and maintenance of foraging have been documented for immature green turtles at several subtropical foraging grounds (Mendonca, 1983; Read et al., 1996; Seminoff, 2000). At Heron Island, Australia, immature green turtles remain active throughout the year, but exhibit significant seasonal shifts in diving patterns; dive durations during the winter ($\bar{X}$ $T_W = 21.4°C$) were twice as long as dive durations during summer ($\bar{X}$ $T_W = 25.8°C$; Southwood et al., 2003b). Field metabolic rates estimated with the doubly labeled water (DLW) method (see Section 1.2.4) for turtles at this site were 43% lower during winter compared with summer, presumably reflecting the direct effects of temperature, longer dive times, and lower activity levels during the winter (Southwood et al., 2006; see Section 1.2.4). Digestive state (fed or fasted) has a pronounced effect on metabolic rate of animals, and decreases in food intake during the winter may also contribute to the seasonal change in metabolic rate observed for green turtles at Heron Island. Oxygen consumption rates of immature green turtles exposed to laboratory simulations of seasonal changes in temperature ($T_{Wsummer} - T_{Wwinter} = 9°C$) and photoperiod at a subtropical site were only 24%–27% lower during exposure to winter conditions compared with summer conditions (Southwood et al., 2003a). Green turtles were fasted prior to $\dot{V}O_2$ trials and activity levels were controlled for in this laboratory study, which may account for the discrepancy between laboratory and field results. Taken together, the results from laboratory and field studies suggest that multiple factors (e.g., temperature, food availability, and diving patterns) may contribute to seasonal shifts in metabolism of green turtles.

Although conditions at some sites are favorable to support year-round foraging and maintenance of active dive patterns, thermal and/or ecological conditions at other sites favor a drastic decrease in activity, prolonged dives, and sometimes entrance into a dormant state. Accumulated evidence from field and laboratory studies suggests that 15°C may be a thermal "activity threshold" for sea turtles, and body temperatures below this threshold trigger shifts in behavior that promote metabolic downregulation and energy conservation (Davenport et al., 1997; Moon et al., 1997; Seminoff, 2000; Hochscheid et al., 2007). Reports of dormant green turtles buried in the muddy seafloor of the Gulf of California (Felger et al., 1976) and torpid immature loggerhead turtles dredged by shrimp trawls from the seafloor of Cape Canaveral shipping channel off the East coast of Florida (Carr et al., 1980) generated a great deal of interest in the behaviors and physiological adjustments associated with overwintering in sea turtles. Moon et al. (1997) exposed immature green and Kemp's ridley turtles to a temperature regime designed to trigger dormancy, and found that both species exhibited extended dive times (2–3 h), decreased activity, and hypophagia when $T_W$ fell below 15°C. Hochscheid et al. (2004) monitored $\dot{V}O_2$, feeding behavior, and activity levels of captive immature loggerhead turtles exposed to natural seasonal variations in $T_W$ at an aquarium facility in Naples, Italy, and found significant seasonal differences in $\dot{V}O_2$. The $\dot{V}O_2$ for turtles at 15.7°C was 80% lower than $\dot{V}O_2$ at 25.4°C. Turtles in this study were permitted to feed ad libitum and winter ingestion rates were much lower than summer ingestion rates, so seasonal differences in nutritional status and digestive state may have contributed to the seasonal difference in metabolism.

There have been impressive advances in remote-monitoring technology since the first anecdotal observations of overwintering behavior in sea turtles. Satellite-linked data recorders have been used to document dive times and surfacing patterns for loggerhead turtles over the course of a year or more in the Mediterranean Sea (Hochscheid et al., 2007a) and the behavioral data collected in the field provide a useful complement to the physiological data collected in the laboratory. Loggerhead turtles in the Mediterranean Sea experienced $T_W$ as low as 12.5°C during the winter and had maximum dive durations of 4.5–8 h, with 85%–92% of dives lasting longer than 3 h (Hochscheid et al., 2007a). Routine dive durations for active loggerhead turtles at temperatures above 20°C are only 5–53 min (Table 1.2). Loggerhead turtles appear to have relied primarily on aerobic metabolism during these long-duration winter dives, as the median surface interval between dives was only 7.9 min. Overwintering behavior characterized by long dive durations (>90 min) has also been reported for green turtles in the Mediterranean (Godley et al., 2002). Although the duration of winter dives made by sea turtles in the Mediterranean Sea is remarkable, it does not appear that turtles are burying in the seafloor substrate and remaining completely inactive during seasonal cold exposure. The initial observations of dormant sea turtles buried along the seafloor off the coast of Florida and in the Gulf of California have not been revisited. Unfortunately, harvesting practices that take advantage of sea turtles in this vulnerable condition may ultimately prevent us from fully understanding this phenomenon.

The effects of chronic shifts in temperature on metabolic machinery at the cellular and molecular level have not been well studied in sea turtles. Southwood et al. (2003) investigated metabolic enzyme activity in green turtles exposed to a laboratory simulation of seasonal changes in temperature. Although there was no significant difference in activity of the mitochondrial enzyme CS in muscle tissue collected during exposure to summer conditions ($T_W = 26$°C) and after 4 weeks exposure to winter conditions ($T_W = 17$°C), activities of PK and LDH were significantly higher in muscle tissue collected during exposure to winter conditions (Figure 1.3; Southwood et al., 2003). This pattern is typical of thermal acclimation, a process by which the direct effects of temperature on biochemical reaction rates are offset by physiological adjustments during prolonged exposure to a new thermal regime (Somero, 1997; Hochachka and Somero, 2002). For example, an increase in enzyme concentration during prolonged cold exposure may compensate for the decreased kinetic energy of reactant molecules at low temperature. If this is the case, then enzyme activities in tissue collected from cold-acclimated animals should be higher than enzyme activities in tissue collected from warm-acclimated animals when tested at the same assay temperature, as was observed for green turtle PK and LDH activity. The exact mechanisms underlying thermal compensation of anaerobic enzyme activity in green turtle tissues have not been studied, but the potential benefits at the organismal level are clear. Preservation of anaerobic capacity for ATP production via low thermal dependence or thermal acclimation insures that sea turtles retain the capacity for burst locomotion and a rapid escape/defense response over a broad range of temperatures.

### 1.2.3.3 Thermoregulation

As ectotherms, reptiles have low levels of metabolic heat production and, consequently, rely on external sources of heat to warm their tissues. Even so, many reptiles are adept at regulating body temperature by making behavioral adjustments, such as shuttling between sun and shade or air and water, in order to take advantage of thermal microclimates (Huey, 1982). Physiological alterations, particularly changes in blood flow and distribution, may modify rates of heat exchange between the animal and its environment such that heat gain is facilitated during basking episodes and heat loss is slowed down during non-basking periods. Sea turtles, being aquatic, have fewer opportunities for behavioral thermoregulation than their terrestrial counterparts. Nevertheless, sea turtles can select thermal habitats by undertaking seasonal migrations; they may also alter their immediate thermal environment by diving, as $T_W$ varies predictably with depth (Southwood et al., 2005). The high thermal conductivity of water places limitations on the degree to which sea turtles may elevate $T_B$ above ambient $T_W$, but larger turtles are capable of retaining metabolically generated heat and

**TABLE 1.2**

**Typical Dive Depths and Durations for Active Sea Turtles (Overwintering Turtles Are Not Included)**

| Species | Routine Dive Depth (m) | Maximum Dive Depth (m) | Routine Dive Duration (min) | Maximum Dive Duration (min) |
|---|---|---|---|---|
| Leatherback turtle | | | | |
| (*Dermochelys coriacea*) | | | | |
| Hatchling[a] | 5 | 17 | 2.1 | 6.0 |
| Immature[b] | — | — | 7.7 | — |
| Adult[c,d,e,f] | 16–90 | 1230 | 7.1–28.6 | 86.1 |
| Green turtle (*Chelonia mydas*) | | | | |
| Hatchling[a] | 2.5 | 9.3 | 1.4 | 4.3 |
| Immature[g,h] | 2.9–5.6 | 7.9 | 13.1 | — |
| Adult[i–k] | 1.2–17.1 | >45 | 10.2–39.7 | >60 |
| Hawksbill turtle | | | | |
| (*Erytmochelys imbricata*) | | | | |
| Hatchling[l,m] | 2.7–9.1 | 72 | 7.5–37.1 | 82 |
| Adult[n] | 8.8–27.7 | ~50 | ~5–90 | ~115 |
| Flatback turtle (*Natador depressus*) | | | | |
| Hatchling[o] | 4 | 12 | 1.7 | ~6 |
| Loggerhead turtle (*Caretta caretta*) | | | | |
| Immature[p] | 1–15 | — | 2–30 | — |
| Adult[q,r] | 3.0–61.0 | 233 | 5.5–53.6 | ~95 |
| Olive ridley turtle | | | | |
| (*Lepidochelys olivacea*) | | | | |
| Immature[s] | — | — | — | ~150 |
| Adult[t] | 20.0–46.7 | 200 | 24.5–48.0 | 115 |
| Kemp's ridley turtle | | | | |
| (*Lepidochelys kempii*) | | | | |
| Immature[u] | 2.1–2.6 | 5.3 | 4.8–6.4 | 22.1 |

References are provided for values that encompass the range reported in the literature.

[a] Salmon et al. (2004).
[b] Standora et al. (1984).
[c] Hays et al. (2004).
[d] Reina et al. (2005).
[e] Lopex-Mendilaharsu et al. (2008).
[f] Casey et al. (2011).
[g] Southwood et al. (2003b).
[h] Makowski et al. (2006).
[i] Godley et al. (2002).
[j] Hays et al. (2004).
[k] Sato et al. (1998).
[l] van Dam and Diez (1996).
[m] Witt et al. (2010).
[n] Storch et al. (2005).
[o] Salmon et al. (2010).
[p] Howell et al. (2010).
[q] Sakamoto et al. (1990).
[r] Minamikawa et al. (1997).
[s] Polovina et al. (2004).
[t] McMahon et al. (2007).
[u] Sasso and Witzell (2006).

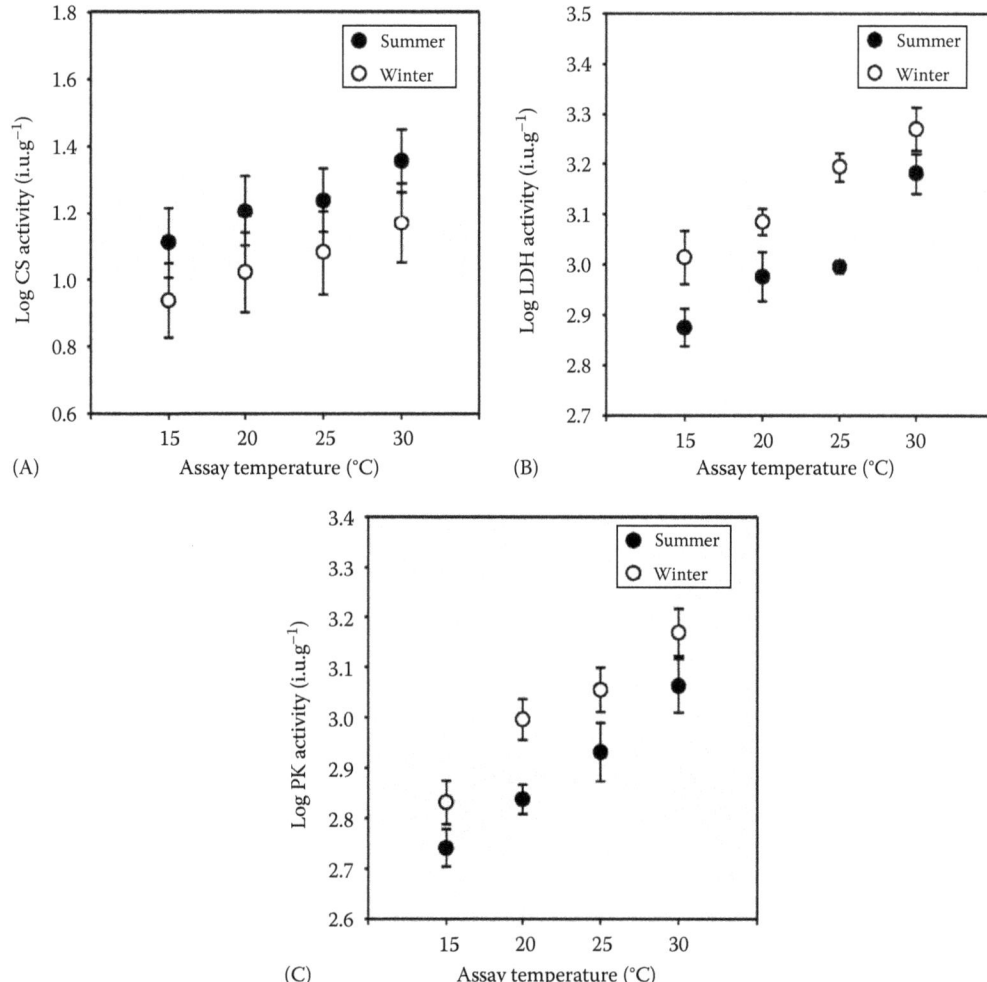

FIGURE 1.3   Temperature effects on the activity of (A) citrate synthase (CS), (B) lactate dehydrogenase, and (C) pyruvate kinase (PK) for immature green turtles acclimated to 26°C (summer) and 17°C (winter) under laboratory conditions. The $Q_{10}$ values for enzymes ranged between 1.44 and 1.69. Enzymes involved in glycolysis (PK) and lactic acid fermentation (LDH) show evidence of thermal acclimation to prolonged cold exposure. (From Southwood, A.L. et al., *J. Exp. Biol.*, 206, 4521, 2003a; Southwood, A.L. et al., *Can. J. Zool.*, 81, 1014, 2003b.)

heat obtained from the environment due to a small surface area to volume ratio and redistribution of blood flow (Heath and McGinnis, 1980; Standora et al., 1982; Paladino et al., 1990; Hochscheid et al., 2002). Even so, the average gradient between internal $T_B$ and ambient $T_W$ in freely swimming adult sea turtles is typically limited to 1°C–2°C (Sato et al., 1994, 1995, 1998). The giant leatherback turtle (typically 300–500 kg as an adult) is exceptional with regard to its thermoregulatory abilities and may establish thermal gradients up to 17°C (Frair et al., 1972).

Observational data from scientists and fishermen in the 1960s and 1970s led to the recognition that leatherback turtles routinely migrate through cold waters off the East Coast of Canada and New England, and that foraging at high latitude is an integral part of their life history (Bleakney, 1965; Lazell, 1980). The ability of leatherback turtles to maintain $T_B$ higher than ambient $T_W$ is thought to be a critical component of their biology that allows this aquatic reptile to exploit cold water habitats. In addition to the thermal benefits of large body size, leatherbacks also possess specific anatomical features that facilitate heat retention. Countercurrent heat exchangers at the base of the front flippers help warm blood returning to the core from the extremities, and a unique arrangement of densely packed blood vessels lining the trachea of adult leatherbacks may serve to warm inspired air as it travels into the core (Greer et al., 1973; Davenport et al., 2009a). Furthermore, leatherbacks have extensive deposits of insulative fat along the carapace and plastron, and within the head and neck region (Frair et al., 1972; Goff and Stenson, 1988; Davenport et al., 2009b).

While the morphological features that promote heat retention are fairly well understood for leatherbacks, there is still some debate as to the role of metabolic heat generation. Comparative analyses of allometric relationships between metabolic rate and mass show no significant difference in resting metabolic intensity between leatherback turtles and green turtles or reptiles in general (Figure 1.2; Wallace and Jones, 2008). In other words, the thermal gradients maintained by leatherback turtles cannot be explained by exceptionally high metabolic rates in comparison with other reptiles. Paladino et al. (1990) constructed a "core-shell" model to predict the magnitude of the gradient between core $T_B$ of a 400 kg leatherback and ambient $T_W$ using metabolic rate and degree of peripheral blood flow as inputs into the model. They concluded that even at low metabolic rates, leatherbacks could maintain high, stable $T_B$ in cold $T_W$ by making simple adjustments in blood flow and using peripheral tissues as insulation. Bostrom and Jones (2007) also used a modeling approach to investigate the significance of metabolic heat generated during locomotion for maintenance of thermal gradients, and concluded that metabolic heat generated by skeletal muscles could play an important role in the thermal strategy of leatherbacks. Experiments with captive immature leatherback turtles (16–37 kg) exposed to an acute decrease in temperature showed that an increase in metabolic heat production due to an increase in activity and a decrease in heat exchange across the skin of the flippers contribute to maintenance of a modest thermal gradient (1°C–2.3°C) in $T_W$ as low as 16°C (Bostrom et al., 2010). Thus far, consideration of the role of metabolic heat in generating large thermal gradients between leatherback turtles and ambient $T_W$ has focused on heat produced in skeletal muscle as a result of locomotion. The contribution of the metabolic heat generated through digestive processes in leatherbacks has not been investigated, but might play a significant role in the thermal strategy of leatherback turtles at their high-latitude foraging grounds.

A very different problem for leatherback turtle thermoregulation is how to prevent overheating during the breeding season in tropical seas. According to the Paladino et al. (1990) model, increased peripheral blood flow would be an effective means to facilitate heat transfer and regulate body temperature in warm environments. Behavioral mechanisms of thermoregulation, such as changes in dive patterns, may act in tandem with circulatory adjustments to insure efficient heat transfer. Southwood et al. (2005) noted a significant negative correlation between dive depth and $T_B$ for female leatherback turtles during the internesting interval offshore from Playa Grande, Costa Rica; $T_B$ during prolonged periods of dives to cool waters at depths of 40–60 m was lower than $T_B$ during prolonged periods of dives to warm waters at shallow depths (Figure 1.4). Incidences of relatively rapid decreases in $T_B$ of leatherback turtles that were resting on the seafloor were noted in this study (Southwood et al., 2005; Bostrom et al., 2006).

Data derived from laboratory studies, field observations, and modeling approaches have greatly expanded our knowledge of thermoregulation in leatherback turtles. The many exciting advances in this field have demonstrated that leatherback turtles use a combination of physiological adjustments, morphological features, and behavioral modifications to stay warm in cool water and prevent overheating in the tropics.

### 1.2.4 FIELD MEASUREMENTS OF METABOLIC RATE

The study of energetics plays a central role in understanding how physiological functioning on the organismal level translates into impacts at the population level. For example, energy acquisition and allocation to various physiological tasks play a critical role in determining growth rates and reproductive outputs, life history traits that have clear implications for demography and population dynamics.

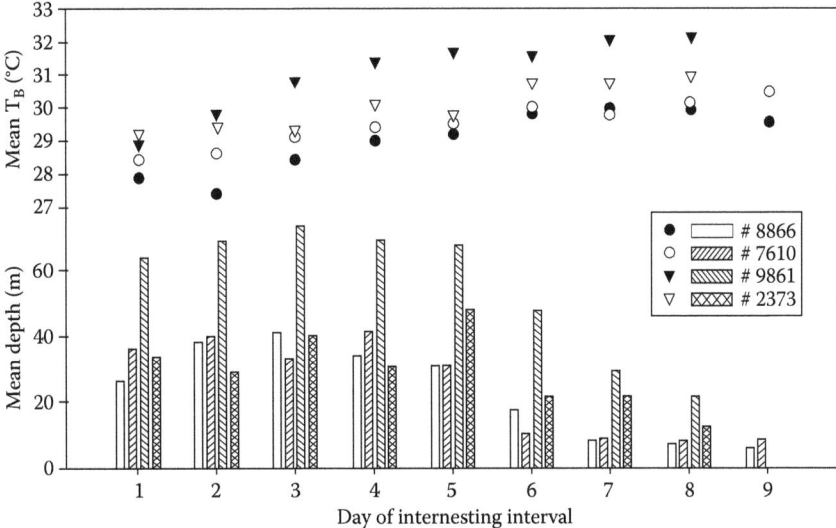

**FIGURE 1.4**   Mean dive depth and mean body temperature ($T_B$) for each day of the internesting interval for four leatherback turtles. There was a significant relationship between mean daily $T_B$ and dive depth for each turtle. Changes in dive behavior may play a role in thermoregulation for leatherback turtles nesting in tropical seas. (From Southwood, A.L. et al., *Physiol. Biochem. Zool.*, 78(2), 291, 2005.)

Figure 1.1 demonstrates the connections between environmental factors, the acquisition and allocation of chemical energy, and animal behavior. As awareness of the interplay between energetics of individuals and population-level phenomena has increased, there has been a push to expand investigations of metabolic physiology to include sea turtles in their natural environment.

Doubly labeled water has been used to estimate field metabolic rates and address questions regarding energy allocation and environmental impacts on sea turtle energetics at various stages of life history. This technique involves injecting the study animal with water enriched with stable isotopes of hydrogen ($^2H$ or $^3H$) and oxygen ($^{18}O$), and subsequently monitoring the decrease in the amount of these stable isotopes in the animal's body water over time. The hydrogen isotope is lost due to water turnover (e.g., respiration, evaporation, excretion of waste products), whereas the oxygen isotope is lost due to water turnover and incorporation into metabolically produced $CO_2$. The difference in washout curves of the two isotopes provides a measure of $CO_2$ production and, therefore, metabolic rate (Speakman, 1997).

Clusella Trullas et al. (2006) employed DLW to investigate differences in the energetic cost of various locomotory activities associated with hatchling dispersal, and documented that the energetic costs of digging and crawling on land were four to five times higher than RMR. Surprisingly, given the low cost of aquatic locomotion compared with terrestrial locomotion, the FMR for swimming hatchlings was seven times higher than RMR.

The DLW technique has also been used to document seasonal shifts in at-sea FMR in immature green turtles (9.8–23.8 kg) resident at Heron Island, Australia (Southwood et al., 2006). Although the mean winter FMR (1.00 W kg$^{-1}$ at $\overline{X}T_W = 21.4°C$) was 43% lower than mean summer FMR (1.70 W kg$^{-1}$ at $\overline{X}T_W = 25.8°C$) at this site, the difference in FMR between seasons was not statistically significant. There was no significant difference in water flux between seasons, which suggests that turtles were still foraging during the relatively mild winter experienced during the course of this study. Turtles significantly altered dive patterns between seasons, so the mean percentage of time spent in shallow water along the reef flats in winter (64.3%) was double that in summer (34.5%) (Southwood et al., 2003, 2006). Increased utilization of reef flat habitats in the winter may reflect a thermal preference, a shift in foraging habitat, or a means of avoiding predators given the turtle's limited metabolic capacity at colder temperatures. Integration of FMR data with dive records and measurements of temperature effects on metabolic enzyme activity led Southwood et al. (2006) to conclude that multiple biotic and abiotic environmental factors act in conjunction with temperature to elicit seasonal changes in energetics of sea turtles, and that temperature alone cannot fully explain seasonal shifts in metabolism.

Wallace et al. (2005) used DLW to investigate energetics of female leatherback turtles during the internesting interval. When compared over a similar range of temperatures, FMRs for adult female leatherback turtles at sea (0.40 W kg$^{-1}$) were comparable to metabolic rates for turtles resting on the beach (0.36–0.40 W kg$^{-1}$; Paladino et al., 1990; Lutcavage et al., 1992), and considerably lower than metabolic rates for turtles engaged in crawling or nest-covering behavior on the beach (1.12–1.51 W kg$^{-1}$; Paladino et al., 1990, 1996). Dive records for leatherback turtles injected with DLW illustrated a pattern of short, shallow dives interspersed with deeper U-dives to the seafloor, presumably to rest in cooler waters. Wallace et al. (2005) concluded that low levels of activity and utilization of cool waters at depth allowed leatherbacks to conserve energy and allocate a high proportion of their energy budget to reproductive activities during the internesting period. Efficient use of energy reserves is particularly important for nesting female sea turtles as they are typically capital breeders (but see Hochscheid et al., 1999; Rosette et al., 2008) and foraging opportunities are limited during breeding season in tropical waters (Casey et al., 2010).

The use of DLW to estimate field metabolic rates of sea turtles has some significant drawbacks, including the high cost of isotopes, potential for error in estimates due to high rates of water turn-over in aquatic species, and the fact that FMR estimates provide just a single mean value for metabolic rate over a given period of time instead of detailed information on metabolic rates associated with specific activities or physiological states (Speakman, 1997; Jones et al., 2009). A promising alternative to DLW is the use of data derived from tri-axial acceleration data loggers to estimate energy expenditure for freely swimming sea turtles. Laboratory studies have demonstrated a strong correlation between $\dot{V}O_2$ and overall dynamic body acceleration in hatchling loggerhead turtles (Halsey et al., 2011) and immature and adult green turtles (Enstipp et al., 2011; Halsey et al., 2011). These laboratory validations of accelerometry as a means to estimate metabolic rate open the door to more detailed studies of the energetics of sea turtles in their natural environment.

## 1.3  PHYSIOLOGY OF DIVING

Given the central importance of aerobic metabolism for routine and sustained activity in sea turtles, it is worth considering how they manage and utilize $O_2$ stores while diving. Instruments to remotely monitor dive behavior (e.g., data loggers and satellite telemeters) have been deployed on all seven species of sea turtles, and the data gained through these studies demonstrate that sea turtles may spend over 90% of time at sea submerged below the water surface with no access to air (Lutcavage and Lutz, 1997). A summary of typical dive durations and dive depths for freely swimming, active sea turtles of various age classes is provided in Table 1.2. Data are limited to turtles exhibiting an active dive pattern (i.e., overwintering turtles are not included), and are meant to give a broad representation of diving behavior and a framework in which to consider the physiological traits associated with diving in sea turtles.

Air-breathing diving vertebrates initiate dives with a finite store of $O_2$ in their system that must last for the entire duration of submergence if the animal is to sustain aerobic metabolism. A shift to heavy reliance on anaerobic metabolism could lead to accumulation of lactate and disruptions in dive patterns, as the animal may need to spend extended periods of time at the surface between dives to clear lactate from its system (Kooyman et al., 1980). Foraging, mating, and other intra- and interspecific interactions take place underwater, and preferential use of aerobic metabolic pathways allows diving animals to maximize the amount of time spent submerged. For aerobic dives, the amount of $O_2$ stored in the lungs and tissues and the rate of $O_2$ utilization (i.e., the metabolic rate) during the dive are important determinants of dive duration, and these aspects of diving physiology have been studied in several species of sea turtle. Mass-specific $O_2$ storage capacity in sea turtles

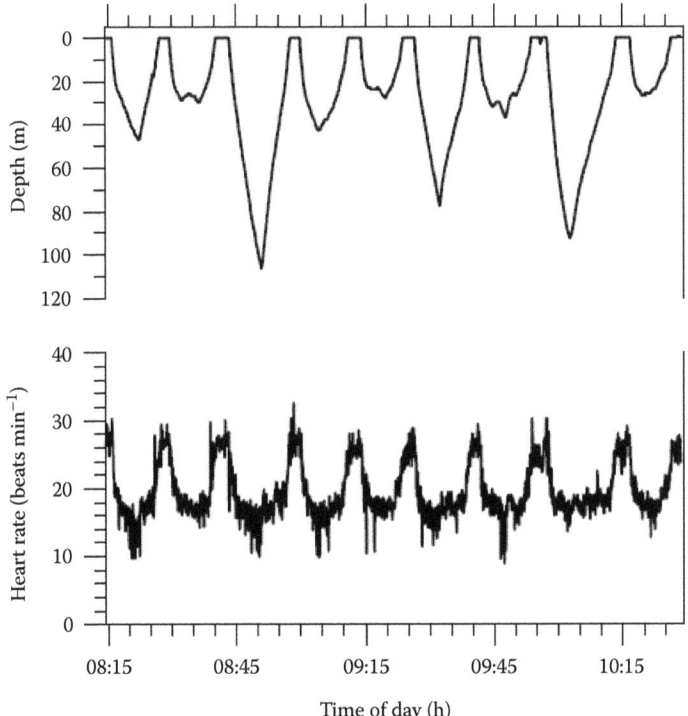

**FIGURE 1.5** Dive trace and corresponding heart rate trace for a leatherback turtle freely swimming at sea during the internesting interval. Dive heart rates are approximately 25%–30% lower than surface heart rates. Cardiovascular adjustments during diving play a role in management and efficient use of on-board $O_2$ stores. (From Southwood, A.L. et al., *J. Exp. Biol.*, 202, 1115, 1999.)

is similar to that observed in other species of reptiles, and cardiovascular adjustments made during submergence allow sea turtles to use these $O_2$ stores efficiently (Lutcavage and Lutz, 1997). Like many other species of air-breathing diving vertebrates, dive heart rates are lower than surface heart rates in sea turtles. Adjustments in heart rate signify alterations in blood flow to manage and conserve $O_2$ stores (Figure 1.5; Davenport et al., 1982; West et al., 1982; Butler et al., 1984; Southwood et al., 1999, 2003). Lutcavage and Lutz (1997) provide a detailed review of cardiorespiratory adaptations of sea turtles that facilitate $O_2$ uptake at the surface and delivery of $O_2$ to tissues during dives. This chapter will provide an update on the current state of knowledge regarding diving capacity and impacts of enforced submergence on sea turtle physiology.

### 1.3.1 Aerobic Dive Limit and Dive Capacity

The term "aerobic dive limit" (ADL) refers to the maximum amount of time that an animal may remain submerged without an increase in blood lactate levels (Kooyman, 1983). Given the difficulty of obtaining blood samples from freely diving animals, ADL has rarely been measured under field conditions. A more common approach is to estimate the maximum dive duration that could be supported by aerobic metabolism based on the total usable $O_2$ stores in the lungs, blood, and muscle tissue and the rate of $O_2$ utilization during the dive ($\dot{V}O_2$). The calculated aerobic dive limit (cADL = $O_2$ stores/$\dot{V}O_2$) is often used to evaluate the theoretical maximum aerobic dive time for a given species (Butler and Jones, 1997), and assumes that a dive must be terminated at the point when $O_2$ stores are fully depleted as $O_2$-sensitive tissues, such as the heart and brain, cannot tolerate prolonged anoxia. Estimates of cADL clearly have limitations in applicability as field recordings of dive durations in many species of birds and mammals demonstrate that the cADL is routinely exceeded (Kooyman et al., 1980; Ponganis et al., 1997; Costa et al., 2001; Butler, 2006). This is likely due to the fact that the cADL does not account for the effects of preferential perfusion to $O_2$-sensitive tissues during a dive and regional reliance on anaerobic metabolism in tissues that are less perfused. In other words, diving animals may utilize both aerobic and anaerobic metabolism to varying degrees in different tissues during the dive, and it is not simply a matter of switching from complete reliance on aerobic pathways to complete reliance on anaerobic pathways when $O_2$ stores run out. This is an important point when considering the use of cADL to predict the diving capabilities of sea turtles, given their extraordinary capacity for anaerobic metabolism and ability to tolerate hypoxia (Berkson, 1966; Lutz et al., 1980; see Section 1.2.2.2).

Given this caveat, values for cADLs have been reported for leatherback and loggerhead turtles, and an equation relating cADL to body mass for loggerhead, flatback, and hawksbill turtles has been published (cADL = 6.956 $M^{0.593}$; Hochscheid et al., 2007). The cADLs reported for adult female leatherback turtles (200–300 kg) all use beach-derived measurements of $O_2$ stores in the numerator (Lutcavage et al., 1990, 1992), but greatly vary depending on what measure of metabolic rate is used in the denominator. Estimates of cADL as low as 5 min and as high as 70 min have been reported for this species, but most estimates fall within the middle of this range (Lutcavage et al., 1992; Southwood et al., 1999; Wallace et al., 2005; Bradshaw et al., 2007). Calculated aerobic dive limits derived from estimates of $\dot{V}O_2$ based on heart rate of diving leatherback turtles range from 33 to 67 min (Southwood et al., 1999), whereas cADL estimates derived from field metabolic rates of turtles over the course of the internesting interval range from 11.7 to 44.3 min (Wallace et al., 2005). Bradshaw et al. (2007) used estimates of $\dot{V}O_2$ based on dive times and lung capacities to derive a cADL of 19.2–48.1 min. As seen with diving birds and mammals, cases in which dive duration exceeds the cADL have been documented for leatherback turtles. Maximum dive durations of 83.3–86.5 min have been reported for adult female leatherback turtles (Fossette et al., 2008; Lopez-Mendilaharsu et al., 2008); these dive durations are well in excess of the maximum cADL estimate of 70 min for this species. Without measurements of blood lactate, it is difficult to assess the degree to which turtles resorted to anaerobic metabolism during these extended dives.

Estimates of cADL for loggerhead turtles vary depending on environmental temperature. Using literature values for $O_2$ stores and calculations of $\dot{V}O_2$ based on a previously derived equation that accounts for temperature and mass effects (Hochscheid et al., 2004), Hochscheid et al. (2005) determined that the cADL for a 52 kg loggerhead turtle increased by almost sevenfold as average sea surface temperature in the Mediterranean Sea decreased from a high of 26°C in July (cADL = 63 min) to low of 15°C in February (cADL = 427 min) (Figure 1.6). Dive durations recorded from this turtle using a satellite-relayed data logger were generally less than the cADL throughout the monitoring period, but there were isolated instances in August and October when the cADL was exceeded (Figure 1.6; Hochscheid et al., 2005). Results from this study highlight how environmental effects on metabolism and physiology translate into behavioral adjustments at the whole animal level.

In addition to temperature, body size may also have an effect on the diving capacity of sea turtles. As animals get larger, the capacity for $O_2$ storage increases to a greater degree than does metabolic rate (Schreer and Kovacs, 1997; Halsey et al., 2006a,b). This is illustrated by analyses of the relationships between maximum lung volumes ($V_L$ in mL) and mass and metabolic rate and mass in sea turtles. The allometric relationship between $V_L$ and mass has an exponent of 0.92 (Hochscheid et al., 2007), whereas estimates of the mass scaling exponent for whole animal metabolic rates are considerably lower (0.793–0.831, derived from mass-specific metabolic rates reported in Wallace and Jones, 2008). This trend has been noted in a wide variety of endothermic taxa, and is thought to contribute to increased dive capacity in larger animals compared with smaller animals. In theory, $O_2$ stores in larger animals will last longer, thus permitting longer dive durations and dives to deeper depth. Sea turtle appear to follow this general trend, although the relationship between dive durations and body size in sea turtles, and in reptiles in general, is much weaker than that observed in endothermic divers (Brischoux et al., 2007; Hochscheid et al., 2007). The assumption that there should be a strong relationship between dive duration and mass

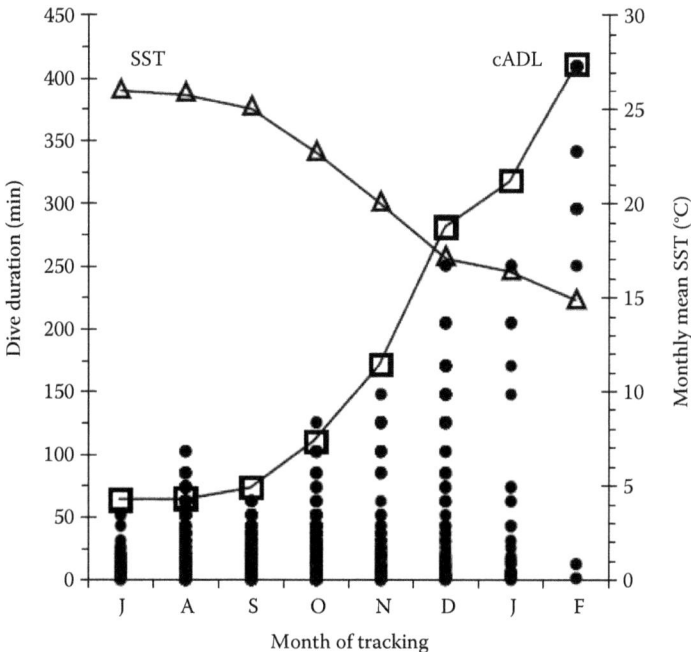

**FIGURE 1.6** An illustration of how calculated aerobic dive limit (cADL, open squares) changes with seasonal changes in sea surface temperature (SST, open triangles) for loggerhead turtles. The filled circles are dive durations recorded from a freely swimming loggerhead turtle. Dive durations rarely exceeded the cADL during this study. (From Hochscheid, S. et al., *Biol. Lett.*, 1, 82, 2005.)

is based on the idea that dive duration is dictated primarily by $O_2$ stores and $O_2$ utilization. Perhaps the weak relationship between these variables observed in reptiles is due in part to their enhanced capacity for anaerobic glycolysis and high tolerance for anaerobic byproducts (Berkson, 1966; Jackson, 2000; Brischoux et al., 2007).

The balance between aerobic and anaerobic metabolism during submergence and the levels of blood and tissue lactate that sea turtles tolerate without altering dive patterns are topics worthy of further investigation. Turtles, in general, have a high blood and tissue buffering capacity (Jackson, 2000), so it is not unreasonable to assume that sea turtles could maintain a consistent dive pattern with accumulation of low to moderate levels of lactate. Furthermore, while $O_2$ is necessary to metabolize lactate, either through oxidation or by conversion to glycogen in the liver or muscle, it is possible that sea turtles could use their on-board $O_2$ stores to metabolize lactate while they are submerged rather than being restricted to the surface. This assertion is supported by studies with painted turtles (*Chrysemys picta bellii*) in which the rate of lactate metabolism following a bout of anoxia was no different for intermittently breathing turtles engaged in sustained exercise than for resting turtles (Warren and Jackson, 2004). A similar principle may apply to metabolism of lactate accumulated during dives made by sea turtles.

### 1.3.2 FORCED SUBMERGENCE

Although sea turtles rely primarily on aerobic metabolism during routine dives under natural conditions, forced (i.e., involuntary) submergence generally elicits a shift to anaerobic metabolism. The early experiments of Berkson (1966) showed that green turtles (size not specified, but presumably adults) could withstand up to 5 h of forced submergence at 18°C–23°C and fully recover upon 15 h exposure to air. Measurements of tracheal air during prolonged dives showed that $O_2$ levels dropped below 1%, and arterial blood $O_2$ dropped to near zero during a 6 h breath hold. Blood lactic acid during prolonged dives increased slowly during the first 60 min, suggesting sequestration of lactic acid in ischemic tissue, and spiked to approximately 18 mmol $L^{-1}$ upon emergence from the dive. Immature loggerhead turtles ($\leq 20$ kg) force dived for up to 90 min at 22°C depleted blood $O_2$ stores by the end of the dive and showed a steady increase in blood lactate throughout the dive and during recovery (Lutz and Bentley, 1985) to a maximum of 20–25 mmol $L^{-1}$ 30 min post-dive. Blood lactate returned to pre-dive levels within 24 h of emergence.

These forced dive studies demonstrate an amazing capacity for anoxia tolerance in green and loggerhead turtles, but the artificial conditions used in these experiments do not necessarily translate into natural diving behavior. The most likely scenario in which sea turtles would experience prolonged anoxia under natural conditions would be during the purported lengthy dives of days to weeks associated with seasonal burial in the seafloor substrate. Unfortunately, the anecdotal reports of winter dormancy in these species (Felger et al., 1976; Carr et al., 1980) have not yet been substantiated by at-sea recordings of overwintering behavior (Godley et al., 2002; Hochscheid et al., 2005, 2007), nor have investigators been able to trigger dormant behavior under laboratory conditions in order to study metabolic downregulation (Moon et al., 1997; Southwood et al., 2003; Hochscheid et al., 2004).

Situations in which wild sea turtles are forcibly submerged due to entrapment in fishing gear suggest that the behavioral and physiological responses are drastically different from what has been recorded under controlled laboratory conditions. Blood biochemistry profiles for immature and adult Kemp's ridley turtles subjected to capture in experimental shrimp trawls at 27°C show significant disruptions in blood biochemistry and blood acidosis in just 2–7 min (Stabenau et al., 1991a). Blood lactate values were 10.2 mmol $L^{-1}$ for Kemp's ridley turtles post-trawl compared with pre-trawl values of 1.7 mmol $L^{-1}$. Similarly, capture in shrimp trawls with tow time up to 30 min in $T_W$ of 25°C–30°C resulted in blood lactate levels of 8.5–20.0 mmol $L^{-1}$ in immature and adult loggerhead turtles (Harms et al., 2003). The average blood lactate level in immature green turtles entangled for 30–240 min in shallow set gillnets ($T_W$ of 26°C–32°C) was 30.6 mmol $L^{-1}$, and the maximum blood lactate observed was 50.6 mmol $L^{-1}$,over twice as high as any lactate values reported for trawl-captured turtles or even turtles subjected to a laboratory forced dive protocol (Berkson, 1966; Lutz and Bentley, 1985; Lutz and Dunbar-Cooper, 1987; Harms et al., 2003; Stabenau and Vietti, 2003; Snoddy et al., 2008). The point at which blood lactate accumulation begins to affect diving patterns is not well understood, but evidence from satellite telemetry studies of green turtles and Kemp's ridley turtles released from gillnets suggests that turtles spent more time at the surface during the first 24 h post-release (Snoddy et al., 2010). Data from these studies suggest that entanglement in fishing gear results in physiological disruptions, the severity of which depends on the length of time that the turtle is forcibly submerged and the degree to which the turtle struggles while entangled. Even if turtles are released alive from nets, disturbances in blood biochemistry may cause alterations in post-release behavior and potentially impact survival (see Section 1.5).

## 1.4   PHYSIOLOGY OF WATER AND SALT BALANCE

Among the reptiles, a very limited number of species exclusively inhabit the marine realm. Those that do are faced with the challenge of maintaining a body fluid composition that is appropriate for optimal physiological functioning despite the desiccating effects of their salty environment. Sea turtle body fluids are hyposmotic to seawater, meaning the concentration of solutes in their body fluids are lower than the concentration of solutes in seawater. Consequently, they must invest energy and make regulatory adjustments to minimize water loss and combat excessive influx of salts if they are to maintain a steady state with regard to body fluid composition.

### 1.4.1   OSMOTIC AND IONIC HOMEOSTASIS

The osmotic concentration of sea turtle body fluids (~300 mosmol $kg^{-1}$) is about 1/3 the osmotic concentration of seawater (Nicolson and Lutz, 1989; Reina and Cooper, 2000). The structure and composition of sea turtle integument is such that the rate of cutaneous water loss is likely very low, and respiratory water loss is assumed to be negligible given the high humidity of air at the water surface (Bentley, 1976; Lillywhite and Maderson, 1982). The primary osmoregulatory challenge for sea turtles is presented by the large salt loads incurred while feeding. The diet of most sea turtles includes marine animals, plants or algae that are isosmotic with the surrounding seawater, and they may also ingest seawater while foraging. The salt loads ingested during feeding or by direct ingestion of seawater must be excreted in order to maintain osmotic homeostasis of body fluids.

The reptile kidney plays an important role in ionic and acid–base regulation, but it is incapable of generating hyperosmotic urine to excrete excess salts (Dantzler, 1976). As is the case with marine birds and teleost fish, marine reptiles have extra-renal mechanisms for salt excretion that contribute to regulation of body fluid osmolality. Cephalic salt glands have arisen independently in diverse reptiles that exploit high-salinity environments (Schmidt-Nielsen and Fänge, 1958). Salt excretion is facilitated in marine iguanas (*Amblyrhynchus cristatus*) by means of nasal salt glands (Shoemaker and Nagy, 1984), in marine and estuarine snakes (*Pelamis* spp., *Laticauda* spp., *Acrochordus granulatus*) by means of a sublingual gland (Dunson, 1979), in estuarine crocodiles (*Crocodylus porosus*) by means of lingual salt glands (Taplin and Grigg, 1981), and in estuarine and marine turtles by means of modified lachrymal glands (Holmes and McBean, 1964; Dunson, 1969).

## 1.4.2 Salt Gland Function

The integral importance of salt glands for exploitation of the marine environment by turtles is indicated in the fossil record, with anatomical evidence of salt glands pre-dating the presence of paddle-like flippers in early marine turtles (Hirayama, 1998). The large, lobular salt glands of sea turtles lie adjacent to the eye orbits and are capable of secreting tears with osmotic concentration as high as 2000 mosmol·kg$^{-1}$, or twice the concentration of seawater (Nicolson and Lutz, 1989). The impressive secretory capacity of sea turtle salt glands makes possible the excretion of excess salts ingested with food, and also permits sea turtles to obtain osmotically free water by directly drinking seawater (Holmes and McBean, 1964; Bennett et al., 1986; Marshall and Cooper, 1988). Evidence for the latter is provided by observations that unfed hatchling sea turtles placed in seawater gain weight whereas unfed hatchlings that are kept dry lose weight (Bentley et al., 1986; Marshall and Cooper, 1988; Reina et al., 2002), and the high rates of water turnover documented using isotopic techniques with freely swimming immature and adult sea turtles (Wallace et al., 2005; Southwood et al., 2006).

Variation in diet of different species of sea turtles may be reflected in secretory capacity of the salt glands. For example, leatherback turtles eat large quantities of jellyfish that are isosmotic with seawater and incur large salt loads relative to other sea turtle species. The osmolality of leatherback turtle (100–300 g) salt gland secretions is 136% higher than secretions of loggerhead (100 g) and hawksbill turtles (2300 g) and 70% higher than secretions of green turtles (450 g) (Hudson and Lutz, 1986).

Activation of salt glands results in an increase in both tear production and the osmotic concentration of tears (Marshall and Cooper, 1988; Nicolson and Lutz, 1989; Reina and Cooper, 2000). The major ionic components of tears are Na$^+$ and Cl$^-$, with other ions (K$^+$, Ca$^{2+}$, Mg$^{2+}$) present in smaller, but significant, amounts (Nicolson and Lutz, 1989). The ionic composition of tears in relation to seawater provides evidence that the salt gland plays an important role in both ionic and osmotic regulation in sea turtles (Lutz, 1997). Salt glands are activated specifically in response to increased levels of plasma sodium (Na$^+$) (Marshall and Cooper, 1988; Reina and Cooper, 2000), rather than an increase in blood volume or osmotic pressure. The mechanisms underlying Na$^+$ detection and the signal transduction pathways leading to salt gland activation have not been thoroughly investigated in sea turtles, but presumably there are similarities with avian salt glands and salt glands of other marine reptiles. In birds, an increase in plasma Na$^+$ concentration triggers cephalic osmoreceptors (Gertsberger et al., 1984), which leads to stimulation of the salt glands via the autonomic nervous system (Fange et al., 1958). Methacholine, a cholinergic agonist, has been shown to stimulate salt gland secretions in estuarine terrapins (*Malaclemys terrapin*) and crocodiles (*Crocodylus porosus*) (Dunson, 1970; Taplin et al., 1982), but conflicting reports of cholinergic effects on sea turtle salt gland activity exist in the literature. Methacholine stimulated tear production in the loggerhead turtle (Schmidt-Nielsen and Fange, 1958), but Reina and Cooper (2000) found an inhibitory effect of methacholine in hatchling green turtles and Reina et al. (2002) found highly variable effects of methacholine on salt gland secretion in leatherback turtles. Variation in study results may be due to ontogenetic or interspecific differences. Holmes and McBean (1964) provided indirect evidence that corticosterone may stimulate salt gland secretion by simulating adrenalectomy with amphenone B, but the results of the study were not entirely conclusive and the role of hormones in salt gland regulation requires further study.

The mechanisms underlying generation of highly concentrated salt gland secretions in sea turtles have not been described in detail, but many aspects of the secretory process may be inferred from comparative anatomical studies and the composition of the secreted fluid. Sea turtle salt glands consist of numerous lobules, each containing a series of blind-ended secretory tubules that drain into a central canal. The central canals of the lobules drain into secondary ducts, which then drain into a main duct that terminates at the posterior corner of the eye (Marshall and Saddlier, 1989). Studies by Marshall (1989) indicate that secretions in sea turtle lachrymal glands are generated using cellular mechanisms similar to those observed in shark rectal glands. Measurements of ion concentrations in cells lining the secretory tubules suggest that active transport mechanisms, likely involving $Na^+$–$K^+$ ATPase pumps and $Na^+$–$2Cl^-$–$K^+$ co-transporters, are employed to secrete ions into the lumen of the tubules during salt gland activation (Marshall, 1989). The primary secretion in the secretory tubules is isosmotic with blood, and the secretion may be concentrated, depending on salt burden and osmotic needs, as it travels through the duct system of the glands. The cells lining the central canals have a high concentration of mitochondria and complex interdigitations between adjacent cells (Marshall and Saddlier 1989), characteristics that suggest these cells may be involved in ATP-dependent ion transport processes and play a role in tear modification. Further modification of tears may occur in the secondary ducts and main ducts, which are well vascularized (Schmidt-Nielsen and Fange, 1958; Marshall and Saddlier, 1989). The metabolic costs associated with salt gland operation are not well studied, but this is a good topic for future research as it would provide valuable information that could be used to refine energy budgets of sea turtles.

The salt gland plays a critical role in maintenance of osmotic homeostasis, and damage to or malfunction of the salt gland can have severe repercussions for sea turtles. Oros et al. (2011) documented high plasma NaCl concentrations and severe salt gland lesions, characterized by inflammation and bacterial cultures, in nine sea turtles that stranded along the coast of the Canary Islands and subsequently died after 2–12 days at a rehabilitation facility. No other gross or histological lesions were observed in these turtles, and death was attributed to salt gland adenitis. Future studies of mechanistic and metabolic aspects of salt gland function should provide useful data for assessing the implications of osmotic imbalance for the health status of sea turtles.

## 1.5 ROLE OF PHYSIOLOGY IN SEA TURTLE CONSERVATION

Physiological studies provide insight into the resiliency of individuals and populations to changes in environmental conditions (Carey, 2005; Tracy et al., 2006; Wikelski and Cooke, 2006; Chown and Gaston, 2008). Sea turtles have a long evolutionary history (180–150 Mya; Poloczanska et al., 2009), yet there has been a dramatic decline in sea turtle populations over the past several decades and all seven species of sea turtles are currently listed as threatened or endangered (www.iucnredlist.org). This chapter highlights aspects of sea turtle physiology relevant to their ability to exploit the marine realm, where they face numerous threats such as incidental capture in fisheries, oil pollution, marine debris, and alterations in abiotic and biotic environmental conditions due to climate change. The following case studies highlight the importance of integrating physiological data with behavioral, ecological, and demographic data to provide the most comprehensive picture of how sea turtles are impacted by alterations to their environment. Data collected through integrative studies may be used by biologists and regulatory agencies to manage and conserve sea turtle populations.

A topic of considerable concern in sea turtle management and conservation is mortality of sea turtles due to incidental capture in fishing gear, and the impact of fisheries-related mortalities on sea turtle populations. As discussed in Section 1.3.2, entanglement in fishing gear has significant effects on the physiology of sea turtles (Lutz and Dunbar-Cooper, 1987; Harms et al., 2003; Stabenau and Vietti, 2003; Snoddy et al., 2008, 2010). The long-term effects of capture-related metabolic disruption on behavior and, ultimately, survivability of sea turtles could be substantial, but to date there has been little effort to integrate physiological data into estimates of mortality. Studies that combine information on physiological status, post-release behavior, and observer and stranding data may be useful in refining estimates of post-release mortality for sea turtles captured in fishing gear. This approach has been used with some success to infer the fate of sharks incidentally captured in longline fishing gear (Moyes et al., 2006; Hight et al., 2007), and shows promise for use with sea turtles (Snoddy et al., 2010). Accurate data on both in-net and post-release mortality are crucially important for refining the current mortality estimates used to govern management decisions with far-reaching conservation, economic, and social consequences.

The Deepwater Horizon oil spill in the Gulf of Mexico on April 20, 2010, dramatically underscored the issue of ocean pollution and its impacts on the health of humans and animals. Approximately 4.9 million barrels (~205 million gallons) of oil leaked into the Gulf of Mexico after the fire and subsequent explosion on the Deepwater Horizon oil drilling platform (Antonio et al., 2011). Over the course of 170 days following the oil spill, a total of 1135 sea turtles stranded or were directly captured as part of the mitigation response to the spill (http://www.nmfs.noaa.gov/pr/health/oilspill/turtles.htm). Fifty-three percent of these turtles were dead, and 42% of all turtles combined (alive and dead) showed direct evidence of oil contamination. Of the turtles that were found alive, 85% were oiled upon discovery. The numerous physiological impacts of oil exposure on sea turtles were reviewed by Lutcavage and Lutz (1997). Exposure of skin to oil may cause dermatitis, irritation of mucous membranes, lesions that could lead to secondary infections, and blockage of tear ducts that may lead to interference with salt gland function (Lutcavage et al., 1995; Lutcavage and Lutz, 1997). Furthermore, oil that is ingested may interfere with digestion and inhalation of petroleum vapors may cause respiratory and metabolic disruptions (Lutcavage and Lutz, 1997). Knowledge of the impacts of oil exposure on sea turtle physiology and behavior is critical for efforts to rehabilitate oiled turtles. Further research into the physiological effects of exposure to the dispersants used to break up surface oil slicks is also warranted, given the wide-scale use of these chemicals in response to the Deepwater Horizon spill.

The damaging effects of marine debris ingestion by sea turtles have been recognized for many years (Carr, 1987; Bjorndal et al., 1994), but this topic has received more attention recently due to increasing awareness of the magnitude of regional debris accumulation in the oceans (Kaiser, 2010). Plastic, nondegradable debris ingested by sea turtles may decrease gut function through mechanical blockage or damage to tissues, and may impact nutritional status of turtles through dietary dilution (Bjorndal et al., 1994; McCauley and Bjorndal, 1999). There is also concern that absorption and accumulation of polychlorinated biphenyls (PCBs) from ingested plastics could lead to sublethal effects on metabolic and endocrine function that could affect reproductive behaviors and success of marine vertebrates (Derraik, 2002). This aspect of marine debris ingestion is a worthy topic of future research, as alterations in reproductive parameters may have important consequences for population demography.

The continuing rise of average global temperature is predicted to impact sea surface temperatures, sea level, surface current patterns, and ocean pH (Hansen et al., 2006; IPCC 2007; Rahmstorf et al., 2007), and these alterations in the oceanic and coastal environments may affect sea turtle distribution, migration patterns, reproductive behaviors, and energetics (Hawkes et al., 2009; Poloczanska et al., 2009; Witt et al., 2010). Consideration of temperature effects on metabolism and performance in sea turtles may allow us to better predict the responses of sea turtles to climate change. Additionally, information on thermal physiology may contribute to rehabilitation and treatment of cold-stunned sea turtles. Cold stunning is a phenomenon in which sea turtles exposed to rapid declines in $T_W$ become inactive, cease swimming and diving, and often strand. Blood biochemistry analyses indicate that cold-stunned immature Kemp's ridley turtles present with metabolic and respiratory acidosis, hypocalcemia, and hypermagnesemia (Innis et al., 2007). Cold-stunned immature green turtles also showed significant alterations in blood biochemistry, particularly with regard to glucose, cations, proteins, uric acid, and blood urea nitrogen (Anderson et al., 2011). Additional information on metabolic status of live stranded cold-stunned sea turtles could help clarify the effects of acute cold exposure on sea turtle physiology and behavior, and identify the mechanistic explanations for cold-stunning behavior. This information is important for refining care provided to cold-stunned turtles during rehabilitation.

## 1.6  CONCLUSIONS

Studies of metabolism and physiology in sea turtles have yielded important insight regarding specializations for the marine environment and responses to changing environmental conditions. Function at the organismal level has implications for population-level processes, such as birth rate, mortality rate, immigration, and emigration (Dunham et al., 1989), which influence the demographic characteristics of a population. For threatened and endangered species of sea turtles, an integrative understanding of their biology that encompasses physiology, behavior, and ecology is critical for crafting effective management and conservation strategies.

## ACKNOWLEDGMENTS

Special thanks to the dedicated laboratory and field researchers who have greatly expanded our knowledge of sea turtle physiology over the past several decades, and to those who will build on this knowledge in the future. I also thank Dr. Kenneth Lohmann and an anonymous reviewer for helpful comments that improved this manuscript.

## REFERENCES

Anderson, E.T., Harms, C.A., Stringer, E.M., Cluse, W.M. 2011. Evaluation of hematology and serum biochemistry of cold-stunned green sea turtles (*Chelonia mydas*) in North Carolina, USA. *J. Zoo. Wildlife Med.* 42, 247–255.

Antonio, F.J., Mendes, R.S., Thomaz, S.M. 2011. Identifying and modeling patterns of tetrapod vertebrate mortality rates in the Gulf of Mexico oil spill. *Aquat. Toxicol.* 105, 177–179.

Baldwin, J., Gyuris, E., Mortimer, K., Patak, A. 1989. Anaerobic metabolism during dispersal of green and loggerhead turtle hatchlings. *Comp. Biochem. Physiol.* 94A, 663–665.

Bennett, A.F. 1972. A comparison of the activities of metabolic enzymes in lizards and rats. *Comp. Biochem. Physiol.* 42B, 637–647.

Bennett, A.F. 1982. The energetics of reptilian activity, in: C. Gans, F.H. Pough (Eds.), *Biology of the Reptilia.* Academic Press, New York, pp. 155–199.

Bennett, A.F., Dawson, W.R. 1976. Metabolism, in: C. Gans, W.R. Dawson (Eds.), *The Biology of Reptilia.* Academic Press, New York, pp. 127–223.

Bennett, J.M., Taplin, L.E., Grigg, G.C. 1986. Seawater drinking as a homeostatic response to dehydration in hatchling loggerhead turtles *Caretta caretta. Comp. Biochem. Physiol.* 83A, 507–513.

Bentley, P. 1976. Osmoregulation, in: C. Gans, W. Dawson (Eds.), *Biology of the Reptilia.* Academic Press, New York, pp. 365–412.

Berkson, H. 1966. Physiological adjustments to prolonged diving in the Pacific green turtle (*Chelonia mydas agassizii*). *Comp. Biochem. Physiol.* 18, 101–119.

Bishop, C.M. 1999. The maximum oxygen consumption and aerobic scope of birds and mammals: Getting to the heart of the matter. *Proc. R. Soc. Lond. B.* 266, 2275–2281.

Bjorndal, K.A., Bolten, A.B., Lagueux, C.J. 1994. Ingestion of marine debris by juvenile green sea turtles in coastal Florida habitats. *Mar. Poll. Bull.* 28(3), 154–158.

Bleakney, J.S. 1965. Reports of marine turtles from New England and eastern Canada. *Can. Field Nat.* 79, 953–972.

Bostrom, B.L., Jones, D.R. 2007. Exercise warms adult leatherback turtles. *Comp. Biochem. Physiol. A* 147, 323–331.

Bostrom, B.L., Jones, T.T., Hastings, M., Jones, D.R. 2010. Behaviour and physiology: The thermal strategy of leatherback turtles. *PLoS One* 5, 1–9.

Bradshaw, C.J.A., McMahon, C.R., Hays, G.C. 2007. Behavioral inference of diving metabolic rate in free-ranging leatherback turtles. *Physiol. Biochem. Zool.* 80, 209–219.

Brischoux, F., Bonnet, X., Cook, T.R., Shine, R. 2007. Allometry of diving capacities: Ectothermy vs. endothermy. *J. Evol. Biol.* 21, 324–329.

Burggren, W., Farrell, A., Lillywhite, H. 1997. Vertebrate cardiovascular systems, in: W.H. Dantzler (Ed.), *Handbook of Physiology*, Section 13: Comparative Physiology. Oxford University Press, New York, pp. 215–307.

Butler, P.J. 2006. Aerobic dive limit. What is it and is it always used appropriately? *Comp. Biochem. Physiol. A* 145, 1–6.

Butler, P.J., Jones, D.R. 1997. Physiology of diving of birds and mammals. *Physiol. Rev.* 77, 837–899.

Butler, P.J., Milsom, W.K., Woakes, A.J. 1984. Respiratory, cardiovascular and metabolic adjustments during steady state swimming in the green turtle, *Chelonia mydas. J. Comp. Physiol. B* 154, 167–174.

Carey, C. 2005. How physiological methods and concepts can be useful in conservation biology. *Integr. Comp. Biol.* 45, 4–11.

Carr, A. 1987. Impact of nondegradable marine debris on the ecology and survival outlook of sea turtles. *Mar. Pollut. Bull.* 18(6), 352–356.

Carr, A., Ogren, L. 1960. The ecology and migrations of sea turtles. 4: The green turtle in the Caribbean Sea. *Bull. Am. Mus. Nat. Hist.* 121(1), 6–48.

Carr, A., Ogren, L., McVea, C. 1980. Apparent hibernation by the Atlantic loggerhead turtle *Caretta caretta* off Cape Canaveral, Florida. *Biol. Conserv.* 19, 7–14.

Casey, J.P., Garner, J., Garner, S., Williard, A.S. 2010. Diel foraging behavior of gravid leatherback sea turtles in deep waters of the Caribbean Sea. *J. Exp. Biol.* 213, 3961–3971.

Chown, S.L., Gaston, K.J. 2008. Macrophysiology for a changing world. *Proc. R. Soc. B* 275, 1469–1478.

Clusella Trullas, S., Spotila, J.R., Paladino, F.V. 2006. Energetics during hatchling dispersal of the olive ridley turtle *Lepidochelys olivacea* using doubly labeled water. *Physiol. Biochem. Zool.* 79, 389–399.

Costa, D.P., Gales, N.J., Goebel, M.E. 2001. Aerobic dive limit: How often does it occur in nature? *Comp. Biochem. Physiol. A* 129, 771–783.

Dantzler, W.H. 1976. Renal function (with special emphasis on nitrogen excretion), in: C. Gans, W. Dawson (Eds.), *Biology of the Reptilia*. Academic Press, New York, pp. 447–503.

Davenport, J., De Verteuil, N., Magill, S.H. 1997. The effects of current velocity and temperature upon swimming in juvenile green turtles *Chelonia mydas. Herpetol. J.* 7, 143–147.

Davenport, J., Fraher, J., Fitzgerald, E., McLaughlin, P., Doyle, T., Harman, L., Cuffe, T. 2009a. Fat head: An analysis of head and neck insulation in the leatherback turtle (*Dermochelys coriacea*). *J. Exp. Biol.* 212, 2753–2759.

Davenport, J., Fraher, J., Fitzgerald, E., McLaughlin, P., Doyle, T., Harman, L., Cuffe, T., Dockery, P. 2009b. Ontogenetic changes in tracheal structure facilitate deep dives and cold water foraging in adult leatherback sea turtles. *J. Exp. Biol.* 212, 3440–3447.

Davenport, J., Ingle, G., Hughes, A.K. 1982. Oxygen uptake and heart rate in young green turtles (*Chelonia mydas*). *J. Zool., Lond.* 198, 399–412.

Derraik, J.G.B. 2002. The pollution of the marine environment by plastic debris: A review. *Mar. Pollut. Bull.* 44, 842–852.

Dunham, A.E., Grant, B.W., Overall, K.L. 1989. Interfaces between biophysical and physiological ecology and the population ecology of terrestrial vertebrate ectotherms. *Physiol. Zool.* 62, 335–355.

Dunson, W.A. 1969. Reptilian salt glands, Exocrine glands. *Proceedings of a Satellite Symposium of the XXIV International Congress of Physiological Sciences*, Philadelphia, PA, pp. 83–103.

Dunson, W.A. 1970. Some aspects of electrolyte and water balance in three estuarine reptiles, the diamondback terrapin, American and "salt water" crocodile. *Comp. Biochem. Physiol.* 32, 161–174.

Dunson, W.A. 1979. Control mechanisms in reptiles, in: R. Gilles (Ed.), *Mechanisms of Osmoregulation in Animals: Maintenance of Cell Volume*. John Wiley & Sons, New York, pp. 273–322.

Else, P.L., Hulbert, J.A. 1981. Comparison of the "mammal machine" and the "reptile machine": Energy production. *Am. J. Physiol.* 240, R3–R9.

Enstipp, M.R., Ciccione, S., Gineste, B., Milbergue, M., Ballorain, K., Ropert-Coudert, Y., Kato, A., Plot, V., George, J-Y. 2011. Energy expenditure of freely swimming adult green turtles (*Chelonia mydas*) and its link with body acceleration. *J. Exp. Biol.* 214, 4010–4020.

Fange, R., Schmidt-Nielsen, K., Osaki, H. 1958. The salt gland of the herring gull. *Biol. Bull.* 115: 162–171.

Felger, R.S., Cliffton, K., Regal, P.J. 1976. Winter dormancy in sea turtles: Independent discovery and exploitation in the Gulf of California by two local cultures. *Science* 191, 283–285.

Fossette, S., Gaspar, P., Handrich, Y., LeMaho, Y., Georges, J.Y. 2008. Dive and beak movement patterns in leatherback turtles *Dermochelys coriacea* during internesting intervals in French Guiana. *J. Anim. Ecol.* 77, 236–246.

Frair, W., Ackman, R.G., Mrosovsky, N. 1972. Body temperature of *Dermochelys coriacea*: Warm turtle from cold water. *Science* 177, 791–793.

Fry, F.E.J. 1947. Effects of the environment on animal activity. University of Toronto Studies, Biological Series 55. Publication of the Ontario Fisheries Research Laboratory 68, pp. 1–62.

Gatz, R.N., Glass, M.L., Wood, S.C. 1987. Pulmonary function of the green sea turtle, *Chelonia mydas*. *J. Appl. Physiol.* 62, 459–463.

Gerstberger, R., Oppermann, S., Kaul, R. 1984. Cephalic osmoreceptor control of salt gland activation and inhibition in the salt adapted duck. *J. Comp. Phys. B* 154, 449–456.

Godley, B.J., Richardson, S., Broderick, A.C., Coyne, M.S., Glen, F., Hays, G.C. 2002. Long-term satellite telemetry of the movements and habitat utilisation by green turtles in the Mediterranean. *Ecography* 25, 352–362.

Goff, G.P., Stenson, G.B. 1988. Brown adipose tissue in leatherback sea turtles: A thermogenic organ in an endothermic reptile? *Copeia* 1988, 1071–1075.

Greer, A.E., Lazell, J.D., Wright, R.M. 1973. Anatomical evidence for a countercurrent heat exchanger in the leatherback turtle (*Dermochelys coriacea*). *Nature* 244, 181.

Guderley, H., Seebacher, F. 2011. Thermal acclimation, mitochondrial capacities and organ metabolic profiles in a reptile (*Alligator mississippiensis*). *J. Comp. Phys. B* 181, 53–64.

Guderley, H., St-Pierre, J. 2002. Going with the flow or life in the fast lane: Contrasting mitochondrial responses to thermal change. *J. Exp. Biol.* 205, 2237–2249.

Halsey, L.G., Blackburn, T.M., Butler, P.J. 2006a. A comparative analysis of the diving behavior of birds and mammals. *Funct. Ecol.* 20, 889–899.

Halsey, L.G., Butler, P.J., Blackburn, T.M. 2006b. A phylogenetic analysis of the allometry of diving. *Am. Nat.* 167, 276–287.

Halsey, L.G., Jones, T.T., Jones, D.R., Liebsch, N., Booth, D.T., 2011. Measuring energy expenditure in sub-adult and hatchling sea turtles via accelerometry. *PLoS One* 6, 1–9.

Hansen, J., Sato, M., Ruedy, R., Lo, K., Lea, D.W., Medina-Elizade, M. 2006. Global temperature change. *Proc. Natl Acad. Sci. USA* 103, 14288–14239.

Harms, C.A., Mallo, K.M., Ross, P.M., Segars, A. 2003. Venous blood gases and lactates of wild loggerhead sea turtles (*Caretta caretta*) following two capture techniques. *J. Wildlife Dis.* 39, 366–374.

Hawkes, L.A., Broderick, A.C., Godfrey, M.H., Godley, B.J. 2009. Climate change and marine turtles. *Endanger. Species Res.* 7, 137–154.

Hays, G.C., Houghton, J.D., Isaacs, C., King, R.S., Lloyd, C., Lovell, P. 2004a. First records of oceanic dive profiles for leatherback turtles, *Dermochelys coriacea*, indicate behavioural plasticity associated with long-distance migration. *Anim. Behav.* 67, 733–743.

Hays, G.C., Houghton, J.D.R., Myers, A.E., 2004b. Pan-Atlantic leatherback turtle movements. *Nature* 429, 522.

Hays, G.C., Metcalfe, J.D., Walne, A.W. 2004c. The implications of lung-regulated buoyancy control for dive depth and duration. *Ecology* 85, 1137–1145.

Hays, G.C., Metcalfe, J.D., Walne, A.W., Wilson, R.P. 2004d. First records of flipper beat frequency during sea turtle diving. *J. Exp. Mar. Biol. Ecol.* 303, 243–260.

Heath, M.E., McGinnis, S.M. 1980. Body temperature and heat transfer in the green sea turtle, *Chelonia mydas*. *Copeia* (4), 767–773.

Hicks, J.W., Wang, T. 1996. Functional role of cardiac shunts in reptiles. *J. Exp. Zool.* 275, 204–216.

Hight, B.V., Holts, D., Graham, J.B., Kennedy, B.P., Taylo, V., Sepulveda, C.A., Bernal, D., Ramon, D., Rasmussen, R., Lai, N.C. 2007. Plasma catecholamine levels as indicators of the post-release survivorship of juvenile sharks caught on experimental drift longlines in the Southern California Bight. *Mar. Freshwater Res.* 58, 145–151.

Hirayama, R. 1998. Oldest known sea turtle. *Nature* 392, 705–708.

Hochachka, P.W., Somero, G.N. 2002. *Biochemical Adaptation: Mechanisms and Process in Physiological Evolution*. Oxford University Press, New York.

Hochscheid, S., Bentivegna, F., Bradai, M.N., Hays, G.C. 2007a. Overwintering behaviour in sea turtles: Dormancy is optional. *Mar. Ecol. Prog. Ser.* 340, 287–298.

Hochscheid, S., Bentivegna, F., Hays, G.C. 2005. First records of dive durations for a hibernating sea turtle. *Biol. Lett.* 1, 82–86.

Hochscheid, S., Bentivegna, F., Speakman, J.R. 2002. Regional blood flow in sea turtles: Implications for heat exchange in an aquatic ectotherm. *Physiol. Biochem. Zool.* 75(1), 66–76.

Hochscheid, S., Bentivegna, F., Speakman, J.R. 2004. Long-term cold acclimation leads to high Q10 effects on oxygen consumption of loggerhead sea turtles *Caretta caretta*. *Physiol. Biochem. Zool.* 77, 209–222.

Hochscheid, S., Godley, B.J., Broderick, A.C., Wilson, R.P. 1999. Reptilian diving: Highly variable dive patterns in the green turtle *Chelonia mydas*. *Mar. Ecol. Prog. Ser.* 185, 101–112.

Hochscheid, S., McMahon, C.R., Bradshaw, C.J.A., Maffucci, F., Bentivegna, F., Hays, G.C. 2007b. Allometric scaling of lung volume and its consequences for marine turtle diving performance. *Comp. Biochem. Physiol. A* 148, 360–367.

Holmes, W.N., Mcbean, R.L. 1964. Some aspects of electrolyte excretion in the green turtle, *Chelonia mydas mydas*. *J. Exp. Biol.* 41, 81–90.

Howell, E.A., Dutton, P.H., Polovina, J.J., Bailey, H., Parker, D.M., Balazs, G.H. 2010. Oceanographic influences on the dive behavior of juvenile loggerhead turtles (*Caretta caretta*) in the North Pacific Ocean. *Mar. Biol.* 157, 1011–1026.

Hudson, D.M., Lutz, P.L. 1986. Salt gland function in the leatherback sea turtle, *Dermochelys coriacea*. *Copeia* (1), 247–249.

Huey, R. 1982. Temperature, physiology, and the ecology of reptiles, in: C. Gans, F.H. Pough (Eds.), *Biology of the Reptilia*. Academic Press, New York, pp. 25–91.

Hulbert, A.J., Else, P.L. 1981. Comparison of the "mammal machine" and the "reptile machine": Energy use and thyroid activity. *Am. J. Physiol.* 241, R350–R356.

Innis, C.J., Tlusty, M., Merigo, C., Weber, E.S. 2007. Metabolic and respiratory status of cold-stunned Kemp's ridley sea turtles (*Lepidochelys kempii*). *J. Comp. Physiol. B* 177, 623–630.

IPCC, 2007: Climate Change 2007: Synthesis Report. Contribution of working groups I, II, and III to the Fourth Assessment Report of the intergovernmental Panel on Climate Change [core writing team, Pachauri, R.K. and Reisinger, A. (eds)]. IPCC, Geneva, Switzerland, p.104.

Jackson, D.C. 2000. Living without oxygen: Lessons from the freshwater turtle. *Comp. Biochem. Physiol. A* 125, 299–315.

Jackson, D.C., Prange, H.D. 1979. Ventilation and gas exchange during rest and exercise in adult green sea turtles. *J. Comp. Physiol. B* 134, 315–319.

James, M.C., Mrosovsky, N. 2004. Body temperatures of leatherback turtles (*Dermochelys coriacea*) in temperate waters off Nova Scotia, Canada. *Can. J. Zool.* 82, 1302–1306.

Jones, T.T., Hastings, M.D., Bostrom, B.L., Andrews, R.D., Jones, D.R. 2009. Validation of the use of doubly labeled water for estimating metabolic rate in the green turtle (*Chelonia mydas* L.): A word of caution. *J. Exp. Biol.* 212, 2635–2644.

Jones, T.T., Reina, R.D., Darveau, C.A., Lutz, P.L. 2007. Ontogeny of energetics in leatherback (*Dermochelys coriacea*) and olive ridley (*Lepidochelys olivacea*) sea turtle hatchlings. *Comp. Biochem. Physiol. A* 147, 313–322.

Kaiser, J. 2010. The dirt on ocean garbage patches. *Science* 328, 1506.

Kooyman, G.L., Wahrenbrock, E.A., Castellini, M.A., Davis, R.W., Sinnett, E.E. 1980. Aerobic and anaerobic metabolism during voluntary diving in Weddell seals: Evidence of preferred pathways from blood chemistry and behavior. *J. Comp. Physiol. B* 138, 335–346.

Kraus, D.R., Jackson, D.C. 1980. Temperature effects on ventilation and acid-base balance of the green turtle. *Am. J. Physiol.* 239, R254–R258.

Lazell, J.D. 1980. New England waters: Critical habitat for marine turtles. *Copeia* 1980, 290–295.

Lillywhite, H.B., Maderson, P.F.A. 1982. Skin structure and permeability, in: C. Gans, F.H. Pough (Eds.), *Biology of the Reptilia*. Academic Press, New York, pp. 397–442.

Lopez-Mendilaharsu, M., Rocha, C.F.D., Domingo, A., Wallace, B.P., Miller, P. 2008. Prolonged, deep dives by the leatherback turtle *Dermochelys coriacea*: Pushing their aerobic dive limits. JMBA2 Biodiversity Records, http://www.mba.ac.uk/jmba/jmba2biodiversityrecords.php

Lutcavage, M.E., Bushnell, P.G., Jones, D.R. 1990. Oxygen transport in leatherback sea turtle *Dermochelys coriacea*. *Physiol. Zool.* 63, 1012–1024.

Lutcavage, M.E., Bushnell, P.G., Jones, D.R. 1992. Oxygen stores and aerobic metabolism in the leatherback sea turtle. *Can. J. Zool.* 70, 348–351.

Lutcavage, M.L., Lutz, P.L. 1997. Diving physiology, in: P.L. Lutz, J.A. Musick (Eds.), *The Biology of Sea Turtles.* CRC Press, Boca Raton, FL, pp. 277–296.

Lutcavage, M.E., Lutz, P.L., Baier, H. 1989. Respiratory mechanics of the loggerhead turtle, *Caretta caretta. Resp. Physiol.* 76, 13–24.

Lutcavage, M.E., Lutz, P.L., Bossart, G.D., Hudson, D.M. 1995. Physiologic and clinicopathologic effects of crude oil on loggerhead sea turtles. *Arch. Environ. Contam. Toxicol.* 28, 417–422.

Lutz, P.L. 1997. Salt, water, and pH balance in the sea turtle, in: P.L. Lutz, J.A. Musick (Eds.), *The Biology of Sea Turtles.* CRC Press, Boca Raton, FL, pp. 343–361.

Lutz, P.L., Bentley, T.B. 1985. Respiratory physiology of diving in the sea turtle. *Copeia* 3, 671–679.

Lutz, P.L., Bergey, A., Bergey, M. 1989. Effects of temperature on gas exchange and acid-base balance in the sea turtle *Caretta caretta* at rest and during routine activity. *J. Exp. Biol.* 144, 155–169.

Lutz, P.L., Dunbar-Cooper, A. 1987. Variations in the blood chemistry of the loggerhead sea turtle, *Caretta caretta. Fish. Bull.* 85, 37–43.

Lutz, P.L., LaManna, J.C., Adams, M.R., Rosenthal, M. 1980. Cerebral resistance to anoxia in the marine turtle. *Resp. Physiol.* 41, 241–251.

Makowski, C., Seminoff, J.A., Salmon, M. 2006. Home range and habitat use of juvenile Atlantic green turtles (*Chelonia mydas* L.) on shallow reef habitats in Palm Beach, Florida, USA. *Mar. Biol.* 148, 1167–1179.

Marshall, A.T. 1989. Intracellular and luminal ion concentrations in sea turtle salt glands determined by x-ray microanalysis. *J. Comp. Physiol. B* 159, 609–616.

Marshall, A.T., Cooper, P.D. 1988. Secretory capacity of the lachrymal salt gland of hatchling sea turtles, *Chelonia mydas. J. Comp. Physiol. B* 157, 821–827.

Marshall, A.T., Saddlier, S.R. 1989. The duct system of the lachrymal salt gland of the green sea turtle, *Chelonia mydas. Cell Tissue Res.* 257, 399–404.

McCauley, S.J., Bjorndal, K.A. 1999. Conservation implications of dietary dilution from debris ingestion: Sublethal effects in post-hatchling loggerhead sea turtles. *Conserv. Biol.* 13, 925–929.

McMahon, C.R., Bradshaw, C.J.A., Hays, G.C. 2007. Satellite tracking reveals unusual diving characteristics for a marine reptile, the olive ridley turtle *Lepidochelys olivacea. Mar. Ecol. Prog. Ser.* 329, 239–252.

Mendonca, M. 1983. Movements and feeding ecology of immature green turtles (*Chelonia mydas*) in a Florida Lagoon. *Copeia* (4), 1013–1023.

Minamikawa, S., Naito, Y., Uchida, I. 1997. Buoyancy control in diving behavior of the loggerhead turtle, *Caretta caretta. J. Ethol.* 15, 109–118.

Moon, D., Mackenzie, D.S., Owens, D.W. 1997. Simulated hibernation of sea turtles in the laboratory: I. feeding, breathing frequency, blood pH, and blood gases. *J. Exp. Zool.* 278, 372–380.

Moyes, C., Fragoso, N., Musyl, M., Brill, R. 2006. Predicting postrelease survival in large pelagic fish. *Trans. Am. Fish. Soc.* 135, 1389–1397.

Nicolson, S.W., Lutz, P.L. 1989. Salt gland function in the green sea turtle (*Chelonia mydas*). *J. Exp. Biol.* 144, 171–184.

Oros, J., Camacho, M., Calabuig, P., Arencibia, A. 2011. Salt gland adenitis as only cause of stranding of loggerhead sea turtles *Caretta caretta. Dis. Aquat. Organ.* 95, 163–166.

Paladino, F.V., O'Connor, P., Spotila, J.R. 1990. Metabolism of leatherback turtles, gigantothermy, and thermoregulation of dinosaurs. *Nature* 344, 858–860.

Paladino, F.V., Spotila, J.R., O'Connor, M.P., Gatten, R.E. 1996. Respiratory physiology of adult leatherback turtles (*Dermochelys coriacea*) while nesting on land. *Chelonian Conserv. Biol.* 2, 223–229.

Penick, D.N., Paladino, F.V., Steyermark, A.C., Spotila, J.R. 1996. Thermal dependence of tissue metabolism in the green turtle, *Chelonia mydas. Comp. Biochem. Physiol. A* 113, 293–296.

Penick, D.N., Spotila, J.R., O'Connor, M.P., Steyermark, A.C., George, R.H., Salice, C.H., Paladino, F.V. 1998. Thermal independence of muscle tissue metabolism in the leatherback turtle, *Dermochelys coriacea. Comp. Biochem. Physiol. A* 120, 399–403.

Perry, S.F. 1983. Reptilian lungs: Functional anatomy and evolution. *Adv. Anat. Embryol. Cell. Biol.* 79, 1–81.

Poloczanska, E., Limpus, C., Hays, G., 2009. Vulnerability of marine turtles to climate change, in: D. Sims (Ed.), *Advances in Marine Biology.* Academic Press, Burlington, VT, pp. 151–211.

Polovina, J.J., Balazs, G.H., Howell, E.A., Parker, D.M., Seki, M.P., Dutton, P.H. 2004. Forage and migration habitat of loggerhead (*Caretta caretta*) and olive ridley (*Lepidochelys olivacea*) sea turtles in the central North Pacific Ocean. *Fish. Oceanogr.* 13, 36–51.

Ponganis, P.J., Kooyman, G.L., Starke, L.N., Kooyman, C.A., Kooyman, T.G. 1997. Post-dive lactate concentrations in emperor penguins, *Aptenodytes forsteri. J. Exp. Biol.* 200, 1623–1626.

Prange, H.D. 1976. Energetics of swimming of a sea turtle. *J. Exp. Biol.* 64, 1–12.

Prange, H.D., Jackson, D.C. 1976. Energetics of swimming of a sea turtle. *Resp. Physiol.* 27, 369–377.

Rahmstorf, S., Cazenave, A., Church, J.A., Hansen, J.E., Keeling, R.F., Parker, D.E., Somerville, R.C.J. 2007. Recent climate observations compared to projections. *Science* 316, 709.

Read, M.A., Grigg, G.C., Limpus, C.J. 1996. Body temperatures and winter feeding in immature green turtles, *Chelonia mydas*, in Moreton Bay, Southeastern Queensland. *J. Herpetol.* 30, 262–265.

Reina, R.D., Abernathy, K.J., Marshall, G.J., Spotila, J.R. 2005. Respiratory frequency, dive behaviour, and social interactions of leatherback turtles, *Dermochelys coriacea*, during the inter-nesting interval. *J. Exp. Mar. Biol. Ecol.* 316, 1–16.

Reina, R.D., Cooper, P.D. 2000. Control of salt gland activity in the hatchling green sea turtle, *Chelonia mydas*. *J. Comp. Physiol. B* 170, 27–35.

Reina, R.D., Jones, T.T., Spotila, J.R. 2002. Salt and water regulation by the leatherback sea turtle *Dermochelys coriacea*. *J. Exp. Biol.* 205, 1853–1860.

Sakamoto, W., Uchida, I., Naito, Y., Kureha, K., Tujimura, M., Sato, K. 1990. Deep diving behavior of the loggerhead turtle near the frontal zone. *Nippon Suisan Gakk.* 56, 1435–1443.

Salmon, M., Hamann, M., Wyneken, J. 2010. The development of early diving behavior by juvenile flatback sea turtles. *Chelonian Conserv. Biol.* 9, 8–17.

Salmon, M., Jones, T.T., Horch, K.W. 2004. Ontogeny of diving and feeding behavior in juvenile sea turtles: Leatherback sea turtles (*Dermochelys coriacea*) and green sea turtles (*Chelonia mydas* L) in the Florida current. *J. Herpetol.* 38, 36–43.

Sasso, C.R., Witzell, W.N. 2006. Diving behaviour of an immature Kemp's ridley turtle (*Lepidochelys kempii*) from Gullivan Bay, Ten Thousand Islands, south-west Florida. *J. Mar. Biol. Assoc. U.K.* 86, 919–925.

Sato, K., Matsuzawa, Y., Tanaka, H., Bando, T., Minamikawa, S., Sakamoto, W., Naito, Y. 1998. Internesting intervals for loggerhead turtles, *Caretta caretta*, and green turtles, *Chelonia mydas*, are affected by temperature. *Can. J. Zool.* 76, 1651–1662.

Sato, K., Sakamoto, W., Matsuzawa, Y., Tanaka, H., Minamikawa, S., Naito, Y. 1995. Body temperature independence of solar radiation in free-ranging loggerhead turtles, *Caretta caretta*, during internesting periods. *Mar. Biol.* 123, 197–205.

Sato, K., Sakamoto, W., Matsuzawa, Y., Tanaka, H., Naito, Y. 1994. Correlation between stomach temperatures and ambient water temperatures in free-ranging loggerhead turtles, *Caretta caretta*. *Mar. Biol.* 118, 343–351.

Schmidt-Nielsen, K. 1984. *Scaling: Why Is Size So Important?* Cambridge University Press, Cambridge, U.K.

Schmidt-Nielsen, K., Fange, R. 1958. Salt glands in marine reptiles. *Nature* 182, 783–785.

Schmidt-Nielsen, K. 1972. Locomotion: Energy cost of swimming, flying, and running. *Science* 177, 222–228.

Schreer, J.F., Kovacs, K.M. 1997. Allometry of diving capacity in air-breathing vertebrates. *Can. J. Zool.* 75, 339–358.

Seebacher, F. 2005. A review of thermoregulation and physiological performance in reptiles: What is the role of phenotypic flexibility? *J. Comp. Physiol. B* 175, 453–461.

Seebacher, F., Guderley, H., Elsey, R.M., Trosclair, P.L. 2003. Seasonal acclimatisation of muscle metabolic enzymes in a reptile (*Alligator mississippiensis*). *J. Exp. Biol.* 206, 1193–1200.

Seebacher, F., Sparrow, J., Thompson, M.B. 2004. Turtles (*Chelodina longicollis*) regulate muscle metabolic enzyme activity in response to seasonal variation in body temperature. *J. Comp. Physiol. B* 174, 205–210.

Seminoff, J.A. 2000. *Biology of the East Pacific Green Turtle, Chelonia mydas agassizii, at a Warm Temperate Feeding Area in the Gulf of California, Mexico.* The University of Arizona, Tuscon, AZ, p. 249.

Shoemaker, V.H., Nagy, K.A. 1984. Osmoregulation in the Galapagos marine iguana, *Amblyrhynchus cristatus*. *Physiol. Zool.* 57, 291–300.

Snoddy, J.E., Landon, M.L., Blanvillain, G., Southwood, A. 2008. Blood biochemistry of sea turtles released from gillnets in the lower Cape Fear River, North Carolina, USA. *Endanger. Species Res.* 12, 235–247.

Snoddy, J.E., Southwood, A. 2010. Movements and post-release mortality of juvenile sea turtles captured in gillnets in the lower Cape Fear River, North Carolina, USA. *Endanger. Species Res.* 12, 235–247.

Somero, G.N. 1997. Temperature relationships: From molecules to biogeography, in: W.H. Dantzler (Ed.), *Handbook of Comparative Physiology*. Oxford University Press, New York, pp. 1391–1444.

Southwood, A.L. 2002. *The Effects of Seasonal Cold Exposure on Metabolism and Behaviour of Juvenile Green Sea Turtles (*Chelonia mydas*), Zoology.* University of British Columbia, Vancouver, BC, Canada, p. 117.

Southwood, A.L., Andrews, R.D., Lutcavage, M.E., Paladino, F.V., West, N.H., George, R.H., Jones, D.R. 1999. Heart rates and diving behavior of leatherback sea turtles in the Eastern Pacific Ocean. *J. Exp. Biol.* 202, 1115–1125.

Southwood, A.L., Andrews, R.D., Paladino, F.V., Jones, D.R. 2005. Effects of diving and swimming behavior on body temperatures of Pacific leatherback turtles in tropical seas. *Physiol. Biochem. Zool.* 78, 285–297.

Southwood, A.L., Darveau, C.A., Jones, D.R. 2003a. Metabolic and cardiovascular adjustments of juvenile green turtles to seasonal changes in temperature and photoperiod. *J. Exp. Biol.* 206, 4521–4531.

Southwood Williard, A., Harden, L.A. 2011. Seasonal changes in thermal environment and metabolic enzyme activity in the diamondback terrapin (*Malaclemys terrapin*). *Comp. Biochem. Physiol. A* 158, 477–484.

Southwood, A.L., Reina, R.D., Jones, V.S., Jones, D.R. 2003b. Seasonal diving patterns and body temperatures of juvenile green turtles at Heron Island, Australia. *Can. J. Zool.* 81, 1014–1024.

Southwood, A., Reina, R., Jones, V., Speakman, J., Jones, D. 2006. Seasonal metabolism of juvenile green turtles (*Chelonia mydas*) at Heron Island, Australia. *Can. J. Zool.* 84, 125–135.

Speakman, J.R. 1997. *Doubly Labelled Water: Theory and Practice.* Chapman and Hall, London, U.K.

Stabenau, E.K., Heming, T.A., Mitchell, J.F. 1991a. Respiratory, acid-base and ionic status of Kemp's ridley sea turtles (*Lepidochelys kempi*) subjected to trawling. *Comp. Biochem. Physiol.* 99A, 107–111.

Stabenau, E.K., Vanoye, C.G., Heming, T.A. 1991b. Characteristics of the anion transport system in sea turtle erythrocytes. *Am. J. Physiol.* 261, R1218–R1225.

Stabenau, E.K., Vietti, K.R.N. 2003. The physiological effects of multiple forced submergences in loggerhead sea turtles (*Caretta caretta*). *Fish. Bull.* 101, 889–899.

Standora, E.A., Spotila, J.R., Foley, R.E. 1982. Regional endothermy in the sea turtle, *Chelonia mydas. J. Therm. Biol.* 7, 159–165.

Standora, E.A., Spotila, J.R., Keinath, J.A., Shoop, C.R. 1984. Body temperatures, diving cycles, and movement of a subadult leatherback turtle, *Dermochelys coriacea. Herpetologica* 40, 169–176.

Storch, S., Wilson, R.P., Hillis-Starr, Z., Adelung, D. 2005. Cold-blooded divers: Temperature-dependent dive performance in the wild hawksbill turtle *Eretmochelys imbricata. Mar. Ecol. Prog. Ser.* 293, 263–271.

Taplin, L.E., Grigg, G.C. 1981. Salt glands in the tongue of the estuarine crocodile. *Science* 212, 1045–1047.

Taplin, L.E., Grigg, G.C., Harlow, P., Ellis, T.M., Dunson, W.A. 1982. Lingual salt glands in *Crocodylus acutus* and *C. johnstoni* and their absence from *Alligator mississipiensis* and Caiman crocodiles. *J. Comp. Physiol.* 149, 43–47.

Tenney, S.M., Bartlett, D., Farber, J.P., Remmers, J.E. 1974. Mechanics of the respiratory cycle in the green turtle (*Chelonia mydas*). *Resp. Physiol.* 22, 361–368.

Tracy, C.R., Nussear, K.E., Esque, T.C., Dean-Bradley, K., Tracy, C.R., DeFalco, L.A., Castle, K.T., Zimmerman, L.C., Espinoza, R.E., Barber, A.M. 2006. The importance of physiological ecology in conservation biology. *Integr. Comp. Biol.* 46, 1191–1205.

Tsuji, J.S. 1988a. Seasonal profiles of standard metabolic rate of lizards (*Sceloporus occidentalis*) in relation to latitude. *Physiol. Zool.* 61, 230–240.

Tsuji, J.S. 1988b. Thermal acclimation of metabolism in Sceloporus lizards from different latitudes. *Physiol. Zool.* 61, 241–253.

Tucker, V.A. 1970. Energetic cost of locomotion in animals. *Comp. Biochem. Physiol.* 34, 841–846.

vanDam, R.P., Diez, C.E. 1996. Diving behavior of immature hawksbills (*Eretmochelys imbricata*) in a Caribbean cliff-wall habitat. *Mar. Biol.* 127, 171–178.

Wallace, B.P., Jones T.T. 2008. What makes marine turtles go: A review of metabolic rates and their consequences. *J. Exp. Mar. Biol. Ecol.* 356, 8–24.

Wallace, B.P., Williams, C.L., Paladino, F.V., Morreale, S.J., Lindstrom, R.T., Spotila, J.R. 2005. Bioenergetics and diving activity of internesting leatherback turtles *Dermochelys coriacea* at Parque Nacional Marinos Las Baulas, Costa Rica. *J. Exp. Biol.* 208, 3873–3884.

Warren, D.E., Jackson, D.C. 2004. Effects of swimming on metabolic recovery from anoxia in the painted turtle. *J. Exp. Biol.* 207, 2705–2713.

West, N.H., Butler, P.J., Bevan, R.M. 1982. Pulmonary blood flow at rest and during swimming in the green turtle, *Chelonia mydas. Physiol. Zool.* 65, 287–310.

Wikelski, M., Cooke, S.J. 2006. Conservation physiology. *Trends Ecol. Evol.* 21, 38–46.

Witt, M., Hawkes, L., Godfrey, M., Godley, B., Broderick, A. 2010. Predicting the impacts of climate change on a globally distributed species: The case of the loggerhead turtle. *J. Exp. Biol.* 213, 901–911.

Wyneken, J., 1997. Sea turtle locomotion: Mechanisms, behavior, and energetics, in: P.L. Lutz, J.A. Musick (Eds.), *The Biology of Sea Turtles.* CRC Press, Boca Raton, FL, pp. 165–198.

# 2 Vision

*Kerstin A. Fritsches and Eric J. Warrant*

## CONTENTS

## 2.1   INTRODUCTION

The visual system of sea turtles is exposed to some of the most varied visual habitats found in the animal kingdom. Aquatic vision in clear oceanic waters and more turbid coastal habitats is presumably of predominant importance for sea turtles when feeding, avoiding predators, and finding a mate. Being air breathers, sea turtle eyes are also frequently exposed to visual scenes above water. And for females and hatchling sea turtles, visual orientation on nesting beaches and sea finding is a major visual task (see Lohmann et al., 1997). With sea finding being a largely nocturnal activity and feeding largely diurnal, sea turtles are also exposed to a wide range of light intensities. Lastly, sea turtles are thought to have evolved from a terrestrial predecessor (Pritchard, 1997), bringing with them the constraints of a terrestrial eye design.

On the other hand, sea turtles, together with other reptiles and birds, are part of the vertebrate group in which vision evolved to the highest levels of complexity (Walls, 1942). This suggests that when sea turtles returned to the aquatic realm, they did so with a great "toolbox" of visual system features, to be adapted to a predominantly aquatic lifestyle. As a "natural experiment," sea turtle vision is therefore of particular interest to vision science.

Given the protected nature of sea turtles and the often invasive nature of vision research, knowledge of sea turtle visual capabilities is relatively limited. The "Biology of Sea Turtle" series has covered the topic of vision in the context of sea finding (Lohmann et al., 1997) and with respect to general sensory abilities (Bartol and Musick, 2002). This review aims to update these previous reviews with the most recent studies and to explore the significance of the currently available data for visual tasks such as finding food or mates and avoiding predators and orienting in water and on land. The vast majority of studies on visual capabilities have concentrated on the green turtle (*Chelonia mydas*), the loggerhead sea turtle (*Caretta caretta*), and the leatherback sea turtle (*Dermochelys coriacea*), and therefore these three species are the focus of this review.

## 2.2   LIGHT ENVIRONMENT

All sea turtle species begin their life on land. However, this initial phase is very short, as hatchlings reach the waters within minutes after emerging from the nest. For the majority of hatchlings, this first visual experience takes place at night, when ambient temperatures are lower (Hendrickson, 1958; Bustard, 1967; Witherington et al., 1990) and animals experience light conditions ranging from relatively bright full moonlight to new moon conditions, as well as dark, overcast nights. In their adult years, usually only female turtles spend any time on land, mainly for laying eggs, and this also predominately occurs at night. However, since sea turtles are air breathers, they do spend a significant proportion of their time on the surface (as much as 19%–26%: Lutcavage and Lutz, 1997), and are able to view the visual scene above water throughout their lifetime.

Once entering the water, the hatchlings commence their "lost years" (Carr, 1987), in which they are thought to inhabit the surface layers of the clear open ocean. Juveniles and adults of green, loggerhead, and leatherback sea turtles are found in clear open ocean water, as well as the coastal waters of clear coral reefs or more temperate shallow benthic habitats with varying visibility. Green and loggerhead sea turtles appear to be predominantly diurnal in their activity patterns (Odgen et al., 1983; Wyneken and Salmon, 1992), although some nocturnal activity has been observed in green turtles (Jessop et al., 2002). Leatherback turtles, on the other hand, are active throughout the 24 h cycle both in the early stages of their lives (Odgen et al., 1983; Wyneken and Salmon, 1992) and as adults (Eckert et al., 1986, 1989; Hays et al., 2004a). This species is also known to dive to significant depths, with maximum diving depths reported to exceed 1200 m (Hays et al., 2004b; Doyle et al., 2008), although the vast majority of dives are within the top 200 m of the water column and are relatively short (Hays et al., 2004b). Both green and loggerhead sea turtles show routine dives to 20 m or shallower, with maximum depths recorded at 211–233 m for the loggerhead sea turtle (Lutcavage and Lutz, 1997).

Due to the sea turtle's varied exposure to terrestrial light (both day and night) and marine light (both shallow and deep), a fuller description of natural daylight in these environments is justified.

### 2.2.1  SPECTRUM OF NATURAL DAYLIGHT

Sunlight—the major source of light on earth—illuminates either directly, as during the day, or indirectly by reflection from the moon at night. The intensity of daylight at the surface of the sea can vary by up to 10 orders of magnitude from day to night (depending on the presence or absence of the moon or clouds: see Warrant, 2008). Its irradiance spectrum also varies during the course of the day (Figure 2.1A), undergoing its most significant changes around dusk and dawn (Johnsen et al., 2006; Sweeney et al., 2011). When the sun is high in the sky, the spectrum of daylight is dominated by longer wavelengths (Moon, 1940; Lythgoe, 1979). As the sun sets, the spectrum becomes increasingly neutral across wavelengths; and then as the sun's disk sinks below the horizon, daylight becomes distinctly blue, with a broad peak centered around 450 nm (Johnsen et al., 2006; Sweeney et al., 2011).

Once the sun disappears well below the horizon, the disk of the moon can also strongly affect the spectrum of nocturnal light. The irradiance spectrum of moonlight is essentially the same as that of full sunlight since the moon is a spectrally flat reflector of the sunlight that strikes it (Figure 2.1A), although its intensity (at full moon) is around 6 orders of magnitude dimmer. When the moon is absent, the stars are the only source of illumination. The irradiance spectrum of starlight is significantly red-shifted (Johnsen et al., 2006), and shows four prominent peaks at 560, 590, 630, and 685 nm (Figure 2.1A).

### 2.2.2  OPTICAL PROPERTIES OF WATER

As light propagates through water it is absorbed and scattered, and this has a marked effect on its quality at different depths in the sea. The intensity of light, as well as its color, degree of polarization, and contrast, is significantly altered as it travels through the ocean.

The intensity of light, even in the clearest oceans and lakes, reduces by approximately 1.5 orders of magnitude for every 100 m of depth. By 500–700 m during the day, light intensities reach starlight levels (Clarke and Denton, 1962). Almost no daylight remains upon entrance to the bathypelagic zone at 1000 m, and it is very unlikely that marine animals can see daylight below this depth (Denton, 1990).

In a clear ocean, light also becomes progressively bluer with depth (Figure 2.1B), with the orange-red part of the spectrum (beyond 550 nm) almost completely absorbed within the first 100 m (Tyler and Smith, 1970). Ultraviolet light is also absorbed, but not quite as effectively, with biologically relevant intensities remaining to at least 200 m (Frank and Widder, 1996; Losey et al., 1999). Below 200 m, the down-welling daylight is almost monochromatic (475 nm).

In the upper depths of the ocean, daylight is visible in all directions around the animal. This is because particles suspended in the water scatter daylight in every direction. Thus, an animal swimming near the surface will see light coming from below, from the sides and of course also from above. The greater the density of particles suspended in the water, the greater the scatter and the brighter this so-called space light. Thus, in very clear water, the space light intensity is lower than in cloudier water. However, with increasing depth, the intensity of the space light seen below and to the sides declines, and the available daylight comes increasingly from above (Figure 2.1C). This radiance distribution is dominated by the position of the sun in shallower water, but this dominance declines with depth, disappearing altogether below the so-called asymptotic depth. Below this depth—which is about 400 m in clearest ocean water—the radiance distribution is vertically symmetric (Jerlov, 1976): light originating laterally and from below is respectively about 40 times and 300 times dimmer than light originating directly from above.

No matter what its intensity, the space light has a detrimental effect on the contrast and visibility of objects in the sea. The scattering of the down-welling daylight by suspended particles in the

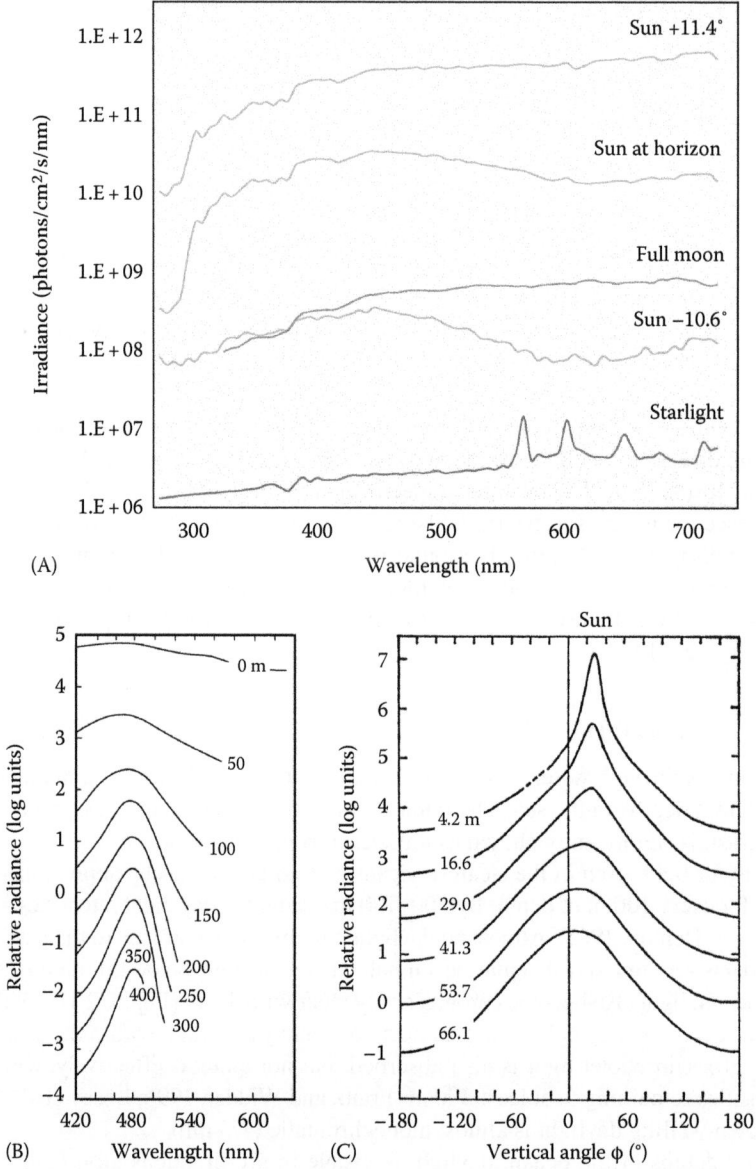

**FIGURE 2.1** The spatial and spectral properties of daylight in terrestrial and marine habitats. (A) The irradiance spectra of sunlight (green curves), moonlight (blue curve), and starlight (red curve) in a terrestrial habitat (spectra were measured on a near-cloudless night with minimal anthropogenic light pollution). Sunlight spectra are shown just prior to sunset (sun elevation +11.4°), at sunset (sun at horizon), and just after sunset (sun elevation −10.6°). Note that even though the shapes of these curves will remain the same, the presence of clouds will cause them to shift downward on the irradiance axis by up to 1–2 orders of magnitude. (Adapted from Johnsen, S. et al., *J. Exp. Biol.*, 209, 789, 2006. With permission.) (B) The relationship between depth (shown in m) and the spectrum of downward irradiance in the Golfe du Lion. (C) The change in the radiance distribution of green light with depth (shown in m) in Lake Pend Oreille. f is the angle relative to vertical (0° = vertical, ±180° = horizontal). The distribution is skewed in the direction of the sun near the surface, but becomes more symmetric with increasing depth. In Lake Pend Oreille, it becomes perfectly symmetric (asymptotic) at approximately 100 m. (Adapted from Jerlov, N.G., *Marine Optics*, Elsevier, Amsterdam, the Netherlands, 1976. With permission.)

interposing body of water will create a veiling "haze" of space light within which the object may disappear from sight. Such scattering greatly reduces the apparent contrasts of objects, and limits the furthest distance at which they can be reliably detected (Lythgoe, 1979, 1988; Jagger and Muntz, 1993). The further away an object, the larger the volume of water between it and the viewer and the greater the number of suspended particles that scatter light into the viewer's line of sight. As the object moves further away, this scattered light eventually becomes brighter than the light that reaches the eye from the object, finally veiling it completely. At this point the visual contrast of the object falls to zero and disappears from view. For a dark object in the brightest and clearest ocean water, this occurs at a range of about 40 m (Lythgoe, 1979).

### 2.2.3  BIOLUMINESCENCE

Bioluminescence is a major light source in the sea, with many ecological meanings (Herring, 1978). Its contrast and visibility—just as for objects illuminated by daylight—are limited by the intensity of the surrounding space light. During the day in the brighter depths above 100 m, most bioluminescent signals are probably not visible (Denton, 1990). But below this depth, as the space light becomes progressively weaker, bioluminescent signals become increasingly visible. In the blackness of the bathypelagic zone they reach their greatest contrast.

Bioluminescent signals are usually point-source flashes, although in some cases they can be somewhat extended, as in the tunicate *Pyrosoma* which may grow to the size of a bus (Herring, 2000). The length of a flash may vary from hundreds of milliseconds to several seconds, and their frequency in the sea can vary from 1 to 160 flashes from each steradian (solid angle) of water per minute. Below 1000 m flash frequency drops considerably, and becomes very infrequent below 2000 m (Clarke and Hubbard, 1959). The intensity and color of these flashes can vary considerably (Herring, 1978), but a typical flash is blue and contains between $10^7$ and $10^{13}$ photons, no doubt a highly visible stimulus in the darkness of the deep sea.

### 2.2.4  CHANGING NATURE OF VISUAL SCENES WITH DEPTH

The optical properties of water and the bioluminescent signals of aquatic organisms together create the light environment that animals see, and this environment changes dramatically with depth (Warrant, 2000; Warrant and Locket, 2004). In the brighter upper depths, the visual scenes viewed by animals are *extended,* that is, light reaches the eye from objects (or suspended particles) located in all directions within the visual scene. Scattering produces an even blue space light, and in coastal waters the sea floor may even be visible. But at greater depths, where the space light is diminished, bioluminescent point sources also begin to appear, especially from below where the space light is dimmest. Upward, and even frontward, the scene is still extended. But downward the scene begins to be dominated by point sources. At still deeper levels, bioluminescent point sources can be seen in all directions. Below 1000 m, beyond the penetration of daylight, the visual scene is entirely *point-like.*

## 2.3  VISUAL HARDWARE

### 2.3.1  SEA TURTLE EYE

The main function of the optical apparatus of the vertebrate eye is to provide the support and structure to transmit and focus incoming light to be processed by the retina, the nervous tissue lining the back of the eye, which in turn transmits the light information to the higher visual centers in the brain. Surrounding the retina and forming the outer layers of the eyeball is the vascularized choroid layer and the sclera. The eyeball itself is filled with fluid, with the aqueous and vitreous humor providing a clear medium that maintains eyeball shape and keeps the retina in the plane of the focused image. In sea turtles, as in all vertebrates, the eyeball is moved by six extra-ocular muscles (Wyneken, 2001), providing the ability to shift gaze independently of head movements.

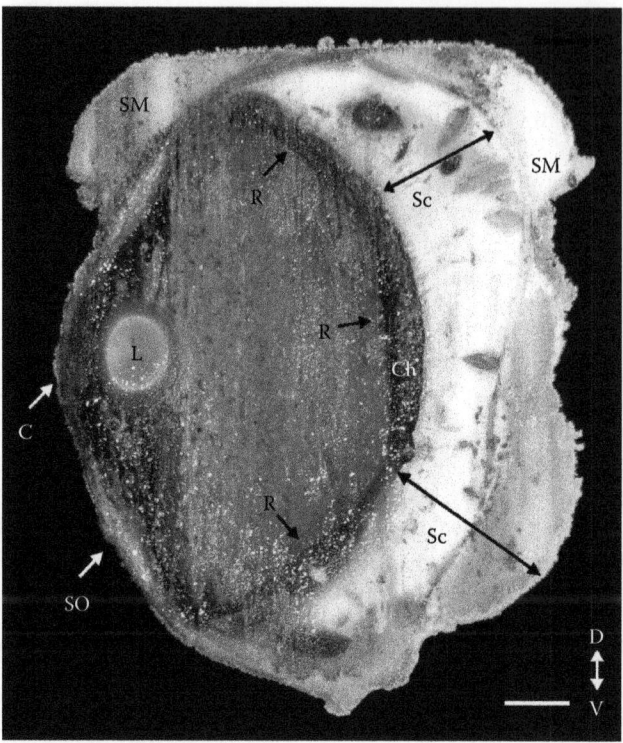

FIGURE 2.2   Cross-section of a frozen leatherback sea turtle eye. C, cornea; Ch, choroid; L, lens; R, retina; Sc, sclera; SO, scleral ossicle; SM, frozen supporting media; scale bar 5 mm. (From Brudenall, D.K. et al., *Vet. Ophthalmol.*, 11, 99, 2008. With permission.)

In adult sea turtles scleral thickness appears to increase as the eye becomes larger (Brudenall et al., 2008; Figure 2.2), presumably to provide added rigidity that prevents deformation of the eye (Brudenall et al., 2008). Sea turtles possess a third eyelid, the nictitating membrane, which protects and moistens the eye. In leatherback sea turtles, unlike other sea turtles studied, the nictitating membrane has extensive plications or folds, which provide a drastically enlarged surface area for mucous secretion (Burne, 1905; Brudenall et al., 2008). This is thought to protect the leatherback eye from the salt secretions which are more highly concentrated than in other species (Hudson and Lutz, 1986).

### 2.3.2   OPTICS AND ACCOMMODATION

Light enters the eye through the cornea (Figure 2.2), which is a clear, specialized part of the sclera. In air, the cornea can achieve high refractive power as its curved outer surface acts as the interface between air and the watery medium inside the eye. However, once the eye is submerged in water, the cornea loses its refractive power, explaining the incorrectly focused image humans experience when opening their eyes under water. In aquatic animals such as fish, the cornea acts purely as a protective barrier and all refractive power is contained in the spherical lens.

Animals adapted to life in both air and aquatic environment have adapted their corneas and lenses to varying degrees, according to their predominant habitat. The freshwater turtle *Pseudemys scripta elegans* (red-eared slider) has a curved cornea and a relatively flat lens (Northmore and Granda, 1991), reflecting a lifestyle where vision in air is more common. In contrast, sea turtles have nearly spherical lenses and corneas with little curvature, reflecting a lifestyle that is primarily aquatic (Beer, 1898; Northmore and Granda, 1991).

As the cornea does not aid in focusing the light under water, all refractive power rests within the lens. Accommodation changes optical power to allow the formation of a clear image of objects at different distances. If and how accommodation is achieved in sea turtles has not been clearly established. Early ophthalmological studies by Beer (1898) and later studies by Ehrenfeld and Koch (1967) found green and loggerhead sea turtles to be short-sighted (myopic) in air while appropriately accommodated (emmetropic) to mildly far-sighted (hyperopic) under water. This result was made plausible by the general consensus that sea turtles lack the anatomical ability to accommodate in the way freshwater turtles do (Walls, 1942; Ehrenfeld and Koch, 1967). Freshwater turtles, like many other reptiles, accommodate by deforming the lens with the help of musculature in the ciliary body and the iris. Small bony plates (scleral ossicles) located at the base of the cornea act as a stabilizing structure to prevent deformation of the eye ball (Duke-Elder, 1958; Figure 2.2).

In their detailed optical study, Northmore and Granda (1991) reached the conclusion that sea turtles are emmetropic in air and far-sighted (hyperopic) in sea water and suggested that sea turtles are capable of accommodation. The authors' explanation for the discrepancy in the literature was that the sea turtles tested in the earlier study might have accommodated incorrectly (Northmore and Granda, 1991). Given these different findings between studies, it is unclear if sea turtles are appropriately accommodated in air. However, some further evidence discussed in Section 2.3 (under visual acuity) suggests that they are not (Bartol et al., 2002).

A recent anatomical study on leatherback sea turtle eyes (Brudenall et al., 2008) has suggested a possible accommodative mechanism in sea turtles, different to the one found in freshwater turtles but shared with marine mammals such as the West Indian manatee (*Trichechus manatus*) and the short-finned pilot whale (*Globicephala macrorhynchus*; (Hatfield et al., 2003). Brudenall et al. (2008) found that leatherback sea turtles have weak musculature in the ciliary body, but a high level of vascularization. The ciliary body has a number of functions, including anchoring the lens in place and producing the aqueous humor, the fluid in the anterior chamber between the lens and the cornea. It has been suggested that aqueous humor production could allow the lens to move by changing the depth of the anterior chamber, as well as reshaping the ciliary body, which holds the lens in place (Hatfield et al., 2003). The extensive vascularization of the sea turtle ciliary body points toward enhanced aqueous humor production to achieve accommodation in this way. Interestingly, similarly to sea turtles, it has been difficult to establish whether the West Indian manatee accommodates in air (Hartman, 1979), pointing to a possible common limitation of this type of accommodation.

Leatherback sea turtles also have well-developed iris sphincter and iris dilator muscles, suggesting that the iris can change size (Brudenall et al., 2008). It is therefore possible that the iris could aid accommodation by squeezing the anterior part of the lens, as has been reported in other turtles (Walls, 1942; Duke-Elder, 1958), but thought to be absent in the green turtle and the hawksbill turtle (*Eretmochelys imbricata*; Ehrenfeld and Koch, 1967). Granda and Dvorak (1977) have also reported that the pupil of turtles in general is somewhat responsive to light, suggesting that beyond the purely accommodative function, the iris might have a further, but limited, role in regulating the amount of light entering the eye.

### 2.3.3 RETINA

Sea turtles have the standard vertebrate arrangement of a retina divided in seven layers (Bartol and Musick, 2001; Brudenall et al., 2008). The retina is inverted with the light-absorbing photoreceptors facing away from the incoming light, a consequence of vertebrate developmental constraints. The outermost layer, adjacent to the choroid, is the pigment epithelium. The photoreceptor layer in sea turtles contain both rods, for dim light vision, and cones, for bright light vision, which are similar in size (Bartol and Musick, 2001; Brudenall et al., 2008). Vertebrate photoreceptors are divided into two sections: the outer segment containing the visual pigments that absorb the light and the inner segment that contains all the supporting organelles for the cell. Sea turtles, as well as other reptiles, birds, and primitive fishes, have oil droplets in between the two segments, intercepting the incoming light before it reaches the outer segments. At the base of the inner segment is the outer nuclear layer, which contains the nuclei of the photoreceptors in a single layer (Bartol and Musick, 2001). The outer plexiform layer houses the synaptic connections between the photoreceptors and the interneurons (bipolar, amacrine, and horizontal cells), the nuclei of which make up the inner nuclear layer. With the inner plexiform layer in between, the interneurons then connect to the ganglion cells in the ganglion cell layer, which represent the last processing step for the neuronal signal within the retina, before further processing occurs in the brain. While the functional anatomy and electrophysiology of the retina have been extensively studied in the freshwater turtle *Pseudemys* (for a comprehensive review see Granda and Dvorak, 1977), very little is known about sea turtles.

### 2.3.4 FURTHER PROCESSING—RETINAL TARGETS IN THE SEA TURTLE BRAIN

The axons of the ganglion cells form the second cranial nerve, the optic nerve, which connects the eyes to the higher centers in the central nervous system (Figure 2.3). In reptiles in general, primary visual targets are found in both the forebrain (diencephalon) and the midbrain (mesencephalon). Within these brain areas, primary visual centers can be found in the hypothalamus, the thalamus, pretectum, superficial layers of the optic tectum (also called optic lobes), as well as the tegmentum (Wyneken, 2007). The contralateral projection targets, originating in one eye and crossing over to the opposite side of the bilateral brain (Figure 2.3A), appear consistent and well preserved among chelonian species, as extensive studies of 21 species from 9 families have shown (Hergueta et al., 1995). The number of primary visual targets that have been identified varies, depending upon the study; however, this appears to be primarily an issue of nomenclature (Papez, 1935; Bass and Northcutt, 1981; Hergueta et al., 1995). Interestingly, the extent and targets of the ipsilateral projections (those that project from the eye to the same side of the brain) differs among chelonians, including the two sea turtle species studied (Hergueta et al., 1995). While both the green turtle and the leatherback sea turtle show ipsilateral projections to the majority of primary visual centers, the extent of the projections is significantly larger in the leatherback sea turtle. The functional significance of this difference in projections is as yet unclear, as there was no consistent explanation based on taxonomy, lifestyle, or the level of binocular overlap of the eyes (Hergueta et al., 1992, 1995).

With the exception of the optic tectum, the function of the different primary visual centers has not been investigated in sea turtles. Lesion studies of the tectum of green turtle hatchlings showed that animals tended to circle when both sides of the tectum were damaged asymmetrically, suggesting that the turning mechanism might be guided by specific locations in both hemispheres (Mrosovsky et al., 1979).

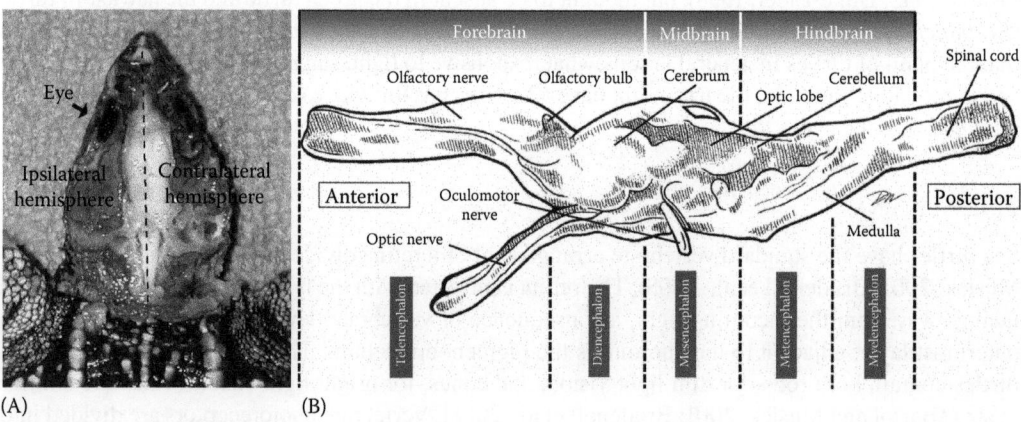

(A)                                        (B)

**FIGURE 2.3** Brain anatomy of the leatherback sea turtle (A) Exposed brain of a hatchling, illustrating the nomenclature of the two brain hemispheres in relation to the eye. (B) Schematic drawing of brain morphology, showing major regions, principal divisions, and landmark structures. (Courtesy of J. Wyneken.)

Extensive but unilateral damage to the tectum did not result in a noticeable change in behavior during sea finding. The authors suggested that the mechanism of sea finding is anatomically very robust, allowing the hatchling to sustain considerable damage while still being able to find the water.

## 2.4   SEA TURTLE VISUAL CAPABILITIES

Visual capabilities are usually divided into a number of key sensitivities. Sensitivity to light in general determines an animal's ability to detect dim light, while spectral sensitivity indicates which wavelengths of light the visual system can detect. Color vision refers to the ability to distinguish between these different wavelengths. Spatial and temporal resolutions are further parameters that determine how well the animal is able to detect detail in both space and time. Polarization sensitivity is another area of interest in sea turtle research. There have been a number of studies testing visual capabilities of sea turtles.

### 2.4.1   SENSITIVITY TO LIGHT

The visual system's sensitivity to dim light is determined by the light-gathering abilities of the optical apparatus, the sensitivity of the photoreceptors and strategies such as pooling of input within the retina or higher centers within the brain. Combining the current data available for sea turtles suggests that this group is not well adapted for dim light conditions.

Sea turtles, in common with most reptiles, have relatively small eyes compared to their body size (Howland et al., 2004). The pupil and lens are small in relation to the size of the eye (Northmore and Granda, 1991; Brudenall et al., 2008), resulting in a small aperture, limiting the level of illumination on the retina (Figure 2.4). The leatherback sea turtles measured by Brudenall et al. (2008) had an F-number (focal length/pupil diameter) of 4.3 with the pupil diameter as measured in the isolated eye after death or an F-number of 2.3 presuming a pupil fully dilated to the diameter of the lens. Northmore and Granda (1991) found similar F-numbers in their study of the green turtle, suggesting an eye design adapted to a diurnal lifestyle (Hughes, 1977).

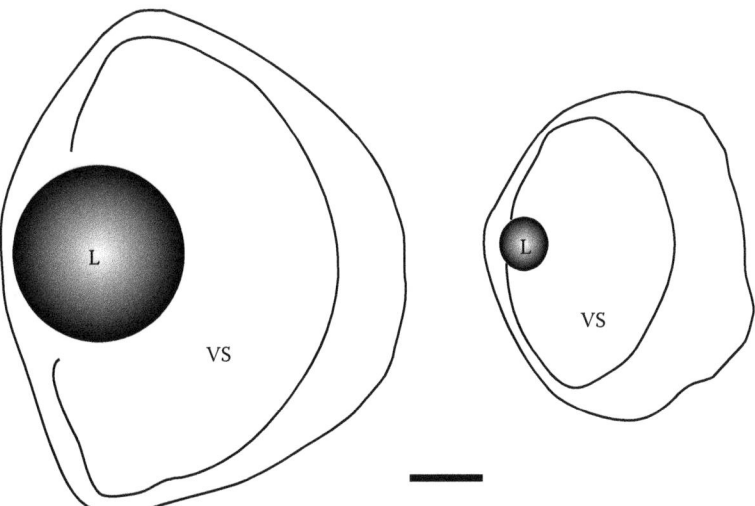

FIGURE 2.4   Simplified schematic drawing of cross-sections of the eye of the bigeye tuna (*Thunnus obesus*, left, body length 152 cm; Fritsches and Warrant, unpublished results) and adult leatherback sea turtle (right, curved carapace length 145 cm) (Modified from Brudenall, D. et al., *Vet. Ophthalmol.*, 11, 99, 2008.). Note the small size of eye and the lens (L) of the sea turtle compared to the similarly sized tuna, even though both species have a similar habitat and depth distribution. VS, vitreous space, fluid-filled space between the lens and the retina; scale bar 1 cm (to scale).

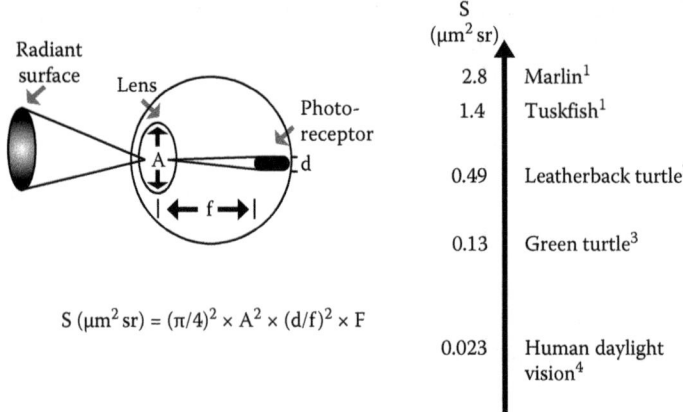

$$S\ (\mu m^2\, sr) = (\pi/4)^2 \times A^2 \times (d/f)^2 \times F$$

**FIGURE 2.5** Schematic drawing illustrating the optical sensitivity S, a measure of an eye's sensitivity to an extended field of light (in units of $\mu m^2$ sr). The optical sensitivity accounts for the optics of the eye (pupil diameter A and focal length f), the dimensions (diameter d), and the fraction of light F (between 0 and 1) absorbed by the photoreceptors. The larger the sensitivity S, the more sensitive the eyes are to low light intensities. Sea turtles have relatively low optical sensitivity compared, for instance, to fish. (1: Fritsches, K.A. et al., *J. Fish Biol.*, 63, 1347, 2003; 2: Brudenall, D. et al., *Vet. Ophthalmol.*, 11, 99, 2008; 3: Mäthger, L.M. et al., *Copeia*, 1, 169, 2007; 4: Land, M.F., Optics and vision in invertebrates, In *Handbook of Sensory Physiology,* ed. F. Crescitelli, Springer Verlag, Berlin, Germany, 1981, pp. 471–592.)

Sea turtles have a duplex retina, containing both rods and cones (loggerhead sea turtle: Bartol and Musick, 2001; green turtle: Mäthger et al., 2007; leatherback sea turtle: Brudenall et al., 2008), which provide specialized functionality for both dim and bright light conditions. Bartol and Musick (2001) found rod photoreceptors relatively evenly distributed throughout the retina in juvenile loggerhead sea turtles, and in similar densities to cone photoreceptors in most areas. Rod and cone numbers were approximately twice that of the ganglion cells in most regions. Animals adapted to vision in dim light tend to have rod-dominated retinas and rod numbers far exceeding those of the ganglion cells.

When calculating optical sensitivity, a measure that considers the optics of the eye as well as photoreceptor dimensions (Figure 2.5), loggerhead sea turtles (Mäthger et al., 2007) as well as leatherback sea turtles (Brudenall et al., 2008) did not possess an eye design particularly well adapted for vision in dim light. In fact, the leatherback sea turtle eye was found to have an optical sensitivity that was five times lower than that of the blue marlin eye. Even the blue tuskfish (*Choerodon albigena*), which shares the reef habitat with sea turtles, has an optical sensitivity that is nearly 3 times higher than the leatherback sea turtle and 10 times higher than the green turtle (Mäthger et al., 2007; Brudenall et al., 2008). While the aforementioned calculations were derived from cone dimensions, sea turtle rods are very similar in size (Bartol and Musick, 2001) and optical sensitivity of rods is likely to be similar.

Neural strategies to improve dim light vision have not been investigated in sea turtles. However, numerous experiments aimed at understanding the visual cues for sea finding have shown that sea turtles use dim light cues to orient on land (Lohmann et al., 1997). Section 2.4 outlines further how the sea turtle visual system might achieve this task.

### 2.4.2 SPECTRAL SENSITIVITY AND COLOR VISION

#### 2.4.2.1 Photoreceptor Spectral Sensitivity and Oil Droplets

One of the prerequisites for color perception is the presence of two or more photoreceptor classes each tuned to the detection of light within a specific wavelength band. A technique called micro-spectrophotometry (MSP) provides direct measurements of the absorbance spectrum of the photopigment in individual photoreceptors by shining a light beam through the light-absorbing outer segment of the individual receptor cell and recording the spectrum of the absorbed light. MSP thus gives a direct measurement of the preferred wavelength ($\lambda_{max}$) of the photopigment of different photoreceptors in a given retina. Liebman and Granda (1971) measured photoreceptor populations of the green turtle and identified four classes of photoreceptors that differed in their $\lambda_{max}$. The rod photoreceptor absorbed maximally at 500–505 nm, while three cone photoreceptor subgroups where found with $\lambda_{max}$ of 440, 502, and 562 nm. There is a strong suggestion that sea turtles possess a fourth visual pigment, absorbing in the UV waveband (Mäthger et al., 2007), but this has not as yet been identified using the MSP method.

In sea turtles only the rod photoreceptor and the accessory cone in the double cone pair do not contain oil droplets. The remaining cone types possess either an orange, yellow, or one of the two types of clear oil droplets (Walls, 1942; Granda and Haden, 1970; Liebman and Granda, 1971, 1975; Mäthger et al., 2007; Figure 2.6A). The oil droplets that appear clear in light microscopy can be differentiated into a type that fluoresces under ultraviolet light and a type that does not (Mäthger et al., 2007; Figure 2.6B). Oil droplets are distributed throughout the retina (Granda and Haden, 1970; Mäthger et al., 2007), with higher distributions in the central and temporal retina found by Mäthger et al. (2007).

Oil droplets are positioned so that most of the light passes through them before entering the outer segment of the photoreceptor (Wortel and Nuboer, 1986). Therefore, the pigmentation of the oil droplet affects the spectral sensitivity of cones by acting as a cut-off filter, usually shifting the sensitivity further toward the red end of the spectrum (Granda and O'Shea, 1972; Neumeyer and Jäger, 1985). This narrowing of the receptor's spectral sensitivity has been shown to enhance color discrimination by reducing the spectral sensitivity overlap between different photoreceptors (Vorobyev, 2003). Different pairings of photoreceptor spectral types and oil droplets further increases the number of possible spectral detection channels. In the freshwater turtle *Trachemys scripta elegans*, seven types of cones could be identified based on their oil droplet color and visual pigment absorbance (Loew and Govardovskii, 2001; Table 2.1). The authors described this as the most complex cone system of any vertebrate studied so far.

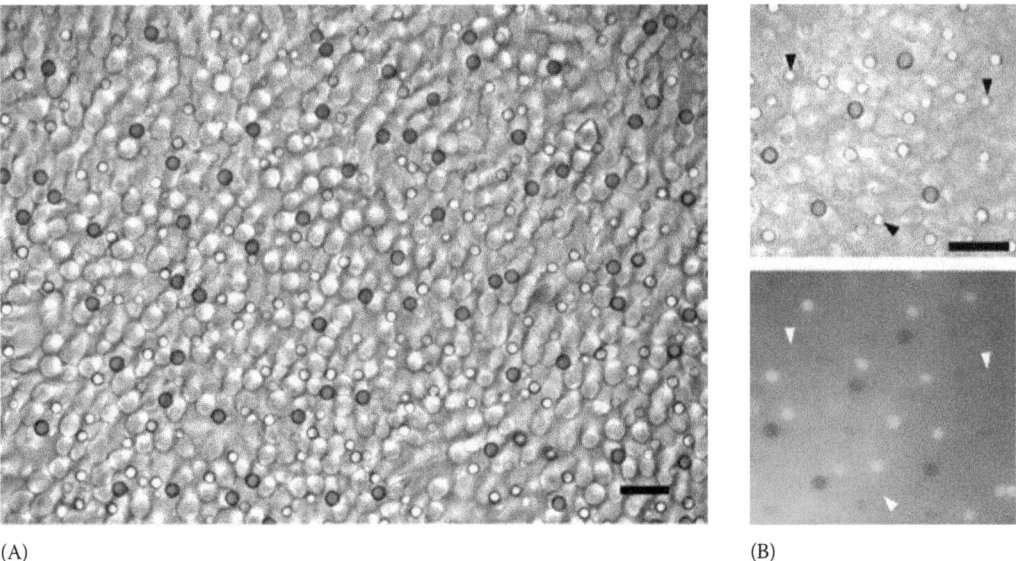

(A)                                                                                    (B)

FIGURE 2.6   Oil droplets in the retina of the green turtle. (A) Image of oil droplets of a whole-mounted retina (brightfield illumination). (B) Images of the same area of retina under brightfield illumination (top) and ultraviolet (UV) illumination (bottom). These photographs show that while most oil droplets fluoresce under UV light, a small number do not, suggesting that this group is transparent in the UV waveband. Scale bars 25 μm. (Modified from Mäthger L.M. et al., *Copeia*, 1, 165, 2007. With permission.)

## TABLE 2.1
## Potential Combinations of Visual Pigments and Oil Droplets in Two Species of Turtles

| | | Double cone | | | | | | |
|---|---|---|---|---|---|---|---|---|
| | Rod | Accessory member | Principal member | Cone | Cone | Cone | Cone | Cone |
| Red-eared turtle | -<br>518 | -<br>617 | Orange<br>617 | Red<br>617 | Orange<br>617 | Yellow<br>515 | Colourless<br>458 | Transparent<br>372 |
| Green turtle | -<br>502 | -<br>502 | ? | Orange<br>562 | Yellow<br>562 | Yellow<br>502 | Colourless<br>440 | Transparent<br>UV (?) |

The schematic drawing (top left) illustrates typical turtle photoreceptors (modified from Walls, 1942), with the inner segments shown in light gray and outer segments in dark gray. Only the single cone and the principal member of the double cone (to the right in the double cone) contain oil droplets (white circle). The table displays combinations of the oil droplet color and the visual pigment $\lambda_{max}$ (nm). The red-eared turtle data have been confirmed (Loew and Govardovskii, 2001), while the green turtle data are based on published (Liebman and Granda,1971; Granda and O'Shea, 1972) and unpublished results (Granda and Liebman in Granda and Dvorak, 1977), as well as strong indications that the green turtle retina contains UV cones (Mäthger et al., 2007). It has not as yet been identified which of the proposed oil droplet/visual pigment combinations is present in the principal member of the green turtle double cone.

In the green turtle, Granda and O'Shea (1972) outlined pairings of oil droplets and visual pigments based on their unpublished data. Combining this and suggestions from other studies, and extrapolating from the situation found in *T. scripta*, the green turtle has the potential for up to six different cone types (Table 2.1). Interestingly, Granda and Dvorak (1977) revealed further detail of Granda and Liebman's unpublished data, showing additional pairings of all three visual pigments with clear oil droplets. Whether this is a common occurrence in the sea turtle retina is still unknown.

### 2.4.2.2   Electrophysiological Studies on Spectral Sensitivity
While MSP data are only available for the green turtle, spectral sensitivity of several different sea turtle species has been investigated using electrophysiological means. Recordings of the electroretinogram (ERG) reveal the massed potential generated by the retina in response to a light stimulus. Hence the spectral sensitivities derived by this method give a view of the overall spectral range the animal is sensitive to, as well as the relative sensitivity to each part of the spectrum. These studies have shown that loggerhead and green turtles appear to have similar spectral sensitivities (Granda and O'Shea, 1972; Levenson et al., 2004), with the loggerhead sea turtle slightly less sensitive to short wavelengths of light. The leatherback sea turtle's spectral sensitivity appears to be different (Crognale et al., 2008; Horch et al., 2008), although the difference was not consistent between studies. Horch et al. (2008) reported little difference in the short-wavelength sensitivities of hatchlings, recorded to be as low as 340 nm, while the leatherback sea turtle was significantly less sensitive to wavelengths above 520 nm compared to the loggerhead sea turtle. Crognale et al. (2008) found that hatchling and adult leatherback sea turtles had an increased sensitivity in the short wavelength region of the spectrum, recorded at around 400 nm, but suggested a similar long wavelength pigment in leatherbacks as has been reported in green and loggerhead sea turtles, using the same methodology (Levenson et al., 2004). The two studies on leatherback sea turtles used different sample preparations, stimulus paradigms, and analyses, which could explain the discrepancies. Given the difference in lifestyle and diving depths between loggerhead and green sea turtles compared to the leatherback, differences in spectral sensitivities are to be expected and this area requires further investigation.

From ERG recordings it is also possible to draw conclusions about the photoreceptor complements, as well as the spectral shift, caused by the oil droplets. Granda and O'Shea (1972) identified three peaks using electrical responses in the green turtle, with the main peak at 520 nm and secondary peaks at 600 and 460 nm. Compared to cone visual pigments with $\lambda_{max}$ identified at 440, 502, and 562 nm using the MSP technique (Liebman and Granda, 1971), these peaks are indicative of the actual spectral sensitivity of the visual pigment and the oil droplets combined. Levenson et al. (2004) identified similar peaks in their electrophysiological study of green and loggerhead sea turtles. However, the long wavelength peak was found around 580 nm, whereas the short-wavelength peak was much less clearly defined. Neither study investigated wavelengths below 400 nm. Also for leatherbacks, the spectral sensitivities based on ERG recordings are highly indicative of the presence of several visual pigments (Crognale et al., 2008; Horch et al., 2008), including one with a peak sensitivity shorter than 450 nm (Crognale et al., 2008).

### 2.4.2.3  Sensitivity to Ultraviolet Light

Hatchlings of green, loggerhead, and leatherback sea turtles respond to ultraviolet light (wavelengths below 400 nm, which are not visible to humans; Witherington and Bjorndal, 1991b; Fritsches, unpublished observation, Figure 2.7), and so do adult green turtles (Ehrenfeld, 1968). Recent electrophysiological data from loggerhead and leatherback sea turtle hatchlings also show responses in the UV band (Horch et al., 2008). Unlike in many other large marine animals, the green turtle's lens transmits light in the UV band, allowing light with wavelengths as short as 300 nm to enter the eye (Mäthger et al., 2007).

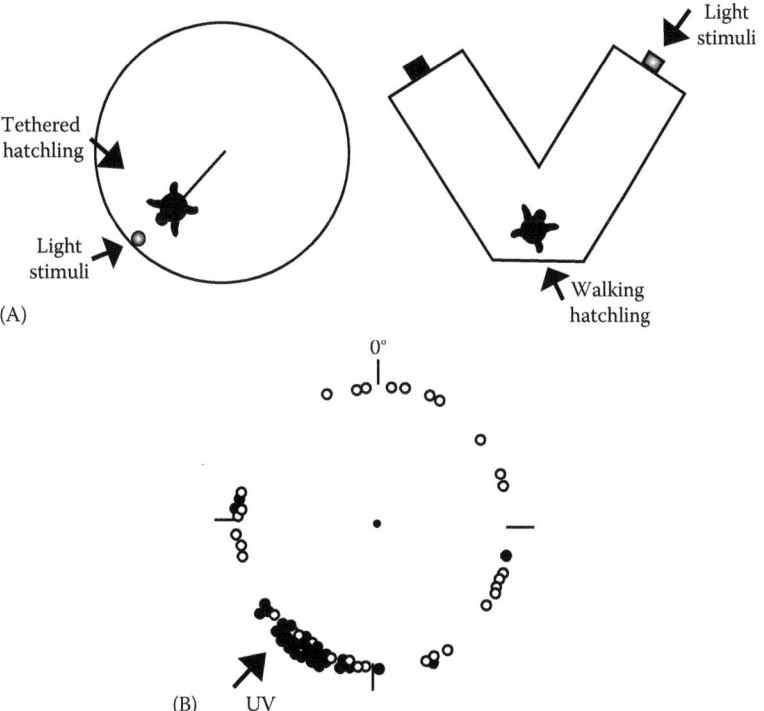

FIGURE 2.7  Behavioral light orientation experiments undertaken with sea turtle hatchlings. (A) Illustrations of two behavioral paradigms: tracking the orientation of a tethered swimming hatchling in relation to a light stimulus (left) or recording the choices of a freely walking hatchling in a Y-maze set-up (right). (B) Unpublished data (KA Fritsches) showing the orientation of a tethered loggerhead sea turtle hatchling to a UV LED light source (365 nm). The black dots show the orientation of the animal recorded (every 20 s over 10 min) while the UV LED is on. The white dots show the random positions of the same animal with the UV stimulus turned off.

While sea turtles have been shown to respond to UV light, it is as yet unknown if they can differentiate UV colors from other wavelengths. Most visual pigments have a second absorption peak in the ultraviolet (the ß-band), allowing UV detection as long as the UV light reaches the retina through a lens transparent to the UV waveband, as is the case in the green turtle (Mäthger et al., 2007) and many reef fish (Siebeck and Marshall, 2001). In order for sea turtles to differentiate the ultraviolet stimulus from other wavelengths, it requires a photoreceptor that solely absorbs in the UV band. The necessary MSP evidence for such a visual pigment is still missing in sea turtles. Liebman and Granda (1971) did not record spectra below 400 nm in their study, probably due to technical constraints. However, the presence of nonfluorescent oil droplets in the green turtle suggests the presence of a UV cone (Mäthger et al., 2007). In *T. scripta* nonfluorescent oil droplets have no selective absorbance at wavelengths greater than 325 nm and are paired with a visual pigment with a $\lambda_{max}$ in the UV spectral region (Loew and Govardovskii, 2001). It is therefore very likely that at least the green turtle can differentiate UV from other wavelengths, expanding its spectral range for color detection.

### 2.4.2.4 Behavioral Studies on Spectral Preferences

Due to the importance of visual stimuli for sea finding in both hatchling and adult sea turtles, there have been a number of studies investigating the behavioral preferences of sea turtles to light of different wavelengths. A common finding was that hatchlings preferred shorter wavelengths (Mrosovsky and Carr, 1967; Ehrenfeld, 1968; Mrosovsky and Shettleworth, 1968). Adult green turtles also showed higher orientation scores when wearing short wavelength spectacles compared to long wavelength ones, suggesting a higher sensitivity to shorter wavelengths (Ehrenfeld and Carr, 1967; Ehrenfeld, 1968).

Witherington and Bjorndal (1991b) found a surprising avoidance behavior to yellow light in loggerhead sea turtle hatchlings, but not in the other species tested (hatchlings of green turtle, olive ridley turtle [*Lepidochelys olivacea*], and hawksbill turtle [*Eretmochelys imbricata*]). In a y-maze choice experiment (Figure 2.7A) the loggerhead hatchlings strongly oriented away from high-intensity yellow light (a "xanthophobia" response to wavelengths between 550 and 600 nm), while other wavelengths produced a positive response. Dim yellow light did not produce any avoidance behavior (Witherington, 1992; Lohmann et al., 1997). It has been hypothesized that this response to yellow light helps negate the potentially confusing influence of celestial bodies for navigation (Witherington and Bjorndal, 1991b). However, it has since been shown that orientation improves with increasing moonlight, regardless of the position of the moon (Salmon and Witherington, 1995). Interestingly, a recent study (Fritsches, 2012) could not replicate avoidance behavior to yellow light in an Australian population of loggerhead sea turtle hatchlings. The hatchlings oriented toward yellow light at all intensities tested, with the experiment closely modeled on those undertaken by Witherington and Bjorndal (1991b) in Florida. This result raises the possibility that different populations of sea turtles may differ in their visual behavior (Fritsches, 2012).

Recent studies in juveniles of green, loggerhead, and leatherback sea turtles failed to show any wavelengths-specific orientation responses (Wang et al., 2007; Gless et al., 2008). The animals' orientation while swimming was studied, using chemical light sticks as stimuli in a darkened laboratory setting. However, a clear species difference in behavior became evident. Green and loggerhead sea turtles orientated toward all lightsticks presented (Wang et al., 2007), while the leatherback juveniles either ignored or oriented away from the same stimuli (Gless et al., 2008). Both studies were motivated by the question of whether chemical lightsticks used in longline fisheries can attract sea turtles. The results were unexpected as all three species are caught in longline fisheries (Lewison et al., 2004), with the more night-active and deep-diving leatherback sea turtle more likely to be exposed to these light sticks. While the practical applications of these findings are unclear, it is interesting to note that preferences for visual stimuli, at least in laboratory conditions, appear to be different for different species of sea turtles.

Given the extensive range of spectral sensitivities identified in the retina, it would be very interesting to establish behaviorally the extent to which sea turtles can discriminate wavelengths. Fehring (1972) trained juvenile loggerhead sea turtles to discriminate between broadband hues to obtain food, establishing that the behavioral training techniques required for establishing color vision are possible in this group. In a recent study, Young et al. (2012) showed that loggerhead sea turtles can distinguish between color targets of blue (450 nm), green (500 nm), and yellow (580 nm) irrespective of the stimulus intensity. This is the first conclusive evidence of color discrimination in sea turtles based on the stimulus wavelength alone. Testing a wider range of wavelengths, Neumeyer and Arnold (1989) could establish that the freshwater turtle *P. scripta* has true tetrachromatic color vision, which includes the UV cone as a fully functional component of a color vision system, one that is significantly more complex than that of humans.

## 2.4.3 SPATIAL RESOLUTION

### 2.4.3.1 Acuity in Water

Most visual scenes contain a variety of small details, and visual systems tend to have evolved to detect detail and important features in the most efficient way for the given species. Limiting factors for detecting spatial detail are the optics of the eye, the density and distribution of both the photoreceptors and ganglion cells of the retina (that determine the array of sampling stations that analyze the visual field), and how the visual information is processed in higher centers.

In sea turtles, visual acuity has been investigated using both electrophysiological and behavioral techniques. For their electrophysiological recordings, Bartol et al. (2002) used visually evoked potentials (VEPs), which are noninvasive recordings of compound field potentials generated by any neural tissue in the visual pathway beyond the retina (Riggs and Wooten, 1972). As response amplitude was correlated with the size of the black and white stripes of the stimulus, recording VEPs in juvenile loggerhead sea turtle revealed a visual resolving threshold of 0.13–0.215 (Bartol et al., 2002). This translates into an average visual angle of 5.4 min of arc, or 11 cycles of black and white stripes per degree of visual angle. Using an operant conditioning paradigm, Bartol (1999) trained juvenile loggerhead sea turtles to distinguish between a gray and a striped panel. During the experiment, acuity thresholds were determined by increasing the number of stripes on the panel until the sea turtle could no longer distinguish the gray from the striped panel. Under these conditions, loggerhead sea turtles showed an acuity threshold of 0.078, translating in a visual angle of 12.9 min of arc or 5 cycles of black and white stripes per degree of visual angle. To put these results in context, human behavioral visual threshold, using a similar paradigm, is far higher at about 1 min of arc or 60 cycles/degree (Riggs, 1965). On the other hand, small tuna with similar visual habitats to sea turtles showed similar grating acuities at 6–7 min/arc or 8–10 cycles/degree when tested in a behavioral paradigm (Nakamura, 1968).

### 2.4.3.2 Acuity in Air

Both the electrophysiological and the behavioral studies by Bartol et al. used visual conditions in water to derive visual acuity thresholds. For vision on land, on the other hand, the few studies done on sea turtles suggest that spatial resolution is impaired. The ability to find the sea does not seem to be affected when the animals are fitted with goggles containing diffusing filters (Ehrenfeld and Carr, 1967), suggesting that this task does not require the resolution of fine detail. Further studies have established that hatchlings integrate visual input over a large part of the visual field, extending over about 180° (Verheijen and Wildschut, 1973; Witherington, 1992). Interestingly, Bartol et al. (2002) recorded good electrophysiological responses when turtles were tested for their ability to detect fine spatial detail in air but they viewed the stimulus with water-filled goggles, recreating the visual optics experienced in water. When the same stimuli were presented to turtles without goggles, the authors reported unreliable responses and no detection threshold could be established, even using the largest stripes available. As only the optical properties of the turtles' vision were changed in this experiment, this suggests that sea turtles do not accommodate appropriately in air, as was suggested in previous studies (Beer, 1898; Ehrenfeld and Koch, 1967), resulting in reduced spatial resolution. So while an accommodative process is available in sea turtles (Brudenall et al., 2008), it might only be sufficient for accurate accommodation under water.

This does not mean that sea turtles are not responsive to shapes on land. While investigating the visual cues guiding sea finding, many studies in hatchlings have shown clear responses to different shapes and silhouettes (Mrosovsky and Shettleworth, 1968; Limpus, 1971; van Rhijn and van Gorkom, 1983; Salmon et al., 1992). However, these cues do not require high acuity but will be equally effective when viewed at low spatial resolution.

### 2.4.3.3 Best Area of Vision

In most animals the sampling array of neural elements in the retina is not homogeneously distributed. Peak spatial resolving power is usually only achieved by a small part of the retina, while the majority of the retinal area contains lower densities of cells. The topography of this cell distribution is highly correlated with the typical visual surrounds of an animal, following the terrain theory of vision (Hughes, 1977). Among sea turtles, the distribution of retinal cells has been studied in green, loggerhead, and leatherback sea turtle hatchlings (Oliver et al., 2000; Bartol and Musick, 2001; Figure 2.8). The retinae of all species contain a horizontal streak, which is a band-like increase in cell density, oriented horizontally along the visual axis. This allows the hatchlings to sample the visual horizon in greater detail, with cell densities approximately twice as high within the streak compared to the periphery (Oliver et al., 2000).

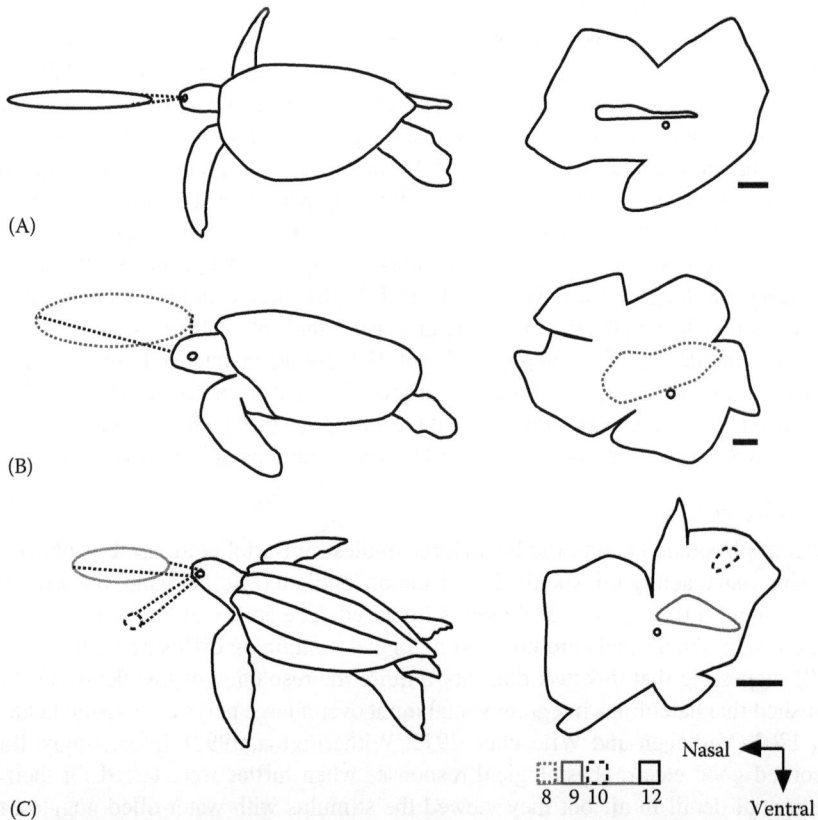

(A)

(B)

(C)

8  9 10     12

Nasal

Ventral

**FIGURE 2.8** Topography of ganglion cells in the sea turtle retina and the approximate corresponding areas in the visual field that are seen at the highest resolution. Estimated best areas of vision (left) in (A) green turtle, (B) loggerhead turtle, and (C) leatherback sea turtles, based on ganglion cell density maps of hatchlings of all three species (right, whole-mounted retinal contour maps). (Modified from Oliver, L. et al., *Mar. Freshwater Behav. Physiol.*, 33, 233, 2000.) Only the outline of the areas of highest ganglion cells densities ($\times 10^3/\text{mm}^2$) found in each retina are shown; open circles in the center of the retinal maps show the location of the optic nerve; scale bars = 1 mm.

Beyond this commonality, which was presumed to be present in adults also, Oliver et al. (2000) found some species-specific differences in the ganglion cell density. The visual streak was best developed in the green turtle, with the highest ganglion cell numbers concentrated within the streak, which was clearly visible as a narrowly defined band (Figure 2.8A). Loggerhead turtles had a lower ganglion cell density per unit area as well as a broader defined band (Figure 2.8B). Leatherback sea turtles had a weakly defined streak but showed a second specialization, an *area temporalis,* which is a small circular area of high cell density. The *area temporalis* of the leatherback sea turtle was located in the dorso-temporal retina, which images the visual field in front of the turtle, below the head (Figure 2.8C). Olivier et al. (2000) suggested that the stronger visual streaks found in the green and loggerhead sea turtle reflect the more surface and shallow water lifestyles of these species, where a horizon plays a larger role. Similar specializations have been found in reef-oriented fish that have a clear view of the sand-water horizon (Collin and Pettigrew, 1988). Leatherback sea turtles routinely dive deeper and have a more open visual habitat, which appears to be reflected in a weak visual streak and a circular area temporalis.

A horizontal streak is most effective if the animal maintains compensatory head or eye movements to maintain the horizontal visual axis, regardless of where the body is in space. Oliver et al. (2000) tested the head compensation of their hatchlings and found that both green and loggerhead sea turtles compensated tilt of their body well, keeping their head at a near-horizontal position at all times. Leatherback hatchlings, on the other hand, showed next to no compensation. The authors speculated that leatherback sea turtles compensate changes of body position using eye movements instead of head movements, which could not be verified due to the dark pigmentation and small size of the eye (Oliver et al., 2000). The ability for eye movements would certainly suit the eye design of the leatherback sea turtle well, as it allows the animal to point its *area temporalis*, the area of highest acuity, toward points of interest located anywhere in the visual field (Walls, 1962).

### 2.4.4 TEMPORAL RESOLUTION

While spatial resolving power defines the ability to see fine spatial detail, temporal resolution reflects the speed at which the visual system can process stimuli that vary in time. The stimulus most often used to identify temporal resolution is a flickering light, where the light changes its intensity over time. As the flicker frequency of the stimulus increases, the response of the visual system to the stimulus decreases. Using this type of flickering stimulus while recording the ERG (retinal mass potential) has been used to characterize the temporal resolution of sea turtles.

Comparisons between studies are difficult, as the criteria for establishing the detection threshold varies. Within studies, however, differences between species can be more easily detected. For instance, Horch and Salmon (2009) determined the detection threshold for leatherback sea turtle hatchlings at 10 Hz, while loggerhead sea turtle hatchlings responses extended to 15 Hz. In leatherback sea turtles Crognale et al. (2008) found that hatchlings show low but detectable responses at stimulus frequencies of 30 Hz, while adults of the same species reached the same low-response amplitude already at 20 Hz. Adult green and loggerhead sea turtles ceased to respond to flickering stimuli at 40 Hz at the light intensities tested (Levenson et al., 2004), with responses in isolated eyes of adult green turtles found as high as 57 Hz (Fritsches, unpublished results).

The pattern that emerges is that leatherback sea turtles are likely to have a lower temporal resolution than both green and loggerhead sea turtles. This suggests an adaptation of this species to its activity pattern at night and feeding at deeper and therefore dimmer depths. This correlation of a low temporal resolution and a habitat preference for deeper and thus dimmer waters has been shown in a range of marine animals groups, from pelagic predatory fishes (Fritsches et al., 2005) to deep sea crustaceans (Frank, 2000).

Little information is available on sea turtles' behavioral response to stimuli that flicker. Mrosovsky (1978) showed that green turtle hatchlings did not distinguish between flashing and steady light sources as long as both sources were matched for intensity. Flashing lights were presented at frequencies between 1 and 14 Hz, which is well within the flicker detection threshold determined by electrophysiology in hatchlings of the closely related loggerhead sea turtle (Horch and Salmon, 2009). Mrosovsky suggested that hatchlings may integrate the visual information over a timeframe of at least 1 s (based on Mrosovsky's lowest flash rate of 1 Hz) before responding. A recent study expanded the range of frequencies tested to include a very slow flicker frequencies of 0.1 Hz (1 sinusoidal flicker over 10 s) and the hatchlings still randomly responded to either the steady or the flickering light, as long as the stimuli were matched in intensity (Fritsches, 2012). As the animals did not appear to show slower responses at slow flicker frequencies, the most likely explanations is that the hatchlings decide on a crawling direction without using a long integration time, but choose the stimulus that appears brighter at the point in time when the decision is made (Fritsches, 2012).

### 2.4.5 POLARIZATION VISION

Natural light is often highly polarized, both in the atmosphere and underwater and many animals (particularly invertebrates) have the capacity to see it. A distinct circular pattern of polarized light is visible in the sky, centered on the disk of the sun or moon, and this pattern changes its position with the sun or the moon's movement during the day or night. Many animals use polarization cues for orientation and navigation (Horváth and Varjú, 2004). Furthermore, marine animals such as cephalopods use polarization sensitivity to aid detection of silvery fish and transparent prey, which are well camouflaged to the nonpolarization-sensitive observer, but appear highly visible against the background to a predator capable of detecting polarized light (Shashar et al., 1998, 2000).

Polarization vision, therefore, has a number of features that would be an advantage to sea turtles; but so far, the evidence points toward sea turtles not using this cue. Ehrenfeld and Carr (1967) tested if polarization vision aids sea finding in adult green turtles on the beach. They attached depolarizing goggles to the animals but found no evidence of disorientation, with the experimental animals finding the sea without problems. Mäthger et al. (2011) tested green turtle hatchlings for their ability to maintain their swimming direction when polarized light was the only directional cue. Tethered hatchlings, with their magnetic sense temporarily disabled (Irwin and Lohmann, 2003), maintained a steady course when presented with a point light source and a polarized light field above them. When the light source was switched off, the hatchlings did not maintain swimming direction, even though a strong polarized light pattern that could guide them remained above them.

One obvious interpretation is that the hatchlings cannot actually detect polarized light and were therefore not able to use this directional cue. Another interpretation, however, is that polarized light cues are not used in this early stage of the swimming frenzy, a possible conclusion based on the very strong hierarchy of sensory cues used in hatchlings in the early hours of their travels (Salmon and Wyneken, 1994; Lohmann and Lohmann, 1996; Lohmann et al., 1997). In birds, for example, it has been shown that polarized light is only used as a calibration reference at sunrise and sunset, but not at other times (Muheim, 2011). The light intensity and spectral composition of the light used might also have affected the results (Mäthger et al., 2011).

In vertebrates there is no clear anatomical correlate of polarization sensitivity, making it as yet impossible to identify that capability based upon eye or retinal design. It will therefore be necessary to undertake further behavioral studies, ideally in a context closely simulating natural conditions, to confirm or refute the hypothesis that sea turtles use polarized light as a cue for orientation, navigation, or prey detection.

## 2.5 VISUAL TASKS

The previous sections of this review outlined the visual environment encountered by sea turtles, as well as their visual capabilities. This section aims to provide interpretations of these findings for the visual tasks that sea turtles encounter throughout their lives and the likely adaptations that are required.

### 2.5.1 VISUAL TASKS ON LAND

#### 2.5.1.1 Adapting a Diurnal Visual System to a Dim Light Task

Sea turtle hatchlings appear to use a preprogrammed sequence of sensory cues in the first hours after emerging from the nest that leads them to the open ocean (Salmon and Wyneken, 1994; Lohmann et al., 1997). Sea turtles without any visual input typically do not orient seaward (Daniel and Smith, 1947; Carr and Ogren, 1960; Ehrenfeld and Carr, 1967; Figure 2.9), indicating that the visual sense is the key for successful orientation on land. It is therefore surprising that given the essential nature of visual orientation in this crucial phase of sea turtle survival, the eye is not well adapted to operate in dim light or detect spatial detail. Evidence suggests that visual acuity is low for terrestrial vision (Ehrenfeld and Koch, 1967; Bartol et al., 2002), and optical design is also not well suited to dim light (Mäthger et al., 2007; Brudenall et al., 2008). Furthermore, the presence of a range of cones with different visual pigments, and the presence of highly pigmented oil droplets (Liebman and Granda, 1971, 1975), also strongly favors bright light conditions.

It is therefore interesting to note that the detection of cues needed to find the sea require little acuity: these are typically spatially broad cues such as silhouettes of dunes and foliage or brightness cues along the horizon (for reviews, see Bartol and Musick, 2002).

Even though the small eye and high F-number of the sea turtle eye suggests an optical construction better suited for use in bright light, the lower than expected spatial and temporal resolution of the visual system may indicate the presence of spatial and temporal summation. In dim light, summation mechanisms can potentially compensate for limitations in optical sensitivity and increase the reliability of vision (Warrant, 1999). If present, summation could potentially allow

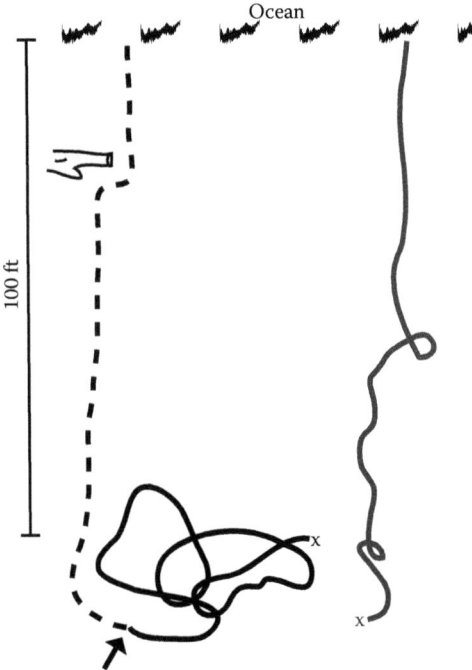

FIGURE 2.9  Beach orientation tracks of adult green turtles when returning to the ocean after nesting. The left track was recorded from a turtle wearing opaque filters that removed all visual input. The animal is clearly not able to locate the ocean. The arrow indicates the point when the blind-fold was removed, at which point the animal oriented correctly to the waterfront. The turtle leaving the right track wore diffusing filters, which removed any fine details as well as polarized light. The track shows some circling but ultimately successful sea finding. (Modified from Ehrenfeld, E.W. and Carr, A., *Animal Behav.*, 15, 25, 1967. With permission.)

sea turtles to reliably discriminate coarse and distant landmarks such as dunes and foliage during sea finding at night.

### 2.5.1.2   Effect of Light Pollution on Sea Finding

It is well established that artificial light disrupts effective sea finding in hatchlings, with the animals showing disorientation either by crawling toward the artificial light source or by crawling in circles (Verheijen, 1958; McFarlane, 1963; Philibosian, 1976; van Rhijn, 1979; Peters and Verhoeven, 1994; Salmon and Witherington, 1995). Sufficient ambient moonlight restores sea finding capability again, allowing the hatchlings to detect the appropriate guidance cues (Salmon and Witherington, 1995). The proposed mechanism of the disrupted sea-finding behavior is "light trapping," which is also found in birds and other animals (Verheijen, 1958). Unlike moonlight or sunlight, artificial light causes light trapping because it tends to be highly directional and provides little illumination of surrounding features. It has been suggested that hatchlings exposed to artificial lights on dark nights are not able to detect other directional cues such as elevation or shape cues, and hence fail to find the sea (Salmon and Witherington, 1995).

A number of strategies have been suggested to reduce at least some of the detrimental effect of artificial lighting on nesting beaches. Using light of specific wavelength (Witherington and Bjorndal, 1991a), lowering and shielding light, turning unnecessary lights off, dune restoration (Tuxbury and Salmon, 2005), as well as dimmer lights embedded in roadways have been suggested (Bertolotti and Salmon, 2005). However, the effectiveness of the different strategies requires careful study in order to choose the best available technology for reducing impact (Witherington and Martin, 2003).

### 2.5.2   Visual Tasks in the Oceanic Phase

Once swimming, loggerhead sea turtle hatchlings cease to use the horizon as a visual cue for orientation (Salmon and Wyneken, 1990) and appear to largely use nonvisual cues for maintaining their swimming direction after entering the sea (Salmon and Lohmann, 1989; Lohmann et al., 1990, 1997). Juvenile turtles maintain heading using multiple sensory input, including vision (Avens and Lohmann, 2003; Mott and Salmon, 2011).

Given the clear water of their pelagic habitat, the visual sense is likely to play a major role for turtle sensory perception. The main visual tasks facing sea turtles in oceanic waters are those of prey detection and predator avoidance. Studies in leatherback post-hatchlings have shown that visual cues elicit a strong prey capture response, increasing stroke rate, diving behavior, orientation toward prey items, and biting (Constantino and Salmon, 2003). Both prey detection and predator avoidance require the visual system to optimize the detection of objects against a background. Sea turtle visual pigments are well adapted to the predominant blue light in the open ocean, with spectral sensitivities shifted toward shorter wavelengths, compared to freshwater turtles (Granda and Dvorak, 1977; Granda, 1979).

### 2.5.2.1   What Is the Function of UV Detection in Sea Turtles?

A common strategy to increase contrast, and therefore object detection, under water is to reduce glare. UV light is strongly scattered in water, and therefore contributes significantly to glare, reducing both the contrast and the quality of the image (Losey et al., 1999). Furthermore, UV light is known to cause damage to retinal tissue. Thus, many animals, including many tropical marine fishes (Siebeck and Marshall, 2001) that share the same habitat as sea turtles, filter UV light with their lens or cornea, preventing short wavelengths entering the eye. Sea turtles, on the other hand, have UV-transparent optical media that do not cut out short-wavelength glare (Mäthger et al., 2007), although oil droplets have been suggested to do so (Walls and Judd, 1933). Nonfluorescent oil droplets, which suggest the presence of UV cones and are likely to be transparent to UV light, are relatively sparse (Mäthger et al., 2007). All other cones, with the exception of the accessory member of the double cone, contain oil droplets acting as cut-off filters against short-wavelength glare. Hence glare protection in sea turtles might be achieved at the level of the photoreceptors, rather than at the lens, while maintaining the ability to detect UV lights. How sea turtle retinas cope with the damaging impact of UV radiation is unknown.

But what is the benefit of UV sensitivity for sea turtles? The most likely function is that of increased prey detection. Johnsen and Widder (2001) showed that surface-oriented planktonic species are transparent at wavelengths above 400 nm but appear opaque in the UV band. This is due to increased scatter at the shorter wavelengths but also because planktonic organisms contain UV-protective pigments to protect themselves against tissue damage (Johnsen and Widder, 2001). Juvenile green and loggerhead sea turtles feed on planktonic prey, while leatherback sea turtles maintain a diet of transparent organisms throughout their lives (for a review see Bjorndal, 1997). Therefore, UV vision might drastically increase the ability to detect such prey in the pelagic environment.

Other functions for UV suggested for other animals with UV sensitivity are less likely, such as mate choice for instance. Many birds and fish show strong UV body patterning and inter- and intraspecific communication in the UV band (for review see Bennett and Cuthill, 1994; Losey et al., 1999). No such patterns or communication have been reported in sea turtles. UV sensitivity has also been linked to polarization vision with the possibility of improved navigation and orientation in sea turtles (Mäthger et al., 2011).

### 2.5.2.2 Spatial Resolution in the Aquatic Environment

Feeding on small planktonic prey requires a certain degree of spatial resolution in order to detect such prey. Sea turtle spatial resolution (between 5 and 11 cycles/degree, depending on method used: Bartol and Musick, 2002) is comparable to other aquatic animals such as most reef fishes (e.g., Collin and Pettigrew, 1989). As mentioned earlier, even in clear water light is scattered, creating a haze that degrades contrast with increasing distance (Lythgoe, 1979; Jagger and Muntz, 1993). For animals with largely aquatic vision, high visual acuity is therefore unlikely to be advantageous, due to the limited viewing distances and highly attenuated spatial frequencies (Guthrie and Muntz, 1993). In contrast, areal predators such as the eagle have been shown to have spatial visual acuity exceeding 130 cycles/degree (Reymond, 1985), a reflection of the animal's clear visual habitat and hunting strategy.

### 2.5.2.3 Dim Light Vision

Sea turtle eyes are adapted for bright light vision but detecting predators at night is a crucial visual task in the open ocean, where there are few places to hide within the water column. Loggerhead and green turtle hatchlings are often associated with sargassum mats and other flotsam in their open ocean phase (Carr, 1987), providing shelter from predation both during the day and night.

Leatherback sea turtles, on the other hand, appear to be active and feed both during the day and night, inhabiting the open waters for most of their lives, but they also feed closer to shore. They also dive to great depths where there is little or no downwelling light. Given this lifestyle, it is surprising that the leatherback optics do not appear to show many adaptations to improve vision in dim light. For instance, the eye of the leatherback sea turtle does not contain a tapetum, a reflective layer behind the retina that improves light sensitivity (Brudenall et al., 2008). In addition, the relatively small size of the lens and pupil (Northmore and Granda, 1991; Brudenall et al., 2008) limits the amount of light that enters the eye. Hence, based on current knowledge, leatherback sea turtles are not considered dim light specialists. In comparison, relative eye and lens size in fishes tends to increase with habitat depth within the top 1000 m of the water column in order to improve the visibility of objects viewed in the dim extended downwelling daylight (Warrant, 2000).

However, a large eye is not necessary to see bright high-contrast pinpoints of bioluminescence. The reliable detection of point light sources only requires a pupil of sufficient size to intercept the required numbers of photons radiating from source and a sufficient spatial resolution to accurately localize their origin (Warrant, 2000; Warrant and Locket, 2004). The eye of the sea turtle, which has a pupil several times the area of the pupils of many deep sea fishes that rely on the detection of bioluminescence to survive, would have no problem in seeing bioluminescence in dark water. Hence when diving deep, below several hundred meters during the day where bioluminescence becomes a stronger stimulus, or at night at any depth, leatherback sea turtles will be able to detect bioluminescent prey. It has been suggested that the relatively consistent distribution of luminous prey in deep water may be one of the main drivers for the leatherback's deep dives (Davenport, 1988).

### 2.5.3 COASTAL HABITAT

Green, loggerhead, and leatherback sea turtles are all known to feed in coastal waters. After their oceanic phase, green turtles switch to a predominately herbivorous diet, dominated by sea grass and algae, while loggerhead sea turtles appear to be highly opportunistic, feeding throughout the water column largely on invertebrates. Leatherback sea turtles maintain a diet of gelatinous organisms throughout their life, both in the open ocean and near the shore (for a review, see Bjorndal, 1997).

Coastal habitats are spectrally diverse with coral reefs among the most colorful habitats on earth. Inhabiting such a habitat, and having a varied diet, might go some way to explain why sea turtles have such a range of spectral sensitivities and most likely an exquisite ability for color discrimination. This is in stark contrast to whales and other mammalian marine carnivores, which appear to be monochromats, possibly for ontogenetic reasons and, at least in some species, as an adaptation to diving into deep, dark waters (Peichl et al., 2001; Levenson and Dizon, 2002; Levenson et al., 2006).

For many animals the major driver for a highly developed color vision system is color communication for mate choice and territorial disputes. For example, the stomatopod shrimp, a crustacean also known as mantis shrimp, has the highest diversity of visual pigments and spectral classes in the animal kingdom (Cronin and Marshall, 1989) and use bright species-specific coloration during territorial displays (Caldwell and Dingle, 1975). This is unlikely to be the case for sea turtles, where conspecifics do not posses highly colorful and spectrally diverse features. However, little is known about whether sea turtles do communicate with visual signals or produce any kind of visual display.

### 2.5.3.1 Ontogenetic Differences in Visual Capabilities

While our knowledge of sea turtle visual capabilities is not extensive among different age groups, the question arises whether there are changes of visual capabilities with age. Similar to many vertebrates, the retina continues to grow throughout the sea turtle's lifetime, increasing the retinal area, as well as the dimensions of cells such as the photoreceptors and oil droplets (Mäthger et al., 2007). This also usually increases the spatial resolving power as the animal's eye increases in size and retinal cells are added, as has been shown in many fish such as the planktivorous sunfish (Hairston et al., 1982).

Ocular changes with age at short wavelengths were suggested for the loggerhead sea turtle and may also possibly occur in the green turtle (Levenson et al., 2004). Levenson et al. (2004) suggested that the lack of a defined peak at shorter wavelengths found in their study might have been caused by an age-related reduction in the transmission of the optical components in that part of the spectrum. However, no such degeneration of optical media was reported in two green turtles estimated to be between 10–30 years old (Mäthger et al., 2007).

It has been shown in birds that spectral discrimination is influenced by both diet (Bowmaker et al., 1993) and by the predominant light intensities in the environment (Hart et al., 2006). This is caused by changes in oil droplet pigmentation and can be manifested within weeks. It is therefore possible (though not yet documented) that changes in the habitats and feeding patterns of sea turtles might affect the spectral sensitivity of the animals at different periods of their lives.

## 2.6  CONCLUSION

Sea turtle eyes have retained many of their reptilian features, such as relatively small eye and pupil size, but at the same time show adaptations suited to a predominantly aquatic lifestyle, such as a flat cornea and a likely inability to appropriately accommodate in air. Sea turtles have the potential for exquisite color vision; however, the full extent of their color discrimination abilities remains to be tested. Their spectral range extends into the UV waveband, which is likely to aid detection of planktonic and transparent prey. Exactly why these animals have retained such a variety of visual pigments and oil droplets is as yet unknown. For instance, little information is available on the role color discrimination may play in behaviors such as feeding or mate choice. There is certainly a need for more extensive and species-specific work on sea turtle behavior in their natural habitat as well as more lab-based and therefore controlled behavioral experiments on color vision in sea turtles to help solve this puzzle.

Given their nocturnal activities on land and, at least in case of the leatherback sea turtle, deep-diving and nocturnal feeding behavior, sea turtle eyes are surprisingly specialized for bright light vision. However, orientation cues used on land at night appear to be based on broad spatial features for which a light-sensitive eye or high acuity is not required. In addition, while the optics of the eyes are not optimized for deep, dimly lit waters, the pupil size is sufficiently large to see bioluminescent point-source light, allowing species such as the leatherback sea turtle to detect bioluminescent prey at depth and at night. The limitations of this form of prey detection and how vision may interact with the other senses are still largely unknown in sea turtles, but expanding our knowledge of this group's visual ecology will aid our understanding of habitat use and feeding strategies in these ocean travelers.

## REFERENCES

Avens, L. and K. J. Lohmann. 2003. Use of multiple orientation cues by juvenile loggerhead sea turtles *Caretta caretta*. *Journal of Experimental Biology* 206: 4317–4325.

Bartol, S. M. 1999. Morphological, electrophysiological and behavioural investigation of the visual acuity of the juvenile loggerhead sea turtle (*Caretta caretta*). PhD thesis. Faculty of the School of Marine Science, College of William and Mary, Williamsburg, VA.

Bartol, S. M. and J. A. Musick. 2001. Morphology and topographically organization of the retina of juvenile loggerhead sea turtles (*Caretta caretta*). *Copeia* 3: 718–725.

Bartol, S. M. and J. A. Musick. 2002. Sensory biology of sea turtles. In *The Biology of Sea Turtles II*, eds. P. L. Lutz, J. A. Musick, and J. Wyneken. pp. 79–102. Boca Raton, FL: CRC Press.

Bartol, S., Musick, J., and A. Ochs. 2002. Visual acuity thresholds of juvenile loggerhead sea turtles (*Caretta caretta*): An electrophysiological approach. *Journal of Comparative Physiology A: Sensory, Neural, and Behavioral Physiology* 187: 953–960.

Bass, A. H. and R. G. Northcutt. 1981. Primary retinal targets in the Atlantic loggerhead sea turtle, *Caretta caretta*. *Cell and Tissue Research* 218: 253–264.

Beer, T. 1898. Die Accommodation des Auges bei den Reptilien. *Pflügers Archiv für die gesammte Physiologie des Menschen und der Thiere* 69: 507–568.

Bennett, A. T. D. and I. C. Cuthill. 1994. Ultraviolet vision in birds; what is its function. *Vision Research* 34: 1471–1478.

Bertolotti, L. and M. Salmon. 2005. Do embedded roadway lights protect sea turtles? *Environmental Management* 36: 702–710.

Bjorndal, K. 1997. Foraging ecology and nutrition in sea turtles. In *The Biology of Sea Turtles,* eds. P. L. Lutz and J. A. Musick. pp. 199–231. Boca Raton, FL: CRC Press.

Bowmaker, J. K., Kovach, J. K., Whitmore, A. V., and E. R. Loew. 1993. Visual pigments and oil droplets in genetically manipulated and carotenoid deprived quail: A microspectrophotometric study. *Vision Research Supplement* 33: 571–578.

Brudenall, D., Schwab, I. R., and K. A. Fritsches. 2008. Ocular morphology of the leatherback sea turtle (*Dermochelys coriacea*). *Veterinary Ophthalmology* 11: 99–110.

Burne, R. H. 1905. Notes on the muscular and visceral anatomy of the leathery turtle (*Dermochelys coriacea*). *Proceedings of the Zoological Society of London* 73: 219–324.

Bustard, H. R. 1967. Mechanism of nocturnal emergence from the nest in green turtle hatchlings. *Nature* 214: 317.

Caldwell, R. L. and H. Dingle. 1975. Ecology and evolution of agonistic behavior in stomatopods. *Naturwissenschaften* 62: 214–222.

Carr, A. F. 1987. New perspectives on the pelagic stage of sea turtle development. *Conservation Biology* 1: 103–121.

Carr, A. L. and L. Ogren. 1960. The ecology and migrations of sea turtles. IV. The green turtle in the Caribbean Sea. *Bulletin of the American Museum of Natural History* 121: 6–45.

Clarke, G. L. and E. J. Denton. 1962. Light and animal life. In *The Sea*, ed. M. N. Hill, pp. 456–468. London, U.K.: Wiley-Interscience.

Clarke, G. L. and C. J. Hubbard. 1959. Quantitative records of the luminescent flashing of oceanic animals at great depths. *Limnology and Oceanography* 4: 163–180.

Collin, S. P. and J. D. Pettigrew. 1988. Retinal topography in reef teleosts. II. Some species with prominent horizontal streaks and high-density areae. *Brain, Behaviour and Evolution* 31: 283–295.

Collin, S. P. and J. D. Pettigrew. 1989. Quantitative comparison of the limits on visual spatial resolution set by the ganglion cell layer in twelve species of reef teleosts. *Brain, Behaviour and Evolution* 34: 184–192.

Constantino, M. A. and M. Salmon. 2003. Role of chemical and visual cues in food recognition by leatherback posthatchlings (*Dermochelys coriacea* L). *Zoology* 106: 173–181.

Crognale, M. A., Eckert, S. A., Levenson, D. H., and C. A. Harms. 2008. Leatherback sea turtle *Dermochelys coriacea* visual capacities and potential reduction of bycatch by pelagic longline fisheries. *Endangered Species Research* 5: 249–256.

Cronin, T. W. and N. J. Marshall. 1989. A retina with at least ten spectral types of photoreceptors in a mantis shrimp. *Nature* 339: 137–140.

Daniel, R. S. and K. U. Smith. 1947. The sea-approaching behaviour of the neonate loggerhead turtle (*Caretta caretta*). *Journal of Comparative Physiology* 40: 413.

Davenport, J. 1988. Do diving leatherbacks pursue glowing jelly? *British Herpetological Society Bulletin* 24: 20–21.

Denton, E. J. 1990. Light and vision at depths greater than 200 metres. In *Light and Life in the Sea*, eds. P. Herring, A. Campbell, M. Whitfield, and L. Maddock. pp. 127–148. Cambridge, U.K.: Cambridge University Press.

Doyle, T. K., Houghton, J. D., Suilleabhain, P. F. et al. 2008. Leatherback turtles satellite-tagged in European waters. *Endangered Species Research* 4: 23–31.

Duke-Elder, W. S. 1958. *The Eye in Evolution*. London, U.K.: Henry Kimpton.

Eckert, S. A., Eckert, K. L., Ponganis, P., and G. L. Kooyman. 1989. Diving and foraging behavior of leatherback sea turtles (*Dermochelys coriacea*). *Canadian Journal of Zoology* 67: 2834–2840.

Eckert, S. A., Nellis, D. W., Eckert, K. L., and G. L. Kooyman. 1986. Diving pattern of two leatherback sea turtles (*Dermochelys coriacea*) during internesting intervals at Sandy Point, St Croix, US Virgin Islands. *Herpetologica* 42: 381–388.

Ehrenfeld, E. W. 1968. The role of vision in the sea finding orientation of the green turtle (*Chelonia mydas*). 2. Orientation mechanism and range of spectral sensitivity. *Animal Behaviour* 16: 281–287.

Ehrenfeld, E. W. and A. Carr. 1967. The role of vision in the sea-finding orientation of the green turtle (*Chelonia mydas*). *Animal Behaviour* 15: 25–36.

Ehrenfeld, E. W. and A. L. Koch. 1967. Visual accommodation in the green turtle. *Science* 155: 827–828.

Fehring, W. K. 1972. Hue discrimination in hatchling loggerhead turtles (*Caretta caretta caretta*). *Animal Behaviour* 20: 632–636.

Frank, T. M. 2000. Temporal resolution in mesopelagic crustaceans. *Philosophical Transactions of the Royal Society of London B* 355: 1195–1198.

Frank, T. M. and E. A. Widder. 1996. UV light in the deep sea: in situ measurements of downdwelling irradiance in relation to the visual threshold sensitivity of UV-sensitive crustaceans. *Marine and Freshwater Behavioural Physiology* 27: 189–197.

Fritsches, K. A. 2012. Australian loggerhead sea turtle hatchlings do not avoid yellow. *Marine and Freshwater Behaviour and Physiology* 45(2):79–89.

Fritsches, K. A., Brill, R. W., and E. J. Warrant. 2005. Warm eyes provide superior vision in swordfishes. *Current Biology* 15: 55–58.

Fritsches, K. A., Litherland, L., Thomas, N., and J. Shand. 2003. Cone visual pigments and retinal mosaics in the striped marlin. *Journal of Fish Biology* 63: 1347–1351.

Gless, J. M., Salmon, M., and J. Wyneken. 2008. Behavioral responses of juvenile leatherbacks *Dermochelys coriacea* to lights used in the longline fishery. *Endangered Species Research* 5: 239–247.

Granda, A. M. 1979. Eyes and their sensitivity to light of different wavelengths. In *Turtles: Perspectives and Research*, eds. M. Harless and H. Morlock. pp. 247–265. New York: John Wiley & Sons.

Granda, A. M. and C. A. Dvorak. 1977. Vision in turtles. In *The Visual System in Vertebrates,* ed. F. Crescitelli. pp. 451–495. Berlin, Germany: Springer Verlag.

Granda, A. M. and K. W. Haden. 1970. Retinal oil globule counts and distributions in two species of turtles: *Pseudemys scripta elegans* (Wied) and *Chelonia mydas mydas* (Linnaeus). *Vision Research* 10: 79–84.

Granda, A. M. and P. O'Shea. 1972. Spectral sensitivity of the green turtle (*Chelonia mydas mydas*) determined by electrical responses to heterochromatic light. *Brain, Behaviour and Evolution* 5: 143–154.

Guthrie, D. M. and W. R. A. Muntz. 1993. Role of vision in fish behaviour. In *Behaviour of Teleost Fishes,* ed. T. J. Pitcher. pp. 87–128. London, U.K.: Chapman & Hall.

Hairston, N. G. J., Li, K. T., and S. S. J. Easter. 1982. Fish vision and the detection of planktonic prey. *Science* 218: 1240–1242.

Hart, N. S., Lisney, T. J., and S. P. Collin. 2006. Cone photoreceptor oil droplet pigmentation is affected by ambient light intensity. *Journal of Experimental Biology* 209: 4776–4787.

Hartman, D. S. 1979. Ecology and behavior of the manatee (*Trichechus manatus*) in Florida. Special Publication No 5, American Society of Mammalogists, Lawrence, KS, 1–153.

Hatfield, J. R., Samuelson, D. A., Lewis, P. A., and M. Chisholm. 2003. Structure and presumptive function of the iridocorneal angle of the West Indian manatee (*Trichechus manatus*), short-finned pilot whale (*Globicephala macrorhynchus*), hippopotamus (*Hippopotamus amphibius*), and African elephant (*Loxodonta africana*). *Veterinary Ophthalmology* 6: 35–43.

Hays, G. C., Houghton, J. D. R., Isaacs, C., King, R. S., Lloyd, C., and P. Lovell. 2004a. First records of oceanic dive profiles for leatherback turtles, *Dermochelys coriacea*, indicate behavioural plasticity associated with long-distance migration. *Animal Behaviour* 67: 733–743.

Hays, G. C., Houghton, J. D. R., and A. E. Myers. 2004b. Pan-Atlantic leatherback turtle movements. *Nature* 429: 6991.

Hendrickson, J. R. 1958. The green sea turtle, *Chelonia mydas* (Linn.) in Malaysia and Sarawak. *Proceedings of the Zoological Society of London* 130: 455–535.

Hergueta, S., Lemire, M., Ward, R., Repérant, J., Rio, J. P., and C. Weidner. 1995. Interspecific variation in the chelonian primary visual system. *Brain Research* 36: 171–193.

Hergueta, S., Ward, R., Lemire, M., Rio, J. P., Repérant, J., and C. Weidner. 1992. Overlapping visual fields and ipsilateral retinal projections in turtles. *Brain Research Bulletin* 29: 427–433.

Herring, P. J. 1978. Bioluminescence in invertebrates other than insects. In *Bioluminescence in Action,* ed. P. J. Herring. pp. 199–240. London, U.K.: Academic Press.

Herring, P. J. 2000. Species abundance, sexual encounter and bioluminescent signalling in the deep sea. *Philosophical Transactions of the Royal Society of London B* 355: 1273–1276.

Horch, K. W., Gocke, J. P., Salmon, M., and R. B. Forward. 2008. Visual spectral sensitivity of hatchling loggerhead (*Caretta caretta* L.) and leatherback (*Dermochelys coriacea* L.) sea turtles, as determined by single-flash electroretinography. *Marine and Freshwater Behaviour and Physiology* 41: 79–91.

Horch, K. W. and M. Salmon. 2009. Frequency response characteristics of isolated retinas from hatchling leatherback (*Dermochelys coriacea* L.) and loggerhead (*Caretta caretta* L.) sea turtles. *Journal of Neuroscience Methods* 178: 276–283.

Horváth, G. and D. Varjú. 2004. *Polarized Light in Animal Vision: Polarization Patterns in Nature.* Berlin, Germany: Springer Verlag.

Howland, H. C., Merola, S., and Basarab, J. R. 2004. The allometry and scaling of the size of the vertebrate eye. *Vision Research* 44: 2043–2065.

Hudson, D. M. and P. L. Lutz. 1986. Salt gland function in the leatherback sea turtle, *Dermochelys coriacea. Copeia* 1: 247–249.

Hughes, A. 1977. The topography of vision in mammals of contrasting life style: Comparative optics and retinal organisation. In *The Visual System in Vertebrates*, ed. F. Crescitelli. pp. 613–756. Berlin, Germany: Springer Verlag.

Irwin, W. P. and K. J. Lohmann. 2003. Magnet-induced disorientation in hatchling loggerhead sea turtles. *Journal of Experimental Biology* 206: 497–501.

Jagger, W. S. and W. R. A. Muntz. 1993. Aquatic vision and the modulation transfer properties of unlighted and diffusely lighted natural waters. *Vision Research* 33: 1755–1763.

Jerlov, N. G. 1976. *Marine Optics,* Amsterdam, the Netherlands: Elsevier.

Jessop, T. S., Limpus, C. J., and J. M. Whittier. 2002. Nocturnal activity in the green sea turtle alters daily profiles of melatonin and corticosterone. *Hormones and Behavior* 41: 357–365.

Johnsen, S., Kelber, A., Warrant, E. J. et al. 2006. Twilight and nocturnal illumination and its effects on color perception by the nocturnal hawkmoth *Deilephila elpenor. Journal of Experimental Biology* 209: 789–800.

Johnsen, S. and E. A. Widder. 2001. Ultraviolet absorption in transparent zooplankton and its implications for depth distribution and visual predation. *Marine Biology* 138: 717–730.

Land, M. F. 1981. Optics and vision in invertebrates. In *Handbook of Sensory Physiology,* ed. F. Crescitelli. pp. 471–592. Berlin, Germany: Springer Verlag.

Levenson, D. H. and A. Dizon. 2002. Genetic evidence for the ancestral loss of short-wavelength-sensitive cone pigments in mysticete and odontocete cetaceans. *Proceedings of the Royal Society London B* 270: 673–679.

Levenson, D., Eckert, S., Crognale, M., Deegan, J. I., and G. Jacobs. 2004. Photopic spectral sensitivity of green and loggerhead sea turtles. *Copeia* 4: 908–911.

Levenson D.H., Ponganis P.J., Crognale M.A., Deegan, J.F.II, Dizon, A., and G. H. Jacobs. 2006. Visual pigments of marine carnivores: Pinnipeds, polar bear, and sea otter. *Journal of Comparative Physiology A* 192(8): 833–843.

Lewison, R. L., Freeman, S. A., and L. B. Crowder. 2004. Quantifying the effects of fisheries on threatened species: The impact of pelagic longlines on loggerhead and leatherback sea turtles. *Ecology Letters* 7: 221–231.

Liebman, P. A. and A. M. Granda. 1971. Microspectrophotometric measurements of visual pigments in two species of turtle, *Pseudemys scripta* and *Chelonia mydas. Vision Research* 11: 105–114.

Liebman, P. A. and A. M. Granda. 1975. Super dense carotenoid spectra resolved in single cone oil droplets. *Nature* 253: 370–372.

Limpus, C. J. 1971. Sea turtle ocean finding behaviour. *Search* 2: 385–387.

Loew, E. R. and V. I. Govardovskii. 2001. Photoreceptors and visual pigments in the red-eared turtle, *Trachemys scripta elegans. Visual Neuroscience* 18: 753–757.

Lohmann, K. J. and C. M. F. Lohmann. 1996. Orientation and open-sea navigation in sea turtles. *Journal of Experimental Biology* 199: 73–81.

Lohmann, K. J., Salmon, M., and J. Wyneken. 1990. Functional autonomy of land and sea orientation systems in sea turtle hatchlings. *Biological Bulletin* 179: 214–218.

Lohmann, K. J., Witherington, B. E., Lohmann, C. M. F., and M. Salmon. 1997. Orientation, navigation, and natal beach homing in sea turtles. In *The Biology of Sea Turtles,* eds. P. L. Lutz and J. A. Musick. pp. 107–135. Boca Raton, FL: CRC Press.

Losey, G. S., Cronin, T. W., Goldsmith, T. H., Hyde, D., Marshall, N. J., and W. N. Mcfarland. 1999. The UV visual world of fishes: A review. *Journal of Fish Biology* 54: 921–943.

Lutcavage, M. E. and P. L. Lutz. 1997. Diving physiology. In *The Biology of Sea Turtles,* eds. P. L. Lutz and J. A. Musick. pp. 277–296. Boca Raton, FL: CRC Press.

Lythgoe, J. N. 1979. *The Ecology of Vision.* Oxford, U.K.: Clarendon Press.

Lythgoe, J. N. 1988. Light and vision in the aquatic environment. In *Sensory Biology of Aquatic Animals,* ed. J. Atema. pp. 57–82. New York: Springer Verlag.

Mäthger, L. M., Litherland, L., and K. A. Fritsches. 2007. An anatomical study of the visual capabilities of the green turtle, *Chelonia mydas. Copeia* 1: 169–179.

Mäthger, L. M., Lohmann, K. J., Limpus, C. J., and K. A. Fritsches. 2011. An unsuccessful attempt to elicit orientation responses to linearly polarized light in hatchling loggerhead sea turtles (*Caretta caretta*). *Philosophical Transactions of the Royal Society of London B* 366: 757–762.

McFarlane, R. 1963. Disorientation of loggerhead hatchlings by artificial road lighting. *Copeia* 1: 153.

Moon, P. 1940. Proposed standard solar-radiation curves for engineering use. *Journal of the Franklin Institute* 230: 583–617.

Mott, C. R. and M. Salmon. 2011. Sun compass orientation by juvenile green sea turtles (Chelonia mydas). *Chelonian Conservation and Biology* 10(1): 73–81.

Mrosovsky, N. 1978. Effects of flashing lights on sea-finding behaviour of green turtles. *Behavioral Biology* 22: 85–91.

Mrosovsky, N. and A. Carr. 1967. Preference for light of short wavelengths in hatchling green turtles, *Chelonia mydas,* tested at their natural nesting beaches. *Behavior* 28: 217–231.

Mrosovsky, N., Granda, A. M., and T. Hay. 1979. Seaward orientation of hatchling turtles: Turning systems in the optic tectum. *Brain, Behaviour and Evolution* 16: 203–221.

Mrosovsky, N. and S. Shettleworth. 1968. Wavelength preferences and brightness cues in the water-finding behavior of sea turtles. *Behavior* 32: 211–257.

Muheim, R. 2011. Behavioural and physiological mechanisms of polarized light sensitivity in birds. *Philosophical Transactions of the Royal Society of London B* 366: 763–771.

Nakamura, E. L. 1968. Visual acuity of two tunas, *Katsuwonus pelamis* and *Euthynnus affinis*. *Copeia* 1: 41–49.

Neumeyer, C. and K. Arnold. 1989. Tetrachromatic colour vision in goldfish and turtle. In *Seeing Contour and Colour*, eds. J. J. Kulikowski, C. M. Dickinson, and I. J. Murray. Oxford, Pergamon Press, pp. 617–631.

Neumeyer, C. and J. Jäger. 1985. Spectral sensitivity of the freshwater turtle *Pseudemys scripta elegans*: Evidence for the filter-effect of colored oil droplets. *Vision Research* 25: 833–838.

Northmore, D. P. M. and A. M. Granda. 1991. Ocular dimensions and schematic eyes of freshwater and sea turtles. *Visual Neuroscience* 7: 627–635.

Odgen, J. C., Robinson, L., Whitlock, K., Daganhardt, H., and R. Cebula. 1983. Diel foraging patterns in juvenile green turtles (*Chelonia mydas* L.) in St. Croix United States Virgin Islands. *Journal of Experimental Marine Biology and Ecology* 66: 199–205.

Oliver, L., Salmon, M., Wyneken, J., Hueter, R., and T. Cronin. 2000. Retinal anatomy of hatchling sea turtles: Anatomical specializations and behavioural correlates. *Marine and Freshwater Behaviour and Physiology* 33: 233–248.

Papez, J. W. 1935. Thalamus of turtles and thalamic evolution. *Journal of Comparative Neurology* 61: 433–475.

Peichl, L., Behrmann, G., and R. H. H. Kroger. 2001. For whales and seals the ocean is not blue: A visual pigment loss in marine mammals. *European Journal of Neuroscience* 13: 1520–1528.

Peters, A. and J. Verhoeven. 1994. Impact of artificial lighting on the seaward orientation of hatchling loggerhead turtles. *Journal of Herpetology* 28: 112–114.

Philibosian, R. 1976. Disorientation of hawksbill turtle hatchlings, *Eretmochelys imbricata*, by stadium lights. *Copeia* 1976: 824.

Pritchard, P. C. H. 1997. Evolution, phylogeny, and current status. In *The Biology of Sea Turtles,* eds. P. L. Lutz and J. A. Musick. pp. 1–28. Boca Raton, FL: CRC Press.

Reymond, L. 1985. Spatial visual acuity of the eagle *Aquila audax*: A behavioural, optical and anatomical investigation. *Vision Research* 25: 1477–1491.

van Rhijn, F. V. 1979. Optic orientation in hatchlings of the sea turtle, *Chelonia mydas*. I. Brightness: Not the only optic cue in sea-finding orientation. *Marine Behaviour and Physiology* 6: 105–121.

van Rhijn, F. V. and J. van Gorkom. 1983. Optic orientation in hatchlings of the sea turtle, *Chelonia mydas* III. Sea-finding behaviour: The role of photic and visual orientation in animals walking on the spot under laboratory conditions. *Marine Behaviour and Physiology* 9: 211–228.

Riggs, L. A. 1965. Visual acuity. In *Vision and Visual Perception,* ed. C. H. Graham. pp. 321–349. New York: John Wiley & Sons.

Riggs, L. A. and B. R. Wooten. 1972. Electrical measures and psychophysical data on human vision. In *Visual Psychophysics*, eds. D. Jameson and L. M. Hurvich. pp. 690–731. *Handbook of Sensory Physiology*, Vol. VII/4. Berlin, Germany: Springer-Verlag.

Salmon, M. and K. J. Lohmann. 1989. Orientation cues used by hatchling loggerhead sea turtles (*Caretta caretta* L.) during their offshore migration. *Ethology* 83: 215–228.

Salmon, M. and B. Witherington. 1995. Artificial lighting and seafinding by loggerhead hatchlings: Evidence for lunar modulation. *Copeia* 4: 931–938.

Salmon, M. and J. Wyneken. 1990. Do swimming loggerhead sea turtles (*Caretta caretta* L.) using light cues for offshore orientation. *Marine Biology & Physiology* 17: 233–246.

Salmon, M. and J. Wyneken. 1994. Orientation by hatchling sea turtles: Mechanisms and implications. *Herpetological Natural History* 2: 13–24.

Salmon, M., Wyneken, J., Fritz, E., and M. Lucas. 1992. Seafinding by hatchling sea turtles: Role of brightness, silhouette and beach slope as orientation cues. *Behaviour* 122: 56–77.

Shashar, N., Hagan, R., Boal, J. G., and R. T. Hanlon. 2000. Cuttlefish use polarization sensitivity in predation on silvery fish. *Vision Research* 40: 71–75.

Shashar, N., Hanlon, R. T., and A. deM. Petz. 1998. Polarization vision helps detect transparent prey. *Nature* 393: 222–223.

Siebeck, U. E. and N. J. Marshall. 2001. Ocular media transmission of coral reef fish—Can coral reef fish see ultraviolet light? *Vision Research* 41: 133–149.

Sweeney, A. M., Boch, C. A., Johnsen, S., and D. E. Morse. 2011. Twilight spectral dynamics and the coral reef invertebrate spawning response. *Journal of Experimental Biology* 214: 770–777.

Tuxbury, S. M. and M. Salmon. 2005. Competitive interactions between artificial lighting and natural cues during seafinding by hatchling marine turtles. *Biological Conservation* 121: 311–316.

Tyler, J. E. and R. C. Smith. 1970. *Measurements of Spectral Irradiance Underwater*. New York: Gordon and Breach Science Publishers.

Verheijen, F. J. 1958. The trapping effect of artificial light sources upon animals. *Archives Neerlandaises de Zoologie* 13: 1–107.

Verheijen, F. J. and J. T. Wildschut. 1973. The photic orientation of hatchling sea turtles during water finding behaviour. *Netherlands Journal of Sea Research* 7: 53–67.

Vorobyev, M. 2003. Coloured oil droplets enhance colour discrimination. *Philosophical Transactions of the Royal Society of London B* 270: 1255–1261.

Walls, G. L. 1942. *The Vertebrate Eye and Its Adaptive Radiation*. New York: Hafner.

Walls, G. L. 1962. The evolutionary history of eye movements. *Vision Research* 2: 69–80.

Walls, G. L. and H. D. Judd. 1933. The intra-ocular colour-filters of vertebrates. *British Journal of Opthalmology* 17: 641–675.

Wang, J. H., Boles, L. C. B., Higgins, B., and K. J. Lohmann. 2007. Behavioral responses of sea turtles to lightsticks used in longline fisheries. *Animal Conservation* 10: 176–182.

Warrant, E. J. 1999. Seeing better at night: Life style, eye design and the optimum strategy of spatial and temporal summation. *Vision Research* 39: 1611–1630.

Warrant, E. J. 2000. The eyes of seep-sea fishes and the changing nature of visual scenes with depth. *Philosophical Transactions of the Royal Society of London B* 355: 1155–1159.

Warrant, E. J. 2008. Nocturnal vision. In *The Senses: A Comprehensive Reference,* eds. T. Albright and R. H. Masland. pp. 53–86. Oxford, U.K.: Academic Press.

Warrant, E. J. and N. A. Locket. 2004. Vision in the deep sea. *Biological Reviews* 79: 671–712.

Witherington, B. E. 1992. Sea-finding behaviour and the use of photic orientation cues by hatchling sea turtles. PhD thesis. Gainesville, FL: University of Florida.

Witherington, B. E. and K. A. Bjorndal. 1991a. Influences of artificial lighting on the seaward orientation of hatchling loggerhead turtles *Caretta caretta*. *Biological Conservation* 55: 139–149.

Witherington, B. E. and K. A. Bjorndal. 1991b. Influences of wavelength and intensity on hatchling sea turtle phototaxis: implications for sea-finding behaviour. *Copeia* 4: 1060–1069.

Witherington, B. E., Bjorndal, K. A., and C. Mccabe. 1990. Temporal pattern of nocturnal emergence of loggerhead turtle hatchlings from natural nests. *Copeia* 4: 1165–1168.

Witherington, B. E. and R. E. Martin. 2003. *Understanding, Assessing, and Resolving Light-Pollution Problems on Sea Turtle Nesting Beaches*. Florida Marine Research Institute. http://www.fws.gov/caribbean/es/PDF/Library%20Items/LightingManual-Florida.pdf (accessed March 31, 2012).

Wortel, J. and J. Nuboer. 1986. The spectral sensitivity of blue-sensitive pigeon cones: Evidence for complete screening by the visual pigment by the oil-droplet. *Vision Research* 26: 885–886.

Wyneken, J. 2001. The anatomy of sea turtles, NOAA Technical Memorandum NMFS-SEFSC-470, U.S. Department of Commerce, Washington, DC.

Wyneken, J. 2007. Reptilian neurology: Anatomy and function. *Veterinary Clinics of North America: Exotic Animal Practice* 10: 837–853.

Wyneken, J. and M. Salmon. 1992. Frenzy and postfrenzy swimming activity in loggerhead, green, and leatherback hatchling sea turtles. *Copeia* 2: 478–484.

Young, M., Salmon M., and R. Forward. 2012. Visual wavelength discrimination by the loggerhead turtle, *Caretta caretta*. *Biological Bulletin* 222: 46–55.

# 3 Natal Homing and Imprinting in Sea Turtles

*Kenneth J. Lohmann, Catherine M.F. Lohmann,*
*J. Roger Brothers, and Nathan F. Putman*

## CONTENTS

## 3.1 INTRODUCTION

One of the most remarkable and mysterious elements of sea turtle biology is the ability of turtles to return to nest in the same geographic area from which they originated. This behavior, often referred to as natal homing, is particularly astonishing because many sea turtles migrate long distances away from their home areas before returning. Explaining how turtles leave a beach as hatchlings and then, years later, locate the same area of coastline after traveling immense distances through the open sea has posed a daunting challenge for biologists who have long struggled to explain the phenomenon without invoking magic.

The first indication that turtles might return to nest on or near their natal beaches came from results of early tagging programs in the 1950s and 1960s (Carr, 1967; Mrosovsky, 1983). These studies revealed that some female green turtles nest in the same areas year after year, which in turn fueled speculation that the nesting locations chosen by adult turtles might be the same beaches where they themselves began life as hatchlings. Initially, there was no way to test the idea because no suitable method for marking turtles existed; the only tags small enough to be placed on hatchlings detached long before turtles grew to maturity. The development of molecular techniques in the early 1990s, however, triggered an explosion of genetic evidence consistent with natal homing (e.g., Meylan et al., 1990; Bowen et al., 1993, 1994; Bowen and Avise, 1995). It is now known that most, if not all, sea turtles, display some degree of natal homing, although the precision of the homing may vary considerably among different populations and species (Bowen and Karl, 2007; Lohmann et al., 2008c).

Although it is now clear that natal homing occurs in sea turtles, little is known about how it is accomplished. For purposes of discussion, it is helpful to consider the process as being composed of two distinct elements. First, a turtle needs to be able to distinguish its natal beach or region from others, a process that might, in principle, involve information about the target area that the turtle has either learned or inherited. Second, a turtle must be able to navigate to the target area from a considerable distance away. The guidance mechanisms used might be unique to natal homing, or they might instead be the same ones that turtles use whenever they travel over long distances, regardless of the destination.

The question of how a turtle identifies its natal beach or region has prompted considerable speculation. It has been widely assumed, but never demonstrated, that the process of natal homing is linked to a special form of learning known as imprinting. Although precise definitions of imprinting differ (e.g., Hasler and Scholz, 1983; Alcock, 2009; Goodenough et al., 2010; Zupanc, 2010), the hallmarks of imprinting are that the learning occurs during a specific, critical period (usually early in the life of the animal), the effects are long-lasting, and the learning cannot be easily modified. For natal homing, the concept is that sea turtles imprint on some characteristic of their natal beach as hatchlings and then use this information to locate the beach years later as adults.

From a scientific perspective, the best way to study whether imprinting occurs in sea turtles would be to raise turtles under conditions in which various potential imprinting cues are manipulated, release the turtles into the ocean to mature, and allow them to return to nest as adults. This approach provides a powerful way to investigate which elements of early experience, if any, affect the behavior of adults. A similar methodology was used to demonstrate that young salmon imprint on the "chemical signature" of the water in their home rivers and use this information, as adults, to relocate their natal tributary when it is time for them to spawn (reviewed by Hasler and Scholz, 1983; Dittman and Quinn, 1996; Lohmann et al., 2008a; Zupanc, 2010).

For sea turtles, experimentation of this type is challenging for several reasons. All sea turtle species are threatened or endangered, so that limited numbers are available for experimental manipulations. In addition, sea turtles have an extremely long maturation period, with most populations and species requiring one or more decades to reach sexual maturity. Given these constraints, it is not surprising that little is known about imprinting in sea turtles, or even whether it truly occurs.

In this chapter, we begin by briefly summarizing the evidence for natal homing and the likely reasons that natal homing evolved in sea turtles. We then discuss the hypothesis of natal-beach imprinting in sea turtles, with an emphasis on chemical and geomagnetic cues, the two types of sensory information upon which turtles have been proposed to imprint. In addition, we summarize the limited experimental and correlational evidence consistent with each idea. Finally, given that clear evidence for imprinting does not yet exist in sea turtles, we discuss whether it is necessary to invoke imprinting to explain natal homing.

## 3.2 EVIDENCE FOR NATAL HOMING

Natal homing generates a testable prediction about the genetic partitioning of nesting populations. If females return faithfully to their rookery of origin, then each nesting population should possess a unique genetic signature in mitochondrial DNA (mtDNA), which is passed directly from females to their offspring (Bowen and Karl, 2007). At present, nearly all evidence for natal homing has come from genetic studies of this type.

An impressive amount of genetic data consistent with natal homing has been obtained during the past two decades. For example, green turtles that nest in Suriname, South America, have mtDNA haplotypes distinct from those of green turtles that nest at Ascension Island, even though the two populations feed in the same areas along the Brazilian coast (Bowen et al., 1992). Similarly, green turtles that nest in different regions of the Great Barrier Reef have different mtDNA haplotypes, even though these groups also share common feeding areas (Norman et al., 1994; Dethmers et al., 2006).

Genetic evidence for natal homing has now been acquired for loggerheads (Bowen et al., 1993; Encalada et al., 1998; Carreras et al., 2006, 2011), hawksbills (Bass et al., 1996), leatherbacks (Dutton et al., 1999), and olive ridleys (Bowen et al., 1998; Shanker et al., 2004). In addition, the Kemp's ridley turtle nests only in a limited area of coastline in the western Gulf of Mexico, a behavior that provides an impressive demonstration of natal homing in and of itself.

Although it is now evident that female turtles typically nest in the vicinity of their natal beaches, the geographic precision of natal homing remains to be determined. In particular, it is not known whether turtles return to nest on or very near to their exact natal sites, or instead home only to general regions. Growing evidence suggests that homing to regions several hundred kilometers in length is common (Bowen et al., 1992; Bowen and Avise, 1995; Bowen and Karl, 2007; Bourjea et al., 2007; Lohmann et al., 2008c). The absence of precise natal homing in many turtle populations, or at least the willingness of turtles to select nest sites over a considerable expanse of coastline (e.g., Bjorndal et al., 1983), may be adaptive because particular nesting areas can be destroyed rapidly by storms, erosion, and flooding. Nevertheless, greater precision may exist in some cases (Peare and Parker, 1996; Lee et al., 2007), and further research is clearly needed.

At first glance, astonishingly precise natal homing appears to exist in at least some situations. Carr (1967) highlighted the remarkable migration of green turtles from the coast of Brazil to Ascension Island, a tiny isolated island in the South Atlantic located approximately 2000 km from the South American coast. The ability of turtles to locate the island seems to argue for pinpoint natal homing and navigational accuracy. An interesting consideration, however, is that Ascension turtles may, in effect, be forced to locate the island because no alternative exists. It is plausible that turtles guide themselves into the general vicinity of the island and would willingly nest anywhere nearby but, confronted with the lack of any other nesting area, they must search extensively using local cues (Lohmann et al., 1999, 2008b) until they either find Ascension or abandon efforts to nest. Thus, the ability of turtles to locate small islands, or other restricted nesting areas (e.g., small sandy beaches on otherwise rocky coastlines) where no alternatives exist, may provide a misleading picture of the accuracy of natal homing.

## 3.3   WHY DID NATAL HOMING EVOLVE?

Migrating hundreds or thousands of kilometers to nest in a particular geographic location carries with it considerable costs. For such a pattern to evolve, the benefits must be correspondingly high. In evolutionary terms, natal homing presumably arose in sea turtles because individuals that returned to their natal areas to nest produced more surviving offspring than those that tried to nest elsewhere.

In all likelihood, the structure of the environment has been a major factor in shaping natal homing, inasmuch as successful reproduction in sea turtles requires a specialized set of environmental conditions that exist only in limited and highly specific geographic areas (Lohmann et al., 1999). For sea turtles, the need to lay eggs on land restricts possible nesting locations to a tiny fraction of the environment in which the animals live. Indeed, even most coastal areas are unsatisfactory because the beach must consist of sand rather than rock or mud and the sand must possess specific qualities favorable for nest construction. In addition, the beach must be free of steep inclines and obstacles that block access from the sea (Hendrickson, 1958; Mortimer, 1995); it must also have suitable incubation temperatures (Carthy et al., 2003) and low densities of egg predators (Mortimer, 1995). Finally, even if optimal conditions exist on the beach, some areas of coastline are much more favorable to hatchling survival than others because of proximity to ocean currents that can help transport hatchlings to suitable developmental habitats (Putman et al., 2010a,b, 2012; Schillinger et al., 2012).

A seemingly irrational feature of natal homing is that turtles sometimes forego reproducing in suitable nearby areas in favor of migrating back to their natal regions. For example, some turtles feed in areas adjacent to nesting beaches used by their own species, but nevertheless migrate vast distances to nest elsewhere (Limpus et al., 1992). For a turtle, however, assessing the suitability of an unfamiliar area for reproduction is likely to be very difficult. A turtle crawling out of the sea to nest at night probably cannot determine that the temperature of the sand during the day is lethally hot, or that a dense population of raccoons is likely to consume all eggs deposited in the area, or that strong seasonal onshore currents will impede the efforts of her hatchlings to migrate offshore.

Under such conditions, in which suitable reproductive habitat is scarce and reproductive output can be strongly affected by factors that are difficult to assess, it is perhaps not surprising that natural selection has favored individuals that return to their natal areas to reproduce. In effect, the very existence of an adult animal confirms that its natal area has the attributes needed for successful reproduction—an assurance that no other location can provide.

## 3.4   WHAT ENVIRONMENTAL CUES MIGHT BE USED IN IMPRINTING?

If we accept the view that sea turtles imprint on their natal beaches or regions, then precisely what information is used? In principle, the sensory information that is exploited needs, at a minimum, to enable a turtle to distinguish its own natal beach, region, or both from alternative areas. Ideally, the information should also provide the turtle with some means for navigating toward the appropriate location from a considerable distance away.

Nearly all discussions of possible imprinting mechanisms in sea turtles have focused on two possibilities: that turtles imprint on distinctive chemical cues associated with their natal beach (Owens et al., 1982; Grassman et al., 1984), or that turtles imprint on the magnetic field of the beach (Lohmann et al., 1999, 2008c). The two ideas are not mutually exclusive, inasmuch as turtles might hypothetically use both types of information together. For example, they might first use magnetic information to arrive in the general region of the beach, and then use chemical cues to localize a more specific nesting site (Lohmann et al., 2008a,b; Putman and Lohmann, 2008).

### 3.4.1   CHEMICAL IMPRINTING HYPOTHESIS

Historically, the first hypothesis proposed to explain natal homing in sea turtles was that turtles imprint on chemical cues unique to their natal beach and use this information as adults to return to that same beach for nesting and mating (Carr, 1979; Owens et al., 1982; Mrosovsky, 1983). This idea was inspired by the discovery that salmon imprint on the chemical cues of their home tributaries and use this information as adults to recognize the streams and rivers in which they hatched (Hasler and Wisby, 1951; Hasler et al., 1978; Hasler and Scholz, 1983).

Relatively little evidence has been obtained to either support or refute the chemical imprinting hypothesis in turtles. Sea turtles possess numerous functional olfactory genes and can thus probably detect diverse olfactory cues (Kishida et al., 2007). Behavioral experiments have revealed that sea turtles can detect chemicals dissolved in water (Manton et al., 1972a,b; Grassman and Owens, 1982; Southwood et al., 2008) as well as airborne odorants (Endres et al., 2009). The latter ability might allow turtles to detect odors carried by winds over considerable distances. Whether aquatic or aerial chemicals play a role in natal homing and beach recognition, however, remains unclear.

#### 3.4.1.1   Chemical Cues and Recognition of the Natal Beach Region

A question at the heart of the chemical imprinting hypothesis is whether turtles can recognize a specific nesting area on the basis of distinctive chemical cues. Some limited evidence suggests that this might be possible, but results are not yet conclusive.

In one study (Grassman et al., 1984), eggs of Kemp's ridley turtles were incubated in sand from Padre Island, Texas. After emerging from their eggs, hatchlings were permitted to crawl across the Padre Island beach and swim through the surf. The turtles were then held in captivity for 4 months, after which they were tested in a water-filled arena consisting of four compartments. One contained a solution made from Padre Island sand and sea water, a second contained a similar solution made from sand and sea water from a different location (Galveston, Texas), and two others contained untreated sea water. The time that turtles spent in each compartment after entering was monitored.

Turtles were found to spend significantly more time per entry in the Padre Island compartment than in any of the others.

Although these results were suggestive, all turtles tested in this study had been "imprinted" to Padre Island sand and water; no experiments were done with turtles that had been similarly "imprinted" to water from another location. Thus, the possibility that turtles preferred Padre Island water for reasons unrelated to early experience (e.g., the Padre Island odors might have smelled more like food than the Galveston water) could not be excluded. A second experiment was therefore carried out to determine whether turtles that had been "imprinted" to either water from Padre Island or water from northern Mexico subsequently preferred the water that they had been exposed to previously (Grassman and Owens, 1989). No such preferences could be discerned, but poor health of the turtles may have adversely affected their performance (Grassman and Owens, 1989).

In an additional experiment, Grassman and Owens (1987) incubated green turtle eggs in sand that was scented with one of two chemicals (morpholine or 2-phenylethanol) that do not exist in the natural habitat. After the eggs hatched, each turtle was held for 3 months in water containing the same chemical that had been present in the sand during incubation. After two additional months without exposure to either chemical, the turtles were tested in a compartmentalized arena containing solutions of morpholine, 2-phenylethanol, and untreated sea water. Turtles that had previously been exposed to morpholine preferred morpholine to 2-phenylethanol, whereas the opposite was true for turtles that had been exposed to the 2-phenylethanol. Interestingly, however, additional groups of turtles that had been exposed to the chemicals only while in the nest (for about 2 months) or only after emerging from the nest (i.e., while living in the water of the holding tank for 3 months) failed to show these preferences.

These results provide additional evidence that turtles can detect chemical cues. Moreover, they demonstrate that, during at least certain developmental periods, turtles can retain the ability to recognize a chemical for at least 2 months. Although this outcome is arguably consistent with chemical imprinting, the significance of the findings is not yet clear. Only turtles exposed to the chemicals for about 5 months duration (approximately 2 months in the nest followed by 3 months in water) acquired the preference. Under natural conditions, hatchlings migrate beyond the waters of their natal beach within a few hours after emerging from their nests (Frick, 1976; Ireland et al., 1978; Salmon and Wyneken, 1987). Thus, if exposure to the chemical for 3 months after hatching is essential for the response to develop, then it is difficult to envision how such a process could occur during the offshore migration. Nevertheless, these initial results are intriguing, and additional studies are needed.

### 3.4.1.2 Limitations of the Chemical Imprinting Hypothesis

At the time that the chemical imprinting hypothesis was first proposed for sea turtles, almost nothing was known about how sea turtles navigate. Chemical (olfactory) imprinting, which had been described in salmon, was the only known mechanism that seemed able to explain how a turtle might recognize its home beach. Largely overlooked, however, was the fact that chemical cues are generally thought to guide salmon to their spawning grounds only after the fish reach the vicinity of the target river (Hasler et al., 1978; Hasler and Scholz, 1983). Under favorable conditions, chemical cues from rivers might extend some distance from a river mouth (Døving et al., 1985), but it appears impossible for such cues to extend across a thousand or more kilometers of ocean, the distance over which some populations of salmon migrate (Dittman and Quinn, 1996). For this reason, olfactory navigation in salmon is generally assumed to function only over short distances, whereas salmon navigation in the open sea is thought to rely on a different set of mechanisms that are not olfactory (Hasler and Scholz, 1983; Quinn, 2005; Lohmann et al., 2008a).

Similarly, if chemical imprinting occurs in sea turtles, then chemical cues associated with the natal beach can probably only be detected by turtles over a limited geographic range close to the natal beach. Odor plumes from a specific beach presumably dissipate rapidly with distance and are carried downcurrent from the source. Turtles approaching from directions other than downcurrent are thus presumably unable to exploit the chemical signature of the beach to reach the beach.

Nevertheless, turtles in many geographic areas appear to approach their nesting areas from many different directions, including some that are upcurrent of the destination (Lohmann et al., 1999).

Two examples vividly illustrate this point: (1) loggerhead turtles that nest at Melbourne Beach, Florida (which has one of the highest densities of loggerhead nests in North America); (2) the Kemp's ridley turtle, which nests in Rancho Nuevo, Mexico. In both cases, oceanographic analyses indicate that waterborne odorants originating at the nesting grounds are carried north by ocean currents (Figure 3.1). Despite this, many loggerheads that nest at Melbourne Beach migrate there from the south (Schroeder et al., 2003). Similarly, although odorants from the Kemp's ridley nesting area at Rancho Nuevo, Mexico, are carried north by currents (Figure 3.1), nesting turtles converge on the site from diverse directions (Morreale et al., 2007). Equivalent situations exist at Tortuguero, Costa Rica, where green turtles apparently converge on the nesting area from feeding grounds that are both upcurrent and downcurrent (Carr, 1967), as well as in South Africa (Hughes, 1995) and along the Great Barrier Reef (Limpus et al., 1992).

These findings imply that turtles are unlikely to use chemical cues to guide themselves to nesting beaches over long distances. Instead, if some type of chemical imprinting does occur in sea turtles, then recognition of the natal beach area is likely to occur only after turtles have already arrived in the general vicinity of the target using navigational processes that are not dependent on olfaction. How might sea turtles find a natal region when starting from locations that vary in distance and direction from the home area? One possibility is described next.

### 3.4.2 Geomagnetic Imprinting Hypothesis

A different hypothesis about imprinting and natal homing emerged from research on how sea turtles navigate. An extensive series of experiments, conducted over the past two decades, has demonstrated that much of a turtle's navigational repertoire depends upon a well-developed magnetic sense (Lohmann and Lohmann, 1996, 2003; Luschi et al., 2007). Several species of sea turtles are known to have the biological equivalent of magnetic compasses, which enable them to determine their magnetic headings (Lohmann, 1991; Lohmann and Lohmann, 1993). In addition, turtles are able to detect subtle differences in various magnetic field features that vary geographically (Lohmann and Lohmann, 1994, 1996; Lohmann et al., 2001, 2012). To a turtle, the magnetic field of south Florida, for example, is distinguishable from the field that exists farther north along the coast (Lohmann et al., 2004). The geomagnetic imprinting hypothesis proposes that turtles imprint on the magnetic field of their natal beach and use this information to return as adults (Lohmann et al., 1999, 2008c).

To understand how turtles might exploit the Earth's magnetic field in natal homing, it is helpful to start by discussing the Earth's magnetic field. To a first approximation, the Earth's field resembles the dipole field of a giant bar magnet (Figure 3.2). Field lines leave the southern hemisphere and curve around the globe before reentering the planet in the northern hemisphere. Several geomagnetic elements vary predictably across the surface of the Earth (Figure 3.2). For example, at each location on the globe, the magnetic field lines intersect the Earth's surface at a specific angle of inclination. At the magnetic equator, the field lines are parallel to the ground and the inclination angle is said to be 0°. The field lines become progressively steeper as one moves toward the magnetic poles; at the poles themselves, the field lines are perpendicular to the Earth's surface. Thus, inclination angle varies predictably with latitude, and an animal able to detect this field element may be able to determine if it is north or south of a particular area. In addition to inclination angle, at least three other magnetic field elements related to intensity (i.e., the intensity of the total field, horizontal field, and vertical field) vary across the Earth's surface in ways that make them suitable for use in position-finding (Lohmann et al., 2007; Figure 3.2).

### 3.4.2.1 How Beaches Can Be Uniquely Identified by Magnetic Parameters

Most major sea turtle rookeries are located on continental coastlines (e.g., Mexico, Costa Rica, the southeastern United States, Africa, and Australia) that are aligned approximately north to south. Thus,

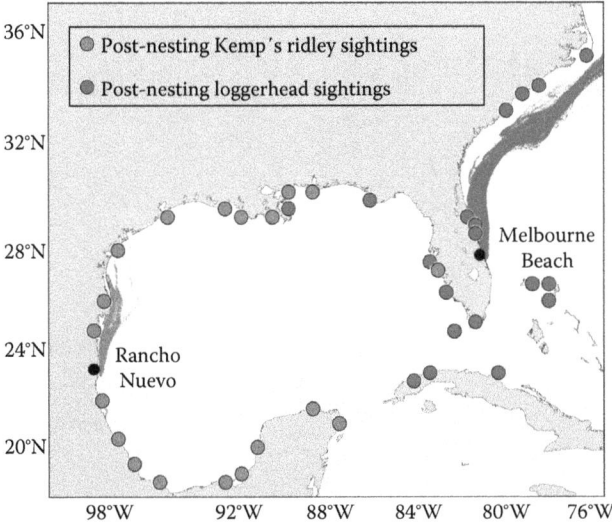

FIGURE 3.1 Dispersion of odorant plumes from Rancho Nuevo in Tamaulipas, Mexico (the main nesting area used by Kemp's ridley sea turtles), and Melbourne Beach, Florida (a major nesting beach of loggerhead sea turtles), based on ocean circulation models (see later). The plot shows hypothetical odor plumes originating at Rancho Nuevo (the green plume emanating from the black dot labeled "Rancho Nuevo") and at Melbourne Beach (the blue plume emanating from the black dot labeled "Melbourne Beach"). In each case, coastal currents are expected to carry odorants primarily northward. Nevertheless, turtles appear to migrate to the nesting area from geographically diverse locations, many of which are situated upcurrent in places that odorants from the nesting beach cannot reach (Schroeder et al., 2003; Morreale et al., 2007). On the diagram, green dots indicate locations where Kemp's ridleys known to nest at Rancho Nuevo have been captured or sighted (Marquez, 1994); blue dots indicate locations where loggerheads known to nest at Melbourne Beach have been captured or sighted (Meylan et al., 1983). At least some of these areas are presumably foraging grounds and thus represent starting locations for the migration to the nesting area (Meylan et al., 1983; Marquez, 1994). The geographically wide distribution of such locations implies that turtles are unlikely to use odorants from the natal beach to guide themselves to the beach over long distances, but do not rule out the possibility that turtles might use such cues to recognize the natal beach once other navigational mechanisms have guided them into the vicinity. The simulations are based on hindcast output from the South Atlantic Bight-Gulf of Mexico ocean circulation model (SABGOM). SABGOM output has a spatial resolution of 5 km and hourly snapshots of velocity. For open boundary conditions, SABGOM is one-way nested inside the 0.08° data assimilative North Atlantic Hybrid Coordinate Ocean Model (Hyun and He, 2010). Major coastal river input in the SABGOM was included in the hindcast using daily runoff data observed by USGS river gauges along the coast. For both momentum and buoyancy forcing at the model surface, three hourly wind data from the North American Regional Reanalysis (NARR, www.cdc.noaa.gov) were incorporated. This model has successfully characterized anomalous coastal conditions in the South Atlantic Bight (Hyun and He, 2010) and predicted the movement of oil from the Deepwater Horizon oil spill in the Gulf of Mexico (North et al., 2011). To generate the predicted odorant plumes, virtual particles (simulated odorants) were released immediately offshore of Rancho Nuevo, Mexico, and Melbourne Beach, Florida, the United States. One hundred particles were released daily from April 20 to June 20, 2010 into SABGOM output using ICHTHYOP (v. 2) particle tracking software (Lett et al., 2008). For advection of particles through the surface SABGOM velocity fields, ICHTHYOP implemented a Runge-Kutta fourth order time-stepping method in which particle position was calculated every half hour and the location of each particle was tracked for 30 days.

during reproductive migrations, the essence of the navigational task that most sea turtles confront is to travel, either through the open sea or along the shoreline of a continent, to a particular coastal area.

How might an animal arrive reliably at a particular region of coastline from a considerable distance away? An interesting possibility is that geomagnetic parameters could be used to identify particular coastal areas. The southeast coast of North America illustrates the basic principle.

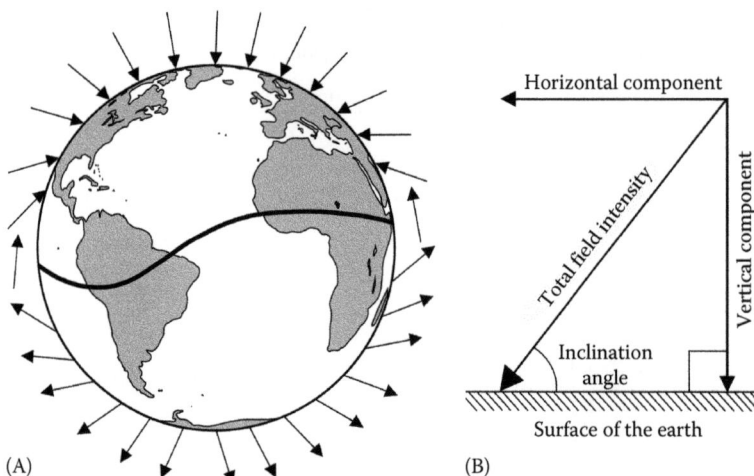

(A)                                                                    (B)

FIGURE 3.2    (A) Diagrammatic representation of the Earth's magnetic field illustrating how field lines (represented by arrows) intersect the Earth's surface, and how inclination angle (the angle formed between the field lines and the Earth) varies with latitude. At the magnetic equator (the curving line across the Earth), field lines are parallel to the Earth's surface. The field lines become progressively steeper as one travels north toward the magnetic pole, where the field lines are directed straight down into the Earth and the inclination angle is 90°. (B) Diagram illustrating four elements of geomagnetic field vectors that might, in principle, provide turtles with positional information. The field present at each location on Earth can be described in terms of total field intensity and inclination angle. The total intensity of the field can be resolved into two vector components: the horizontal field intensity and the vertical field intensity. (Whether turtles or other animals can resolve the total field into vector components is not known.)

The coastline is aligned approximately north–south, whereas isolines of inclination (lines along which inclination angle is constant) trend east–west. As a consequence, every area of coastline is marked by a different inclination angle (Figure 3.3). Similarly, isolines of total field intensity run approximately east–west in this geographic area and different coastal locations are thus marked by different intensities (Figure 3.4). In effect, different coastal areas have unique "magnetic signatures" that might, in principle, be used to identify a natal region and distinguish it from all other locations along the same coast. The same is true along nearly all continental coastlines used by sea turtles for nesting worldwide (Lohmann et al., 1999).

Several variants of the geomagnetic imprinting hypothesis are possible. The simplest is that turtles imprint on a single element of the field (e.g., either inclination angle or intensity). To locate the area later in life, the turtle would need only to find the coastline, and then swim north or south along it to reach the target region; alternatively, a turtle in the open ocean might adjust its position until it arrives at the correct isoline and then swim along the isoline until arriving at the coast and natal area. In either case, a turtle could hypothetically determine whether it is north or south of the goal by assessing whether the inclination angle or intensity at a given location is greater or lesser than the value at the natal area.

More complex scenarios are also possible. For example, turtles might imprint on both inclination and intensity, and use both elements as redundant markers of the natal area upon return.

### 3.4.2.2    Detection of Magnetic Parameters

Sea turtles have the sensory abilities needed for geomagnetic imprinting. Hatchling loggerheads can perceive both magnetic inclination angle (Lohmann and Lohmann, 1994) and magnetic field intensity (Lohmann and Lohmann, 1996). Furthermore, when hatchlings were subjected to magnetic fields that exist at widely separated locations along their open-sea migratory pathway, they responded by swimming in directions that would, in each case, facilitate movement along the migratory route

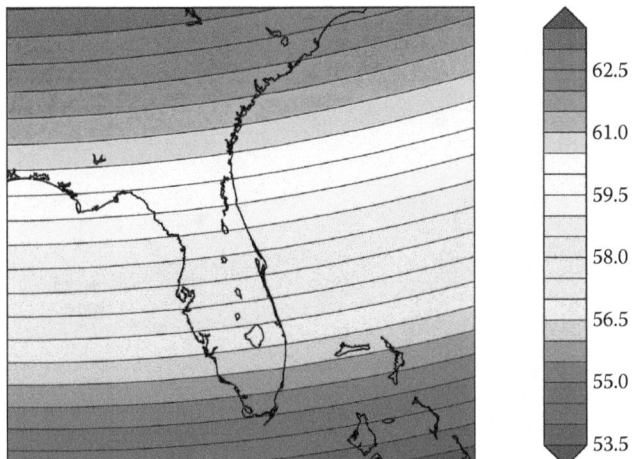

**FIGURE 3.3** Variation in magnetic inclination angle along the southeastern U.S. coastline. The color code (right) indicates the value of the inclination angle in degrees; higher values indicate more steeply inclined field lines. Black isolines bordering each color on the map indicate increments of 0.5°. Because the isolines trend east–west while the east coast of the U.S. trends north–south, every area along the Atlantic coast has a different inclination angle associated with it. In principle, turtles might be able to exploit these unique "magnetic signatures" to locate specific coastal nesting areas (Lohmann et al., 2008b). Isolines were derived from the International Geomagnetic Reference Field (IGRF) model for the year 2012.

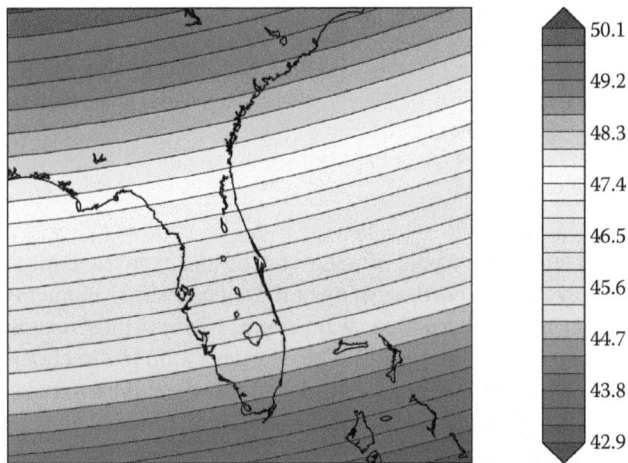

**FIGURE 3.4** Variation in magnetic intensity along the southeastern U.S. coastline. The color code (right) indicates the intensity of the field in microTesla. Black isolines bordering each colored stripe on the map indicate increments of 30 μT. Because the isolines trend east–west while the east coast of the U.S. trends north–south, every area along the Atlantic coast has a different magnetic field intensity associated with it. In principle, turtles might be able to exploit these differences, either alone or in combination with inclination angle (Figure 3.3), to locate specific coastal nesting areas. Isolines were derived from the International Geomagnetic Reference Field (IGRF) model for the year 2012.

(Lohmann et al., 2001, 2012; Fuxjager et al., 2011; Putman et al., 2011). These results demonstrate that turtles can distinguish among magnetic fields that exist in different geographic locations.

Additional work has demonstrated that older turtles also use magnetic positional information to facilitate navigation toward specific geographic goals along coastlines (Lohmann et al., 2004, 2007). Juvenile green turtles captured in feeding grounds along the east coast of Florida were

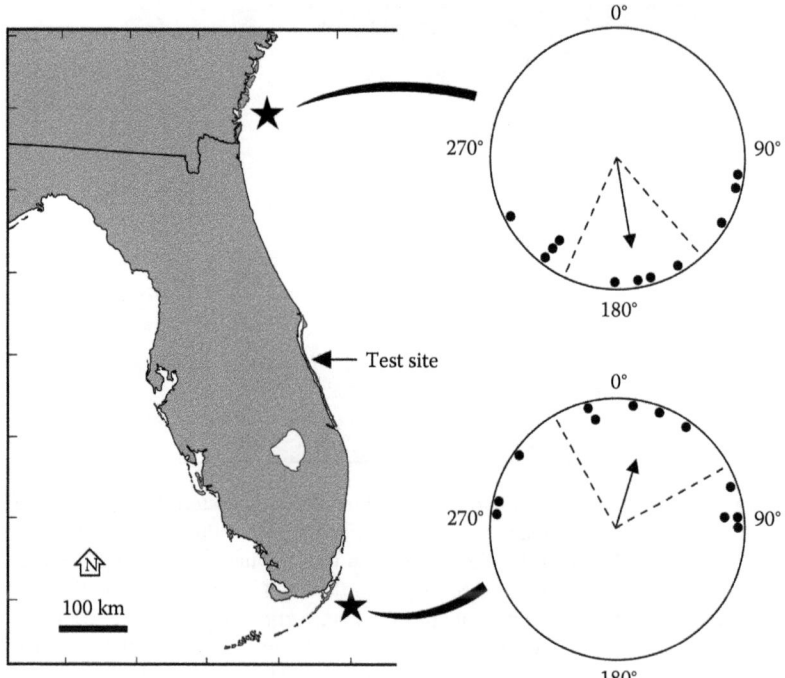

**FIGURE 3.5** Evidence for a magnetic map in juvenile green turtles. Juvenile turtles were captured in feeding grounds near the test site in Melbourne Beach, Florida, the United States. Each turtle was exposed to a magnetic field that exists at one of two distant locations (represented by stars along the coastline). Turtles exposed to the field from the northern site swam approximately southward, whereas those exposed to the field from the southern site swam approximately north. In the orientation diagrams, each dot represents the mean angle of a single turtle. The arrow in the center of each circle represents the mean angle of the group. Dashed lines represent the 95% confidence interval for the mean angle. (Modified from Lohmann, K. J. et al., *Nature*, 428, 909, 2004. See text for discussion.)

tethered to a tracking system inside a pool of water on land and exposed to magnetic fields that exist at locations approximately 340 km north or south of the capture site (Lohmann et al., 2004). Turtles exposed to the field from the northern area swam south, whereas those exposed to the field from the southern location swam north (Figure 3.5). Thus, turtles swam in directions that would have led them home had they actually been displaced to the locations where the two fields exist. These results imply that, well before the turtles mature, they have already acquired a "magnetic map" (Lohmann et al., 2007) and the skills needed to navigate toward specific coastal areas.

Although the geomagnetic imprinting hypothesis appears to be compatible with the known navigational mechanisms of sea turtles, no direct evidence supporting or refuting the idea has yet been obtained. Circumstantial evidence consistent with the hypothesis, however, has come from genetic analyses of loggerhead turtles nesting in Florida (Shamblin et al., 2011). Genetic data have revealed a surprising mirror-image pattern of haplotype frequencies on the east and west coasts of Florida. In other words, some groups of turtles with comparable haplotype frequencies were found nesting on opposite sides of the Florida peninsula at similar latitudes (Figure 3.6). The authors point out that the magnetic fields that exist at similar latitudes on opposite sides of Florida are similar, and that turtles from the Gulf sometimes enter the Atlantic to forage and vice versa. If turtles imprint on the field of their natal beach, but some inadvertently travel along the "wrong" coast of Florida when seeking a nesting site with the appropriate magnetic signature, then the pattern of haplotype frequencies can easily be explained (Shamblin et al., 2011; Figure 3.6). Interestingly, such a pattern cannot easily be explained by chemical imprinting, inasmuch as beaches on the Gulf and Atlantic

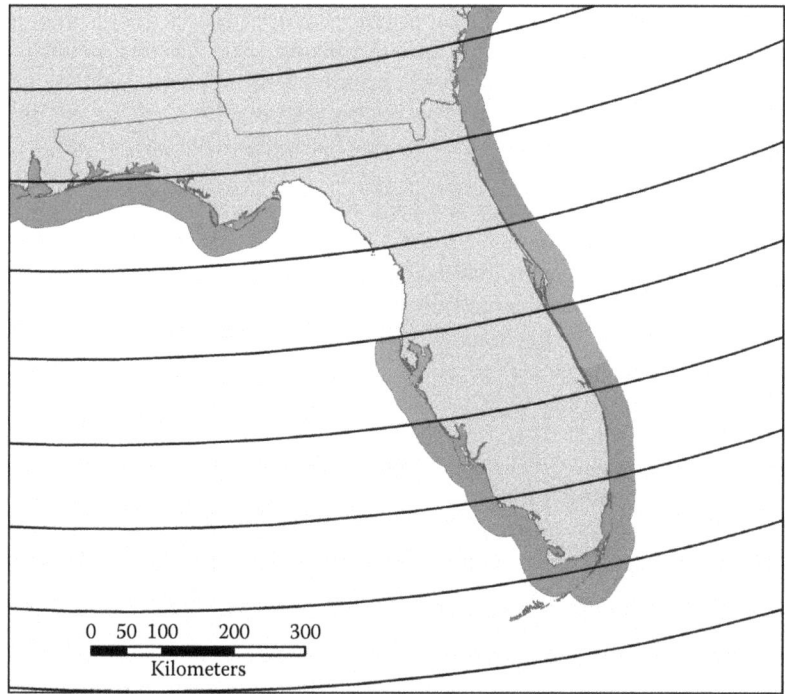

**FIGURE 3.6** Diagrammatic representation of a possible relationship between haplotype frequencies of nesting loggerheads and magnetic parameters on the east and west coasts of Florida, as suggested by Shamblin et al. (2011). Matching colors on the east and west coasts of Florida indicate geographic regions in which one or more groups of turtles on opposite coasts have indistinguishable haplotype frequencies. The diagram illustrates that, when groups of turtles with similar haplotype frequencies are found on opposite sides of the peninsula, they are found at similar latitudes, and in areas with similar magnetic signatures. The diagram is intended only to illustrate the concept and omits considerable complexity. For example, not all groups of turtles within the shaded areas have haplotype frequencies matching those in the corresponding region on the opposite coast (see Shamblin et al., 2011 for details); moreover, the exact boundaries of the colored regions are not yet known and have been drawn somewhat arbitrarily. Isolines on the map indicate increments of 1° of magnetic inclination angle. See text for discussion.

sides of Florida differ in numerous environmental factors (e.g., water temperature, water currents, coastal geology) and are thus unlikely to have similar chemical signatures.

### 3.4.2.3 Limitations of the Geomagnetic Imprinting Hypothesis

Unlike chemical cues dispersing from a home beach, the Earth's magnetic field is well suited for use in guiding long-distance migrations because it is present everywhere in the ocean and varies predictably across the globe (Lohmann et al., 2007). Young loggerheads exploit the Earth's field in following complex, transoceanic routes (Lohmann et al., 2012; Putman et al., 2012). A limitation of geomagnetic imprinting, however, is that it is better suited to bringing turtles into the general area of a target than it is in guiding turtles to a specific location with pinpoint accuracy (Lohmann et al., 2008c; Putman and Lohmann, 2008).

One reason is that the Earth's magnetic field and its constituent elements (such as inclination and intensity) change gradually over time (Skiles, 1985; Lohmann et al., 2008c). This change in magnetic field elements, known as secular variation, poses a potential difficulty for the geomagnetic imprinting hypothesis because field changes that occur at the natal site during a turtle's absence might cause navigational errors during return migrations (Lohmann et al., 1999, 2008c; Putman and Lohmann, 2008).

Several simple modeling exercises, however, imply that geomagnetic imprinting is compatible with present and recent rates of secular variation (Lohmann et al., 2008c; Putman and Lohmann, 2008). For example, an analysis of navigational errors that would hypothetically occur at three major, widely separated continental nesting beaches suggests that simple strategies of geomagnetic imprinting can return turtles to an appropriate geographic region, even after an absence of a decade or more (Lohmann et al., 2008c).

The nesting area of the Kemp's ridley has been studied particularly thoroughly in this regard (Putman and Lohmann, 2008; Figure 3.7). Simulations indicate that, for the Kemp's ridley, a strategy of returning to a coastal area marked by a specific inclination angle would be effective, inasmuch as isolines of inclination seldom shift more than a few kilometers along the coast during a year. A turtle imprinting on inclination and returning after a decade would, on average, arrive approximately 23 km from its precise natal site (Putman and Lohmann, 2008), a distance that typically puts it well within the area of beach used for nesting, or at least close enough to locate the area using additional, local cues.

If turtles update their knowledge of the magnetic field at the nesting beach each time they visit, then subsequent return trips would likely involve much smaller navigational errors attributable to secular variation. Females of most species return to nest every 2–4 years once they begin nesting (Lohmann et al., 1999), an interval considerably shorter than the 1–3 decades that some populations and species require to mature.

In some cases, turtles are not absent from the natal beach region for the entire maturation period. For example, juvenile loggerheads show natal homing on a very broad, regional scale well before their first reproductive migration (Bowen et al., 2004). When these turtles leave the open ocean to establish coastal feeding sites, they choose foraging grounds within their general natal region more often than would be expected by chance, although the sites chosen are often a considerable distance from the natal beach (Bowen et al., 2004; Bowen and Karl, 2007). This regional homing raises the interesting possibility that loggerheads diminish effects of secular variation on natal homing accuracy by updating their knowledge of the field in their natal region long before their first reproductive migration. If so, then navigational errors due to secular variation might be considerably smaller than some estimates (Lohmann et al., 2008c; Putman and Lohmann, 2008).

Another intriguing possibility is that geomagnetic imprinting might simply bring turtles into the general vicinity of their natal site, close enough that other local cues (e.g., chemical cues, sounds of breaking waves) can lead turtles to suitable nesting areas (Lohmann et al., 2008b,c; Putman and Lohmann, 2008). If so, then turtles might either imprint on particular local cues (such as the chemical signature of the beach) or instead simply recognize features of appropriate nesting areas and select any suitable beach within a large target area.

### 3.4.2.4  Island-Nesting Turtles and Geomagnetic Imprinting

Although most major sea turtle rookeries are located along continental coastlines, some populations nest on islands. Island-nesting populations are thought to be derived, evolutionarily, from populations that nest on continents (Bowen and Karl, 2007), but whether the two groups locate their natal regions in the same way is not known. It is possible, for example, that different strategies of navigation and imprinting evolved as an adaptation for island nesting (Lohmann et al., 2008c). Experimental evidence indicates that adult turtles use magnetic cues when navigating to islands, although in what exact way is not known (Lohmann, 2007; Luschi et al., 2007).

In principle, finding an island using a single magnetic element such as inclination or intensity is possible, inasmuch as a turtle might follow an isoline that intersects the island or passes nearby (Lohmann and Lohmann, 2006; Lohmann et al., 2007). Thus, a magnetic imprinting process similar to that outlined previously for continental nesting sites might suffice in some cases. Alternatively or additionally, a more complex strategy, such as imprinting on two elements of the field and using some form of bicoordinate magnetic navigation to return, might also be feasible in some situations (Lohmann et al., 1999, 2008c).

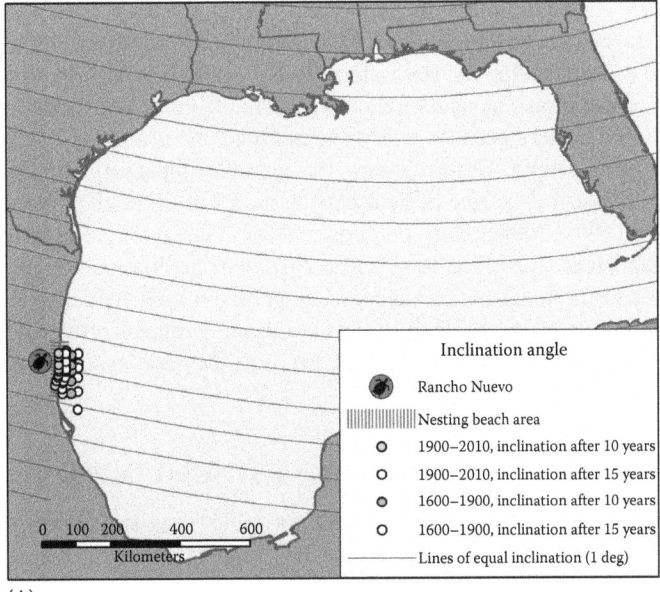

(A)

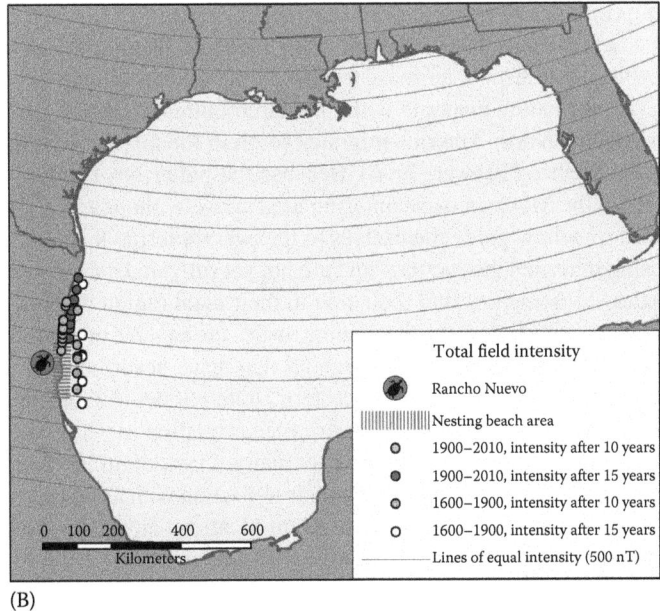

(B)

FIGURE 3.7 Map of the Gulf of Mexico illustrating the nesting area of the Kemp's ridley turtle and the locations to which turtles would hypothetically return under two simple magnetic imprinting strategies. The red hatched lines indicate the region of coastline (approximately 160 km) in which 98% of nests of the species are deposited (Marquez et al., 2001; Plotkin, 2007). The turtle symbol indicates the location of Rancho Nuevo, Mexico (23.20°N, 97.77°W), the site of peak nesting density (Marquez et al., 2001; Plotkin, 2007). Isolines in (A) indicate 1° increments of magnetic inclination; in (B) they indicate 500 nT increments of total intensity. (A) Predicted locations of returns if turtles imprint on the magnetic inclination angle at Rancho Nuevo and then return 10 or 15 years later to the coastal location with the same inclination angle. Each colored dot indicates the return location for a turtle leaving the coast in a specific year (1900, 1905, and so on). Because some return locations are nearly identical, not all dots are visible. (B) Predicted locations of returns if turtles imprint on the magnetic intensity at Rancho Nuevo. Conventions are as before. (Modified from Putman, N. F. and Lohmann, K. J., *Curr. Biol.*, 18, R596, 2008.)

In considering the possibility of geomagnetic imprinting on islands, an important factor is that an island typically represents a smaller target than a continental coastline. Along coastlines, small navigational errors due to secular variation are unlikely to prevent turtles from finding a suitable nesting site nearby. By contrast, navigational errors arising from secular variation potentially pose a greater risk for island-nesters because a turtle might miss the island entirely and thus be unable to nest (Lohmann et al., 1999). Unfortunately, the possible impact of secular variation in such settings is difficult to assess. The rate of field change varies in different geographic areas (Skiles, 1985; Lohmann et al., 1999, 2008c); thus, potential navigational errors are greatly influenced by the exact location of the island, how long turtles in a particular population are absent from the area, whether they return to the natal region (at least temporarily) as juveniles, and so on (Lohmann et al., 1999). In addition, estimates of likely effects are strongly influenced by assumptions about the precise navigational strategy that turtles use (whether, e.g., they use a single magnetic element or a bicoordinate magnetic map).

## 3.5   CAN NATAL HOMING BE EXPLAINED WITHOUT IMPRINTING?

Given that no compelling experimental evidence of imprinting yet exists in sea turtles, it is reasonable to ask whether imprinting occurs at all. Natal homing requires a turtle to somehow navigate back to a particular geographic location, or at least to a particular geographic region, from a considerable distance away. The question, then, is whether it is possible to explain natal homing in sea turtles without invoking imprinting.

Research with other animals has revealed that, under at least some conditions, guidance mechanisms can evolve to transport animals to specific locations through processes that do not involve imprinting. A fascinating example is the monarch butterfly (*Danaus plexippus*). Much of the population that inhabits North America migrates south in the fall to overwinter in a particular, limited area in central Mexico (Brower, 1996). Because the butterflies live less than a year, all of the monarchs that reach the Mexican overwintering area arrive without any prior experience in the area. Thus, butterflies somehow guide themselves to the overwintering location without imprinting on it, presumably using inherited instructions that are not yet fully understood (Reppert et al., 2010).

Whether sea turtles can somehow find their way to their natal region without imprinting on it is unclear. What is known, however, is that at least some turtles are capable of guiding themselves along complex, transoceanic migratory routes, even though they have never been to these areas before. Hatchling loggerhead sea turtles from the southeastern United States, for example, enter the sea for the first time and immediately embark on a long-distance migration in which they follow a circular migratory route around the Sargasso Sea before eventually returning to the North American coast (Carr, 1986; Bolten et al., 1998). The turtles begin their migration with an inherited "magnetic map," in which regional magnetic fields that exist at different locations along the route elicit directional swimming, which helps steer turtles along the pathway (Lohmann et al., 2012; Putman et al., 2012). Given that such a navigational system exists in loggerheads, it is at least conceivable that similar, inherited responses to regional magnetic fields play a role in guiding turtles back toward a particular geographic region.

Nevertheless, reasons exist to favor the prevailing hypothesis that turtles imprint on some aspect of their natal beaches or the nearshore environment. For example, during the period from 1978 to 2000, an effort was made to establish a nesting area for the Kemp's ridley turtle in Padre Island, Texas, in the hope of providing an additional nesting colony in the event that the population nesting at Rancho Nuevo went extinct (Shaver and Wibbels, 2007). Procedures were modified as the project progressed, but in many cases, eggs from Rancho Nuevo were incubated in sand at Padre Island. Hatchlings were then allowed to run across the beach at Padre Island and into the surf, after which turtles were "head-started" (raised for some months in captivity) at a Galveston facility before being released into the western Gulf of Mexico (Shaver and Wibbels, 2007). At least some of these turtles survived and returned to Texas to nest (Shaver and Caillouet, 1998). Although it is impossible to determine what aspect or aspects of egg-rearing, head-starting, and release might have affected the selection of eventual nesting site, or what environmental feature(s) turtles use to recognize this new natal area, the results at least suggest that early experience can establish a preference for nesting in a particular geographic area.

An additional reason to favor imprinting over an inherited program is that coastal areas constantly change. Storms rapidly destroy some nesting beaches and create new ones; sea levels rise and fall over time, and climate and ocean currents shift. For these reasons, a flexible system based on imprinting may be advantageous, inasmuch as it potentially allows a turtle to target a nesting area that existed at the start of its own life, rather than an ancestral nesting area that might have ceased to exist long before. Also, different nesting turtles within a population often begin their migrations back to the natal beach from widely dispersed locations (Carr, 1967; Schroeder et al., 2003). Because the migratory route that each turtle must follow to reach its home area appears to vary greatly among individuals in a population, it is difficult to imagine how an inherited or "hard-wired" set of responses can reliably guide a turtle back to the region of its natal beach. On the other hand, it is conceivable that only the final destination is inherited and that each turtle discovers the route to that location independently. Thus, despite reasons to favor the imprinting hypothesis, the possibility that alternative mechanisms underlie natal homing cannot yet be excluded.

## 3.6   SUMMARY AND FUTURE DIRECTIONS

Natal homing is a well-documented event in the life history of most or all species of sea turtles. It probably functions to return adult turtles to nesting beaches that possess a relatively rare combination of environmental factors favorable for nest construction, egg incubation, and the survival of hatchlings. The precision of natal homing is not well understood, but probably varies among different populations and species. Despite its prevalence, natal homing has not been studied extensively and the mechanisms that underlie it are unknown. In principle, turtles might recognize their natal beach or region using information about the target area that has either been learned or inherited. Although definitive evidence is lacking, it is widely assumed that turtles imprint on key features of their nesting beach region or nearshore environment during development or as hatchlings, and then use this information to return as adults. Sensory cues that might function in imprinting include the chemical profile of the beach or surrounding waters, or the magnetic signature of the area. Limited evidence exists for the involvement of both types of cues in different parts of natal homing, but numerous questions remain.

The long generation time of sea turtles, combined with their status as threatened species, make it challenging to conduct the kinds of experiments (e.g., Hasler and Scholz, 1983) that were instrumental in unraveling olfactory imprinting in salmon. Nonetheless, long-term studies of chemical and geomagnetic imprinting are theoretically possible and need to be done.

Even in the absence of direct tests of imprinting hypotheses, progress may still be possible through more indirect means. A promising avenue of research, as suggested by Shamblin et al. (2011), is to combine genetic studies with analyses of geomagnetic parameters, with a view toward determining whether patterns of variation in mtDNA among different nesting beaches can be explained by the geomagnetic imprinting hypothesis. Such questions appear to be fertile grounds for future inquiry.

## REFERENCES

Alcock, J. 2009. *Animal Behavior: An Evolutionary Approach*, 9th edn. Sunderland, MA: Sinauer Associates.

Bass, A. L., D. A. Good, K. A. Bjorndal et al. 1996. Testing models of female reproductive migratory behaviour and population structure in the Caribbean hawksbill turtle, *Eretmochelys imbricata*, with mtDNA sequences. *Molecular Ecology* 5:321–328.

Bjorndal, K. A., A. B. Meylan, and B. J. Turner. 1983. Sea turtle nesting at Melbourne Beach, Florida, I. Size, growth, and reproductive biology. *Biological Conservation* 26:65–77.

Bolten, A. B., K. A. Bjorndal, H. R. Martins et al. 1998. Transatlantic developmental migrations of loggerhead sea turtles demonstrated by mtDNA sequence analysis. *Ecological Applications* 8:1–7.

Bourjea, J., S. Lapegue, L. Gagnevin et al. 2007. Phylogeography of the green turtle, *Chelonia mydas*, in the southwest Indian Ocean. *Molecular Ecology* 16:175–186.

Bowen, B. W. and J. C. Avise. 1995. Conservation genetics of marine turtles. In *Conservation Genetics: Case Histories from Nature*, eds. J. C. Avise and J. L. Hamrick, pp. 190–237. New York: Chapman and Hall.

Bowen, B. W., J. C. Avise, J. I. Richardson, A. B. Meylan, D. Margaritoulis, and S. R. Hopkins-Murphy. 1993. Population structure of loggerhead turtles (*Caretta caretta*) in the northwestern Atlantic Ocean and Mediterranean Sea. *Conservation Biology* 7:834–844.

Bowen, B. W., A. L. Bass, S. Chow et al. 2004. Natal homing in juvenile loggerhead turtles (*Caretta caretta*). *Molecular Ecology* 13:3797–3808.

Bowen, B. W., A. M. Clark, F. A. Abreu-Grobois, A. Chaves, H. A. Reichart, and R. J. Ferl. 1998. Global phylogeography of the ridley sea turtles (*Lepidochelys* spp.) as inferred from mitochondrial DNA sequences. *Genetica* 101:179–189.

Bowen, B. W., N. Kamezaki, and C. J. Limpus. 1994. Global phylogeography of the loggerhead turtle (*Caretta caretta*) as indicated by mitochondrial-DNA haplotypes. *Evolution* 48:1820–1828.

Bowen, B. W. and S. A. Karl. 2007. Population genetics and phylogeography of sea turtles. *Molecular Ecology* 16:4886–4907.

Bowen, B. W., A. B. Meylan, J. P. Ross, C. J. Limpus, G. H. Balazs, and J. C. Avise, 1992. Global population structure and natural history of the green turtle (*Chelonia mydas*) in terms of matriarchal phylogeny. *Evolution* 46:865–881.

Brower, L. 1996. Monarch butterfly orientation: Missing pieces of a magnificent puzzle. *Journal of Experimental Biology* 199:93–103.

Carr, A. F. 1967. *So Excellent a Fishe: A Natural History of Sea Turtles*. New York: Scribner.

Carr, A. F. 1979. *The Windward Road: Adventures of a Naturalist on Remote Caribbean Shores*. Tallahassee, FL: University Presses of Florida.

Carr, A. F. 1986. Rips, FADS and little loggerheads. *Bioscience* 36:92–100.

Carreras, C., M. Pascual, L. Cardona et al. 2011. Living together but remaining apart: Atlantic and Mediterranean loggerhead sea turtles (*Caretta caretta*) in shared feeding grounds. *Journal of Heredity* 102:666–677.

Carreras, C., S. Pont, F. Maffucci et al. 2006. Genetic structuring of immature loggerhead sea turtles (*Caretta caretta*) in the Mediterranean Sea reflects water circulation patterns. *Marine Biology* 149:1269–1279.

Carthy, R. R., A. M. Foley, and Y. Matsuzawa. 2003. Incubation environment of loggerhead nests: Effects on hatching success and hatchling characteristics. In *Loggerhead Sea Turtles*, eds. A. B. Bolten and B. E. Witherington, pp. 144–153. Washington, DC: Smithsonian Books.

Dethmers, K. E. M., D. Broderick, C. Moritz et al. 2006 The genetic structure of Australasian green turtles (*Chelonia mydas*): Exploring the geographical scale of genetic exchange. *Molecular Ecology* 15:3931–3946.

Dittman, A. H. and T. P. Quinn. 1996. Homing in Pacific salmon: Mechanisms and ecological basis. *Journal of Experimental Biology* 199:83–91.

Døving, K. B., H. Westerberg, and P. B. Johnsen. 1985. Role of olfaction in the behavioral and neuronal responses of Atlantic salmon, *Salmo salar*, to hydrographic stratification. *Canadian Journal of Fisheries and Aquatic Sciences* 42:1658–1667.

Dutton, P. H., B. W. Bowen, D. W. Owens, A. Barragan, and S. K. Davis. 1999. Global phylogeography of the leatherback turtle (*Dermochelys coriacea*). *Journal of Zoology* 248:397–409.

Encalada, S. E., K. A. Bjorndal, A. B. Bolten et al. 1998. Population structure of loggerhead turtle (*Caretta caretta*) nesting colonies in the Atlantic and Mediterranean as inferred from mitochondrial DNA control region sequences. *Marine Biology* 130:567–575.

Endres, C. S., N. F. Putman, and K. J. Lohmann. 2009. Perception of airborne odors by loggerhead sea turtles. *Journal of Experimental Biology* 212:3823–3827.

Frick, J. 1976. Orientation and behavior of hatchling green turtles (*Chelonia mydas*) in the sea. *Animal Behaviour* 24:849–857.

Fuxjager, M. J., B. S. Eastwood, and K. J. Lohmann. 2011. Orientation of hatchling loggerhead sea turtles to regional magnetic fields along a transoceanic migratory pathway. *Journal of Experimental Biology* 214:2504–2508.

Goodenough, J., B. McGuire, and E. M. Jakob. 2010. *Perspectives on Animal Behavior*, 3rd edn. Hoboken, NJ: John Wiley & Sons.

Grassman, M. A. and D. W. Owens. 1982. Development and extinction of food preferences in the loggerhead sea turtle, *Caretta caretta. Copeia* 4:965–969.

Grassman, M. and D. W. Owens. 1987. Chemosensory imprinting in juvenile green sea turtles, *Chelonia mydas*. *Animal Behaviour* 35:929–931.

Grassman, M. A. and D. W. Owens. 1989. A further evaluation of imprinting in Kemp's ridley sea turtle. In *Proceedings of the 1st International Symposium on Kemp's Ridley Sea Turtle Biology, Conservation and Management*, eds. C. W. Caillouet Jr. and A. M. Landry Jr., pp. 90–105. Texas A&M University Sea Grant College Program, College Station, TX.

Grassman, M. A., D. W. Owens, J. P. Mcvey et al. 1984. Olfactory-based orientation in artificially imprinted sea turtles. *Science* 224:83–84.

Hasler, A. D. and A. T. Scholz. 1983. *Olfactory Imprinting and Homing in Salmon*. Berlin, Germany: Springer-Verlag.

Hasler, A. D., A. T. Scholz, and R. M. Horrall. 1978. Olfactory imprinting and homing in salmon. *American Scientist* 66:347–355.

Hasler, A. D. and W. J. Wisby. 1951. Discrimination of stream odors by fishes and its relation to parent stream behavior. *American Naturalist* 85:223–238.

Hendrickson, J. R. 1958. The green sea turtle, *Chelonia mydas* (Linn.) in Malaya and Sarawak. *Proceedings of the Zoological Society of London* 130:455–535.

Hughes, G. R. 1995. Conservation of sea turtles in the southern Africa region. In *Biology and Conservation of Sea Turtles*, ed. K. A. Bjorndal, pp. 397–404. Washington, DC: Smithsonian Institute Press.

Hyun, K. H. and R. He. 2010. Coastal upwelling in the South Atlantic Bight: A revisit of the 2003 cold event using long term observations and model hindcast solutions. *Journal of Marine Systems* 83:1–13.

Ireland, L. C., Frick, J. A., and Wingate, D. B. 1978. Nighttime orientation of hatchling green turtles (*Chelonia mydas*) in open ocean. In *Animal Migration: Navigation and Homing*, eds. K. Schmidt-Koenig and W. T. Keeton, pp. 420–429. New York: Springer-Verlag.

Kishida, T., S. Kubota, Y. Shirayama, and H. Fukami. 2007. The olfactory receptor gene repertoires in secondary-adapted marine vertebrates: Evidence for reduction of the functional proportions in cetaceans. *Biology Letters* 3:428–430.

Lee, P. L. M., P. Luschi, and G. C. Hays. 2007. Detecting female precise natal philopatry in green turtles using assignment methods. *Molecular Ecology* 16:61–74.

Lett, C., P. Verley, P. Mullon et al. 2008. A Lagrangian tool for modelling ichthyoplankton dynamics. *Environmental Modelling and Software* 23:1210–1214.

Limpus, C. J., J. D. Miller, C. J. Paramenter, D. Reimer, N. McLachlan, and R. Webb. 1992. Migration of green (*Chelonia mydas*) and loggerhead (*Caretta caretta*) turtles to and from eastern Australian rookeries. *Wildlife Research* 19:347–358.

Lohmann, K. J. 1991. Magnetic orientation by hatchling loggerhead sea turtles (*Caretta caretta*). *Journal of Experimental Biology* 155:37–49.

Lohmann, K. J. 2007. Sea turtles: Navigating with magnetism. *Current Biology* 17:R102–R104.

Lohmann, K. J., S. D. Cain, S. A. Dodge, and C. M. F. Lohmann. 2001. Regional magnetic fields as navigational markers for sea turtles. *Science* 294:364–366.

Lohmann, K. J., J. T. Hester, and C. M. F. Lohmann. 1999. Long-distance navigation in sea turtles. *Ethology Ecology and Evolution* 11:1–23.

Lohmann, K. J. and C. M. F. Lohmann. 1993. A light-independent magnetic compass in the leatherback sea turtle. *Biological Bulletin* 185:149–151.

Lohmann, K. J. and C. M. F. Lohmann. 1994. Detection of magnetic inclination angle by sea turtles: A possible mechanism for determining latitude. *Journal of Experimental Biology* 194:23–32.

Lohmann, K. J. and C. M. F. Lohmann. 1996. Detection of magnetic field intensity by sea turtles. *Nature* 380:59–61.

Lohmann, K. J. and C. M. F. Lohmann. 2003. Orientation mechanisms of hatchling loggerheads. In *Loggerhead Sea Turtles*, eds. A. Bolten and B. Witherington, pp. 44–62. Washington, DC: Smithsonian Institution Press.

Lohmann, K. J. and C. M. F. Lohmann. 2006. Sea turtles, lobsters, and oceanic magnetic maps. *Marine and Freshwater Behaviour and Physiology* 39:49–64.

Lohmann, K. J., C. M. F. Lohmann, L. M. Ehrhart, D. A. Bagley, and T. Swing. 2004. Geomagnetic map used in sea-turtle navigation. *Nature* 428:909–910.

Lohmann, K. J., C. M. F. Lohmann, and C. S. Endres. 2008a. The sensory ecology of ocean navigation. *Journal of Experimental Biology* 211:1719–1728.

Lohmann, K. J., C. M. F. Lohmann, and N. F. Putman. 2007. Magnetic maps in animals: Nature's GPS. *Journal of Experimental Biology* 210:3697–3705.

Lohmann, K. J., P. Luschi, and G. C. Hays. 2008b. Goal navigation and island-finding in sea turtles. *Journal of Experimental Marine Biology and Ecology* 356:83–95.

Lohmann, K. J., N. F. Putman, and C. M. F. Lohmann. 2008c. Geomagnetic imprinting: A unifying hypothesis of long-distance natal homing in salmon and sea turtles. *Proceedings of the National Academy of Sciences of the United States of America* 105:19096–19101.

Lohmann, K. J., N. F. Putman, and C. M. F. Lohmann. 2012. The magnetic map of hatchling loggerhead sea turtles. *Current Opinion in Neurobiology* 22:336–342.

Luschi, P., S. Benhamou, C. Girard et al. 2007. Marine turtles use geomagnetic cues during open-sea homing. *Current Biology* 17:126–133.

Manton, M. L., D. W. Ehrenfeld, and A. Karr. 1972a. Operant method for study of chemoreception in green turtle, *Chelonia mydas*. *Brain Behavior and Evolution* 5:188–201.

Manton, M., A. Karr, and D. W. Ehrenfeld, 1972b. Chemoreception in migratory sea turtle, *Chelonia mydas*. *Biological Bulletin* 143:184–195.

Marquez, M. R. 1994. Synopsis of biological data on the Kemp's ridley turtle, *Lepidochelys kempi* (Garman, 1880). *NOAA Technical Memorandum NMFS-SEFSC*-343:1–91.

Marquez, R., P. Burchfield, M. A. Carrasco et al. 2001. Update on the Kemp's Ridley turtle nesting in Mexico. *Marine Turtle Newsletter* 92:2–4.

Meylan, A. B., K. A. Bjorndal, and B. J. Turner. 1983. Sea turtles nesting at Melbourne Beach, Florida, II. Post-nesting movements of *Caretta caretta*. *Biological Conservation* 26:79–90.

Meylan, A. B., B. W. Bowen, and J. C. Avise. 1990. A genetic test of the natal homing versus social facilitation models for green turtle migration. *Science* 248:724–727.

Morreale, S. J., P. T. Plotkin, D. J. Shaver, and H. J. Kalb. 2007. Adult migration and habitat utilization: Ridley turtles in their element. In *Biology and Conservation of Ridley Sea Turtles*, ed. P. T. Plotkin, pp. 213–229. Baltimore, MD: Johns Hopkins.

Mortimer, J. A. 1995. Factors influencing beach selection by nesting sea turtles. In *Biology and Conservation of Sea Turtles*, ed. K. A. Bjorndal, pp. 45–51. Washington, DC: Smithsonian Institute Press.

Mrosovsky, N. 1983. *Conserving Sea Turtles*. London, U.K.: The British Herpetological Society c/o The Zoological Society of London Regent's Park.

Norman, J. A., C. Moritz, and C. J. Limpus. 1994. Mitochondrial DNA control region polymorphisms: Genetic markers for ecological studies of marine turtles. *Molecular Ecology* 3:363–373.

North, E. W., E. E. Adams, Z. Schlag, C. R. Sherwood, R. He, K. H. Hyun, and S. A. Socolofsky. 2011. Simulating oil droplet dispersal from the *Deepwater Horizon* spill with a Lagrangian approach. In *Monitoring and Modeling the Deepwater Horizon Oil Spill: A Record-Breaking Enterprise, Geophysical Monograph Series*, Vol. 195, eds. Y. Liu, A. MacFadyen, Z.-G. Ji, and R. H. Weisberg, pp. 217–226. Washington, DC: American Geophysical Union.

Owens, D. W., M. A. Grassman, and J. R. Hendrickson. 1982. The imprinting hypothesis and sea turtle reproduction. *Herpetologica* 38:124–135.

Peare, T. and P. G. Parker. 1996. Local genetic structure within two rookeries of *Chelonia mydas* (the green turtle). *Heredity* 77:619–628.

Plotkin, P. 2007. *Biology and Conservation of Ridley Sea Turtles*. Baltimore, MD: The Johns Hopkins University Press.

Putman, N. F., J. M. Bane, and K. J. Lohmann. 2010a. Sea turtle nesting distributions and oceanographic constraints on hatchling migration. *Proceedings of the Royal Society B* 277:3631–3637.

Putman, N. F., C. S. Endres, C. M. F. Lohmann, and K. J. Lohmann. 2011. Longitude perception and bicoordinate magnetic maps in sea turtles. *Current Biology* 21:463–466.

Putman, N. F. and K. J. Lohmann. 2008. Compatibility of magnetic imprinting and secular variation. *Current Biology* 18:R596–R597.

Putman, N. F., T. J. Shay, and K. J. Lohmann. 2010b. Is the geographic distribution of nesting in the Kemp's ridley turtle shaped by the migratory needs of offspring? *Integrative and Comparative Biology* 50:305–314.

Putman, N. F., P. Verley, T. J. Shay, and K. J. Lohmann. 2012. Simulating transoceanic migrations of young loggerhead sea turtles: Merging magnetic navigation behavior with an ocean circulation model. *Journal of Experimental Biology* 215:1863–1870.

Quinn, T. P. 2005. *The Behavior and Ecology of Pacific Salmon and Trout*. Seattle, WA: The Washington University Press.

Reppert, S. M., R. J. Gegear, and C. Merlin. 2010. Navigational mechanisms of migrating monarch butterflies. *Trends in Neurosciences* 33:399–406.

Salmon, M. and J. Wyneken. 1987. Orientation and swimming behavior of hatchling loggerhead turtles *Caretta caretta* L. during their offshore migration. *Journal of Experimental Marine Biology and Ecology* 109:137–153.

Schroeder, B. A., A. M. Foley, and D. A. Bagley. 2003. Nesting patterns, reproductive migrations, and adult foraging areas of loggerhead turtles. In *Loggerhead Sea Turtles*, eds. A. B. Bolten and B. E. Witherington, pp. 114–124. Washington, DC: Smithsonian Books.

Shamblin, B. M., M. G. Dodd, D. A. Bagley et al. 2011. Genetic structure of the southeastern United States loggerhead turtle nesting aggregation: Evidence of additional structure within the peninsular Florida recovery unit. *Marine Biology* 158:571–587.

Shanker, K., J. Ramadevi, and B. C. Choudhury. 2004. Phylogeography of olive ridley turtles (*Lepidochelys olivacea*) on the east coast of India: Implications for conservation theory. *Molecular Ecology* 13:1899–1909.

Shaver, D. J. and C. W. Caillouet, Jr. 1998. More Kemp's ridley turtles return to south Texas to nest. *Marine Turtle Newsletter* 82:1–5.

Shaver, D. J. and T. Wibbels. 2007. Head starting the Kemp's ridley sea turtles. In *Biology and Conservation of Ridley Sea Turtles*, ed. P. T. Plotkin, pp. 297–324. Baltimore, MD: Johns Hopkins.

Shillinger, G. L., E. Di Lorenzo, H. Luo et al. 2012. On the dispersal of leatherback turtle hatchlings from Mesoamerican nesting beaches. *Proceedings of the Royal Society B* 279:2391–2395.

Skiles, D. D. 1985. The geomagnetic field: Its nature, history, and biological relevance. In *Magnetite Biomineralization and Magnetoreception in Organisms: A New Biomagnetism*, eds. J. L. Kirschvink, D. S. Jones, and B. J. MacFadden, pp. 43–102. New York: Plenum Press.

Southwood, A., K. Fritsches, R. Brill, and Y. Swimmer. 2008. Sound, chemical, and light detection in sea turtles and pelagic fishes: Sensory-based approaches to bycatch reduction in longline fisheries. *Endangered Species Research* 5:225–238.

Zupanc, G. K. H. 2010. *Behavioral Neurobiology: An Integrative Approach*. New York: Oxford University Press.

# 4 The Skeleton
## An In Vivo View of Structure

*Jeanette Wyneken*

## CONTENTS

## 4.1 INTRODUCTION

Sea turtles, like all other turtles, have a unique body design that is characterized by the rib cage surrounding the shoulders and hips so the limbs appear to emerge from within the armored body. The anatomy of the sea turtle is what physically interacts with the environment. The bony skeleton gives sea turtle species their unique shape and size. The skeleton is dynamic and functions in many roles, ranging from serving as the system of levers upon which contracting muscles act to providing protective armor. The skeleton also serves as a mineral reserve and provides a record of the animal's growth as well as a history of trophic level or food consumption (Biasatti, 2004; Avens and Snover, Chapter 5; Jones and Seminoff, Chapter 9). Bone is strong, dynamic, and persists long after death—long past the time when soft tissues have become food for many other organisms. When fossilized, the skeleton can teach us about the evolutionary history of the animals (see Zangerl, 1980).

For vertebrate biologists, paleontologists, species managers, and veterinarians, the sea turtle skeleton is a system of landmarks and key characteristics. The skeleton also provides many of the characteristics that we use to identify species. Through its formation as a rigid shell, the skeleton also can provide a convenient platform for the attachment of biologging devices, leading to the understanding of behavior that would otherwise be intractable. While the skeleton provides bony armor, it is vulnerable to injury or disease. Very often though, the sea turtle skeleton is an inconvenient "black box" providing such "good protection" that access is limited along with our understanding and ability to treat disease of the structures within.

Perhaps, the most common inquiries I receive, as a sea turtle anatomist/biologist, are as follows: "What is this"?; "What part of body is it from"?; "What species did this bone come from and how do you know"?; and from clinicians, "Can you take a look at the radiograph (or CT) and tell me if this is normal"? One goal of this chapter is to provide the foundation for identifying bones and understanding where they occur in the sea turtle's body. Species identification from the bones is beyond the scope of this chapter and is addressed in detail elsewhere (see Wyneken, 2001). While interest in skeletons has a very long history (see Romer, 1956), past approaches have not been able to describe the skeleton as it occurs in living animals. Advances in increasingly effective and available in vivo imaging techniques (digital radiography and CT imaging) allow us to view the skeleton as it occurs in life. Discussion of bones and their uses in understanding the biology of sea turtles are scattered in chapters throughout this book. Together, these approaches and opportunities prompted an updated view of the sea turtle skeleton.

## 4.2 COMPONENTS OF THE SKELETON

The skeleton of any animal can be described from a number of perspectives or scales including (1) the tissues that form its elements, (2) evolutionary history, (3) its origins and embryological development, and (4) the distribution and functions of its elements in the body.

### 4.2.1 TISSUES

The skeleton is composed of bones and cartilages. These are composite tissues formed of collagen fibers, cells, and a matrix of minerals and water. Cartilage tends to contain more water than bone, and its matrix is glycoprotein-based. Young cheloniid turtles and leatherbacks tend to possess much hyaline cartilage, which is a resilient tissue. Sea turtle bone has a mineralized matrix of calcium phosphate and limited calcium carbonate. The bones may form by intermembranous (sesamoid or dermal) or endochondral (cartilage replacement) processes. Sea turtles retain cartilage throughout life, particularly as articular surfaces, supporting the bases of the great vessels, and the medial aspect of the eye's sclera. Cartilage forms a much greater portion of the skeleton in hatchlings and juveniles than it does in adults. But, in leatherbacks, it persists as a major component of the skeleton throughout life.

In cheloniids, bone formation is by cartilage replacement, deposition of periosteal bone around the cortical compact bone (see Avens and Snover, Chapter 5 the volume; Snover and Rhodin, 2008), and by intramembranous condensations of bone (Gilbert et al., 2008). In *Dermochelys*, the long bone growth, such as is described for the humerus (Snover and Rhodin, 2008), is primarily by chondroosseus formation. Extensive articular cartilages and subchondral articular bone are rich in blood vessels that support chrondroosseus bone growth in *Dermochelys*. The abundant cartilages are not seen in radiographs, which cause the skeleton of leatherbacks to appear loosely articulated (Figure 4.1). In fact, their limb skeletal structures are quite strong, and the limb segments are fairly stiff except in the youngest animals.

### 4.2.2 BONY GROUPS

Typically, the skeleton is described by the embryonic and evolutionary origins of its main parts: skull and hyoid apparatus, axial skeleton, and appendicular skeleton (Figure 4.2). In sea turtles, each of these bony groups is a composite of several structures. For example, the skull and hyoid (Figure 4.3) includes the cranium (often broadly termed the braincase), jaws, and the hyoid apparatus (throat skeleton). In life, the hyoid apparatus (the hyoid body and paired ceratohyal bones and cartilages) is associated with the lower jaw and receives tongue and throat muscles. The hyoid apparatus is located between two rami of the lower jaw, and its ceratohyal bones and cartilages extend around the back of the skull. Sea turtles like birds and many fish also have a ring of endochondral bones in each eye (scleral ossicles) and hyaline cartilage within the sclera supporting the back of each eyeball. The division of the skeleton is described and illustrated next.

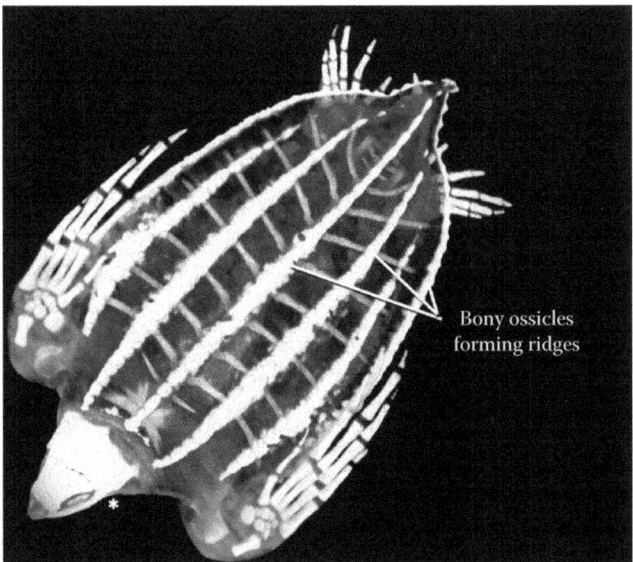

FIGURE 4.1    Three-dimensional (3D) CT scan of a juvenile leatherback (*Dermochelys coriacea*). The dermal ossicles of the carapace are in the initial stages of ossifying so that the characteristic longitudinal ridges are present. This specimen has a depressed fracture at the frontal–parietal suture (*). Because leatherbacks have extensive articular cartilages, most bones appear to be widely separated. The nuchal bone, nine ribs, and large transverse processes of the sacral vertebrae are seen clearly.

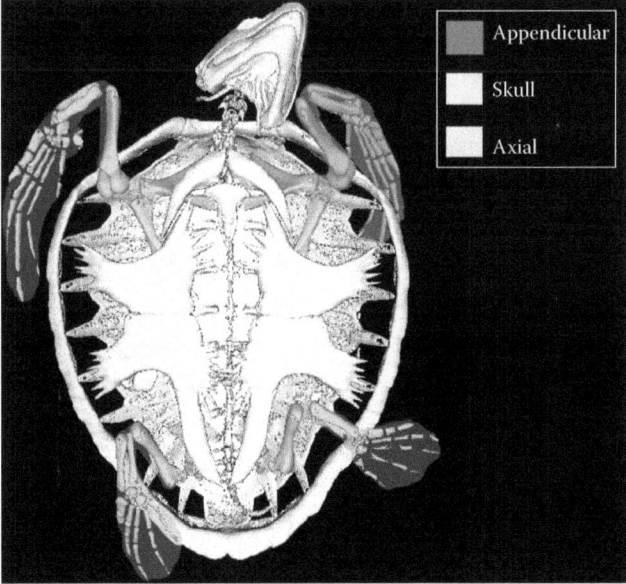

FIGURE 4.2    Ventral view of a 3D CT scan of a neritic stage juvenile loggerhead (*C. caretta*). The three skeletal regions are color-coded. Note that the shoulder and pelvic skeletal elements are within the rib cage, a characteristic of turtles.

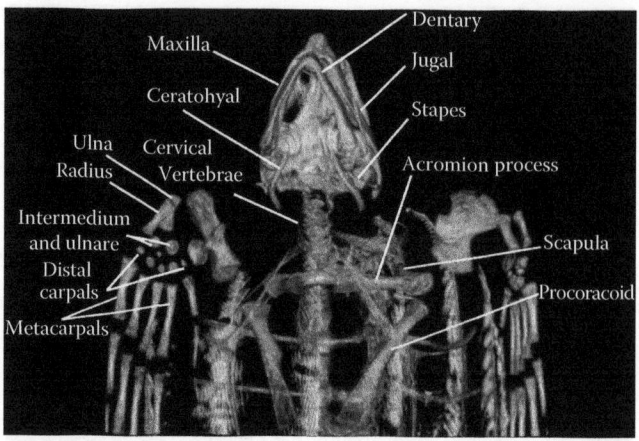

**FIGURE 4.3** Ventral view of a 3D CT scan of a juvenile leatherback focusing on the cranial, hyoid, and pectoral appendage skeletal elements.

The division of the skeleton is described and illustrated next.

### 4.2.2.1 Axial Skeleton

The axial skeleton is composed of the vertebrae, the ribs and their derivatives, and other bones that form the carapace and plastron (Figure 4.2). The cheloniid carapace includes a nuchal bone, marginals, pleurals, neurals, and a suprapygal bone that may be a single structure or divided as two (Gilbert et al., 2008; Pritchard, 2008) (Figure 4.4). The neural bones are ankylosed as modified neural spines to the vertebral bodies, and, together, they enclose the spinal cord in the trunk region.

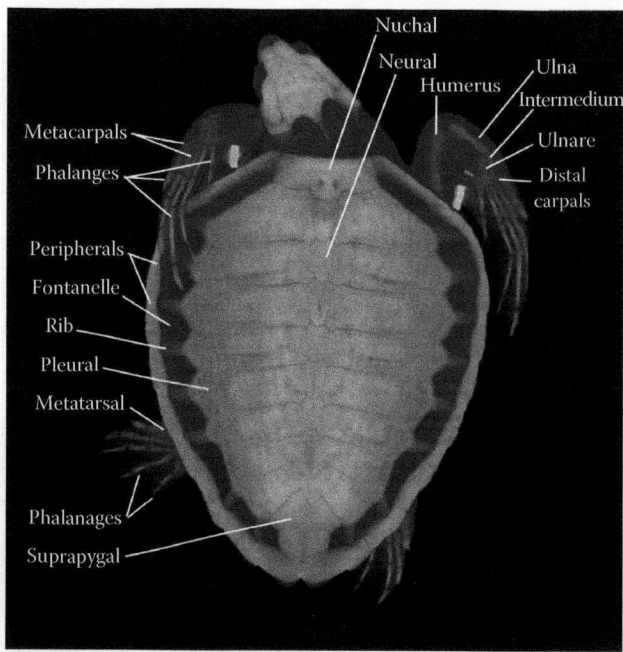

**FIGURE 4.4** Dorsal view, CT scan 3D reconstruction of neritic stage juvenile *C. caretta*. This is an MIP volume reconstruction that shows some bone and the soft tissue for reference. The major appendicular and dorsal carapacial bones are noted here and discussed in the text. This turtle had metal flipper tags in both flippers (bright rectangles) and a pit tag (white line) near the right pisiform bone.

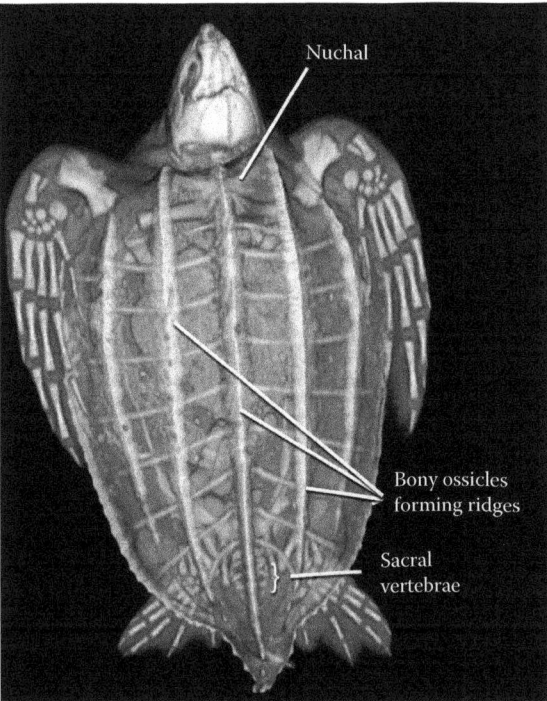

FIGURE 4.5 Dorsal view, CT scan 3D reconstruction of juvenile *D. coriacea*. This is a volume reconstruction that shows some bone and the soft tissue including the lungs and dorsally positioned parts of the intestines. The scapulae are visible deep to and extending and laterally from the nuchal bone. In a healthy animal, these bony processes would be more dorsoventrally positioned. The sacral vertebrae of leatherbacks loosely articulate with cartilaginous ends of the ilia. They are more mobile than those of cheloniid sea turtles.

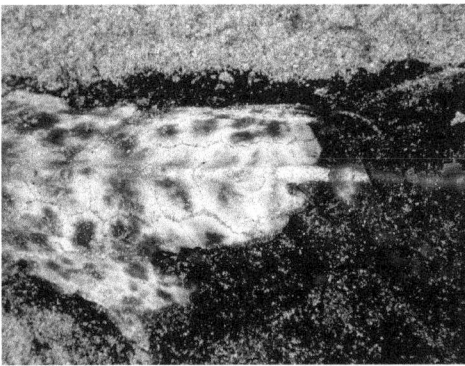

FIGURE 4.6 Dermal ossicles (white bone with irregular sutures) are just deep to the skin. These bones were exposed on a beached *D. coriacea* carcass so the black skin has sand scattered across much of the surface.

In the posterior carapace, the vertebrae are free from the neurals starting in the sacral region. The sacral and proximal caudal vertebrae are within the carapace. Posterior to the suprapygal bone, the caudal vertebrae form the tail.

The axial skeleton also includes ribs that are modified as pleural bones in the cheloniid carapace (Figure 4.4). In *D. coriacea*, the carapace is formed by ribs, a nuchal bone, and vertebrae (Figures 4.1 and 4.5) covered by a thick blubber layer and a layer of interlocking dermal ossicles (Figure 4.6). These bony ossicles are embedded in peripheral blubber and are covered by waxy skin.

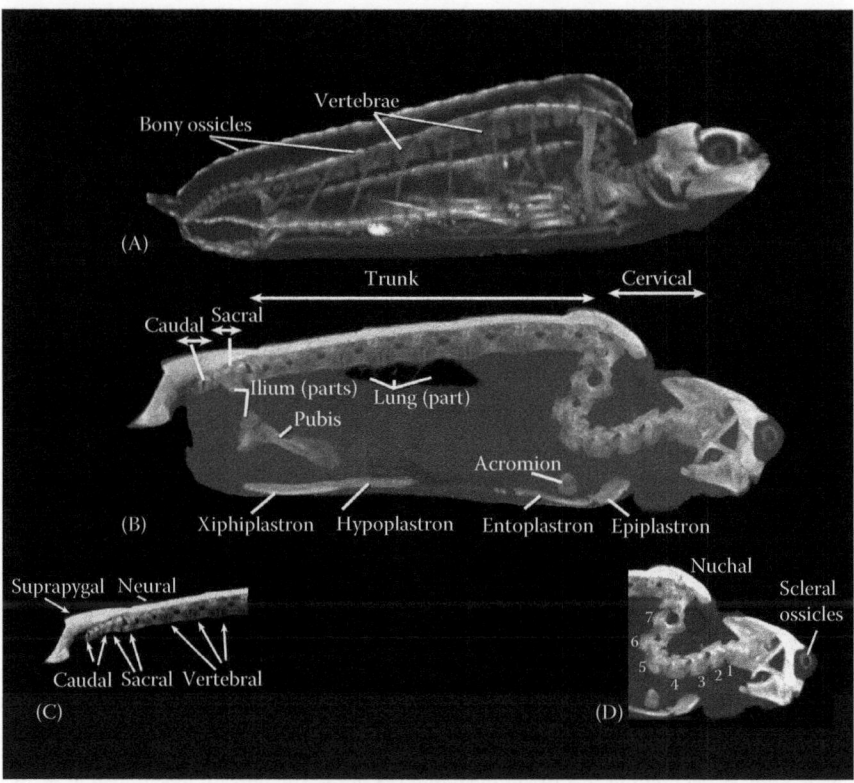

**FIGURE 4.7** CT scan. Parasagittal view of a juvenile (A) *D. coriacea* and (B) *C. caretta* showing the regionally specialized vertebrae (cervical, trunk, sacral, and caudal) as well as the relationship of vertebrals to the overlying dermal ossicles layer (A) or neurals (B). (C) Parasagittal view of the posterior vertebrae medial to ilia. (D) Parasagittal view of the cervical vertebrae and nuchal vertebra.

Like all turtles, sea turtles have seven mobile cervical vertebrae including a three-part atlas and axis. The eighth vertebra transitions between cervical and trunk as it articulates with a convex articular surface on the ventral side of the nuchal, the most anterior vertebra of the carapace (Figure 4.7). Eight trunk (thoracic) vertebrae follow and articulate with rib heads. The single-headed ribs align at the junction of two vertebral bodies (Pritchard, 1988). The rib heads are ventral to the expanded pleurals and neurals. There are usually eight to nine elongated neural bones. However, variation is common (Pritchard, 1988) so that some neurals are divided in two by transverse sutures. Supernumerary neural bones are common in some species, particularly *Lepidochelys olivacea*.

In cheloniids, pleural bones comprise the ribs, and their intercostal bony expansions form much of the dorsal carapace. The distal ribs articulate with the peripheral bones along the margin of the carapace (Figure 4.4). The anterior-most carapace bone is the nuchal, which appears to develop from neural crest tissue (Gilbert et al., 2008). The posterior-most peripheral bone, the pygal, develops as a dermal bone. Between the last neural bone and the pygal is the suprapygal, which lacks articulation with any vertebral components (Figure 4.4). In *D. coriacea*, the carapace skeleton is formed by the nuchal, vertebrae formed by vertebral bodies with incomplete neural arches and eight ribs. These bones are covered with blubber and dermal ossicles (Figure 4.1 and 4.8). There are no neural bones, and the ribs remain in neotenic form, never expanding as pleurals. *Dermochelys* lacks a suprapygal, pygal, and peripheral bones.

All sea turtles have two to three sacral vertebrae with long transverse processes that articulate with the medial ilium. Twelve or more caudal vertebrae follow. The caudal vertebrae of females are

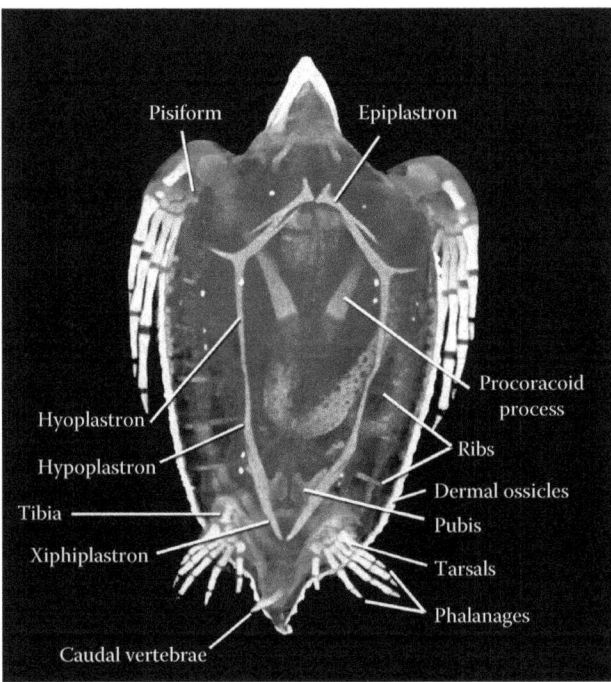

**FIGURE 4.8** Ventral view of a juvenile *D. coriacea*, 3D reconstruction of CT scan. The plastron skeletal elements are reduced. The lateral ridges formed of dermal ossicles are found superficially in blubber layer. Components of the pectoral and pelvic girdles and appendages can be identified in the view. They are discussed in the text.

short and decrease in size distally. The caudal vertebrae of males grow in depth, width, and length during puberty so that mature males have long prehensile tails equipped with robust lateral and dorsal processes (see Wyneken, 2001). It is not known if the number of tail vertebrae increases in males as they approach puberty. In hatchlings, there appear to be no differences between males and females in the number of tail vertebrae.

The ontogeny of the shell differs between cheloniids and *D. coriacea*. In hatchling cheloniids, the carapace skeleton is composed of ribs, a nuchal, and vertebrae. The shell skeleton becomes increasingly ossified with age. The spaces between ribs fill with intermembranous bone that grows distally from the neurals to form the pleural bones; these make up much of the carapace's bony armor (Figure 4.9). The ribs articulate with the peripheral bones, which ossify after hatching. Ossification proceeds in cranial to caudal fashion in freshwater turtles (Gilbert et al., 2008); this process appears to be similar in at least *Caretta caretta, Eretmochelys imbricata*, and *Chelonia mydas*. In *Lepidochelys kempii*, the peripheral bones also widen as the plurals grow distally. The fontanelles (spaces bordered by the rib ends, distal aspects of the plueral bones, and the peripherals) are filled with a fibrous connective tissue membrane underlying the scutes (Figure 4.4). The fontanelles are closed in some adult *L. kempii, L. olivacea*, and *C. caretta*, but often are retained in *C. mydas, E. olivacea*, and some *N. depressus* (Pritchard, 1979; Wyneken, 2001). In contrast, *D. coriacea* never develops plurals or peripheral bones, and the plastron bones that it has in common with other sea turtles are reduced. Instead the *D. coriacea* shell gains strength and protection form a layer in interlocking bony ossicles. The bony ossicles are displaced from the other bones by blubber layer so they are just beneath the skin of the shell (Figures 4.5 and 4.7). The ossicles form within the scales that cover the bodies of hatchlings, appearing in the ridges of posthatchings when the scales have been shed.

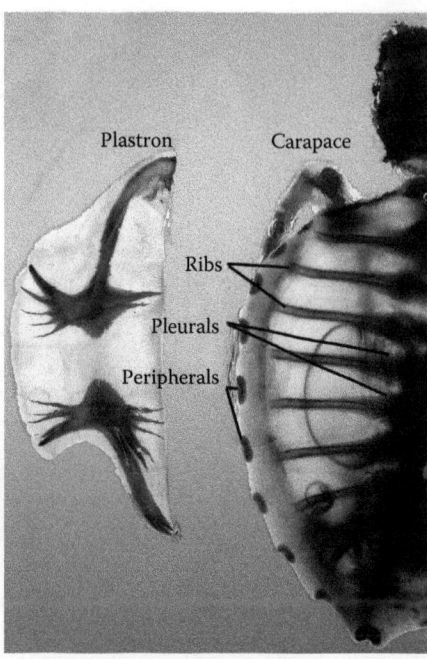

**FIGURE 4.9** Cleared-stained posthatchling green turtle, *C. mydas*. Bone stains red and cartilage stains blue. The left half of the plastron was removed showing the four left plastral bones. The peripherals started ossification individually; the pleurals started to ossify near the neurals. Neither is ossified at hatching.

The plastron of all sea turtles is a composite of four paired bony elements (epiplastra, hypoplastra, hyoplastra, and xiphiplastra). Cheloniids also have one unpaired bony element, the entoplastron, located medially in the anterior half of the plastron (Figure 4.10). The epiplastral bones are homologous to the clavicles, and the entoplastron is thought to be homologous to the interclavicle (Zangerl, 1939, 1969; Rieppel, 1996; see review by Gilbert et al., 2008). Zangerl (1939) suggests that the hypoplastra, hyoplastra, and xiphiplastra may be homologous to tetrapod gastralia (abdominal ribs). In *D. coriacea*, the paired plastron bones are very reduced (Figure 4.8), forming an incomplete ring of bone around the periphery of the plastron (Deraniyagala, 1939; Pritchard, 1979).

### 4.2.2.2 Appendicular Skeleton

The appendicular skeleton, perhaps, is the feature that is most distinctive in marine turtles. Appendicular bones support the flippers (forelimbs), hind limbs, shoulders and hips (the pectoral and pelvic girdles, respectively). The pectoral and pelvic girdles are located within the shell, an arrangement that is characteristic of all turtles (Figures 4.1 and 4.2). In sea turtles, the forelimb articulates with the pectoral girdle just inside the axial margins of the shell. The shell is reduced compared to that of other turtles, making the shoulder joint accessible, but externally, it is cryptic in appearance and position. Additionally, the heads of both the humerus and the femur are offset. It is easy to palpate the large medial process of the humerus and the minor trochanter of the femur, which are frequently mistaken for heads of these long bones (see below and Avens and Snover, Chapter 5. This volume).

#### *4.2.2.2.1  Pectoral Girdles and Limbs*

Sea turtle forelimbs are modified from the typical tetrapod limbs pattern by elongation, widening, and flattening of the wrist and hand. As the forelimb forms, the head of the humerus is displaced

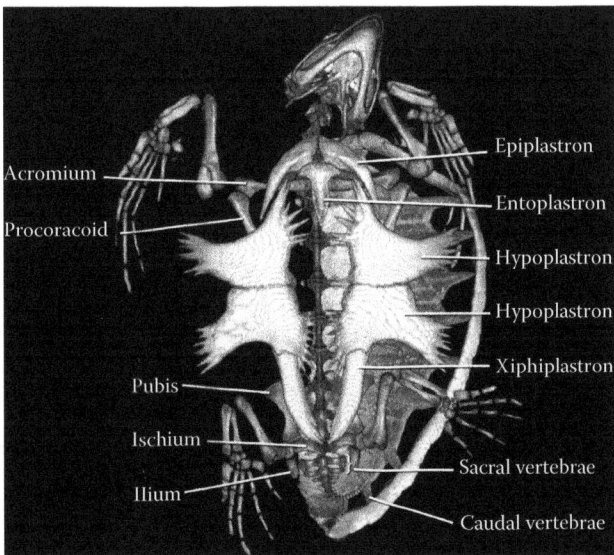

FIGURE 4.10   *C. caretta*, ventral view with right carapace, digitally removed to expose the appendages and their girdles. The plastral elements are well ossified but do not fill the entire plastron. The vertebrals can be seen through the persistent plastral fontanelle. The rib heads articulate with the junction of two vertebral bodies. The sacral vertebrae closely articulate with the ilia; however, the cartilaginous and fibrous ligaments are not detected in the CT bone reconstruction.

distally. The humerus has undergone torsion as it develops and so morphological dorsal surface of the humerus faces anteroventrally, and the elbow moves the forearm from an anterolateral position to a ventromedial position (Figure 4.2).

Each forelimb articulates with its pectoral girdle, which is a prominent triradiate structure composed of two bones, the scapula and the procoracoid (=coracoid) (Figure 4.10). The shoulder girdle forms the medial aspect of the shoulder joint and serves as a major site for the attachment of the swimming musculature. The procoracoid, a ventrally positioned bone that lies parallel to the plastron, is flat and widens at its distal end. The procoracoids each terminate in a crescent-shaped procoracoid cartilage that supports major muscle attachments. Each scapula is oriented dorsoventrally and attaches to the carapace at the nuchal bone and adjacent to the first trunk vertebra (Figure 4.11). The acromion process extends medially at an angle of ~110° angle (cheloniids) or ~130° angle (*D. coriacea*) from the ventral part of the scapula. The medial end of acromion process articulates with the plastron's fibrous connective tissue via ligaments just posterior to the medial extent of the epiplastral bones in *D. coriacea* (Figure 4.3). In cheloniids, they attach to both the cranial end of the entoplastron (Figures 4.7 and 4.10) and the posterior edges of the epiplastra. The scapula and procoracoid join laterally to form the shoulder joint (glenoid fossa). Two of the three major of the flipper retractor and adductor muscles (Supracoracoideus and Coracobrachialis, but not Pectoralis) originate from the procoracoid, acromion processes, and acromiocoracoid ligament (not seen in the CT scans) (Walker, 1973; Wyneken, 1997; Rivera et al., 2011).

The forelimb is composed of humerus, radius and ulna, carpals, metacarpals, and phalanges of five digits (Figures 4.3 and 4.4). The flipper blade is formed by widening and flattening of the wrist bones and elongation of the digits. The humeral shaft is robust and somewhat flattened; its head is offset from the bone's shaft by about 20°. The humerus has a large medial process that extends beyond the head. The humeral head and distal articulations to the radius and ulna are largely cartilaginous in *D. coriacea*. The medial process is bony in cheloniids (Figure 4.12) but remains cartilaginous in *D. coriacea*. Just distal to the head, along ventral surface is the

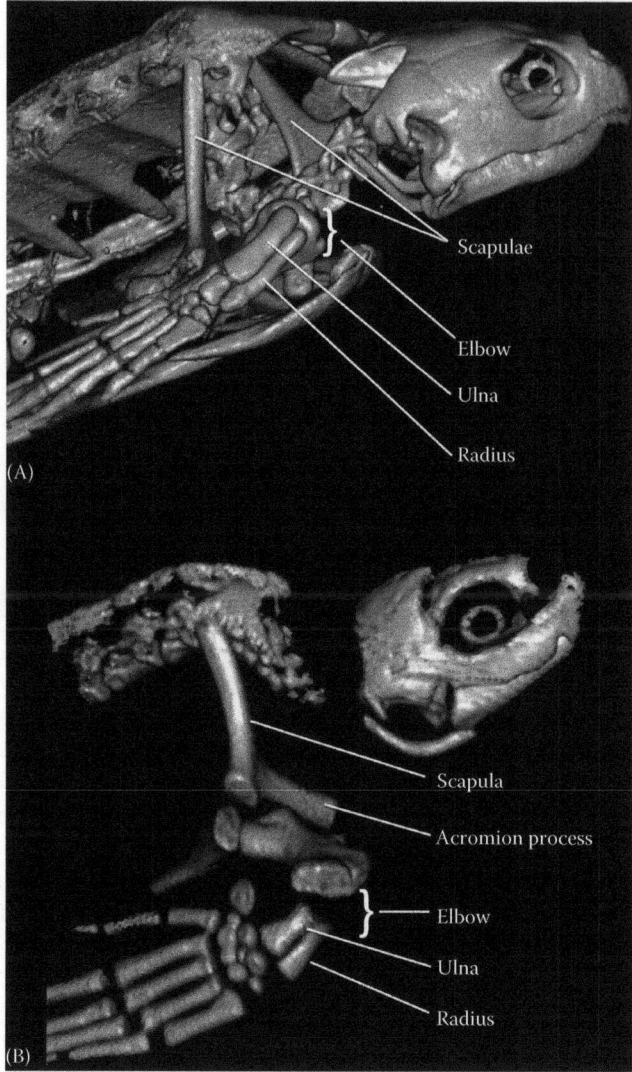

**FIGURE 4.11** CT scans. (A) *C. caretta* and (B) *D. coriacea* right lateral views of "digital dissections" from CT bone reconstructions. The relationship of the scapula to the nuchal can be seen using this method of image exploration. Each turtle's flipper is flexed at the elbow so the radius and ulna are in clear view. The greater spaces between bones in the leatherback are due to the high proportion of cartilage in this species' skeleton. The skull shows the scleral ossicles of the left eye in both and the nasolachrymal foramen as a white spot in the anterior ventral orbit of *C. caretta*.

lateral (=pectorodeltoid) process that appears as if it is transverse to the shaft (Figure 4.10). In *D. coriacea*, the lateral process is displaced nearly half way down the shaft (Figure 4.3). These two processes are the insertion sites for most of the major muscles that move the flipper during swimming, particularly during powerstroking.

The flipper blade extends from the elbow distally. Orienting the bones of the flipper in a radiograph or CT can be daunting (errors are common in even current publications). However, there are several landmarks that help define the major axes (termed the preaxial–postaxial axis or radial–ulnar axis). The first digit has just two phalanges and a large flat metacarpal. (In cheloniids, there is usually a claw present on the terminal phalanx of Digit I). These components mark the radial

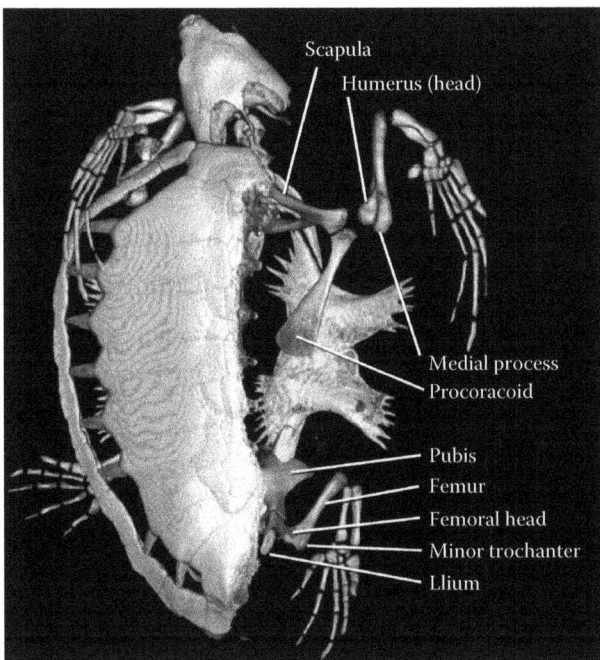

**FIGURE 4.12** *C. caretta* dorsal view with right carapace digitally removed (same reconstruction and digital dissection as in Figure 4.10). This view exposes pectoral and pelvic girdles and their appendages. The major muscle attachment sites on the humerus and femur are noted as they are important landmarks for the two joints.

side of the flipper, which is the leading-edge side. The ulnar side is easily identified by the large flat pisiform bone (rectangular in cheloniids and nearly round in dorsoventral outline in *D. coriacea*); this is the trailing-edge side of the flipper blade (Figures 4.3 through 4.5).

The flipper skeleton is composed of a short, mostly straight radius, and a shorter slightly curved ulna, with 9–10 wide and flat wrist elements (Figures 4.4 and 4.8) and five digits (metacarpals and phalanges). The radius articulates with the radiale, a bone that is quite small. The ulna articulates with the ulnare and intermedium. The distal wrist is bounded by a row of distal carpals that articulate with one another, in order, and the fifth articulates with an expanded and flattened pisiform bone. A centrale is distal to the intermedium and articulated with the second, third, and fourth distal carpals. The pisiform and the distal carpals demark the distal wrist. The five elongated metacarpals and their phalanges form the largest part of the flipper blade. The forearm bones (radius and ulna) are short in sea turtles and, in adults, become functionally fused along their shafts by fibrous connective tissue. Similarly, a network of collagen-rich ligaments tightly attaches the wrist elements to one another. In cheloniids, Digit I has two phalanges, and Digits II–V have three phalanges each (Figures 4.4, 4.5, and 4.8). In *D. coriacea*, Digits I and V have two phalanges, and Digits II–IV have three phalanges each. The phalanges often are nearly circular in cross-section. *Chelonia mydas* and *Natator depressus* typically have one claw located on Digit I on each limb; *C. caretta*, *E. imbricata*, and both *Lepidochelys* species have claws on Digits I and II on each limb. *Dermochelys* lacks claws.

### 4.2.2.2.2   Pelvic Girdles and Limbs

The pelvis is comprised of three pairs of bones (pubis, ischium, and ilium) joined by cartilage in neonate cheloniids and throughout life in *D. coriacea*. The pubic bones and the ischia form the more ventrally positioned part of the pelvis (Figure 4.10). The ilia are oriented dorsoventrally and articulated with the sacral vertebrae dorsally. The ilia also attach the pelvis to the carapace via ligaments. All three bones form the acetabulum (hip socket) ventrolaterally.

The head of the femur is strongly offset from the relatively straight shaft and articulates with the acetabulum, which has a deep cup form. A bony process, the major trochanter, is oriented dorsal and posterior to the head (Figure 4.12); this process is the attachment site for the thigh retractors. The minor trochanter faces ventrally and receives the attachment of the thigh adductors (Walker, 1973). The distal femur articulates with the tibia and fibula of the shank. There is no patella. The hindfoot is a foot but not a flipper because of its shape and function.

The shank and tarsal bones are often as challenging to orient as those of the flipper because of their gross symmetry. In both cheloniids and *D. coriacea*, the first digit has one large tapering metatarsal and two phalanges (Figure 4.16). In cheloniids, the tibial side of the limb is that with the claw on the first digit. The fibular side of the hind limb has a short flat "hook-shaped" metatarsal, and the digit has two phalanges (Figure 4.16). The tibia tends to be slightly more stout that the fibula and has a triangular articular surface at its proximal end. The fibula has convex articular surfaces at both ends, and the shaft is concave toward the tibia (Romer, 1956). The tibia articulates with the first distal tarsal. The fibula articulates with the fibulare and intermedium, which often is a single bone, the astragalocalcaneum (after Sánchez-Villagra et al., 2007). There are four smaller distal tarsals completing the ankle. These articulate, sequentially, with metatarsals I–IV. The fifth ("hook-shaped") metatarsal articulates with the outer (anatomically medial) edge of the fourth distal tarsal. Metatarsals II–V resemble phalanges except their proximal articular surfaces somewhat irregular. Digit I has two phalanges; Digits II–IV have three phalanges each. In leatherbacks and some cheloniid individuals, a third distal-most cartilaginous phalanx sometimes is found in Digit V.

### 4.2.2.3   Skull

The same skull bones occur in all marine turtle species. However, their specific form, particularly in the palatal bones and some articulations, differs with species. Skull shape and the patterns of bones of the palate are diagnostic for species identification (see Wyneken, 2001, 2005).

The skull is often envisioned as a single unit, yet it is composed of three parts (the chondrocranium, dermatocranium, and splanchnocranium) that have different phylogenetic, positional, and developmental origins. Chondrocranial bones are endochondral in origin. They encase much of the brain and form the posterior skull including the parietal bones, whose form can be diagnostic. Bones of the dermatocranium form as intramembranous bone, often from neural crest; they are often flat and make up the outer casing and roof of the skull. The braincase is a composite of parts of the chondrocranium roofed by dermatocranial bones (Kardong, 2012). Most of the endochondral skull bones are deep within the skull housing the brain and inner ear. Bones of the back of the skull are endochondral, including the occipital condyles, the occipital series, and supraoccipital (Figures 4.13 and 4.14). So are the scleral ossicles (Figure 4.7B) and the articular of the lower jaw (Figure 4.15).

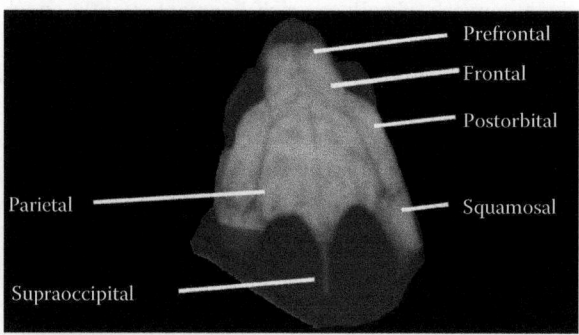

**FIGURE 4.13**   Dorsal close-up of a *C. caretta* head showing the bony suture detail that is possible with high-resolution (1.5 mm voxels) clinical CT imaging.

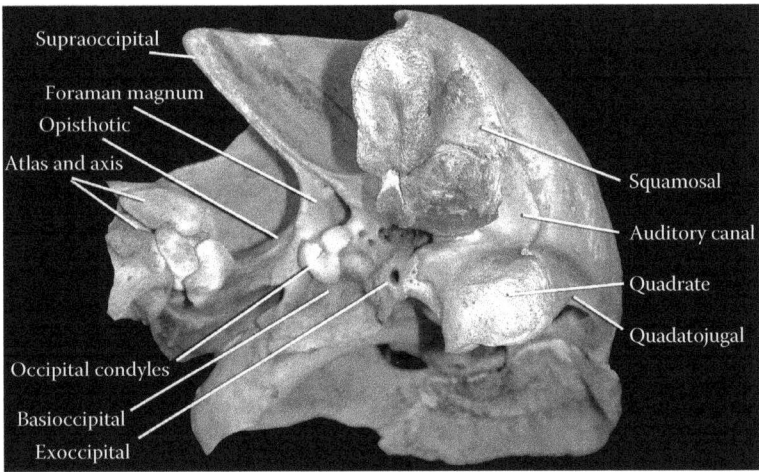

Supraoccipital

Foraman magnum

Opisthotic

Atlas and axis

Squamosal

Auditory canal

Quadrate

Quadatojugal

Occipital condyles

Basioccipital

Exoccipital

FIGURE 4.14 Posterior view of a *C. caretta* skull and the atlas and axis showing the cervical-to-skull articulation. The occipital condyles have three components that allow limited head rotation and less-limited head "nodding" movements. The bones that form the back of the skull and the more internal braincase are all endochondral bones from the occipital, otic, and sphenoid bone series.

Most of the superficial bones of the face are dermatocranial bones: premaxillae, maxillae, postorbitals, prefrontals, parietals, jugals, quadratojugals, and squamosals. The bones of the lower jaw (mandible) are dermatocranial and endochondral (Figure 4.15). These include the large dentary, surangular, angular, and splenial. The bones of the palate are also dermatocranial, including the buccal surfaces of the premaxillae and maxillae, the vomer, palatines, and pterygoids. These composites that make up the palate are important in species identification, and they form the partial secondary palate and primary palate, in cheloniids, adjacent to the braincase and often are sites of injury when sea turtles are caught on fishing hooks (Figure 4.16). The sutures between dermatocranial bones are often difficult to see even with high-resolution digital CT or radiographs, yet they are sometimes apparent following trauma to the skull.

The elements of the splanchnocranium are the same in all species. However, their specific form differs slightly with age and among species. They include the cartilaginous part of the mandible. The splanchnocranium forms the jaws (both the mandibles) and skeletal housing of the sense organs. By the time of hatching, upper and lower jaws are composites of several dermatocranial bones and the splanchnocranial elements reduced. The remaining splanchnocranium includes the hyoid apparatus, which is the mobile throat skeleton. The hyoid, which has both boney and cartilaginous parts, serves as a site for muscle attachments in the jaws, throat, and tongue (Schumacher, 1973). Each ear has one bone, a stapes (=columella), which is also part of the splanchnocranium (Figure 4.17).

## 4.3 OVERVIEW AND PERSPECTIVE

New imaging technologies have provided opportunities to explore the structure of sea turtles in vivo. Why does this new perspective and the description of the parts and pieces of skeletal anatomy matter? In addition to species identification, applications extend to paleontology, functional morphology, ecology, developmental biology, veterinary medicine, and conservation biology. Additional reasons for understanding sea turtle anatomy arise continually. For example, when considering where internal tags, flipper tags, or tag attachment systems should be placed (all important for mark and recapture studies), a basic understanding of skeletal anatomy, skeletal stability, and growth improve tag retention and minimize tag impacts (e.g., Epperly et al.,

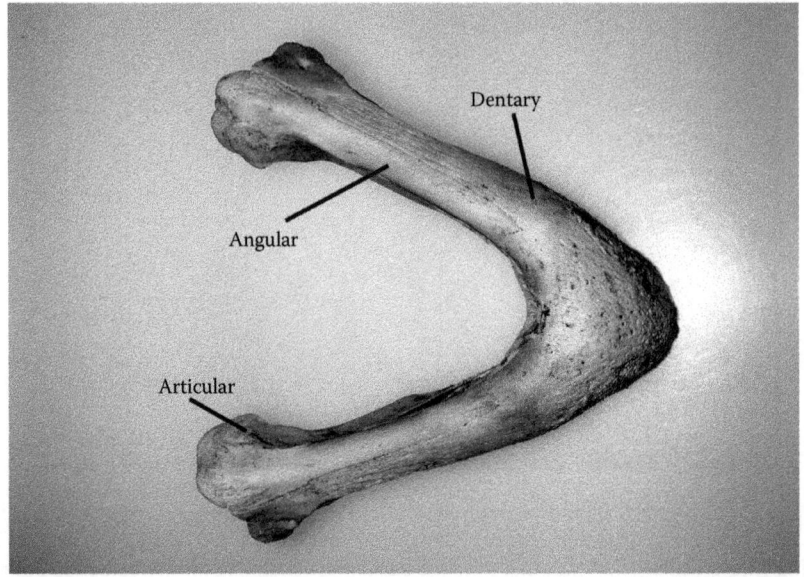

(A)

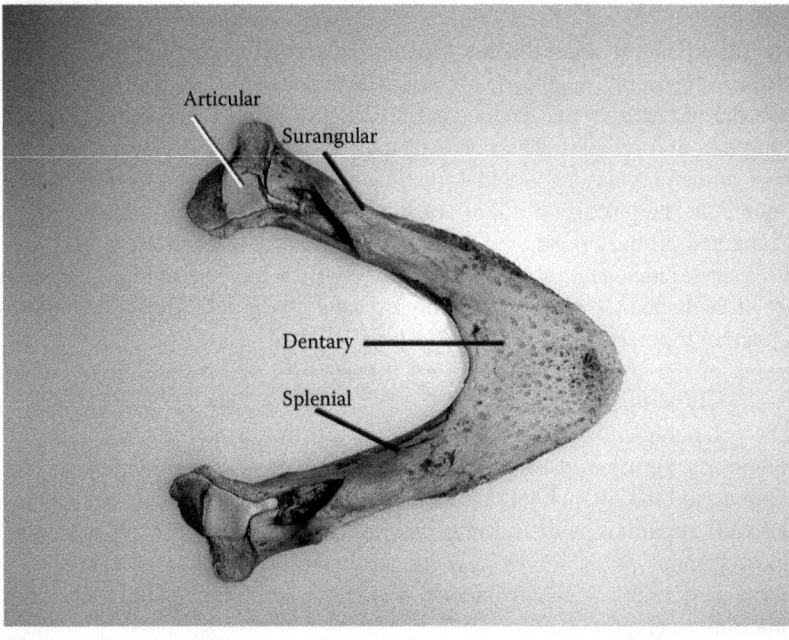

(B)

FIGURE 4.15    (A) Ventral view of a *C. caretta* mandible. (B) Dorsal (buccal) view of the same mandible. The lower jaw has a cartilaginous core (Meckel's cartilage) that is encased in dermal bones. The only remaining endochondral bone is the articular. In life, the articular is part of the jaw joint; it articulates with the quadrate.

2007; Wyneken et al., 2010; Mansfield et al., 2012). Understanding the relationships of bones to critical or sensitive structures, bone shapes, and limits imposed by skeletal architecture helps innovative tagging systems without harming the animal (Lutcavage et al., 2001; Casey et al., 2010). Finding and identifying bones with particular developmental characteristics has been fundamental to understanding bone growth and advancing skeletochronologic aging techniques

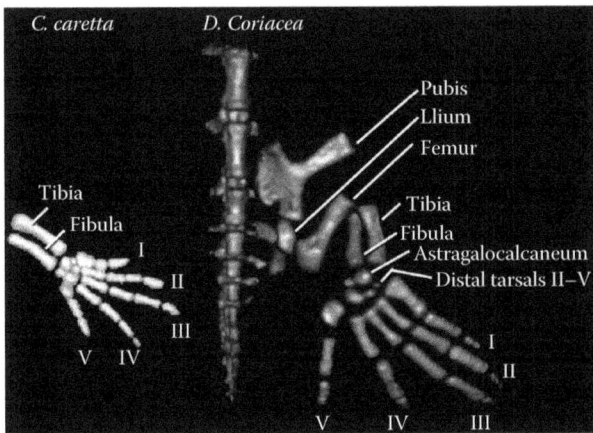

FIGURE 4.16   Dorsal view of the hind limb skeletons of *C. caretta* and *D. coriacea*. These views show the hind limb skeleton for both including the shank and (pes) tarsal bones, metatarsals, and phalanges.

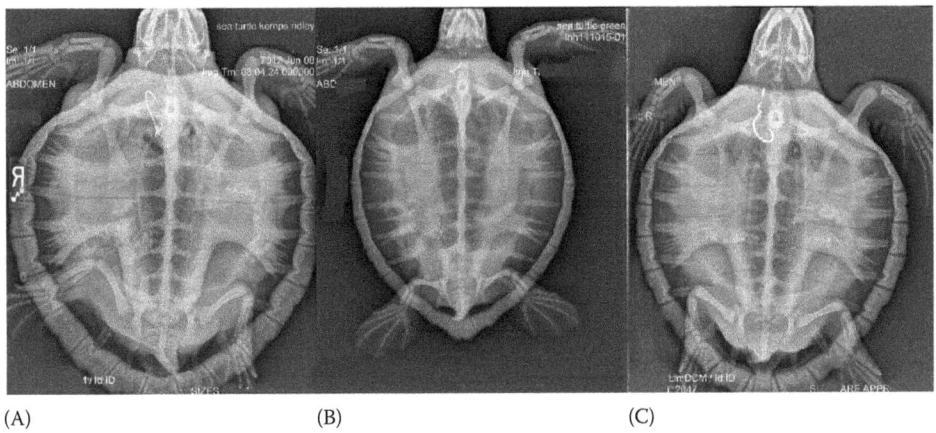

(A)                                    (B)                                    (C)

FIGURE 4.17   Digital radiographs of *L. kempii* (A,C) and *C. mydas* (B) that ate fishing bait with hooks. The position of the hooks (A,C) within the shell caudal to the acromial processes makes them difficult to reach. However, the flexibility and normal mobility of the esophagus makes the position of the hooks relative to the skeleton possible to manipulate. When the hooks are outside of the shell, access is less problematic. In all cases, however, if the hooks perforate the esophagus or stomach, their removal is more challenging, and the risk of infection and damage to other critical structures is increased. (Courtesy of J. Flanagan.)

(Snover and Hohn, 2004; Avens and Goshe, 2007; Snover and Rhodin, 2008; Avens et al., 2009; Casale et al., 2011; Chaper 5).

   In other contexts, understanding the skull bones, including the composite nature of the many bones that make up the skull, their diversity, loose versus tight articulations, and their arrangements is important to many aspects of biology. Particularly, knowing the form of the palate, jaw joint, and mandibles is important for the removal of hooks (a common injury associated with bycatch) and treating hook-related injuries. Estimating the impact of hook location on the survival probabilities of bycaught turtles depends upon an understanding of the skeleton as well as associated soft tissues (Chaloupka et al., 2004; Ryder et al., 2006; see Chapter 12). The impacts of other kinds of fishing and dredging gear (e.g., Upite, 2011) require knowledge of skeletal anatomy so that trauma can be identified and assessed. The knowledge of the skeleton is very relevant to the treatment of injured or sick sea turtles. It often delineates what can best be treated surgically and what must be treated medically (e.g., Wyneken et al., 2005).

The list enumerating why the parts and pieces of skeletal anatomy matter likely is a long one that will continue to grow as innovation and inspiration to better understand the biology of sea turtles develops.

## ACKNOWLEDGMENTS

I am indebted to the technicians and Fred Steinberg, MD of University MRI for teaching me about the utility of modern medical imaging in basic research. They generously provided access to their high-resolution GE LightSpeed CT scanner and software and enthusiastically tolerated sea turtles, live and dead at the end of long work days. The images would not have been obtained without the superb technical expertise of G. Boykin, A. Kaufman, and S. Rubel. D. Wilke taught me many of the techniques that I use today in image processing. I enthusiastically thank A. Rivera and M. Salmon who helped improved this chapter through their very constructive reviews.

## REFERENCES

Avens, L. and L. R. Goshe. 2007. Comparative skeletochronological analysis of Kemp's ridley (*Lepidochelys kempii*) and loggerhead (*Caretta caretta*) humeri and sclera ossicles. *Marine Biology* 152:1309–1317.
Avens, L., J. C. Taylor, L. R. Goshe, T. T. Jones, and M. Hastings. 2009. Use of skeletochronological analysis to estimate the age of leatherback sea turtles *Dermochelys coriacea* in the western North Atlantic. *Endangered Species Research* 8:165–177.
Biasatti, D. M. 2004. Stable carbon isotopic profiles of seaturtle humeri: Implications for ecology and physiology. *Palaeogeography, Palaeoclimatology, Palaeoecology* 206(3–4):203–216.
Casale, P., N. Conte, D. Freggi, C. Cioni, and R. Argano. 2011. Age and growth determination by skeletochronology in loggerhead sea turtles (*Caretta caretta*) from the Mediterranean Sea. *Scientia Marina* 75:197–203.
Casey, J., J. Garner, S. Garner, and A. Southwood Williard. 2010. Diel foraging behavior of gravid leatherback sea turtles in deep waters of the Caribbean Sea. *Journal of Experimental Biology* 213:3961–3971.
Chaloupka, M., D. Parker, and G. H. Balazs. 2004. Modeling postrelease mortality of pelagic loggerhead sea turtles exposed to the Hawaii-based longline fishery. *Marine Ecology Progress Series* 280:285–293.
Deraniyagala, P. E. P. 1939. The tetrapod reptiles of Ceylon. *Ceylon Journal of Science (Colombo Museum Natural History Series)* 1:1–412.
Epperly, S. P., J. Wyneken, J. P. Flanagan, C. A. Harms, and B. Higgins. 2007. Attachment of popup archival transmitting tags to loggerhead sea turtles (*Caretta caretta*). *Herpetological Review* 38(4):419–425.
Gilbert, S. F., J. A. Cebra-Thomas, and A. C. Burke. 2008. How the turtle gets its shell. In J. Wyneken, M. H. Godfrey, and V. Bels (Eds.), *Biology of Turtles: From Structures to Strategies of Life*, pp. 1–16. CRC Press, Boca Raton, FL.
Kardong, K. V. 2012. *Vertebrates: Comparative Anatomy, Function, and Evolution*, 3rd edn. McGraw-Hill, New York, 794pp.
Lutcavage, M., A. G. J. Rhodin, S. S. Sadove, and C. R. Conroy. 2001. Direct carapacial attachment of satellite tags using orthopedic bioabsorbable mini-anchor screws on leatherback turtles in Culebra, Puerto Rico. *Marine Turtle Newsletter* 95:9–12.
Mansfield, K. L., J. Wyneken, D. Rittschof, M. Walsh, C. W. Lim, P. Richards. 2012. Satellite tag attachment methods for tracking neonate sea turtles. *Marine Ecology Progress Series* 457:181–192. doi: 10.3354/meps09485.
Pritchard 1979. Encyclopedia of Turtles. T.F.H. Publications. Jersey City, NJ.
Pritchard, P. C. H. 1988. A survey of neural bone variation among recent chelonian species, with functional interpretations. *Acta Zoologica Cracoviensia* 31(26), 625–686.
Pritchard, P. C. H. 2008. Evolution and structure of the turtle shell. In J. Wyneken, M. H. Godfrey, and V. Bels (Eds.), *Biology of Turtles: From Structures to Strategies of Life*, pp. 45–84. CRC Press, Boca Raton, FL.
Rieppel, O. 1996. Testing the homology by congruence: The pectoral girdle of turtles. *Proceedings of the Royal Society of London Series B* 263:1395.
Rivera, A. R. V., J. Wyneken, and R. W. Blob. 2011. Forelimb kinematics and motor patterns of swimming loggerhead sea turtles (*Caretta caretta*): Are motor patterns conserved in the evolution of new locomotor strategies? *Journal of Experimental Biology* 214:3314–3323.

Romer, A. S. 1956. *Osteology of the Reptiles*. University of Chicago Press, Chicago, 722pp.

Ryder, C. E., T. A. Conant, and B. A. Schroeder. 2006. Report of the workshop on marine turtle longline post-interaction mortality. U.S. Department of Commerce, NOAA Technical Memorandum NMFS-F/OPR-29, 36p.

Sánchez-Villagra, M. R., C. Mitgutsch, H. Nagashima, and S. Kuratani. 2007. Autopodial development in the sea turtles *Chelonia mydas* and *Caretta caretta*. *Zoological Science* 24:257–263.

Schumacher, G. H. 1973. The head muscles and hyolaryngeal skeleton of turtles and crocodilians. In C. Gans and T. Parsons (Eds.), *Biology of the Reptilia*, Vol. 4 (Morphology D), pp. 101–199. Academic Press, New York.

Snover, M. L. and A. A. Hohn. 2004. Validation and interpretation of annual skeletal marks in loggerhead (*Caretta caretta*) and Kemp's ridley *Lepidochelys kempii*) sea turtles. *Fishery Bulletin* 102:682–692.

Snover, M. L. and A. G. J. Rhodin. 2008. Comparative ontogenetic and phylogenetic aspects of chelonian chondro-osseous growth and skeletochronology. In J. Wyneken, M. H. Godfrey, and V. Bels (Eds.), *Biology of Turtles*, pp. 17–43. CRC Press, Boca Raton, FL.

Upite, C. 2011. Evaluating sea turtle injuries in northeast fishing gear. U.S. Department of Commerce, Northeast Fisheries Science Center Reference Document 11-10.

Walker, Jr., W. F. 1973. The locomotor apparatus of Testudines. In C. Gans and T. S. Parsons, *The Biology of the Reptilia*, Vol. 3, Chap. 1, pp. 1–100. Academic Press, New York.

Wyneken, J. 1997. Sea turtle locomotion: Mechanisms, behavior, and energetics. In: P. Lutz and J. Musick (Eds.), *The Biology of Sea Turtles*. pp 168–198. CRC Press, Inc. Boca Raton, FL.

Wyneken, J. 2001. Guide to the anatomy of sea turtles. NMFS Technical Publication. NOAA Technical Memorandum NMFS-SEFSC-470. 172pp.

Wyneken, J. 2005. Computed tomography and magnetic resonance imaging of reptile anatomy through CT and MRI imaging. In D. R. Mader (Ed.), *Reptile Medicine and Surgery*, pp. 1088–1095. Elsevier Press, Philadelphia, PA.

Wyneken, J., S. P. Epperly, B. Higgins, E. McMichael, C. Merigo, and J. P. Flanagan. 2010. PIT tag migration in seaturtle flippers. *Herpetological Review* 41(4):448–454.

Wyneken, J., D. R. Mader, S. Weber, and C. Merigo. 2005. Medical care of seaturtles. In D. R. Mader (Ed.), *Reptile Medicine and Surgery*, pp. 972–1007. Elsevier Press, Philadelphia, PA.

Zangerl, R. 1939. The homology of the shell elements in turtles. *Journal of Morphology* 65:383.

Zangerl, R. 1969. The turtle shell. In C. Gans and A. d'A. Bellairs (Eds.), *The Biology of the Reptilia*, Vol. 1, pp. 311–319. Academic Press, New York.

Zangerl, R. 1980. Patterns of phylogenetic differentiation in the toxochelyid and cheloniid sea turtles. *American Zoologist* 20:585–596.

# 5 Age and Age Estimation in Sea Turtles

*Larisa Avens and Melissa L. Snover*

## CONTENTS

## 5.1  INTRODUCTION

While the specific life histories of different species and populations vary to some extent, sea turtles typically are long-lived, slow-growing, and wide-ranging, occupying multiple habitats over the course of their development (Musick and Limpus, 1997; Heppell et al., 2002). Each of these habitats is likely to present varying threats that will impact survival probabilities (Lewison et al., 2004; Moore et al., 2009; Bolten et al., 2011). Knowledge of how long each developmental area is occupied is therefore critical to understanding how habitat-specific survival rates may influence the proportion of the population that survives to reach Age at Sexual Maturation (ASM).

ASM is a key component of population dynamics, and understanding its mean, variation, and any potential temporal trend in the mean within a population is essential for assessments of long-term population trajectories (Heppell et al., 2003). Characterizations of population dynamics can be highly sensitive to variations in stage duration and ASM values, as minor parameter changes may result in diverging predictions (SEFSC, 2009). However, ASM within populations cannot be assumed to remain constant, as growth rates and stage durations are potentially influenced by a number of different factors. For example, fisheries population assessments have revealed numerous examples of possible harvest impacts, with ASM often decreasing in response to size-specific fishing pressure (e.g., Law and Grey, 1989; Law, 2000; Olsen et al., 2004; Swain et al., 2007). Sea turtle populations have also been harvested, either directly or indirectly through fisheries bycatch, to fractions of their historic sizes, yet many of those same populations are now receiving conservation efforts and some appear to be rebounding (Jackson et al., 2001; Kamezaki et al., 2003; Balazs and Chaloupka, 2004a; Heppell et al., 2005; Chaloupka et al., 2008). As a result, it is reasonable to assume that ASM may have changed over time, and any such variability must be quantified in order to successfully manage threatened and endangered sea turtle populations.

Furthermore, growth rates can be influenced by the physical and biological characteristics of the environment (Suryan et al., 2009; Hatase et al., 2010; Snover et al., 2010), and information regarding the nature and magnitude of these effects is needed to better understand ASM variability. For example, as populations recover, density-dependent factors can be expected to influence growth potential in juvenile foraging areas (Bjorndal et al., 2000a; Balazs and Chaloupka, 2004b). Also, climate processes have recently been demonstrated to influence sea turtle nesting remigration intervals (Saba et al., 2007; Mazaris et al., 2009; del Monte-Luna et al., 2012), and it is likely that they influence the productivity of foraging habitats as well (Snover, 2008; Waycott et al., 2009). Insight into how these long-term phenomena may be affecting growth rates, stage durations, and ASM is essential not only to characterizing overall population dynamics but also to being able to recognize the impacts of acute events, such as the recent 2010 Deepwater Horizon MC-252 oil spill in the Gulf of Mexico (Kerr et al., 2010), on these population parameters.

Although some static "snapshots" of age and growth parameters are available for a small number of sea turtle populations, isolated assessments are insufficient, as values are likely to change over time in response to natural and anthropogenic influences on habitat and sea turtle mortality. As a result, effective management of sea turtle species will require determination of temporal and spatial trends in age and growth parameters, the scope of variation, and the drivers underlying observed differences (NRC, 2010). In this chapter, we review current methods available for age estimation in sea turtles, including the benefits and drawbacks to each, and also present a summary of stage duration and ASM estimates that have been made using the different techniques. In addition, we review other age-estimation methodologies for their potential application to sea turtle populations.

## 5.2 APPROACHES TO AGE ESTIMATION

### 5.2.1 MARK–RECAPTURE

The earliest estimates of age-at-size and age at first reproduction for wild sea turtles were made using growth data from mark–recapture studies (e.g., Mendonça, 1981; Frazer and Ehrhart, 1985; Frazer and Ladner, 1986). This method remains one of the most common approaches currently used to estimate somatic growth rates, stage durations, and ASM in sea turtles (Tables 5.1 through 5.3). As the name implies, mark–recapture for the purpose of obtaining growth rates involves capturing wild sea turtles, taking morphometric measurements, tagging each turtle with a durable identification mark, and releasing. Growth rates in natural habitats are then obtained when tagged turtles are recaptured during subsequent capture events and new morphometric measurements taken.

A variety of methods have been used to capture sea turtles, and approaches are often specifically adapted to target particular size-classes or habitats. The most common methods include some form of netting, such as scoop or entanglement nets (e.g., Kubis et al., 2009; Lopez-Castro et al., 2010). Another common method is hand-capture, either using snorkel or SCUBA, or diving from a boat platform (e.g., Limpus and Limpus, 2003; Balazs and Chaloupka, 2004b; and see Ehrhart and Ogren, 1999, for a full review of techniques). Some turtles are also acquired from nonlethal interactions with fisheries, such as pound nets (e.g., Braun-McNeill et al., 2008). Once captured, turtles are measured and/or, less commonly, weighed. Most mark–recapture studies report size in carapace length, measured either as a straight-line carapace length (SCL) using calipers or as a curved carapace length (CCL) using a flexible tape measure (see Wyneken, 2001, for a full review of measurement types and methods). Prior to release, turtles are tagged in front or rear flippers either internally, using passive-integrated transponder tags or wire tags, or externally, typically using metal Monel or Inconel tags (see Balazs, 1999, for a full discussion of tagging techniques). Some success has also been found with the use of living tags where tissue from the lighter-colored plastron is transplanted onto the carapace (Hendrickson and Hendrickson, 1981). This technique is commonly used for

TABLE 5.1

## Summary of Stage Duration and ASM Estimates for Loggerhead Sea Turtles (C. caretta; Cc)

| Spp | Pop | Sex | Pelagic Stage Duration (year) | SAS (cm) | Benthic Stage Duration (year) | ASM (year) | SSM (cm) | Tech | Source |
|---|---|---|---|---|---|---|---|---|---|
| Cc | NWA | C | — | 50 SCL | 5 | 10–15 | 75 SCL | MR | Mendonça (1981) |
| Cc | NWA | C | — | — | — | 12–30 | 74, 92 SCL | MR | Frazer and Ehrhart (1985) |
| Cc | NWA | C | — | — | — | 13–15 | 86 CCL | Sk | Zug et al. (1986) |
| Cc | NWA | C | — | — | — | 22–26 | 92.5 CCL | Sk | Klinger and Musick (1995) |
| Cc | NWA | C | — | — | — | 25–30 | 92 SCL | Sk | Parham and Zug (1997) |
| Cc | NWA | C | 6.5–11.5 | 46–64 CCL | — | — | — | LF | Bjorndal et al. (2000b) |
| Cc | NWA | C | — | 46 CCL | 20 | — | 87 CCL | LF | Bjorndal et al. (2001) |
| Cc | NWA | C | — | 42 SCL | 24 | 30 | 83 SCL | MR | NMFS (2001) |
| Cc | NWA | C | — | 49 SCL | 32 | 39 | 90 SCL | MR | NMFS (2001) |
| Cc | NWA | C | 9–24 | 48.5–51.1 SCL | 17 | 24.3–38.9 | 87 CCL | Sk | Snover (2002) |
| Cc | NWA | C | 7 | 46 CCL | — | — | — | Sk | Bjorndal et al. (2003) |
| Cc | NWA | C | — | 50 SCL | >17.4 (50–80 cm SCL) | — | — | MR | Braun-McNeill et al. (2008) |
| Cc | NWA | C | — | — | — | 23.8–37.7 | 90 SCL | Sk | Vaughan (2009) |
| Cc | NWA (Med)[a] | C | — | — | — | 38 | 80 CCL | Sk | Piovano et al. (2011) |
| Cc | Med | C | 4 | 30 CCL | — | — | — | LF | Casale et al. (2009b) |
| Cc | Med | C | — | 35 CCL | 9.5–22.3 | 16–28 | 66.5–84.7 CCL | MR | Casale et al. (2009a) |
| Cc | Med | C | — | 30 CCL | 19.5–25.3 | 23.4–29.3 | 80 CCL | LF | Casale et al. (2011a) |
| Cc | Med | C | — | — | — | 14.9–28.5 | 66.4–87.7 CCL | Sk | Casale et al. (2011b) |
| Cc | Med | C | — | — | — | 24 | 69 CCL | Sk | Piovano et al. (2011) |
| Cc | SWA | C | 8–19 | 47.0–65.5 CCL | — | 32 | 102.5 CCL | Sk | Petitet et al. (2012) |
| Cc | NPac | C | >8.9 (42 SCL) | — | — | — | — | Sk | Zug et al. (1995) |
| Cc | SPac/Aus | F | — | 77–87 CCL | 9–23 | — | 90.5–101.5 CCL | MR | Limpus and Limpus (2003) |

Populations (Pop) include the Northwestern Atlantic (NWA), Mediterranean (Med), North Pacific (NPac), South Pacific/Australia (SPac/Aus). Sex is specified as males (M), females (F), or both sexes combined (C). The size representing the shift from pelagic to benthic habitats (SAS) and the size representing size at sexual maturity (SSM) are given as either straight (SCL) or curved (CCL) carapace length. When only partial benthic stage lengths were reported, the size-classes over which stage length was estimated are reported in parenthesis after the stage length. Techniques (Tech) used to make the stage length and age estimates are mark–recapture (MR), length frequency (LF), skeletochronology (Sk). ASM, age at sexual maturation.

a Estimated ASM for loggerheads with genetic origins in the NWA but recovered from the Med.

## TABLE 5.2
### Summary of Stage Duration and ASM Estimates for Green Sea Turtles (*Chelonia mydas*; Cm)

| Spp | Pop | Sex | Pelagic Stage Duration (year) | SAS (cm) | Benthic Stage Duration (year) | ASM (year) | SSM (cm) | Tech | Source |
|---|---|---|---|---|---|---|---|---|---|
| Cm | NWA | C | — | 30 SCL | 22 | 25–30 | — | MR | Mendonça (1981) |
| Cm | NWA | C | — | — | — | 18–27 | 88, 99 SCL | MR | Frazer and Ehrhart (1985) |
| Cm | NWA | C | — | — | — | 19–24 | 104–177 kg | MR | Ehrhart and Witham (1992) |
| Cm | NWA | C | — | — | >11 years (28–74 cm SCL) | — | — | Sk | Zug and Glor (1998) |
| Cm | NWA | M | 1–7 (mean = 3) | — | — | 35.5–50 | 84.8–94.9 SCL | Sk | Goshe et al. (2010) |
| Cm | NWA | F | 1–7 (mean = 3) | — | — | 42–44 | 99.5–101.5 SCL | Sk | Goshe et al. (2010) |
| Cm | NWA | C | — | — | 17–20 (20–78.5 cm SCL) | — | — | Sk | Avens et al. In Press |
| Cm | Car | C | — | — | — | 27–33 | 105.75–111.75 CCL | MR | Frazer and Ladner (1985) |
| Cm | Car | C | — | 30 SCL | >11–13.5 (30–70 cm SCL) | — | — | LF | Bjorndal et al. (1995) |
| Cm | Car | C | — | 30 SCL | >17 (30–75 cm S) | — | — | MR | Bjorndal et al. (2000a) |
| Cm | Car | M | — | — | — | 19 | 108 CCL | KA | Bell et al. (2005) |
| Cm | Car | F | — | — | — | 15–17 | 108 CCL | KA | Bell et al. (2005) |
| Cm | EPac | C | — | 46.6 SCL | 9.6–21.9 | — | 60–77.3 SCL | MR | Seminoff et al. (2002) |
| Cm | Pac/HI | C | 4–10 | 35–37 SCL | >20 | >30 | 92 SCL | Sk | Zug et al. (2002) |
| Cm | Pac/HI | C | — | 35 SCL | — | 35–50 | 80 SCL | MR | Balazs and Chaloupka (2004) |
| Cm | SPac/Aus | M | — | 35 CCL | 35 | >40 | 95 CCL | MR | Limpus and Chaloupka (1997) |
| Cm | SPac/Aus | F | — | 35 CCL | 35 | >40 | 95 CCL | MR | Limpus and Chaloupka (1997) |
| Cm | SPac/Aus | C | — | — | — | 25–50 | — | MR | Chaloupka et al. (2004) |
| Cm | WIO | C | — | — | — | 28.8 | 101.7 CCL | MR | Watson (2006) |

Populations (Pop) include the Northwestern Atlantic (NWA), Caribbean (Car), Gulf of Mexico (GoM), East Pacific (EPac), Pacific/Hawaii (Pac/HI), South Pacific/Australia (SPac/Aus), and the West Indian Ocean (WIO). Sex is specified as males (M), females (F), or both sexes combined (C). The size representing the shift from pelagic to benthic habitats (SAS) and the size representing size at sexual maturity (SSM) are given as either straight (SCL) or curved (CCL) carapace length. When only partial benthic stage lengths were reported, the size classes over which stage length was estimated are reported in parenthesis after the stage length. Techniques (Tech) used to make the stage length and age estimates are mark–recapture (MR), length frequency (LF), skeletochronology (Sk), known-aged individuals that were tagged, released, and recovered (KA). ASM, age at sexual maturation.

TABLE 5.3

**Summary of Stage Duration and ASM Estimates for Kemp's Ridleys (*L. kempii*; Lk) Olive Ridleys (*L. olivacea*; Lo), Hawksbills (*E. imbricata*; Ei), and Leatherbacks (*D. coriacea*; Dc)**

| Spp | Pop | Sex | Pelagic Stage Duration (year) | SAS (cm) | Benthic Stage Duration (year) | ASM (year) | SSM (cm) | Tech | Source |
|---|---|---|---|---|---|---|---|---|---|
| Lk | GoM | C | — | — | — | 11–16 | 65 SCL | Sk | Zug et al. (1997) |
| Lk | GoM | C | — | — | — | 15–20 | 63 SCL | Sk | Chaloupka and Zug (1997) |
| Lk | GoM | C | — | — | 8–9 | 8–13 | 56, 64.2 SCL | MR | Schmid and Witzell (1997) |
| Lk | GoM | F | — | — | — | 10–18 | 58.1 S - 67.5 CCL | KA | Shaver and Wibbels (2007) |
| Lk | GoM | C | 1 | 21 SCL | 8.9–15.7 | 9.9–16.7 | 60 SCL | Sk | Snover et al. (2007b) |
| Lk[a] | GoM | F | — | — | — | 9.7–22.8 | 58.1–65.8 SCL | KA | Caillouet et al. (2011) |
| Lo | NPac | C | — | — | — | 13 | 60 SCL | Sk | Zug et al. (2006) |
| Ei | Car | C | — | 21.4 SCL | 16.5–19.3 | — | 78.8–88.7 SCL | MR | Boulon (1994) |
| Ei | Car | C | — | 23 SCL | >4.8–16.1 (23–61 cm SCL) | — | — | MR | Diez and van Dam (2002) |
| Ei | Pac/HI | C | — | — | — | 14–20 | 78.6 SCL | Sk | Snover et al. (2012) |
| Dc | NPac | C | — | — | — | 13–14 | 144.5 CCL | Sk | Zug and Parham (1996) |
| Dc | Car | F | — | — | — | 12–14 | — | DNA | Dutton et al. (2005) |
| Dc | NWA | C | — | — | — | 24.5–29 16–22 | 145 CCL 125 CCL | Sk | Avens et al. (2009) |

Populations (Pop) include the Northwestern Atlantic (NWA), Caribbean (Car), Gulf of Mexico (GoM), North Pacific (NPac), Pacific/Hawaii (Pac/HI). Sex is specified as males (M), females (F), or both sexes combined (C). The size representing the shift from pelagic to benthic habitats (SAS) and the size representing size at sexual maturity (SSM) are given as either straight (SCL) or curved (CCL) carapace length. When only partial benthic stage lengths were reported, the size-classes over which stage length was estimated are reported in parenthesis after the stage length. Techniques (Tech) used to make the stage length and age estimates are mark–recapture (MR), length frequency (LF), skeletochronology (Sk), known-aged individuals that were tagged, released, and recovered (KA), and DNA fingerprinting (DNA).

[a] Reported ages and sizes of nesters include all documented reports of nesters and are not necessarily first-time nesters. ASM, age at sexual maturation.

marking large numbers of hatchlings or yearlings prior to release, and the location of the transplant on the carapace indicates the year-class (Bell et al., 2005; Shaver and Wibbels, 2007). Initial work with microsatellite DNA analysis has shown promise for a method to identify not only individuals but also mother–daughter relationships, providing information on ASM as inferred through genera-tion time (Dutton et al., 2005).

Numerous studies have highlighted the importance of recapture interval to the validity of estimating annual growth rates from mark–recapture data (Chaloupka and Limpus, 1997; Chaloupka and Musick, 1997; Limpus and Chaloupka, 1997; Braun-McNeill et al., 2008). As ectotherms, metabolic processes in sea turtles are tightly related to environmental temperature, and, as a result, growth rates will vary throughout the season (Spotila et al., 1997), and a partial year of growth can result in under- or overestimating annual growth rates. In general, most recent studies limit datasets to recapture intervals >10–11 months (e.g., Chaloupka and Limpus, 1997; Limpus and Chaloupka, 1997; Braun-McNeill et al., 2008; Kubis et al., 2009).

Mark–recapture studies result in empirical measurements of absolute growth rates or amount of growth measured over the actual time at large. Irrespective of the precise recapture interval, these data (with the limitations discussed earlier) are usually extrapolated to annual growth rates for individuals within the size range captured in the study as a direct ratio of change in carapace length between captures divided by the recapture interval in years. Mean and variance of these growth rates are then calculated for binned size-classes, or the data are modeled statistically, com-monly with a Generalized Additive Modeling (GAM) approach to achieve size-specific growth rates inferred from a smoothing spline (Hastie and Tibshirani, 1990; see Chaloupka and Limpus, 1997, for more information on the application of GAM models to mark–recapture data). Both of these methods result in carapace length versus change in carapace length $\times$ year$^{-1}$. This relationship can be integrated over the size range of turtles in the study to yield an age-at-size relationship, and from this, stage durations, or at least time to grow over the size range of captured turtles, can be esti-mated (e.g., Chaloupka and Limpus, 1997; Diez and van Dam, 2002; Seminoff et al., 2002). Growth models, such as von Bertalanffy, Gompertz, or logistic, can also be fitted to growth data, resulting in fitted parameters for asymptotic lengths and growth coefficients, generating an age-at-size curve (e.g., Frazer and Ehrhart, 1985; Schmid and Witzell, 1997; Casale et al., 2009b). However, caution is needed when developing growth curves from mark–recapture data and interpreting the results. First, different functional forms of growth models (i.e., von Bertalanffy or Gompertz) may describe growth pattern better for one species or sex than another (e.g., Goshe et al., 2010), and many stud-ies do not evaluate best model fits when describing the relationship between age and size. Second, by their nature, asymptotic growth models assume a universal abrupt slowing of growth rates as asymptotic sizes are approached, which may or may not be representative of actual growth and which can impact inferences about ASM, especially if the fitted or assumed asymptotic growth rate is close to the assumed size at maturity. Finally, it is not possible to use growth models to estimate growth rates, stage durations, or ASM when not all of these size or stage classes were within the range of turtles actually measured in the study.

Factors that limit the feasibility of sea turtle age estimation through mark–recapture studies include small sample sizes, partial year measurements of growth, incomplete information for all sizes, and the need for long-term, multiyear efforts that can be costly. Studies have shown that initial or mean carapace length, year, sex, and habitat can have significant impacts on individual growth rates (Chaloupka and Limpus, 1997; Limpus and Chaloupka, 1997; Balazs and Chaloupka, 2004b; Kubis et al., 2009); therefore, 1 year of growth from a few individuals in different years may not be representative of "typical" growth. Sea turtle growth rates are highly variable, gener-ally with a lognormal distribution that results in a long tail of occasional high growth rates but with lower mean growth rates (e.g., Braun-McNeill et al., 2008). The importance of the growth that occurs in these tails may be critical to a full understanding of stage duration and ASM, as turtles may use compensatory growth throughout their life histories to achieve their large sizes at first reproduction (Bjorndal et al., 2003). However, these occasional and exceptional growth rates may go undetected in traditional mark–recapture studies, especially those with shorter durations and limited geographic scope, biasing estimates of ASM derived from these data upward.

Some mark–recapture studies involve tagging and releasing large numbers of hatchlings and/or captive-reared yearling sea turtles (Bell et al., 2005; Shaver and Wibbels, 2007; Caillouet et al., 2011). Recaptured or resighted turtles from these studies yield individuals of known age and are valuable for verifying our understanding of juvenile growth rates and ASM. However, it should be noted that for any given population, there is likely a range of ASM, and while individuals maturing the earliest are the most likely to be observed, their ages may not be representative of mean ASM for the population. The number of tagged individuals from any cohort will decline over time due to mortality, reducing the likelihood of intercepting tagged turtles that mature later. Hence, the reported ASMs of those turtles that are first to return may not be representative of the norm for a population and should be interpreted with care (Bell et al., 2005).

Somewhat related to mark–recapture studies are length-frequency studies, which attempt to detect cohorts from length distributions and estimate growth rates from the differences among peaks within those distributions (Pauly and Morgan, 1987; Fournier et al., 1990). Although these analyses can incorporate mark–recapture data (e.g., Bjorndal et al., 1995), recapture of individuals is not essential, and therefore data sources also have included strandings (Bjorndal et al., 2001; Casale et al., 2009a, 2011b) and fishery bycatch (Bjorndal et al., 2000b; Casale et al., 2011b). However, while individuals do not need to be resampled for length-frequency analyses, it is necessary for populations to be sampled more than once to improve accuracy of estimation of intrinsic growth rates (von Bertalanffy k) as well as the number of age classes in the population (Fournier et al., 1990). For the technique to be applied effectively, a number of assumptions must be met: (1) the size structure of the subsample must represent the population as a whole (i.e., all size-classes must be sampled), (2) the study organism must exhibit annual recruitment to the sampling location, (3) growth patterns must approximate a von Bertalanffy-type curve, and (4) lengths must be normally distributed within each cohort (Pauly and Morgan, 1987; Fournier et al., 1990). In sea turtles, it appears that the technique may be useful for identifying earlier age classes (e.g., Bjorndal et al., 1995, 2000b; Casale et al., 2009a). However, given the observed variability in sea turtle growth rates (e.g., Braun-McNeill et al., 2008; Casale et al., 2009b), potentially resulting from multiple influences (e.g., climatic factors, habitat quality, sex, genetic origin, health, migratory movements), von Bertalanffy growth cannot be assumed (Chaloupka and Musick, 1997; Chaloupka and Zug, 1997). As variability in growth rates can also obscure cohort distinctions, it is necessary to validate designation of the number of age classes in a population through alternate means (Fournier et al., 1990). Furthermore, due to the rapid decrease, or even cessation, of somatic growth upon maturation observed for sea turtles, the applicability of this technique to the entire range of size-classes is unclear (Bjorndal et al., 2001).

*Loggerheads (Caretta caretta)*: Mark–recapture studies have been used to estimate juvenile stage durations and/or ASM for loggerheads in the western North Atlantic (Mendonça, 1981; Frazer and Ehrhart, 1985; Schmid, 1995, 1998; NMFS SEFSC, 2001; Braun-McNeill et al., 2008), the Caribbean (Bjorndal and Bolten, 1988a), the Mediterranean (Casale et al., 2009b), and the western South Pacific (Limpus and Limpus, 2003; Table 5.1). In addition, length-frequency analyses have been conducted for the western North Atlantic (Bjorndal et al., 2000b; 2001) and the Mediterranean (Casale et al., 2009a, 2011b).

For loggerheads in the Northwestern Atlantic, only one study has estimated the length of the oceanic stage (Bjorndal et al., 2000b; Table 5.1). Based on a length-frequency analysis, Bjorndal et al. (2000b) estimated 6.5–11.5 years from hatching to recruitment to near-shore, benthic habitats, with the duration depending on size at recruitment. For the same population, several mark–recapture and length-frequency studies have estimated the duration of the neritic stage and ASM (Table 5.1). Estimates for neritic stage duration range from 5 to 32 years and estimates of ASM range from 10 to 39 years. However, more recent studies have demonstrated that the lower end of this range, estimated by Mendonça (1981), is not probable and, more likely, the result of short recapture intervals overestimating annual growth rates.

*Kemp's (Lepidochelys kempii) and Olive (Lepidochelys olivacea) Ridleys*: While the number of mark–recapture studies dedicated to Kemp's ridleys has been limited (however, see Schmid and Witzell, 1997; Schmid, 1998), much has been learned about Kemp's ridley growth and ASM from the large numbers of marked hatchlings (Higgins et al., 1997) and 1-year-old headstart turtles (Shaver and Wibbels, 2007) released into the wild and subsequently recaptured. From these data, minimum ASM is understood to be as low as 10 years. This number is consistent with the results of a mark–recapture study of wild Kemp's ridleys, which estimated ASM of 8–13 years (Schmid and Witzell, 1997; Table 5.3). Olive ridleys have a markedly different life history that does not lend itself to mark–recapture studies. This species remains in oceanic habitats after hatching and throughout their lives, except to mate and nest (Bolten, 2003). To date, the difficulties in detecting aggregations of juvenile turtles in these habitats large enough to warrant mark–recapture efforts have inhibited attempts to apply mark–recapture techniques to this species.

*Greens (Chelonia mydas)*: Long-term mark–recapture studies of green turtles have resulted in estimates of ASM at 18–30 years for juveniles using habitat in Florida (Mendonça, 1981; Frazer and Ehrhart, 1985; Ehrhardt and Witham, 1992), 35–50 years in Hawaii (Balazs and Chaloupka, 2004b), and 25–50 years in the Great Barrier Reef, Australia (Limpus and Chaloupka, 1997; Chaloupka et al., 2004). In addition, marked and released hatchlings and yearlings from the Cayman Turtle Farm have been recorded nesting or mating at 15–19 years (Bell et al., 2005), indicating a minimum ASM for this species in the Caribbean. Additional studies have estimated growth rates and durations of juvenile stages in the Caribbean (Bjorndal and Bolten, 1988b; Bjorndal et al., 1995, 2000a), the East Indian Ocean (Watson, 2006), the West Pacific (Pilcher, 2010), and the East Pacific (Seminoff et al., 2002; Table 5.2).

*Hawksbills (Eretmochelys imbricata)*: Only a few studies on growth rates of juvenile hawksbills have been conducted (Table 5.3) and those have occurred in Australia (Chaloupka and Limpus, 1997), the Caribbean (Boulon, 1994; Diez and van Dam, 2002; Bjorndal and Bolten, 2010), and the East Indian Ocean (Watson, 2006); however, the latter did not yield enough information to estimate stage duration or ASM. The southern Great Barrier Reef dataset presented by Chaloupka and Limpus (1997) did not contain growth rates of mature individuals, and therefore, ASM could not be estimated; however, based on observed juvenile growth rates, they predicted that decades would be required for a hatchling hawksbill to reach maturity. For Caribbean hawksbills, Boulon (1994) estimated that it would take 16.5–19.3 years for 21.4 cm SCL hawksbills in St. Thomas, U.S. Virgin Islands, to mature, depending on size at maturation. Similarly, for hawksbills in Puerto Rico, Diez and van Dam (2002) estimate a range of 4.8–16.1 years to grow from 23 to 61 cm SCL, with duration depending on habitat type. For hawksbills inhabiting forage areas producing the highest growth rates, Diez and van Dam (2002) estimate 14.7 years to grow from 23 cm SCL to adult size.

*Flatbacks (Natator depressus)*: This species also has a distinctly different life history from most of those previously discussed in that a pelagic stage is absent, and individuals remain in near-shore habitats throughout their lives (Walker and Parmenter, 1990; Bolten, 2003). Few age and growth parameters are available for the flatback turtle, although marking of hatchlings was conducted from 1974 to 1982, and as of 2004, one of these turtles had returned to nest at the age of 21 years (Limpus, 2007). However, no other mark–recapture or length-frequency studies have been conducted to date on this species.

*Leatherbacks (Dermochelys coriacea)*: Like olive ridleys, leatherbacks also remain in pelagic habitats throughout most of their life history (Bolten, 2003), and no mark–recapture studies have been conducted on juveniles. However, Dutton et al. (2005) analyzed nesting trends at St. Croix, U.S. Virgin Islands, following the implementation of conservation efforts that increased hatchling production. A surge in nesting trends detected 12–14 years following the start of conservation efforts was attributed to those initial hatchlings maturing and returning to nest. Furthermore, preliminary results of microsatellite DNA analysis that detected probable mother–daughter pairs support an estimate of 12–14 years ASM (Dutton et al., 2005). However, as with information yielded by returns of tagged hatchlings or head-started yearlings, this result would likely represent the younger ASMs over the range that will be typical for this population.

## 5.2.2 CAPTIVE GROWTH

The study of sea turtles under natural conditions is challenging due to ontogenetic habitat shifts (Musick and Limpus, 1997), migratory movements (Plotkin, 2003), and general inaccessibility away from nesting beaches (NRC, 2010). As a result, a number of age and growth studies have focused instead on turtles held captive for varying periods of time. In contrast to mark–recapture, during captive studies, known-identity individual turtles are readily available for measurement on a regular schedule, allowing calculation of standardized growth rates. However, these benefits are counterbalanced at least to some extent by the costs of rearing facilities (e.g., tanks, equipment, food) and labor required to maintain captive turtles over long periods of time, which in turn constrain sample sizes for most captive studies (Zug et al., 1986).

Captive studies of small numbers of loggerheads kept under favorable conditions for periods of several years yielded growth trajectories that, if continued unchanged, would have allowed the turtles to attain adult size in as little as 6 or 7 years (Caldwell, 1962; Uchida, 1967). In contrast, loggerheads captive reared in small enclosures in a temperate location under more natural temperature and salinity regimes, but still regularly fed to satiation, exhibited growth rates consistent with 19–20 years ASM at 92.5 cm SCL (Frazer and Schwartz, 1984). At the Cayman Turtle Farm, first egg production for female Kemp's ridleys was observed as early as 5–7 years (summarized in Caillouet et al., 2011), and green turtles reached maturity at a minimum of 8–10 years (Wood and Wood, 1980). This observed range for green turtle ASM was slightly greater than those extrapolated from some earlier captive growth studies, which estimated 4–6 years (Hendrickson, 1958), 5 years (Carr, 1968), and 8 years (Bustard, 1976), but was at the center of the 4–13 years range presented by Hirth (1971). Growth data for a small number of Western Samoan hawksbills kept in captivity for slightly longer than 2 years indicated that maturity might be reached even more rapidly for this species, with turtles attaining 50 cm at 3.5 years of age (Witzell, 1980). Yet even more remarkable were growth results for leatherbacks maintained in captivity for periods of several weeks to just over 3 years, which suggested that this largest sea turtle species might attain adult size in as little as 2, 3, or 6 years (Deraniyagala, 1952; Birkenmeier, 1971; Bels et al., 1988). However, fitting growth curves to a combination of (1) growth rate data yielded by a recent leatherback captive-rearing effort spanning 2 years, during which juveniles were fed to satiation ≥3 times daily, and (2) adult length-at-age data generated by other studies have produced a higher ASM estimate of approximately 16 years (Jones et al., 2011).

Captive studies have provided a great deal of insight into the shapes of growth curves, the influence of various factors on growth rates, and growth potential (Frazer and Schwartz, 1984; Swingle et al., 1993; Jones et al., 2011). However, at least for Cheloniid sea turtle species for which mark–recapture data are available, stage duration and ASM estimates yielded by growth observations of wild turtles have been greater than those based on captive growth rates (see Tables 5.1 through 5.3 and *Species-Specific Information in section earlier*). A number of different factors associated with captive studies may contribute to these differences (summarized in Frazer and Ladner, 1986). Captive care often entails optimal temperature and feeding regimes as well as small enclosures that limit physical activity (Frazer and Schwartz, 1984; Swingle et al., 1993), which can maximize growth potential and decrease relevance of resulting growth rates to migratory, wild populations that experience variability in environmental conditions and forage availability. Also, although captive growth rates are typically collected over short time frames, length- or weight-at-age relationships have been extrapolated beyond the available data range with some frequency. If growth curves that are not representative of a species' growth patterns are applied, or if models neglect to take into account ontogenetic decay in growth trends, this may generate downward biases for resulting stage duration and ASM estimates.

### 5.2.3 SKELETOCHRONOLOGY

Skeletochronology is a subdiscipline of the broader field of study called sclerochronology, which is the investigation of temporally associated life history characteristics through analysis of hard or calcified structures in organisms. Sclerochronology is quite similar to the study of dendrochronology (the study of tree rings) in that the techniques are based on the premise that the structure being analyzed exhibits alternating, recurring cycles of fast and slow growth manifesting as increments that can be distinguished by their appearance and composition. This field not only encompasses the study of bones comprising the axial and appendicular skeleton or skeletochronology (Klevezal, 1996) but also includes analyses of scales (scalimetry) (Panfili et al., 2002), fish otoliths (otolithometry) (Secor et al., 1995; Panfili et al., 2002), mammalian teeth (dental chronology) (Klevezal, 1996; Hohn, 2009), claws (Thomas et al., 1997), cephalopod (Rodhouse and Hatfield, 1990), gastropod (Chatzinikolaou and Richardson, 2007), statoliths, bivalve shells (Richardson, 2001), echinoderm tests (Gage, 1992), and corals (Marschal et al., 2004), among others. A systematic approach to this course of investigation into age and growth was initiated in the late-1800s, primarily in the form of scalimetric fish studies (Carlander, 1987). However, it was not until the mid-1900s that researchers began to apply the technique in earnest, resulting in a proliferation of sclerochronologic age-estimation studies involving diverse taxa that continues to the present day.

Although the specifics of sclerochronological studies may differ according to study organism and research question, to apply the technique with confidence, four main concerns must be addressed. First, care must be taken that the structure selected for analysis exhibits the greatest possible clarity and retention of growth increments appropriate for the time frame being investigated (i.e., daily, monthly, and/or yearly) (Klevezal, 1996). In conjunction with this aspect, many different methods for sclerochronology sample preparation and visualization of growth increments have been developed and applied. Although the details of these approaches exceed the scope of this chapter, they range anywhere from simple visual inspection of the surface of an untreated sample (Scheffer and Myrick, 1980) to complex histological processing (Klevezal, 1996) and examination using microradiography (Hohn, 1980) or electron microscopy (Brothers et al., 1976). Conspicuousness of increments can vary according to preparation technique, as well as the study organism and structure, making it necessary to also assess the efficacy of the chosen sample preparation method prior to analysis.

Before using increments within calcified structures as indicators of age, the second issue to consider is the need to determine how frequently the increments are deposited; ideally, this is assessed across all age groups, as deposition may vary depending upon life stage (Campana, 2001). For vertebrates at least, annual growth increment deposition appears to be driven mainly by endogenous, physiological cycles (Schauble, 1972; Beamish and McFarlane, 1987; Simmons, 1992) synchronized with local environmental cues (Beamish and McFarlane, 1987; Castanet et al., 1993). However, supplemental, nonannual increments may occur as the result of other factors that influence activity and growth (Jakob et al., 2002; Olgun et al., 2005). Validation of deposition frequency is typically accomplished by analyzing samples collected from organisms whose age is established, either through captive-rearing, or when known-age young are marked prior to release into the wild and subsequently recovered (Castanet et al., 1993; Klevezal, 1996; Campana, 2001). Characterization of deposition frequency can also be accomplished through analysis of samples from organisms whose hard structures have been marked in some manner at a known time in their lives (Klevezal, 1996; Campana, 2001). A number of different compounds can be used as bone markers including calcein, alizarin red, lead acetate, and procion dyes, although their toxicity complicates their use (Klevezal, 1996). Perhaps the most frequently applied marker is oxytetracycline, which is commonly used as an antibiotic, but at higher, species-specific dosages is incorporated into growing skeletal structures as a layer that fluoresces when viewed under ultraviolet light (Lipinski, 1986; Gage, 1992; Klevezal, 1996; Campana, 2001; Goldman et al., 2006). Analysis of each of these sample types allows calculation of deposition rate over a known time frame, that is, the total

lifespan or the time between bone-marking and death. Another approach is that of marginal increment analysis (MIA), which involves characterization of the amount of bone growth external to the
previous increment relative to time of year, with the assumption that the deposition cycle will reflect
deposition frequency (Campana, 2001; Panfili et al., 2002; Caillet et al., 2006). Finally, validation
of growth increment interpretation can also be assessed through comparison of age estimates generated using sclerochronology and results from radiometric analyses, which involve characterization
of lead–radium disequilibria ($^{210}$Pb:$^{226}$Ra ratios) in the structure being analyzed (Andrews et al.,
2007). Recent advances in this field have increased the potential for use of the technique as an
independent age estimator as well (Andrews et al., 2009).

The third issue to consider is that, despite their mechanical strength, hard and calcified structures
are often modified or damaged over the course of an organism's life. These changes may decrease
the clarity and interpretability of the growth increments within, making it essential to develop
analytical methods to compensate for the loss of information (Klevezal, 1996). For example, scales
and teeth can experience breakage and wear or cessation of increment deposition (Beamish and
McFarlane, 1987; Klevezal, 1996), and molluscan shells are damaged by abrasion, crushing, and
boring organisms (Zuschin et al., 2003). Although skeletal bone can also break, the main imped-
ance to age estimation in these structures is a phenomenon termed "resorption," resulting from bone
remodeling (Klevezal, 1996; Castanet, 2006). Examination of a cross section of skeletal bone reveals
spongy, cancellous bone toward the center, which is encircled by compact bone, where formation
of skeletal growth increments or growth marks (GMs) occurs. Within the compact bone, early GMs
are present toward the center, while accretion of new GMs occurs at the outer margin in the peri-
osteum (Figure 5.1; Zug et al., 1986; Klevezal, 1996). As organisms grow and age, the cancellous
bone expands and early, inner GMs may be destroyed, or "resorbed" (Figure 5.1; Klevezal, 1996;
Castanet, 2006), with the extent of resorption varying among species and individuals. However,
characterization of early GM deposition patterns can make it possible to develop correction factors
that allow the estimation of the number of GMs lost when resorption has occurred (Klevezal, 1996;
Parham and Zug, 1997; Curtin et al., 2008). The estimated number of lost GMs is then added to the
observed number of GMs to yield an age estimate.

Finally, sclerochronological analyses are sometimes applied to back-calculate longitudinal age-
at-size relationships and somatic growth rates through conversion of sequential growth increment
measurements within an individual structure to estimates of body size (Campana, 1990; Francis,
1990; Gonzaléz et al., 1996; Vigliola et al., 2000; Chatzinikolaou and Richardson, 2007). However,
for this technique to yield accurate results, it is necessary not only to validate the frequency of increment deposition (see earlier) but also to demonstrate a predictable, proportional relationship between
increment size and the somatic measure of interest (Campana, 1990; Francis, 1990; Vigliola et al.,
2000). Although characterization of this relationship can be accomplished to some extent through
collection of an ontogenetic series of sclerochronological samples and body sizes for a population cross section, this approach can be confounded by individual variability (Hare and Cowen,
1995; Vigliola et al., 2000). Alternatively, validation might be carried out through comparison of
back-calculated, individual growth trajectories to those observed in captivity or mark–recapture
(Hare and Cowen, 1995; Chatzinikolaou and Richardson, 2007). Application of growth models to
back-calculated growth data can be used as an alternative means to direct age estimation to assess
length-at-age, stage durations, and age at maturation (e.g., Fabens, 1965; Frazer et al., 1990).

## 5.2.3.1 Skeletochronology and Cheloniid Sea Turtles

Perhaps the most familiar approach to sclerochronological age estimation in turtles is the counting
of growth increments that manifest as ridges on the scutes of the carapace and/or plastron in many
terrestrial and semiaquatic species (Cagle, 1946; Galbraith and Brooks, 1987; Lagarde et al., 2001).
Although there is some indication that the scutes of hawksbill sea turtles (*Eretmochelys imbricata*) may retain age-related information (Kobayashi, 2000; Tucker et al., 2001; Palaniappan, 2007),
the external surfaces of sea turtle scutes and scales are typically smooth, negating the possibility

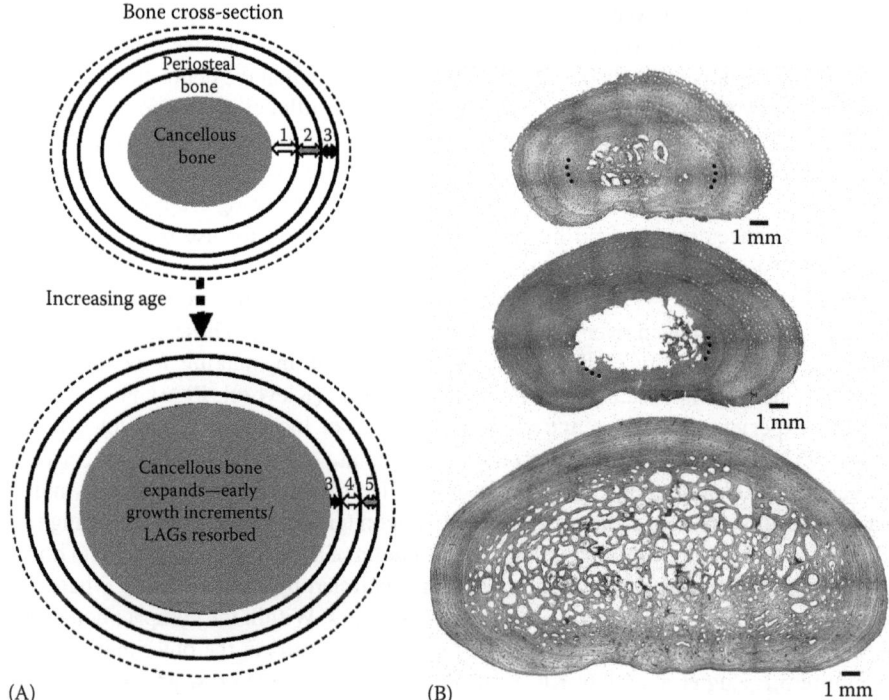

FIGURE 5.1   (A) Schematic representation of bone resorption. As organisms age and grow, the proportion of cancellous, porous bone in the interior may increase, destroying early growth increments and the lines of arrested growth (LAGs) that mark their outer margins toward the center of the bone. Black lines denote LAGs; arrows and numbers represent successive growth increments. (B) Humerus cross sections from three Kemp's ridley (*L. kempii*) sea turtles increasing in size from top to bottom, demonstrating the progression of bone resorption. The top section is from the humerus of a known-age Kemp's ridley that was tagged as a posthatchling, released, and stranded at the age of 3.0 years at 33.2 cm SCL; the annulus, or diffuse LAG denoting the outer margin of the first growth increment, is intact, and its lateral boundaries are marked with the dotted black lines. The middle section is from the humerus of another known-age Kemp's ridley tagged as a posthatchling, released, and later stranded dead at the age of 4.25 years at 43.0 cm SCL, and in this section, the annulus has been partly resorbed. The section at the bottom was collected from a wild, stranded Kemp's ridley 62.0 cm SCL, and the annulus and additional early LAGs are completely resorbed.

of applying this traditional approach. As a result, the most common sclerochronological study of sea turtles to date has been skeletochronological analysis of the lines of arrested growth (LAGs; Castanet et al., 1993) that denote the outer edges of skeletal GMs in periosteal or compact bone. Early on in this field of study, Zug established that in the Cheloniid sea turtles, the humerus bone (Figure 5.2) exhibited the least amount of resorption relative to other skeletal elements, particularly in the narrowest part of the diaphyseal shaft (Zug et al., 1986; Zug, 1990). Due to this initial finding and additional verification (Avens and Goshe, 2007), subsequent skeletochronological analyses of the Cheloniid sea turtle species have focused on this skeletal element (Goshe et al., 2009). While some studies have attempted to analyze LAGs in humerus biopsies collected from live turtles (Klinger and Musick, 1992; Bjorndal et al., 1998), results have been mixed, as LAG visibility and clarity vary around the circumference of the humerus (Snover et al., 2011). Consequently, thorough skeletochronological analysis of this bone requires examination of full cross sections (Snover et al., 2011) and is therefore restricted to samples collected from dead, stranded turtles.

During humerus collection and preparation for analysis, several factors must be taken into account to ensure optimal results. When initially removing the front flipper from a dead, stranded turtle, it is necessary to take particular note of the location of the joint where the humerus articulates

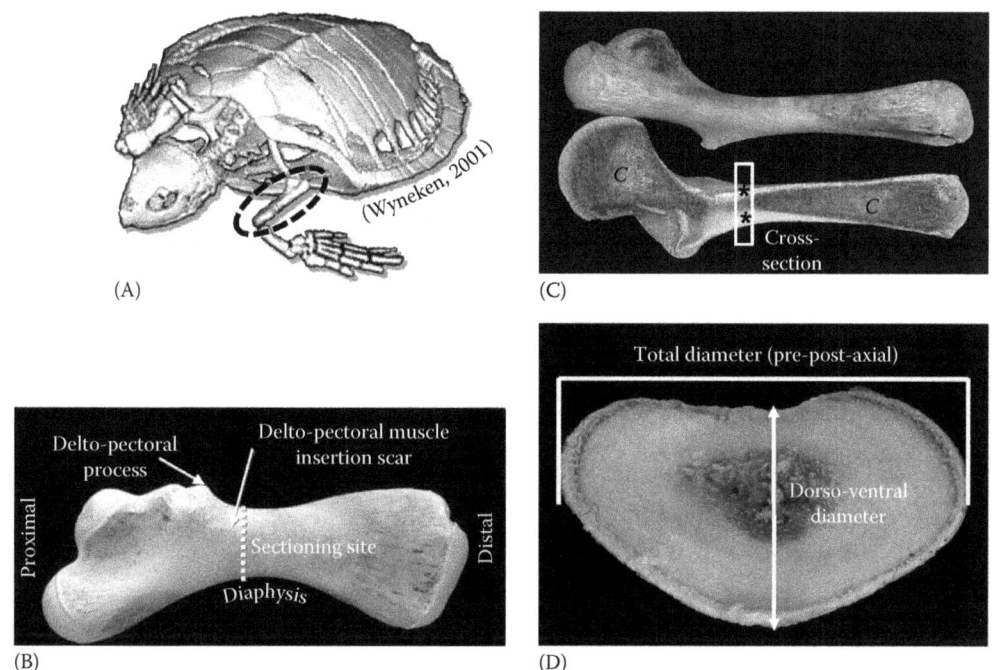

**FIGURE 5.2** (A) CT scan of a juvenile loggerhead sea turtle (*C. caretta*; Wyneken 2001) with left humerus bone circled to show its relative size, position, and form. (B) Ventral view of a juvenile Kemp's ridley (*L. kempii*) left humerus, showing recommended site for cross-sectioning for skeletochronological analysis at the distal end of the insertion scar for the deltopectoral muscle, just distal to the deltopectoral crest. (C) Longitudinally sectioned loggerhead humerus highlighting distribution of cancellous bone (C) and periosteal bone (*). White rectangle denotes sectioning site, where the proportion of periosteal bone is greatest, yet muscle insertion scars at the periphery do not impinge on the outermost growth increments. (D) Cross section of Kemp's ridley (*L. kempii*) humerus with measurement axes labeled.

with the pectoral girdle (Figure 5.2; Wyneken, 2001), as attempting to cut the flipper distal to the joint may sever the humerus, compromising its usefulness for analysis. Dissection of tissue away from the humerus is best done using a knife, not a scalpel, as the latter can easily cut away exterior layers of periosteal bone, impeding later GM analyses. After this initial dissection, the humerus can be stored frozen for future use; although preservation in ethanol is also acceptable, long-term storage of bone tissue in formaldehyde is discouraged, as it can result in unregulated bone decalcification. Alternatively, any soft tissue attached to the bones can be removed by cooking the humeri in water at moderate temperatures (at high heat, the bones may crack) and then drying them in the sun for several weeks, after which they can be stored dry or vacuum-sealed to prevent insect infestation.

Preparation of sea turtle humeri for analysis of skeletal GMs has generally involved examination of cross sections either left untreated or that have undergone extensive histological processing (reviewed in Goshe et al., 2009). The first step in each of these approaches is to cut a cross section at the distal end of the deltopectoral muscle insertion scar (Figure 5.2) and perpendicular to the long axis of the bone, often using a low-speed saw with a diamond-coated blade immersed in water (Snover and Hohn, 2004; Goshe et al., 2009). Any additional sections for complementary analyses (e.g., oxytetracycline; also, see stable isotopes in *Loggerhead* and *Kemp's Ridley* sections and *Bomb-Radiocarbon* section) should be collected consecutive to the skeletochronology section. Particular care must be taken with humerus alignment during this initial step, as even small deviations in sectioning location or axis will result in skewed humerus section and GM measurements. For analysis of untreated bone, the cross section is cut to be 0.5–0.8 mm thick and is immersed in a 4:6 solution of glycerin:ethanol for examination under a dissecting microscope

(Parham and Zug, 1997), but otherwise can be stored dry. If sample preparation is to involve further histology, a 2–3 mm thick initial cross section is cut, which is then fixed and decalcified using a combination of formaldehyde and a dilute hydrochloric, nitric, or formic acid solution (Snover and Hohn, 2004; Goshe et al., 2009; Piovano et al., 2011; Avens et al., 2012). After decalcification, a microtome is used to cut 25 μm thick "thin" sections, which are then stained using hematoxylin, to highlight the LAGs within the bone (Figure 5.3; Snover and Hohn, 2004). Although different types of hematoxylin have been used, Erhlich's modified hematoxylin (recipe in Frazier, 1982) seems to perform particularly well with humerus sections (Frazier, 1982). Thin sections are mounted in glycerin on microscope slides, which can be examined using transmitted light under a dissecting or compound microscope. As significant fading of stained sections has been observed over fairly short time frames (i.e., several months), high-magnification images of stained sections should be acquired and retained both for analysis and archiving. Comparison of the two preparation methods

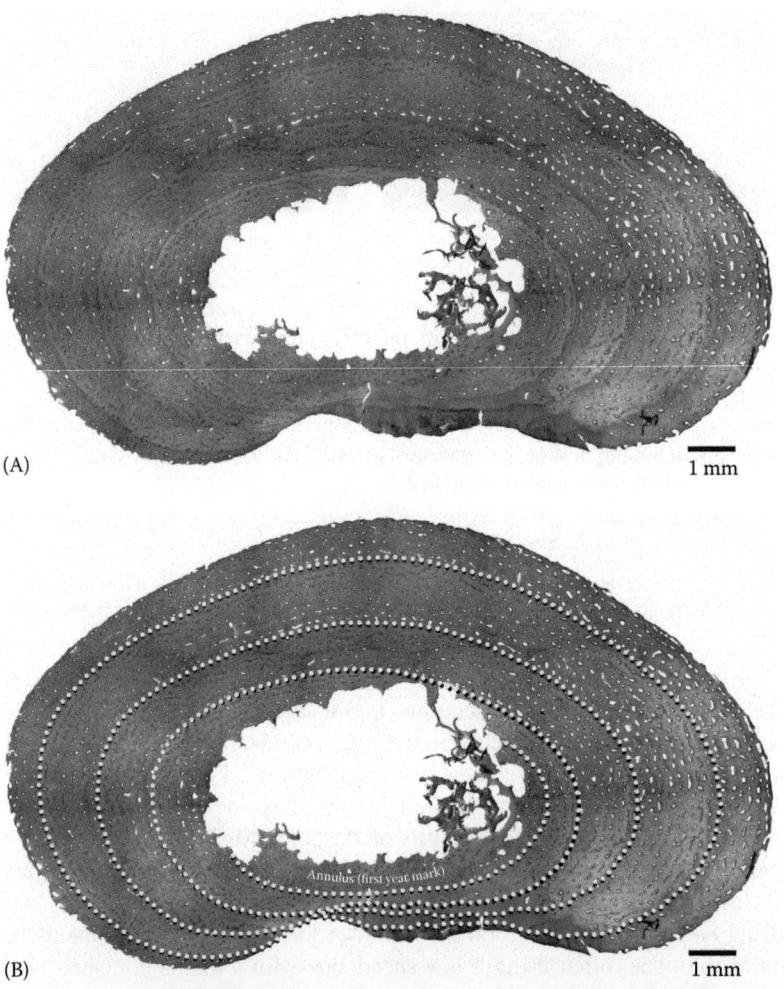

(A)

1 mm

(B)

1 mm

FIGURE 5.3   (A) High-magnification image of Kemp's ridley (*L. kempii*) humerus cross section taken from a known-age juvenile tagged as posthatchling, released, and later stranded dead at the age of 4.25 years and 43.0 cm SCL. The humerus section was decalcified, microtomed, and stained with modified Ehrlich's hematoxylin (see text for additional processing details) to highlight the lines of arrested growth (LAGs) within the bone. LAGs manifest as the darker-stained, narrow lines, with the exception of the annulus, or first year mark, which appears darker, but more diffuse. (B) The same humerus section as in (A), but with the LAGs marked with yellow, dotted lines to specifically denote their location.

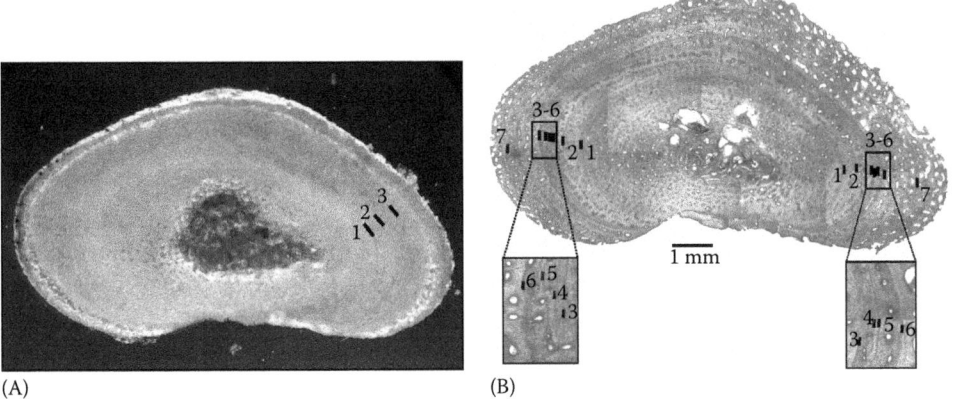

(A)                                                    (B)

FIGURE 5.4  Comparison of the appearance of lines of arrested growth (LAGs) that mark the outer edges of skeletal growth increments observed in unstained and stained sections taken from the same 31.7 cm SCL juvenile Kemp's ridley (*L. kempii*) humerus. (A) Unstained section with three darker-shaded lines visible (the margin appears darker due to refraction of light at the bone's edge). (B) Stained section with 7 LAGs. Part of the discrepancy in LAG numbers between the two section types is due to closely spaced LAGs (3–6 in expanded boxes).

has demonstrated that LAG visibility and readability are improved with histological processing (Figure 5.4; Goshe et al., 2009). As a result, approaches incorporating decalcification, microtoming, and hematoxylin staining are recommended not only to maximize the information yielded by each bone but also to ensure comparability among studies (Goshe et al., 2009).

For the purpose of relating bone and somatic growth, measurements of humerus sections and LAGs have been taken along both the prepostaxial (humerus section width) and dorsoventral axis (humerus section thickness) of bone cross sections (Zug et al., 1986) (Figure 5.2D). Along each of these axes, growth increment measurements have been obtained by measuring distance between each pair of LAGs (Zug et al., 1986) or by measuring full diameters (Snover, 2002; Snover and Hohn, 2004; Goshe et al., 2010; Piovano et al., 2011) or radii (Parham and Zug, 1997; Bjorndal et al., 2003) for each LAG and then calculating the difference between pairs of sequential measurements. LAG spacing and retention increase along the ventral edge of the humerus, improving readability and interpretability (Parham and Zug, 1997), suggesting that perhaps measurement along this axis might be most appropriate. However, an insertion scar on the dorsal aspect often compresses and/or interrupts LAGs, impeding diameter measurements along the dorsoventral axis. Also, humerus thickness along the dorsoventral axis is not quite as strongly correlated with carapace length as is pre–postaxial humerus width (Snover, 2002); therefore, incorporating the former into analyses may decrease the accuracy of length and somatic growth back-calculation. Radius measurements are additionally confounded by the amorphous and inconsistent nature of resorption cores, which make it difficult to define a standardized point from which to begin measurements that is applicable and repeatable across all humeri (Snover et al., 2007a; Piovano et al., 2011). This difficulty can be avoided through measurement of total humerus and LAG diameters along the axis parallel to the dorsal edge of the bone (Snover and Hohn, 2004; Snover et al., 2007a; Goshe et al., 2010; Piovano et al., 2011; see "total diameter," Figure 5.2D).

*Loggerheads* (*C. caretta*): The extent to which skeletochronology has been applied varies among the Cheloniid sea turtle species (Tables 5.1 through 5.3), as does the degree to which mark deposition frequency, early LAG resorption, and back-calculation validation have been characterized. Loggerheads in the North Atlantic Ocean have received perhaps the greatest attention (Table 5.1), beginning with the first systematic application of the technique to sea turtles by Zug et al. (1986). Skeletochronological analysis of humeri collected from known-age (Snover and Hohn, 2004) and oxytetracycline bone-marked (Klinger and Musick, 1992; Coles et al., 2001) loggerheads has supported the hypothesis of annual mark deposition, although occurrence of supplemental marks due to unusual life events has been documented (e.g., release following extended captivity [Snover and Hohn, 2004; Snover et al., 2011]). Resorption of early LAGs in loggerhead humeri does occur, and various modeling approaches have been applied to attempt to account for this lost information (Zug et al., 1986; Klinger and Musick, 1992; Parham and Zug, 1997; Snover, 2002). Although characterization of early LAG deposition patterns is needed to increase the accuracy of these models (Parham and Zug, 1997), the relative inaccessibility of oceanic juvenile loggerheads has made it difficult to address this issue. However, a combination of data from oceanic (Bjorndal et al., 2003) and neritic juvenile (Snover et al., 2007a) loggerheads in the North Atlantic indicates that patterns change among life stages (Snover et al., 2007a), which is an important consideration for future studies. The potential for accurate back-calculation of length-at-age and growth rates for the species has been indicated by a strong relationship between humerus (skeletochronology section) diameter and SCL (Snover, 2002; Snover and Hohn, 2004) and tested using humeri collected from neritic juveniles that were tagged and at large ≥1 year prior to stranding dead (Snover, 2002; Snover et al., 2007a). Back-calculated estimates of SCL based on LAG diameter thought to be deposited closest to the time of tagging were not significantly different from actual SCL measurements at tagging, validating the efficacy of this approach (Snover et al., 2007a).

Minimum loggerhead oceanic stage duration for the northwestern Atlantic as estimated through skeletochronology (7.0 years to 46 cm CCL; Bjorndal et al., 2003) corresponded well with that yielded by length-frequency analysis (6.5 years to reach 46.0 cm CCL; Bjorndal et al., 2000b). Estimates of mean duration for the stage extrapolated backward from linear growth trajectories of neritic juveniles characterized using skeletochronology were 14.8 ± 3.3 years, with age at transition into neritic habitat ranging from 9 to 24 years at 48.5–51.1 cm SCL (Snover, 2002). Mean neritic stage duration (49–90 cm SCL) in the northwestern Atlantic has been estimated as 17 years (Snover, 2002). Early analyses of skeletochronological age and growth data for loggerheads using simple regressions produced lower estimates of ASM of 13–15 years at 86 cm CCL (Zug et al., 1986). However, subsequent refinement of techniques, application of growth curves, and assumption of 92 cm SCL at maturity have produced mean loggerhead ASM estimates for the northwestern Atlantic of around 22–26 years (Klinger and Musick, 1995), 25–30 years (Parham and Zug, 1997), 24.3–38.8 years (Snover, 2002), and 23.8–37.7 years (Vaughan, 2009), comparable to predictions of models based on mark–recapture data (Frazer and Ehrhart, 1985; Braun-McNeill et al., 2008; Table 5.1). Similarly, in the southwestern Atlantic, skeletochronological analysis of Brazilian loggerheads indicates a mean oceanic stage duration of 11.5 years (range 8–19 years), with a mean ASM of 32 years (Petitet et al., 2012). Skeletochronological analysis of loggerheads inhabiting the Mediterranean Sea yielded estimates of 14.9–28.4 years to reach maturity at sizes ranging from 66.5 to 84.7 cm CCL (Casale et al., 2011a), which is similar to the 16–28 years predicted by mark–recapture data (Casale et al., 2009b), but somewhat lower than the 15.4–34.9 years yielded by length-frequency analysis (Casale et al., 2011b) for the same population. A subsequent study that incorporated potential influence of genetic origin yielded a similar age range of 19–25 years for Mediterranean loggerheads 66.5 cm CCL (Piovano et al., 2011).

A few studies have applied skeletochronology to loggerhead sea turtles in the Pacific Ocean as well (Table 5.1). Analysis of humeri collected from oceanic juveniles 13.0–42.0 cm CCL by-caught in driftnets indicated that these turtles ranged from 1.4 to 8.9 years of age (Zug et al., 1995). Age-specific growth rates yielded by fitting a growth curve to these length-at-age data yielded an estimate of 10 years to grow from hatchling to 46.5 cm SCL (Zug et al., 1995). More recently, skeletochronological analysis of humeri collected from neritic loggerheads 49–90 cm CCL stranded in Baja California Sur, Mexico, yielded age estimates of 6–31 years. The age distribution of the sample was bimodal, with a peak at 12 years and another at 24 years, perhaps due to size/age-specific differences in mortality for turtles in this region (Bickerman, 2011).

Skeletochronology in loggerheads has also been complemented with stable isotope analyses to investigate a significant increase in growth observed in the humeri of neritic juveniles in the northwestern Atlantic (Snover et al., 2010). Stable isotope signatures in animal tissues reflect those of the environment at the time the tissue was formed (Wallace et al., 2009), and, as retention of those signatures is dependent upon the turnover rate of a given tissue, the relatively inert nature of compact bone can provide a long-term isotopic record (Snover et al., 2010). Stable isotope ratios of nitrogen ($\delta^{15}N$) are typically used to assess trophic level, and in the marine environment, carbon ($\delta^{13}C$) can offer insight into foraging location (nearshore vs. offshore) (Wallace et al., 2009). Using skeletochronology, Snover et al. (2010) identified significant increases in loggerhead growth rates (as inferred through skeletal growth increment width) that were proposed to correspond with an ontogenetic change in foraging habitat and/or prey preference. Analysis of $\delta^{15}N$ and $\delta^{13}C$ in bone samples collected to either side of the growth transition demonstrated that this increased growth did coincide with a shift from oceanic to neritic habitat (Snover et al., 2010).

*Kemp's (L. kempii) and Olive (L. olivacea) Ridleys*: Skeletochronological analysis of ridleys has focused most on the Kemp's ridley (Table 5.3), due at least in part to the need for life history data to inform management of this highly endangered species (Snover et al., 2007b). Analysis of humerus samples from known-age head-started and coded wire tagged (CWT) Kemp's ridleys allowed validation of annual LAG deposition for the species (Snover and Hohn, 2004; Snover et al., 2007b). A strong proportional relationship between humerus section diameter and SCL comparable to that observed for loggerheads (earlier) provided a basis for back-calculation of length-at-age and growth rates (Snover and Hohn, 2004; Snover et al., 2007b), although validation is still needed. Also similar to loggerheads (*earlier*), changes in stable isotope ratios of $\delta^{15}N$ and $\delta^{13}C$ within the humerus reflected a transition from oceanic to neritic habitat, although for Kemp's ridleys, this occurred as early as 1 year of age (Snover, 2002). Sex-specific analyses suggest that the polyphasic growth observed for Kemp's ridleys (Chaloupka and Zug, 1997) might differ between males and females and reflect transition of juveniles among neritic habitats (Snover, 2002; Snover et al., 2007b). ASM estimates are based mostly on Atlantic strandings and range from 11 to 16 years at 65 cm SCL (Zug et al., 1997) and 9.9–16.7 years at 60 cm SCL (Snover et al., 2007b), which are consistent with estimates obtained through mark–recapture and observed ages of 10–18 years for head-started turtles when first detected nesting at 58.1–67.5 cm SCL (Shaver and Wibbels, 2007). However, as growth rates appear to differ between the Atlantic and the Gulf of Mexico (Schmid and Witzell, 1997; Zug et al., 1997) and may also differ between males and females (Snover, 2002; Snover et al., 2007b), additional characterization is needed. Only one skeletochronological study of olive ridleys has been published to date, involving the species in the North Pacific, and these data predict a median ASM of 13 years at 60 cm SCL, which falls within the range of estimates for Kemp's ridleys (Zug et al., 2006; Table 5.3).

*Green Turtles (Chelonia mydas)*: Annual LAG deposition for green turtles has been validated through oxytetracycline bone-marking (Snover et al., 2011) and comparison of back-calculated SCL to SCL at tagging for humeri recovered from known-age and tagged, stranded turtles (Goshe et al., 2010; Avens et al., 2012). However, additional characterization of LAG deposition frequency for adult green turtles is recommended (Goshe et al., 2010). Oceanic stage duration in the northwestern Atlantic has been estimated using skeletochronology at 3–6 years (Zug and Glor, 1998) and 1–7 years (Goshe et al., 2010; Table 1.2), which is similar to the range predicted using stable isotope analysis (3–5 years; Reich et al., 2007). Neritic juveniles, 28–74 cm SCL from the Indian River Lagoon on the east coast of Florida, were predicted to be 3.3–13.6 years of age (Zug and Glor, 1998), which falls within the 2–22 years estimated for juveniles 18.1–78.5 cm SCL on the Gulf coast of Florida, in St. Joseph Bay (Avens et al., 2012). Although their sample did not include adult individuals, Zug and Glor (1998) cautiously estimated ASM for green turtles from the Indian River Lagoon as 34 years at 101.5 cm SCL. In a subsequent analysis involving Atlantic green turtles stranded along the entire U.S. Atlantic coast, ASM of females was estimated to range from 30 to 44 years, depending upon the size at maturation and the source populations (i.e., Florida, Costa Rica, and Mexico), with a minimum of 28 years to maturity (Goshe et al., 2010). Males exhibited slower growth, which might bias estimates of female ASM upward in combined models (Goshe et al., 2010). The estimates presented by Goshe et al. (2010) are somewhat greater than those predicted from mark–recapture data in this region (26–36 years) (Mendonça, 1981; Frazer and Ehrhart, 1985; Frazer and Ladner, 1986). Discrepancies among these studies are likely due in some part to differences in sample size and time frames of the studies and intervals used to calculate growth rates as well as growth rate variability among foraging grounds (Goshe et al., 2010). This last factor may also be influential in the Caribbean, where initial returns of adult green turtles live-tagged as hatchlings and head-started yearlings have generated much lower minimum ASM estimates; females released as hatchlings and yearlings returned to nest at 17 and 15 years, respectively, while males were observed mating at 19 years of age after release as hatchlings and at 15 years of age for those released as yearlings (Bell et al., 2005). Large-scale tagging projects of this type (*see also Head-Started and CWT Kemp's Ridleys*) hold great potential to offer insight into age and growth parameters, once sufficient time has passed to allow characterization of the full age distribution at maturation.

In the Pacific, oceanic stage duration predicted through skeletochronology for Hawaiian green turtles was 4–10 years (Zug et al., 2002), slightly longer than that predicted for the northwestern Atlantic. Hawaiian green turtle growth rates estimated using skeletochronology were fairly comparable to those observed through mark–recapture, and ASM was estimated at ≥30 years (Zug et al., 2002), with mark–recapture data suggesting 35–40 or even ≥50 years for the same population (Balazs and Chaloupka, 2004b).

*Hawksbills (Eretmochelys imbricata)*: Only one skeletochronological analysis of hawksbills has been completed to date (Snover et al., 2012; Table 5.3), and it was based on a small nesting population in the main Hawaiian Islands. Although sample size was small (n = 30), validation of mark deposition frequency and age estimates was provided by two sources: (1) MIA and (2) comparison of results to growth data from a hawksbill that was tagged as a small juvenile (32.9 cm SCL) and subsequently observed nesting 20 years later at 76.4 cm SCL. Results of this study estimated ASM for the population between 17 and 22 years, which is comparable to 14.7 years time from 23 cm to ASM estimated by Diez and van Dam (2002) for Caribbean hawksbills. The tagged turtle reported in the study was a minimum of 20 years old and estimated to be 22–23 years old based on probable age at first capture. As it is possible that the turtle had nested prior to being observed, this age may not necessarily be the age at first reproduction.

*Flatbacks (Natator depressus)*: To date, no published skeletochronological studies for the species are available. As noted earlier (*Mark–Recapture: Flatbacks*), one flatback turtle tagged and released as a hatchling returned to nest at the age of 21 years (Limpus, 2007).

### 5.2.3.2 Skeletochronology and Leatherbacks (*D. coriacea*)

Due to the high degree of vascularization and remodeling present in leatherback skeletal elements (reviewed by Snover and Rhodin, 2008), the humerus exhibits a great deal of resorption and is not suitable for skeletochronological analysis (Figure 5.5A; Zug and Parham, 1996). Instead, studies have focused on the scleral ossicles, which are the small bones encircling the pupil of the eye (Figure 5.5B; Zug and Parham, 1996), as these structures exhibit less remodeling relative to other bones and contain marks whose appearance is consistent with that of LAGs (Zug and Parham, 1996; Avens et al., 2009). Histological preparation of these bones is quite similar to that of humeri, with the exception that the full structure is decalcified, and Mayer's hematoxylin (formula in Frazier, 1982) appears to yield better results than the Ehrlich's modified hematoxylin stain often used for humeri. Analysis of marks at the lateral edges of ossicles collected from leatherbacks in the southeastern Pacific suggested a mean ASM for the species in this region of 13–14 years (Table 5.3); however, the nature of the marks in the ossicles could not be verified (Zug and Parham, 1996). The exceptionally migratory behavior of leatherbacks (e.g., Fossette et al., 2010), crypticity of small

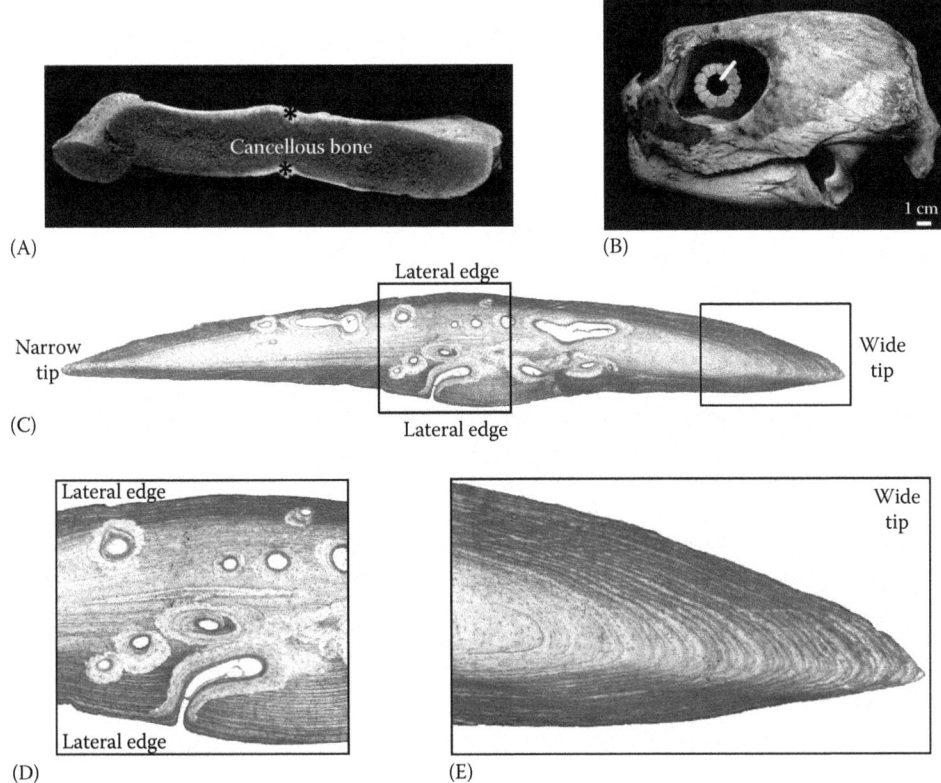

(A)                                                                    (B)

Narrow
tip

Wide
tip

(C)

Lateral edge

Lateral edge

Wide
tip

(D)                                          (E)

**FIGURE 5.5** Leatherback (*D. coriacea*) (A) Longitudinally sectioned humerus showing the predominance of cancellous bone and lack of periosteal bone (*) retaining growth increments that could be used for skeletochronological analysis. (B) Skull and scleral ossicle ring that surrounds the pupil of the eye. White line denotes the plane of sectioning for skeletochronology. (C) Section of an entire scleral ossicle decalcified, microtomed, and stained with Mayer's hematoxylin. (D) Magnified view of central ossicle section showing LAG compression, particularly at the top lateral edge. C denotes the "core mark" or LAG deposited at hatching (Avens et al. 2009) (E) Magnified view of ossicle section wide tip showing increased ossicle spacing, when compared to the lateral edges.

juveniles (Eckert, 2002), and rarity of strandings retaining eyes to allow ossicle collection combine to represent a major impediment to validation studies for the species. As a result, indirect validation of deposition rate was attempted by a proxy comparison of LAGs in the ossicles and humeri of individual Kemp's ridleys (Avens and Goshe, 2007), a species for which annual mark deposition in the humerus had been demonstrated (Snover and Hohn, 2004). The number of LAGs in the ossicles and humeri was equivalent, suggesting that the marks in the ossicles represent annual growth cycles and could be used for skeletochronological analysis (Avens and Goshe, 2007). However, these analyses also revealed that because of axial differences in ossicle growth patterns, as well as mark compression and resorption, fewer marks are retained at the lateral edges than in the wide tip of the bone (Figure 5.5C through E), suggesting that LAG counts may have been underestimated in the previous leatherback study (Zug and Parham, 1996). Subsequent analysis of LAGs in the wide tips of leatherback ossicles from the western North Atlantic partially validated (i.e., for the LAG denoting the end of the first year of growth) with ossicles from captive-reared juveniles did yield increased ASM estimates of 16–22 years at 125 cm CCL and 24.5–29.0 years at145 cm CCL (Avens et al., 2009). These values contrast sharply with the results of the initial Zug and Parham (1996) study, as well as ASM estimates inferred through anatomic features (3–6 years, reviewed by Snover and Rhodin, 2008), captive growth (2–6 years in Deraniyagala, 1952; Birkenmeier, 1971; Bels et al., 1988; 16 years in

Jones et al., 2011), and microsatellite DNA analysis (12–14 years Dutton et al., 2005). Despite the associated difficulty, additional investigation into wild leatherback growth rates and size at age, as well as validation of ossicle LAG deposition through traditional and/or other means (e.g., see Section 5.2.4), is needed to help resolve these discrepancies.

### 5.2.3.3 Closing Remarks

In summary, although skeletochronology can be a useful tool for assessing sea turtle age and growth, the technique can only be applied with confidence when assumptions relating to (1) suitability of analytical structure and preparatory method, (2) frequency of mark deposition, (3) estimation of resorbed LAGs, and (4) back-calculation of growth and length-at-age are addressed. One drawback to this method is that analysis is typically restricted to dead, stranded turtles whose cause of death is unknown, potentially resulting in parameter bias due to atypical growth and stage durations for compromised individuals or decreased longevity when fishery interactions occur. Also, given that resorption prohibits the direct determination of absolute age for individual turtles, skeletochronology is best-suited for providing broader, population-wide characterizations. Despite these limitations, when appropriately applied and validated, skeletochronology can be an effective means of collecting sea turtle age and growth data relatively rapidly when compared to traditional methods, particularly with respect to characterization of individual length-at-age and growth trajectories. Annual LAG deposition ensures consistency in the time frames for which growth rates are calculated, minimizing potential biases that arise when recapture intervals are variable (Snover et al., 2007b). If necropsies and genetic analyses are conducted in conjunction with skeletochronology, results will offer insight into sex- and stock-specific differences in length-at-age and growth rates (e.g., Piovano et al., 2011; Avens et al., 2012). Furthermore, when complemented with techniques such as growth increment-specific stable isotope and trace element analyses, skeletochronology should also allow characterization of the influence of factors such as foraging location, trophic ecology, and contaminant exposure on age and growth parameters.

### 5.2.4 BOMB-RADIOCARBON (BOMB-$^{14}$C)

Radiocarbon ($^{14}$C) levels in the environment were low prior to the mid-1900s, partly due to depletion resulting from burning of fossil fuels since the beginning of industrialization (Druffel, 1980). However, large pulses of the radioisotope were introduced during atmospheric testing of thermonuclear devices that began in the 1950s, peaked ~1961–1962, and then was greatly reduced following the Partial Test Ban Treaty in 1963 (Nydal and Gislefoss, 1996). The advent and later decrease in numbers of these detonations yielded corresponding fluctuations in $^{14}$C quantities in the atmosphere, with highest levels occurring during the early- to mid-1960s (Broecker et al., 1985; Nydal and Gislefoss, 1996); following a lag of several years, this pattern was also repeated in freshwater systems (Peng and Broecker, 1980; Campana et al., 2008). After an additional delay, bomb-derived $^{14}$C became widely incorporated into oceanic waters as well, although at much lower levels (Broecker et al., 1985), through a combination of diffusion (equilibration between atmospheric and oceanic $CO_2$), precipitation, and riverine input (Campana et al., 2008). Bomb-$^{14}$C became not only integrated into surface waters in the form of dissolved inorganic carbon (DIC) (Druffel, 1980) but also entered oceanic food webs through photosynthetic activity of phytoplankton (Roark et al., 2006). Regional differences in $^{14}$C levels may occur due to varying contributions of high-$^{14}$C freshwater to estuarine systems (Campana and Jones, 1998; Vadopalas et al., 2011). Also, temperature fluctuations can affect gas diffusivity and solubility, as well as water viscosity, all of which influence the incorporation of $CO_2$-containing bomb-$^{14}$C into surface waters (Broecker et al., 1985). Furthermore, deep ocean waters remain depleted in bomb-$^{14}$C relative to the surface, and, as a result, oceanographic processes that result in mixing, such as wind, currents, upwelling, and downwelling, can alter $^{14}$C levels (Nydal et al., 1984;

Broecker et al., 1985; Nydal and Gislefoss, 1996; Kerr et al., 2005). Despite these factors, the overall timing of bomb-[14]C incorporation has been relatively synchronous across the world's surface ocean waters (<200 m), with an initial increase in the late 1950s and a peak spanning the late 1960s and early 1970s, followed by a gradual, continued decrease (Broecker et al., 1985; Nydal and Gislefoss, 1996).

The introduction and subsequent distribution of bomb-[14]C into the global environment have been well described, and this characterization has proven informative for oceanographic and forensic science (Kalish et al., 2001; Reimer et al., 2004). Furthermore, this rapid, massive influx of bomb-[14]C inadvertently laid the foundation for global-scale chemical marking that would subsequently facilitate age estimation and validation studies (Campana, 2001). As the chemical composition of animal tissues reflects that of the environment at the time of formation (Peterson and Fry, 1987), bomb-[14]C levels in organisms worldwide also peaked around the time of atmospheric nuclear testing (e.g., Druffel and Linick, 1978; Weidman and Jones, 1993; Kalish, 1995a; Campana, 1997). This in turn generated a time-specific marker in organismal tissues (Campana, 2001), reflecting bomb-[14]C levels specific to particular environments and time scales, as influenced by the distributional factors mentioned earlier (e.g., Campana and Jones, 1998; Kerr et al., 2005). Retention of the bomb-[14]C marker within organisms is dependent upon tissue stability, with calcified structures being perhaps the most widespread relatively metabolically inert tissues among vertebrate and invertebrate taxa. Although analysis of calcified structures has been most prevalent in bomb-[14]C age-estimation studies to date, vertebrate eye lens nuclei retain embryonic cells throughout life (Bloemendal, 1977) and as a result can retain long-term bomb-[14]C exposure records as well (Lynnerup et al., 2008). Over the past several decades, bomb-[14]C levels in tissues have been predominantly measured using accelerator mass spectrometry (e.g., Kalish, 1993; Campana, 1997; Kerr et al., 2005; Andrews et al., 2011). Results are typically reported as $\Delta^{14}C$, the deviation of the measured value in per mil (‰) from the oxalic acid radiocarbon reference standard (age-corrected [14]C activity in tree rings from 1890, corrected for decay prior to 1950), normalized for isotopic fractionation of $\delta^{13}C$ relative to the PDB (Pee Dee Belemnite) standard (Stuiver and Polach, 1977).

Through comparison with appropriate reference chronologies, bomb-[14]C analysis can be used as an independent age estimator (Kalish, 1995b; Campana, 1997; Andrews et al., 2007), but has more often been applied as a means of validating other age-estimation methods such as sclerochronology (Kalish, 1995a,b; Campana et al., 2002; Andrews et al., 2005, 2011; Ardizzone et al., 2006; Stewart et al., 2006; Armsworthy and Campana, 2010). However, for the latter approach to succeed the technique being validated must yield age estimates that meet the assumption of normal error distribution (Piner et al., 2005; Armsworthy and Campana, 2010), and particular care must be taken to eliminate contamination of tissue used for [14]C analysis by more recent material (Campana, 1997; Campana and Jones, 1998). As long term, direct DIC [14]C measurements at specific locations are frequently limited (Kalish, 1995a,b), extended reference chronologies were initially established using [14]C values from annual growth increments primarily in hermatypic corals (Druffel and Linick, 1978; Druffel, 1980; Kalish, 1993), but also in bivalves (Weidman and Jones, 1993), whose calcium carbonate composition directly reflects DIC [14]C values. Although these chronologies represented significant advances, their application was constrained, as regional differences in bomb-[14]C trends made inference beyond the limited geographic distribution of hermatypic coral species problematic (Campana, 1997; Kerr et al., 2004). However, the finding that 70%–90% of fish otolith aragonite is derived from DIC (Kalish, 1991), combined with availability of long-term, multispecies otolith archives originally collected for sclerochronological analysis, allowed development of additional chronologies with wider geographic scope (Campana, 1997; Kalish et al., 2001; Kerr et al., 2004; Piner and Wischniowski, 2004; Campana et al., 2008). The most valuable samples for developing chronologies, as well as for age estimation and validation, are those that encompass the time frame spanning and extending slightly beyond the sharp increase in [14]C in the marine environment ~1958–1965 (Kalish, 1995a; Campana, 1997; Andrews et al., 2007). Measured $\Delta^{14}C$ values from

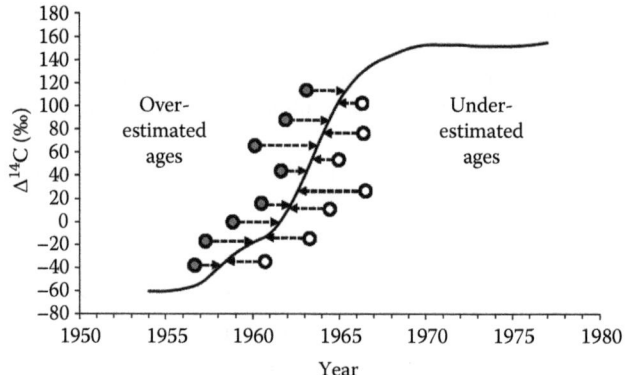

**FIGURE 5.6** Sample oceanic bomb radiocarbon (bomb-[14]C) chronology (solid line) obtained from hermatypic corals showing sharp increase in the late 1950s and 1960s, after which levels of bomb-[14]C are stabilized. Validation of ages generated using other methods, such as sclerochronology, can be accomplished through comparison with age estimates obtained through bomb-[14]C analysis. If ages are underestimated using the alternate method(s), they will be shifted to the right of the bomb-[14]C chronology, whereas if they are overestimated, they will be shifted to the left. (Adapted from Druffel, E.M., *Radiocarbon*, 22, 363, 1980.)

samples are plotted against reference chronology $\Delta^{14}$C curves and visually compared to infer age and also to validate ages generated using other methods (e.g., Figure 5.6; Campana, 1997; Campana et al., 2002, 2006; Piner et al., 2006; Andrews et al., 2011). In the latter case, overestimating ages will result in values shifted to the left of the chronology curve, while underestimation will yield values shifted to the right (Kalish, 1993; Kilada et al., 2007) (Figure 5.6). Detection of prebomb [14]C values can only provide an estimate of minimum longevity, as the lack of a trend in oceanic [14]C chronologies prior to 1958 does not allow inference of year-specific values prior to that time (Kalish, 1995a,b; Campana, 1997).

Due to the physicochemical, temporal, and spatial considerations described earlier, current use of bomb-[14]C age estimation is most straightforward when (1) it is applied to long-lived organisms that do not range widely and (2) involves analysis of samples whose calcium carbonate structure directly reflects surface (<200 m) DIC [14]C levels. Consequently, results have been perhaps most clear-cut when studies have focused on relatively sedentary osteichthyid fish species in continental shelf waters (e.g., Kalish, 1993, 1995b; Baker and Wilson, 2001; Andrews et al., 2005, 2007, 2011; Kerr et al., 2005; Piner et al., 2005, 2006; Armsworthy and Campana, 2010), rhodoliths (Frantz et al., 2000), coralline sponges (Fallon and Guilderson, 2005), bivalves (Kilada et al., 2007, 2009; Vadopalas et al., 2011), and various coral species (Roark et al., 2005, 2006; Sherwood et al., 2005). Interpretation of [14]C measurements becomes more difficult when organisms are highly migratory, as uncertainty regarding relevance of reference chronologies arises (Kalish, 1995b; Ardizzone et al., 2006). This can be particularly problematic when organisms transition between estuarine and oceanic environments (Campana and Jones, 1998; Campana et al., 2008) or when movements span entire ocean basins (Kalish et al., 1996; Kerr et al., 2006). When carbon in calcified structures is primarily drawn from dietary sources instead of DIC, as in the case of shark vertebrae (e.g., Kerr et al., 2006) and beluga whale teeth (*Delphinapterus leucas*; Stewart et al., 2006), $\delta^{13}$C measurements are necessary to assess the source (Campana et al., 2002). Dietary influence can be especially significant if deep-water prey or long-lived prey representing an integrated bomb-[14]C signal are consumed, as both of these can artificially deplete measured [14]C values (Ardizzone et al., 2006; Kerr et al., 2006). Finally, additional questions regarding interpretability arise if there is a possibility that a calcified structure has been remodeled to some extent throughout life, which may also depress [14]C values through the introduction of recently formed tissue (Campana et al., 2002; Ardizzone et al., 2006; Kerr et al., 2006).

Because available evidence indicates that sea turtles are generally slow to mature and (barring adverse anthropogenic interactions) long-lived, bomb-radiocarbon may prove to be a valuable tool for estimating age and validating other age-estimation techniques in sea turtles. However, sea turtles are also generally highly migratory throughout their lives, in some cases, moving repeatedly between estuarine and oceanic environments (e.g., McClellan and Read, 2007; Mansfield et al., 2009). As a result, it may be challenging to establish appropriate reference chronologies that allow accurate interpretation of sea turtle bomb-$^{14}$C measurements. For Cheloniid sea turtle species, bone archives in museums or other institutions may contain humeri collected, or retaining growth increments deposited, during the time of greatest $^{14}$C increase in the world's oceans. Alternatively, comparison of ages estimated through bomb-$^{14}$C analysis of eye lens nuclei and skeletochronology of humeri for recently stranded, large loggerheads, or green turtles that may have hatched during that same time period may be informative. Finally, although the amount of resorption in leatherback ossicles varies widely, some do retain the earliest growth increments (Avens et al., 2009), which might allow comparison of early $^{14}$C values to appropriate reference chronologies (A. Andrews, unpublished data).

### 5.2.5 TELOMERE SHORTENING

The chromosomes of eukaryotic cells are capped by repeating, noncoding nucleotide sequences called telomeres, with the sequence $(TTAGGG)_n$ being highly conserved in all vertebrate species (Meyne et al., 1989; Blackburn, 1991). During each round of somatic cell replication, the failure of DNA polymerase to completely replicate the $3'$ end of the linear DNA strand results in a shortening of the telomere (Watson, 1972). Telomeres therefore appear to serve the purpose of protecting the coding DNA from attrition, but also facilitate chromosome sorting and help prevent breakage and fusion (Haussmann and Vleck, 2002). When telomere length reaches a minimum threshold, cell replication ceases, followed by senescence and possibly cell death (apoptosis) (Blackburn, 1991). In some frequently replicating cells (i.e., stem or germ cells), telomere length may be maintained by the ribonucleoprotein enzyme telomerase, which adds the necessary nucleotide repeats to the telomere ends (Greider, 1996). However, attrition of the telomeric sequence does occur in most cells due to the downregulation or absence of telomerase activity, which is thought to minimize the potential for cell immortalization and tumor formation (Heidinger et al., 2011). Because the extent of the telomeric decrease is related to the number of times a cell has divided, it has been hypothesized that this phenomenon might serve as a "molecular clock," allowing age and perhaps even longevity to be assessed (Haussmann and Vleck, 2002). The relationship between telomeres and aging has been mainly investigated in the context of human health (e.g., Epel et al., 2004) and forensic science (e.g., Takasaki et al., 2003). However, it has also been proposed that the technique may hold promise for the estimation of ages and lifespans of wild animal populations for which collection of this information is typically challenging, if not prohibitively difficult (Haussmann and Vleck, 2002; Nakagawa et al., 2004).

Telomere measurements have been obtained from numerous tissue types for age-related studies, including blood (Haussmann and Vleck, 2002; Haussmann et al., 2003b; Epel et al., 2004), skin (Friedrich et al., 2000; Hatase et al., 2008), and synovial fluid (Friedrich et al., 2000). However, blood cells have frequently been the tissue of choice (leucocytes in mammals, nucleated red blood cells in fish, birds, and reptiles), as it is thought that because the bone marrow cells that generate blood cells exhibit high levels of mitotic activity, the probability of observing telomeric change is increased (Haussmann and Vleck, 2002; Nakagawa et al., 2004). Furthermore, blood samples are often collected in the course of animal population field studies for other purposes and might therefore be readily available for analysis (Haussmann and Vleck, 2002).

Irrespective of tissue type, a number of different analytical methods have been applied to measure telomere length (Nakagawa et al., 2004; Haussmann and Mauck, 2008). Perhaps the most common are various telomeric restriction fragment (TRF) assays, which use restriction enzymes to dissociate telomeres from chromosomes along with differing lengths of nontelomeric DNA, depending on the location of the restriction site (Juola et al., 2006). Telomere length is then taken as the average of the range of fragment sizes (Nakagawa et al., 2004; Haussmann and Mauck, 2008). Quantitative polymerase chain reaction (Q-PCR) assays provide a measure of the relative ratio (T/S ratio) of telomere repeats (T) to an arbitrary full-genome reference standard (S), which minimizes the variability of the measurement relative to TRF assays. This method does not yield absolute telomere length, but allows within-species, interindividual comparisons and evaluation of telomere length rate of change (Nakagawa et al., 2004). However, Q-PCR can be confounded by the presence of interstitial telomeric repeat sequences in the centromeric region (as in birds; Nakagawa et al., 2004) or in microsatellite DNA (as in cetaceans; Dunshea et al., 2011), which are also amplified during the assay. More recently, fluorescent markers have been hybridized to telomeres in situ in metaphase cells (quantitative fluorescence in situ hybridization) and lengths inferred through measurement of fluorescent light intensity (Haussmann and Mauck, 2008; Kim et al., 2011), but the technical expertise and specialized equipment needed for the method have limited its use.

Over the past decade, wildlife telomere studies have often focused on bird species, perhaps because their comparative longevity relative to mammals of similar size has been of particular interest in aging research (Nakagawa et al., 2004). The majority of these studies involved a cross-sectional population sampling regime and demonstrated a significant overall decrease in telomere length with age for species with varying lifespans (Haussmann and Vleck, 2002; Haussmann et al., 2003a,b, 2007; Hall et al., 2004; Juola et al., 2006; Bize et al., 2009). One notable exception was the long-lived Leach's storm petrel (*Oceanodrama leucorha*), which exhibited an *increase* in telomere length with age of a magnitude similar to the decrease observed in other species (Haussmann et al., 2003b). While longitudinal studies of individuals have also indicated an overall decrease in telomere length over time, they have offered insight into factors that confound the telomere-age relationship as well, such as significant interindividual variability due to heredity and experience as well as non-linear change in shortening rates over time (Hall et al., 2004; Bize et al., 2009; Heidinger et al., 2011). Furthermore, despite the significance of age-telomere length correlations, a lack of predictive power has been acknowledged (Hall et al., 2004; Juola et al., 2006), as the range of ages associated with particular telomere lengths in different species might span as much as 10–30 years (Haussmann et al., 2003b; Hall et al., 2004; Juola et al., 2006; Bize et al., 2009).

Although telomere research has not been as extensive in reptilian systems, a small number of cross-sectional studies have been conducted. In male California garter snakes (*Thamnophis elegans*), erythrocyte TRF length decreased significantly with age as determined through skeletochronology, with the age for a particular TRF length spanning up to 5 years (Bronikowski, 2008). Similarly, erythrocyte TRF length significantly decreased with size in American alligators (*Alligator mississippiensis*), potentially allowing general age classification (Scott et al., 2006). However, as individuals with similar TRFs differed by as much as 230 cm total body length (Scott et al., 2006), telomere length did not appear to suffice as an independent indicator of age. European freshwater turtles (*Emys orbicularis*) exhibited no decrease in telomere length between embryos and adults (Girondot and Garcia, 1999). Hatase et al. (2008) analyzed skin and blood from captive loggerhead sea turtles and observed no significant correlation between estimated age and relative T/S ratios, although overall skin T/S ratios were lower in older than in younger turtles. A visual inspection of the data from Hatase et al. (2008) reveals that the skin samples taken from turtles aged 0–2 years appeared to exhibit higher T/S ratios than turtles with ages $\geq$10 years (Figure 1 in Hatase et al., 2008). The difference is almost, but not quite, significant (one-tailed, independent samples t-test; $T(18) = 1.69$; $P = 0.0539$), and there is a great deal of overlap in the data (0–2 years range = 0.716–2.011; $\geq$10 years range = 0.348–1.283). Furthermore, if the 10–12 years turtles are compared to the 35–36 years turtles, there is no detectable difference in the T/S ratios ($T(7) = 0.10$; $P = 0.46$; 10–12 years range = 0.348–1.283; 35–36 years range = 0.642–1.159).

Hatase et al. (2010) attempted to apply the results of Hatase et al. (2008) to a study of nesting loggerheads at Yakushima Island, Japan. Initial comparison of stable isotope ratios ($\delta^{13}C$ and $\delta^{15}N$) in egg yolks from loggerhead nests was used to assign nesting females to one of two foraging groups, oceanic or neritic. As the oceanic turtles were significantly smaller than the neritic turtles, telomere length (T) to control gene length (C), or T/C ratios, in skin samples of turtles from the two foraging groups were also compared to estimate age. No significant difference in T/C ratios was found between the groups, and, as a result, the conclusion was drawn that the ages of oceanic and neritic foragers did not differ. However, from the results of Hatase et al. (2008), one would have predicted no significant difference in T/C ratios even if turtle ages were as much as 25 years apart. Although the detection of two different foraging groups with significantly different sizes at Yakushima is a compelling result, the determination of age structure based on application of T/C ratios remains inconclusive.

The possibility that age might be determined through its relationship with telomere length is appealing; as sample collection for this technique is fairly noninvasive, analytical results can be produced relatively rapidly when compared to traditional mark–recapture and, unlike skeletochronology, the method can be applied to live animals. Nevertheless, despite the general relationship between chronological age and telomere length in a number of different taxa (approximately 53% of telomere length explained by age; Dunshea et al., 2011), the wide age ranges typically associated with specific telomere lengths limit predictive ability (Dunshea et al., 2011). Part of this variability is likely the result of differences in tissue sample type, analytical approach, and the presence of interstitial telomere repeats, which might be resolved to some extent through continued refinement and standardization of methods (e.g., Haussmann and Mauck, 2008). However, controlling for the significant influences of heredity (Heidinger et al., 2011) and individual exposure to myriad factors that elevate oxidative stress and increase telomere attrition (Von Zglinicki, 2002; Dunshea et al., 2011) on telomere length is far less feasible. Assessing the full extent of variability in telomere length, the rate of change over time, and underlying causes through longitudinal population study would be impeded by constraints similar to those for traditional mark–recapture methods (see *Mark–Recapture*). Although methodological advances may increase the viability of this approach in the future, the factors outlined earlier, combined with the lack of significant results from turtle telomere studies conducted to date, suggest that, currently, the technique may not be applicable to sea turtle species.

### 5.2.6 AMINO ACID RACEMIZATION

In all organisms, the amino acids comprising proteins are potentially configured as one of two optical isomers or enantiomers: *L* (levorotatory) or *D* (dextrorotatory) (Masters-Helfman and Bada, 1975). Almost without exception, amino acids are initially synthesized as *L*-enantiomers; however, because this uniform state is thermodynamically unstable, over time, *L*-amino acids "racemize" or change their configuration to become *D*-enantiomers, eventually yielding a racemic mixture with a *D/L* ratio of 1.0 (Masters-Helfman and Bada, 1975). The rate at which this occurs is partly dependent upon amino acid type, as structural differences can affect racemization rate (Kvenvolden et al., 1973); for example, whereas racemization half-life at 20°C is 15,000 years for aspartic acid, for isoleucine, it is 100,000 years (Bada et al., 1974). Furthermore, ambient pH can also have an effect, as racemization is facilitated when hydrolysis of peptide bonds occurs (Kvenvolden et al., 1973). Finally, and perhaps most significantly, racemization rates are strongly influenced by thermal environment, progressing more rapidly with increasing temperature (Bada, 1985). For example, aspartic acid exhibits a half-life of a few days at ~100°C, several thousand years at 26°C–27°C, and up to 10,000 years at 18°C, suggesting that substantial differences can result from even relatively small temperature variations (Bada, 1985).

The comparatively high racemization rate of aspartic acid relative to other amino acids makes it most appropriate for assessing age in more recent biological tissues. *D/L* ratios for aspartic acid, as well as other amino acids, are typically measured through comparison of the peak heights and/or peak areas for the two enantiomers yielded by chromatographic analysis (Bada and Protsch, 1973;

Kvenvolden et al., 1973; Ohtani and Yamamoto, 2005). To maximize the probability of observing a change in the $D/L$ ratio, study organisms must exhibit long lifespans and high body temperatures (Masters-Helfman and Bada, 1975). The probability of detecting racemization in a tissue is also increased greatly if it exhibits low metabolic activity and remains structurally stable throughout the organism's life (Masters-Helfman and Bada, 1975). Tissues in teeth display these characteristics (Yekkala et al., 2006), and aspartic acid racemization rates in teeth calibrated to body temperature have been used to estimate age in humans (Masters-Helfman and Bada, 1975; Ohtani and Yamamoto, 2005) and narwhals (*Monodon monoceros*; Bada et al., 1983). The latter study analyzed both teeth internal to the mouth, as well as the long male "tusk" that erupts externally, and results highlighted the importance of temperature history. Whereas racemization for the internal teeth experiencing fairly constant body temperature was measurable and reasonably consistent with age as estimated through tooth growth layer group (GLG) counts, virtually no racemization was observed in the erupted tusk constantly exposed to cold water temperatures (Bada et al., 1983). Statistical correspondence between known age and that estimated through calibrated aspartic acid racemization in teeth has been fairly good (summarized in Ohtani and Yamamoto, 2005); however, discrepancies between actual and estimated age may be as high as 10–15 years (Masters-Helfman and Bada, 1975).

More relevant to sea turtles, the lens of the vertebrate eye, whose development and structure is highly conserved among taxa, is another tissue with characteristics suitable for amino acid racemization study (Bloemendal, 1977). Lens formation begins early in embryonic development, and, while the first cells formed (i.e., the lens placode) comprise a foundation for accretion of additional cell layers, the lens nucleus itself remains isolated and inert, retaining embryonic cells throughout life (Bloemendal, 1977). Aspartic acid $D/L$ ratios in human eye lenses showed a high degree of correlation with known age; however, age ranges corresponding with very similar $D/L$ measurements could potentially span up to 30 years (Masters et al., 1977). Fin whale (*Balaenoptera physalus*) ages estimated through aspartic acid racemization within eye lenses were strongly correlated to counts of ear plug laminations thought to relate to age (Nerini, 1983). Due to apparent similarity between the human and fin whale lens racemization rates, these values were combined to estimate age from bowhead whales (*Balaena mysticeta*) eye lens nuclei (George et al., 1999). Although age estimates increased with body length, standard error did as well; interestingly, in some cases, ages estimated for bowhead whales of similar lengths differed by >100 years, and the highest age estimate was $211 \pm 35$ years (George et al., 1999). Later analyses using aspartic acid racemization rates specific to bowhead whales yielded somewhat lower age estimates for adults, with a maximum estimate of $145.7 \pm 23.2$ years (Rosa et al., 2004). Fin whale racemization rates were also applied to minke whales (*Balaenoptera acutorostra*) (Olsen and Sunde, 2002) and used to supplement narwhal racemization data (Garde et al., 2007) to estimate age and longevity for the two species. However, concerns were raised regarding measurement repeatability, potential for age underestimation, and significant errors in age estimation due to cross species application of racemization rates (Olsen and Sunde, 2002; Garde et al., 2010). For harp seals (*Pagophilus groenlandicus*), ages estimated through calibrated, species-specific lens aspartic acid racemization and GLG counts corresponded fairly closely (Garde et al., 2010).

For sea turtles, it has been proposed that characterization of aspartic acid racemization in the nucleus of the eye lens might provide insight into age parameters in a manner similar to that observed in mammal studies (Frazier, 1982). As in other vertebrates, the embryonic cells retained in sea turtle lens nuclei would be suitable for this type of analysis. Furthermore, longevity for some sea turtle species, as estimated through other means such as mark–recapture and skeletochronology, is sufficient to allow for the possibility that racemization might be observed. However, as sea turtles are predominantly ectothermic (Spotila et al., 1997) and yet highly migratory (Musick and Limpus, 1997; Plotkin, 2003), with large interindividual differences in movements and habitat use, thermal exposure patterns would be highly variable. Given the significant influence of temperature on racemization rates and the improbability of characterizing thermal histories of individual lenses, it is unlikely that this technique can be effectively applied to sea turtle species.

## 5.3  SUMMARY

As described throughout the chapter, a number of age-estimation techniques have been proposed for sea turtles, and while some have the potential to provide and/or validate absolute age estimates (e.g., mark–recapture, skeletochronology, bomb-radiocarbon), others may only yield relative ages (e.g., telomere shortening, amino acid racemization). The information presented herein suggests that no single, optimal age-estimation technique for sea turtles is currently available. Growth data yielded by mark–recapture studies are, of course, reflective of wild populations. However, differences in recapture intervals combined with intra- and interindividual variability in growth intervals, growth rates, and habitat use can confound population-wide inference of stage durations and/or ages. This is of particular concern when sample sizes are small due to the low recapture rates that are often characteristic of sea turtle mark–recapture studies. Recovery of adult sea turtles marked as hatchlings or head-started juveniles can potentially be very informative, provided that such tagging studies are of sufficient duration to allow characterization of the entire age distribution at maturation, not just the early returns. Although captive growth studies increase accessibility of sea turtles for growth rate measurements, rearing often occurs under optimal growth conditions with respect to temperature, food quantity, and activity level, yielding data that are not necessarily representative of wild turtles. Both mark–recapture and captive studies require large investments of time, logistical support, and funding, in order to realize results. In contrast, skeletochronology offers the potential for relatively rapid, broad characterization of population-wide age and growth parameters, but only when the necessary assumptions are addressed for appropriate application of the technique. Even when the method is correctly applied, the issue of potential parameter bias produced by the sampling of stranded turtles dead of unknown causes remains a concern. While length-frequency analysis can also yield information more quickly than direct growth observations, for sea turtles, the approach is confounded by a priori assumption of the number of age classes in the population and the requirement of von Bertalanffy-type growth. In addition, the highly variable growth rates relative to size and age and considerable decrease in adult growth observed for sea turtles may obscure association between size and cohort. The current state of telomere length age-estimation research suggests that its potential use is contingent upon further, significant refinement of the method, and the dependence of aspartic acid racemization technique on constant temperature history invalidates the use of this method for highly migratory, ectothermic organisms, such as sea turtles. As demonstrated previously for various fish and invertebrate species, analyses of bomb-radiocarbon (bomb-$^{14}$C) in sea turtles may prove to be very informative for independent age estimation as well as validation of skeletochronological techniques. However, success of the method depends on access to archived and/or recent samples retaining growth increments deposited during the 1950s and 1960s, as well as significant financial support, as these analyses are quite expensive (i.e., U.S. \$275 to \$1000 per sample; Campana and Jones, 1998; Baker and Wilson, 2001; Kerr et al., 2005; Piner et al., 2005). In time, perhaps, new technologies or analytical approaches (e.g., age estimation from T-cell DNA rearrangements; Zubakov et al., 2010) may offer additional insight into sea turtle age and longevity. However, in the interim, given that the greatest confidence in a result is generated when multiple lines of inquiry converge upon the same solution, collection of age and growth data through mark–recapture and skeletochronology should be continued and supplemented with bomb-$^{14}$C analyses.

## ACKNOWLEDGMENTS

The authors gratefully acknowledge L. Goshe and A. Hohn, as well as anonymous reviewers, for their constructive comments and suggestions to improve this manuscript.

# REFERENCES

Andrews, A. H., E. J. Burton, L. A. Kerr et al. 2005. Bomb radiocarbon and lead-radium disequilibria in otoliths of bocaccio rockfish (*Sebastes paucispnins*): A determination of age and longevity for a difficult-to-age fish. *Marine and Freshwater Research* 56:517–528.

Andrews, A. H., J. M. Kalish, S. J. Newman, and J. M. Johnston. 2011. Bomb radiocarbon dating of three important reef-fish species using Indo-Pacific $\Delta^{14}C$ chronologies. *Marine and Freshwater Research* 62:1259–1269.

Andrews, A. H., L. A. Kerr, G. M. Cailliet, T. A. Brown, C. C. Lundstrom, and R. D. Stanley. 2007. Age validation of canary rockfish (*Sebastes pinniger*) using two independent otolith techniques: Lead-radium and bomb radiocarbon dating. *Marine and Freshwater Research* 58:531–541.

Andrews, A. H., D. M. Tracey, and M. R. Dunn. 2009. Lead-radium dating of orange roughy (*Hoplostethus atlanticus*): Validation of a centenarian life span. *Canadian Journal of Fisheries and Aquatic Sciences* 66:1130–1140.

Ardizzone, D., G. M. Cailliet, L. J. Natanson, A. H. Andrews, L. A. Kerr, and T. A. Brown. 2006. Application of bomb radiocarbon chronologies to shortfin mako (*Isurus oxyrinchus*) age validation. *Environmental Biology of Fishes* 77:355–366.

Armsworthy, S. L. and S. E. Campana. 2010. Age determination, bomb-radiocarbon validation and growth of Atlantic halibut (*Hippoglossus hippoglossus*) from the Northwest Atlantic. *Environmental Biology of Fishes* 89:279–295.

Avens, L. and L. R. Goshe. 2007. Comparative skeletochronological analysis of Kemp's ridley (*Lepidochelys kempii*) and loggerhead (*Caretta caretta*) humeri and sclera ossicles. *Marine Biology* 152:1309–1317.

Avens, L., L. R. Goshe, C. A. Harms et al. 2012. Population characteristics, age structure, and growth dynamics of neritic juvenile green turtles (*Chelonia mydas*) in the northeastern Gulf of Mexico. *Marine Ecology Progress Series* 458:213–229.

Avens, L., J. C. Taylor, L. R. Goshe, T. T. Jones, and M. Hastings. 2009. Use of skeletochronological analysis to estimate the age of leatherback sea turtles *Dermochelys coriacea* in the western North Atlantic. *Endangered Species Research* 8:165–177.

Bada, J. L. 1985. Amino acid racemization dating of fossil bones. *Annual Review of Earth and Planetary Sciences* 13:241–268.

Bada, J. L., E. Mitchell, and B. Kemper. 1983. Aspartic acid racemization in narwhal teeth. *Nature* 303:418–420.

Bada, J. L. and R. Protsch. 1973. Racemization reaction of aspartic acid and its use in dating fossil bones. *Proceedings of the National Academy of Sciences of the United States of America* 70:1331–1334.

Bada, J. L., R. A. Schroeder, R. Protsch, and R. Berger. 1974. Concordance of collagen-based radiocarbon and aspartic-acid racemization ages. *Proceedings of the National Academy of Sciences of the United States of America* 71:914–917.

Baker, M. S., Jr. and C. A. Wilson. 2001. Use of bomb radiocarbon to validate otolith section ages of red snapper *Lutjanus campechanus* from the northern Gulf of Mexico. *Limnology and Oceanography* 46:1819–1824.

Balazs, G. H. 1999. Factors to consider in the tagging of sea turtles. In *Research and Management Techniques for the Conservation of Sea Turtles*, eds. K. L. Eckert, K. A. Bjorndal, F. A. Abreu-Grobois, and M. Donnelly, pp. 101–109. Washington, DC: IUCN/SSC Marine Turtle Specialist Group Publication No. 4.

Balazs, G. H. and M. Chaloupka. 2004a. Thirty-year recovery trend in the once depleted Hawaiian green sea turtle stock. *Biological Conservation* 117:491–498.

Balazs, G. H. and M. Chaloupka. 2004b. Spatial and temporal variability in somatic growth of green sea turtles (*Chelonia mydas*) resident in the Hawaiian Archipelago. *Marine Biology* 145:1043–1059.

Beamish, R. J. and G. A. McFarlane. 1987. Current trends in age determination methodology. In *Age and Growth of Fish*, eds. R. C. Summerfelt and G. E. Hall, pp. 15–42. Ames, IA: Iowa State University Press.

Bell, C. D. L., J. Parsons, R. J. Austin, A. C. Broderick, G. Ebanks-Petrie, and B. J. Godley. 2005. Some of them came home: The Cayman Turtle Farm headstarting project for the green turtle *Chelonia mydas*. *Oryx* 39:137–148.

Bels, B., F. Rimbolt-Baly, and J. Lescure. 1988. Croissance et maintien en captivité, de la tortue luth *Dermochelys coriacea* (Vandelli, 1761). *Revue Francaise d'Aquariologie* 15:59–64.

Bickerman, K. 2011. A skeletochronological analysis of a stranded population of loggerhead (*Caretta caretta*) sea turtles in Baja California Sur, Mexico. Masters thesis, Columbia University, New York.

Birkenmeier, E. 1971. Juvenile leathery turtles, *Dermochelys coriacea* (Linnaeus), in captivity. *Brunei Museum Journal* 2:160–172.

Bize, P., F. Criscuolo, N. B. Metcalfe, L. Nasir, and P. Monaghan. 2009. Telomere dynamics rather than age predict life expectancy in the wild. *Proceedings of the Royal Society of London, B* 276:1679–1683.

Bjorndal, K. A. and A. B. Bolten. 1988a. Growth rates of juvenile loggerheads, *Caretta caretta*, in the southern Bahamas. *Journal of Herpetology* 22:480–482.

Bjorndal, K. A. and A. B. Bolten. 1988b. Growth rates of immature green turtles, *Chelonia mydas*, on feeding grounds in the southern Bahamas. *Copeia* 1988:555–564.

Bjorndal, K. A. and A. B. Bolten. 2010. Hawksbill sea turtles in seagrass pastures: Success in a peripheral habitat. *Marine Biology* 157:135–145.

Bjorndal, K. A., A. B. Bolten, R. A. Bennett, E. R. Jacobson, R. J. Wronski, J. J. Valeski, and P. J. Eliazar. 1998. Age and growth in sea turtles: Limitations of skeletochronology for demographic studies. *Copeia* 1998:23–30.

Bjorndal, K. A., A. B. Bolten, and M. Y. Chaloupka. 2000a. Green turtle somatic growth model: Evidence for density dependence. *Ecological Applications* 10:269–282.

Bjorndal, K. A., A. B. Bolten, A. L. Coan Jr., and P. Kleiber. 1995. Estimation of green turtle (*Chelonia mydas*) growth rates from length-frequency analysis. *Copeia* 1995:71–77.

Bjorndal, K. A., A. B. Bolten, T. Dellinger, C. Delgado, and H. R. Martins. 2003. Compensatory growth in oceanic loggerhead sea turtles: Response to a stochastic environment. *Ecology* 84:1237–1249.

Bjorndal, K. A., A. B. Bolten, B. Koike, B. A. Schroeder, D. J. Shaver, W. G. Teas, and W. N. Witzell. 2001. Somatic growth function for immature loggerhead sea turtles, *Caretta caretta*, in southeastern U.S. waters. *Fishery Bulletin* 99:240–246.

Bjorndal, K. A., A. B. Bolten, and H. R. Martins. 2000b. Somatic growth model of juvenile loggerhead sea turtles *Caretta caretta*: Duration of pelagic stage. *Marine Ecology Progress Series* 202:265–272.

Blackburn, E. H. 1991. Structure and function of telomeres. *Nature* 350:569–573.

Bloemendal, H. 1977. The vertebrate eye lens. *Science* 197:127–138.

Bolten, A. B. 2003. Variation in sea turtle life history patterns: Neritic vs. oceanic developmental stages. In *The Biology of Sea Turtles*, Vol. II, eds. P. L. Lutz, J. A. Musick, and J. Wyneken, pp. 243–257. Boca Raton, FL: CRC Press.

Bolten, A. B., L. B. Crowder, M. G. Dodd et al. 2011. Quantifying multiple threats to endangered species: An example from loggerhead sea turtles. *Frontiers in Ecology and the Environment* 9:295–301.

Boulon, R. H., Jr. 1994. Growth rates of wild juvenile hawksbill turtles, *Eretmochelys imbricata*, in St. Thomas, United States Virgin Islands. *Copeia* 1994:811–814.

Braun-McNeill, J., S. P. Epperly, L. Avens, M. L. Snover, and J. C. Taylor. 2008. Growth rates of loggerhead sea turtles (*Caretta caretta*) from the western North Atlantic. *Herpetological Conservation and Biology* 3:273–281.

Broecker, W. S., T.-H. Peng, G. Ostlund, and M. Stuiver. 1985. The distribution of bomb radiocarbon in the ocean. *Journal of Geophysical Research* 90:6953–6970.

Bronikowski, A. M. 2008. The evolution of aging phenotypes in snakes: A review and synthesis with new data. *Age* 30:169–176.

Brothers, E. B., C. P. Mathews, and R. Lasker. 1976. Daily growth increments in otoliths from larval and adult fishes. *Fishery Bulletin* 74:1–8.

Bustard, R. 1976. Turtles of coral reefs and coral islands. In *Biology and Geology of Coral Reefs*, Vol. III, eds. O. A. Jones and R. Endean, pp. 343–368. New York: Academic Press.

Cagle, F. R. 1946. The growth of the slider turtle, *Pseudemys scripta elegans*. *American Midland Naturalist* 36:685–729.

Cailliet, G. M., W. D. Smith, H. F. Mollet, and K. J. Goldman. 2006. Age and growth studies of chrondrichthyan fishes: The need for consistency in terminology, verification, validation, and growth function fitting. *Environmental Biology of Fishes* 77:211–228.

Caillouet, C. W., Jr., D. J. Shaver, A. M. Landry, D. W. Owens, and C. H. Pritchard. 2011. Kemp's ridley sea turtle (*Lepidochelys kempii*) age at first nesting. *Chelonian Conservation and Biology* 10:288–293.

Caldwell, D. K. 1962. Growth measurements of young captive Atlantic sea turtles in temperate waters. *Los Angeles County Museum Contributions in Science* 50:1–8.

Campana, S. E. 1990. How reliable are growth back-calculations based on otoliths? *Canadian Journal of Fisheries and Aquatic Sciences* 47:2219–2227.

Campana, S. E. 1997. Use of radiocarbon from nuclear fallout as a dated marker in the otoliths of haddock *Melanogrammus aeglefinus*. *Marine Ecology Progress Series* 150:49–56.

Campana, S. E. 2001. Accuracy, precision and quality control in age determination, including a review of the use and abuse of age validation methods. *Journal of Fish Biology* 59:197–242.

Campana, S. E., J. M. Casselman, and C. M. Jones. 2008. Bomb radiocarbon chronologies in the Arctic, with implications for the age validation of lake trout (*Salvelinus namaycush*) and other Arctic species. *Canadian Journal of Fisheries and Aquatic Sciences* 65:733–743.

Campana, S. E. and C. M. Jones. 1998. Radiocarbon from nuclear testing applied to age validation of black drum, *Pogonias cromis. Fishery Bulletin* 96:185–192.

Campana, S. E., C. Jones, G. A. McFarlane, and S. Myklevoll. 2006. Bomb dating and age validation using the spines of spiny dogfish (*Squalus acanthus*). *Environmental Biology of Fishes* 77:327–336.

Campana, S. E., L. J. Natanson, and S. Myklevoll. 2002. Bomb dating and age determination of large pelagic sharks. *Canadian Journal of Fisheries and Aquatic Sciences* 59:450–455.

Carlander, K. D. 1987. A history of scale age and growth studies of North American freshwater fish. In *Age and Growth of Fish*, eds. R. C. Summerfelt and G. E. Hall, pp. 3–14. Ames, IA: Iowa State University Press.

Carr, A. 1968. *The Turtle: A Natural History*. London, U.K.: Cassell and Company.

Casale, P., P. P. d'Atore, and R. Argano. 2009a. Age at size and growth rates of early juvenile loggerhead sea turtles (*Caretta caretta*) in the Mediterranean based on length frequency analysis. *Herpetological Journal* 19:29–33.

Casale, P., N. Conte, D. Freggi, C. Cioni, and R. Argano. 2011a. Age and growth determination by skeletochronology in loggerhead sea turtles (*Caretta caretta*) from the Mediterranean Sea. *Scientia Marina* 75:197–203.

Casale, P., A. D. Mazaris, and D. Freggi. 2011b. Estimation of age at maturity of loggerhead sea turtles *Caretta caretta* in the Mediterranean using length-frequency data. *Endangered Species Research* 13:123–129.

Casale, P., A. D. Mazaris, D. Freggi, C. Vallini, and R. Argano. 2009b. Growth rates and age at adult size of loggerhead sea turtles (*Caretta caretta*) in the Mediterranean Sea, estimated through capture-mark-recapture. *Scientia Marina* 73:589–595.

Castanet, J. 2006. Time recording in bone microstructures of endothermic animals: Functional relationships. *General Palaeontology* 5:629–636.

Castanet, J., H. Francillon-Viellot, F. J. Meunier, and A. DeRicqles. 1993. Bone and individual aging. In *Bone. Bone Growth – B*, Vol. 7, ed. B. K. Hall, pp. 245–283. Boca Raton, FL: CRC Press.

Chaloupka, M., K. A. Bjorndal, G. H. Balazs et al. 2008. Encouraging outlook for recovery of a once severely exploited marine megaherbivore. *Global Ecology and Biogeography* 17:297–304.

Chaloupka, M. and C. Limpus. 1997. Robust statistical modeling of hawksbill sea turtle growth rates (southern Great Barrier Reef). *Marine Ecology Progress Series* 146:1–8.

Chaloupka, M., C. Limpus, and J. Miller. 2004. Green turtle somatic growth dynamics in a spatially disjunct Great Barrier Reef metapopulation. *Coral Reefs* 23:325–335.

Chaloupka, M. Y. and J. A. Musick. 1997. Age, growth and population dynamics. In *The Biology of Sea Turtles*, eds. P. L. Lutz and J. A. Musick, pp. 233–276. Boca Raton, FL: CRC Press.

Chaloupka, M. and G. R. Zug. 1997. A polyphasic growth function for the endangered Kemp's ridley sea turtle, *Lepidochelys kempii. Fishery Bulletin* 95:849–856.

Chatzinikolaou, E. and C. A. Richardson. 2007. Evaluating growth and age of netted whelk *Nassarius reticulates* (Gastropoda: Nassariidae) using statolith growth rings. *Marine Ecology Progress Series* 342:163–176.

Coles, W. C., J. A. Musick, and L. A. Williamson. 2001. Skeletochronology validation from an adult loggerhead (*Caretta caretta*). *Copeia* 2001:240–242.

Curtin, A. J., G. R. Zug, P. A. Medica, and J. R. Spotila. 2008. Assessing age in the desert tortoise *Gopherus agassizii*: Testing skeletochronology with individuals of known age. *Endangered Species Research* 5:21–27.

Deraniyagala, P. E. P. 1952. *A Colored Atlas of Some Vertebrates from Ceylon. Vol. I (Fishes)*. Colombo, Sri Lanka: The Ceylon Government Press.

Diez, C. E. and R. P. van Dam. 2002. Habitat effect on hawksbill turtle growth rates on feeding grounds at Mona and Monito Islands, Puerto Rico. *Marine Ecology Progress Series* 234:301–309.

Druffel, E. M. 1980. Radiocarbon in annual coral rings of Belize and Florida. *Radiocarbon* 22:363–371.

Druffel, E. M. and T. W. Linick. 1978. Radiocarbon in annual coral rings of Florida. *Geophysical Research Letters* 5:913–916.

Dunshea, G., D. Duffield, N. Gales, M. Hindell, R. S. Wells, and S. N. Jarman. 2011. Telomeres as age markers in vertebrate molecular ecology. *Molecular Ecology Resources* 11:225–235.

Dutton, D. L., P. H. Dutton, M. Chaloupka, and R. H. Boulon. 2005. Increase of a Caribbean leatherback turtle *Dermochelys coriacea* nesting population linked to long-term nest protection. *Biological Conservation* 126:186–194.

Eckert, S. A. 2002. Distribution of juvenile leatherback sea turtle *Dermochelys coriacea* sightings. *Marine Ecology Progress Series* 230:289–293.

Ehrhardt, L. M. and R. Witham. 1992. Analysis of growth of the green sea turtle (*Chelonia mydas*) in the western central Atlantic. *Bulletin of Marine Science* 50:275–281.

Ehrhart, L. M. and L. H. Ogren. 1999. Studies in foraging habitats: Capturing and handling turtles. In *Research and Management Techniques for the Conservation of Sea Turtles*, eds. K. L. Eckert, K. A. Bjorndal, F. A. Abreu-Grobois, and M. Donnelly, pp. 61–64. Washington, DC: IUCN/SSC Marine Turtle Specialist Group Publication No. 4.

Epel, E. S., E. H. Blackburn, J. Lin, F. S. Dhabhar, N. E. Adler, J. D. Morrow, and R. Cawthon. 2004. Accelerated telomere shortening in response to life stress. *Proceedings of the National Academy of Sciences of the United States of America* 101:17312–17315.

Fabens, A. J. 1965. Properties and fitting of the von Bertalanffy growth curve. *Growth* 29:265–289.

Fallon, S. J. and T. P. Guilderson. 2005. Extracting growth rates from the nonlaminated coralline sponge *Astrosclera willeyana* using bomb radiocarbon. *Limnology and Oceanography: Methods* 3:455–461.

Fossette, S., C. Girard, M. López-Mendilaharsu et al. 2010. Atlantic leatherback migratory paths and temporary residence areas. *PLoS One* 5:1–12.

Fournier, D. A., J. R. Sibert, J. Majkowski, and J. Hampton. 1990. MULTIFAN a likelihood-based method for estimating growth parameters and age composition from multiple length frequency data sets illustrated using data for southern bluefin tuna (*Thunnus maccoyii*). *Canadian Journal of Fisheries and Aquatic Science* 47:301–317.

Francis, R. I. C. C. 1990. Back-calculation of fish length: A critical review. *Journal of Fish Biology* 36:883–902.

Frantz, B. R., M. Kashgarian, K. H. Coale, and M. S. Foster. 2000. Growth rate and potential climate record from a rhodolith using $^{14}$C accelerator mass spectrometry. *Limnology and Oceanography* 45:1773–1777.

Frazer, N. G. and L. M. Ehrhart. 1985. Preliminary growth models for green, *Chelonia mydas*, and loggerhead, *Caretta caretta*, turtles in the wild. *Copeia* 1985:73–79.

Frazer, N. B., J. W. Gibbons, and J. L. Greene. 1990. Exploring Fabens' growth interval model with data on a long-lived vertebrate, *Trachemys scripta* (Reptilia: Testudinata). *Copeia* 1990:112–118.

Frazer, N. B. and R. C. Ladner. 1986. A growth curve for green sea turtles, *Chelonia mydas*, in the U.S. Virgin Islands 1913–1914. *Copeia* 1986:798–802.

Frazer, N. B. and F. J. Schwartz. 1984. Growth curves for captive loggerhead turtles, *Caretta caretta*, in North Carolina, USA. *Bulletin of Marine Science* 34:485–489.

Frazier, J. 1982. Age determination studies in marine turtles. Final report for contracts NA 81-GA-C-00018 and NA 81-GF-A-184 with National Marine Fisheries Service (Southeast Fisheries Science Center) and U.S. Fish and Wildlife Service (Region 2).

Friedrich, U., E.-U. Griese, M. Schwab, P. Fritz, K. P. Thon, and U. Kotz. 2000. Telomere length in different tissues of elderly patients. *Mechanisms of Ageing and Development* 119:89–99.

Gage, J. D. 1992. Growth bands in the sea urchin *Echinus esculentus*: Results from tetracycline-mark-recapture. *Journal of the Marine Biological Association of the United Kingdom* 72:257–260.

Galbraith, D. A. and R. J. Brooks. 1987. Addition of annual growth lines in adult snapping turtles *Chelydra serpentina. Journal of Herpetology* 21:359–363.

Garde, E., A. K. Frie, G. Dunshea, S. H. Hansen, K. M. Kovacs, and C. Lydersen. 2010. Harp seal ageing techniques—Teeth, aspartic acid racemization, and telomere sequence analysis. *Journal of Mammalogy* 91:1365–1374.

Garde, E., M. P. Heide-Jørgensen, S. H. Hansen, G. Nachman, and M. C. Forchammer. 2007. Age-specific growth and remarkable longevity in narwhals (*Monodon monoceros*) from west Greenland as estimated by aspartic acid racemization. *Journal of Mammalogy* 88:49–58.

George, J. C., J. Bada, J. Zeh, L. Scott, S. E. Brown, T. O'Hara, and R. Suydam. 1999. Age and growth estimates of bowhead whales (*Balaena mysticetus*) via aspartic acid racemization. *Canadian Journal of Zoology* 77:571–580.

Girondot, M. and J. Garcia. 1999. Senescence and longevity in turtles: What telomeres tell us. In *Proceedings of the 9th Ordinary General Meeting of the Societas Europaea Herpetologica in Current Studies in Herpetology*, eds. R. Guyétant and C. Miaud, pp. 133–137. Le Bourget du Lac, France: Université de Savoie.

Goldman, K. J., S. Branstetter, and J. A. Musick. 2006. A re-examination of the age and growth of sand tiger sharks, *Carcharias taurus*, in the western North Atlantic: The importance of ageing protocols and use of multiple back-calculation techniques. *Environmental Biology of Fishes* 77:241–252.

Gonzaléz, A. F., B. G. Castro, and A. Guerra. 1996. Age and growth of the short-finned squid *Illex coindetii* in Galician waters (NW Spain) based on statolith analysis. *ICES Journal of Marine Science* 53:802–810.

Goshe, L. R., L. Avens, J. Bybee, and A. A. Hohn. 2009. An evaluation of histological techniques used in skeletochronological age estimation of sea turtles. *Chelonian Conservation and Biology* 8:217–222.

Goshe, L. R., L. Avens, F. S. Scharf, and A. L. Southwood. 2010. Estimation of age at maturation and growth of Atlantic green turtles (*Chelonia mydas*) using skeletochronology. *Marine Biology* 157:1725–1740.

Greider, C. W. 1996. Telomere length regulation. *Annual Review of Biochemistry* 65:337–365.

Hall, M. E., L. Nasir, F. Daunt et al. 2004. Telomere loss in relation to age and early environment in long-lived birds. *The Proceedings of the Royal Society of London, B* 271:1571–1576.

Hare, J. A. and R. A. Cowen. 1995. Effect of age, growth rate, and ontogeny on the otolith size-fish size relationship in bluefish, *Pomatomus saltatrix*, and the implications for back-calculation of size in fish early life history stages. *Canadian Journal of Fisheries and Aquatic Sciences* 52:1909–1922.

Hastie, T. J. and R. J. Tibshirani. 1990. Generalized additive models. In *Monographs on Statistics and Applied Probability*, p. 43. London, U.K.: Chapman and Hall.

Hatase, H., K. Omuta, and K. Tsukamoto. 2010. Oceanic residents, neritic migrants: A possible mechanism underlying foraging dichotomy in adult female loggerhead turtles (*Caretta caretta*). *Marine Biology* 157:1337–1342.

Hatase, H., R. Rudo, K. K. Watanabe et al. 2008. Shorter telomere length with age in the loggerhead turtle: A new hope for live sea turtle age estimation. *Genes and Genetic Systems* 83:423–426.

Haussmann, M. F. and R. A. Mauck. 2008. New strategies for telomere-based age estimation. *Molecular Ecology Resources* 8:264–274.

Haussmann, M. F. and C. M. Vleck. 2002. Telomere length provides a new technique for aging animals. *Oecologia* 130:325–328.

Haussmann, M. F., C. M. Vleck, and I. C. T. Nisbet. 2003a. Calibrating the telomere clock in common terns, *Sterna hirundo*. *Experimental Gerontology* 38:787–789.

Haussmann, M. F., D. W. Winkler, C. E. Huntington, I. C. T. Nisbet, and C. M. Vleck. 2007. Telomerase activity is maintained throughout the lifespan of long-lived birds. *Experimental Gerontology* 42:610–618.

Haussmann, M. F., D. W. Winkler, K. M. O'Reilly, C. E. Huntington, I. C. T. Nisbet, and C. M. Vleck. 2003b. Telomeres shorten more slowly in long-lived birds and mammals than in short-lived ones. *Proceedings of the Royal Society of London, B* 270:1387–1392.

Heidinger, B. J., J. D. Blount, W. Boner, K. Griffiths, N. B. Metcalfe, and P. Monaghan. 2011. Telomere length in early life predicts lifespan. *Proceedings of the National Academy of Sciences* 109:1743–1748.

Hendrickson, J. R. 1958. The green sea turtle, *Chelonia mydas* (Linn.) in Malaya and Sarawak. *Proceedings of the Zoological Society of London* 130:455–535.

Hendrickson, L. P. and J. R. Hendrickson. 1981. A new method for marking sea turtles? *Marine Turtle Newsletter* 19:6–7.

Heppell, S. S., D. T. Crouse, L. B. Crowder et al. 2005. A population model to estimate recovery time, population size, and management impacts on Kemp's ridley sea turtles. *Chelonian Conservation and Biology* 4:767–773.

Heppell, S. S., L. B. Crowder, D. T. Crouse, S. P. Epperly, and N. B. Frazer. 2003. Population models for Atlantic loggerheads: Past, present and future. In *Loggerhead Sea Turtles*, eds. A. B. Bolton and B. E. Witherington, pp. 255–273. Washington, DC: Smithsonian Books.

Heppell, S. S., M. L. Snover, and L. B. Crowder. 2002. Sea turtle population ecology. In *The Biology of Sea Turtles*, Vol. II, eds. P. Lutz, J. A. Musick, and J. Wyneken, pp. 275–306. Boca Raton, FL: CRC Press.

Higgins, B. M., B. A. Robertson, and T. D. Williams. 1997. Manual for mass wire tagging of hatchling sea turtles and the detection of internal wire tags. NOAA Technical Memorandum NMFS-SEFSC-402.

Hirth, H. F. 1971. *Synopsis of Biological Data on the Green Turtle, Chelonia mydas (Linnaeus) 1758*. FAO Fisheries Synopsis No. 85, Rome, Italy.

Hohn, A. A. 1980. Analysis of growth layers in the teeth of *Tursiops truncatus*, using light microscopy, microradiography, and SEM. Report of the International Whaling Commission (Special Issue No. 3:155–160).

Hohn, A. A. 2009. Age estimation. In *Encyclopedia of Marine Mammals*, 2nd edn., eds. W. F. Perrin, B. Würsig, and J. G. M. Thewissen, pp. 11–17. San Diego, CA: Academic Press.

Jackson, J. B. C., M. X. Kirby, W. H. Berger et al. 2001. Historical overfishing and the recent collapse of coastal ecosystems. *Science* 293:629–638.

Jakob, C., A. Seitz, A. J. Criveeli, and C. Miaud. 2002. Growth cycle of the marbled newt (*Triturus marmoratus*) in the Mediterranean region assessed by skeletochronology. *Amphibia-Reptilia* 23:407–418.

Jones, T. T., M. D. Hastings, B. L. Bostrom, D. Pauly, and D. R. Jones. 2011. Growth of captive leatherback turtles, *Dermochelys coriacea*, with inferences on growth in the wild: Implications for population decline and recovery. *Journal of Experimental Marine Biology and Ecology* 399:84–92.

Juola, F. A., M. F. Haussmann, D. C. Dearborn, and C. M. Vleck. 2006. Telomere shortening in a long-lived marine bird: Cross-sectional analysis and test of an aging tool. *The Auk* 123:775–783.

Kalish, J. M. 1991. C-13 and O-18 isotopic disequilibria in fish otoliths: Metabolic and kinetic effects. *Marine Ecology Progress Series* 75:191–203.

Kalish, J. M. 1993. Pre- and post-bomb radiocarbon in fish otoliths. *Earth and Planetary Science Letters* 114:549–554.

Kalish, J. M. 1995a. Radiocarbon and fish biology. In *Recent Developments in Fish Otolith Research*, eds. D. H. Secor, J. M. Dean, and S. E. Campana, pp. 637–653. Columbia, SC: University of South Carolina Press.

Kalish, J. M. 1995b. Application of the bomb radiocarbon chronometer to the validation of redfish *Centroberyx affinis* age. *Canadian Journal of Fisheries and Aquatic Sciences* 52:1399–1405.

Kalish, J. M., J. M. Johnston, J. S. Gunn, and N. P. Clear. 1996. Use of the bomb radiocarbon chronometer to determine age of southern bluefin tuna *Thunnus maccoyii*. *Marine Ecology Progress Series* 143:1–8.

Kalish, J. M., R. Nydal, K. H. Nedreaas, G. S. Burr, and G. L. Eine. 2001. A time history of pre- and post-bomb radiocarbon in the Barents Sea derived from Arcto-Norwegian cod otoliths. *Radiocarbon* 43:843–855.

Kamezaki, N., Y. Matsuzawa, O. Abe et al. 2003. Loggerhead turtles nesting in Japan. In *Loggerhead Sea Turtles*, eds. A. B. Bolten and B. E. Witherington, p. 319. Washington, DC: Smithsonian Institution Press.

Kerr, L. A., A. H. Andrews, G. M. Cailliet, T. A. Brown, and K. H. Coale. 2006. Investigations of $\Delta^{14}C$, $\delta^{13}C$, and of $\delta^{15}N$ in vertebrae of white shark (*Carcharodon carcharias*) from the eastern North Pacific Ocean. *Environmental Biology of Fishes* 77:337–353.

Kerr, L. A., A. H. Andrews, B. R. Frantz, K. H. Coale, T. A. Brown, and G. M. Cailliet. 2004. Radiocarbon in otoliths of yelloweye rockfish (*Sebastes ruberrimus*): A reference time series for the coastal waters of southeast Alaska. *Canadian Journal of Fisheries and Aquatic Sciences* 61:443–451.

Kerr, L. A., A. H. Andrews, K. Munk et al. 2005. Age validation of quillback rockfish (*Sebastes maliger*) using bomb radiocarbon. *Fishery Bulletin* 103:97–107.

Kerr, R., E. Kintisch, and E. Stokstrad. 2010. Will *Deepwater Horizon* set a new standard for catastrophe? *Science* 328:674–675.

Kilada, R. W., S. E. Campana, and D. Roddick. 2007. Validated age, growth, and mortality estimates of the ocean quahog (*Arctica islandica*) in the western Atlantic. *ICES Journal of Marine Science* 64:31–38.

Kilada, R. W., S. E. Campana, and D. Roddick. 2009. Growth and sexual maturity of the northern propellerclam (*Cyrtodaria siliqua*) in eastern Canada, with bomb radiocarbon age validation. *Marine Biology* 156:1029–1037.

Kim, Y. J., V. K. Subramani, and S. H. Sohn. 2011. Age prediction in the chickens using telomere quantity by quantitative fluorescence in situ hybridization technique. *Asian-Australian Journal of Animal Science* 24:603–609.

Klevezal, G. A. 1996. *Recording Structures of Mammals: Determination of Age and Reconstruction of Life History* (M. V. Mina and A. V. Oreshkin, translators). Brookfield, VT: A. A. Balkema Publishers.

Klinger, R. C. and J. A. Musick. 1992. Annular growth layers in juvenile loggerhead turtles (*Caretta caretta*). *Bulletin of Marine Science* 51:224–230.

Klinger, R. C. and J. A. Musick. 1995. Age and growth of loggerhead turtles (*Caretta caretta*) from Chesapeake Bay. *Copeia* 1995:204–209.

Kobayashi, M. 2000. An analysis of the growth based on the size and age distributions of the hawksbill sea turtle inhabiting Cuban waters. *Japanese Journal of Veterinary Research* 48:129–135.

Kubis, S., M. Chaloupka, L. Ehrhart, and M. Bresette. 2009. Growth rates of juvenile green turtles *Chelonia mydas* from three ecologically distinct foraging habitats along the east central coast of Florida, USA. *Marine Ecology Progress Series* 389:257–269.

Kvenvolden, K. A., E. Peterson, J. Wehmiller, and P. E. Hare. 1973. Racemization of amino acids in marine sediments determined by gas chromatography. *Geochimica et Cosmochimica Acta* 37:2215–2225.

Lagarde, F., X. Bonnet, B. T. Henen, J. Corbin, K. A. Nagy, and G. Naulleau. 2001. Sexual size dimorphism in steppe tortoises (*Testudo horsfieldi*): Growth, maturity, and individual variation. *Canadian Journal of Zoology* 79:1433–1441.

Law, R. 2000. Fishing, selection, and phenotypic evolution. *ICES Journal of Marine Science* 57:659–668.

Law, R. and D. R. Grey. 1989. Evolution of yields from populations with age-specific cropping. *Evolutionary Ecology* 3:343–359.

Lewison, R. L., S. A. Freeman, and L. B. Crowder. 2004. Quantifying the effects of fisheries on threatened species: The impact of pelagic longlines on loggerhead and leatherback sea turtles. *Ecology Letters* 7:221–231.

Limpus, C. J. 2007. *A Biological Review of Australian Marine Turtle Species. 5. Flatback Turtle, Natator depressus (Garman)*. The State of Queensland, Australia: Environmental Protection Agency.

Limpus, C. J. and M. Y. Chaloupka. 1997. Nonparametric regression modelling of green sea turtle growth rates (southern Great Barrier Reef). *Marine Ecology Progress Series* 149:23–34.

Limpus, C. J. and D. J. Limpus. 2003. Biology of the loggerhead turtle in western South Pacific Ocean foraging areas. In *Loggerhead Sea Turtles*, eds. A. B. Bolton and B. E. Witherington, pp. 93–113. Washington, DC: Smithsonian Books.

Lipinski, M. 1986. Methods for the validation of squid age from statoliths. *Journal of the Marine Biological Association of the United Kingdom* 66:505–526.

Lopez-Castro, M. C., V. Koch, A. Mariscal-Loza, and W. J. Nichols. 2010. Long-term monitoring of black turtles *Chelonia mydas* at coastal foraging areas off the Baja California Peninsula. *Endangered Species Research* 11:35–45.

Lynnerup, N., H. Kjeldsen, S. Heegaard, C. Jacobsen, and J. Heinimeier. 2008. Radiocarbon dating of the human eye lens crystallines reveal proteins without carbon turnover throughout life. *PLoS One* 3: e1529. doi: 10.1371/journal.pone.0001529.

Mansfield, K. L., V. S. Saba, J. A. Keinath, and J. A. Musick. 2009. Satellite tracking reveals a dichotomy in migration strategies among juvenile loggerhead turtles in the Northwest Atlantic. *Marine Biology* 156:2555–2570.

Marschal, C., J. Garrabou, J. G. Harmelin, and M. Pichon. 2004. A new method for measuring growth and age in the precious red coral *Corallium rubrum* (L.). *Coral Reefs* 23:423–432.

Masters, P. M., J. L. Bada, and J. S. Zigler, Jr. 1977. Aspartic acid racemisation in the human lens during ageing and in cataract formation. *Nature* 268:71–73.

Masters-Helfman, P. and J. L. Bada. 1975. Aspartic acid racemization in tooth enamel from living humans. *Proceedings of the National Academy of Sciences of the United States of America* 72:2891–2894.

Mazaris, A. D., A. S. Kallimanis, J. Tzanopoulos, S. P. Sgardelis, and J. D. Pantis. 2009. Sea surface temperature variations in core foraging grounds drive nesting trends and phenology of loggerhead turtles in the Mediterranean Sea. *Journal of Experimental Marine Biology and Ecology* 379:23–27.

McClellan, C. M. and A. J. Read. 2007. Complexity and variation in loggerhead sea turtle life history. *Biology Letters* 3(6):592–594. doi:10.1098/rsbl.2007.0355.

Mendonça, M. T. 1981. Comparative growth rates of wild immature *Chelonia mydas* and *Caretta caretta* in Florida. *Journal of Herpetology* 15:447–451.

Meyne, J., R. L. Tarliff, and R. K. Moyzis. 1989. Conservation of the human telomere sequence $(TTAGGG)_n$ among vertebrates. *Proceedings of the National Academy of Sciences of the United States of America* 86:7049–7053.

del Monte-Luna, P., V. Guzman-Hernandez, E. A. Cuevas, F. Arreguin-Sanchez, and D. Lluch-Belda. 2012. Effect of North Atlantic climate variability on hawksbill turtles in the southern Gulf of Mexico. *Journal of Experimental Marine Biology and Ecology* 412:103–109.

Moore, J. E., B. P. Wallace, R. L. Lewison, R. Žydelis, T. M. Cox, and L. B. Crowder. 2009. A review of marine mammal, sea turtle and seabird bycatch in USA fisheries and the role of policy in shaping management. *Marine Policy* 33:435–451.

Musick, J. A. and C. J. Limpus. 1997. Habitat utilization and migration in juvenile sea turtles. In *The Biology of Sea Turtles*, eds. P. L. Lutz and J. A. Musick, pp. 137–164. Boca Raton, FL: CRC Press.

Nakagawa, S., N. J. Gemmell, and T. Burke. 2004. Measuring vertebrate telomeres: Applications and limitations. *Molecular Ecology* 13:2523–2533.

National Marine Fisheries Service Southeast Fisheries Science Center (NMFS SEFSC). 2001. Stock assessments of loggerhead and leatherback sea turtles and an assessment of the impact of the pelagic longline fishery on the loggerhead and leatherback sea turtles of the Western North Atlantic. U.S. Department of Commerce NOAA Technical Memorandum NMFS-SEFSC-455.

National Research Council (NRC). 2010. *Assessment of Sea-Turtle Status and Trends: Integrating Demography and Abundance*. Washington, DC: The National Academies Press.

Nerini, N. K. 1983. Age determination of fin whales (*Balaenoptera physalus*) based upon aspartic acid racemization in the lens nucleus. Report of the International Whaling Commission 33, pp. 447–448.

Nydal, R. and J. S. Gislefoss. 1996. Further application of bomb $^{14}C$ as a tracer in the atmosphere and ocean. *Radiocarbon* 38:389–406.

Nydal, R., S. Gulliksen, K. Løvseth, and F. Skogseth. 1984. Bomb $^{14}C$ in the ocean surface 1966–1981. *Radiocarbon* 26:7–45.

Ohtani, S. and T. Yamamoto. 2005. Strategy for the estimation of chronological age using the aspartic acid racemization method with special reference to coefficient of correlation between D/L ratios and ages. *Journal of Forensic Science* 50:1–8.

Olgun, K., N. Uzum, A. Avci, and C. Miaud. 2005. Age, size, and growth of the southern crested newt *Triturus karelinii* (Strauch 1870) in a population from Bozdag (Western Turkey). *Amphibia-Reptilia* 26:223–230.

Olsen, E. M., M. Heino, G. R. Lilly et al. 2004. Maturation trends indicative of rapid evolution preceded the collapse of northern cod. *Nature* 428:932–935.

Olsen, E. and J. Sunde. 2002. Age determination of minke whales (*Balaenoptera acutorostrata*) using the aspartic acid racemization technique. *Sarsia* 87:1–8.

Palaniappan, P. M. 2007. The carapacial scutes of hawksbill turtles (*Eretmochelys imbricate*): Development, growth dynamics, and utility as an age indicator. PhD thesis (unpublished), Charles Darwin University, Alice Springs, Australia.

Panfili, J., H. de Pontual, H. Troadec, and P. J. Wright, eds. 2002. *Manual of Fish Sclerochronology*. Brest, France: Ifremer-IRD Coedition.

Parham, J. F. and G. R. Zug. 1997. Age and growth of loggerhead sea turtles (*Caretta caretta*) of coastal Georgia: An assessment of skeletochronological age-estimates. *Bulletin of Marine Science* 61:287–304.

Pauly, D. and G. R. Morgan. 1987. Length-based methods in fisheries research. *ICLARM Conference Proceedings*, International Center for Living Aquatic Resources Management, Manila, Phillipines, and Kuwait Institute for Scientific Research, Safat, Kuwait, Vol. 13, 468 pp.

Peng, T.-H. and W. Broecker. 1980. Gas exchange rates for three closed-basin lakes. *Limnology and Oceanography* 25:789–796.

Peterson, B. J. and B. Fry. 1987. Stable isotopes in ecosystem studies. *Annual Review of Ecology and Systematics* 18:293–320.

Petitet, R., E. R. Secchi, L. Avens, and P. G. Kinas. 2012. Age and growth of loggerhead sea turtle (*Caretta caretta*) in southern Brazil. *Marine Ecology Progress Series* 456:255–268.

Pilcher, N. 2010. Population structure and growth of immature green turtles at Mantanani, Sabah, Malaysia. *Journal of Herpetology* 44:168–171.

Piner, K. R., O. S. Hamel, J. L. Menkel, J. R. Wallace, and C. E. Hutchinson. 2005. Age validation of canary rockfish (*Sebastes pinniger*) from off the Oregon coast (USA) using the bomb radiocarbon method. *Canadian Journal of Fisheries and Aquatic Sciences* 62:1060–1066.

Piner, K. R., J. R. Wallace, O. Hamel, and R. Mikus. 2006. Evaluation of ageing accuracy of bocaccio (*Sebastes paucispinis*) rockfish using bomb radiocarbon. *Fisheries Research* 77:200–206.

Piner, K. R. and S. G. Wischniowski. 2004. Pacific halibut chronology of bomb radiocarbon in otoliths from 1944 to 1981 and a validation of ageing methods. *Journal of Fish Biology* 64:1060–1071.

Piovano, S., M. Clusa, C. Carreras, C. Giacoma, M. Pascual, and L. Cardona. 2011. Different growth rates between loggerhead sea turtles (*Caretta caretta*) of Mediterranean and Atlantic origin in the Mediterranean Sea. *Marine Biology* 158:2577–2587.

Plotkin, P. 2003. Adult migrations and habitat use. In *The Biology of Sea Turtles*, Vol. II, eds. P. Lutz, J. A. Musick, and J. Wyneken, pp. 225–242. Boca Raton, FL: CRC Press.

Reich, K. J., K. A. Bjorndal, and A. B. Bolten. 2007. The "lost years" of green turtles: Using stable isotopes to study cryptic lifestages. *Biology Letters* 3:712–714.

Reimer, P. J., T. A. Brown, and R. W. Reimer. 2004. Discussion: Reporting and calibration of post-bomb $^{14}$C data. *Radiocarbon* 46:1299–1304.

Richardson, C. R. 2001. Molluscs as archives of environmental change. *Oceanography and Marine Biology An Annual Review* 39:103–164.

Roark, E. B., T. P. Guilderson, R. B. Dunbar, and B. L. Ingram. 2006. Radiocarbon-based ages and growth rates of Hawaiian deep-sea corals. *Marine Ecology Progress Series* 327:1–14.

Roark, E. B., T. P. Guilderson, S. Flood-Page et al. 2005. Radiocarbon-based ages and growth rates of bamboo corals from the Gulf of Alaska. *Geophysical Research Letters* 32:L04606. doi:04610.01029/02004FL021919.

Rodhouse, P. G. and E. M. C. Hatfield. 1990. Age determination in squid using statolith growth increments. *Fisheries Research* 8:323–334.

Rosa, C., J. C. George, J. Zeh, T. M. O'Hara, O. Botta, and J. L. Bada. 2004. Update on age estimation of bowhead whales (*Balaena mysticetus*) using aspartic acid racemization. IWC/SC/56/BRG6. Cambridge, U.K.: International Whaling Commission.

Saba, V. S., P. Santidrián-Tomillo, R. D. Reina et al. 2007. The effect of the El Niño Southern Oscillation on the reproductive frequency of eastern Pacific leatherback turtles. *Journal of Applied Ecology* 44:395–404.

Schauble, M. K. 1972. Seasonal variation of newt forelimb regeneration under controlled environmental conditions. *Journal of Experimental Zoology* 181:281–286.

Scheffer, V. B. and A. C. Myrick. 1980. A review of studies to 1970 or growth layers in the teeth of marine mammals. *Report of the International Whaling Commission (Special Issue)* 3:51–63.

Schmid, J. R. 1995. Marine turtle populations on the east-central coast of Florida: Results of tagging studies at the Cape Canaveral, Florida, 1986–1991. *Fishery Bulletin* 93:139–151.

Schmid, J. 1998. Marine turtle populations on the west-central coast of Florida: Results of tagging studies at the Cedar Keys, Florida, 1986–1995. *Fishery Bulletin* 96:589–602.

Schmid, J. R. and W. N. Witzell. 1997. Age and growth of wild Kemp's ridley turtles (*Lepidochelys kempi*): Cumulative results from tagging studies in Florida. *Chelonian Conservation Biology* 2:532–537.

Scott, N. M., M. F. Haussmann, R. M. Elsey, P. L. Trosclair III, and C. M. Vleck. 2006. Telomere length shortens with body length in *Alligator mississippiensis*. *Southeastern Naturalist* 5:685–692.

Secor, D. H., J. M. Dean, and S. E. Campana (eds.), and Miller, A. B. (assoc. ed.). 1995. *Recent Developments in Otolith Research*. The Belle W. Baruch Library in Marine Science Number 19. Columbia, SC: University of South Carolina Press.

Southeast Fisheries Science Center (SEFSC). 2009. An assessment of loggerhead sea turtles to estimate impacts of mortality reductions on population dynamics. NMFS SEFSC Contribution PRD-08/09-14, July 2009.

Seminoff, J. A., A. Resendiz, W. J. Nichols, and T. T. Jones. 2002. Growth rates of wild green turtles (*Chelonia mydas*) at a temperate foraging area in the Gulf of California, Mexico. *Copeia* 2002:610–617.

Shaver, D. J. and T. Wibbels. 2007. Head-starting ridley sea turtles. In *Synopsis of the Biology and Conservation of the Ridley Sea Turtle*, ed. P. Plotkin, pp. 297–323. Washington, DC: Smithsonian Institute Press.

Sherwood, O. A., D. B. Scott, M. J. Risk, and T. P. Guilderson. 2005. Radiocarbon evidence for annual growth rings in the deep-sea octocoral *Primnoa resedaeformis*. *Marine Ecology Progress Series* 301:129–134.

Simmons, D. J. 1992. Circadian aspects of bone biology. In *Bone: Bone Growth-A*, Vol. 6, ed. B. K. Hall, pp. 91–128. Boca Raton, FL: CRC Press.

Snover, M. L. 2002. Growth and ontogeny of sea turtles using skeletochronology: Methods, validation and application to conservation. PhD dissertation, Duke University, Durham, U.K.

Snover, M. L. 2008. Ontogenetic habitat shifts in marine organisms: Influencing factors and the impact of climate variability. *Bulletin of Marine Science* 83:53–67.

Snover, M. L., L. Avens, and A. A. Hohn. 2007a. Back-calculating length from skeletal growth marks in loggerhead sea turtles *Caretta caretta*. *Endangered Species Research* 3:95–104.

Snover, M. L., G. H. Balazs, S. K. K. Murakawa, S. K. Hargrove, M. R. Rice, and W. A. Sietz. 2012. Age and growth rates of Hawaiian hawksbill turtles (*Eretmochelys imbricata*) using skeletochronology. *Marine Biology*.

Snover, M. L. and A. A. Hohn. 2004. Validation and interpretation of annual skeletal marks in loggerhead (*Caretta caretta*) and Kemp's ridley (*Lepidochelys kempii*) sea turtles. *Fishery Bulletin* 102:682–692.

Snover, M. L., A. A. Hohn, L. B. Crowder, and S. S. Heppell. 2007b. Age and growth in Kemp's ridley sea turtles: Evidence from mark recapture and skeletochronology. In *Synopsis of the Biology and Conservation of the Ridley Sea Turtle*, ed. P. Plotkin, pp. 89–105. Washington, DC: Smithsonian Institute Press.

Snover, M. L., A. A. Hohn, L. B. Crowder, and S. A. Macko. 2010. Combining stable isotopes and skeletal growth marks to detect habitat shifts in juvenile loggerhead sea turtles *Caretta caretta*. *Endangered Species Research* 13:25–31.

Snover, M. L., A. A. Hohn, L. R. Goshe, and G. H. Balazs. 2011. Validation of annual skeletal marks in green sea turtles *Chelonia mydas* using tetracycline labeling. *Aquatic Biology* 12:197–204.

Snover, M. L. and A. G. J. Rhodin. 2008. Comparative ontogenetic and phylogenetic aspects of chelonian chondro-osseous growth and skeletochronology. In *Biology of Turtles*, eds. J. Wyneken, M. H. Godfrey, and V. Bels, pp. 17–43. Boca Raton, FL: CRC Press.

Spotila, J. R., M. P. O'Connor, and F. V. Paladino. 1997. Thermal biology. In *The Biology of Sea Turtles*, eds. P. L. Lutz and J. A. Musick, pp. 297–314. Boca Raton, FL: CRC Press.

Stewart, R. E. A., S. E. Campana, C. M. Jones, and B. E. Stewart. 2006. Bomb radiocarbon dating calibrates beluga (*Delphinapterus leucas*) age estimates. *Canadian Journal of Zoology* 84:1840–1852.

Stuiver, M. and H. A. Polach. 1977. Reporting of $^{14}C$ data. *Radiocarbon* 19:355–363.

Suryan, R. M., V. S. Saba, B. P. Wallace, S. A. Hatch, M. Frederiksen, and S. Wanless. 2009. Environmental forcing on life history strategies: Evidence for multi-trophic level responses at ocean basin scales. *Progress in Oceanography* 81:1–4.

Swain, D. P., S. F. Sinclair, and J. M. Hanson. 2007. Evolutionary response to size-selective mortality in an exploited fish population. *Proceedings of the Royal Society of London, B* 274:1015–1022.

Swingle, W. M., D. I. Warmolts, J. A. Keinath, and J. A. Musick. 1993. Exceptional growth rates of captive loggerhead sea turtles, *Caretta caretta*. *Zoo Biology* 12:491–497.

Takasaki, T., A. Tsufi, N. Ikeda, and M. Ohishi. 2003. Age estimation in dental pulp DNA based on human telomeres. *International Journal of Legal Medicine* 117:232–234.

Thomas, R. B., D. W. Beckman, K. Thompson, K. A. Buhlmann, J. W. Gibbons, and D. L. Moll. 1997. Estimation of age for *Trachemys scripta* and *Deirochelys reticularia* by counting annual growth layers in claws. *Copeia* 1994:842–845.

Tucker, A. D., D. Broderick, and L. Kampe. 2001. Age estimation of *Eretmochelys imbricata* by schlerochronology of carapacial scutes. *Chelonian Conservation and Biology* 4:219–222.

Uchida, I. 1967. On the growth of the loggerhead turtle, *Caretta caretta*, under rearing conditions. *Bulletin of the Japanese Society of Scientific Fisheries* 33:497–506.

Vadopalas, B., C. Weidman, and E. K. Cronin. 2011. Validation of age estimation in geoduck clams using the bomb radiocarbon signal. *Journal of Shellfish Research* 30:303–307.

Van Houtan, K. S. and J. M. Halley. 2011. Long-term climate forcing in loggerhead sea turtle nesting. *PLoS One* 6:1–7.

Vaughan, J. R. 2009. Evaluation of length distributions and growth variance to improve assessment of the loggerhead sea turtle (*Caretta caretta*). Masters thesis, Oregon State University, Corvallis, OR.

Vigliola, L., M. Harmelin-Vivien, and M. G. Meekan. 2000. Comparison of techniques of back-calculation of growth and settlement marks from the otoliths of 3 species of *Diplodus* from the Mediterranean Sea. *Canadian Journal of Fisheries and Aquatic Sciences* 57:1291–1299.

Von Zglinicki, T. 2002. Oxidative stress shortens telomeres. *Trends in Biochemical Sciences* 27:339–344.

Wallace, B. P., L. Avens, J. Braun-McNeill, and C. McClellan. 2009. The diet composition of immature loggerheads: Insights on trophic niche, growth rates, and fisheries interactions. *Journal of Experimental Marine Biology and Ecology* 373:50–57.

Walker, T. A. and C. J. Parmenter. 1990. Absence of a pelagic phase in the life cycle of the flatback turtle, *Natator depressa* (Garman). *Journal of Biogeography* 17:275–278.

Watson, J. 1972. Origin of concatameric T4 DNA. *Nature* 239:197–201.

Watson, D. M. 2006. Growth rates of sea turtles in Watamu, Kenya. *Earth & Environment* 2:29–53.

Waycott, M., C. M. Duarte, T. J. B. Carruthers et al. 2009. Accelerating loss of seagrasses across the globe threatens coastal ecosystems. *Proceedings of the National Academy of Sciences of the United States of America* 106:12377–12381.

Weidman, C. R. and G. A. Jones. 1993. A shell-derived time history of bomb-C-14 on Georges Bank and its Labrador Sea implications. *Journal of Geophysical Research* 98:14577–14588.

Witzell, W. N. 1980. Growth of captive hawksbill turtles, *Eretmochelys imbricata*, in Western Samoa. *Bulletin of Marine Science* 30:909–912.

Wood, J. R. and F. E. Wood. 1980. Reproductive biology of captive green turtle *Chelonia mydas*. *American Zoologist* 20:499–505.

Wyneken, J. 2001. The anatomy of sea turtles. NOAA Technical Memorandum NMFS-SEFSC-470.

Yekkala, R., C. Meers, A. Van Schepdael, J. Hoogmartens, I. Lambrichts, and G. Willems. 2006. Racemization of aspartic acid from human dentin in the estimation of chronological age. *Forensic Science International* 156S:89–94.

Zubakov, D., F. Liu, M. C. van Zelm et al. 2010. Estimating human age from T-cell DNA rearrangements. *Current Biology* 20:970–971.

Zug, G. R. 1990. Age determination of long-lived reptiles: Some techniques for seaturtles. *Annales des Sciences Naturelles, Zoologie et Biologie Animale* 13:219–222.

Zug, G. R., G. H. Balazs, and J. A. Wetherall. 1995. Growth in juvenile loggerhead sea turtles (*Caretta caretta*) in the north Pacific pelagic habitat. *Copeia* 1995:484–487.

Zug, G. R., G. H. Balazs, J. A. Wetherall, D. M. Parker, and S. K. K. Murakawa. 2002. Age and growth of Hawaiian green sea turtles (*Chelonia mydas*): An analysis based on skeletochronology. *Fishery Bulletin* 100:117–127.

Zug, G. R., M. Chaloupka, and G. H. Balazs. 2006. Age and growth in olive ridley sea turtles (*Lepidochelys olivacea*) from the North-central Pacific: A skeletochronological analysis. *Marine Ecology* 27:263–270.

Zug, G. R. and R. E. Glor. 1998. Estimates of age and growth in a population of green sea turtles (*Chelonia mydas*) from the Indian River lagoon system, Florida: A skeletochronological analysis. *Canadian Journal of Zoology* 76:1497–1506.

Zug, G. R., J. H. Kalb, and S. J. Luzar. 1997. Age and growth in wild Kemp's ridley sea turtles *Lepidochelys kempii* from skeletochronological data. *Biological Conservation* 80:261–268.

Zug, G. R. and J. F. Parham. 1996. Age and growth in leatherback turtles, *Dermochelys coriacea* (Testudines: Dermochelyidae): A skeletochronological analysis. *Chelonian Conservation and Biology* 2:244–249.

Zug, G. R., A. H. Wynn, and C. Ruckdeschel. 1986. Age determination of loggerhead sea turtles, *Caretta caretta*, by incremental growth marks in the skeleton. *Smithsonian Contributions to Zoology* 427:1–34.

Zuschin, M., M. Stachowitsch, and R. J. Stanton Jr. 2003. Patterns and processes of shell fragmentation in modern and ancient marine environments. *Earth-Science Reviews* 63:33–82.

# 6 Molecular Genetics of Sea Turtles

*Michael P. Jensen, Nancy N. FitzSimmons,
and Peter H. Dutton*

## CONTENTS

## 6.1   INTRODUCTION

Since the first volume of *The Biology of Sea Turtles* (Lutz and Musick, 1997), studies using molecular techniques to address a variety of questions about sea turtle biology and life history have grown rapidly. In the late 1980s researchers had just begun using mitochondrial (mt) DNA to investigate how sea turtle rookeries are genetically linked through female dispersal and set a benchmark when providing compelling evidence of female natal homing. The growing popularity of using molecular techniques in sea turtle research is illustrated by the number of genetic papers presented at the *Annual Symposium on Sea Turtle Conservation and Biology* over the past two decades (Figure 6.1). The rapid progress in DNA sequencing and genotyping technology has expanded the scope of molecular genetics, and the symposia presentations include diverse topics ranging from mating systems and kinship among individuals to relationships among populations and species.

Over the past 20 years molecular genetics has come to play a central role in addressing questions that are directly relevant to the conservation of sea turtles (Table 6.1). Most studies to date have used maternally inherited mtDNA (control region) and nuclear microsatellites as the markers of choice,

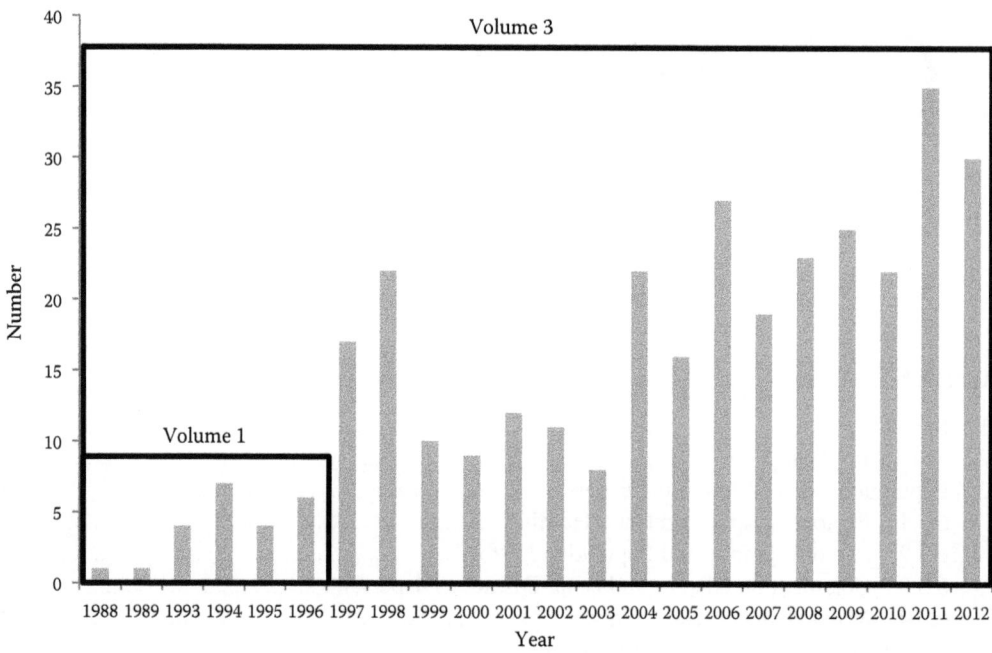

**FIGURE 6.1**   The number of genetics presentations at the Annual Symposium on Sea Turtle Biology and Conservation from 1988 to 2012. The small box highlights the number of genetics presentations prior to Volume 1 of the *Biology of Sea Turtles* being published in 1997.

---

**TABLE 6.1**

**Molecular Genetic Landmarks for Sea Turtle Biology and Conservation**

**Contribution**

Confirming natal homing in sea turtles

Demonstrating that multiple paternity exists in many sea turtle populations

Connecting foraging areas to rookery origins

Identifying populations of concern

Determining population genetic structure

Resolving taxonomic uncertainty

Defining management units within species

Establishing parentage; pedigree analysis

Detecting hybridization

Understanding population connectivity

---

as these regions are the most highly variable markers which makes them ideal for population-level analysis. In a study of Atlantic green turtles, Meylan et al. (1990) provided the first evidence of natal homing by showing that turtles nesting on Aves Island and those nesting at Tortuguero, Costa Rica, had fixed haplotype differences for mtDNA, despite these two populations mixing at foraging grounds in the Caribbean. Since then, numerous studies have shown that this pattern of mtDNA differentiation among rookeries is a common feature among all species of marine turtles, albeit to varying degree.

    The genetic differentiation of rookeries led to another important advance for the use of molecular tools, the ability to determine the origin of turtles sampled away from nesting beaches. The value of this was demonstrated when Bowen et al. (1995) were able to trace loggerhead turtles from developmental habitats in the central and eastern Pacific to nesting beaches in Japan, thereby providing evidence for the trans-Pacific migration of loggerhead turtles. Since then, detailed studies using mixed stock analysis (MSA) have aimed to explain the processes that generate the composition of turtles at mixed foraging areas. This is one of the most active fields in sea turtle genetics as researchers seek to generate more precise and reliable estimates that may be used for threat assessment. Genetic tools have also proven useful in answering questions about the reproductive behavior of sea turtles. Since the first study that used protein isozyme polymorphisms to document multiple paternity in loggerhead turtles (Harry and Briscoe 1988), researchers have used dozens of microsatellite loci to understand mating systems in all species of sea turtles, and even so, questions remain regarding sperm storage and possible fitness benefits of different mating strategies. Furthermore, a combination of mtDNA and microsatellite data has been used to document several cases of hybridization in sea turtles. Molecular genetic studies of sea turtles exist within broader genetic fields that are constantly evolving. New tools and techniques such as single nucleotide polymorphisms (SNPs) and mitogenomics are important additions to the molecular toolbox that promise to overcome some of the limitations of past studies. In this chapter we review the current role and scope of molecular genetics in sea turtle research. Empirical examples are used to highlight some key findings that best describe the patterns and processes that genetic studies of sea turtles aim to unravel.

## 6.2   SEA TURTLE PHYLOGENY

Several recent studies have advanced our knowledge of the relationship among sea turtle lineages and their placement within the Testudines. Sea turtles are placed within the superfamily Chelonioidea (containing the families Cheloniidae and Dermochelyidae), which forms a monophyletic group most closely related to freshwater mud turtles (Kinosternoidea) and snapping turtles (Chelydridae) based on sequence data from 14 nuclear genes (Barley et al., 2010). Within the Chelonioidea, *Dermochelys coriacea* has a basal position as the older lineage relative to the other marine turtles (Bowen et al., 1993; Dutton et al., 1996; Naro-Maciel et al., 2008), which split into two subfamilies, the Chelonini (*Chelonia mydas* and *Natator depressus*) and the Carettini (*Lepidochelys olivacea*, *Lepidochelys kempi*, *Caretta caretta*, and *Eretmochelys imbricata*) about 63 MYA (Figure 6.2) (Naro-Maciel et al., 2008). Early phylogenetic studies of sea turtles based on sequencing of the mtDNA control region (d-loop), ND4, and Cyt*b* placed flatback turtles as a sister species to the Carettini (Bowen et al., 1993; Dutton et al., 1996), but using sequence data (7340 base pair [bp]) from the mtDNA genes 12S and 16S and four nuclear genes, *N. depressus* is grouped with the *C. mydas* lineages (Naro-Maciel et al., 2008). Estimated divergence times among species are 34 MYA (95% HPD: 14.1–60.1) between flatback and green turtles, and 29 MYA (95% HPD: 16.5–44.3) between hawksbill turtles and the combined loggerhead and ridley lineages (Naro-Maciel et al., 2008).

The advent of new sequencing technologies has recently allowed the entire mtDNA genome of sea turtles to be sequenced (Duchene et al., 2012; Frey and Dutton 2012; Morin Shamblin et al., 2012b). This expands the data from earlier studies by Dutton et al. (1996), which used ~1433 bp sequences (from three regions) to over 16,000 bp. The new studies detect additional genetic variation in the chelonids and will provide new insights into the evolutionary relationships among species (Duchene et al., 2012). Also, recent whole-mitogenome data support previous hypotheses (Avise et al., 1992) that the Testudines mitochondrial clock is slower than the conventional $2 \times 10^{-2}$ rate estimated for other animal lineages (Duchene et al., 2012). Whole mitogenome analyses support previous nuclear-mtDNA topology (Naro-Maciel et al., 2008; see also Thomson and Shaffer, 2010), placing *N. depressus* as a sister taxon to *Chelonia* and provide more precise divergence times. For example, the estimated divergence time between Pacific and Atlantic *C. mydas* lineages was estimated to be 3.09 MYA (1.76–4.87) in comparison to 7 MYA (1.92–13.47 HPD) in the Naro-Maciel et al. (2008) study (Duchene et al., 2012). However, this may in part reflect incomplete sampling distributions if samples from different lineages were used. Preliminary analysis of all 16,281 bp

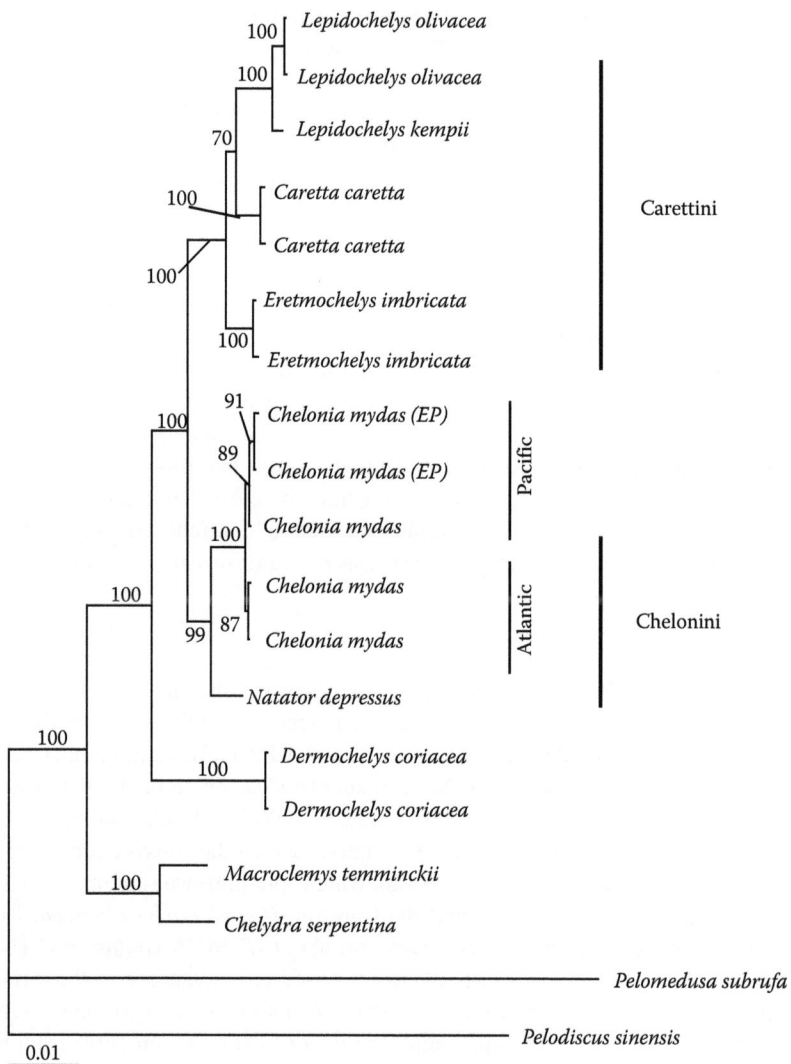

**FIGURE 6.2** Sea turtle phylogeny based on maximum parsimony (MP) and Bayesian analyses sequences from a combined 980 bp of mtDNA (12S and 16S) and 6350 bp from four nuclear genes (BDNF, Cmos, R35, Rag1, and Rag2). The numbers above the branches are MP bootstrap value. All posterior probability values (PP) from combined and mixed—model Bayesian analyses were 100%. (Adapted from Naro-Maciel, E. et al., *Mol. Phylogenet. Evol.*, 49, 659, 2008.)

of each mitochondrial genome sequence for leatherbacks from representative rookeries around the world has revealed surprisingly little additional variation (Dutton et al., unpublished data) and rules out a lack of resolution in the genetic marker as an alternative explanation of the low diversity found in *Dermochelys* by earlier studies (Dutton et al., 1996, 1999).

## 6.3  POPULATIONS, GENE FLOW, AND DISPERSAL

The value of using mtDNA markers became apparent from the initial use of DNA analysis to study sea turtles because the female-to-offspring mode of inheritance allows for tracking of rookery history and the relationships between rookeries (Avise and Bowen, 1994). By investigating the genetic structure among rookeries, it was possible to confirm the operation of natal homing by females in their choice of nesting regions. In addition, through the comparison of the relationships among mtDNA genetic lineages to their geographic locations, it was also possible to address questions of rookery history, colonization, and long-distance dispersal. However, relying solely on mtDNA markers to elucidate population histories and boundaries is not sufficient, given the importance of understanding marine turtle behavior for conservation purposes and how turtles have responded to past climatic and sea-level changes. Nuclear- and genome-wide markers allow assessment of male-mediated gene flow, and variations in their mutation rates provide insights into population processes at variable time scales.

### 6.3.1 PHYLOGEOGRAPHY AND NATAL HOMING

Many genetic studies have focused on understanding the history of marine turtle populations using phylogeographic techniques that compare the relationship among genetic lineages to the geographic locations where those lineages are found (Avise, 2009). These studies also provide insights into the extent of genetic structure among rookeries and the operation of natal homing by adults in their selection of breeding locations. Early genetic studies of green turtles uncovered genetic structure among distant rookeries, thus confirming that natal homing occurs when a female returns to nest (Meylan et al., 1990). This demonstrated the usefulness of genetics to understand marine turtle behavior and population dynamics. Natal homing behavior was also revealed for male green turtles in their choice of breeding grounds, although this did not preclude male-mediated gene flow from occurring, likely through opportunistic mating by males during migration (FitzSimmons et al., 1997a). Population genetic studies have now been conducted on all species of marine turtles in many regions, and it is evident that natal homing behavior is shared among all species, though there is considerable variation in the extent of genetic structure among populations and the implied extent of natal homing, when compared both across and within species.

Evidence of strong natal homing is indicated in several studies, but it is far from being a predictable phenomenon. In green turtles, strong natal homing is evidenced by significant genetic differentiation between island rookeries at Ashmore and Scott reefs off northwestern Australia (Dethmers et al., 2006; Jensen, 2010); these rookeries are located only 225 km apart. This contrasts with the nearby Northwest Shelf population that spans over 1,000 km of coastline and offshore islands (Dethmers et al., 2006; Jensen, 2010). Genetic differentiation was also found for rookeries on two islands in Taiwan that are separated by ~250 km (Cheng et al., 2008) and between locations in Japan that are <60 km apart (Nishizawa et al., 2011). In hawksbill turtles, rookeries in Iran (FitzSimmons, 2010) and Barbados (Browne et al., 2009) <50 km from each other have also displayed significant genetic differences. Some loggerhead rookeries <100 km apart are genetically distinct in the Mediterranean (Garofalo et al., 2009) and in Florida (Shamblin et al., 2011a), but this is not consistent among rookeries. Interesting support for strong natal homing comes from the green turtle rookery in Tortuguero, Costa Rica, where there was a negative correlation between genetic relatedness (using DNA minisatellite fingerprinting) and distance among individual nesting turtles over a 2 year period, indicating natal precision even within a rookery along the nesting beach. However, this was not found for green turtles nesting at Melbourne beach in Florida (Peare and Parker, 1996). Leatherback turtles appear to have the least strict natal homing overall, with some rookeries located over 2000 km apart not showing significant genetic differentiation (Dutton et al., 1999, 2007).

### 6.3.2 PHYLOGEOGRAPHY AND POPULATION HISTORY

Within species there is considerable variation in the estimated divergence times among haplotypes, and observations of shallow divergence and limited genetic variation have led to suggestions of genetic bottlenecks within ocean basins, or even globally. In contrast to its basal position as the oldest lineage in the marine turtle phylogeny, leatherback turtles were found to have a shallow phylogeny and low genetic diversity, suggesting a relatively recent global radiation for this oldest of lineages (Dutton et al., 1999). The most divergent (1.4%) leatherback haplotypes are estimated to have diverged <1 MYA. One explanation supporting this pattern is that leatherback turtles went through population declines caused by repeated glaciations during the Pleistocene, followed by subsequent population expansion (Dutton et al., 1999, but see Rivalan et al., 2006). Shallow phylogenies are also observed in the geographically isolated flatback and Kemps ridley turtles, with maximum divergence among haplotypes estimated at p=0.7% (Pittard, 2010) and 0.9% (Bowen et al., 1998), respectively for each species, which is also suggestive of past genetic bottlenecks (Shanker et al., 2004; Pittard, 2010). In leatherbacks, only 11 mtDNA haplotypes were identified from 281 samples in 12 populations (Dutton et al., 1999, 2007). A small number of haplotypes (12 in 274 samples) have also been found in flatback turtles (Pittard, 2010) for which, similar to leatherbacks, there is a predominant, presumed ancestral haplotype that is found in most regions (Pittard, 2010). But in contrast to leatherbacks, some flatback rookeries less than 300 km apart show genetic differentiation (Pittard, 2010). Comparisons between the two species suggest that both went through recent bottlenecks, from which they expanded into new areas, but that variation in the extent of natal homing has resulted in several functionally independent rookeries in flatback turtles, in contrast to regionally broader metapopulations for leatherbacks (Dutton et al., 1999, 2007).

Loggerhead turtles have phylogenies that reflect strong phylogeographic structure within and between the Atlantic-Mediterranean and Indo-Pacific ocean basins in which levels of genetic diversity and structure vary between regions. The two main divergent genetic lineages in loggerhead turtles are separated by a maximum sequence divergence of 6.3%, with an estimated time of divergence during the Pliocene of ∼3MYA (Bowen, 2003). Within ocean basins, loggerhead turtles have considerably greater genetic diversity and regional structure among Atlantic populations in comparison to Indian Ocean and western Pacific populations. In the Atlantic and Mediterranean, at least 43 haplotypes (based on 380 bp sequences) have been published (Bolten et al., 1998; Laurent et al., 1998; Bass et al., 2004; Bowen et al., 2004; Roberts et al., 2005; Carreras et al., 2006; Reece et al., 2006; Casale et al., 2008; Monzón-Argüello et al., 2009, 2010b; Reis et al., 2010a; Shamblin et al., 2011a; Yilmaz et al., 2011), with at least another 16 known haplotypes yet to be published (http://accstr.ufl.edu/ccmtdna.html). These haplotypes represent two different clades, separated by an average of 5.1% sequence divergence (Encalada et al., 1998), with one clade found in the United States and Brazil and the other found in the United States, Mexico, and Mediterranean rookeries. The distribution and relationship among mtDNA haplotypes has led to an hypothesis that during the Pleistocene, loggerhead populations contracted to nesting locations closer to the equator (southern Florida and Mexico) and later colonized into their northern range in the United States and Mediterranean (Encalada et al., 1998) and south to Brazil (Reis et al., 2010b). Tests for genetic bottlenecks support a scenario of population expansion after bottlenecks for the Florida (Reece et al., 2005), Brazil, and Mediterranean populations (Reis et al., 2010b), possibly as early as the late Pliocene in response to the closing of the Isthmus of Panama (Reece et al., 2005). Among the relatively smaller Pacific loggerhead populations, only four haplotypes (among 362 samples) have been observed, and there is little sequence diversity (Hatase et al., 2002; Boyle et al., 2009). This is indicative of a strong ocean-wide population bottleneck, with haplotypes estimated to have diverged 500–700,000 years ago (Hatase et al., 2002).

In contrast, both green and hawksbill turtle populations display high levels of genetic diversity and phylogeographic structure in both the Indo-Pacific and the Atlantic and Mediterranean basins. Among 27 green turtle rookeries in the Indo-Pacific, 25 haplotypes were observed, with sequence divergences of up to 8.4% (Dethmers et al., 2006). Among the Atlantic and Mediterranean rookeries, sequence divergence is lower (maximum p=3.3%; estimated from Bjorndal et al., 2005), but haplotype diversity is high. A total of 47 haplotypes have been published (Allard et al., 1994; Lahanas et al., 1994; Encalada et al., 1996; Bass and Witzell, 2000; Bass et al., 2006; Bjorndal et al., 2006; Formia et al., 2006, 2007; Naro-Maciel, 2006; Foley et al., 2007; Ruiz-Urquiola et al., 2010; Bagda et al., 2012) and another 18 haplotypes are yet to be published (http://accstr.ufl.edu/cmmtdna.html). Evidence of population bottlenecks is restricted to particular rookeries, as based on specific tests for bottlenecks (Reece et al., 2005; Formia et al., 2006) or from low estimates of historic effective population size (e.g., Dethmers et al., 2006). Similar to the case for green turtles, high levels of genetic diversity characterize hawksbill turtles in the Atlantic and Caribbean (26 haplotypes in 12 rookeries; Monzón-Argüello et al., 2011) as well as those in the Indo-Pacific (48 haplotypes in 8 populations, FitzSimmons, 2010).

### 6.3.3 COLONIZATION HISTORY AND LONG-DISTANCE DISPERSAL

Two aspects are apparent in the genetic structure of sea turtle populations: a disconnect between current population size and genetic diversity, and limited correlations between genetic distance and geographic distance among rookeries. Apparently, genetic structure and levels of genetic diversity in marine turtle populations are complex due to varied colonization histories that range from single colonization events to multiple colonizations from diverse populations. Thus the extent of genetic diversity is not necessarily indicative of population size (e.g., Lahanas et al., 1994; Bjorndal et al., 2006; Dethmers et al., 2006), as would be expected from classic population genetic theory. Additionally, there is considerable evidence that marine turtle rookeries have undergone major geographic shifts in response to climate change, with many rookeries being established within the past 10,000 years. For example, green, flatback, and olive ridley turtles nest within the Gulf of Carpentaria, Australia, on beaches that did not exist until ~8000 years ago (Jones and Torgersen, 1988). This represents around 200–250 turtle generations and provides a test for how long it takes for rookeries to become genetically differentiated once colonized. All three species indicate significant genetic divergence from the nearest rookeries, though the level of divergence varies ($F_{ST} = 0.07$–$0.80$; Jensen et al., unpublished data; Pittard, 2010), suggesting either a low level of ongoing gene flow with neighboring rookeries or, equally likely, that not enough generations have elapsed to observe a more defined genetic separation. Complex colonization patterns are also reflected in several nonsignificant results from genetic tests for isolation by distance among rookeries (Bass et al., 1996; Carreras et al., 2006; Bourjea et al., 2007; Garofalo et al., 2009). Although patterns of isolation by distance have been observed (Figure 6.3), it is often relatively weak ($r^2 < 0.3$), or dependent upon certain geographic boundaries (Reece et al., 2005; Dethmers et al., 2006; Bourjea et al., 2007; Reis et al., 2010b; Pittard, 2010; LeRoux 2012). From a conservation perspective, the long generation times of marine turtles may act as a buffer against the loss of genetic diversity when population size is reduced, as evidenced by the diversity observed in Kemps ridleys (Bowen et al., 1991, 1998; Kichler et al., 1999). However, several populations have been observed with no mtDNA diversity (Dutton et al., 1999; Formia et al., 2006; Carreras et al., 2007; Browne et al., 2009; Reis et al., 2010b; Monzón-Argüello et al., 2011), and although this may have occurred through founder effects of small colonization events, long-term genetic bottlenecks may also have contributed to the lack of diversity.

Extant marine turtles have adapted to changing climate, sea levels, and oceanic current patterns throughout their evolution of over 110 MYA (Hirayama, 1998), including periods of glaciation and sea-level changes of >200 m (Haq et al., 1987). Genetic evidence suggests that this likely resulted in a series of regional colonization and extinction events, in which strict natal homing would not allow for such adaptation (e.g., Reece et al., 2005). Apart from a need for relaxed natal homing that might allow turtle populations to shift rookery locations over hundreds of kilometers, occasional long-distance dispersal has also been a feature of sea turtle evolution.

Many sea turtle species show evidence of long-distant dispersal in the past, as seen in the widespread distribution of some mtDNA haplotypes. Olive ridley populations in India were found to have a low frequency of individuals with a haplotype that is found in Malaysia and Australia, and another haplotype was shared with the eastern Pacific olive ridleys in Costa Rica. This evidence and the observation that the mtDNA haplotypes of Kemps ridley are more closely related to haplotypes that predominate along the east coast of India led to a hypothesis that the Indo-western Pacific population is the ancestral source for other olive ridley populations (Bowen et al., 1998; Shanker

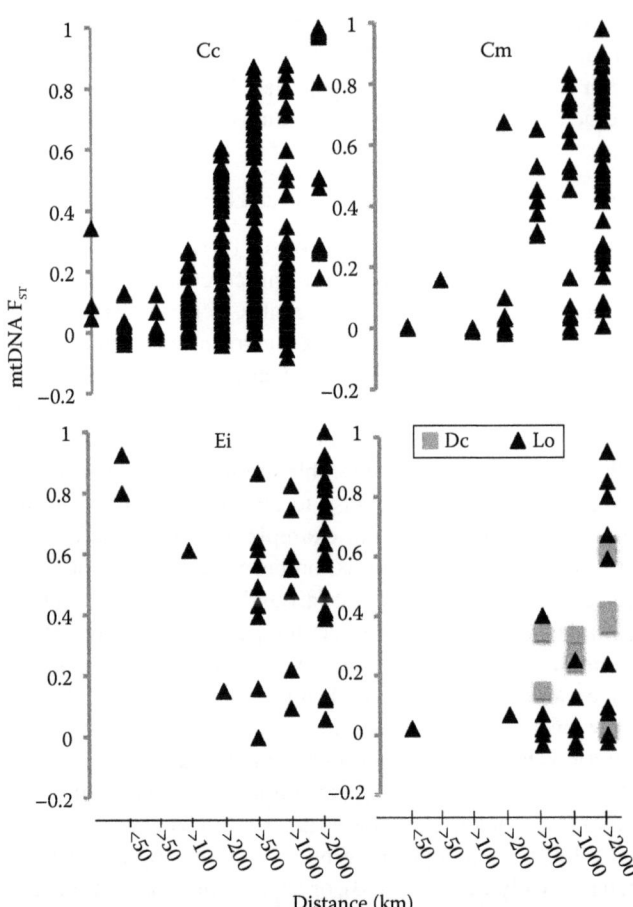

**FIGURE 6.3** Estimates of $F_{ST}$ derived from mtDNA control region data from pairwise comparisons between rookeries versus the distance between rookeries. Distances are based on distance categories of <50, >50, >100, >200, >500, >1000, and >2000 km. Species abbreviations are as follows: *Caretta caretta* (Cc), *Chelonia mydas* (Cm), *Eretmochelys imbricata* (Ei), *Dermochelys coriacea* (Dc), and *Lepidochelys olivacea* (Lo). (Data taken from Lahanas, P.N. et al., *Genetica*, 94, 57, 1994; FitzSimmons, N.N. et al., *Genetics*, 147, 1843, 1997b; Dutton, P.H. et al., *J. Zool.*, 248, 397, 1999; Hatase, H. et al., *Mar. Biol.*, 141, 299, 2002; Chassin-Noria, O. et al., *Genetica*, 121, 195, 2004; Shanker, K. et al., *Mol. Ecol.*, 13, 1899, 2004; Bjorndal, K.A. et al., *Mar. Biol.*, 147, 1449, 2005; Bjorndal, K.A. et al., *Chelonian Conserv. Biol.*, 5, 262, 2006; López-Castro, M.C. and Rocha-Olivares, A., *Mol. Ecol.*, 14, 3325, 2005; Formia, A. et al., *Conserv. Genet.*, 7, 353, 2006; Camacho-Mosquera, L. et al., *Investig. Mar.*, 37, 77, 2008; Cheng, I.J. et al., *J. Zool.*, 276, 375, 2008; Boyle, M.C. et al., *Proc. R. Soc. [Lond.]*, 276, 1993, 2009; Browne, D.C. et al., *Conserv. Genet.*, 11, 1541, 2009; Garofalo, L. et al., *Mar. Biol.*, 156, 2085, 2009; Reis, E.C. et al., *Conserv. Genet.*, 11, 1467, 2010b; Shamblin, B.M. et al., *Mar. Biol.*, 158, 571, 2011a; Monzón-Argüello, C. et al., *J. Exp. Mar. Biol. Ecol.*, 407, 345, 2011; Nishizawa, H. et al., *Endanger. Species Res.*, 14(2), 141, 2011; Yilmaz, C. et al., *Biochem. Syst. Ecol.*, 39, 266, 2011; Saied, A. et al., *Mar. Ecol. Prog. Ser.*, 450, 207, 2012.)

et al., 2004). Under this scenario, the Indo-western Pacific and eastern Pacific populations were established by an eastward trans-oceanic dispersal, in contrast to a proposed westward colonization from the eastern Pacific (Pritchard, 1969). Long-distance dispersal is also implied in hawksbill turtles where the most common haplotype found in rookeries in the Persian Gulf off Iran was also found in a western Pacific rookery in the Solomon Islands (FitzSimmons, 2010). Likewise, a common green turtle haplotype in Micronesia was found in Australian rookeries in both the Pacific and Indian Oceans (Dethmers et al., 2006).

Genetic evidence has supported previous hypotheses that turtles may have travelled around the Cape of Good Hope, allowing for gene flow from the Indian Ocean to the Atlantic. This would be most likely during periods of warmer oceanic temperatures, which may allow an increased flow of the relatively warm water of the Agulhas Current southwest along the coast of South Africa and into the Atlantic (Bard and Rickaby, 2009). An hypothesis of colonization of the Atlantic Ocean by olive ridley migrants from the Indian Ocean (Pritchard, 1969) is supported by the occurrence of mtDNA haplotypes in the Atlantic that are closely related to haplotypes found in the Indian Ocean (Bowen et al., 1998). The phylogeographic structure of loggerhead turtles indicates at least two long-distance dispersals, one around the Cape of Good Hope, as evidenced by the grouping of the only haplotype found in Oman with haplotypes from the North Atlantic and the presence of the only haplotype found in South Africa from rookeries in the North Atlantic and Mediterranean (Bowen, 2003). In green turtles, Atlantic haplotypes were found in high proportions in two rookeries in the southwest Indian Ocean to the west of Madagascar, but not in rookeries 500 km to the north (Bourjea et al., 2007). The presence of only a single Atlantic haplotype in the Indo-Pacific rookeries led to the hypothesis that gene flow was not extensive, but that it was relatively recent (given the lack of new mutations to the Atlantic haplotype), and that the dispersal may have been from the Atlantic into the Indo-Pacific as has been observed in hammerhead sharks (Duncan et al., 2006; Bourjea et al., 2007).

Whether these various genetic data represent long-distance dispersal events by a few individuals, a series of step-wise dispersals, or if they reflect genetic relicts from a large, widespread ancestral population is largely unknown. Additionally, it is not known whether these events happen as a result of long-distance dispersal by posthatchling turtles that never make it back to their natal areas, or are due to displacement by breeding females. Studies of foraging-ground turtles also provide evidence of long-distance dispersal. Among hawksbill turtles, a previously unidentified haplotype from an Indian Ocean foraging ground (Okayama et al., 1999) was found to be the only haplotype observed among 20 nesting turtles at Principe in the eastern Atlantic (Monzón-Argüello et al., 2011). At an Atlantic foraging ground in Brazil, 12% of loggerhead turtles had the same haplotype as commonly observed in Australian rookeries (Reis et al., 2010b). The migratory limits within the life cycles of turtles from most populations are poorly known, though the trans-oceanic voyages of loggerheads (Bowen et al., 1995; Laurent et al., 1998; Boyle et al., 2009; Monzón-Argüello et al., 2011) and leatherbacks (Dutton et al., 2000) are good examples of what is possible.

### 6.3.4  MALE-MEDIATED GENE FLOW

Several published studies have compared the genetic structure observed with mtDNA to that with nuclear DNA, and it is apparent that a priori predictions cannot be made about the extent of male-mediated gene flow. Since the development of the first microsatellite loci for marine turtles (FitzSimmons et al., 1995), a proliferation of loci have been developed for loggerhead (Shamblin et al., 2007, 2009; Monzón-Argüello, 2008), green (Dutton and Frey, 2009; Shamblin et al., 2012a), hawksbill (Lin et al., 2008; Miro-Herrans et al., 2008), olive ridley (Aggarwal et al., 2004, 2008), and leatherback (Alstad et al., 2011; Roden and Dutton, 2011) turtles for studies of genetic structure and mating systems. All studies have found evidence of male-mediated gene flow, but there is considerable variation in results (Table 6.2). In several recent studies, less than half of the pairwise tests between populations indicate male-mediated gene flow, and in several of these there are examples of less gene flow estimated using microsatellite loci than by mtDNA sequencing (Table 6.2). Such results may be interpreted as a lack of male-mediated gene flow and evidence of how higher mutation rates at microsatellite loci may lead to genetic divergence between populations that is not always observed within the mtDNA.

One concern related to studies of male-mediated gene flow is the need for better knowledge of the spatial distribution of populations, especially the extent of overlapping feeding grounds, and how this could provide avenues for opportunistic mating during breeding migrations. Future tests of male-mediated gene flow need to be structured to incorporate appropriate geographic scales that consider

## TABLE 6.2
## Studies That Have Compared Genetic Structure at Nuclear Microsatellites to mtDNA and Evidence for Male-Mediated Gene Flow

| Species | Region | #Loci | Sample Size (# Rookeries) | Evidence of Male-Mediated Gene Flow | Reference |
|---------|--------|-------|---------------------------|-------------------------------------|-----------|
| Cc | Western Atlantic | 5 | 459 (9) | 64 of 72 tests[a] | Bowen et al. (2005) |
| Cc | Mediterranean | 7 | 112 (7) | 5 of 11 tests[b] | Carreras et al. (2007) |
| Cc | Turkey | 6 | 256 (18) | 10 of 10 tests[b] | Yilmaz et al. (2011) |
| Cm | Australia | 4 | 275 (9) | 6 of 6 tests[b] | FitzSimmons et al. (1997b) |
| Cm | Pacific Mexico | 3 | 123 (4) | 3 of 5 tests[b] | Chassin-Noria et al. (2004) |
| Cm | Global | 4 | 337 (16) | Ocean basins[c] | Roberts et al. (2004) |
| Cm | Japan | 4 | 67 (3) | 1 of 1 test[b] | Nishizawa et al. (2011) |
| Dc | Atlantic | 16 | 1417 (9) | 1 of 36 tests[b] | Dutton et al. (2013) |
| Ei | Indian Ocean | 5 | 64 (2) | Not tested | Zolgharnein et al. (2011) |
| Lo | French Guiana | 11 | 46 (1) | Not tested | Plot et al. (2011) |
| Nd | Australia | 11 | 370 (11) | 22 of 59 tests[b] | Pittard (2010) |

[a] Estimates based on MIGRATE (Beerli, 2002).
[b] Estimates based on $F_{ST}$ and $F_{ST}$ analogs.
[c] Pairwise values not shown.

the distribution of feeding grounds used by the populations. Microsatellite studies may be affected by homoplasy (mutations in different lineages that create identical alleles) occurring among distinct populations (Roberts et al., 2004). This may lead to erroneously concluding that male-mediated gene flow has occurred between populations, particularly if sampling designs are not appropriate.

Comparisons of genetic structure observed at microsatellite versus mtDNA markers provide important insights about population-wide diversity but afford only a limited understanding of male behavior. To understand whether male-mediated gene flow is due to "relaxed" natal philopatry in males, or whether it is due to opportunistic matings by males as they migrate through breeding grounds en route to their natal areas, requires sampling males at breeding grounds. This allows for comparisons of the mtDNA haplotype frequencies of males versus females at breeding grounds, and it would be a true test of male natal philopatry. The first study to investigate this found that in three Australian populations, green turtle males, like females, have strong natal philopatry and that male-mediated gene flow is opportunistic and depends upon the timing of breeding and the geographic locations of feeding grounds and mating grounds (FitzSimmons et al., 1997a). In contrast, weak but significant haplotype differences were observed between breeding male and female hawksbill turtles in Puerto Rico, and there was evidence that some males had originated from different rookeries (e.g., Costa Rica) (Velez-Zuazo et al., 2008).

## 6.4  FEEDING GROUNDS AND MIGRATORY BEHAVIOR

Most sea turtle species have a circumglobal distribution across tropical and subtropical waters, with hundreds of nesting beaches and foraging grounds making up a complex network of migratory routes. After hatching from tropical and subtropical beaches, posthatchling sea turtles spend years at the mercy of the prevailing currents (Musick and Limpus, 1997). Here the turtles grow larger and as they reach approximately 20–40 cm in curved carapace length (CCL), some species (e.g., green and hawksbill turtles) settle into neritic benthic habitats (Bjorndal, 1980; Balazs, 1982; Musick and Limpus, 1997) while other species, such as leatherbacks (and to some extent ridleys) stay in deeper pelagic waters. Some take up permanent residency and show strong fidelity to a chosen foraging area, while others undertake further developmental migrations with temporary settlement in developmental areas before finally settling in a specific area or using seasonal habitats. As mature adults, females migrate periodically between breeding and foraging grounds during breeding seasons, in some cases travelling several thousand kilometers (Limpus, 2007, 2009; Benson et al., 2011). The ability to link turtles at feeding grounds, or those encountered along migratory routes, back to their breeding habitat is challenging, but it is a fundamental component of effective management and conservation. Both mark-recapture and satellite telemetry studies have connected rookeries to foraging habitats for many populations of sea turtles (Bentivegna, 2002; Godley et al., 2002, 2003; Shillinger et al., 2008). However, these techniques cannot yet be used to connect the non-adult portion of the population to their natal rookery. Molecular techniques have opened up new possibilities to assess the connectivity between nesting and foraging areas, especially for immature sea turtles.

## 6.4.1 Mixed Stock Analysis

When mtDNA haplotypes exhibit significant frequency shifts among rookeries, they can be used to infer the natal origin of turtles captured along migration corridors and in feeding habitats. Mixed stock analysis (MSA) was first developed to detect the proportion of genetically differentiated salmon stocks from different rivers to mixed stocks of salmon caught in oceanic fisheries (Pella and Milner, 1987; Grant et al., 1980). Salmon and sea turtles share the life history traits of natal homing that results in breeding stocks that are genetically differentiated, coupled with highly migratory life history stages where stocks mix in foraging habitats. Since the early 1990s, researchers have used MSA methods to identify the rookery origins of sea turtles in the pelagic stage (Bowen et al., 1995; Bolten et al., 1998), in juvenile benthic foraging grounds (Bass and Witzell, 2000; Engstrom et al., 2002; Velez-Zuazo et al., 2008), in adult foraging grounds (Bass et al., 1998; Velez-Zuazo et al., 2008), in fisheries bycatch (Bowen et al., 1995; Laurent et al., 1998; Prosdocimi et al., 2011), and in strandings (Rankin-Baransky et al., 2001; Maffucci et al., 2006; Prosdocimi et al., 2011).

MSA studies have demonstrated the complexity of sea turtle migratory patterns that differ not only among species but also among populations within the same species, and every study reveals a unique scenario. There are regions where both green and loggerhead turtles demonstrate strong fidelity to their neritic foraging area from early recruitment (Limpus et al., 1992), while in other regions turtles switch between different developmental habitats before settling in an adult foraging ground upon reaching sexual maturity (Bjorndal et al., 2003; Godley et al., 2003; Pilcher, 2010). A recurring theme in MSA of sea turtles is the attempt to determine the mechanisms that generate the composition of turtles at mixed foraging grounds. Several hypotheses have been proposed to quantify the roles that rookery size, distance between rookeries and foraging grounds, juvenile natal homing behavior, and ocean currents play in shaping the mixture of turtles in foraging aggregations.

## 6.4.2 Factors Shaping the Composition of Foraging Grounds

The idea that larger rookeries in a region contribute more turtles to associated feeding grounds is intuitive. Early studies using MSA showed that juvenile loggerhead turtles found in oceanic foraging aggregations around the Azores and Madeira in the eastern Atlantic originated from nesting beaches in Mexico (~10%), south Florida (~70%), and northern Florida to North Carolina (~20%) (Bolten et al., 1998). Some of these eastern Atlantic turtles also pass through the Strait of Gibraltar and enter the western Mediterranean. Here 50% (or more) of loggerhead turtles caught in pelagic drift longline fisheries have been found to originate from western Atlantic rookeries (Laurent et al., 1998; Carreras et al., 2006). Despite the long distances involved, the contributions of turtles are roughly proportional to the size of the rookeries they came from. Similar to loggerhead turtles in the Pacific (Bowen et al., 1995), immature turtles from western Atlantic rookeries forage in the eastern Atlantic and Mediterranean but eventually traverse back across the Atlantic where they recruit into coastal areas along the eastern seaboard of the United States (Bolten et al., 1998; Laurent et al., 1998). This is supported by the findings that most foraging loggerhead turtles in neritic habitats throughout the

Mediterranean originate from Mediterranean rookeries (Laurent et al., 1998; Maffucci et al., 2006), while those recruiting into neritic habitats of the southeastern United States are from local rookeries (Bass et al., 2004; Bowen et al., 2004; Reece et al., 2006). After entering neritic foraging aggregations, the stock contributions are no longer proportional to the size of the rookeries alone. Instead, foraging areas share similar haplotype profiles to nearby rookeries, suggesting that immature loggerhead turtles tend to choose foraging areas near their natal origin (Bowen et al., 2004), thus disputing the idea of random mixing (Sears et al., 1995; Witzell et al., 2002; Reece et al., 2006).

While the model of random recruitment explains how some oceanic aggregations are formed (Bolten et al., 1998), there are many studies showing contrasting patterns of dispersal. For example, juvenile green turtles foraging in east-central Florida are significantly differentiated from green turtle foraging in the Bahamas only 350 km away (Bass and Witzell, 2000). Likewise, green turtle foraging grounds along the Great Barrier Reef (GBR) in Australia show a gradual shift in foraging ground composition along a north–south transect (Jensen, 2010). Foraging areas in the southern GBR (sGBR)are dominated by turtles from nearby sGBR rookeries and northern GBR (nGBR) foraging areas are dominated by turtles with a nGBR origin. This may reflect juvenile natal homing. However, it may be more a function of geography, as it appears that posthatchling turtles do not mix in the pelagic stage to the same extent as Atlantic loggerhead turtles due to varied oceanic currents affecting the two regions (Boyle, 2007).

While rookery size and distance might explain how marine turtles are distributed across foraging grounds, the results are somewhat ambiguous as disproportionately large or small contributions from some rookeries cannot be explained by size and distance alone. Green turtles foraging around Barbados in the West Indies showed large (25%) contributions of turtles from Ascension Island, more than 5,500 km away, and substantial contributions (19%) came from the much larger rookery at Tortuguero in Costa Rica, located "only" 2,600 km away. There was also a substantial contribution (18.5%) from the distant and much smaller south Florida rookery (Luke et al., 2004). While neither distance nor size plays a major role in recruitment to the Barbados foraging aggregation, ocean currents might partly explain this scenario. Barbados is located where the North and South Equatorial Currents meet, and turtles from both Ascension Island and south Florida rookeries feed into these two major Atlantic current systems. Costa Rica, on the other hand, is affected by smaller and more local current systems that would bring fewer posthatchling turtles toward Barbados (Luke et al., 2004). Similarly, foraging loggerhead turtles in the western Mediterranean Sea are mainly derived from western Atlantic rookeries, whereas turtles in the eastern Mediterranean mainly originate from Mediterranean rookeries, thus providing a strong association between location and ocean current systems (Carreras et al., 2006). Likewise, as the South Equatorial Current approaches the east coast of Australia, it splits into the southward East Australian Current and the northward North Queensland Current, and this pattern possibly influences the strong partitioning of foraging green turtles between the nGBR and the sGBR (Jensen, 2010). The use of high-resolution ocean current data to model the movement of passively dispersing (or modeled swimming behavior) of turtles is increasing (e.g., Blumenthal et al, 2009a; Godley et al., 2010; Proietti et al., 2012). For example, a recent study showed a significant correlation between foraging compositions generated by ocean current models and those from MSA for a number of hawksbill turtle foraging aggregations throughout the Caribbean (Blumenthal et al., 2009a), highlighting the important role of ocean currents in shaping the composition of foraging areas.

### 6.4.3 DIFFERENCES BETWEEN TIME, SIZE, AND GENDER

Temporal variation in the composition of turtles at foraging grounds should be considered, given that foraging aggregations are potentially highly dynamic when composed of turtles from multiple rookeries. Seasonal movement is common in both green and loggerhead turtles along the east coast of the United States (Avens and Lohmann, 2004). Developmental migrations from strictly juvenile to adult foraging grounds is common in loggerhead turtles (Bolten et al., 1998; Bjorndal et al., 2000; McClellan and Read, 2007) but has also been reported for green (Godley et al., 2003;

Bjorndal et al., 2005; Pilcher, 2010) and hawksbill turtles (Whiting and Koch, 2006; Grossman et al., 2007; Blumenthal et al., 2009b). In other areas, juvenile and adult turtles share foraging grounds and juveniles show strong fidelity to the same area throughout their life (Limpus et al., 1992, 1994; Broderick et al., 1994). The extent to which these different patterns in the use of foraging grounds, or the specific locations of foraging grounds, are related to temporal variation in the stock composition of foraging aggregations is not well understood. Bass et al. (2004) found no temporal variation in haplotype frequency for immature loggerhead turtles at a North Carolina foraging aggregation sampled over three consecutive years. Jensen (2010) found no temporal variation in adult green turtle foraging grounds on the GBR and neither did Naro-Maciel et al. (2007) for green turtles in Brazil. Velez-Zuazo et al. (2008) found no evidence of temporal variation in a 5 year study of hawksbill turtles from Puerto Rico. The only study to report temporal variation in foraging grounds is a 12 year study from a highly dynamic foraging ground for immature green turtles in the Bahamas where haplotype frequencies from a single year was found to be significantly different from other years (Bjorndal and Bolten, 2008). However, marine turtle foraging populations are unlikely to be static. The recruitment of juveniles from several rookeries is a complex process that is affected by variation in output from rookeries, which is caused by variation in nesting numbers, natural catastrophes, predation, and human impacts as well as varying ocean currents. These changes at rookeries or in ocean currents are likely to be reflected in foraging ground compositions. Temporal variation in the composition of foraging aggregations is expected if they are comprised off turtles from a large number of rookeries, and for highly dynamic foraging aggregations where juveniles stay for a short amount of time, such as in the Bahamas (Bjorndal and Bolten, 2008).

A recent study of green turtle aggregations at six major foraging grounds, spanning a north–south transect along the entire length (~2,300 km) of the GBR, combined MSA with data from more than 30 years of mark-recapture efforts (Jensen, 2010). Overall, the MSA estimates were in agreement with estimates derived from tag returns and provided confidence in relying on point estimates from MSA. Interestingly, there were significant shifts in haplotype frequencies between juveniles and adults at the most northern foraging ground (Torres Strait), resulting in major shifts in the estimated stock contributions. Here, fewer juveniles (53%) originated from the nGBR stock in comparison to adults (89%). This trend was apparent in the four most northern foraging grounds. The observed patterns at the various foraging grounds likely resulted from several causes, the mostly likely of which were that (1) juveniles have shifted foraging grounds as they mature, especially those from distant nesting regions; or that (2) reduced hatching success from the main nGBR rookery at Raine Island for well over a decade (Limpus et al., 2000; Limpus, 2007) has resulted in reduced recruitment into the nGBR foraging ground. The latter possibility suggests a need to take action to conserve the nGBR population and highlights the direct conservation and management values of monitoring foraging areas using genetic techniques. The combined strength of data derived from mark-recapture studies, demographic studies to determine sex, maturity, and breeding status of the turtles, genetic studies to determine stock composition, and satellite telemetry, are needed to provide informed assessments of foraging populations necessary for guiding sustainable management of marine turtles.

Another confounding factor is that, foraging areas where turtles from rookeries that are female biased due to warmer incubation temperature mix with turtles from cooler more male-producing rookeries would be expected to generate different MSA estimates between males and females (see Jensen, 2010). Bass et al. (1998) found a small difference in the contribution between males and females from different rookeries at a green turtle foraging ground in Nicaragua. However, sample sizes were small (30 for each sex) and the results remain inconclusive. Sex-based dispersal remains poorly understood in marine turtles. Because marine turtles lack obvious morphological sex characteristics prior to maturity, the gonads of immature must be examined using laparoscopy (Miller and Limpus, 2003), or hormonal assays performed to determine sex (Diez and Van Dam, 2003). This compounds the logistical difficulties in sampling a sufficiently large number of both males and females, especially if sex ratios are highly skewed. As a result of these challenges most studies have been unable to analyze foraging composition by sex.

### 6.4.4 LIMITATIONS OF MSA

MSA has provided valuable new insights into the distribution of marine turtle populations, but in many cases the estimates are affected by large uncertainty, often due to the haplotype composition of the source populations. Ideally, mtDNA haplotype frequencies would show highly significant shifts among rookeries, and the presence of unique haplotypes would make it straightforward to assign individuals to their natal rookery. However, this is typically not the case, and the occurrence of common mtDNA haplotypes that are shared among rookeries may lead to unreliable MSA results with large confidence intervals. Examples of this include the common loggerhead turtle haplotypes CC-A1 and CC-A2, that are found across western Atlantic and Mediterranean rookeries (Bowen et al., 2004; Carreras et al., 2007; Shamblin et al., 2011a), haplotypes C1 and C3 that are shared among green turtle rookeries in the Indo-Pacific (Dethmers et al., 2010), and the A and F haplotypes that dominate the Caribbean hawksbill turtle rookeries (Velez-Zuazo et al., 2008). As a result, MSA estimates may not reflect the true mixture of sea turtles in the foraging areas. One way to address this issue is to look for more resolution in the genetic markers used. As sequencing techniques have become cheaper, and more efficient, researchers are starting to sequence a longer segment of the mtDNA control region hoping to increase the resolution of the genetic marker and thereby the power of the MSA. Another important criteria for a successful MSA is the sampling of all (or most) possible source rookeries, especially when populations share widespread haplotypes. Recently, efforts have been made to expand geographic sampling and to add resolution to genetic analyses for Caribbean hawksbills by re-sequencing samples using a longer (740 bp) segment of the mtDNA control region. By doing this, rookeries that were previously indistinguishable based on old 384 bp sequences may now be differentiated (Velez-Zuazo et al., 2008; LeRoux et al., 2012).

The number of "orphan" haplotypes, those not observed at the rookeries but seen in foraging grounds, is a good indication of inadequate sampling of source populations. Medium frequencies of orphan haplotypes are often indicative of an unsampled source, while low frequencies of orphan haplotypes are indicative of either an unsampled source or insufficient sampling of already sampled rookeries. This is highlighted by a recent study of juvenile hawksbill turtles foraging around the Cape Verde Islands (Monzón-Argüello et al., 2010a). Here, all three haplotypes found (n = 28) were orphan haplotypes not found at any rookery, highlighting obvious gaps in sampling of key rookeries. However, as more rookeries are characterized for mtDNA variation, the number of orphan haplotypes seen in foraging aggregations should decrease. These examples accentuate the importance of being critical when using MSA. Ideally, the interpretation of MSA results should use an integrated approach considering demographic, ocean current, stable isotope, mark-recapture, and/or satellite tracking data if these are available, in order to draw conclusions that are biologically meaningful.

## 6.5  CONSERVATION AND MANAGEMENT IMPLICATIONS: A POPULATION PERSPECTIVE

One aim of many genetic studies is to inform management decisions to aid in effective conservation. This includes knowledge about which rookeries should be considered part of the same breeding population, and which function as separate populations, the amount of genetic exchange among populations, the extent of genetic variability and insights into the dynamics of population history and colonization. To focus management decisions at a population level, the term "Management Unit" has been used to signify functionally independent populations in which a loss of individuals in one population is not likely to be replaced from animals in another population within time frames relevant to management (Moritz, 1994). For example, Management Units (MUs) have been defined for green turtles (Dethmers et al., 2006; Formia et al., 2006; Bourjea et al., 2007), loggerhead turtles in the Atlantic and Mediterranean (Encalada et al., 1998; Shamblin et al., 2011a;

Yilmaz et al., 2011), leatherback turtles in the Pacific and Atlantic (Dutton et al., 2007; unpublished data), hawksbill turtles in the Indo-Pacific and Caribbean (FitzSimmons, 2010; LeRoux et al., 2012), and flatback turtles in Australia (Pittard, 2010).

Typically, the identification of MUs has been based upon significant genetic differentiation of mtDNA haplotypes (based on $F_{ST}$ values) among rookeries (or groups of rookeries), though this approach has limitations. It is possible to have relatively low gene flow between two populations that is sufficient to prevent genetic divergence, yet low enough that the populations function as demographically independent populations. In this context, Palsbøll et al. (2007) suggest setting a level of <10% migration per generation to define MUs, which could be assessed by genetic studies or through tagging data. Genetic studies may have inherent limitations though. For example some rookeries may be functioning independently, but because of recent colonization, not yet appear differentiated based on the genetic markers being used. In such cases it becomes more important to have field data to identify populations, for example, having tagging data that show a lack of exchange of individuals between rookeries over decades, or data on differences in the timing of nesting (summer and winter), as were used to identify separate hawksbill populations in Australia (Limpus, 2009). Additionally, there is some evidence of temporal variation in the mtDNA haplotype frequencies of turtles nesting in different years in some populations studied (Shamblin et al., 2011a) but not in others (Hatase et al., 2002; Bjorndal et al., 2005; Formia et al., 2007; Velez-Zuazo et al., 2008; Jensen, 2010), thus robust sampling designs may need to include samples collected across years for the identification of MUs.

The importance of male-mediated gene flow is limited when defining MUs for sea turtles. While the nuclear exchange of genes is crucial to genetic diversity in a population, no amount of male-mediated gene flow will bring back a breeding population if the rookeries go extinct. Thus, male-mediated gene flow needs to be considered relative to the extent of genetic divergence among populations as indicated by mtDNA markers. For example, the indication of substantial male-mediated gene flow between northern and southern GBR green turtle populations is less relevant given a high degree of mtDNA differentiation ($F_{ST}=0.8$; FitzSimmons et al., 1997b) which demonstrates little exchange of females between the rookeries in this region. In contrast, mtDNA and microsatellite data on loggerhead populations in Turkey (Yilmaz et al., 2011) suggest a metapopulation structure among some rookeries due to inconsistent mtDNA haplotype differentiation among pairs of rookeries, and strong male-mediated gene flow at microsatellite loci among all areas.

Because sea turtles migrate long distances at various times throughout their life, they often occupy habitats under the authority of multiple countries and may spend a considerable amount of time in international waters. Nations that host sea turtle populations at either nesting and/or foraging habitats have legal jurisdiction over animals that also spend parts of their lives within the borders of other nations. The use of MSA is therefore an extremely important tool for providing information that can help provide information for setting up international agreements for effective management of sea turtles, taking into account the trans-boundary nature of populations (Dutton and Squires, 2011). From a management perspective, MSA provides an important tool for identifying threatened sea turtle populations away from the breeding grounds. For example, MSA has been used to show that 50% of loggerhead turtles caught in some Mediterranean fisheries originated from rookeries in the southeastern United States (Laurent et al., 1998). In the North Pacific, MSA studies have shown that loggerheads encountered as fisheries bycatch on the high seas and foraging grounds off the coast of Baja California, Mexico, all originate from the rookeries in Japan (Bowen et al., 1995; Dutton et al., unpublished data). The Caribbean highlights the complexity of management, because turtles reside and migrate through habitats within multiple countries. MSA studies of foraging ground composition show that green turtles (Luke et al., 2004; Bjorndal and Bolten, 2008), loggerhead turtles (Engstrom et al., 2002) and hawksbill turtles (Bass, 1999; Velez-Zuazo et al., 2008; Browne et al., 2009) all cross international borders when migrating between foraging and nesting grounds in this region.

As MSA estimates get more precise, they may provide an effective means of monitoring trends at oceanic and coastal foraging grounds for all size classes and genders. Comparing the origin of adult turtles to that of juvenile turtles that have recently recruited into benthic foraging areas will make it possible to detect early signs of changing contributions which may indicate population decline or increase at the nesting beaches. In recent years, the potential effects of climate change on sea turtle populations have become an increasing concern (Hamann et al., 2007; Fuentes et al., 2009). Climate change might vary the carrying capacity of foraging grounds, alter the currents that transport juveniles to those foraging grounds (Fuentes et al., 2009), and will likely affect sex ratios in some turtle populations (Hamann et al., 2007; Fuentes et al., 2010). In all cases, long-term monitoring of the composition of foraging grounds may provide an effective way of detecting significant population changes as well as identifying female- and/or male-producing rookeries. Overall, the growing impact of conservation genetics will allow for more precise conservation decisions to be made at both regional and global scales for sea turtles.

## 6.6 MATING SYSTEMS

Mating systems influence demographic processes but are difficult to observe for sea turtles and this is one area where genetic approaches have been particularly informative. Multiple paternity has been examined using microsatellite markers and has been documented in loggerhead (e.g., Zbinden et al., 2007), olive ridley (Hoekert et al., 2002; Jensen et al., 2006), Kemp's ridley (Kichler et al., 1999), green (e.g., FitzSimmons, 1998; Ireland et al., 2003), leatherback (Crim et al., 2002; Stewart and Dutton, 2011), hawksbill (Joseph and Shaw, 2010), and flatback (Theissinger et al., 2009) turtles. The extent of multiple paternity has ranged from very low values (FitzSimmons, 1998) to over 90% (Jensen et al., 2006; Zbinden et al., 2007) in some populations. The extent to which multiple fathers contribute to clutches varies considerably among studies, with low levels of contributions from secondary males (FitzSimmons, 1998; Hoekert et al., 2002; Lee and Hays, 2004) to equal contributions (Zbinden et al., 2007). However, in several studies, primary and secondary fathers contributed to all clutches within a season for a particular female, indicating that sperm storage had occurred regardless of the proportions of sperm (Stewart and Dutton, 2011). One study investigated paternity relative to the order of egg deposition in two multiply-sired clutches of green turtles (Lara-De La Cruz et al., 2010), and the data suggest that sperm from different males is mixed within the oviduct and that fertilization may function as a raffle system. It has been proposed that marine turtles, unlike most birds, may have a first male sperm precedence for fertilization (FitzSimmons, 1998), but this has not yet been tested. There is genetic evidence to support field observations (Limpus, 1993) that the marine turtle mating system is promiscuous, as (inferred) individual male genotypes have been observed in the offspring of more than one female (Crim et al., 2002). Studies that analyzed successive clutches have not found evidence of successful mating by "new" males between clutches (Stewart and Dutton, 2011), although variation in male success across clutches does occur (Theissinger et al., 2009).

While there are many theoretical explanations for multiple paternity (such as increased offspring fitness, ensuring fertilization, male coercion), Lee and Hays (2004) suggest that it may be driven by male density and avoidance of aggressive mating behavior. In fact, Jensen et al. (2006) found higher levels of multiple paternity in mass-nesting olive ridley populations (90%) than in solitary nesters (30%), indicating the role of density and/or adult sex ratio. Few studies have been able to test for a relationship between the extent of multiple paternity and female characteristics (e.g., size), clutch size, and hatching success or hatchling fitness. A positive correlation was found between the number of fathers and female body size among clutches of 15 loggerhead females, and limited evidence supported a relationship between hatching success and the level of multiple paternity (Zbinden et al., 2007). But this was not found in green turtles (18 females), nor was there any relationship between the presence or absence of multiple paternity to clutch size or clutch success (Lee and Hays, 2004). Several studies suffer from small sample sizes, in terms of the number of females, number of offspring, or the number of loci analyzed. The importance of designing an experimental assay that has sufficient power to detect multiple paternity was illustrated when initial results based on small sample sizes that failed to detect multiple paternity in leatherbacks, were later overturned by a study of the same population that found 42% of clutches had multiple paternity when over 1000 hatchlings from successive clutches of 12 known nesting leatherbacks were analyzed at 7 microsatellite loci (Stewart and Dutton, 2011).

## 6.7   POPULATION VITAL PARAMETERS

Vital parameters such as age to maturity, survival, sex ratios, and population size (including males) are still lacking for most sea turtle populations, and this has made it difficult to conduct meaningful population risk assessments (NRC, 2010). Although vital parameters are difficult to observe directly, genetic analysis provides a practical approach to understand these processes. The ever-increasing number of informative microsatellite loci and improvements in field sampling methods will facilitate expansion of paternity studies and form the basis for other new areas of study. For instance, by comparing genotypes of hatchlings with that of the mothers, it is possible to infer the male genotypes in the breeding population and make progress on tackling some previously elusive population vital rates.

### 6.7.1   Sex Ratios of Breeding Populations

Little is known about adult breeding sex ratios, known as the operational sex ratio (OSR), in sea turtles (Hays et al., 2010). Most research on sex ratios has focused on hatchling and juvenile stages in sea turtles, and there has been concern, prompted by general findings of female-biased hatchling sex ratios, that populations of turtles may become entirely feminized due to warming climate trends and temperature-dependent sex determination. Stewart and Dutton (2011) used kinship analysis to obtain the genotypes of successfully breeding males in a leatherback population without ever encountering them in the field. They assessed hatchlings belonging to 46 female leatherbacks and found that 47 different males had mated with those females (Stewart and Dutton, 2011). One male had mated with three different females, and several others had mated with two females. Using a similar approach, Wright et al. (2012) found that for a green turtle population, despite having a 95% female hatchling sex ratio, there were at least 1.4 reproductive males to every breeding female. These studies show that OSRs may not necessarily be female biased as feared and that breeding males may outnumber breeding females in encounter rates at breeding grounds. Stewart and Dutton (2011) identified one male that had been actively breeding in both 2009 and 2010 (with different females), providing evidence that some males may breed yearly, as also observed in green turtles (Limpus, 1993). Expansion of these studies across multiple years to account for male breeding behavior will be required to accurately estimate the number of breeding males in the population.

### 6.7.2   Age to First Reproduction

The age at first reproduction is one of the most important vital parameters for demographic modeling, and is uncertain for many sea turtle populations, because it has not been possible to easily tag hatchlings and monitor at what age they reach maturity. In leatherbacks for instance, estimates from chondro-osseous morphology, skeletochronology, and growth rate modeling have suggested a range from 3 to 29 years for the age of first reproduction (Rhodin, 1985; Zug and Parham, 1996; Avens et al., 2009; Jones et al., 2011). Dutton et al. (2005) inferred age of first reproduction at around 12–15 years from analysis of demographic trend data, generally corroborating the more recent estimates of 13–16 years proposed by Jones et al. (2011). Genetic fingerprinting was also used to show that first-time nesters in the 1990s were closely related and possibly the genetic offspring of leatherbacks nesting in the 1980s (Dutton et al., 2005) (Figure 6.4).

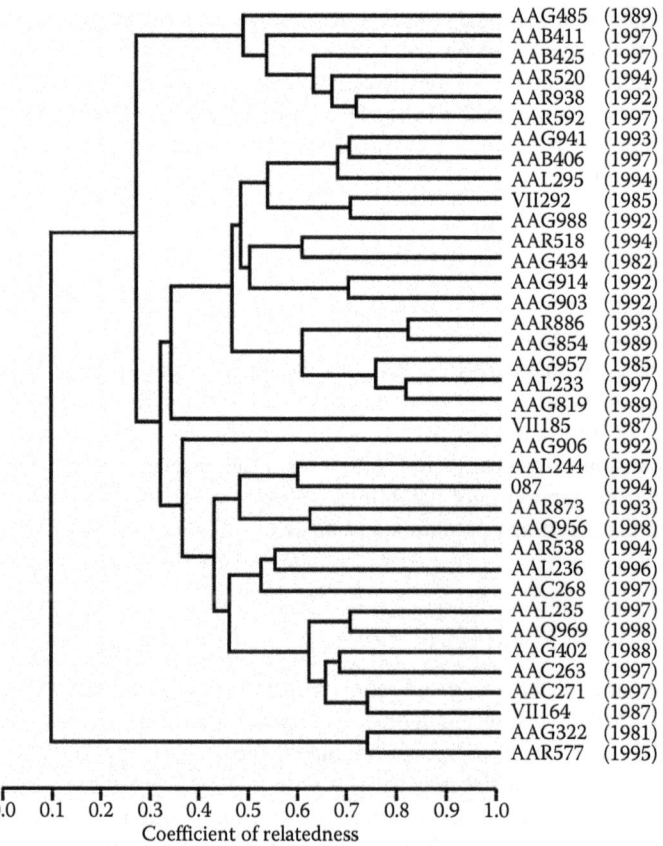

**FIGURE 6.4** Family groups identified among 37 St. Croix leatherback nesters based on relatedness determined with microsatellite genotyping and mtDNA sequencing. The year the turtle was first observed to nest is given in brackets; old-timers, such as AAG322 (identified in 1981) and AAG434 (identified in 1982) are most likely mothers of recent (post 1993) first-time nesters such as AAR577 (1995) and AAR518 (1994), respectively. (From Dutton, D.L. et al., *Biol. Conserv.*, 126, 186, 2005.)

A new approach uses genetic analyses to "tag" hatchlings for a long-term capture-mark-recapture study, using non-injurious sampling methods established for collecting hatchling DNA, which will be used to create a genetic fingerprint or "tag" to identify individual turtles throughout their lifetime (Dutton et al., 2008; Stewart and Dutton, 2012). Genetic samples routinely collected from first-time nesters in future years will be analyzed and compared to the stored hatchling genotypes to identify the individuals that were originally "tagged" at birth and directly determine age at first reproduction and juvenile survival rates for this population by following a cohort of hatchlings to adulthood. Age-specific vital rates of adult females, such as birth and death rates, also may also be estimated by monitoring these cohorts through their lifetimes, providing crucial information for future studies of the species. Given the rapid advances occurring in biotechnology and information management systems, it should be possible to expand the use of genetic fingerprinting in a broad range of Capture-Mark-Recapture applications in the future.

## 6.8 FUTURE DIRECTIONS

There is a growing need for genetic tools to test for finer resolution in genetic structure, based on what is known from field data. Recent efforts to uncover additional genetic structure among rookeries using mtDNA have taken two contrasting directions. In one approach, Shamblin et al. (2012b) sequenced a majority (16,350 bp) of the mtDNA genome to uncover genetic structure among green turtle populations in the Greater Caribbean that were dominated by a common haplotype (CM-A5). This was done by selectively sequencing 20 individuals with the CM-A5 haplotype to determine if there were sequence variants among them. Four variants of CM-A5 were found that were geographically structured, resulting in higher $F_{ST}$ values for three eastern Caribbean rookeries. In contrast, Tikochinski et al. (2012) developed primers to amplify a known microsatellite locus found on the 3′ end of the mtDNA control region to look for cryptic genetic structure in green turtle rookeries in the Mediterranean, in which there was little variation. Microsatellite repeats are a common feature of the mtDNA control region in several species and have been used in phylogenetic and population genetic studies, although it can be difficult to obtain reliable results (Lunt et al., 1998). Within the nesting and stranded green turtles studied (n = 289), sequencing of the microsatellite region yielded 33 haplotypes and repeated sampling of the same individuals (n = 20) gave identical sequences. Both approaches offer promise as techniques to investigate whether there is phylogeographic structure among individuals that share widespread haplotypes.

Rapidly evolving techniques to develop genome-wide markers are likely to lead to a future shift in the genetic markers of choice for some studies. SNPs have been identified within the genome of green turtles (Roden et al., 2009a,b). These SNPs are useful for detecting population structure in green turtles (Roden, 2009), although initial cross-species tests have had limited success, suggesting that SNP markers will need to be developed for each species of sea turtles (Dutton et al., unpublished data; Quinzin et al., unpublished data). Nevertheless, SNPs show great promise for ultimately replacing microsatellites due to their higher data quality and genotyping efficiency (Morin et al., 2004). Currently, next-generation sequencing (NGS) platforms are being used for sea turtle nuclear SNP discovery and have the potential for identifying large numbers of new SNPs as the process becomes significantly more automated and less labor-intensive than traditional PCR and Sanger sequencing (Nielsen et al., 2011). One of the main challenges of these advances in the application of NGS technologies will be dealing with the vast quantity of data generated. Informatics and statistical methods for managing and applying results have not kept pace with the rapidly evolving technologies, and some of the basic analytical approaches for determining stock structure and other aspects of conservation genetics will need to be further developed.

Advances in sampling techniques have also improved the capacity for genetic studies. For studies of hatchlings, sampling of DNA from a sliver off the carapace of hatchlings (e.g., Theissinger et al., 2009) rather than taking blood samples has allowed a less invasive, easier and much quicker technique. To individually identify females nesting at beaches that are not monitored at night, Shamblin et al. (2011b) developed the technique of getting the mothers' DNA from the eggshells of recently (<15 h) deposited eggs to conduct microsatellite analyses. This could be an important tool for determining the number of clutches being laid by females, or the total number of nesting females at beaches where it is not possible to encounter females while nesting.

Although the number of genetic studies has increased dramatically over the past three decades (Figure 6.1), and some may be inclined to think there are a plethora of studies, there remains a great deal of genetic research to be done on sea turtles to contribute to their conservation. This includes genetic studies of turtle diseases (Quackenbush et al., 2001) or commensals (Rawson et al., 2003), mechanisms of sex determination (Torres Maldonado et al., 2002), or molecular evolution of turtle DNA (Russell and Beckenbach, 2008) as well as the topics discussed above. Ultimately, the contribution of genetic studies to our understanding of marine turtles will continue well into the future.

## ACKNOWLEDGMENTS

Thanks to E. Naro-Maciel for providing an adapted figure from Naro-Maciel et al. (2008) and to B. Shamblin for helpful reviewer comments. Thanks to Kelly Stewart and Camryn Allen for comments on the manuscript. Also, thanks to William Perrin and Elizabeth Whitman for editorial assistance.

## REFERENCES

Aggarwal, R. K., A. Lalremruata, T. Velavan, A. P. Sowjanya, and L. Singh. 2008. Development and characterization of ten novel microsatellite markers from olive ridley sea turtle (*Lepidochelys olivacea*). *Conserv. Genet.* 9: 981–984.

Aggarwal, R. K., T. P. Velavan, D. Udaykumar et al. 2004. Development and characterization of novel microsatellite markers from the olive ridley sea turtle (*Lepidochelys olivacea*). *Mol. Ecol. Notes* 4: 77–79.

Allard, M. W., M. M. Miyamoto, K. A. Bjorndal, A. B. Bolten, and B. W. Bowen. 1994. Support for natal homing in green turtles from mitochondrial DNA sequences. *Copeia.* 1: 34–41.

Alstad, T., B. Shamblin, D. Bagley, L. Ehrhart, and C. Nairn. 2011. Isolation and characterization of tetranucleotide microsatellites from the leatherback turtle (*Dermochelys coriacea*). *Conserv. Genet. Resour.* 3: 457–460.

Avens, L. and K. J. Lohmann. 2004. Navigation and seasonal migratory orientation in juvenile sea turtles. *J. Exp. Biol.* 207: 1771–1778.

Avens, L., J. C. Taylor, L. R. Goshe, T. T. Jones, and M. Hastings. 2009. Use of skeletochronological analysis to estimate the age of leatherback sea turtles *Dermochelys coriacea* in the western North Atlantic. *Endanger. Species Res.* 8(3): 165–177.

Avise, J. C. 2009. Phylogeography: Retrospect and prospect. *J. Biogeogr.* 36: 3–15.

Avise, J. C. and B. W. Bowen. 1994. Investigating sea turtle migration using DNA markers. *Curr. Opin. Genet. Dev.* 4: 882–886.

Avise, J. C., B. W. Bowen, T. Lamb, A. B. Meylan, and E. Bermingham. 1992. Mitochondrial DNA evolution at a turtle's pace: Evidence for low genetic variability and reduced microevolutionary rate in the testudines. *Mol. Biol. Evol.* 9: 457–473.

Bagda, E., F. Bardakci, and O. Turkozan. 2012. Lower genetic structuring in mitochondrial DNA than nuclear DNA among the nesting colonies of green turtles in the Mediterranean. *Biochem. Syst. Ecol.* 43: 192–199.

Balazs, G. H. 1982. Growth rates of immature green turtles in the Hawaiian archipelago. In *Biology and Conservation of Sea Turtles*, ed. K. A. Bjorndal, pp. 117–125. Washington, DC: Smithsonian Institution Press.

Bard, E. and R. E. M. Rickaby. 2009. Migration of the subtropical front as a modulator of glacial climate. *Nature* 460: 380–383.

Barley, A. J., P. Q. Spinks, R. C. Thomson, and H. B. Shaffer. 2010. Fourteen nuclear genes provide phylogenetic resolution for difficult nodes in the turtle tree of life. *Mol. Phylogenet. Evol.* 55: 1189–1194.

Bass, A. L. 1999. Genetic analysis to elucidate the natural history and behavior of hawksbill turtles (*Eretmochelys imbricata*) in the wider Caribbean: A review and re-analysis. *Chelonian Conserv. Biol.* 3: 195–199.

Bass, A. L., S. P. Epperly, and J. Braun-McNeill. 2004. Multi-year analysis of stock composition of a loggerhead turtle (*Caretta caretta*) foraging habitat using maximum likelihood and Bayesian methods. *Conserv. Genet.* 5: 783–796.

Bass, A. L., S. P. Epperly, and J. Braun-McNeill. 2006. Green turtle (*Chelonia mydas*) foraging and nesting aggregations in the Caribbean and Atlantic: Impact of currents and behavior on dispersal. *J. Hered.* 94: 346–354.

Bass, A. L., D. A. Good, K. A. Bjorndal et al. 1996. Testing models of female reproductive migratory behaviour and population structure in the Caribbean hawksbill turtle, *Eretmochelys imbricata*, with mtDNA sequences. *Mol. Ecol.* 5: 321–328.

Bass, A. L., C. J. Lagueux, and B. W. Bowen. 1998. Origin of green turtles, *Chelonia mydas*, at 'Sleeping Rocks' off the northeast coast of Nicaragua. *Copeia* 1998: 1064–1069.

Bass, A. L. and W. N. Witzell. 2000. Demographic composition of immature green turtles (*Chelonia mydas*) from the east central Florida coast: Evidence from mtDNA markers. *Herpetologica.* 56: 357–367.

Beerli, P. 2002. Migrate version 1.7.6.1 no longer available. Migrate 3.3.1 Available at http://popgen.sc.fsu.edu/Migrate/Migrate-n.html (accessed October 5, 2012).

Benson, S. R., T. Eguchi, D. G. Foley et al. 2011. Large-scale movements and high-use areas of western Pacific leatherback turtles, *Dermochelys coriacea. Ecosphere* 2(7): 1–27.

Bentivegna, F. 2002. Intra-mediterranean migrations of loggerhead sea turtles (*Caretta caretta*) monitored by satellite telemetry. *Mar. Biol.* 141: 795–800.

Bjorndal, K. A. 1980. Nutrition and grazing behavior of the green turtle *Chelonia mydas. Mar. Biol.* 56: 147–154.

Bjorndal, K. A. and A. B. Bolten. 2008. Annual variation in source contributions to a mixed stock: Implications for quantifying connectivity. *Mol. Ecol.* 17: 2185–2193.

Bjorndal, K. A., A. B. Bolten, and M. Y. Chaloupka. 2003. Survival probability estimates for immature green turtles *Chelonia mydas* in the Bahamas. *Mar. Ecol. Prog. Ser.* 252: 273–281.

Bjorndal, K. A., A. B. Bolten, and H. R. Martins. 2000. Somatic growth model of juvenile loggerhead sea turtles *Caretta caretta*: Duration of pelagic stage. *Mar. Ecol. Prog. Ser.* 202: 265–272.

Bjorndal, K. A., A. B. Bolten, L. Moreira, C. Bellini, and M. Marcovaldi. 2006. Population structure and diversity of Brazilian green turtle rookeries based on mitochondrial DNA sequences. *Chelonian Conserv. Biol.* 5: 262–268.

Bjorndal, K. A., A. B. Bolten, and S. Troëng. 2005. Population structure and genetic diversity in green turtles nesting at Tortuguero, Costa Rica, based on mitochondrial DNA control region sequences. *Mar. Biol.* 147: 1449–1457.

Blumenthal, J., F. Abreu-Grobois, T. J. Austin et al. 2009a. Turtle groups or turtle soup: Dispersal patterns of hawksbill turtles in the Caribbean. *Mol. Ecol.* 18: 4841–4853.

Blumenthal, J. M., T. J. Austin, C. D. L. Bell et al. 2009b. Ecology of hawksbill turtles, *Eretmochelys imbricata*, on a western Caribbean foraging ground. *Chelonian Conserv. Biol.* 8: 1–10.

Bolten, A. B., K. A. Bjorndal, H. R. Martins et al. 1998. Transatlantic developmental migrations of loggerhead sea turtles demonstrated by mtDNA sequence analysis. *Ecol. Appl.* 8: 1–7.

Bourjea, J., S. Lapègue, L. Gagnevin et al. 2007. Phylogeography of the green turtle, *Chelonia mydas*, in the Southwest Indian Ocean. *Mol. Ecol.* 16: 175–186.

Bowen, B. W. 2003. What is a loggerhead turtle? The genetic perspective. In *Loggerhead Sea Turtles*, eds. A. B. Bolten and B. E. Witherington, pp. 7–27. Washington, DC: Smithsonian Institution Press.

Bowen, B. W., F. A. Abreu-Grobois, G. H. Balazs, N. Kamezaki, C. J. Limpus, and R. J. Ferl. 1995. Trans-Pacific migrations of the loggerhead turtle (*Caretta caretta*) demonstrated with mitochondrial DNA markers. *Proc. Natl Acad. Sci. U.S.A.* 92: 3731–3734.

Bowen, B. W., A. L. Bass, S. Chow et al. 2004. Natal homing in juvenile loggerhead turtles (*Caretta caretta*). *Mol. Ecol.* 13: 3797–3808.

Bowen, B. W., A. L. Bass, L. Soares, and R. J. Toonen. 2005. Conservation implications of complex population structure: Lessons from the loggerhead turtle (*Caretta caretta*). *Mol. Ecol.* 14: 2389–2402.

Bowen, B. W., A. M. Clark, F. A. Abreu-Grobois, A. Chaves, H. A. Reichart, and R. J. Ferl. 1998. Global phylogeography of the ridley sea turtles (*Lepidochelys* spp.) as inferred from mitochondrial DNA sequences. *Genetica* 101: 179–189.

Bowen, B. W., A. B. Meylan, and J. C. Avise. 1991. Evolutionary distinctiveness of the endangered Kemp's ridley sea turtle. *Nature* 352: 709–711.

Bowen, B. W., W. S. Nelson, and J. C. Avise. 1993. A molecular phylogeny for marine turtles: Trait mapping, rate assessment, and conservation relevance. *Proc. Natl Acad. Sci. U.S.A.* 90: 5574–5577.

Boyle, M. C. 2007. Post-hatchling sea turtle biology. PhD dissertation, James Cook University, Douglas, Queensland, Australia.

Boyle, M. C., N. N. FitzSimmons, C. J. Limpus, S. Kelez, X. Velez-Zuazo, and M. Waycott. 2009. Evidence for transoceanic migrations by loggerhead sea turtles in the southern Pacific Ocean. *Proc. R. Soc. Lond. [Biol.].* 276: 1993–1999.

Broderick, D., C. Moritz, J. D. Miller et al. 1994. Genetic studies of the hawksbill turtle *Eretmochelys imbricata*: Evidence for multiple stocks in Australian waters. *Pac. Conserv. Biol.* 1: 123–131.

Browne, D. C., J. A. Horrocks, and F. A. Abreu-Grobois. 2009. Population subdivision in hawksbill turtles nesting on Barbados, West Indies, determined from mitochondrial DNA control region sequences. *Conserv. Genet.* 11: 1541–1546.

Camacho-Mosquera, L., D. Amorocho, L. M. Mejía-Ladino, J. D. Palacio-Mejía, and F. Rondón-González. 2008. Genetic characterization of the olive ridley sea turtle -*Lepidochelys olivacea* (Eschscholtz, 1829)— In Gorgona National Natural Park (Colombian Pacific) from mitochondrial DNA sequences. *Investig. Mar.* 37: 77–92.

Carreras, C., M. Pascual, L. Cardona et al. 2007. The genetic structure of the loggerhead sea turtle (*Caretta caretta*) in the Mediterranean as revealed by nuclear and mitochondrial DNA and its conservation implications. *Conserv. Genet.* 8: 761–775.

Carreras, C., S. Pont, F. Maffucci et al. 2006. Genetic structuring of immature loggerhead sea turtles (*Caretta caretta*) in the Mediterranean Sea reflects water circulation patterns. *Mar. Biol.* 149: 1269–1279.

Casale, P., D. Freggi, P. Gratton, R. Argano, and M. Oliverio. 2008. Mitochondrial DNA reveals regional and interregional importance of the central Mediterranean African shelf for loggerhead sea turtles (*Caretta caretta*). *Sci. Mar.* 72: 541–548.

Chassin-Noria, O., A. Abreu-Grobois, P. H. Dutton, and K. Oyama. 2004. Conservation genetics of the East Pacific green turtle (*Chelonia mydas*) in Michoacan, Mexico. *Genetica* 121: 195–206.

Cheng, I. J., P. H. Dutton, C. L. Chen, H. C. Chen, Y-H. Chen, and J. W. Shea. 2008. Comparison of the genetics and nesting ecology of two green turtle rookeries. *J. Zool.* 276: 375–384.

Crim, J. L., L. D. Spotila, J. R. Spotila et al. 2002. The leatherback turtle, *Dermochelys coriacea*, exhibits both polyandry and polygyny. *Mol. Ecol.* 11: 2097–2106.

Dethmers, K. E. M., D. Broderick, C. Moritz et al. 2006. The genetic structure of Australasian green turtles (*Chelonia mydas*): Exploring the geographical scale of genetic exchange. *Mol. Ecol.* 15: 3931–3946.

Dethmers, K. E. M., M. P. Jensen, N. N. FitzSimmons, D. Broderick, C. J. Limpus, and C. Moritz. 2010. Migration of green turtles (*Chelonia mydas*) from Australasian feeding grounds inferred from genetic analyses. *Mar. Fresh. Res.* 61: 1376–1387.

Diez, C. E. and R. P. Van Dam. 2003. Sex ratio of an immature hawksbill sea turtle aggregation at Mona Island, Puerto Rico. *J. Herpetol.* 37: 533–537.

Duchene, S., A. Frey, A. Alfaro-Nunez et al. 2012. Marine turtle mitogenome phylogenetics and evolution. *Mol. Phylogenet. Evol.* 6(1): 241–250.

Duncan, K. M., A. P. Martin, B. W. Bowen, and H. G. De Couet. 2006. Global phylogeography of the scalloped hammershark (*Sphyrna lewini*). *Mol. Ecol.* 15: 2239–2251.

Dutton, P. H., G. Balazs, A. Dizon, and A. Barragan. 2000. Genetic stock identification and distribution of leatherbacks in the Pacific: potential effects on declining populations. In the *Proceedings of the Eighteenth International Sea Turtle Symposium*, Abreu-Grobois, F. A., R. Briseño-Dueñas, R. Márquez-Millán and L. Sarti-Martinez, Editors,  U.S. Dept. Commerce. NOAA Tech. Memo., NMFS-SEFSC-436, 38-39, Miami, FL.

Dutton, P. H., B. W. Bowen, D. W. Owens, A. Barragan, and S. K. Davis. 1999. Global phylogeography of the leatherback turtle (*Dermochelys coriacea*). *J. Zool.* 248: 397–409.

Dutton, P. H., S. K. Davis, T. Guerra, and D. Owens. 1996. Molecular phylogeny for marine turtles based on sequences of the ND4-leucine tRNA and control regions of mitochondrial DNA. *Mol. Phylogenet. Evol.* 5: 511–521.

Dutton, D. L., P. H. Dutton, M. Chaloupka, and R. H. Boulon. 2005. Increase of a Caribbean leatherback turtle *Dermochelys coriacea* nesting population linked to long-term nest protection. *Biol. Conserv.* 126(2): 186–194.

Dutton, P. H., T. Eguchi, K. R. Stewart, J. Alexander-Garner, and S. Garner. 2008. Pilot study to develop non-injurious genetic sampling methods to mass tag hatchling leatherback turtles. Final Report No. 41526 to the U.S. Fish and Wildlife Service, Christiansted, St. Croix, USVI.

Dutton, P. H. and A. Frey. 2009. Characterization of polymorphic microsatellite markers for the green turtle (*Chelonia mydas*). *Mol. Ecol. Resour.* 9: 354–356.

Dutton, P. H., C. Hitipeuw, M. Zein, S. R. Benson, G. Petro, J. Pita, V. Rei, L. Ambio, and J. Bakarbessy. 2007. Status and genetic structure of nesting populations of leatherback turtles (*Dermochelys coriacea*) in the western Pacific. *Chelonian Conserv. Biol.* 6: 47–53.

Dutton, P. H., m S. Roden, K. R. Stewart, E. LaCasella, M. Tiwari, A. Formia, J. Thomé, S. R. Livingstone, S. Eckert, D. Chacon-Chaverri, P. Rivalan, and P. Allman. 2013. Population stock structure of leatherback turtles (*Dermochelys coriacea*) in the Atlantic revealed using mtDNA and microsatellite markers. Conservation Genetics, in press.

Dutton, P. H. and D. Squires. 2011. A holistic strategy for Pacific sea turtle conservation. In *Conservation and Sustainable Management of Sea Turtles in the Pacific Ocean*, eds. P. H. Dutton, D. Squires, and A. Mahfuzuddin, pp. 37–59. Honolulu, HI: University of Hawaii Press.

Encalada, S. E., K. A. Bjorndal, A. B. Bolten et al. 1998. Population structure of loggerhead turtle (*Caretta caretta*) nesting colonies in the Atlantic and Mediterranean as inferred from mitochondrial DNA control region sequences. *Mar. Biol.* 130: 567–575.

Encalada, S. E., P. N. Lahanas, K. A. Bjorndal, A. B. Bolten, M. M. Miyamoto, and B. W. Bowen. 1996. Phylogeography and population structure of the Atlantic and Mediterranean green turtle *Chelonia mydas*: A mitochondrial DNA control region sequence assessment. *Mol. Ecol.* 5: 473–483.

Engstrom, T. N., P. A. Meylan, and A. B. Meylan. 2002. Origin of juvenile loggerhead turtles (*Caretta caretta*) in a tropical developmental habitat in Caribbean Panama. *Anim. Conserv.* 5: 125–133.

FitzSimmons, N. N. 1998. Single paternity of clutches and sperm storage in the promiscuous green turtle (*Chelonia mydas*). *Mol. Ecol.* 7: 575–584.

FitzSimmons, N. N. 2010. Population genetic studies in support of conservation and management of hawksbill turtles in the Indian Ocean. Marine Conservation Fund Award 98210–7-G126, Final Report to U.S. Fish and Wildlife Service.

FitzSimmons, N. N., C. J. Limpus, J. A. Norman, A. R. Goldizen, J. D. Miller, and C. Moritz. 1997a. Philopatry of male marine turtles inferred from mitochondrial DNA markers. *Proc. Natl Acad. Sci. U.S.A.* 94: 8912–8917.

FitzSimmons, N. N., C. Moritz, C. J. Limpus, L. Pope, and R. Prince. 1997b. Geographic structure of mito-chondrial and nuclear gene polymorphisms in Australian green turtle populations and male-biased gene flow. *Genetics* 147: 1843–1854.

FitzSimmons, N. N., C. Moritz, and S. S. Moore. 1995. Conservation and dynamics of microsatellite loci over 300 million years of marine turtle evolution. *Mol. Biol. Evol.* 12: 432–440.

Foley, A. M., K. E. Singel, P. H. Dutton, and T. M. Summers. 2007. Characteristics of a green turtle (*Chelonia mydas*) assemblage in northwestern Florida determined during a hypothermic stunning event. *Gulf Mexico Sci.* 2007: 131–143.

Formia, A., A. C. Broderick, F. Glen, B. J. Godley, G. C. Hays, and M. W. Bruford. 2007. Genetic composition of the Ascension Island green turtle rookery based on mitochondrial DNA: Implications for sampling and diversity. *Endanger. Species Res.* 3: 145–158.

Formia, A., B. J. Godley, J. F. Dontaine, and M. W. Bruford. 2006. Mitochondrial DNA diversity and phylogeog-raphy of endangered green turtle (*Chelonia mydas*) populations in Africa. *Conserv. Genet.* 7: 353–369.

Frey, A. and P. H. Dutton. 2012. Whole mitogenomic sequences for further resolution of ubiquitous dloop haplotypes in Pacific Green turtles. In *Proceedings of the Thirty-first Annual Symposium on Sea Turtle Biology and Conservation.* Jones, T. T. and Wallace, B. P., Editors, U.S. Dept. Commerce. NOAA Tech. Memo., NMFS-SEFSC-631, 27, Miami, FL.

Fuentes, M. M. P. B., M. Hamann, and C. J. Limpus. 2010. Past, current and future thermal profiles for green turtle nesting grounds: Implications from climate change. *J. Exp. Mar. Biol. Ecol.* 383: 56–64.

Fuentes, M. M. P. B., M. Hamann, and V. Lukoschek. 2009. Marine reptiles. In *A Marine Climate Change Impacts and Adaptation Report Card for Australia* 2009, eds. E. S. Poloczanska, A. J. Hobday, and A. J. Richardson, National Climate Change Adaptation Research Facility, Gold Coast, Australia, NCCARF Publication 05/09, ISBN 978-1-921609-03-9.

Garofalo, L., T. Mingozzi, A. Micò, and A. Novelletto. 2009. Loggerhead turtle (*Caretta caretta*) matrilines in the Mediterranean: Further evidence of genetic diversity and connectivity. *Mar. Biol.* 156: 2085–2095.

Godley, B. J., C. Barbosa, M. W. Bruford et al. 2010. Unravelling migratory connectivity in marine turtles using multiple methods. *J. Appl. Ecol.* 47: 769–778.

Godley, B. J., E. H. S. M. Lima, S. Åkesson et al. 2003. Movement patterns of green turtles in Brazilian coastal waters described by satellite tracking and flipper tagging. *Mar. Ecol. Prog. Ser.* 253: 279–288.

Godley, B. J., S. Richardson, A. C. Broderick, M. S. Coyne, F. Glen, and G. C. Hays. 2002. Long-term satellite telemetry of the movements and habitat utilisation by green turtles in the Mediterranean. *Ecography* 25: 352–362.

Grant, W. S., G. B. Milner, P. Krasnowski, F. M. Utter. 1980. Use of bio-chemical genetic variants for identification of sockeye salmon (Oncorrhynchus nerka) stocks in Cook Inlet, Alaska. *Canadian Journal of Fisheries and Aquatic Sciences,* 37: 1236–1247.

Grossman, A., C. Bellini, A. Fallabrino et al. 2007. Second TAMAR-tagged hawksbill recaptured in Corisco Bay, West Africa. *Mar. Turtle Newslett.* 116: 26.

Hamann, M., C. J. Limpus, and M. A. Read. 2007. Vulnerability of marine reptiles to climate change in the Great Barrier Reef. In: *Climate Change and the Great Barrier Reef: A Vulnerability Assessment*, pp. 465–497. Johnson, J. E., and Marshall, P. A., Townsville, Queensland, Australia: Great Barrier Reef Marine Park Authority and Australian Greenhouse Office.

Haq, B. U., J. Hardenbol, and P. R. Vail. 1987. Chronology of fluctuating sea levels since the triassic. *Science* 235: 1156–1167.

Harry, J. L. and D. A. Briscoe. 1988. Multiple paternity in the loggerhead turtle (*Caretta caretta*). *J. Hered.* 79: 96–99.

Hatase, H., M. Kinoshita, T. Bando et al. 2002. Population structure of loggerhead turtles, *Caretta caretta*, nest-ing in Japan: Bottlenecks on the Pacific population. *Mar. Biol.* 141: 299–305.

Hays, G. C., S. Fossette, K. A. Katselidis, G. Schofield, and M. B. Gravenor. 2010. Breeding periodicity for male sea turtles, operational sex ratios, and implications in the face of climate change. *Conserv. Biol.* 24(6): 1636–1643.

Hirayama, R. 1998. Oldest known sea turtle. *Nature* 392: 2259–2262.

Hoekert, W. E. J., H. Neufeglise, A. D. Schouten, and S. B. J. Menken. 2002. Multiple paternity and female-biased mutation at a microsatellite locus in the olive ridley sea turtle (*Lepidochelys olivacea*). *Heredity* 89: 107–113.

Ireland, J. S., A. C. Broderick, F. Glen et al. 2003. Multiple paternity assessed using microsatellite markers, in green turtles *Chelonia mydas* (Linnaeus, 1758) of Ascension Island, South Atlantic. *J. Exp. Mar. Biol. Ecol.* 291: 149–160.

Jensen, M. P. 2010. Assessing the composition of green turtle (*Chelonia mydas*) foraging grounds in Australasia using mixed stock analyses. PhD dissertation, University of Canberra, Australia.

Jensen, M. P., F. A. Abreu-Grobois, J. Frydenberg, and V. Loeschcke. 2006. Microsatellites provide insight into contrasting mating patterns in arribada vs. non-arribada olive ridley sea turtle rookeries. *Mol. Ecol.* 15: 2567–2575.

Jones, T. T., M. D. Hastings, B. L. Bostrom, D. Pauly, and D. R. Jones. 2011. Growth of captive leatherback turtles, *Dermochelys coriacea*, with inferences on growth in the wild: Implications for population decline and recovery. *J. Exp. Mar. Biol. Ecol.* 399(1): 84–92.

Jones, M. R. and T. Torgersen. 1988. Late Quaternary evolution of Lake Carpentaria on the Australian–New Guinea continental shelf. *Aust. J. Earth Sci.* 35: 313–324.

Joseph, J. and P. W. Shaw. 2010. Multiple paternity in egg clutches of hawksbill turtles (*Eretmochelys imbricata*). *Conserv. Genet.* 12: 601–605.

Kichler, K., M. T. Holder, S. J. Davis, R. Márquez, and D. W. Owens. 1999. Detection of multiple paternity in the Kemp's ridley sea turtle with limited sampling. *Mol. Ecol.* 8: 819–830.

Lahanas, P. N., M. M. Miyamoto, K. A. Bjorndal, and A. B. Bolten. 1994. Molecular evolution and population genetics of Greater Caribbean green turtles (*Chelonia mydas*) as inferred from mitochondrial DNA control region sequences. *Genetica* 94: 57–66.

Lara-De La Cruz, L. I., K. O. Nakagawa, H. Cano-Camacho et al. 2010. Detecting patterns of fertilization and frequency of multiple paternity in *Chelonia mydas* of Colola (Michoacan, Mexico). *Hidrobiológica* 20: 85–89.

Laurent, L., P. Casale, M. N. Bradai et al. 1998. Molecular resolution of marine turtle stock composition in fishery bycatch: A case study in the Mediterranean. *Mol. Ecol.* 7: 1529–1542.

Lee, P. L. M. and G. C. Hays. 2004. Polyandry in marine turtles: Females make the best of a bad job. *Proc. Natl Acad. Sci. U.S.A.* 101: 6530–6535.

LeRoux, R. A., P. H. Dutton, F. A. Abreu-Grobois. 2012. Re-examination of population structure and phylogeography of hawksbill turtles in the Wider Caribbean using longer mtDNA sequences. *J. Hered.* doi: 10.1093/jhered/ess055

Limpus, C. J. 1993. The green turtle, *Chelonia mydas*, in Queensland: Breeding males in the southern Great Barrier Reef. *Wildlife Res.* 20: 513–523.

Limpus, C. J. 2007. *A Biological Review of Australian Marine Turtles*: 2. *Green Turtle (Chelonia mydas) (Linnaeus)*. State of Queensland, Australia: Environmental Protection Agency. ISBN 978-0-9803613-2-2.

Limpus, C. J. 2009. *A Biological Review of Australian Marine Turtles*: 3. *Hawksbill Turtle (Eretmochelys imbricata) (Linnaeus)*. State of Queensland, Australia: Environmental Protection Agency. ISBN 978–0–9803613–2–2.

Limpus, C. J., P. J. Couper, and M. A. Read. 1994. The green turtle, *Chelonia mydas*, in Queensland: Population structure in a warm temperate feeding area. *Mem. Queensl. Mus.* 35: 139–154.

Limpus, C. J., J. D. Miller, C. J. Parmenter, and D. J. Limpus. 2000. The green turtle, *Chelonia mydas*, population of Raine Island and the northern Great Barrier Reef: 1843–2001. *Mem. Queensl. Mus.* 49: 349–440.

Limpus, C. J., J. D. Miller, C. J. Parmenter, D. Reimer, N. McLachlan, and R. Webb. 1992. Migration of green (*Chelonia mydas*) and loggerhead (*Caretta caretta*) turtles to and from eastern Australian rookeries. *Wildlife Res.* 19(3): 347–358.

Lin, G., A. Chang, H. W. Yap, and G. H. Yue. 2008. Characterization and cross-species amplification of microsatellites from the endangered Hawksbill turtle (*Eretmochelys imbricate*). *Conserv. Genet.* 9: 1071–1073.

López-Castro, M. C. and A. Rocha-Olivares. 2005. The panmixia paradigm of eastern Pacific olive ridley turtles revised: Consequences for their conservation and evolutionary biology. *Mol. Ecol.* 14: 3325–3334.

Luke, K., J. A. Horrocks, R. A. LeRoux, and P. H. Dutton. 2004. Origins of green turtle (*Chelonia mydas*) feeding aggregations around Barbados, West Indies. *Mar. Biol.* 144: 799–805.

Lunt, D. H., L. E. Whipple, and B. C. Hyman. 1998. Mitochondrial DNA variable number tandem repeats (VNTRs): Utility and problems in molecular ecology. *Mol. Ecol.* 7: 1441–1455.

Lutz, P. L. and J. A. Musick. 1997. *The Biology of Sea Turtles*. Boca Raton, FL: CRC Press.

Maffucci, F, W. H. C. F. Kooistra, and F. Bentivegna. 2006. Natal origin of loggerhead turtles, *Caretta caretta*, in the neritic habitat off the Italian coasts, central Mediterranean. *Biol. Conserv.* 127: 183–189.

McClellan, C. M. and A. J. Read. 2007. Complexity and variation in loggerhead sea turtle life history. *Biol. Lett.* 3: 592.

Meylan, A. B., B. W. Bowen, and J. C. Avise. 1990. A genetic test of the natal homing versus social facilitation models for green turtle migration. *Science* 248: 724–727.

Miller, J. D. and C. J. Limpus. 2003. Ontogeny of marine turtle gonads. In *The Biology of Sea Turtles*, eds. P. Lutz, J. Musick, and J. Wyneken, pp. 225–241. Boca Raton, FL: CRC Press.

Miro-Herrans, A. T., X. Velez-Zuazo, J. P. Acevedo, and W. O. McMillan. 2008. Isolation and characterization of novel microsatellites from the critically endangered hawksbill sea turtle (*Eretmochelys imbricata*). *Mol. Ecol. Res.* 8: 1098–1101.

Monzón-Argüello, C., N. S. Loureiro, C. Delgado et al. 2011. Príncipe Island hawksbills: Genetic isolation of an eastern Atlantic stock. *J. Exp. Mar. Biol. Ecol.* 407: 345–354.

Monzón-Argüello, C., J. Muñoz, A. Marco, L. F. López-Jurado, and C. Rico. 2008. Twelve new polymorphic microsatellite markers from the loggerhead sea turtle (*Caretta caretta*) and cross-species amplification on other marine turtle species. *Conserv. Genet.* 9: 1045–1049.

Monzón- Argüello, C., C. Rico, C. Carreras et al. 2009. Variation in spatial distribution of juvenile loggerhead turtles in the eastern Atlantic and western Mediterranean Sea. *J. Exp. Mar. Biol. Ecol.* 373(2): 79–86.

Monzón-Argüello, C., C. Rico, A. Marco, P. Lopez, and L. F. López-Jurado. 2010a. Genetic characterization of eastern Atlantic hawksbill turtles at a foraging group indicates major undiscovered nesting populations in the region. *J. Exp. Mar. Biol. Ecol.* 387: 9–14.

Monzón-Argüello, C., C. Rico, E. Naro-Maciel et al. 2010b. Population structure and conservation implications for the loggerhead sea turtle of the Cape Verde Islands. *Conserv. Genet.* 10: 1871–1884.

Morin, P. A. 2012. Applications of "next generation" of sequencing and SNP genotyping for population genetics and phylogeography studies. In *Proceedings of the Thirty-first Annual Symposium on Sea Turtle Biology and Conservation*. Jones, T. T. and Wallace, B. P., Editors, U.S. Dept. Commerce. NOAA Tech. Memo., NMFS-SEFSC-631, 23, Miami, FL.

Morin, P. A., G. Luikart, R. K. Wayne, and SNP workshop group. 2004. SNPs in ecology, evolution and conservation. *Trends Ecol. Evol.* 19: 208–216.

Moritz, C. 1994. Defining 'evolutionary significant units' for conservation. *Trends Ecol. Evol.* 9: 373–375.

Musick, J. A. and C. J. Limpus. 1997. Habitat utilization and migration in juvenile sea turtles. In *Biology of Sea Turtles*, eds. P. L. Lutz and J. A. Musick, Vol. 1, pp. 137–163. Boca Raton, FL: CRC Press.

Naro-Maciel, E. 2006. Connectivity and structure of Atlantic green sea turtles (*Chelonia mydas*): A genetic perspective. PhD dissertation, Columbia University, New York.

Naro-Maciel, E., J. H. Becker, E. H. S. M. Lima, M. A. Marcovaldi, and R. DeSalle. 2007. Testing dispersal hypotheses in foraging green sea turtles (*Chelonia mydas*) of Brazil. *J. Hered.* 98: 29–39.

Naro-Maciel, E., M. Le, N. N. FitzSimmons, and G. Amato. 2008. Evolutionary relationships of marine turtles: A molecular phylogeny based on nuclear and mitochondrial genes. *Mol. Phylogenet. Evol.* 49: 659–662.

National Research Council Committee on the Review of Sea-Turtle Population Methods. 2010. *Assessment of Sea-Turtle Status and Trends: Integrating Demography and Abundance*. Washington, DC: The National Academies Press.

Nielsen, R., J. S. Paul, A. Albrechtsen, and Y. S. Song. 2011. Genotype and SNP calling from next-generation sequencing data. *Nat. Rev. Genet.* 12: 443–451.

Nishizawa, H., O. Abe, J. Okuyama, M. Kobayashi, and N. Arai. 2011. Population genetic structure and implications for natal philopatry of nesting green turtles *Chelonia mydas* in the Yaeyama Islands, Japan. *Endanger. Species Res.* 14(2): 141–148.

Okayama, T., R. Diaz-Fernandez, Y. Baba et al. 1999. Genetic diversity of the hawksbill turtle in the Indo-Pacific and Caribbean regions. *Chelonian Conserv. Biol.* 3: 362–367.

Palsbøll, P. J., M. Berube, and F. W. Allendorf. 2007. Identification of management units using population genetic data. *Trends Ecol. Evol.* 22: 11–16.

Peare, T. and P. G. Parker. 1996. Local genetic structure within two rookeries of *Chelonia mydas* (the green turtle). *Heredity* 77: 619–628.

Pella, J. J. and G. B. Milner. 1987. Use of genetic markers in stock composition analysis. In *Population Genetics and Fisheries Management*, eds. N. Ryman and F. Utter, pp. 247–275. Seattle, WA: University of Washington Press.

Pilcher, N. J. 2010. Population structure and growth of immature green turtles at Mantanani, Sabah, Malaysia. *J. Herpetol.* 44(1): 168–171.

Pittard, S. D. 2010. Genetic population structure of the flatback turtle (*Natator depressus*): A nuclear and mitochondrial DNA analysis. Honours thesis, University of Canberra, Canberra, Australia.

Plot, V., B. De Thoisy, S. Blanc et al. 2011. Reproductive synchrony in a recovering bottlenecked sea turtle population. *J. Anim. Ecol.* 81: 341–351.

Pritchard, P. H. C. 1969. Studies of the systematics and reproductive cycles of the genus *Lepidochelys*. PhD dissertation, University of Florida, Gainesville, FL.

Proietti, M. C., J. W. Reisser, P. G. Kinas et al. 2012. Green turtle *Chelonia mydas* mixed stocks in the western South Atlantic, as revealed by mtDNA haplotypes and drifter trajectories. *Mar. Ecol. Prog. Ser.* 447: 195–209.

Prosdocimi, L., V. González Carman, D. A. Albareda, and M. I. Remis. 2011. Genetic composition of green turtle feeding grounds in coastal waters of Argentina based on mitochondrial DNA. *J. Exp. Mar. Biol. Ecol.* 412: 37–45.

Quackenbush, S. L., R. N. Casey, R. J. Murcek et al. 2001. Quantitative analysis of herpesvirus sequences from normal tissue and fibropapillomas of marine turtles with real-time PCR. *Virology* 287: 105–111.

Rankin-Baransky, K., C. J. Williams, A. L. Bass, B. W. Bowen, and J. R. Spotila. 2001. Origin of logger-head turtles stranded in the northeastern United States as determined by mitochondrial DNA analysis. *J. Herpetol.* 35: 638–646.

Rawson, P. D., R. Macnamee, M. G. Frick, and K. L. Williams. 2003. Phylogeography of the coronulid barnacle, *Chelonibia testudinaria*, from loggerhead sea turtles, *Caretta caretta*. *Mol. Ecol.* 12: 2697–2706.

Reece, J. S., T. A. Castoe, and C. L. Parkinson. 2005. Historical perspectives on population genetics and conservation of three marine turtle species. *Conserv. Genet.* 6: 235–251.

Reece, J. S., L. M. Ehrhart, and C. L. Parkinson. 2006. Mixed stock analysis of juvenile loggerheads (*Caretta caretta*) in Indian River Lagoon, Florida: Implications for conservation planning. *Conserv. Genet.* 7: 345–352.

Reis, E. C., L. S. Soares, and G. Lôbo-Hajdu. 2010a. Evidence of olive ridley mitochondrial genome introgression into loggerhead turtle rookeries of Sergipe, Brazil. *Conserv. Genet.* 11: 1587–1591.

Reis, E. C., L. S. Soares, S. M. Vargas et al. 2010b. Genetic composition, population structure and phylogeography of the loggerhead sea turtle: Colonization hypothesis for the Brazilian rookeries. *Conserv. Genet.* 11: 1467–1477.

Rhodin, A. G. J. 1985. Comparative chondro-osseous development and growth of marine turtles. *Copeia.* 1985(3): 752–771.

Rivalan, P., P. H. Dutton, E. Baudry, S. E. Roden, and M. Girondot. 2006. Demographic scenario inferred from genetic data in leatherback turtles nesting in French Guiana and Suriname. *Biol. Conserv.* 130: 1–9.

Roberts, M. A., C. J. Anderson, B. Stender et al. 2005. Estimated contribution of Atlantic coastal loggerhead turtle nesting populations to offshore feeding aggregations. *Conserv. Genet.* 6: 133–139.

Roberts, M. A., T. S. Schwartz, and S. A. Karl. 2004. Global population genetic structure and male-mediated gene flow in the green sea turtle (*Chelonia mydas*): Analysis of microsatellite loci. *Genetics* 166: 1857–1870.

Roden, S. E. 2009. Detecting green turtle population structure in the Pacific using single nucleotide polymorphisms (SNPs). Master's thesis, University of San Diego, San Diego, CA, pp. 1–68.

Roden, S. E. and P. H. Dutton. 2011. Isolation and characterization of 14 polymorphic microsatellite loci in the leatherback turtle (*Dermochelys coriacea*) and cross-species amplification. *Conserv. Genet. Resour.* 3: 49–52.

Roden, S. E., P. H. Dutton, and P. A. Morin. 2009a. Characterization of single nucleotide polymorphism markers for the green sea turtle (*Chelonia mydas*). *Mol. Ecol. Resour.* 9: 1055–1060.

Roden, S. E., P. H. Dutton, and P. A. Morin. 2009b. AFLP fragment isolation technique as a method to produce random sequences for single nucleotide polymorphism discovery in the green turtle, *Chelonia mydas*. *J. Hered.* 100: 390–393.

Ruiz-Urquiola, A., F. B. Riverón-Giró, E. Pérez-Bermúdez et al. 2010. Population genetic structure of greater Caribbean green turtles (*Chelonia mydas*) based on mitochondrial DNA sequences, with an emphasis on rookeries from southwestern Cuba. *Revista de Investigaciones Mar.* 31: 33–52.

Russell, R. D. and A. T. Beckenbach. 2008. Recoding of translation in turtle mitochondrial genomes: Programmed frameshift mutations and evidence of a modified genetic code. *J. Mol. Evol.* 67: 682–695.

Saied, A., F. Maffucci, S. Hochscheid et al. 2012. Loggerhead turtles nesting in Libya: An important management unit for the Mediterranean stock. *Mar. Ecol. Prog. Ser.* 450: 207–218.

Sears, C. J., B. W. Bowen, R. W. Chapman, S. B. Galloway, S. R. Hopkins-Murphy, and C. M. Woodley. 1995. Demographic composition of the feeding population of juvenile loggerhead sea turtles (*Caretta caretta*) off Charleston, South Carolina: Evidence from mitochondrial DNA markers. *Mar. Biol.* 123: 869–874.

Shamblin, B., B. Berry, D. Lennon, D. Bagley, L. Ehrhart, and C. Nairn. 2012a. Tetranucleotide microsatellite loci from the endangered green turtle (*Chelonia mydas*). *Conserv. Genet. Resour.* 4: 783–785.

Shamblin, B. M., K. A. Bjorndal, A. B. Bolten et al. 2012b. Mitogenomic sequences better resolve stock structure of southern greater Caribbean green turtle rookeries. *Mol. Ecol.* 21: 2330–2340.

Shamblin, B. M., M. G. Dodd, D. A. Bagley et al. 2011a. Genetic structure of the southeastern United States loggerhead turtle nesting aggregation: Evidence of additional structure within the peninsular Florida recovery unit. *Mar. Biol.* 158: 571–587.

Shamblin, B. M., M. G. Dodd, K. L. Williams, M. G. Frick, R. Bell, and C. J. Nairn. 2011b. Loggerhead turtle eggshells as a source of maternal nuclear genomic DNA for population genetic studies. *Mol. Ecol. Resour.* 11: 110–115.

Shamblin, B. M., B. C. Faircloth, M. Dodd et al. 2007. Tetranucleotide microsatellites from the loggerhead sea turtle (*Caretta caretta*). *Mol. Ecol. Notes* 7: 784–787.

Shamblin, B. M., B. C. Faircloth, M. G. Dodd et al. 2009. Tetranucleotide markers from the loggerhead sea turtle (*Caretta caretta*) and their cross-amplification in other marine turtle species. *Conserv. Genet.* 10: 577–580.

Shanker, K., J. Ramadevi, B. C. Choudhury, L. Singh, and R. K. Aggarwal. 2004. Phylogeography of olive ridley turtles (*Lepidochelys olivacea*) on the east coast of India: Implications for conservation theory. *Mol. Ecol.* 13: 1899–1909.

Shillinger, G. L., D. M. Palacios, H. Bailey et al. 2008. Persistent leatherback turtle migrations present opportunities for conservation. *PLoS Biol.* 6: e171. doi:10.1371/journal.pbio.0060171.

Stewart, K. R. and P. H. Dutton. 2011. Paternal genotype reconstruction reveals multiple paternity and sex ratios in a breeding population of leatherback turtles (*Dermochelys coriacea*). *Conserv. Genet.* 12: 1101–1113.

Stewart, K. R. and P. H. Dutton. 2012. Sea turtle CSI: It's all in the genes. State of the world's sea turtles report, Vol. 7, pp. 12–13.

Theissinger, K., N. N. FitzSimmons, C. J. Limpus, C. J. Parmenter, and A. D. Phillott. 2009. Mating system, multiple paternity and effective population size in the endemic flatback turtle (*Natator depressus*) in Australia. *Conserv. Genet.* 10: 329–346.

Thomson, R. C. and H. B. Shaffer. 2010. Sparse supermatrices for phylogenetic inference: Taxonomy, alignment, rogue taxa, and the phylogeny of living turtles. *Syst. Biol.* 59: 42–58.

Tikochinski, Y., R. Bendelac, A. Barash, A. Daya, Y. Levy, and A. Friedmann. 2012. Mitochondrial DNA STR analysis as a tool for studying the green sea turtle (*Chelonia mydas*) populations: The Mediterranean Sea case study. *Mar. Genomics* 6: 17–24.

Torres Maldonado, L. C., A. Landa Piedra, N. Moreno Mendoza, A. Marmolejo Valencia, A. Meza Martinez, and H. Merchant Larios. 2002. Expression profiles of Dax1, Dmrt1, and Sox9 during temperature sex determination in gonads of the sea turtle *Lepidochelys olivacea*. *Gen. Comp. Endocrinol.* 129: 20–26.

Velez-Zuazo, X., W. D. Ramos, R. P. Van Dam, C. E. Diez, A. Abreu-Grobois, and W. O. McMillan. 2008. Dispersal, recruitment and migratory behaviour in a hawksbill sea turtle aggregation. *Mol. Ecol.* 17: 839–853.

Whiting, S. and A. U. Koch. 2006. Oceanic movement of a benthic foraging juvenile hawksbill turtle from the Cocos (Keeling) Islands. *Mar. Turtle Newslett.* 112: 15–16.

Witzell, W. N., A. L. Bass, M. J. Bresette, D. A. Singewald, and J. C. Gorham. 2002. Origin of immature loggerhead sea turtles (*Caretta caretta*) at Hutchinson Island, Florida: Evidence from mtDNA markers. *Fish. Bull.* 100: 624–631.

Wright, L. I., L. Kimberley Stokes, W. J. Fuller et al. 2012. Turtle mating patterns buffer against disruptive effects of climate change. *Proc. R. Soc. Lond. [Biol.].* 279: 2122–2127.

Yilmaz, C., O. Turkozan, and F. Bardakci. 2011. Genetic structure of loggerhead turtle (*Caretta caretta*) populations in Turkey. *Biochem. Syst. Ecol.* 39: 266–276.

Zbinden, J. A., C. R. Largiadér, F. Leippert, D. Margaritoulis, and R. Arlettaz. 2007. High frequency of multiple paternity in the largest rookery of Mediterranean loggerhead sea turtles. *Mol. Ecol.* 16: 3703–3711.

Zolgharnein, H., M. A. Salari-aliabadi, and A. M. Forougmand. 2011. Genetic population structure of hawksbill turtle (*Eretmochelys imbricta*) using microsatellite analysis. *Agriculture* 9: 56–62.

Zug, G. R. and J. Parham. 1996. Age and growth in leatherback turtles, *Dermochelys coriacea* (Testudines: Dermochelyidae): A skeletochronological analysis. *Chelonian Conserv. Biol.* 2: 244–249.

# 7 Oceanic Habits and Habitats
## Dermochelys coriacea

*Vincent S. Saba*

## CONTENTS

## 7.1 INTRODUCTION

This chapter will focus on the coastal and pelagic ocean habitats of leatherback turtles (*Dermochelys coriacea*) in terms of migration, foraging, physiology, and dive behavior with links to climate and oceanography. Satellite telemetry provides the framework by which scientists can track sea turtles in three dimensions such that crucial data can be acquired on migratory pathways, swim speed, dive behavior, and the ambient water temperature of the turtle's marine environment. Whereas the oceanic habitats of loggerhead turtles are based on satellite-tracking studies of both males and females at multiple life history stages (Chapter 8), those of leatherback turtles are almost entirely based on the internesting and postnesting movements of mature females, with the exception of some regional studies that have tagged in-water adult and subadult males and females. The juvenile life history

stage of leatherbacks (curved carapace length [CCL] < 100 cm) is one of the biggest mysteries of
sea turtle biology and has confounded researchers for decades. Therefore, the majority of the data
presented in this chapter will be based on mature female leatherback studies although some work is
presented on the in-water deployment of satellite tags of male and female subadults (CCL > 100 cm
but not sexually mature), adults, and on the limited dataset of juvenile (CCL < 100 cm) sightings
with inferences on their physiological constraints.

The unique physiology of adult leatherbacks (Chapter 1) facilitates their use of cooler, more
temperate waters. Adults are essentially facultative homeotherms and can thus maintain a core
body temperature above ambient depending on the temperature gradient between the surrounding
water and the turtle. Gigantothermy combined with a facultative heat exchange system (via blood
circulation changes) and the use of peripheral tissues as insulation (Paladino et al., 1990) have
resulted in a physiology that enables the adult latitudinal range to be the largest among all reptiles.
Adult leatherbacks have been observed (alive and well) in near-freezing waters of the Gulf of St.
Lawrence for weeks at a time where floating ice passes by on the surface waters (James et al., 2006).
Most striking is that these wide-ranging reptiles with relatively fast growth rates (Jones et al., 2011b)
forage exclusively on gelatinous zooplankton (Chapter 10), an energy poor group of taxa that are
~95% water (Doyle et al., 2007). The low energy density of leatherback prey suggests that large
juveniles, subadults, and adults require large patches of prey to sustain their metabolisms, fuel
distant migrations, and successfully reproduce.

Capable of basin-wide migrations and extremely deep dives (on occasion), the streamlined,
hydrodynamic body form of leatherbacks renders this species the fastest and deepest diving relative to all other sea turtles. Consequently, adult leatherbacks utilize a wide variety of Large Marine
Ecosystems (LMEs) around the globe where they forage on gelatinous zooplankton. In this chapter,
the common foraging areas will be described based on the associated LME (Sherman, 1991).

Finally, the majority of the satellite-tracking studies presented here utilized the "harness
method" (Eckert and Eckert, 1986) for satellite tag attachment. A new attachment method called
"direct attachment" now appears to be more common due to its substantial reduction in drag
(Jones et al., 2011a). Moreover, there have been multiple studies showing increased swim speed and
longer dive duration among leatherbacks equipped with tags via the "direct attachment" method versus those equipped with tags using the "harness method" (Fossette et al., 2008; Jones et al., 2011a).
However, it is unlikely that the "harness method" affects the overall migratory pathways and general
dive behavior of the mature females that are discussed in this chapter.

This chapter is structured as a biological and physical oceanographic framework for the oceanic
habitat and behavior of postnesting female leatherbacks. It begins with the biological and physical
constraints on gelatinous zooplankton—the common prey of leatherback turtles and likely driving
force governing their postnesting migrations. The paucity of gelatinous zooplankton observations
combined with the wide-ranging nature of leatherbacks requires an alternative proxy for leatherback prey abundance. Here, phytoplankton biomass and large phytoplankton net primary productivity (NPP) are used as proxies for leatherback-foraging hotspots. The primary migratory pathways,
foraging hotspots, and dive behavior of postnesting females from nesting beaches worldwide are
detailed based on satellite-tracking studies along with a review of the limited anecdotal observations of juveniles. Finally, bottom-up and climatic forcing on leatherback phenotype, reproductive
frequency, and population status is discussed followed by closing remarks on future directions.

## 7.2 A BIOPHYSICAL OCEANOGRAPHIC CONTEXT FOR LEATHERBACK PREY: THE JELLYFISH DIET

### 7.2.1 PHYTOPLANKTON: THE BASE OF THE MARINE FOOD WEB

Autotrophic organisms are the base of food webs in terrestrial, freshwater, and marine ecosystems.
Composed of plants, bacteria, and protists, phytoplankton fill this role in the marine environment.

Marine phytoplankton may be responsible for one third of the globe's photosynthetically fixed carbon (Welp et al., 2011). The growth of phytoplankton is limited by sunlight, nutrients (nitrogen, phosphorous, iron, and other trace metals), and grazing rates of zooplankton. Therefore, phytoplankton are typically at their highest biomass during the winter–spring transition just after the winter winds have settled and nutrients are plentiful, day length and sunlight irradiance increase, and zooplankton have not yet achieved significant biomass to graze down the phytoplankton.

Global estimates of marine phytoplankton biomass in the surface waters can be achieved via ocean color satellite sensors. The Sea-viewing Wide Field-of-view Sensor (SeaWiFS) ocean color satellite mission lasted from 1997 to 2010 and is now succeeded by the Moderate-resolution Imaging Spectroradiometer and the Visible and Infrared Imager/Radiometer Suite ocean color missions. Ocean color measured by satellites is run through an algorithm to calculate the ocean surface concentration of the phytoplankton pigment chlorophyll-*a* (chl-*a*) (Figure 7.1). Located in the chloroplasts of plant cells, chl-*a* is the essential molecule that absorbs and transfers light energy in the process of photosynthesis. Therefore, the scientific community typically uses satellite estimates of surface chl-*a* to estimate the ocean's surface biomass of phytoplankton. A major caveat to this approach is that much of the global ocean's maximum concentration occurs at depth (deep chl-*a* max) where the satellite sensors cannot detect ocean color, particularly within the subtropical gyres at around 50–80 m. While not relatively high in phytoplankton biomass, the subtropical gyres make up a large portion of the global ocean and thus account for a substantial proportion of global NPP.

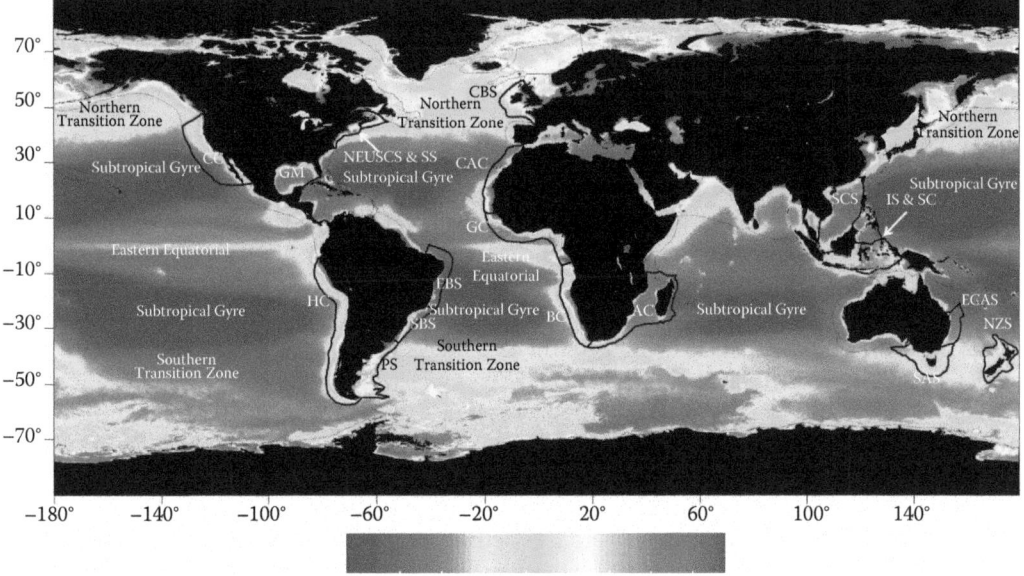

**FIGURE 7.1** Composite of surface chlorophyll-*a* (chl-*a*) measured by the SeaWiFS satellite sensor from 1997 to 2010 (level-3 9 km data; http://oceancolor.gsfc.nasa.gov). The chl-*a* color bar is on a log scale. The coastal LMEs that are known leatherback foraging/migration hotspots are highlighted with solid-bold black lines (Sherman, 1991). The LMEs utilized by Western Atlantic leatherbacks are the Scotian Shelf, Northeast U.S. Continental Shelf, GM, Celtic-Biscay Shelf, and the CAC. The Eastern Atlantic population target the Guinea Current, BC, East Brazil Shelf, South Brazil Shelf, and Patagonian Shelf LMEs. Western Indian leatherbacks are only associated with the AC and the BC LMEs. The Western Pacific population utilizes a diverse suite of LMEs comprised of the SCS, Indonesian Sea, Sulu-Celebes Sea, East Central Australian Shelf, Southeast Australian Shelf, New Zealand Shelf, CC LMEs, and, to a lesser extent, the Humboldt Current LME. The Gulf of Panama and the HC LME are the only coastal areas associated with the migratory pathways of Eastern Pacific leatherbacks.

However, the use of surface chl-*a* to discern regions of high phytoplankton biomass is a valuable tool for animal-tracking studies looking to highlight foraging hotspots of sea turtles, finfish, marine mammals, sea birds, and sharks.

Figure 7.1 is a composite of the entire SeaWiFS mission's estimate of chl-*a* concentration in the surface waters of the global ocean. The relevant LMEs that will be discussed in this chapter, as they relate to leatherback foraging areas, are outlined with solid black lines. The boundaries of LMEs are based on both oceanography and fishery activity (Sherman, 1991) and are all coastal areas. There are, of course, many pelagic and deep-sea fisheries operating outside these LMEs but the majority of global fishery catch derives from these coastal LMEs (Sherman, 1991).

Marine regions with high biomass of marine phytoplankton are continental shelves, eastern boundary coastal upwelling regions, eastern equatorial upwelling regions, and temperate pelagic areas. Coastal areas are the most productive due to (1) their shallow depths allowing the wind-forced mixed-layer to interact with the nutrient-rich sea floor and thus maintain a more than adequate vertical advection and retention of nutrients throughout the euphotic zone (depth range where there is enough sunlight for photosynthesis, ~0–200 m); (2) coastal upwelling along eastern boundary regions (i.e., California Current [CC] and Canary Current [CAC] LMEs in Figure 7.1); (3) close proximity to river delta and land run-off flow that is high in nutrients (i.e., northern Gulf of Mexico [GM] LME); and (4) close proximity to iron dust sources (Figure 7.1). The subtropical gyres are relatively low in surface chl-*a* (Figure 7.1) and are sometimes referred to as "ocean deserts" due to their low productivity. Eastern Equatorial upwelling regions, on the other hand, are very productive compared to the subtropical gyres in the north and south (Figure 7.1) due to a shoaling of the equatorial thermocline in the east and the ensuing vertical advection of nutrient-rich water from equatorial undercurrents. Known as "the equatorial cold tongues" due to the cooler sea surface temperature (SST) from upwelled water, the Eastern Equatorial Pacific and Atlantic are the most productive pelagic waters in the tropics (Figure 7.1).

Northern and Southern Transition Zones are characterized by the high gradients in surface chl-*a* and separate the unproductive subtropical gyres and highly productive subpolar gyres (Figure 7.1). The temperate transition zones have strong winter winds that cause deep mixed layers, ensuring vertical transport of nutrients to the euphotic zone for the ensuing spring phytoplankton bloom. The strong gradient in phytoplankton biomass between the subtropical gyres and subpolar gyres has been shown to be a major hotspot for marine predators (Block et al., 2011), including sea turtles (Polovina et al., 2001; Mansfield et al., 2009). These transition zones attract predators due to their abrupt separation of phytoplankton biomass between gyres and the physical aggregation of plankton and other lower trophic organisms due to physical discontinuities (Polovina et al., 2001).

## 7.2.2 BIOLOGICAL AND PHYSICAL CONTROLS OF GELATINOUS ZOOPLANKTON BIOMASS

Observations of subadult and adult leatherbacks actively foraging on gelatinous zooplankton have been mostly limited to studies in the coastal waters of Nova Scotia, Canada (Heaslip et al., 2012), and central California (Benson et al., 2007; Graham et al., 2010). In both regions, leatherbacks have been observed exclusively foraging on large species of scyphozoans. Off central California, *Chrysaora fuscescens* (sea nettle) is the common leatherback prey (Benson et al., 2007) whereas *Cyanea capillata* (lion's mane) and, to a lesser extent, *Aurelia aurita* (moon jellyfish) are the prey for leatherbacks in the waters of Nova Scotia, Canada (Heaslip et al., 2012). A small number of leatherback feeding observations were also made in the offshore waters of Karujeu Island, Solomon Islands (near a nesting beach) where a mature male foraged on dense patches of the very small scyphozoan *Linuche unguiculata* (thimble jellyfish) (Fossette et al., 2012). Because there is such a paucity of observations of leatherbacks actively foraging on gelatinous zooplankton, it is not valid to draw broad conclusions on the typical species consumed by leatherbacks on a global scale. Other gelatinous zooplankton such as ctenophores, hydrozoans, cubozoans, staurozoans, and salpidae may also be common prey items.

In the oceans, observations of gelatinous zooplankton biomass and distribution, both verti-cally and horizontally in the water column, are very sparse and typically restricted to coastal waters (Purcell, 2012). A global review of the available data on integrated gelatinous zooplankton biomass in the epipelagic zone (0–200 m water depth) reported that biomass decreased with water column depth and that this relationship was driven by bottom-up factors because shallow coastal waters are higher in NPP than offshore pelagic waters (Lilley et al., 2011). These results imply that food resources for gelatinous zooplankton (phytoplankton, crustacean zooplankton, fish larvae) are likely in greater abundance in coastal waters. Another possible explanation, in addition to resource availability, is the presence of hard bottom substrate in coastal regions, which is required for polyp attachment among the majority of scyphozoans. However, there are some scyphozoans that do not have a benthic polyp stage (*Pelagia* spp.) and can be found in the open ocean.

Regional studies have also reported bottom-up control of gelatinous zooplankton biomass. In the Mediterranean waters of Mar Menor, Spain, large gelatinous zooplankton abundance was linked to diatom (large phytoplankton) abundance (Pérez-Ruzafa et al., 2002). In the North Sea, gelatinous zooplankton frequency was positively correlated to the North Atlantic Oscillation (NAO) suggest-ing a possible link to NPP based on continuous plankton recorder data (Attrill et al., 2007). Positive NAO phases result in stronger westerly winds (deeper mixed layer) and thus delay the start of the subpolar spring phytoplankton bloom by 2–3 weeks (Henson et al., 2009). Moreover, in the north-western Mediterranean Sea, the annual standing stocks of gelatinous zooplankton from 1974 to 2003 were constrained by bottom-up and climatic mechanisms such that there was an association between the NAO and winter vertical mixing (García-Comas et al., 2011).

In temperate regions, the largest biomass of gelatinous zooplankton commonly occurs in the sum-mer months after the spring phytoplankton blooms (Sullivan et al., 2001; Pérez-Ruzafa et al., 2002; Lynam et al., 2004; Decker et al., 2007). It is likely that extensive blooms of large scyphozoans occur a few months after the spring bloom due to both biological (ecosystem response to the spring phytoplankton bloom) and physical mechanisms (ocean temperature and stratification changes). The production of ephyrae (small immature scyphozoans) that release from benthic polyps is also positively correlated to resource availability (Purcell et al., 1999) and may also be constrained by the size and timing of the spring phytoplankton bloom (Lynam et al., 2004).

While bottom-up forcing appears to be an essential constraint on gelatinous zooplankton biomass, the physical characteristics of the ocean also control their abundance and distribution. Temperature can be a key constraint on gelatinous zooplankton growth at multiple life history stages (Purcell, 2012). Physical retention areas such as thermal fronts, haloclines, current bound-aries, mesoscale eddies, and shelf breaks are all associated with aggregations of gelatinous zoo-plankton. These retention areas have been shown to be scyphozoan hotspots off the coast of central California (Benson et al., 2007). The accumulation of high-gelatinous zooplankton biomass in such areas, however, is still ultimately subject to constraints imposed by the productivity of the ecosystem (Lilley et al., 2011).

## 7.3  MIGRATION AND FORAGING AREAS

### 7.3.1  Adults

#### 7.3.1.1  Western Atlantic Population

The migratory pathways and foraging areas of the Western Atlantic (also called North Atlantic) population of leatherbacks derive from satellite-tracking studies of postnesting females ($n \approx 40$) throughout the western tropical Atlantic (Ferraroli et al., 2004; Eckert, 2006; Hays et al., 2004; Fossette et al., 2010a,b; Figure 7.2A and C) and from in-water satellite tag deployments offshore of Ireland ($n = 2$; Doyle et al., 2008; Figure 7.2A) and Nova Scotia, Canada ($n = 38$; James et al., 2005a; Figure 7.2B). Satellite tags were deployed on nesting females in Panama, Suriname, French

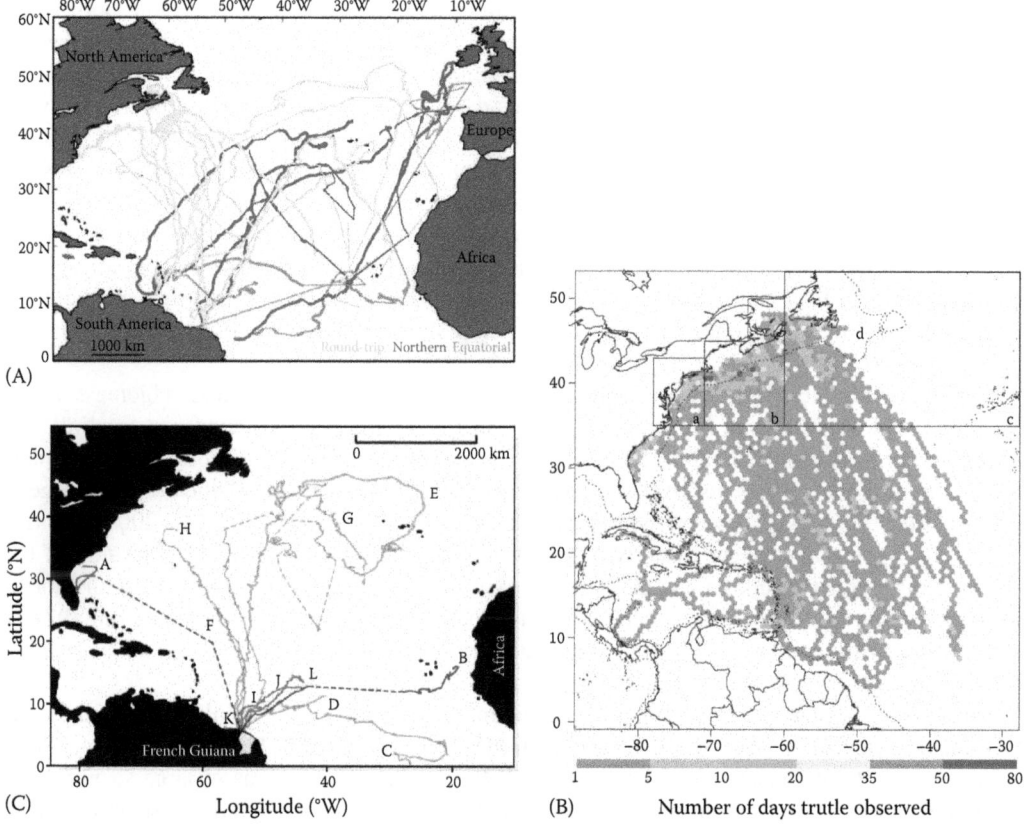

**FIGURE 7.2** Migratory pathways and high-use areas of Western Atlantic leatherbacks. (A) Movements of 6 in-water captured leatherbacks from the waters of Nova Scotia, Canada (*n*=4), and Ireland (*n*=2) and 15 satellite-tracks of postnesting female leatherbacks from French Guiana, Suriname, and Grenada. Green lines denote round-trip migration, blues lines are northern migration, and orange lines are equatorial. (Reproduced from Fossette, S. et al., *J. Mar. Syst.*, 81, 225, 2010b. With permission.) (B) Spatial use of 38 leatherbacks (sub-adult and adult males and females) that were equipped with satellite tags in the waters of Nova Scotia, Canada. Color denotes the number of days the turtles were observed in each hexagon; a=the Mid-Atlantic Bight, b=Northeast Coastal Area, c=Northeast Distant Areas, and d=Grand Banks. (Reproduced from James, M.C. et al., *Ecol. Lett.* 8, 195, 2005a. With permission.) (C) Movements of 12 postnesting female leatherbacks from French Guiana and Suriname in the years in 1999 (blue), 2000 (green), 2001 (orange), and 2002 (red); the dashed line denotes track segments containing uncertain locations due to low transmission frequency. (Reproduced from Ferraroli, S. et al., *Nature* 429, 521, 2004. With permission.)

Guiana, Grenada, and Trinidad. The postnesting migrations from leatherbacks nesting in Trinidad (Eckert, 2006) are not shown in Figure 7.2 because these tracks are very similar to those reported for postnesting females from Suriname, French Guiana, and Grenada (Ferraroli et al., 2004; Fossette et al., 2010b Hays et al., 2004).

The observed migratory pathways of postnesting females appear to be exclusive to the Northern Hemisphere (Figure 7.2). In-water captures of male and female subadults/adults in the waters of Nova Scotia, Canada, and Ireland also show long-term migratory pathways that are also restricted to the Northern Hemisphere (Figure 7.2A and B). Postnesting females migrated to productive coastal LMEs and to pelagic waters in the productive Northern Atlantic Transition Zone and Eastern Equatorial Atlantic (Figures 7.1 and 7.2). In the northwest Atlantic, the most common LMEs that were highly utilized as foraging areas were the Northeast U.S. Continental Shelf and the Scotian Shelf. In the northeast Atlantic, the Celtic-Biscay Shelf and CAC LMEs are common high-use areas where leatherbacks forage. To a lesser extent, the GM LME is also a foraging area for postnesting leatherbacks tagged in Panama (Figure 7.3A). Therefore, this population has three general migration/foraging strategies that are equatorial, coastal, and pelagic with some individuals remaining in tropical regions and others (the majority) moving into temperate waters (Fossette et al., 2010b).

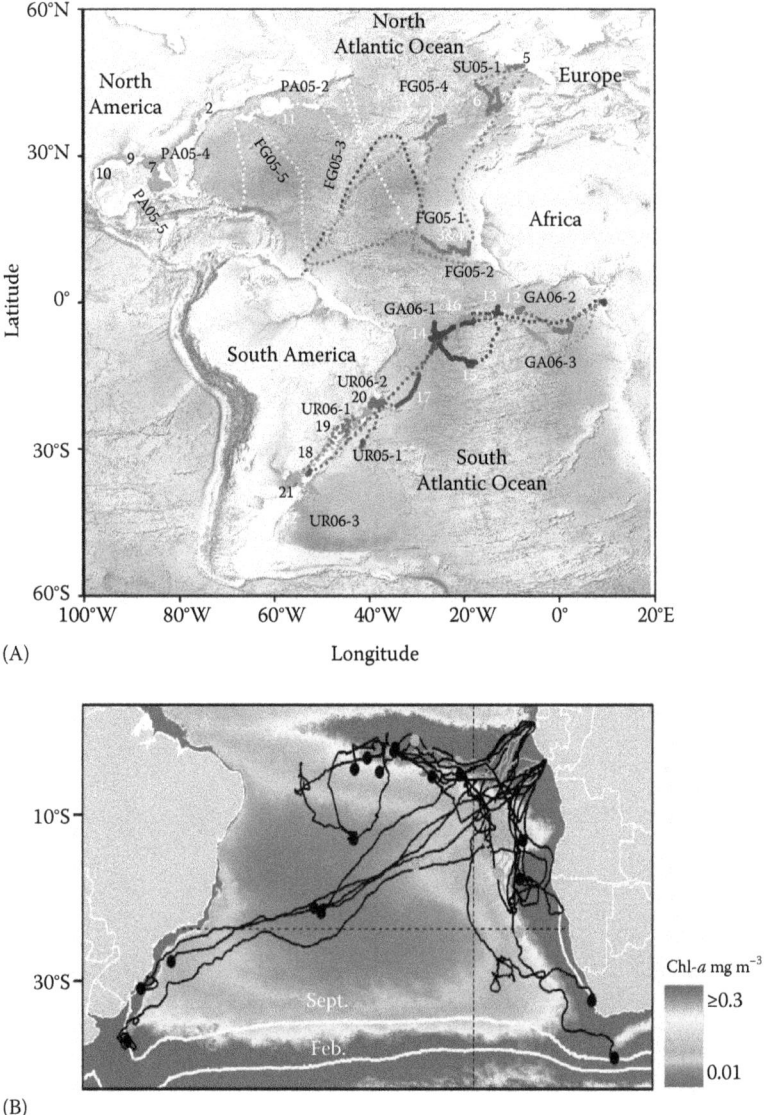

FIGURE 7.3 Movements of Western and Eastern Atlantic leatherbacks. (A) Movements of nine Western Atlantic postnesting females from Panama (*n*=3), French Guiana, and Suriname (*n*=6). Also shown are movements of seven Eastern Atlantic leatherbacks composed of postnesting females from Gabon, Africa (*n*=3), and in-water deployments (*n*=4) among turtles incidentally among Uruguayan fisheries off the coast of South America. Transit and temporary residence areas (TRAs) for each turtle are identified by dotted and solid lines, respectively. Each TRA is identified by a number in black and white. (Reproduced from Fossette, S. et al., *PLoS ONE*, 5, e13908, 2010a. With permission.) (B) Migratory pathways of 25 Eastern Atlantic postnesting females from Gabon, Africa, overlaid on a composite of MODIS surface chl-*a* (4 km; 2006–2009). Mean northerly and southerly position of the 15°C SST isotherm (white contours) derived from annual MODIS Aqua night-time SST (4 km product; 2006–2009). (Reproduced from Witt, M.J. et al., *Proc. R. Soc. B*, 278, 2338, 2011. With permission.)

### 7.3.1.2 Eastern Atlantic Population

Nesting along the coastlines of tropical West Africa, primarily in Gabon, Eastern Atlantic (also referred to as South Atlantic), leatherbacks migrate and forage exclusively in the Southern Hemisphere (Fossette et al., 2010a; Witt et al., 2011; Figure 7.3). These postnesting females ($n=28$) migrated into the productive eastern equatorial Atlantic, the coastlines of southern Africa, and some crossed the South Atlantic Subtropical Gyre to reach the coastal waters of South America (Figure 7.3). In-water deployments off the coast of Uruguay ($n=4$) also showed persistent coastal fidelity to South America waters (Fossette et al., 2010a). These coastal foraging areas are contained within the East Brazil Shelf, South Brazil Shelf, Patagonian Shelf, Guinea Current, and Benguela Current (BC) (Figure 7.1). As with the Western Atlantic population, both tropical and temperate foraging sites in pelagic and coastal waters are observed.

### 7.3.1.3 Western Indian Population

Nesting on the Indian Ocean coastline of South Africa, this small population of leatherbacks has a much smaller migration range than other populations. However, the sample size of satellite-tracked postnesting females is the smallest ($n=9$; Luschi et al., 2006; Lambardi et al., 2008) of all the studies reported in this chapter, and thus the migration area described here may be underrepresented. These postnesting females migrated south of the nesting beach via off-shelf and on-shelf waters (Luschi et al., 2006; Figure 7.4), possibly linked to the Agulhas Current (AC) that runs south along the coast (Lambardi et al., 2008). Some females crossed into the Atlantic Ocean while others

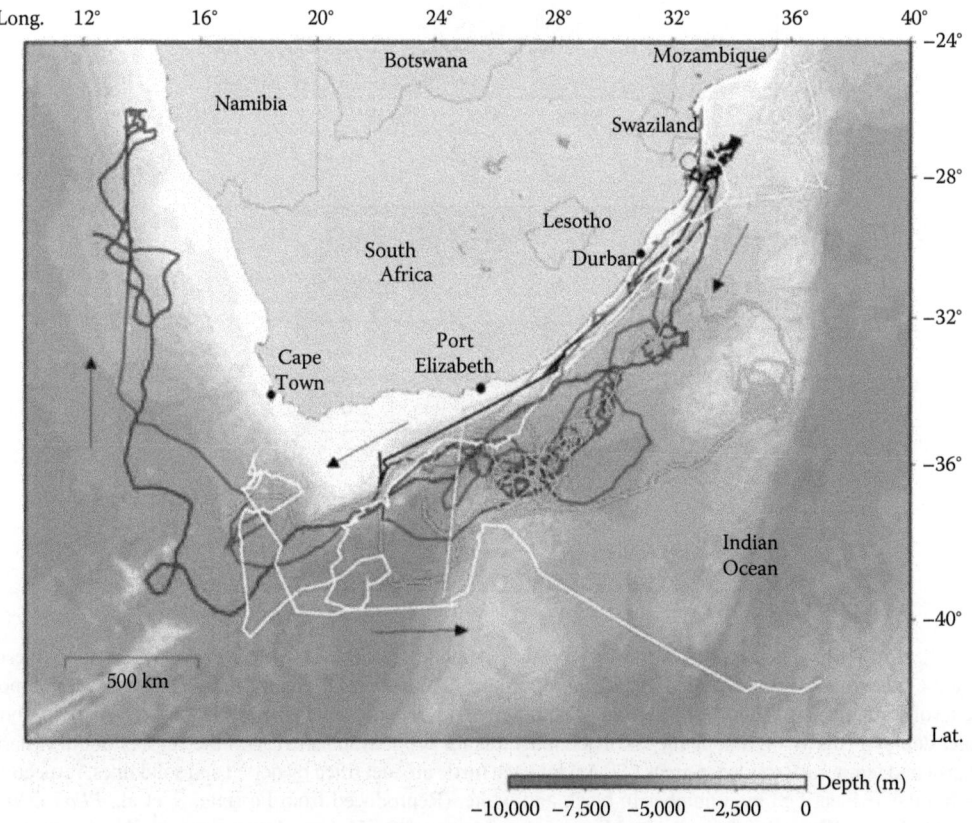

**FIGURE 7.4** Postnesting migratory pathways of six Western Indian leatherbacks tracked by satellite in the years 1996–2003. Yellow dot denotes the nesting beach. (Reproduced from Luschi, P. et al., *S. Afr. J. Sci.*, 102, 51, 2006. With permission.)

remained in the Indian Ocean (Figure 7.4). The only two LMEs associated with this population are the BC and AC (Figure 7.1). The South Atlantic Transition Zone also appears to be the targeted foraging area for this population (Figures 7.1 and 7.4).

### 7.3.1.4 Western Pacific Population

The Western Pacific population represents the largest sample size of satellite-tracked postnesting and in-water captures from a single population ($n = 126$) and migrates to a diverse suite of coastal LMEs and productive pelagic waters throughout the Pacific basin (Benson et al., 2011). Postnesting females were tracked from beaches in Papua Barat, Papua New Guinea, and the Solomon Islands while in-water capture deployments derived from the coastal waters of Central California (Figure 7.5). Postnesting females from beaches in the Western Pacific migrated to both the Northern and Southern Hemisphere, but this was dependent on the time of year such that summer nesters moved into Northern Hemisphere foraging areas and winter nesters moved into Southern Hemisphere areas (Benson et al., 2011). The most common coastal LMEs were the South China Sea (SCS), Indonesian Sea, Sulu-Celebes Sea, East Central Australian Shelf, Southeast Australian Shelf, New Zealand Shelf, and CC (Figures 7.1 and 7.5). Pelagic foraging areas were in the North Pacific Transition Zone and Eastern Equatorial Pacific (Figures 7.1 and 7.5). In-water deployments off the coast of Central California showed migration to the Eastern Equatorial Pacific and across the Pacific basin to waters near the nesting beaches in the Western Pacific (Figure 7.5). Finally, genetic data suggest that a small portion of the leatherbacks caught as bycatch among the fisheries of coastal Peru within the Humboldt Current LME (Figure 7.1) originate from Western Pacific nesting beaches (Dutton et al., 2010).

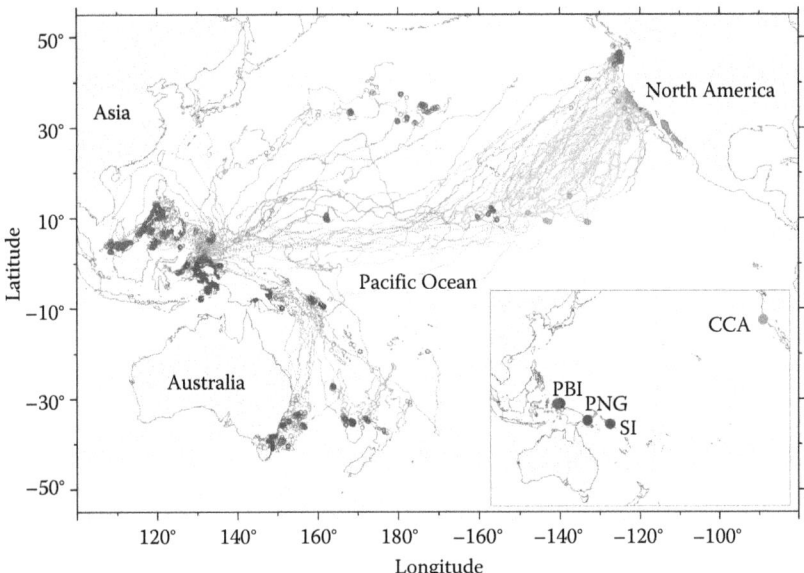

FIGURE 7.5   Migratory pathways and high-use areas of Western Pacific leatherbacks ($n = 126$). Large, darker circles indicate area-restricted search behavior; small, lighter dots indicate transiting behavior. Color of track indicates deployment season: red=summer nesters, blue=winter nesters, green=deployments at central California foraging grounds. Inset shows deployment locations; PBI=Papua Barat, Indonesia, PNG=Papua New Guinea, SI=Solomon Islands, CCA=central California. Black boxes represent ecoregions for which habitat associations were quantitatively examined (refer to Benson et al., 2011): SCS=South China, Sulu and Sulawesi Seas, IND=Indonesian Seas, EAC=East Australia Current Extension, TAS=Tasman Front, KE=Kuroshio Extension, EEP=equatorial eastern Pacific, and CCE=California Current Ecosystem. (Reproduced from Benson, S.R. et al., *Ecosphere*, 2, art84, 2011. With permission.)

The most striking migratory feature of this population is the basin-wide migration between the foraging areas in the CC LME and nesting beaches in the Western Pacific that is among the longest of any marine vertebrate. Second, this population of postnesting females is observed to have highly diversified migratory pathways originating from nesting beaches that are in relatively close proximity to each other. The Western Pacific population appears to be the only genetically distinct breeding population that utilizes waters in both the Northern and Southern Hemisphere (Benson et al., 2011).

### 7.3.1.5 Eastern Pacific Population

The major nesting beaches for the Eastern Pacific population are located in Costa Rica and Mexico. The majority of the postnesting females have been tracked from this population nested at Playa Grande, Costa Rica ($n = 46$; Shillinger et al., 2008, 2011) where they leave the nesting beach on a southwest trajectory toward the Galapagos Islands until they disperse into the South Pacific Subtropical Gyre (Figure 7.6A). A much smaller sample size of postnesting females ($n = 7$) that nested in Mexiquillo, Mexico had similar southwest trajectories shown in Figure 7.6A although one female headed to the coast of Chile while another headed west of 140°W (Eckert and Sarti, 1997).

The internesting movements of females from Playa Grande, Costa Rica have also been well described (Shillinger et al., 2010; Figure 7.6B). At Playa Grande, females produce eight clutches on average with an internesting interval of about 10 days between each clutch (Shillinger et al., 2010). When the turtles are not nesting within their 10-day intervals, they typically move into the productive Gulf of Papagayo, just north of Playa Grande, Costa Rica. Although not yet observed, it is possible that the productive Gulf of Papagayo serves as a foraging area for energy-starved nesting females in between nesting events (Shillinger et al., 2010). Prior studies using swim behavior, dive behavior, and turtle beak movements suggest that nesting leatherbacks forage in between nesting events (Fossette et al., 2008).

Comparing the migratory pathways of the Eastern Pacific population (Figure 7.6A) to all other populations worldwide (Figures 7.2 through 7.5), a clear dichotomy emerges regarding the variability in female dispersal from nesting beaches. With the exception of the Eastern Pacific population, the utilization of coastal LMEs (postnesting) is a common observation among leatherbacks. Of the 53 postnesting females satellite-tracked in the Eastern Pacific from both Mexico and Costa Rica (Eckert and Sarti, 1997; Shillinger et al., 2008, 2011), only three individuals showed any tracks at or near the highly productive, coastal upwelling HC LME off the coastlines of Peru and Chile (Figure 7.1), and only one individual (from Playa Grande, Costa Rica) stayed within the Gulf of Panama for the entire track duration (Figure 7.6A). However, reports of incidental bycatch among the small-scale, artisanal fisheries of coastal Peru indicate that adult leatherbacks are in fact utilizing the HC LME (Alfaro-Shigueto et al., 2007, 2011). It is possible that longer postnesting track durations (>1.5 years) may have shown higher utilization of the HC LME. The mean remigration interval for the eastern Pacific population is ~3.7 years, and thus the tracking data presented here only represent a portion of their movement behavior in between nesting seasons. Alternatively, coastal Peru may be a sink for leatherbacks in the Eastern Pacific due to the widespread nature of the artisanal fisheries operating in this region (Alfaro-Shigueto et al., 2011). Therefore, the pelagic offshore majority described in Figure 7.6A that migrated and foraged throughout the unproductive South Pacific Subtropical Gyre (Figure 7.1) may have been anthropogenically selected due to high mortality rates off the coast of Peru (Saba et al., 2008a). The implications of the Eastern Pacific population primarily utilizing one of the most unproductive pelagic regions will be discussed in Section 7.6.

### 7.3.2 LOST YEARS: NEONATES AND JUVENILES

Upon entering the ocean for the first time as hatchlings (neonates), very little is known as to where these palm-sized leatherbacks go or what motivates their migration. In the Eastern Pacific, currents

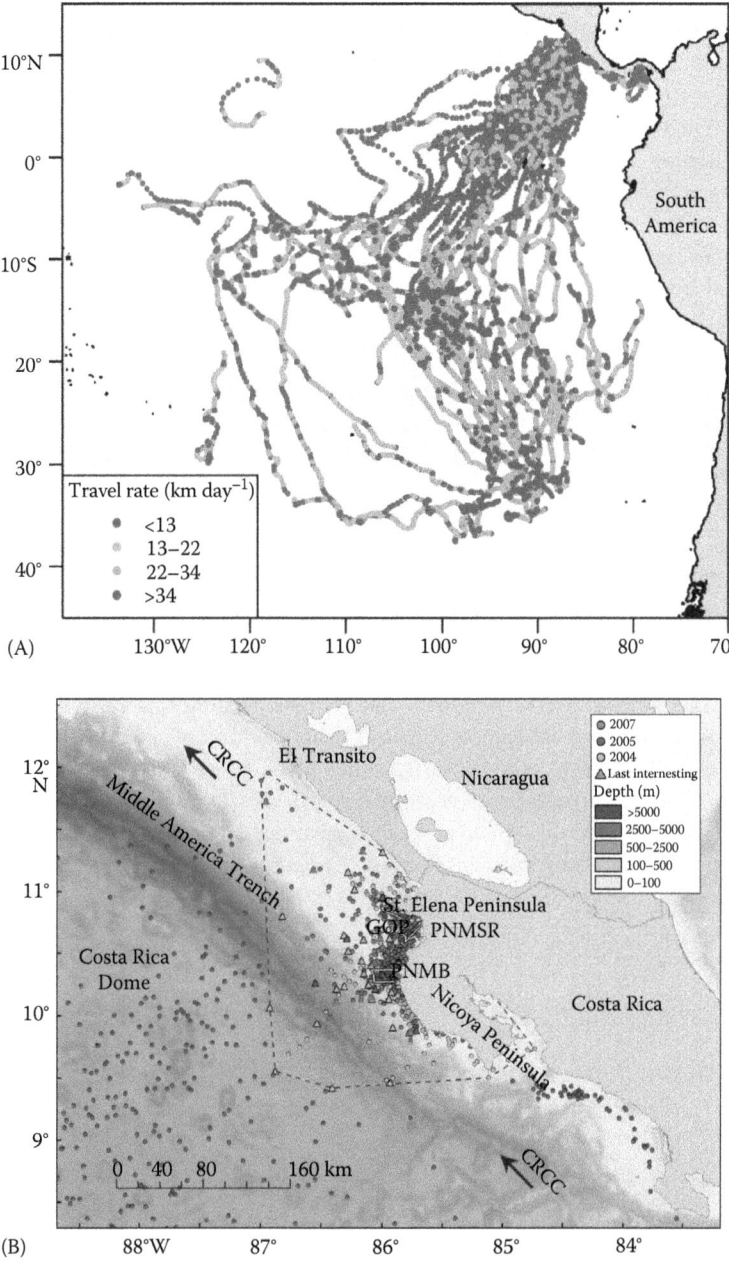

FIGURE 7.6  Postnesting and internesting movements of Eastern Pacific leatherbacks. (A) Switching state-space model-derived daily positions for leatherback turtles tagged at Playa Grande, Costa Rica (*n*=46), color-coded by daily travel rate. (Reproduced from Shillinger, G.L. et al., *Mar. Ecol. Prog. Ser.*, 422, 275, 2011. With permission.) (B) Internesting positions for leatherback turtles during 2004 (yellow), 2005 (pink), and 2007 (green) overlaid bathymetry. Triangles demarcate the last internesting position per platform transmitter terminal within each of the three tracking seasons. Dashed black line represents 95% minimum convex polygon (MCP) for combined tracking seasons. Polygons bordered in white are Playa Grande National Marine Park and Santa Rosa National Marine Park. Circles outside of the MCP illustrate how turtle movements continued postinternesting but have been grayed out as they are not considered in the analysis. GOP=Gulf of Papagayo. Costa Rica Coastal Current is denoted with directionality arrows. (Reproduced from Shillinger, G.L. et al., *Endanger. Species Res.*, 10, 215, 2010. With permission.)

and eddies near Playa Grande, Costa Rica, had a consistent pattern of moving offshore during the peak time when hatchlings enter the ocean (Shillinger et al., 2012). Areas north and south of Playa Grande had significant coastal retention suggesting that Playa Grande evolved to be a major nesting beach for the Eastern Pacific population due to the persistent offshore transport of hatchlings, further away from the high concentration of predators in the shallow coastal waters (Shillinger et al., 2012). There have been other studies suggesting that the near shore currents along nesting beaches transport hatchlings into pelagic waters. These studies will be discussed further and in greater detail in the section that follows but it is at least worth mentioning here given the lack of any observational data regarding leatherback hatchling dispersal from nesting beaches. It is likely that the habitat of leatherback hatchlings within the first 6 months of life may be mostly constrained to pelagic waters due to lower predation rates.

The migration ecology of juvenile leatherbacks (<100 cm CCL) is one of the greatest "unknowns" of sea turtle biology. Referred to as "the lost years," this is the period of time between entering the ocean as a hatchling and the first return to the nesting beach as a mature adult. Although a few subadults (CCL > 100 cm but not sexually mature) have been tracked (James et al., 2005a), there is not a single tracking study of a leatherback < 100 cm CCL worldwide. There was an attempt to track a large juvenile (CCL = 97.5 cm) that was captured off the coast of Liberia, West Africa, but the transmitter failed after deployment (Danish Galthea 3 Expedition, unpublished data).

A recent captive growth study by Jones et al. (2011b) reported that leatherbacks reach a length of 77 cm CCL after 2.23 years of age. From these measured growth rates, leatherbacks may reach sexual maturity at ~16 years of age (Jones et al., 2011b). Jones et al. (2011b) also reviewed leatherback bycatch data among Pacific longlines and gillnets, finding that turtles as small as 83 cm CCL (2.5–3 years old) were caught among tropical fisheries. Moreover, leatherbacks caught among some Peruvian gillnet fisheries were typically either juveniles or subadults (de Paz et al., 2006). A global review of the limited number of juvenile leatherback sightings showed that turtles that have a CCL < 100 cm are restricted to tropical and subtropical waters (<30° absolute latitude; Figure 7.7) or, likewise, ocean temperatures warmer than 26°C (Eckert, 2002).

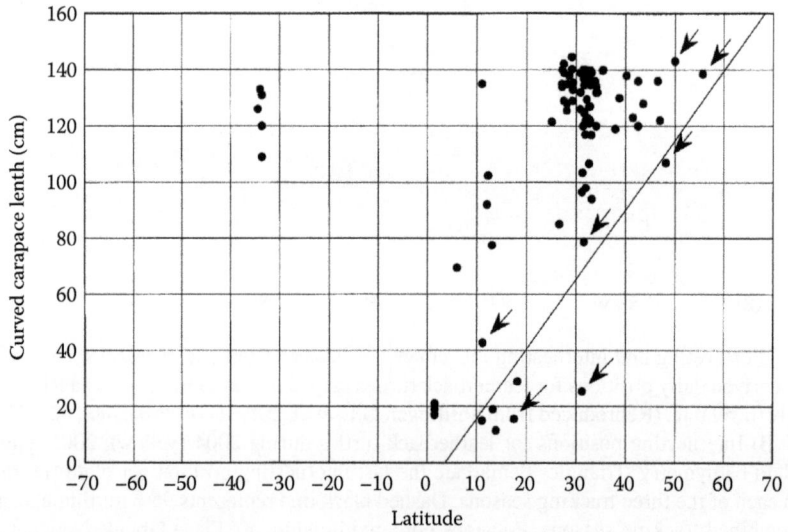

FIGURE 7.7    Scatterplot of juvenile sizes compared with latitude of the sighting or stranding. A linear regression was plotted against the highest latitude sighting within each (20 cm) size class (arrows). (Reproduced from Eckert, S.A., *Mar. Ecol. Prog. Ser.*, 230, 289, 2002. With permission.)

These studies suggest that juvenile leatherbacks cannot tolerate the colder, temperate high latitude regions where larger subadults and adults are commonly observed (Eckert, 2002; Jones et al., 2011b). Whether this apparent size-based limitation is due to either/both the minimum size requirement for gigantothermy (CCL > 100 cm) or/and an ontogenetic limitation of other adaptations such as metabolic heat production and counter-current circulatory heat exchange is unknown.

### 7.3.3 THEORETICAL ONTOGENY OF ADULT MIGRATION/FORAGING AREA FIDELITY

Recent studies investigating the ontogeny of adult sea turtle migration and foraging areas have proposed a common oceanographic mechanism, called the "hatchling drift scenario" (Fossette et al., 2010a; Hays et al., 2010a; Putman et al., 2010) as the primary driving force of the persistent migratory pathways and high-foraging site fidelity of adults. These studies suggest that the strength and direction of near shore currents and eddies that are adjacent to nesting beaches define the adult migratory pathways via hatchling imprinting (Fossette et al., 2010a; Hays et al., 2010a; Putman et al., 2010). Theoretically, hatchlings are entrained and transported in near shore currents and eddies that have similar headings and long-term (years) drift patterns to the postnesting migratory pathways of the nesting females from the respective nesting beach.

In the Atlantic Ocean, Fossette et al. (2010a) reviewed satellite-tracked surface drifter data originating from locations close to the same nesting beaches where females were originally satellite-tagged in Figure 7.3A. These nesting sites were in Panama, Suriname, French Guiana (Western Atlantic population), and in Gabon (Eastern Atlantic population). The majority of the postnesting migratory pathways in Figure 7.3A were similar to some of the satellite-tracked drifters originating near nesting beaches (Fossette et al., 2010a). Off the coast of South Africa in the southwestern Indian Ocean, satellite-tracked surface drifters also had similar trajectories and long-term drift pathways as the postnesting migrations of Western Indian leatherbacks (Sale and Luschi, 2009) shown in Figure 7.4. In the Western Pacific, the Northern/Southern Hemisphere dichotomy in migratory pathways between summer and winter nesting leatherbacks may be due to differential hatching transport from seasonal currents (Gaspar et al., 2012).

The "hatchling drift scenario" theory has also been applied to loggerhead turtles in the Mediterranean Sea (Hays et al., 2010a) and Western Atlantic (Putman et al., 2010; Chapter 8). However, not all of the Western Atlantic leatherback migratory pathways (Figure 7.3A) were associated with the offshore currents. Some postnesting females from French Guiana transited across the North Atlantic Gyre toward the Azores, a pathway not associated with surface drifters originating from this nesting site (Fossette et al., 2010a). A critical component that some of these studies have not considered is the thermal constraints of smaller turtles. As mentioned previously, only subadult and adult leatherbacks are observed in the colder, temperate foraging areas, likely due to the thermal constraints of juveniles. Therefore, if some of these long-term drift trajectories (1–3 years) are thought to carry hatchlings and juveniles to the same temperate foraging areas that are inhabited by subadults and adults, then why is there a lack of observations of smaller turtles in these areas? It may be safer to assume that the "hatchling drift scenario" may only be partially responsible for the ontogeny of adult leatherback foraging area fidelity and that other factors such as magnetic imprinting (Lohmann et al., this book) or Lévy and Brownian movement patterns (Humphries et al., 2010) also play a role.

Examination of the adult leatherback migratory pathways in this chapter raises critical questions regarding the evolution of the remarkable dichotomies among populations that do not seem intuitive from a bioenergetics perspective. For example, postnesting leatherbacks from Mexico and Costa Rica are substantially closer to the highly productive CC LME (Figure 7.1) than those nesting in Papua Barat, Indonesia, yet both genetic and satellite-tracking studies show no occurrence of Eastern Pacific leatherbacks utilizing this LME (Figure 7.6A). Moreover, postnesting females from Papua Barat will transit the entire Pacific basin to reach the CC LME while others from this nesting site will forage much closer in the SCS and IS LMEs (Figure 7.5). If the energetic costs of distant migration can be saved by foraging in productive LMEs that are much closer to nesting beaches, then why do some postnesting females completely neglect these areas whereas others do not? Why are eastern Atlantic leatherbacks only observed in the Southern Hemisphere whereas those from the western Atlantic are only observed in the Northern Hemisphere? These conundrums suggest that ocean currents may in fact play a role in adult foraging site fidelity. The diversity of migratory pathways and foraging areas of the Atlantic and Western Pacific populations may be one way this species has evolved to tolerate instability in the quality of foraging areas.

### 7.3.4  ADULT ROUND-TRIP MIGRATION

Some of the tracking studies reported here observed track durations to and from nesting beaches and highlighted some common aspects of adult round-trip migration. It is safe to assume that mature male leatherbacks frequent nesting beaches just as often, if not more often, than mature females. This assumption is based on both observations of mature males near nesting beaches (Reina et al., 2005; Fossette et al., 2012) and migrations of in-water captured males from foraging areas in the SS LME (James et al., 2005a). Many sea turtle populations may be female-biased (Chapter 13) but male turtles return to nesting beaches more frequently than females, which could compensate for their lower proportion within the population (Hays et al., 2010b). Breeding at or near nesting beaches is common for many species of sea turtle and has been reported for leatherbacks using mounted video cameras on internesting females from Playa Grande, Costa Rica (Reina et al., 2005).

Off the coast of Nova Scotia, Canada, James et al. (2005a) reported eight Western Atlantic leatherbacks that underwent round-trip migration from the temperate foraging waters of the SS LME (Figure 7.1) to tropical waters further south (Figure 7.2A and B). Interestingly, mature males migrated south to coastal tropical waters while subadults and mature females more commonly migrated to pelagic tropical waters (James et al., 2005a). From the coast of Central California, some of the Western Pacific leatherbacks made round-trip migrations southeast to the Eastern Equatorial Pacific (Figure 7.5) before heading back northwest to the CC LME (Benson et al., 2011). There is a consistent seasonal pattern in the round-trip migrations in both of these populations moving into temperate waters in the spring and summer and then back to tropical waters in the fall and winter. These seasonal round-trip migrations are likely being driven by (1) the temperature constraints of large leatherbacks limit them from winter temperatures in temperate regions, even though they are facultative homeotherms; (2) the nesting beaches for leatherbacks are in tropical regions and thus mature males and females need to return to tropical areas for reproduction; (3) gelatinous zooplankton is highest in the summer months following the spring phytoplankton bloom when temperature and the frequency of physical retention zones increase causing favorable conditions for gelatinous zooplankton blooms; and (4) the greater availability of large-sized gelatinous zooplankton in cooler, temperate waters due to enhanced trophic transfer efficiencies from simpler food webs with large phytoplankton at the base of the food chain (discussed further in Section 7.5.1). The mean body weight of leatherbacks upon leaving the waters of the SS LME (Figures 7.1 and 7.2A and B) was ~33% larger than the body weight of nesting females of the same carapace length nesting in St. Croix (James et al., 2005a). This suggests that these cooler, temperate LMEs are critical foraging areas that allow these turtles to bulk up the energy reserves required for their southward migration to tropical waters and nesting beaches. Energy-intake calculations for leatherbacks foraging in the SS LME suggest that these turtles consume approximately 261 Lion's mane jellyfish per day, which is equivalent to about 73% of their body mass per day or three to seven times their daily metabolic requirements (Heaslip et al., 2012).

## 7.4 DIVE BEHAVIOR

### 7.4.1 DIEL VERTICAL MIGRATION

Diel (or diurnal) vertical migration (DVM) is a common daily movement pattern of many marine organisms including plankton, finfish, tuna, sharks, and whales. It is the movement between shallow depths at night and back down to deeper depths during the daylight hours. There are various driving forces that are thought to cause DVM ranging from predator–prey dynamics to physiological and reproductive factors. This behavior is a common observation among leatherbacks in all of the tracking studies that have analyzed dive behavior during migration, foraging, and internesting movements near nesting beaches (James et al., 2005b; Hays et al., 2006; Luschi et al., 2006; Fossette et al., 2007; Shillinger et al., 2010, 2011). Most of these studies suggest that the DVM of gelatinous zooplankton is causing the ensuing behavior of leatherbacks such that they are simply tracking their prey throughout the water column. However, there are not many field observations of DVM among gelatinous zooplankton, and, in some of the few direct observations, DVM was not associated with daily vertical movements patterns (Hays et al., 2011). There may also be visual factors driving the DVM observed with leatherbacks such that during the brighter daylight hours, the turtles can capture prey at deeper depths due to the enhanced light levels.

### 7.4.2 FORAGING

The majority of adult leatherback migration studies discussed here have reported detailed swim speed and dive behavior. In the Atlantic Ocean, there is a persistent pattern of adult movement behavior based on location (James et al., 2005b; Eckert, 2006; Hays et al., 2006; Fossette et al., 2010b). When subadult and adult leatherbacks were in temperate latitudes, their swim speed decreased and their diving became shallower (Figure 7.8). Slower swim speeds are indicative of foraging and digestion (Fossette et al., 2010b). As the turtles moved through the North Atlantic Subtropical Gyre, their swim speed increased, and their dive depths became substantially deeper (James et al., 2005b; Eckert, 2006; Hays et al., 2006; Fossette et al., 2010b). However, once the turtles reached equatorial areas, their swim speed once again decreased and diving became shallower, similar to that of temperate areas. Shallow diving and slow swim speeds were also associated with coastal areas (Figure 7.8) (Fossette et al., 2010b). These studies have associated slow swim speeds and shallow diving with high-success foraging, suggesting that gelatinous zooplankton prey in the coastal LMEs, the North Atlantic Transition Zone, and the Eastern Equatorial Atlantic are more abundant and at shallower depths compared to the North Atlantic Subtropical Gyre turtles where the turtles need to dive deeper. Moreover, the turtles' dive depths appear to track the depths of the thermocline (and nutricline) (Bailey et al., 2012), which are both deeper in the middle of the gyre but then shoal on the edges of the gyre. Phytoplankton biomass as indicated by surface chl-$a$ (Figure 7.1) and zooplankton biomass (Fossette et al., 2010b) derived from satellite ocean color data conform with the high-use patterns described in Figure 7.8 such that prey availability is likely higher in the coastal LMEs, the North Atlantic Transition Zone, and the Eastern Equatorial Atlantic.

Interestingly, the profound spatial-based diving behavior observed among Western Atlantic leatherbacks was not evident among Eastern Pacific leatherbacks (Shillinger et al., 2011; Bailey et al., 2012). The dive depths of Eastern Pacific postnesting females did not significantly increase in the center of the South Pacific Subtropical Gyre, and thus latitude or the depth of the thermocline was not significant factors driving dive depth over their entire postnesting migration area. A possible explanation for this is that the depth of the thermocline of the South Pacific Subtropical Gyre is substantially deeper than that of the North Atlantic Subtropical Gyre (Bailey et al., 2012). Therefore, the turtles do not expend the additional energy required or risk the physiological consequences of frequent deep dives to reach the extremely deep (among the deepest in the global ocean) thermocline in the South Pacific Subtropical Gyre. In other words, it may be a "high risk, small reward" scenario.

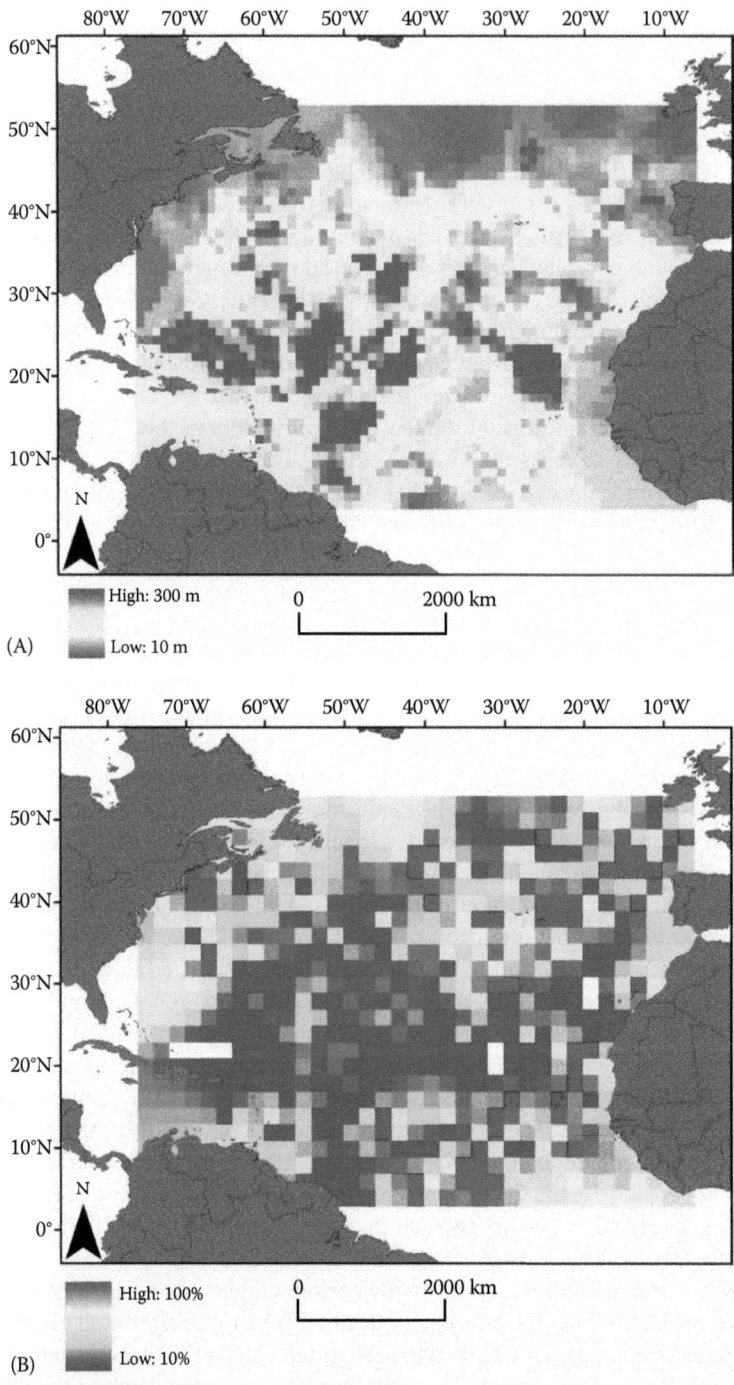

**FIGURE 7.8**  Diving and high-use activity by Western Atlantic leatherbacks. (A) Mean depth of dives recorded in individual 6 h intervals for 20 satellite-tracked leatherback turtles during their migration in the North Atlantic Ocean. The scale indicates the mean depth of the dives performed in each pixel of $1° \times 1°$. (B) High-success foraging areas of 21 satellite-tracked leatherback turtles during their migration in the North Atlantic Ocean (the scale indicates the mean percentage of time spent achieving high-foraging success per pixel of $2° \times 2°$, refer to Fossette et al., 2010b for details). (Reproduced from Fossette, S. et al., *J. Mar. Syst.*, 81, 225, 2010b. With permission.)

### 7.4.3 THERMOREGULATION

Currently, the relationship between swim speed, dive behavior, and location is assumed to be a function of prey distribution. However, one must also consider that dive behavior may also be a function of temperature and thus the turtles' requirement to thermoregulate (Chapter 1). There has been recent debate over the dominating thermoregulatory mechanisms that control leatherback core body temperature (i.e., size vs. movement). Wallace and Jones (2008) extensively reviewed this subject and concluded that leatherback thermoregulation depends on an integrated balance between their unique suite of adaptations for heat production (e.g., metabolic and behavioral modifications) and retention (e.g., large thermal inertia, blood flow adjustments, insulation, and behavioral modifications) to achieve and maintain preferred differentials between body temperature and water temperature in varied thermal environments.

One of the few studies of the relationship between temperature and movement was recently conducted among internesting female leatherbacks from Playa Grande National Marine Park, Costa Rica, where the turtles utilized the Gulf of Papagayo between nesting events (Figure 7.6B; Shillinger et al., 2010). Internesting turtles were tracked during three consecutive nesting years, one of which was anomalously warm due to an El Niño event, and the study found that the turtles dove deeper/longer and swam faster during the warmer year (Shillinger et al., 2010). From a thermoregulatory perspective, one would suspect that this is due to the turtles' need to dive deeper to reach the thermocline and remain in deeper, cooler water longer due to warmer SSTs. However, the increased swim speeds during the warm period would be counterintuitive given the extra heat generated. Conversely, from a foraging perspective, one would suspect that the deeper dives, longer dive duration, and faster swim speeds were due to a higher rate of prey patchiness due to the El Niño event and resulting decreased upwelling in the Gulf of Papagayo. Therefore, there are currently no empirical data that can determine the reason for the marked differences in dive depth and swim speed during the warmer year. One can only surmise that the behavior change could be a function of both thermoregulation and prey distribution.

## 7.5 AN OCEANOGRAPHIC CONTEXT FOR ADULT FORAGING HOTSPOTS

### 7.5.1 LINK BETWEEN DIATOMS, TEMPERATURE, AND LEATHERBACK FORAGING HOTSPOTS

Worldwide, it is evident that cooler, temperate waters are common foraging areas for subadult and adult leatherbacks. In marine food webs, the efficiency of carbon transfer from primary producers (phytoplankton) to upper trophic levels (zooplankton, fish, turtles, birds, whales) is a function of both the magnitude of primary productivity, the community structure of the ecosystem, and the efficiency of trophic linkages (Ryther, 1969). Essentially, the larger the plant cells at the beginning of the food chain, the fewer the trophic levels required to convert the organic matter to a useful form (Ryther, 1969). Therefore, ecosystems that have a high abundance of large phytoplankton (i.e., diatoms, dinoflagellates, cell sizes often >5 μm) relative to small phytoplankton (i.e., cryptophytes, cyanobacteria, cell sizes often <5 μm) will have fewer trophic steps in the food chain up to the top predators and generally higher overall energy transfer efficiencies to higher trophic levels. Large phytoplankton species, such as diatoms, are generally more prevalent in regions with large nutrient inputs such as coastal LMEs and high latitude pelagic regions (Falkowski et al., 2004). Encased within a cell wall made of silica, diatoms are commonly associated with the most productive marine ecosystems around the world.

In situ measurements of marine NPP, let alone large phytoplankton productivity, are extremely sparse both spatially and temporally. Therefore, scientists rely on models based on biogeochemical and physical principles to provide regional and global-scale NPP estimates over annual and decadal time-scales. Figure 7.9 is a satellite-derived ocean color model estimate (Uitz et al., 2010) of the

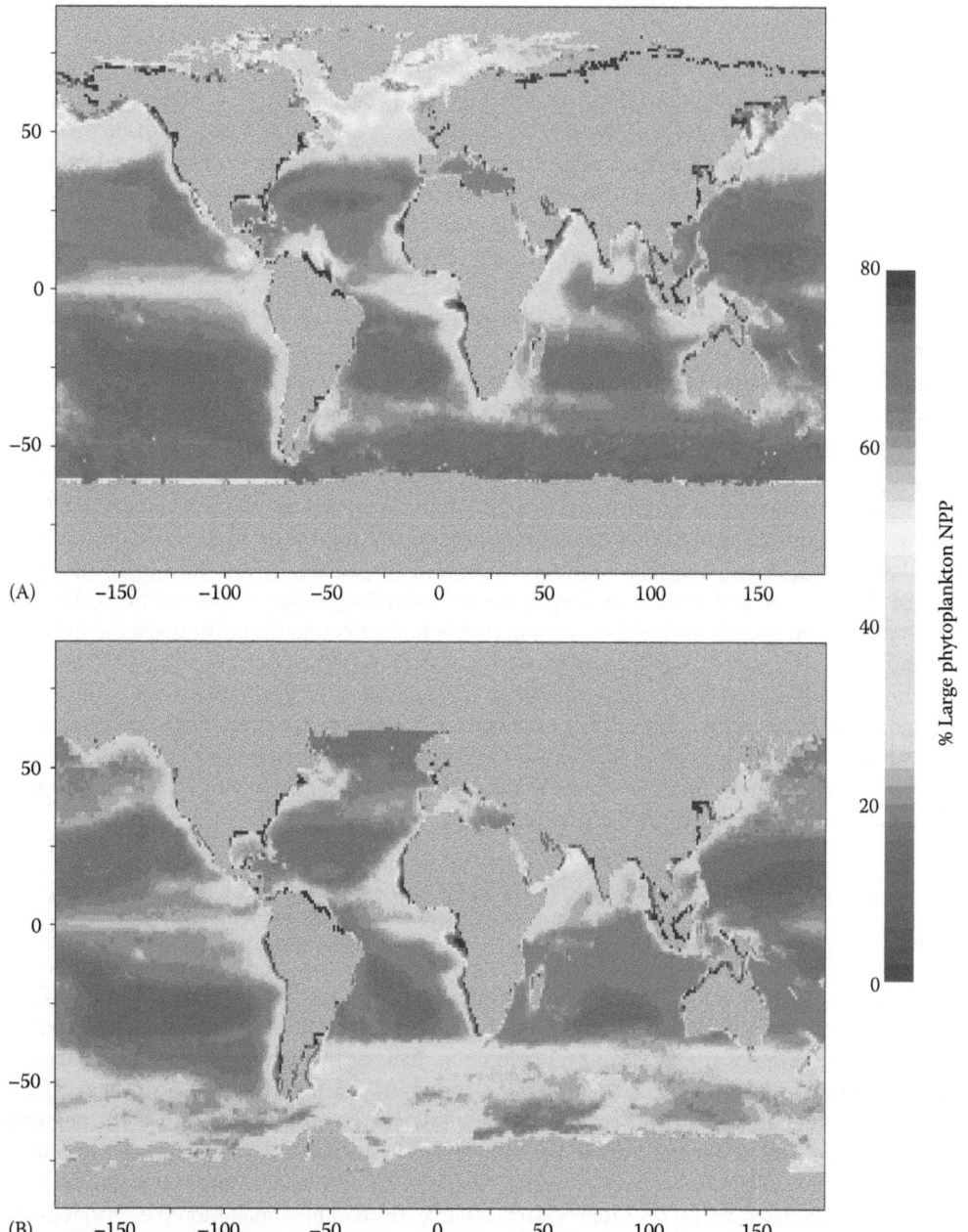

**FIGURE 7.9**  Seasonal climatology (1998–2007) of large phytoplankton (diatoms and dinoflagellates; >20 μm cell size) NPP for the (A) boreal summer/austral winter (June–August) and (B) the boreal winter/austral summer (December–February) based on SeaWiFS ocean color data. (Reproduced from Uitz, J. et al., *Global Biogeochem. Cycles*, 24, GB3016, 2010. With permission.)

mean proportion of large phytoplankton NPP (cell size > 20 μm) for the global ocean during the boreal and austral summer from 1998 to 2007. Comparing Figure 7.9 to Figure 7.1, it is evident that all of the coastal LMEs associated with leatherback foraging areas are high in large phytoplankton NPP. Moreover, the Southern and Northern Transition Zones in temperate waters as well as the Eastern Equatorial tropical waters are also rich in large phytoplankton (Figure 7.9).

It is clear from the tracking studies reviewed in this chapter that most postnesting females are typically foraging in waters that are rich in large phytoplankton, particularly in temperate, coastal LMEs (Figure 7.1). This is consistent with the high-potential trophic transfer efficiency of these ecosystems (Ryther, 1969). Western Pacific leatherbacks derived from nesting beaches in Indonesia migrate across the entire Pacific basin over a 10–12 month period to reach the CC LME (Benson et al., 2011) that is rich in large phytoplankton NPP (Figures 7.1 and 7.9).

Given that a large portion of leatherback migratory pathways worldwide lead to cooler, temperate waters, it is likely these colder waters are more favorable foraging areas due to increased trophic transfer efficiency. Foraging in cooler waters may be beneficial due to a variety of physiological factors such as a greater frequency of larger gelatinous organisms and more efficient food assimilation. Higher energy transfer efficiencies from phytoplankton to upper trophic levels in cooler, high latitude ecosystems may be due to either or both: (1) suppressed rates of microbial herbivory in colder water that increases the proportion of NPP available to zooplankton; (2) the lower maturity of high latitude ecosystems that result in simpler, more efficient food webs (Friedland et al., 2012). Indeed, the largest fishery yields derive from cooler, temperate LMEs, likely driven by higher trophic transfer efficiency associated with cold water systems that are dominated by large phytoplankton (Friedland et al., 2012).

## 7.5.2 Oceanographic Fronts, Mesoscale Eddies, and Currents

Aside from the general patterns of phytoplankton biomass, large phytoplankton productivity, and temperature, smaller mesoscale features and currents also appear to influence leatherback movement and behavior. Among Western Pacific leatherbacks, high-use areas or area-restricted searches occurred in mesoscale eddies, current boundaries (i.e., Transition Zones), frontal regions, and coastal retention areas (Benson et al., 2011). Just outside the CBS LME, the behavior of an adult leatherback was associated with an anticyclonic (rotating clockwise) mesoscale eddy such that the turtle resided 66 days in this feature by looping around the eddy with the same anticyclone rotation (Doyle et al., 2008). The movement behavior of Western Indian leatherbacks has also been associated with mesoscale eddies (Lambardi et al., 2008).

Ocean currents are also suggested to influence the movement behavior of adults. Among the Western Indian population, postnesting females appear to be influenced by currents encountered during their migration such that their trajectories were associated with the respective currents either along or offshore (Lambardi et al., 2008). In the Eastern Pacific, ocean currents had a substantial effect on migratory pathways of postnesting females leaving Playa Grande, Costa Rica (Shillinger et al., 2008). Interannual variability in the location and strength of currents between 12°N and 5°S was associated with interannual variability in the movement patterns of postnesting females leaving Playa Grande, Costa Rica, over a 3 year period. The migration corridor for these leatherbacks was constrained to a migration route of lowest mean kinetic energy, and thus the turtles avoided currents that would push them outside their southwest heading (Shillinger et al., 2008). This "path of least resistance" makes sense in terms of bioenergetics, especially for postnesting females that are energy starved after the resource demanding process of migration and nesting.

## 7.6 BOTTOM-UP FORCING AND CLIMATE VARIABILITY

### 7.6.1 ENSO Influence on Eastern Pacific Leatherbacks

A prime example of the connection between large phytoplankton NPP and leatherback ecology derives from a nesting model for the eastern Pacific leatherback population in Costa Rica (Saba et al., 2007). This model predicts the number of remigrant nesting females based on the phase of the El Niño Southern Oscillation (ENSO). The ENSO is a coupled ocean-atmosphere natural climate oscillation in the tropical Pacific that is associated with global climate variability (i.e., temperature, rainfall). The La Niña phase of ENSO is associated with cooler SST and a shallower thermocline in the Eastern Equatorial Pacific, which is accompanied by shoaling of the nutricline and thus increasing the amount of upwelled nutrients (particularly iron) into the euphotic zone. This results in higher than average NPP in the Eastern Equatorial Pacific from the substantial increase in the relative abundance of large phytoplankton (Chavez et al., 1999). The opposite occurs during the El Niño phase when SST is warmer than average and large phytoplankton NPP is below average. These phases of ENSO can last between 9 months and 2 years (Figure 7.10A, inset).

Saba et al.'s (2007) nesting model estimates the annual number of remigrant nesting females at Playa Grande based on the Multivariate ENSO Index (MEI; Wolter and Timlin, 1998), and satellite-derived sea surface chl-*a* (ocean color) data averaged over the 12 months prior to each nesting season (Figure 7.10A). The MEI is based on six climatic variables over the tropical Pacific. The fit of the model's estimate to the observed was statistically significant over 16 years of remigrant data. When La Niña conditions persist over the year before a nesting season, the probability that previously tagged females will return to nest increases and vice versa for El Niño conditions. In summary, when large phytoplankton are in higher abundance in the Eastern Equatorial Pacific, the foraging conditions for mature females are more suitable such that these turtles have a greater chance of accumulating the threshold level of energy reserve that is required for migration, vitellogenesis, and the nesting process.

Although ENSO-associated variability is quite high in the nesting variability of leatherbacks at Playa Grande, Costa Rica, there is still a paucity of leatherback foraging in the Eastern Equatorial Pacific based on the first year to year and a half of postnesting tracking data (Shillinger et al., 2008, 2011). It is possible that intensive foraging occurs in the Eastern Equatorial Pacific on the way back to Playa Grande or toward the end of the nearly 4 year remigration interval reported for the Eastern Pacific population (Saba et al., 2008a). This component of the migratory cycle was not captured in the tracking study by Shillinger et al. (2008, 2011) and would require tags that last 3–4 years or on in-water captures of mature adults in their southern, temperate foraging areas. The neglect of the Eastern Equatorial Pacific as a foraging area on their way south suggests that postnesting females predominately target cooler, temperate foraging areas such as the South Pacific Transition Zone (Figure 7.1) due to increased trophic transfer efficiency of these waters as discussed in Sections 7.3.4 and 7.5.1. Moreover, leatherback high use of Eastern Equatorial waters just after nesting seasons is not a common observation among the other tracking studies reviewed in the chapter. The extraordinary energy-intake rate (Heaslip et al., 2012) and body weight increase (James et al., 2005a) observed among leatherbacks foraging in the SS LME (Figures 7.1 and 7.2A and B) suggest that cooler, temperate waters produce the largest energy gains.

### 7.6.2 Interpopulation Differences in Phenotype, Fecundity, and Population Status

A review of leatherback-nesting populations worldwide revealed substantial interpopulation differences in mature female body size, fecundity, and nesting trends (Saba et al., 2008b). The review was combined with an analysis of ocean color model NPP in the migration and foraging areas of postnesting females and suggested that differences in the estimated magnitude and variability of NPP among the migration and foraging areas of postnesting females were driving the interpopulation trait differences. Smaller and less-fecund mature females were associated with migration and foraging areas that were lower in NPP (Figure 7.10B) and/or those that had a higher proportion of interannual variability in NPP. The largest and most fecund mature females derived from the nesting beaches in South Africa representing the Western Indian population (Saba et al., 2008b). These females nest in the highest latitude compared to other populations due to the warm AC maintaining warm air temperatures at the adjacent beaches. These postnesting females do not have to travel as far to reach cooler, temperate foraging areas south of the nesting beach (Figure 7.4) and also appear to be conserving energy by passively migrating with currents (Lambardi et al., 2008). Therefore, one can surmise that Western Indian Ocean leatherbacks

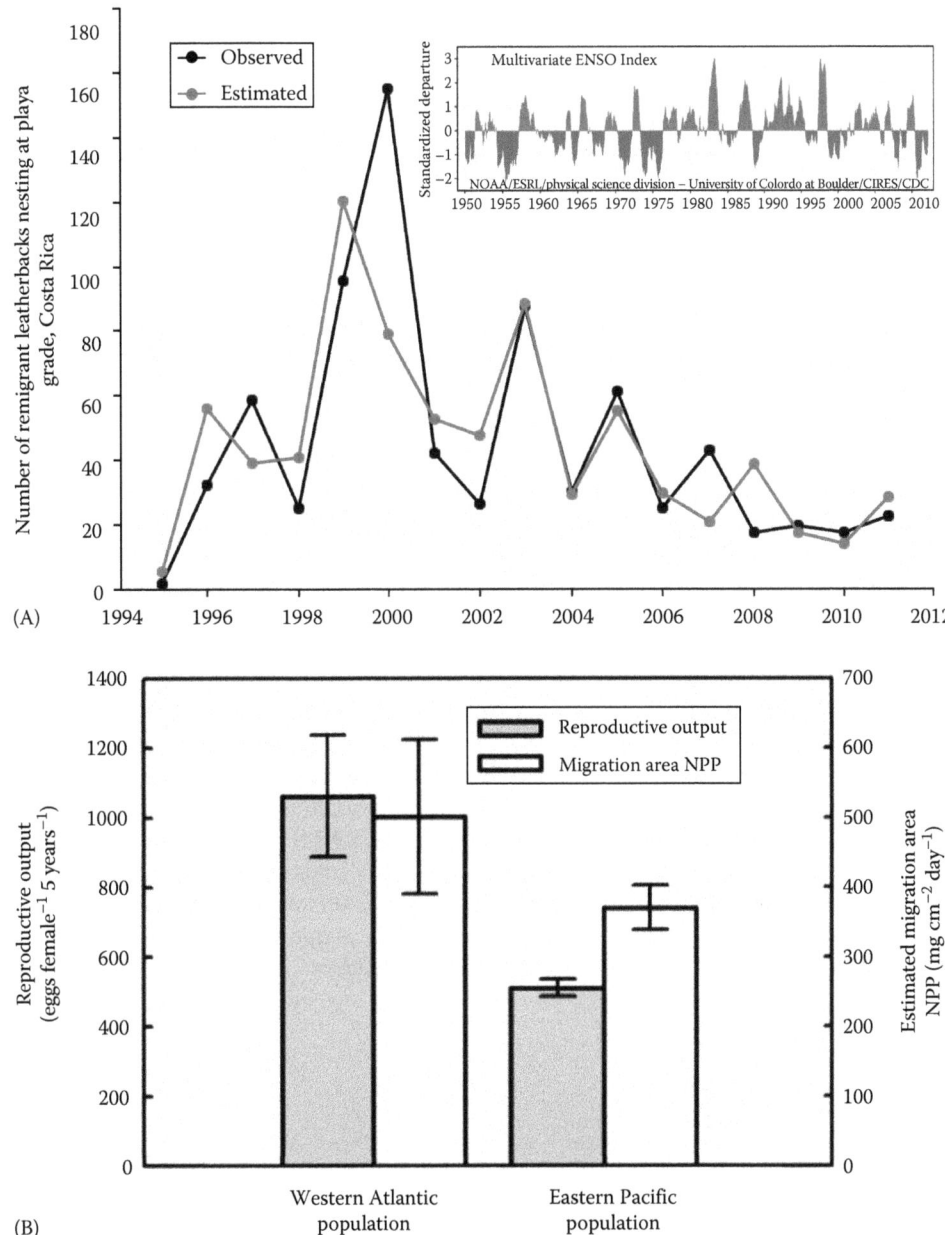

FIGURE 7.10   (A) Observed versus estimated number of Eastern Pacific remigrant nesting females at Playa Grande, Costa Rica. The estimated number of remigrants is based on the updated Saba et al. (2007) model that uses the Multivariate El Niño Southern Oscillation (ENSO) Index (MEI; shown in panel) and SeaWiFS surface chl-*a* in the Eastern Equatorial Pacific as a proxy for ENSO-associated foraging conditions for the Eastern Pacific population nesting at Playa Grande, Costa Rica. (B) Reproductive output of western Atlantic (WA) and eastern Pacific (EP) nesting females and estimated mean monthly NPP at their respective total migration area. Estimates of NPP derived from the VGPM algorithm (refer to Saba et al. [2008b] for details).

nesting in South Africa do not have to expend as much energy on migration as females from other populations and can thus expend more energy on growth and reproduction leading them to be the largest and most fecund females.

The smallest and least fecund females were from the eastern Pacific population (nesting in Mexico and Costa Rica) where NPP variability in the equatorial waters is primarily interannual (associated with ENSO) and does not have the predictable bloom period of foraging areas of other populations. Moreover, the South Pacific Subtropical Gyre where these postnesting females tend to migrate and forage is one of the most unproductive areas in the global ocean (Figures 7.1, 7.9, and 7.10B). Reduced reproductive output of the Eastern Pacific population may increase their sensitivity to anthropogenic mortality (i.e., fisheries bycatch, egg poaching) via lower and more variable recruitment rates and may explain their precipitously declining population trends in Costa Rica and Mexico despite continued beach protection (Saba et al., 2008b; Wallace and Saba, 2009).

As discussed in Section 7.3.1.5, another exclusive trait of the eastern Pacific nesting population is the rare occurrence of migration to coastal waters (Figure 7.6A). This finding is quite extraordinary considering that the highly productive HC LME (Figure 7.1) is relatively close to the nesting beaches in Costa Rica. The smaller, less-fecund females appear to be those that are bypassing the coastal waters after nesting. The single coastal forager observed in the studies by Shillinger et al. (2008, 2011) that remained in the Gulf of Panama was one of the largest and most fecund individuals in the tracking study (Bailey et al., 2012).

Populations in the Atlantic and Indian Oceans are either stable or increasing while those in the Pacific are declining. Whereas nesting data in the western Pacific are limited, the eastern Pacific dataset is quite extensive. Population recovery time in the eastern Pacific may be prolonged, given the continued high mortality in the HC LME (Alfaro-Shigueto et al., 2011). Populations in the Atlantic and Indian Oceans have recovered or are stable due to their recruitment rates that are likely double that of the eastern Pacific population. Populations in the western Pacific also appear to be declining but data are limited regarding their long-term trends, body size, and reproductive output. One might surmise that mature females nesting in Papua Barat that migrate across the Pacific to the CC LME have a reduced reproductive output compared to those migrating to closer foraging areas in the SCS LME and other more local LMEs. Indeed, the mean CCL of winter nesters was significantly greater than the CCL for summer nesters while curved carapace width was significantly greater for both winter nesters and for turtles that moved into temperate waters. These data further support the theory of higher trophic transfer efficiency in colder, temperate ecosystems. Nonetheless, more data are required from the western Pacific nesting populations in order to make any inferences regarding bottom-up forcing and their reproductive output.

## 7.7 FUTURE DIRECTIONS

This chapter highlighted the migratory pathways and high-use areas of adult and subadult leatherbacks worldwide. Fisheries mortality reductions at high-use areas (Chapter 12), specifically within the Pacific Ocean coastal LMEs highlighted in Figure 7.1, should be a management priority to support recovery of both the Eastern and Western Pacific leatherback populations. Acquiring more data on roundtrip migration, especially in the Eastern Pacific where populations have precipitously declined, is essential in order to apply marine spatial-planning management strategies over longtime periods. More data on the juvenile life history stage are also required for effective conservation measures. Sea turtle population growth rates are most sensitive to survival during the late juvenile stage (Crowder et al., 1994), and thus this stage of life history should be critically assessed when designing recovery plans for declining populations. Presently, these data are virtually nonexistent with the exception of some incidental fishery catch reports (de Paz et al., 2006) and a captive growth study (Jones et al., 2011b).

More research is also needed regarding the physical and biological mechanistic underpinnings that are responsible for the changes in leatherback dive behavior as a function of ocean basin, time, space, temperature, prey availability, and thermoregulation. Answering such questions will provide further insight to the potential impacts of climate change in the oceanic habitats of leatherback turtles. What are the consequences of a warming ocean on leatherback habitat and prey availability and what are the bioenergetic implications of living in a warmer ocean? Finally, global and regional estimates of gelatinous zooplankton biomass over time would be a tremendous step-forward in understanding the oceanic habits and habitats of leatherback turtles.

## REFERENCES

Alfaro-Shigueto, J., P. H. Dutton, M. Van Bressem, and J. Mangel. 2007. Interactions between leatherback turtles and Peruvian artisanal fisheries. *Chelonian Conservation and Biology* 6:129–134.

Alfaro-Shigueto, J., J. C. Mangel, F. Bernedo, P. H. Dutton, J. A. Seminoff, and B. J. Godley. 2011. Small-scale fisheries of Peru: A major sink for marine turtles in the Pacific. *Journal of Applied Ecology* 48:1432–1440.

Attrill, M. J., J. Wright, and M. Edwards. 2007. Climate-related increases in jellyfish frequency suggest a more gelatinous future for the North Sea. *Limnology and Oceanography* 52:480–485.

Bailey, H., S. Fossette, S. J. Bograd, G. L. Shillinger, A. M. Swithenback, J. Georges, P. Gaspar. 2012. Movement patterns for a critically endangered species, the leatherback turtle (*Dermochelys coriacea*), linked to foraging success and population status. *PLoS ONE* 7:e36401.

Benson, S. R., T. Eguchi, D. G. Foley, K. A. Forney, H. Bailey, C. Hitipeuw, B. P. Samber, R. F. Tapilatu, V. Rei, P. Ramohia, J. Pita, and P. H. Dutton. 2011. Large-scale movements and high-use areas of western Pacific leatherback turtles, *Dermochelys coriacea*. *Ecosphere* 2:art84.

Benson, S. R., K. A. Forney, J. T. Harvey, J. V. Carretta, and P. H. Dutton. 2007. Abundance, distribution, and habitat of leatherback turtles (*Dermochelys coriacea*) off California, 1990–2003. *Fishery Bulletin* 105:337–347.

Block, B. A., I. D. Jonsen, S. J. Jorgensen, A. J. Winship, S. A. Shaffer, S. J. Bograd, E. L. Hazen, D. G. Foley, G. A. Breed, A. L. Harrison, J. E. Ganong, A. Swithenbank, M. Castleton, H. Dewar, B. R. Mate, G. L. Shillinger, K. M. Schaefer, S. R. Benson, M. J. Weise, R. W. Henry, and D. P. Costa. 2011. Tracking apex marine predator movements in a dynamic ocean. *Nature* 475:86–90.

Chavez, F. P., P. G. Strutton, G. E. Friederich, R. A. Freely, G. C. Feldman, D. G. Foley, and M. J. McPhaden 1999. Biological and chemical response of the Equatorial Pacific Ocean to the 1997–98 El Niño. *Science* 286:2126–2131.

Crowder, L. B., D. T. Crouse, S. S. Heppell, and T. H. Martin. 1994. Predicting the impact of turtle excluder devices on loggerhead sea turtle populations. *Ecological Applications* 4:437–445.

Decker, M. B., C. W. Brown, R. R. Hood, J. E. Purcell, T. F. Gross, J. C. Matanoski, R. O. Bannon, and E. M. Setzler-Hamilton. 2007. Predicting the distribution of the scyphomedusa *Chrysaora quinquecirrha* in Chesapeake Bay. *Marine Ecology Progress Series* 329:99–113.

Doyle, T., J. Houghton, R. McDevitt, J. Davenport, and G. Hays. 2007. The energy density of jellyfish: Estimates from bomb-calorimetry and proximate-composition. *Journal of Experimental Marine Biology and Ecology* 343:239–252.

Doyle, T. K., J. D. Houghton, P. F. O'Súilleabháin, V. J. Hobson, F. Marnell, J. Davenport, and G. C. Hays. 2008. Leatherback turtles satellite-tagged in European waters. *Endangered Species Research* 4:23–31.

Dutton, P. H., E. L. LaCasella, J. Alfaro-Shigueto, and M. Donoso. 2010. Stock origin of leatherback (*Dermochelys coriacea*) foraging in the southeastern Pacific. *Proceedings of the 30th Annual Symposium on Sea Turtle Biology and Conservation*, Goa, India.

Eckert, S. A. 2002. Global distribution of juvenile leatherback turtles, *Dermochelys coriacea*. *Marine Ecology Progress Series* 230:289–293.

Eckert, S. A. 2006. High-use oceanic areas for Atlantic leatherback sea turtles (*Dermochelys coriacea*) as identified using satellite telemetered location and dive information. *Marine Biology* 149:1257–1267.

Eckert, S. A. and K. L. Eckert. 1986. Harnessing leatherbacks. *Marine Turtle Newsletter* 37:1–3.

Eckert, S. A. and L. Sarti. 1997. Distant fisheries implicated in the loss of the world's largest leatherback nesting population. *Marine Turtle Newsletter* 78:2–7.

Falkowski, P. G., M. E. Katz, A. H. Knoll, A. Quigg, J. A. Raven et al. 2004. The evolution of modern eukaryotic phytoplankton. *Science* 305:354–360.

Ferraroli, S., J.-Y. Georges, P. Gaspar, and Y. Le Maho. 2004. Where leatherback turtles meet fisheries. *Nature* 429:521–522.

Fossette, S., H. Corbel, P. Gaspar, Y. Le Maho, and J. Georges. 2008. An alternative technique for the long-term satellite tracking of leatherback turtles. *Endangered Species Research* 4:33–41.

Fossette, S., S. Ferraroli, H. Tanaka et al. 2007. Dispersal and dive patterns in gravid leatherback turtles during the nesting season in French Guiana. *Marine Ecology Progress Series* 338:233–247.

Fossette, S., C. Girard, M. Lopez-Mendilaharsu, P. Miller, A. Domingo, D. Evans, L. Kelle, V. Plot, L. Prosdocimi, S. Verhage, P. Gaspar, and J. Y. Georges. 2010a. Atlantic leatherback migratory paths and temporary residence areas. *PLoS ONE* 5:e13908.

Fossette, S., A. C. Gleiss, J. P. Casey, A. R. Lewis, and G. C. Hays. 2012. Does prey size matter? Novel observations of feeding in the leatherback turtle (*Dermochelys coriacea*) allow a test of predator-prey size relationships. *Biology Letters* 8(3):351–354. doi:10.1098/rsbl.2011.0965.

Fossette, S., V. J. Hobson, C. Girard, B. Calmettes, P. Gaspar, J.-Y. Georges, and G. C. Hays. 2010b. Spatio-temporal foraging patterns of a giant zooplanktivore, the leatherback turtle. *Journal of Marine Systems* 81:225–234.

Friedland, K. D., C. A. Stock, K. F. Drinkwater, J. S. Link, R. T. Leaf, B. V. Shank, J. M. Rose, C. H. Pilskaln, and M. J. Fogarty. 2012. Pathways between primary production and fisheries yields of large marine ecosystems. *PLoS ONE* 7:e28945.

García-Comas, C., L. Stemmann, F. Ibanez, L. Berline, M. G. Mazzocchi, S. Gasparini, M. Picheral, and G. Gorsky. 2011. Zooplankton long-term changes in the NW Mediterranean Sea: Decadal periodicity forced by winter hydrographic conditions related to large-scale atmospheric changes? *Journal of Marine Systems* 87:216–226.

Gaspar P., S. R. Benson, P. H. Dutton, A. Reveillere, G. Jacob, C. Meetoo, A. Dehecq, S. Fossette. 2012. Oceanic dispersal of juvenile leatherback turtles: Going beyond passive drift modeling. *Marine Ecology Progress Series* 457:265–284.

Graham, T. R., J. T. Harvey, S. R. Benson, J. S. Renfree, and D. A. Demer. 2010. Acoustic identification and enumeration of scyphozoan jellies as prey for leatherback sea turtles (*Dermochelys coriacea*) off central California. *ICES Journal of Marine Science* 67:1739–1748.

Hays, G. C., T. Bastian, T. K. Doyle, S. Fossette et al. 2011. High activity and Lévy searches: Jellyfish can search the water column like fish. *Proceedings of the Royal Society B* 279:465–473.

Hays, G. C., S. Fossette, K. A. Katselidis, P. Mariani, and G. Schofield. 2010a. Ontogenetic development of migration: Lagrangian drift trajectories suggest a new paradigm for sea turtles. *Journal of the Royal Society Interface* 7:1319–1327.

Hays, G. C., S. Fossette, K. A. Katselidis, G. Schofield, and M. B. Gravenor. 2010b. Breeding periodicity for male sea turtles, operational sex ratios, and implications in the face of climate change. *Conservation Biology* 24:1636–1643.

Hays, G. C., V. J. Hobson, J. D. Metcalfe, D. Righton, and D. W. Sims. 2006. Flexible foraging movements of leatherback turtles across the North Atlantic Ocean. *Ecology* 87:2647–2656.

Hays, G. C., J. D. R. Houghton, and A. E. Myers. 2004. Pan-Atlantic leatherback turtle movements. *Nature* 429:522.

Heaslip, S. G., S. J. Iverson, W. D. Bowen, and M. C. James. 2012. Jellyfish support high energy intake of leatherback sea turtles (*Dermochelys coriacea*): Video evidence from animal-borne cameras. *PLoS ONE* 7:e33259.

Henson, S. A., J. P. Dunne, and J. L. Sarmiento. 2009. Decadal variability in North Atlantic phytoplankton blooms. *Journal of Geophysical Research* 114:C04013.

Humphries, N. et al. 2010. Environmental context explains Lévy and Brownian movement patterns of marine predators. *Nature* 465:1066–1069.

James, M. C., C. Andrea Ottensmeyer, and R. A. Myers. 2005a. Identification of high-use habitat and threats to leatherback sea turtles in northern waters: New directions for conservation. *Ecology Letters* 8:195–201.

James, M. C., J. Davenport, and G. C. Hays. 2006. Expanded thermal niche for a diving vertebrate: A leatherback turtle diving into near-freezing water. *Journal of Experimental Marine Biology and Ecology* 335:221–226.

James, M. C., R. A. Myers, and C. A. Ottensmeyer. 2005b. Behaviour of leatherback sea turtles, *Dermochelys coriacea*, during the migratory cycle. *Proceedings of the Royal Society B* 272:1547–1555.

Jones, T. T., B. L. Bostrom, M. Carey, B. Imlach et al. 2011a. Determining transmitter drag and best practice attachment procedures for sea turtle biotelemetry studies. NOAA Technical Memorandum NMFS-SWFSC-480.

Jones, T. T., M. D. Hastings, B. L. Bostrom, D. Pauly, and D. R. Jones. 2011b. Growth of captive leatherback turtles, *Dermochelys coriacea*, with inferences on growth in the wild: Implications for population decline and recovery. *Journal of Experimental Marine Biology and Ecology* 399:84–92.

Lambardi, P., J. R. E. Lutjeharms, R. Mencacci, G. C. Hays, and P. Luschi. 2008. Influence of ocean currents on long-distance movement of leatherback sea turtles in the southwest Indian Ocean. *Marine Ecology Progress Series* 353:289–301.

Lilley, M. K. S., S. E. Beggs, T. K. Doyle, V. J. Hobson, K. H. P. Stromberg, and G. C. Hays. 2011. Global patterns of epipelagic gelatinous zooplankton biomass. *Marine Biology* 158:2429–2436.

Luschi, P., J. R. Lutjeharms, P. Lambardi, R. Mencacci, G. R. Hughes, and G. C. Hays. 2006. A review of migratory behaviour of sea turtles off southeastern Africa. *South African Journal of Science* 102:51–58.

Lynam, C. P., S. J. Hay, and A. S. Brierley. 2004. Interannual variability in abundance of North Sea jellyfish and links to the North Atlantic Oscillation. *Limnology and Oceanography* 49:637–643.

Mansfield, K. L., V. S. Saba, J. Keinath, and J. A. Musick. 2009. Satellite telemetry reveals a dichotomy in migration strategies among juvenile loggerhead sea turtles in the northwest Atlantic. *Marine Biology*, 156:2555–2570.

Paladino, F. V., M. P. O'Connor, and J. R. Spotila, 1990. Metabolism of leatherback turtles, gigantothermy, and thermoregulation of dinosaurs. *Nature* 344:858–860.

de Paz, N., J. C. Reyes, M. Ormeñõ, H. A. Anchante, and A. J. Altamirano. 2006. Immature leatherback mortality in coastal gillnet fisheries off San Andre's, Southern Peru. In: Frick, M., Panagopoulou, A., Rees, A. F., Williams, K. (Eds.), *26th Annual Symposium on Sea Turtle Biology and Conservation*, International Sea Turtle Society, Crete, Greece, p. 376.

Pérez-Ruzafa, A., J. Gilabert, J. M. Gutierrez et al. 2002. Evidence of a planktonic food web response to changes in nutrient input dynamics in the Mar Menor coastal lagoon, Spain. *Hydrobiologia* 476:359–369.

Polovina, J. J., E. Howell, D. R. Kobayashi, and M. P. Seki. 2001. The transition zone chlorophyll front, a dynamic global feature defining migration and forage habitat for marine resources. *Progress in Oceanography* 49:469–483.

Purcell, J. 2012. Jellyfish and climate change: Increased invasion and affects on foodwebs. *Annual Review of Marine Science* 4:209–235.

Purcell, J. E., J. R. White, D. A. Nemazie, and D. A. Wright. 1999. Temperature, salinity and food effects on asexual reproduction and abundance of the scyphozoan *Chrysaora quinquecirrha*. *Marine Ecology Progress Series* 180:187–196.

Putman, N. F., J. M. Bane, and K. J. Lohmann. 2010. Sea turtle nesting distributions and oceanographic constraints on hatchling migration. *Proceedings of the Royal Society B* 277:3631–3637.

Reina, R. D., K. J. Abernathy, G. J. Marshall, and J. R. Spotila. 2005. Respiratory frequency, dive behaviour and social interactions of leatherback turtles, *Dermochelys coriacea* during the inter-nesting interval. *Journal of Experimental Marine Biology and Ecology* 316:1–16.

Ryther, J. H. 1969. Photosynthesis and fish production in the sea. *Science* 166:72–76.

Saba, V. S., P. Santidrián-Tomillo, R. D. Reina, J. R. Spotila, J. A. Musick, D. A. Evans, and F. V. Paladino. 2007. The effect of the El Niño Southern Oscillation on the reproductive frequency of eastern Pacific leatherback turtles. *Journal of Applied Ecology* 44:395–404.

Saba, V. S., G. L. Shillinger, A. M. Swithenbank, B. A. Block, J. R. Spotila, J. A. Musick, and F. V. Paladino. 2008a. An oceanographic context for the foraging ecology of eastern Pacific leatherback turtles: Consequences of ENSO. *Deep Sea Research I* 55:646–660.

Saba, V. S., J. R. Spotila, F. P. Chavez, and J. A. Musick. 2008b. Bottom-up and climatic forcing on the worldwide population of leatherback turtles. *Ecology* 89:1414–1427.

Sale, A. and P. Luschi. 2009. Navigational challenges in the oceanic migrations of leatherback sea turtles. *Proceedings of the Royal Society B* 276:3737–3745.

Sherman, K. 1991. The large marine ecosystem concept: Research and management strategy for living marine resources. *Ecological Applications* 1:349–360.

Shillinger, G. L., E. Di Lorenzo, H. Luo, S. J. Bograd, E. L. Hazen, H. Bailey, and J. R. Spotila. 2012. On the dispersal of leatherback turtle hatchlings from Mesoamerican nesting beaches. *Proceedings of the Royal Society B* 279: 2391–2395. doi:10.1098/rspb.2011.2348.

Shillinger, G. L., D. M. Palacios, H. Bailey, S. J. Bograd et al. 2008. Persistent leatherback turtle migrations present opportunities for conservation. *PLoS Biology* 6:e171.

Shillinger, G. L., A. M. Swithenbank, H. Bailey, S. J. Bograd, M. R. Castelton, B. P. Wallace, J. R. Spotila, F. V. Paladino, R. Piedra, and B. A. Block. 2011. Vertical and horizontal habitat preferences of post-nesting leatherback turtles in the South Pacific Ocean. *Marine Ecology Progress Series* 422:275–289.

Shillinger, G. L., A. M. Swithenbank, S. J. Bograd, H. Bailey, M. R. Castelton, W. Bp, J. R. Spotila, F. V. Paladino, R. Piedra, and B. A. Block. 2010. Identification of high-use internesting habitats for eastern Pacific leatherback turtles: Role of the environment and implications for conservation. *Endangered Species Research* 10:215–232.

Sullivan, B. K., D. Van Keuren, and M. Clancy. 2001. Timing and size of blooms of the ctenophore *Mnemiopsis leidyi* in relation to temperature in Narragansett Bay, RI. *Hydrobiologia* 451:113–120.

Uitz, J., H. Claustre, B. Gentili, and D. Stramski. 2010. Phytoplankton class-specific primary production in the world's oceans: Seasonal and interannual variability from satellite observations. *Global Biogeochemical Cycles* 24:GB3016.

Wallace, B. P. and T. T. Jones. 2008. What makes marine turtles go: A review of metabolic rates and their consequences. *Journal of Experimental Marine Biology and Ecology* 356:8–24.

Wallace, B. P. and V. S. Saba. 2009. Environmental and anthropogenic impacts on intra-specific variation in leatherback turtles: Opportunities for targeted research and conservation. *Endangered Species Research* 7:11–21.

Welp, L. R. et al. 2011. Interannual variability in the oxygen isotopes of atmospheric $CO_2$ driven by El Niño. *Nature* 477:579–582.

Witt, M. J., E. Augowet Bonguno, A. C. Broderick, M. S. Coyne, A. Formia, A. Gibudi, G. A. Mounguengui Mounguengui, C. Moussounda, N. S. M, S. Nougessono, R. J. Parnell, G. P. Sounguet, S. Verhage, and B. J. Godley. 2011. Tracking leatherback turtles from the world's largest rookery: Assessing threats across the South Atlantic. *Proceedings of the Royal Society B* 278:2338–2347.

Wolter, K. and M. S. Timlin. 1998. Measuring the strength of ENSO—How does 1997/98 rank? *Weather* 53:315–324.

# 8 Oceanic Habits and Habitats
## Caretta caretta

*Katherine L. Mansfield and Nathan F. Putman*

## CONTENTS

## 8.1 INTRODUCTION

Like other cheloniid sea turtle species, loggerheads (*Caretta caretta*) are long-lived, late-maturing animals. Loggerheads spend the majority of their lives in a marine environment, occurring within both near-shore (neritic) waters and offshore (oceanic) waters. The oceanic habitat loggerheads occupy is broadly defined as seaward of the 200 m isobath (Bolten, 2003a,b; Witherington et al., 2006); waters with depths exceeding 200 m are considered oceanic, shallower waters are considered neritic (Lalli and Parsons, 1993; Tomczak and Godfrey, 1994). Between the ocean surface and floor, loggerheads live within a three-dimensional, dynamic world. As highly migratory, poikilothermic reptiles, loggerhead distribution and behavior are invariably linked to the environment these turtles inhabit—an environment influenced by physical processes and subject to shifting temperatures, currents, habitats, and prey availability.

### 8.1.1 CARETTA LIFE HISTORY AND THE OCEANIC ZONE

Loggerhead life history models proposed by Carr (1986) and reviewed by Musick and Limpus (1997) and Bolten (2003a,b) suggest a generalized oceanic to neritic life history pattern where early juvenile development occurs within the oceanic zone, followed by larger juveniles' development to maturity within the neritic zone. The ontogenetic stages described by these models are defined by distinct geospatial shifts in size distributions. Turtles leave their natal beaches as hatchlings, transitioning through the neritic zone to the oceanic zone. They remain in the oceanic zone until recruiting back to neritic waters as larger, juveniles where they remain through maturity (Carr, 1986; Musick and Limpus, 1997; Bolten, 2003a). The observed mean size at recruitment to neritic waters varies by ocean basin and ranges from 42 to 64 cm curved carapace length (CCL) in the Atlantic to 44 cm straight carapace length (SCL) to 67 cm CCL in the Pacific (Bjorndal et al., 2000, 2001;

Limpus and Limpus, 2003; reviewed by Bolten, 2003a). Using stable isotope and skeletal growth data, Snover et al. (2010) confirmed that an oceanic to neritic habitat and foraging shift likely occurs among northwestern Atlantic loggerheads at a mean length of 45.5 cm SCL.

Yet, the unidirectional pattern assumed by classic life history models is more variable due to resource availability and plasticity in habitat selection among neritic juveniles (Keinath, 1993; Laurent et al., 1998; Witzell, 1999, 2002; McClellan and Read, 2007; Casale et al., 2008; Mansfield et al., 2009; Arendt et al., 2012c) and adults (Hatase et al., 2002a; Hawkes et al., 2006; Reich et al., 2010). Bycatch data and diet analyses from the Mediterranean (Laurent et al., 1998; Casale et al., 2008) and Atlantic (Eckert and Martins, 1989; Bolten et al., 1993; Witzell, 1999, 2002; Watson et al., 2005) indicate that juvenile loggerheads larger than 45.5 cm (or 42–67 cm CCL) SCL occupy and forage within oceanic habitats. Satellite telemetry further confirms that some juveniles (46.7–90.5 cm SCL) originating from near-shore neritic habitats in the northwest Atlantic exhibit a shift in habitat selection from near-shore neritic back to oceanic habitats (Morreale and Standora, 1988; Eckert and Martins, 1989; Keinath, 1993; McClellan and Read, 2007; Mansfield et al., 2009; Arendt et al., 2012c).

Based on size–frequency distributions from the Azores and southeastern United States (Bjorndal et al., 2000, 2001), Bolten (2003a) suggested that almost all Atlantic loggerheads fully recruit to neritic habitats by the time the turtles reach 64 cm SCL. Yet, the juvenile turtles tracked from western Atlantic neritic foraging habitats to the oceanic zone in the northwestern Atlantic maintained an oceanic lifestyle for a minimum of 1–3.5 consecutive years and ranged in mean size from 64.0 SCL (n = 10; range: 52.9–74.2 SCL McClellan and Read, 2007) to 69.5 cm SCL (n = 6; range: 63.5–90.5 SCL Mansfield et al., 2009). This implies that some larger juveniles may return to an oceanic lifestyle for more than transitory migrations or short transitional developmental periods. The turtles tracked by McClellan and Read (2007) and Mansfield et al. (2009) were transported in the Gulf Stream, a major current system that facilitates the oceanic transport of younger stages (Carr, 1986; Witherington, 2002; Putman et al., 2010a). While these older turtles remained entrained within the northern boundary of the North Atlantic Subtropical Gyre, it appears that larger juveniles returning to the oceanic zone do not complete the same transit around the gyre as is hypothesized for the younger, smaller oceanic juveniles. Despite up to 3.5 years of observed tracks, none of these turtles traveled further east than 35°–30° W latitude (Mansfield et al., 2009); none traveled as far eastward as the Azores.

Satellite telemetry also reveals that post-nesting female loggerheads transit through oceanic waters and/or establish post-nesting oceanic foraging habitats off of Japan in the northwestern Pacific (Hatase et al., 2002a), off of Cape Verde in the eastern Atlantic (Hawkes et al., 2007), in the Arabian Sea off of Oman (Rees et al., 2010), and in the Gulf of Mexico (Girard et al., 2009). Stable isotope analyses from female nesters in the western Atlantic/southeastern United States suggest a similar oceanic post-nesting foraging strategy among some individuals (Reich et al., 2010). A phenotypic dichotomy was observed between these adult foragers: smaller females foraged within the oceanic zone or on oceanic prey; larger females were found within neritic waters or associated with coastal prey items (Hatase et al., 2002a; Hawkes et al., 2007; Reich et al., 2010). This phenotypic dichotomy is not reflected among the neritic (or "neritic back to oceanic") juveniles satellite-tracked by McClellan and Read (2007) and Mansfield et al. (2009); however, the spatial scope and sample sizes of these studies may be too small to detect such a pattern.

Bolten (2003a,b) hypothesized that the ontogenetic shift from oceanic to neritic habitats occurs in order to maintain or maximize juvenile growth rates. Yet, why do neritic juveniles return to oceanic habitats? In the western North Atlantic, loggerheads over-winter in proximity to the western edge of the Gulf Stream (Shoop and Kenney, 1992; Keinath, 1993; Epperly et al., 1995; Hawkes et al., 2007; Mansfield et al., 2009; Arendt et al., 2012a). Most of the turtles tracked from neritic back to oceanic waters in the North Atlantic traveled within the Gulf Stream current. It is possible that these turtles were simply transported by prevailing oceanic currents back to the oceanic zone where the turtles forage opportunistically. Why, then, do these turtles remain in an oceanic habitat for extended periods? The answer may still be related to maximizing growth rates—maximizing growth relative to resource availability. It is possible that this life-history strategy is becoming more prevalent; with declines in prey availability within coastal waters (Lipcius and Stockhausen, 2002; Mansfield et al., 2009) and documented changes in diet over time (Seney and Musick, 2006), turtle distributions may be shifting accordingly. More long-term habitat, diet and tracking studies are needed to better understand why some larger juveniles return to the open ocean.

An updated Atlantic loggerhead life history model was proposed by the Turtle Expert Working Group (2009) with five size-based life stages that account for a more "transitional" juvenile stage (e.g., Laurent et al., 1998; Bolten, 2003a; Casale et al., 2008) between the strictly oceanic, smaller juveniles and the predominantly neritic larger turtles (Figure 8.1). This model acknowledges oceanic migrations and oceanic foraging among adults. Stage I, representing Year One (including hatchlings, post-hatchlings, and early neonates or oceanic juveniles), reflects the transition from terrestrial to neritic to oceanic habitats (TEWG, 2009). Stage II represents an exclusively oceanic developmental stage (the first of three juvenile stages), followed by morphologically larger neritic and/or oceanic juvenile Stages III and IV (TEWG, 2009). Adult Stage V may occur within neritic, oceanic, or terrestrial habitats (TEWG, 2009). Of note, this model is based on flexible size distributions indicative of each Stage. Juvenile Stages III and IV reflect neritic size distributions unique to different geographic regions along the eastern United States. Other recent life history models (e.g., NMFS and USFWS, 2008) do not differentiate between these juvenile stages, nor may the specific size distributions be applicable in a broader global context.

It is clear that the oceanic zone provides important habitat for all loggerhead life history stages. The movements between neritic and oceanic habitats by different loggerhead life stages are no

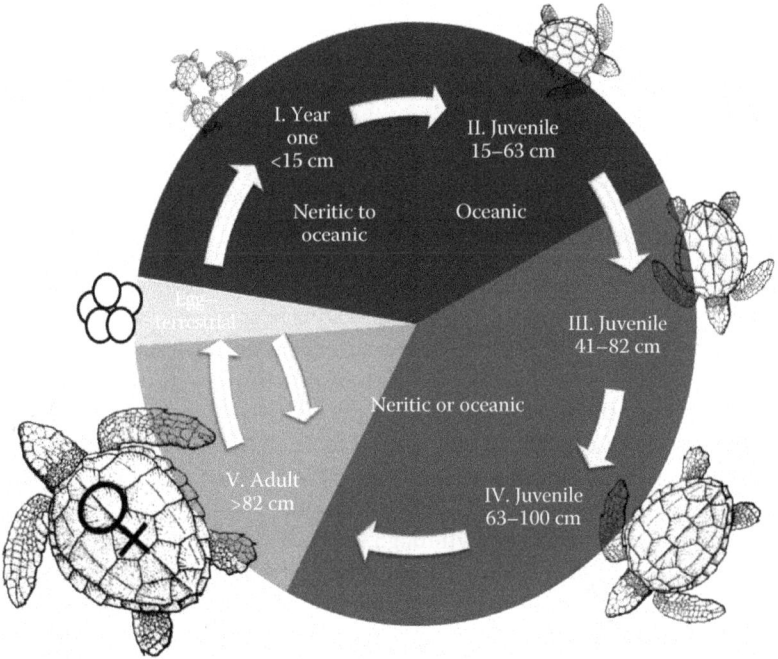

**FIGURE 8.1** Conceptual model of loggerhead size distributions by life stage and habitat type as proposed by TEWG (2009). Dark blue represents Stages I and II, or the more traditional early oceanic life stage where turtles solely occupy oceanic habitats after transitioning through neritic waters from their terrestrial natal beaches. Blue represents juvenile stages III and IV, or more traditional larger, "neritic" juveniles that may occupy both neritic and oceanic habitats. These stages reflect the transitional oceanic-to-neritic (and possibly back to oceanic) habitat selection by morphologically larger juvenile loggerheads. Pale blue represents the adult stage and the oceanic, neritic, and terrestrial habitats occupied by adult female loggerheads. Size distributions derived from Atlantic loggerhead populations. These distributions do not necessarily reflect global populations. (Adapted with permission from a design by Katrina Phillips; turtle artwork by Jack Javech, NOAA.)

longer speculative (McClellan and Read, 2007). Emerging data derived from increasing sample sizes, broader spatial effort, advances in sampling methods, and newer technologies all indicate that life history patterns, habitat use, and connectivity among loggerhead populations are more complex and exciting than previously assumed.

## 8.2   EARLY OCEANIC DISPERSAL

Young loggerhead sea turtles are iconic oceanic migrants. Post-hatchlings leave coastal waters in favor of nursery and developmental habitats in the oceanic zone. Transoceanic migrations are documented among loggerheads in the Pacific and Atlantic (Carr, 1987; Bolten et al., 1992, 1998; Bowen et al., 1995; Resendiz et al., 1998; Nichols et al., 2000; Boyle et al., 2009); loggerheads from some populations traverse the width of entire ocean basins before reaching maturity and returning to their natal beaches to reproduce. The immediate neritic-to-oceanic habitat shift by hatchlings is hypothesized to be driven, in part, by low survival of small turtles over the continental shelf (reviewed by Bolten, 2003a,b). Relative to the oceanic zone, neritic areas provide inhospitable habitat for young turtles due to intense predation from increased predator density (Carr, 1987; Collard and Ogren, 1990). Land-based nutrient runoff promotes high productivity at the base of the marine food-web. This, in turn, supports an abundance of fish and bird species that opportunistically prey on young sea turtles. Additionally, the relative shallowness of the coastal and continental shelf waters reduces the volume over which predatory fish search the water column for food. This may increase the likelihood that predators will encounter turtles restricted to the surface of the ocean. Loggerheads primarily nest on temperate and subtropical coastlines where nearshore waters can become unsuitably cold during winter. Within some temperate regions, prolonged exposure to cold, winter sea temperatures may limit growth, and may cause cold-stunning, metabolic disruptions, or mortality among small turtles remaining in neritic waters (Collard and Ogren, 1990; Morreale et al., 1992). Thus, from a metabolic standpoint, even without increased predation risk, remaining in some coastal waters throughout the year may not be the optimal habitat choice for loggerhead turtles (Putan et al. 2012a).

### 8.2.1   Hatchling Dispersal and Major Ocean Currents

The ecological importance of hatchling loggerheads quickly getting from neritic to oceanic waters is clear when examining the distribution of loggerhead nesting beaches relative to ocean currents; many of the largest nesting assemblages tend to occur near major current systems (Musick and Limpus, 1997). These currents are likely responsible for the transport of young turtles to productive, oceanic feeding grounds. Additionally, variation in nest density across many sea turtle nesting assemblages appears to be influenced by oceanic factors that promote transport of hatchlings (Putman et al., 2010a,b; Okuyama et al., 2011; Shillinger et al., 2012). For example, along the southeastern United States, loggerhead nest density is greatest in regions that are in closest proximity to the Gulf Stream System, the swift, warm current that flows through the Gulf of Mexico (Loop Current), the Florida Straits (Florida Current), and then northward and northeastward over the continental slope off the southeastern United States (Gulf Stream) (Putman et al., 2010a). More than 90% of the variation in loggerhead nest density along this region can be accounted for by the distance that hatchlings must travel before reaching the mean shoreward position of the Gulf Stream System (Putman et al., 2010a). Similarly, beaches in Japan with high nest densities are nearer currents that promote the oceanic dispersal of hatchlings than beaches with lower nest densities (Okuyama et al., 2011). Other loggerhead populations may exhibit similar patterns; numerous major nesting assemblages occur along continental coastlines in close proximity to major ocean currents. Among these are east Australia (East Australian Current), Masirah Island of Oman (Ras al Hadd Jet), Tongaland of South Africa (Agulhas Current), south equatorial Brazil (Brazil Current) and the eastern Yucatan Peninsula of Mexico (Yucatan Current) (Putman et al., 2010a). This biogeographic pattern is likely the result of adult loggerheads nesting near their natal sites (Bowen and Karl, 2007; Lohmann et al., 2008).

Adult female loggerheads are known to exhibit fidelity to specific nesting beaches within and between reproductive seasons (Bowen, 1995; Addison, 1996; Papi et al., 1997). Presumably, nesting beaches that promote oceanic transport of hatchlings enhance early survivorship, thus more turtles are likely to return to these areas to reproduce upon reaching maturity (Putman et al., 2010a,b; Shillinger et al., 2012).

Although loggerhead spatial nesting patterns imply that ocean currents play a key role in moving young turtles from neritic to oceanic waters (Putman et al., 2010a), traversing entire ocean basins is not likely to be solely explained by adults nesting near favorable currents. Studies on the movement of pelagic organisms have found that localized retention typically predominates over long-distance transport (Cowen et al., 2006; Cowen and Sponaugle, 2009). For individuals dispersing from coastal areas (such as hatchling loggerheads from the southeastern United States, east Australia, Japan, Brazil, and South Africa), frontal convergence zones occurring at coastal-oceanic boundaries minimize offshore transport (Belkin et al., 2009; Cowen and Sponaugle, 2009). Neritic and oceanic water properties often differ in salinity, temperature, and thus density. A horizontal pressure gradient is established where the water masses meet (Tomczak and Godfrey, 1994; Belkin et al., 2009). Due in part to Earth's rotation, the resulting frontal boundary becomes part of a geostrophic balance in which flow is directed along the front rather than across it (Tomczak and Godfrey, 1994). Thus, for objects shoreward of such fronts, some offshore velocity, in addition to that of the prevailing currents, is needed to reach oceanic waters.

## 8.2.2   OCEANIC JUVENILES: ACTIVE SWIMMERS—NOT SO PASSIVE DRIFTERS?

This additional velocity might come in the form of storm events or when the geostrophic balance breaks down in regions where high-velocity currents meet topographic obstacles (e.g., as the Gulf Stream flows past the "Charleston Bump" along the eastern U.S. coast [Bane and Dewar, 1988]). Although often characterized as "passive migrants" and "passive drifters" (e.g., Carr, 1986; Hays and Marsh, 1997; Musick and Limpus, 1997; Bolten, 2003a; Luschi et al., 2003), perhaps the most reliable source of additional velocity is from the turtles themselves. Simulations of hatchling loggerhead dispersal in an ocean circulation model indicate that as little as 12 h of post-frenzy swimming directed offshore from southeast Florida increased turtles' probability of survival versus scenarios of passive drift; and swimming for only 48 h significantly increased turtles' chances of reaching the Azores (Putman et al., 2012a). It is worth noting that across the southeastern United States, offshore swimming was most beneficial for simulated turtles from nesting regions where passive drift was most favorable. Thus, the important role of ocean circulation on successful oceanic dispersal of hatchlings should not be minimized; however, caution is warranted in assuming that the movements of even very young turtles are entirely determined by ocean currents.

This is likely to apply equally for turtles after they have successfully entered oceanic waters. Under controlled laboratory conditions, various sensory stimuli elicit oriented movement by young turtles. Odors dissolved in sea water elicit orientation in 5 month old loggerheads (Owens et al., 1982). Loggerheads from days to several years old set consistent headings based on visual cues (Avens and Lohmann, 2003; Lohmann and Lohmann, 2003; Wang et al., 2007). Hatchling detect rotational displacements and orient their swimming direction into approaching waves (Salmon and Lohmann, 1989; Lohmann and Lohmann, 1996b). Hatchlings and juvenile loggerheads also orient their swimming in response to magnetic stimuli (Avens and Lohmann, 2003; Lohmann et al., 2007).

Given the suite of sensory information that young turtles respond to with directed movement, it is somewhat surprising that the dogma that turtles behave as passive drifters in the open sea has persisted. The two chief arguments in favor of the "passive migrant" hypothesis are (1) turtles do not need to expend energy swimming because they can be carried to favorable foraging areas by ocean currents (Carr, 1986; Bjorndal et al., 2003; Luschi et al., 2003); and (2) turtles swim too slowly to progress against prevailing currents even if they were active (e.g., Carr, 1986; Hays and Marsh, 1997; Witherington, 2002; Bjorndal et al., 2003; Bolten, 2003a,b; Okuyama et al., 2011).

On a broad scale, support for these hypotheses primarily comes from opportunistic sightings of young turtles in the open sea "downstream" from nesting beaches (e.g., Florida turtles are found in large numbers around the Azores and Japanese turtles are found in large numbers north of Hawai'i) (Carr, 1986). Early life history distributions are often described within the broad spatial context of gyre circulation, particularly within the Atlantic (reviewed by Musick and Limpus, 1997; Bolten, 2003a,b). At a smaller scale, young turtles are found in association with passive materials such as pelagic algae and flotsam along downwelling fronts (Carr, 1986; Witherington, 2002).

To some degree, these arguments are valid. Classic textbook images of ocean circulation provide generalized, broad-scale insight to gyre systems in the major ocean basins—these gyres are often depicted as continuous circles, moving water from the western basins (e.g., from nesting beaches) to the eastern basins (e.g., to oceanic foraging grounds) and back again. Likewise, satellite imagery of major currents, such as the Gulf Stream, conveys the impression of currents as well-defined bands of water, steadily transporting water in a single, dominant direction (McGrath et al., 2010). Furthering this simplified view of ocean dynamics are several studies simulating turtle movement with ocean circulation model output of monthly averaged velocity fields which do not resolve many oceanic processes (e.g., Hays and Marsh, 1997; Monzón-Argüello et al., 2012; Scott et al., 2012a,b) or that placed simulated passive particles deeper in the water column than hatchlings or young turtles typically encounter (e.g., Hays and Marsh, 1997 [25–75 m]; Shillinger et al., 2012 [0–50 m]). In each case, the energy of currents (and potential for dispersion) is likely underestimated relative to what an actual turtle experiences on a daily basis at the ocean surface.

In reality, passively drifting objects rarely remain entrained in large-scale currents that might otherwise appear to necessarily result in long-distance movements (Lozier and Gawarkiewicz, 2001; McGrath et al., 2010). To blame are turbulent, high-energy processes that occur at the ocean's surface. Meanders in currents induce a centrifugal force that pulls water parcels from the center to the edge of the current where wind effects (e.g., Ekman transport) and submesoscale eddies (i.e., circulating currents 10–40 km in diameter, which are often associated with the boundaries of large, fast-flowing currents) detrain passively drifting objects (Bower and Rossby, 1989; Sponaugle et al., 2005; McGrath et al., 2010). For example, as a result of these processes, of 191 surface drifting buoys (drogued at 10, 15, and 40 m depth) within the Gulf Stream, 95% were expelled into the Sargasso Sea in less than a year (McGrath et al., 2010).

Given the highly dispersive nature of the ocean's surface, it is unlikely that turtles can be guaranteed transport to distant foraging locations by surface currents alone (Figure 8.2). Although some turtles might reach distant foraging areas by drifting passively, many more will not. Recent high-resolution modeling studies simulating early loggerhead dispersal in the Atlantic and Pacific suggest that young turtles dispersing from beaches in the western basin (the United States and Japan, respectively) are unable to reach known foraging areas of juveniles in the eastern basin (the Mediterranean Sea and Baja California, respectively) as passive drifters; they are instead ejected from east-flowing currents toward the gyre's center (Okuyama et al., 2011; Putman et al., 2012b).

Even so, how can the relatively weak swimming capabilities of hatchling and oceanic stage turtles influence their distribution at a large, ocean basin scale? Field and laboratory measurements indicate that hatchling loggerheads can sustain swimming speeds of only 0.15–0.30 m/s (O'Hara, 1980; Salmon and Wyneken, 1987; Witherington, 1991). Oceanic stage juveniles (SCL 26–48 cm) tested in a flow tank sustained swimming speeds ranging from 0.2 to 0.6 m/s (Revelles et al., 2007a). Turtle swimming velocities are indeed much slower than some currents that they encounter (e.g., the Gulf Stream's surface velocities can reach 2 m/s; [Johns et al., 1995]). However, currents exceeding turtle swim speeds by an order of magnitude cover only a relatively small area of the ocean. Over much of the earth's oceanic habitat, swimming behavior might contribute substantially to an oceanic juvenile's net velocity. Moreover, in very fast currents, swimming oriented perpendicular to the direction of flow could have a profound influence on the path any one turtle takes (Scott et al., 2012a).

Support for this hypothesis comes from a number of laboratory experiments examining hatchling loggerheads' orientation behavior in response to magnetic fields (Lohmann et al., 2001;

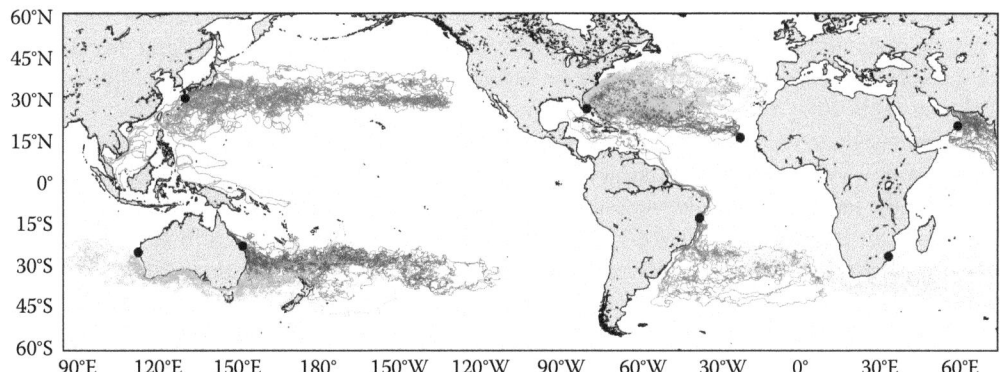

FIGURE 8.2 Map showing predicted dispersal patterns from eight major loggerhead nesting beaches under the assumption of passive drift. Virtual particles were released from each black dot in 2005 during the 3 months of peak hatchling emergence from the corresponding loggerhead rookery and tracked for 6 years within the Global Hybrid Coordinate Ocean Model. Particle-tracking methods used are similar to those described in Putman et al. (2012a,b). Trajectories in red are from Yakushima Island, Japan; purple from Wreck Island, Australia; orange from Dirk Hartog Island, Australia; green from Jupiter Beach, Florida, the United States; navy from Boa Vista, Cape Verde Islands; magenta from Salvador, Brazil; aqua from Tongaland, South Africa; and brown from Masirah Island, Oman.

Füxjager et al., 2011; Putman et al., 2011). Loggerheads detect two parameters of earth's magnetic field: (1) the inclination angle (the angle that field lines intersect the surface of the earth); and (2) the intensity (the strength of the field) (Lohmann et al., 2007). These two parameters form the basis of a crude bicoordinate positional system, from which young loggerheads appear to assess their location (Putman et al., 2011). Hatchling loggerheads from Florida, the United States, exposed to magnetic fields that exist at numerous oceanic regions in the North Atlantic adopt headings which would likely facilitate their transatlantic migration (Lohmann et al., 2012).

Natural selection, it seems, favors turtles that use pairings of inclination angle and field intensity as a proxy to estimate the oceanic conditions typically found at a given location. In fields that exist near narrow, fast-flowing major currents (like the Gulf Stream), turtles choose headings that would lead them into the current. For instance, turtles presented with a field that exists near northeast Florida swam east (a direction that leads into the Gulf Stream) (Lohmann et al., 2001); and, in a field along the north-central limb of the gyre, turtles swam northeast (a direction that leads into the east-flowing Gulf Stream and opposes southward flow that is typical in this region) (Füxjager et al., 2011). In contrast, turtles adopt headings in the same direction as broad, slow-flowing currents. For example, a field that exists near the Cape Verde Islands elicited southwesterly orientation, consistent with the direction of surface flow where the broad Canary Current meets the North Equatorial Current (Putman et al., 2011).

A recent set of modeling experiments simulating the first 5 years of oceanic dispersal in loggerheads suggest that minimal swimming activity in response to three magnetic fields that exist along the northern-limb of the North Atlantic Subtropical Gyre could greatly influence turtles' oceanic distribution (Figure 8.4; Putman et al., 2012b). Simulated turtles that swam for 1–3 h/day at 0.2 m/s upon encountering certain oceanic regions were more likely than passive drifters to remain within warm-water currents favorable for growth and survival, avoid areas on the perimeter of the migratory route where predation risk and thermal conditions pose threats, and successfully return to the open-sea migratory route if carried into coastal areas (Figure 8.4; Putman et al., 2012b). These simulated turtles were also 43%–187% more likely than passive drifters to reach the Azores (Putman et al., 2012b), a known, productive foraging area frequented by Florida loggerheads (Carr, 1987; Bolten, 2003a).

Such swimming behavior may be of profound ecological importance because it provides a mechanism for turtles to target somewhat specific oceanic regions. At the ocean-basin scale, some areas consistently possess more favorable temperature and productivity regimes than others (Polovina et al., 2006; Hawkes et al., 2007; Kobayashi et al., 2008). Thus, natural selection has likely favored orientation biases that increase turtles' chances of reaching areas that favor growth (Peckham et al., 2011; Scott et al., 2012a).

Further insight to early in-water behavior has resulted from recent advances in the use of small-scale solar-powered satellite tags on neonate (<1 year old) loggerheads in the western Atlantic. Until recently, much of our understanding of early dispersal, habitat use, and in-water behavior of oceanic sea turtles has been derived from opportunistic encounters, changes in size distributions and phylogeography across large ocean basins, laboratory studies, numerical models, and inferred analyses using trace elements. Most satellite telemetry has focused on larger oceanic juveniles (>30–40 cm SCL; e.g., Riewald et al., 2000; Bolten, 2003a; Sasso and Epperly, 2007; McCarthy et al., 2010)—turtles large enough to carry heavy, battery-powered satellite tags. Advancements in satellite tag technology and tag attachment methods recently resulted in the first, long-term solar-powered satellite tracks of oceanic stage turtles less than one year of age (3.5–9 months old, 11–18 cm SCL; Mansfield et al., 2012, in review) (Figure 8.3). Seventeen turtles were released from southeast Florida and tracked for 1–7 months. With the brief exception of one turtle, all turtles remained off the continental shelf, in oceanic waters deeper than 200 m (Mansfield et al., in review). All turtles were initially transported by the Gulf Stream; however, over half of the turtles departed the Gulf Stream to the east within 1–2 months, suggesting that this major ocean current is not a primary, long-term mechanism of transport for these young turtles (Mansfield et al., in review). Several of these turtles traveled to the North Atlantic (covering distances over 4000 km; Mansfield

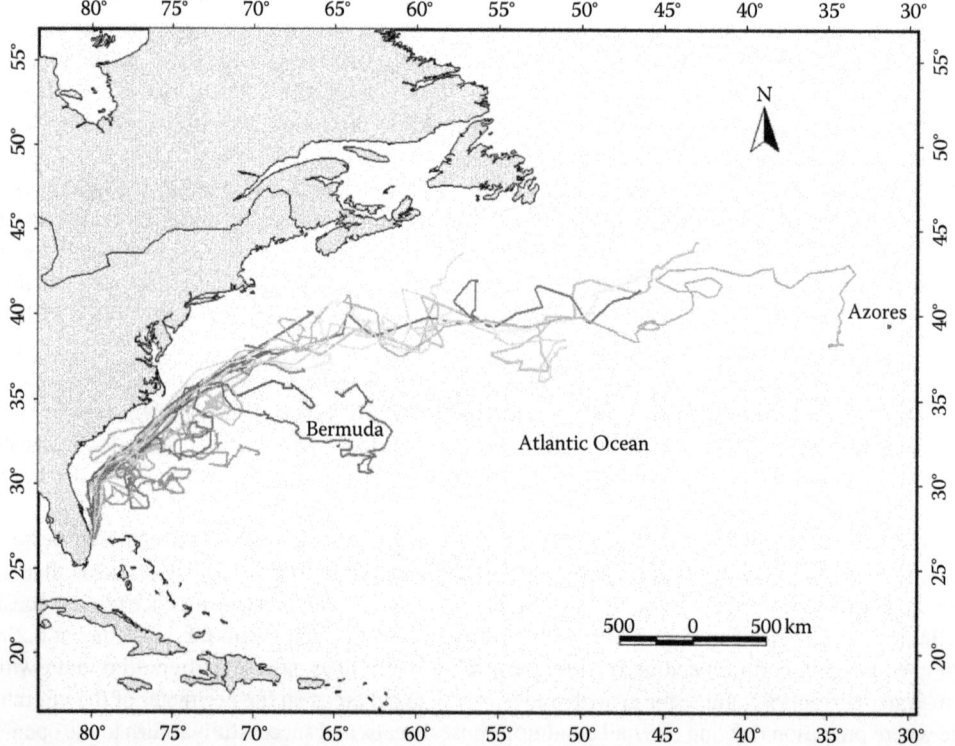

FIGURE 8.3   First satellite tracks of neonate (11–18 cm SCL) loggerhead sea turtles. Seventeen turtles were released from the southeastern United States 2009–2011. Turtles were tracked an average of 85 days (ranging 27–219 days). Each colored line represents an individual turtle. (Adapted from Mansfield, K.L. et al., J. Wyneken, J. Luo, and W. P. Porter. In review. Voyage of the young mariners: First oceanic satellite tracks of neonate loggerhead sea turtles.)

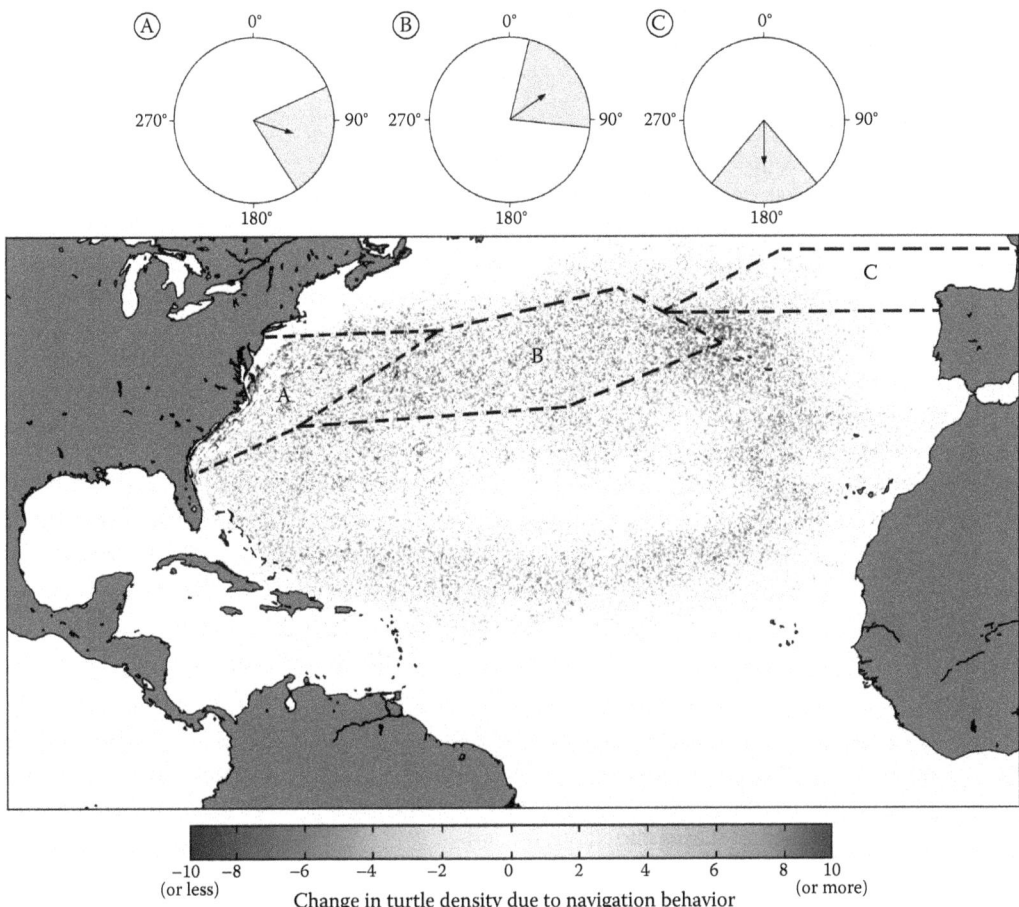

**FIGURE 8.4** Effects of magnetic navigation behavior on simulated spatial patterns of loggerhead turtle abundance in the North Atlantic. (Modified after Putman, N.F. et al., *J. Exp. Biol.*, 215, 1863, 2012.). The map displays the results of a 5 year period in which the distribution of turtles that drifted passively was compared to the distribution of turtles that engaged in 2 h of magnetic navigation behavior per day within the Global Hybrid Coordinate Ocean Model. Regions delineated with hatched black lines (A, B, and C) demarcate three "navigation zones" that encompass regional magnetic fields known to elicit navigational responses in young loggerheads (Lohmann et al., 2001; Füxjager et al., 2011). Zones are irregularly shaped because of how the Earth's field varies across the Atlantic. The direction turtles swam in each navigation zone is shown by the gray-shaded area in the corresponding circular diagram. Each hour per day that a turtle swam, it adopted a new heading randomly chosen from within this gray area. Locations highlighted in blue correspond to areas where navigation behavior led to an increased abundance of turtles relative to passive drift. Locations in red correspond to areas where the same navigation behavior led to a decrease in the abundance of turtles relative to passive drift. Locations with no difference in predicted turtle density between the two scenarios are colored white (indicated by 0 on the color scale at the bottom of the figure). Most blue coloration is distributed within the currents of the North Atlantic Subtropical Gyre (a favorable habitat for young loggerheads), indicating that magnetic navigation increases turtle density in this oceanic area. In particular, a dense cluster of blue can be seen around the Azores (islands off the coast of Portugal), suggesting that responses to regional magnetic fields facilitate migration to this productive foraging area. By contrast, red coloration is mostly concentrated near the margins of the gyre. Thus, relative to passive drift, magnetic navigation apparently diminished the density of turtles in suboptimal locations where predation or wintertime temperatures pose risk, such as along the eastern U.S. coast and in the northeastern corner of the gyre.

et al., in review), yet none traveled as far east as the Azores before tag transmissions ceased. Turtles traveling to the northwestern Atlantic, past South Carolina oriented ENE (Mansfield et al., in review)—consistent with the general eastward orientation that hatchlings adopt when exposed to simulated magnetic fields that correspond to locations north of Florida and west of the Azores (Lohmann and Lohmann, 1996a; Lohmann et al., 2001; Füxjager et al., 2011; Putman et al., 2012b).

Within the North Atlantic, classic hypotheses suggest that turtles entrain within currents associated with the North Atlantic Subtropical Gyre, moving along the gyre's circulatory system in a long, developmental migration. However, due to logistical sampling constraints, few data are available on loggerhead distributions within regions found inside the gyre—a region where areas of favorable growth may exist. Most data are concentrated at points along the gyre (e.g., Azores, Canary Islands) that are relatively accessible for in-water sampling. New neonate tracking data (e.g., Mansfield et al., in review) coupled with modeling work (e.g., Putman et al., 2012b) and magnetic orientation studies (Lohmann et al., 2001, 2012; Füxjager et al., 2011; Putman et al. 2011) suggest that while some turtles may follow the classic circular gyre-based developmental migration route, others may "eject" into the gyres' center. This might also be the case for North Pacific loggerheads; numerous juvenile turtles are caught in the central North Pacific long-line fishery near Hawai'i (Peckham et al. 2011). In the Atlantic and Pacific, some turtles clearly travel eastward of the gyres (e.g., the Mediterranean Sea in the Atlantic and Baja California Peninsula in the Pacific), but the proportion of turtles following these different (though possibly overlapping) trajectories is unknown.

## 8.3  FORAGING OPPORTUNITIES AND MICROHABITATS

Loggerhead high-use areas occur in association with a number of oceanographic features. These include frontal areas and downwelling zones where flotsam and *Sargassum* aggregate (Carr, 1986, 1987; Polovina et al., 2000; Witherington, 2002; Witherington et al., 2012), in areas where primary productivity and chlorophyll are elevated (Polovina et al., 2004, 2006; Mansfield et al., 2009), and in association with the edges of mesoscale eddies (identified using remotely sensed sea surface height) (e.g., Polovina et al., 2006; Revelles et al., 2007b; Mansfield et al., 2009; Kobayashi et al., 2011). These are all areas of resource opportunity—where potential refuge and prey are concentrated.

In the Atlantic, *Sargassum* originates in the Gulf of Mexico and advects over time to the western and northwestern Atlantic via the Gulf Stream System (Gower and King 2011). Using remote sensing, Gower et al. (2006, 2011) illustrate that *Sargassum* distribution spatially varies with season. The greatest seasonal *Sargassum* biomass occurs off the southeastern U.S. coast (Florida to Cape Hatteras, North Carolina) from July to August (Gower and King 2011; Figure 8.5). This roughly coincides with peak hatchling production within the western Atlantic loggerhead populations.

Although, in principle, turtles might be passively carried to these smaller-scale habitats where shelter and foraging opportunities are high (by processes such as Langmuir circulation [Carr, 1986; Witherington, 2002]), it appears that turtles do not leave this up to chance. Experiments conducted in the field and laboratory indicate that post-hatchling loggerheads actively orient toward *Sargassum*, (Smith and Salmon, 2009). Active movement by young loggerheads is almost certainly context-dependent; of turtles observed in the field in association with *Sargassum* mats and other flotsam, approximately 61% were motionless while the remainder was weakly paddling (Witherington, 2002). In waters where food and shelter are less concentrated (or when human observers are not within turtles' sight), behavior might be expected to differ. Oceanic stage loggerheads in the eastern Atlantic displayed relatively straight paths when in waters with low productivity, whereas tracks became more sinuous near frontal zones (McCarthy et al., 2010). Neritic juvenile foragers satellite-tracked by Mansfield et al. (2009) established statistically significant fidelity to discrete coastal foraging areas; however, the turtles tracked from neritic waters back to oceanic habitats did not exhibit fidelity to any specific area. Rather, they were found in association with the edges of mesoscale eddies (Mansfield et al., 2009). This likely reflects the more stochastic foraging

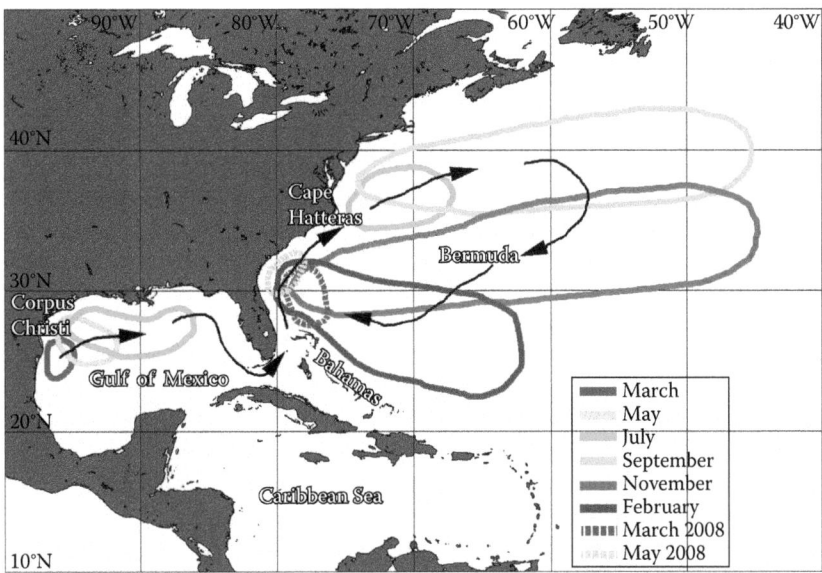

FIGURE 8.5 Simplified outline diagram showing the average seasonal extent of *Sargassum* biomass in March, May, July, September, November, and February, based on monthly medium resolution imaging spectrometer (MERIS) count distributions, 2002–2008. Only in 2008 does MERIS detect significant *Sargassum* in the Atlantic between March and June (dashed outlines). (From Gower, J.F.R. and King, S.A., *Int. J. Remote Sens.*, 32(7), 1917, 2011. With permission.)

environment occurring in the open ocean compared to more concentrated prey availability within neritic waters. Similarly, loggerheads (41–81 cm SCL) tracked by satellite north of Hawai'i moved along downwelling fronts, but appeared to do so by swimming westward against weak east-flowing currents (Polovina et al., 2000).

Thus, the emerging hypothesis is that turtles benefit from taking an active role in their movements throughout all life history stages. This is not to suggest that oceanic stage loggerheads swim continuously, or even very much; nor does this suggest that numerical and conceptual models that assume turtles to be passive drifters are without merit. Much can be learned about the oceanic constraints and factors shaping the ecology of sea turtles using such techniques (e.g., Blumenthal et al., 2009; Putman et al., 2010b; Okuyama et al., 2011; Lohmann et al., 2012). However, it is important to recognize that young, oceanic stage turtles likely rely on strategies beyond passive drift, and that even a minimal level of swimming can have an important function in determining their oceanic distribution as well as influencing their ecology and evolution (Putman et al., 2012a,b).

## 8.4 OCEANIC CONSTRAINTS AND CONNECTIVITY

Loggerhead sea turtles rely on and occupy oceanic habitats during all stages of their life history (Figure 8.1). Determining the oceanic distribution of loggerheads (and the mechanisms that control it) is of primary importance for conservation and management. Understanding large-scale connectivity among nesting beaches, migratory/dispersal routes, and oceanic foraging grounds requires insight into environmental and physiological constraints on loggerhead movement. With the exception of terrestrial activities (nesting, limited basking), early embryonic development, and hatchling emergence, turtles occupy a 3D oceanic space limited only by the sea surface, the ocean floor, and the turtles' diving ability. Within this 3D world, loggerhead distributions are constrained both vertically within the water column and horizontally (in a broad geographic context) by their size or life history stage, environmental conditions, oceanographic features, and resource availability. This, in turn, strongly influences the connectivity among loggerheads in the oceanic environment.

### 8.4.1 Physiological and Environmental Constraints on Distribution

As obligate air-breathers, all sea turtles must spend time at the sea surface. The proportion of time spent occupying different vertical strata within the water column varies among sea turtle life stages. Hatchlings emerge from their nests, enter the ocean and begin a neritic-to-oceanic migration fueled by residual energy from their yolk stores (reviewed by Musick and Limpus, 1997; Bolten, 2003a,b). During their initial dispersal from natal beaches and extending at least into their first year at sea, loggerheads remain predominantly within the upper few meters of the water column (Musick and Limpus, 1997; Witherington, 2002; Bolten, 2003a). *In situ* observations of sea turtles gave rise to early hypotheses of post-hatchling turtles occupying surface habitats, often in association with floating *Sargassum* communities (Brongersma, 1972; Fletmeyer, 1978; Carr, 1986, 1987; Witherington, 2002). Organisms commonly associated with surface-dwelling *Sargassum* communities typically compose the diet of post-hatchling loggerheads (Carr and Meylan, 1980; reviewed by Bjorndal, 1997; Boyle and Limpus, 2008), suggesting that young oceanic juveniles remain near the sea surface. Mansfield et al. (2012, in review) satellite-tracked lab-reared loggerheads 3.5–9 months old (n = 7 and 17; 11–18 cm SCL) in the western Atlantic. These turtles were tracked for periods of 1–6 months, and represent smallest turtles to be satellite-tracked to date (Mansfield et al., 2012). The combination of numerous, high-quality messages received from the tags and the optimal charge rates of the tags' solar cells suggest that the tags (and therefore the turtles) were exposed to air at the sea surface for extended periods (Mansfield et al., 2012).

In the Azores, prey items consumed by oceanic turtles (9.5–56.0 cm CCL) included species found more than 100 m below the surface (Frick et al., 2009). Bolten (2003a) describes satellite telemetry work in the Azores where oceanic stage turtles (~40 to 50 cm SCL, Riewald et al., 2000) spend 75% of their time at the sea surface (<5 m depths). Twenty percent of the turtles' dives ranged as deep as 100 m (Bolten, 2003a). Juveniles tracked using radio, acoustic, and satellite telemetry from neritic waters show that larger, demersally foraging juveniles may spend less than 5% of their time at the sea surface on average dives ranging from a few meters to tens of meters deep (Byles, 1988; Keinath, 1993; Nelson, 1996; reviewed by Lutcavage and Lutz, 1997; Mansfield, 2006). Adult loggerheads may dive to depths exceeding 200 m, spending as little as 5% of their time at the surface (Sakamoto et al., 1990; Lutcavage and Lutz, 1997; Arendt et al., 2012b). This suggests an ontogenetic shift in vertical habitat use from a few meters at the sea surface during a turtle's first year or so at sea, to deeper diving forays as turtles' lungs develop, as turtles encounter different foraging resources, and as turtles require a greater food intake to sustain growth. By maturity, loggerheads are predominantly demersal foragers (reviewed by Bjorndal, 1997), spending much less time at the sea surface.

Loggerheads are poikilothermic reptiles. Thus, their distribution, both vertically within the water column and across broad ocean basins, is constrained by temperature (Bell and Richardson, 1978; Spotila and Standora, 1985; Spotila et al., 1997; Mansfield et al., 2009). In an oceanic environment, it makes evolutionary sense that small cold-blooded turtles remain within the metabolically beneficial thermal environment found at the sea surface. Byles (1988), Nelson (1996), and Mansfield (2006) suggest that there are seasonal differences in dive behavior among acoustically tracked neritic juveniles—turtles may be more likely to remain closer to the sea surface during the cooler seasons. In a stratified water column, turtles off the coast of Virginia remained above the thermocline (~22°C to 23°C) (Mansfield, 2006). Thus, temperature is also likely to constrain loggerheads to surface waters and limit the depth to which turtles forage. The physiological and ecological implications of turtle dive behavior in response to vertical temperature regimes that differ by season and region are worth further investigation.

The horizontal distribution of turtles is also constrained by ocean temperatures. The juveniles tracked by Mansfield et al. (2009) from neritic habitats back to the oceanic zone were constrained by the 10°C–15°C isotherms in the northwestern Atlantic. Similar temperature constraints are documented among all but the youngest life history stages (e.g., Lutcavage and Musick, 1985; Musick and Limpus, 1997; Mansfield, 2006; Hawkes et al., 2007; Kobayashi et al., 2008). These tracking data are consistent with experiments that show that at in-water temperatures below 20°C young log-gerhead feeding behavior is reduced. In temperatures ≤15°C, turtle locomotor abilities are reduced; in temperatures <10°C sea turtles may become "cold-stunned" as evidenced by loss of buoyancy and disrupted metabolic pathways (Schwartz, 1978; Morreale et al., 1992; Davenport, 1997; Spotila et al., 1997). As such, at a global scale, loggerhead distributions are largely constrained by sea sur-face temperature (Witt et al., 2010).

### 8.4.2 ASSESSING CONNECTIVITY: PERSPECTIVES ON GENETIC MARKERS

Assessing the proportion of turtles from a given rookery that encounter different regions, oceanic conditions, and anthropogenic hazards are important inputs for demographic models that predict whether a population is likely to grow or decline (Wallace et al., 2008; Peckham et al., 2011). Of particular concern is determining the influence that mortality from anthropogenic sources, such as in fisheries, is likely to have on different nesting rookeries. For such assessments it is necessary to determine from which rookery bycaught turtles originate. This has primarily been done by comparing a fragment of mtDNA extracted from blood or tissue samples of turtles caught in the ocean with those acquired from turtles at nesting beaches (Bowen and Karl, 2007).

In principle, if unique genetic markers differentiated turtles from each nesting beach it would be possible to unambiguously identify the rookery of origin for any turtle caught at sea. However, the genetic data available from nesting beaches rarely have the resolution needed to assign an individual turtle to a specific rookery. Typically, "genetically distinct" rookeries are identified by differences in haplotype frequencies, not on the basis of completely unique haplotypes (Bowen and Karl, 2007). In most cases, numerous turtles must be sampled within an oceanic region and a statistical estimate is made of the contribution from different nesting rookeries (ideally data are available for all significant nesting sites). This "mixed-stock analysis" relies on the haplotype frequencies of nesting rookeries, the relative population size of the rookeries, and maximum-likelihood or Bayesian algorithms to estimate the contribution of each rookery that provides the best fit to the haplotype frequencies observed in a given oceanic region (Bolker et al., 2007; Bowen and Karl, 2007). These inputs provide a "foraging-ground-centric" perspective that estimates the percentage of turtles in a foraging ground that come from each of the rookery sources for which genetic data are available.

These techniques appear to work well when few nesting populations are involved and the overlap in haplotypes among rookeries is low, such as the case for Pacific loggerheads. Genetic techniques identified the central North Pacific and Baja California foraging areas to comprise turtles primarily from Japan (97% ± 5% s.d. and 92% ± 7% s.d., respectively) with only minimal contributions from Australia (3% ± 5% s.d. and 8% ± 7% s.d, respectively) (Bowen et al., 1995). Given the more recent finding that a small fraction of loggerheads nesting in Japan possess the "Australian" haplotype (Hatase et al., 2002b), it is likely that the contribution from actual Australian turtles is even lower (Bowen and Karl, 2007). Oceanic stage loggerheads found off the coast of Peru and in the eastern South Pacific were identified as coming from the east Australian and New Caledonia rookery (100% ± 0.0% s.d.) (Boyle et al., 2009). These data imply that a certain proportion of loggerheads make developmental migrations from the western Pacific to the eastern Pacific, but that there is minimal connectivity across the equator (see also Figure 8.2).

In the Atlantic basin, published data on population structure of oceanic loggerheads has come from the eastern Atlantic (near the Azores, western France, Madeira, the Mediterranean Sea, and the Canary Islands) and the South Atlantic at a sea mount chain near Brazil (Bolten et al., 1998; Laurent et al., 1998; Monzón-Argüello et al., 2009, 2012; Reis et al., 2010). Although there is some inconsis-tency with the sets of nesting colonies that are included as potential sources in these analyses (some authors include rookeries inside the Mediterranean, others do not), because each foraging area in the eastern Atlantic contains approximately the same haplotype frequency as the basin-wide haplotype frequency of nesting rookeries, the general consensus is that oceanic stage loggerheads have little, if any, population structure (Bolten et al., 1998; Bowen et al., 2005; Bowen and Karl, 2007).

Recent papers challenge the idea that loggerheads are well mixed in their eastern Atlantic foraging grounds by taking a "rookery-centric" perspective of sea turtle distributions (Bolker et al., 2007) to estimate how the turtles originating from particular rookeries are distributed among foraging areas (Monzón-Argüello et al., 2009, 2012). These analyses suggest that turtles from more northerly rookeries are more likely to be found in more northerly foraging grounds and more southerly rookeries are better represented in southerly foraging grounds (Monzón-Argüello et al., 2009). For instance, 30% of the south Florida population was estimated to go to the northerly foraging ground near the Azores, whereas only 13% of the Mexico population was estimated to reach it. In contrast, south Florida contributed only 13% of its population to the southerly foraging ground near the Canary Islands, but Mexico contributed 34% of its population (Monzón-Argüello et al., 2009).

However, these findings should be interpreted with caution; the 95% confidence intervals of the reported means are large. The confidence intervals of estimates made for the south Florida population, for example, indicate that 2%–35% reach the Canary Islands, 4%–47% reach Madeira, 8%–55% reach the Mediterranean, 8%–58% reach the Azores, and 0%–33% go to an unknown location (Monzón-Argüello et al., 2009). Additionally, the implications of these rookery-centric analyses might be misleading when taken literally. For instance, rookery-centric analyses estimate that 4% of the south Florida loggerhead population reaches the west coast of France (Monzón-Argüello et al., 2012). If we assume that 67,000 nests are laid yearly in south Florida (Monzón-Argüello et al., 2009) with 100 eggs per nest (Crouse et al., 1987), that survival of these turtles is no more than 1% (Crouse et al., 1987), and that 4% of the population strands in western France (Monzón-Argüello et al., 2012), this would mean, on average, that 2680 turtles from south Florida strand in France each year (this is high given <100 reported strandings from 1995 to 2009) (Monzón-Argüello et al., 2012).

Another possible complication in applying this analysis to oceanic stage loggerheads in the eastern Atlantic is that foraging grounds are unlikely to be independent of one another. So, stating that 30% of a population goes to the Azores and a different 30% goes to Madeira might be over-interpreting the data. An ocean circulation model indicates fairly high connectivity would be expected among these oceanic regions based on surface currents (Figure 8.6). Indeed, oceanic loggerheads satellite-tracked from Madeira traveled to the Azores and Canary Islands (McCarthy et al., 2010), suggesting that the same individuals make use of foraging areas over much of the eastern Atlantic.

Perhaps the biggest problem with rookery-centric analyses is that many large oceanic regions that might serve as foraging areas are under-sampled (or unsampled). If samples were obtained in the Sargasso Sea or Bahamas, rookery-centric estimates might change substantially. This problem is less pronounced for foraging-centric analyses because in most cases the largest source rookeries have been sampled. However, in the South Atlantic foraging area near Brazil, similar problems are observed with foraging ground-centric analyses. There the estimated contributions are dominated by turtles from nearby Brazil (60%), Australia (29%), and Greece (10%) (Reis et al., 2010). This would suggest that exceptionally long-distance connectivity among loggerheads occurs in the South Atlantic, a finding not observed in other ocean basins. The authors caution that the Australian and Greek haplotypes could originate from unsampled rookeries in West Africa or the Indo-Pacific (Reis et al., 2010). Surface circulation in the South Atlantic, South Pacific, and South Indian Oceans suggest this caution is warranted; connectivity from Africa to Brazil seems more likely than from Australia or the Mediterranean Sea (Figure 8.2). This result highlights a potential complication of many mixed-stock analyses in which not all possible sources are sampled.

The power of both rookery-centric and foraging ground-centric mixed-stock analyses is limited not only by missing sampling locations but also by widely shared haplotypes among rookeries. The south Florida nesting population is an order of magnitude larger than any other rookery in the Atlantic and has two haplotypes (CC-A1 and CC-A2) that constitute 48% and 41% of its population, respectively (Encalada et al., 1998; Monzón-Argüello et al., 2009). Quite inconveniently,

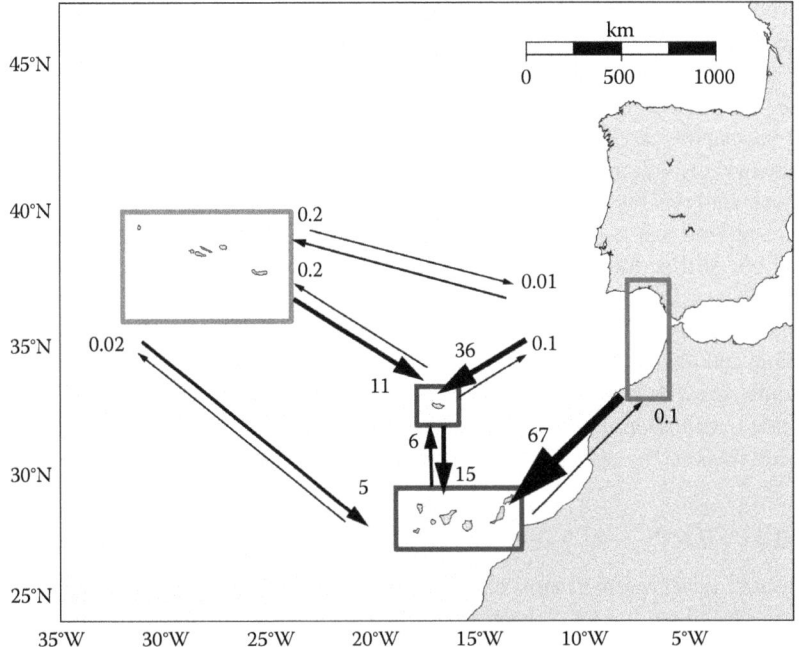

**FIGURE 8.6**  Connectivity among loggerhead foraging grounds in the eastern Atlantic based on modeled surface currents from the Global Hybrid Coordinate Ocean Model. Rectangles served as "release" and "recruitment" zones for passively drifting particles. Green corresponds to the Azores, blue Madeira, magenta the mouth of the Mediterranean Sea, and purple the Canary Islands. Particles were released at 5 day intervals throughout 2007 and tracked backwards for 3 years using Ichthyop (v. 2) particle-tracking software. (Lett et al., 2008). Numbers near each region indicate the percentage of particles entering that area arriving from the corresponding release zone. The width of the arrows is proportional to the contribution from each region. This analysis reveals strong connectivity among some foraging grounds and weak connectivity among others. For example, only 0.2% of the particles from the mouth of the Mediterranean Sea reached the Azores, whereas 36% reached Madeira and 67% reached the Canary Islands.

haplotype CC-A1 is also the primary haplotype for nesting loggerheads from northeast Florida–North Carolina (99%), northwest Florida (78%), and the Cape Verde Islands (68%). Haplotype CC-A2 is the primary haplotype for turtles in the other rookeries: Dry Tortugas (86%), Mexico (55%), Greece (90%), Israel (85%), east Turkey (59%), west Turkey (94%), Cyprus, Lebanon, and Crete (each 100%) (Monzón-Argüello et al., 2009).

Not surprisingly, these two haplotypes account for approximately 80% of turtles sampled in oceanic regions around the Azores, Madeira, the Canary Islands, and at the Strait of Gibraltar (Monzón-Argüello et al., 2009). Thus, the haplotypes found at much lower frequencies (both in nesting populations and in the oceanic region) must be more heavily relied upon to determine the contribution of individual rookeries to a foraging zone. Being rare, these haplotypes might go unobserved in nesting populations and thus limit the robustness with which population structure in the open sea is assessed. As such, difficult-to-explain results are sometimes obtained. For instance, 4.8% (±4.4% s.d.) turtles from the Azores are estimated to be of Cape Verde Island origin (Monzón-Argüello et al., 2009), and yet 26.0% (±8.1% s.d.) of the turtles stranding in western France are estimated to come from Cape Verde (Monzón-Argüello et al., 2012). If these turtles reach western France following the routes proposed by Monzón-Argüello et al. (2012), they would first travel in the vicinity of the Azores. Because Cape Verde does not contribute to the Azores and western France with roughly equal frequency, perhaps the safest conclusion is that the technique is not sensitive enough for such fine-scale comparisons.

The reduced resolution caused by the extensively shared major haplotypes in the Atlantic may be solved by recent developments employing longer mtDNA control region sequences (Abreu-Grobois et al., 2006). For the Cape Verde loggerhead population (from where data are available), they are able to split its common CC-A1 haplotype into three sub-haplotypes that are distinct from the southeastern U.S. sub-haplotype (Monzón-Argüello et al., 2010). If the use of longer sequences provides the same increased resolution across the Atlantic basin, it is likely that significant improvements of the mixed-stock analyses results can be expected.

With the resolution that is presently available to identify source rookeries of oceanic stage loggerheads and our still rudimentary understanding of loggerheads' oceanic movements, the best advice may be that of Bowen and Karl (2007), "values [from mixed-stock analyses] should not be over-interpreted, but instead provide useful qualitative estimates." These estimates can then be used to further refine and test hypotheses about the movement of turtles in the open sea. One promising avenue of future research is to use genetic markers in combination with other tools (such as stable isotopes, in situ oceanographic data, and output from ocean circulation models) to generate more definitive predictions on the population structure of oceanic stage loggerheads.

## 8.5 CONCLUSIONS, DATA NEEDS, AND FUTURE DIRECTIONS

In the past decade, novel research tools have provided exciting new insight to the oceanic habits and habitats of loggerhead sea turtles. Advances in telemetry technologies and tag attachment methods have resulted in smaller turtles being tracked for longer periods (Seney et al., 2010; Mansfield et al., 2012, in review). These data may be used to identify nursery areas and quantify early behavior and dispersal patterns that will help us better understand the horizontal and vertical habitat constraints turtles encounter. Tracking data will also help to better inform programmed "behavior" of turtles within ocean circulation models as well as help predict areas of management and conservation concern. Yet, despite recent advances in telemetry, remote sensing, and ocean modeling, many knowledge gaps remain and many management questions are yet to be answered. Study of ocean-dwelling organisms and management of oceanic habitats remains problematic due to difficult sampling logistics, cross-national or high seas jurisdiction, and spatially variable features and habitats. Oceanic juvenile developmental and foraging habitats are not yet defined for most of the world's oceans. In-water behavior and dispersal of post-hatchlings through early oceanic stage turtles is still not well understood. Oceanic stage survivorship estimates are critically needed for all species of sea turtles. A discussion is also needed to address how we can effectively manage oceanic species and habitats when these habitats are not geographically fixed in time or space but comprise dynamic oceanographic features (e.g., major ocean currents, mesoscale eddies, and frontal systems) or transient resources and habitats (e.g., *Sargassum* communities).

To address some of the persistent questions shadowing our understanding of sea turtle oceanic habits and habitats, advances in technology and, as Bolten (2003b) stresses, multidisciplinary, integrated approaches are still needed. Of primary importance are basic, spatially explicit data on loggerhead activity levels, diving behavior, mortality, energy expenditure, prey items (and the distribution of prey), and energy metabolism. Smaller, lighter satellite tags are needed to better address the challenges associated with tracking small, fast-growing turtles over time and space. Sampling and surveys are needed across a wider area covering the entire oceanic range of turtles. More long-term data are needed on turtles occupying oceanic habitats in waters beyond just the Atlantic (and, to some extent, the North Pacific) in order to better understand the short- and long-term impacts of catastrophic events such as oil spills, climate change, as well as the more pervasive effects of pollution and fisheries bycatch. Integrative research that makes use of recent advances in telemetry (e.g., Mansfield et al., 2012), genetic markers (e.g., Shamblin et al., 2012), and high-resolution models of ocean circulation that incorporate swimming behavior to predict sea turtle movement (e.g., Putman et al., 2012b) will provide deeper insights into the ecology of oceanic loggerheads than the individual use of any one of these techniques. Given the logistical difficulties and expense of studying oceanic species, the use of increasingly sophisticated inferential methods (e.g., remote sensing and numerical models) paired with opportunistic data (e.g., from fisheries bycatch, strandings, and surveys in the vicinity of islands) and in situ observations (e.g., emerging telemetry options) appears to be the trajectory of research on oceanic stage loggerheads for the foreseeable future.

## ACKNOWLEDGMENTS

NFP was supported by funds from the North Carolina State University Initiative for Biological Complexity. KLM was supported in part by the National Academies Research Associateship Program. We thank Alberto Abreu-Grobois and Sheryan Epperly for insightful comments.

## REFERENCES

Abreu-Grobois, F. A., J. Horrocks, A. Formia, R. LeRoux, X. Velez-Zuazo, P. Dutton, L. Soares, P. Meylan, and D. Browne. 2006. New mtDNA Dloop primers which work for a variety of marine turtle species may increase the resolution capacity of mixed stock analysis. In *Book of Abstracts Twenty Sixth Annual Symposium on Sea Turtle Biology and Conservation,* eds. M. Frick, A. Panagopoulou, A. F. Rees, and K. Williams, April 3–8, 2006, p. 179. Athens, Greece: International Sea Turtle Society.

Addison, D. S. 1996. *Caretta caretta* (loggerhead sea turtle) nesting frequency, *Herpetological Review* 27: 76.

Arendt, M. D., A. L. Segars, J. I. Byrd et al. 2012a. Distributional patterns of adult male loggerhead sea turtles (*Caretta caretta*) in the vicinity of Cape Canaveral, Florida, USA during and after major annual breeding aggregation. *Marine Biology* 159: 101–112.

Arendt, M. D., A. L. Segars, J. I. Byrd et al. 2012b. Migration, distribution and diving behavior of adult male loggerhead sea turtles (*Caretta caretta*) following dispersal from a major breeding aggregation in the western North Atlantic. *Marine Biology* 159: 113–125.

Arendt, M. D., A. L. Segars, J. I. Byrd et al. 2012c. Seasonal distribution patterns of juvenile loggerhead sea turtles (*Caretta caretta*) following capture from a shipping channel in the northwest Atlantic. *Marine Biology* 159: 127–139.

Avens, L. and K. J. Lohmann. 2003. Use of multiple orientation cues by juvenile loggerhead sea turtles *Caretta caretta*. *Journal of Experimental Biology* 206: 4317–4325.

Bane, J. M. and W. K. Dewar. 1988. Gulf Stream bimodality and variability downstream of the Charleston Bump. *Journal of Geophysical Research* 93: 6695–6710.

Belkin, I. M., P. C. Cornillon, and K. Sherman. 2009. Fronts in large marine ecosystems. *Progress in Oceanography* 81: 223–236.

Bell, R. and J. I. Richardson. 1978. An analysis of tag recoveries from loggerhead sea turtles (*Caretta caretta*) nesting on Little Cumberland Island, Georgia. *Florida Marine Research Publication* 33: 20–24.

Bjorndal, K. A. 1997. Foraging ecology and nutrition of sea turtles. In *The Biology of Sea Turtles*, eds. P. L. Lutz and J. A. Musick, pp. 199–231. Boca Raton, FL: CRC Press.

Bjorndal, K. A., A. B. Bolten, T. Dellinger, C. Delgado, and H. R. Martins. 2003. Compensatory growth in oceanic loggerhead sea turtles: Response to a stochastic environment. *Ecology* 84(5): 1237–1249.

Bjorndal, K. A., A. B. Bolten, B. Koike et al. 2001. Somatic growth function for immature loggerhead sea turtles in southeastern U.S. waters. *Fishery Bulletin* 99: 240–246.

Bjorndal, K. A., A. B. Bolten, and H. R. Martins. 2000. Somatic growth model for juvenile loggerhead sea turtles *Caretta caretta*: Duration of pelagic stage. *Marine Ecology Progress Series* 202: 265–272.

Blumenthal J. M., F. A. Abreu-Grobois, T. J. Austin et al. 2009. Turtle groups or turtle soup: Dispersal patterns of hawksbill turtles in the Caribbean. *Molecular Ecology* 18: 4841–4853.

Bolker, B. M., T. Okuyama, K. A. Bjorndal, and A. B. Bolten. 2007. Incorporating multiple mixed stocks in mixed stock analysis: 'Many to many' analyses. *Molecular Ecology* 16: 685–695.

Bolten, A. B. 2003a. Active swimmers—Passive drifters: The oceanic juvenile stage of loggerheads in the Atlantic System. In *Loggerhead Sea Turtles*, eds. A. B. Bolten and B. E. Witherington, pp. 63–78. Washington, DC: Smithsonian Institution Press.

Bolten, A. B. 2003b. Variation in sea turtle life history patterns: Neritic vs. oceanic development stages. In *The Biology of Sea Turtles, Volume II,* ed. P. L. Lutz, J. A. Musick, and J. Wyneken, pp. 243–257. Boca Raton, FL: CRC Press.

Bolten, A. B., K. A. Bjorndal, H. R. Martins et al. 1998. Transatlantic developmental migrations of loggerhead sea turtles demonstrated by mtDNA sequence analysis. *Ecological Applications* 8: 1–7.

Bolten, A. B., H. R. Martins, K. A. Bjorndal, M. Cocco, and G. Gerosa. 1992. *Caretta caretta* pelagic movement and growth. *Herpetological Review* 23(4): 116.

Bolten, A. B.H. R. Martins, K. A. Bjorndal, and J. Gordon. 1993. Size distribution of pelagic-stage loggerhead sea turtles (*Caretta caretta*) in the waters around the Azores and Madiera. *Arquipelago: Life and Marine Sciences* 11A: 49–54.

Bowen, B. W. 1995. Voyages of the ancient mariners: Tracking marine turtles with genetic markers. *Bioscience* 45: 528–534.

Bowen, B. W., F. A. Abreu-Grobois, G. H. Balazs, N. Kamezaki, C. J. Limpus, and R. J. Ferl. 1995. Trans-Pacific migrations of the loggerhead turtle (*Caretta caretta*) demonstrated with mitochondrial DNA markers. *Proceedings of the National Academy of Science of the United States of America*. 92: 3731–3734.

Bowen, B. W. and S. A. Karl. 2007. Population genetics and phylogeography of sea turtles. *Molecular Ecology* 16: 4886–4907.

Bower, A. S. and T. Rossby. 1989. Evidence for cross-frontal exchange processes in the Gulf Stream based on isopycnal RAFOS float data. *Journal of Physical Oceanography* 19: 1177–1190.

Boyle, M. C., N. N. FitzSimmons, C. J. Limpus, S. Kelez, X. Velez-Zuazo, and M. Waycott. 2009. Evidence for transoceanic migrations by loggerhead sea turtles in the southern Pacific Ocean. *Proceedings of the Royal Society B* 276: 1993–1999.

Boyle, M. C. and C. J. Limpus. 2008. The stomach contents of post-hatching green and loggerhead sea turtles in the southwest Pacific: An insight into habitat association. *Marine Biology* 155: 233–241.

Brongersma, L. D. 1972. European Atlantic turtles. *Zoologische Verhandelingen* 121: 1–318.

Byles, R. A. 1988. Behavior and ecology of sea turtles from the Chesapeake Bay, PhD dissertation. Virginia Institute of Marine Science, College of William and Mary, Williamsburg, VA.

Carr, A. F. 1986. Rips, FADs and little loggerheads. *Bioscience* 36: 92–100.

Carr, A. F. 1987. New perspectives on the pelagic stage of sea turtle development. *Conservation Biology* 1: 103–121.

Carr, A. F. and A. B. Meylan. 1980. Evidence of passive migration of green turtle hatchlings in *Sargassum*. *Copeia* 1980: 366–368.

Casale, P., G. Abbate, D. Greggi, N. Conte, M. Oliverio, and R. Argano. 2008. Foraging ecology of loggerhead sea turtles *Caretta caretta* in the central Mediterranean Sea: Evidence for a relaxed life history model. *Marine Ecology Progress Series* 372: 265–276.

Collard S. B. and L. H. Ogren. 1990. Dispersal scenarios for pelagic post-hatchling sea turtles. *Bulletin of Marine Science* 47: 233–243.

Cowen, R. K., C. B. Paris, and A. Srinivasan. 2006. Scales of connectivity in marine populations. *Science* 311: 522–527.

Cowen, R. K. and S. Sponaugle. 2009. Larval dispersal and marine population connectivity. *Annual Review of Marine Science* 1: 443–466.

Crouse, D. T., L. B. Crowder, and H. Caswell. 1987. A stage-based population model for loggerhead sea turtles and implications for conservation. *Ecology* 68: 1412–1423.

Davenport, J. 1997. Temperature and the life-history strategies of sea turtles. *Journal of Thermal Biology* 22: 479–488.

Eckert, S. A. and H. R. Martins. 1989. Transatlantic travel by juvenile loggerhead turtle. *Marine Turtle Newsletter* 45: 15.

Encalada, S. E., K. A. Bjorndal, A. B. Bolton et al. 1998. Population structure of loggerhead turtle (*Caretta caretta*) nesting colonies in the Atlantic and Mediterranean as inferred from mitochondrial DNA control region sequences. *Marine Biology* 130: 567–575.

Epperly, S. P., J. Braun, A. J. Chester, F. A. Cross, J. V. Merriner, and P. A. Tester. 1995. Winter distribution of sea turtles in the vicinity of Cape Hatteras and their interactions with the summer Flounder Fishery. *Bulletin of Marine Science* 56(2): 547–568.

Fletmeyer, J. R. 1978. Underwater tracking evidence that neonate loggerhead sea turtles seek shelter in drifting *Sargassum*. *Copeia* 1: 148.

Frick, M. G., K. L. Williams, A. B. Bolten, K. A. Bjorndal, and H. R. Martins. 2009. Foraging ecology of oceanic-stage loggerhead turtles *Caretta caretta*. *Endangered Species Research* 9: 91–97.

Füxjager, M. J., B. S. Eastwood, and K. J. Lohmann. 2011. Orientation of hatchling loggerhead sea turtles to regional magnetic fields along a transoceanic migratory pathway. *Journal of Experimental Biology* 21: 2504–2508.

Girard, C., A. D. Tucker, and B. Calmettes. 2009. Post-nesting migrations of loggerhead sea turtles in the Gulf of Mexico: Dispersal in highly dynamic conditions. *Marine Biology* 156: 1827–1839.

Gower, J., C. Hu, G. Borstad, and S. King. 2006. Ocean color satellites show extensive lines of floating Sargassum in the Gulf of Mexico. *IEEE Transactions on Geoscience and Remote Sensing* 44: 3619–3625.

Gower, J. F. R. and S. A. King. 2011. Distribution of floating Sargassum in the Gulf of Mexico and the Atlantic Ocean mapped using MERIS. *International Journal of Remote Sensing* 32(7): 1917–1929.

Hatase, H., M. Kinoshita, T. Bando et al. 2002b. Population structure of loggerhead turtles, *Caretta caretta*, nesting in Japan: Bottlenecks on the Pacific population. *Marine Biology* 141: 299–305.

Hatase, H., N. Takai, Y. Matsuzawa et al. 2002a. Size-related divergences in feeding habitat use of adult female loggerhead turtles *Caretta caretta* around Japan determined by stable isotope analyses and satellite telemetry. *Marine Ecology Progress Series* 233: 273–281.

Hawkes, L. A., A. C. Broderick, M. S. Coyne et al. 2006. Phenotypically linked dichotomy in sea turtle foraging requires multiple conservation approaches. *Current Biology* 16: 990–995.

Hawkes, L. A., A. C. Broderick, M. S. Coyne, M. H. Godfrey, and B. J. Godley. 2007. Only some like it hot: Quantifying the environmental niche of the loggerhead sea turtle. *Diversity and Distributions* 13(4): 447–457.

Hays, G. C. and R. Marsh. 1997. Estimating the age of juvenile loggerhead sea turtles in the North Atlantic. *Canadian Journal of Zoology* 75: 40–46.

Johns, W. E., T. J. Shay, J. M. Bane, and D. R. Watts. 1995. Gulf Stream structure, transport, and recirculation near 68°W. *Journal of Geophysical Research* 100: 817–838.

Keinath, J. A. 1993. Movements and behavior of wild and head-started sea turtles, PhD dissertation. Virginia Institute of Marine Science, College of William and Mary, Gloucester Point, VA.

Kobayashi, D. R., I.-J. Cheng, D. M. Parker, J. J. Polovina, N. Kamekazi, and G. H. Balazs. 2011. Loggerhead turtle (*Caretta caretta*) movement off the coast of Taiwan: Characterization of a hotspot in the East China Sea and investigation of mesoscale eddies. *ICES Journal of Marine Science* 68: 707–718.

Kobayashi, D. R., J. J. Polovina, D. M. Paker et al. 2008. Pelagic habitat characterization of loggerhead sea turtles, *Caretta caretta*, in the North Pacific Ocean (1997–2006): Insights from satellite tag tracking and remotely sensed data. *Journal of Experimental Marine Biology and Ecology* 356(2008): 96–114.

Lalli, C. M. and T. R. Parsons. 1993. *Biological Oceanography: An Introduction*. New York: Pergamon Press.

Laurent L., P. Casale, M. N. Brandai, et al. 1998. Molecular resolution of marine turtle stock composition in fishery bycatch: A case study in the Mediterranean. *Molecular Ecology* 7: 1529–1542.

Lett, C., P. Verley, P. Mullon et al. 2008. Lagrangian tool for modelling ichthyoplankton dynamics. *Environmental Modeling and Software* 23: 1210–1214.

Limpus, C. J. and D. J. Limpus. 2003. Biology of the loggerhead turtle in Western South Pacific Ocean foraging areas. In *Loggerhead Sea Turtles*, eds. A. B. Bolten and B. E. Witherington, pp. 93–113. Washington, DC: Smithsonian Institution Press.

Lipcius., R. N. and W. T. Stockhausen. 2002. Concurrent decline of the spawning stock, recruitment, larval abundance, and size of blue crab in Chesapeake Bay. *Marine Ecology Progress Series* 226: 45–61.

Lohmann, K. J., S. D. Cain, S. A. Dodge, and C. M. F. Lohmann. 2001. Regional magnetic fields as navigational markers for sea turtles. *Science* 294: 364–366.

Lohmann, K. J. and C. M. F. Lohmann. 1996a. Detection of magnetic field intensity by sea turtles. *Nature* 380: 59–61.

Lohmann, K. J. and C. M. F. Lohmann. 1996b. Orientation and open-sea navigation in sea turtles. *Journal of Experimental Biology* 199: 73–81.

Lohmann, K. J. and C. M. F. Lohmann. 2003. Orientation mechanisms in hatchling sea turtles. In *Loggerhead Sea Turtles*, eds. A. B. Bolten and B. E. Witherington, pp. 44–62. Washington, DC: Smithsonian Institution Press.

Lohmann, K. J., C. M. F. Lohmann, and N. F. Putman. 2007. Magnetic maps in animals: Nature's GPS. *Journal of Experimental Biology* 210: 3697–3705.

Lohmann, K. J., N. F. Putman, and C. M. F. Lohmann. 2008. Geomagnetic imprinting: A unifying hypothesis of long-distance natal homing in salmon and sea turtles. *Proceedings of the National Academy of Sciences of the United States of America* 105: 19096–19101.

Lohmann, K. J., N. F. Putman, and C. M. F. Lohmann. 2012. The magnetic map of hatchling loggerhead sea turtles. *Current Opinion in Neurobiology* 22: 336–342.

Lozier, M. S. and G. Gawarkiewicz. 2001. Cross-frontal exchange in the Middle Atlantic Bight as evidenced by surface drifters. *Journal of Physical Oceanography* 31: 2498–2510.

Luschi, P., G. C. Hays, and F. Papi. 2003. A review of long-distance movements by marine turtles, and the possible role of ocean currents. *Oikos* 103: 293–302.

Lutcavage, M. E. and P. L. Lutz. 1997. Diving physiology. In *The Biology of Sea Turtles*, eds. P. L. Lutz and J. A. Musick, pp. 277–296. Boca Raton, FL: CRC Press.

Lutcavage, M. E. and J. A. Musick. 1985. Aspects of the biology of sea turtles in Virginia. *Copeia* 38(4): 329–336.

Mansfield, K. L. 2006. Sources of mortality, movements and behavior of sea turtles in Virginia, PhD dissertation. Virginia Institute of Marine Science, College of William and Mary, Gloucester Point, VA.

Mansfield, K. L., V. S. Saba, J. Keinath, and J. A. Musick. 2009. Satellite telemetry reveals a dichotomy in migration strategies among juvenile loggerhead sea turtles in the northwest Atlantic. *Marine Biology* 156: 2555–2570.

Mansfield, K. L., J. Wyneken, J. Luo, and W. P. Porter. In review. Voyage of the young mariners: First oceanic satellite tracks of neonate loggerhead sea turtles.

Mansfield, K. L., J. Wyneken, D. Rittschoff, M. Walsh, C. W. Lim, and P. M. Richards. 2012. Satellite tag attachment methods for tracking neonate loggerhead (*Caretta caretta*) sea turtles. *Marine Ecology Progress Series* 457: 181–192.

McCarthy, A. L., S. Heppell, F. Royer, and T. Dellinger. 2010. Identification of likely foraging habitat of pelagic loggerhead sea turtles (*Caretta caretta*) in the North Atlantic through analysis of telemetry track sinuosity. *Progress in Oceanography* 86: 224–231.

McClellan, C. M. and A. J. Read. 2007. Complexity and variation in loggerhead sea turtle life history. *Biology Letters* 3(6): 592–594.

McGrath, G. G., T. Rossby, and J. T. Merrill. 2010. Drifters in the Gulf Stream. *Journal of Marine Research* 68: 699–721.

Monzón-Argüello, C., F. Dell'Amico, P. Moriniere et al. 2012. Lost at sea: Genetic, oceanographic, and meteorological evidence for storm-forced dispersal. *Journal of the Royal Society: Interface* 73: 1725–1732.

Monzón-Argüello C., C. Rico, C. Carreras, P. Calabuig, A. Marco, and L. F. Lopez-Juardo. 2009. Variation in spatial distribution of juvenile loggerhead turtles in the eastern Atlantic and western Mediterranean Sea. *Journal of Experimental Marine Biology and Ecology* 373: 79–86.

Monzón-Argüello C., C. Rico, E. Naro-Maciel et al. 2010. Population structure and conservation implications for the loggerhead sea turtle of the Cape Verde Islands. *Conservation Genetics* 11: 1871–1884.

Morreale S. J., A. B. Meylan, S. S. Sadove, and A. E. Standora. 1992. Annual occurrence and winter mortality of marine turtles in New York waters. *Journal of Herpetology* 26(2): 130–308.

Morreale, S. J. and A. E. Standora. 1988. Early life stage ecology of sea turtles in northeastern U.S. waters. NOAA Technical Memorandum NMFS-SEFSC-413.

Musick J. A. and C. J. Limpus. 1997. Habitat utilization and migration in juvenile sea turtles. In *The Biology of Sea Turtles*, eds. P. L. Lutz and J. A. Musick, pp. 137–163. Boca Raton, FL: CRC Press.

National Marine Fisheries Service and U.S. Fish and Wildlife Service. 2008. *Recovery Plan for the Northwest Atlantic Population of the Loggerhead Sea Turtle (Caretta caretta)*, Second Revision. National Marine Fisheries Service, Silver Spring, MD, http://www.nmfs.noaa.gov/pr/pdfs/recovery/turtle_loggerhead_atlantic.pdf (accessed June 1, 2012).

Nelson, D. A. 1996. Subadult loggerhead sea turtle (*Caretta caretta*) behavior in St. Marys entrance channel, Georgia, U.S.A. PhD dissertation. College of William and Mary, Gloucester Point, VA.

Nichols, W. J., A. Resendiz, J. A. Seminoff, and B. Resendiz. 2000. Transpacific migration of a loggerhead turtle monitored by satellite telemetry. *Bulletin of Marine Science* 67(3): 937–947.

O'Hara, J. 1980. Thermal influences on the swimming speed of loggerhead turtle hatchlings. *Copeia* 773–780.

Okuyama, J., T. Kitagawa, K. Zenimoto et al. 2011. Trans-Pacific dispersal of loggerhead turtle hatchlings inferred from numerical simulation modeling. *Marine Biology* 158: 2055–2063.

Owens, D. W., M. A. Grassman, and J. R. Hendrickson. 1982. The imprinting hypothesis and sea turtle reproduction. *Herpetelogica* 38: 124–135.

Papi, F., P. Luschi, E. Crosio, and G. R. Hughes. 1997. Satellite tracking experiments on the navigational ability and migratory behavior of the loggerhead sea turtle, *Caretta caretta*. *Marine Biology* 129: 215–220.

Peckham, S. H., D. Maldonado-Diaz, Y. Tremblay et al. 2011. Demographic implications of alternative foraging strategies in juvenile loggerhead turtles *Caretta caretta* of the North Pacific Ocean. *Marine Ecology Progress Series* 425: 269–280.

Polovina, J. J., G. H. Balazs, E. A. Howell, D. M. Parker, M. P. Seki, and P. H. Dutton. 2004. Forage and migration habitat of loggerhead (*Caretta caretta*) and olive ridley (*Lepidochelys olivacea*) sea turtles in the central North Pacific Ocean. *Fisheries Oceanography* 13: 36–51.

Polovina, J. J., D. R. Kobayashi, D. Parker, M. P. Seki, and G. H. Balazs. 2000. Turtles on the edge: Movement of loggerhead turtles (*Caretta caretta*) along oceanic fronts, spanning longline fishing grounds in the central North Pacific, 1997–1998. *Fisheries Oceanography* 9(1): 71–82.

Polovina, J., I. Uchida, G. H. Balazs, E. A. Howell, D. Parker, and P. Dutton. 2006. The Kuroshio Extension Bifurcation Region: A pelagic hotspot for juvenile loggerhead sea turtles. *Deep-Sea Research II* 53: 326–339.

Putman, N. F., J. B. Bane, and K. J. Lohmann. 2010a. Sea turtle nesting distributions and oceanographic constraints on hatchling migration. *Proceedings of the Royal Society B* 277: 3631–3637.

Putman, N. F., C. S. Endres, C. M. F. Lohmann, and K. J. Lohmann. 2011. Longitude perception and bicoordinate magnetic maps in sea turtles. *Current Biology* 21: 463–466.

Putman, N. F., R. Scott, P. Verley, R. Marsh, and G. C. Hays. 2012a. Natal site and offshore swimming influence fitness and long-distance ocean transport in young sea turtles. *Marine Biology* 159: 2117–2126.

Putman, N. F., T. J. Shay, and K. J. Lohmann. 2010b. Is the geographic distribution of nesting in the Kemp's ridley turtle shaped by the migratory needs of offspring? *Integrative and Comparative Biology* 50: 305–314.

Putman, N. F., P. Verley, T. J. Shay, and K. J. Lohmann. 2012b. Simulating transoceanic migrations of young loggerhead sea turtles: Merging magnetic navigation behavior with an ocean circulation model. *Journal of Experimental Biology* 215: 1863–1870.

Rees, A. F., A. S. Salim, A. C. Broderick, M. C. Coyne, N. Papathanasopoulou, and B. J. Godley. 2010. Behavioural polymorphism in one of the world's largest population of loggerhead sea turtles *Caretta caretta*. *Marine Ecology Progress Series* 418: 201–212.

Reich, K. J., K. A. Bjorndal, M. J. Frick, B. E. Witherington, C. Johnson, and A. B. Bolten. 2010. Polymodal foraging in adult female loggerheads (*Caretta caretta*). *Marine Biology* 157: 113–121.

Reis, E. C., L. S. Soares, S. M. Vargas et al. 2010. Genetic composition, population structure and phylogeography of the loggerhead sea turtle: Colonization hypothesis for the Brazilian rookeries. *Conservation Genetics* 11: 1467–1477.

Resendiz, A., B. Resendiz, W. J. Nichols, J. A. Seminoff, and N. Kamekazi. 1998. First confirmed east-west transpacific movement of a loggerhead sea turtle, *Caretta caretta*, released in Baja California, Mexico. *Pacific Science* 52(2): 151–153.

Revelles, M., C. Carreras, L. Cardona et al. 2007a. Evidence for an asymmetrical size exchange of loggerhead sea turtles between the Mediterranean and the Atlantic thought the Straits of Gibraltar. *Journal of Experimental Marine Biology and Ecology* 349: 261–271.

Revelles, M., J. Isern-Fontanet, L. Cardona, M. San Felix, C. Carreras, and A. Aguilar. 2007b. Mesoscale eddies, surface circulation and the scale of habitat selection by immature loggerhead sea turtles. *Journal of Experimental Marine Biology and Ecology* 347: 41–57.

Riewald, B., A. B. Bolten, and K. A. Bjorndal. 2000. Use of satellite telemetry to estimate post-hooking behavior and mortality of loggerhead sea turtles in the pelagic longline fishery in the Azores. Final Report to the National Marine Fisheries Service, Honolulu Laboratory, December 2000, Archie Carr Center for Sea Turtle Research University of Florida, Gainesville, FL.

Sakamoto, W., I. Uchida, Y. Naito, K. Kureha, M. Tujimura, and K. Sato. 1990. Deep diving behavior of the loggerhead turtle near the frontal zone. *Nippon Suisan Gakkaishi* 56: 1435–1443.

Salmon, M. and K. J. Lohmann. 1989. Orientation cues used by hatchling loggerhead sea turtles (*Caretta caretta* L.) during their offshore migration. *Ethology* 83: 215–228.

Salmon, M. and J. Wyneken. 1987. Orientation and swimming behavior of hatchling loggerhead turtles *Caretta caretta* L. during their offshore migration. *Journal of Experimental Marine Biology and Ecology* 109: 137–153.

Sasso, C. R. and S. P. Epperly. 2007. Survival of pelagic juvenile loggerhead turtles in the open ocean. *Journal of Wildlife Management* 71(6): 1830–1835.

Schwartz, F. J. 1978. Behavioral and tolerance responses to cold water temperatures by three species of sea turtles (Reptilia, *Cheloniidae*) in North Carolina. In *Proceedings of the Florida Interregional Conference on Sea Turtles*, ed. G. E. Henderson, July 24–25, 1976, Vol. 33, pp. 16–18. Jensen Beach, FL: Florida Marine Research Publication.

Scott, R., R. Marsh, and G. C. Hays. 2012a. A little movement orientated to the geomagnetic field makes a big difference in strong flows. *Marine Biology* 59: 481–488.

Scott, R., R. Marsh, and G. C. Hays. 2012b. Life in the really slow lane: Loggerhead sea turtles mature late relative to other reptiles. *Functional Ecology* 26: 227–235.

Seney, E. E., B. M. Higgins, and A. M. Landry. 2010. Satellite transmitter attachment techniques for small juvenile sea turtles. *Journal of Experimental Marine Biology and Ecology* 384: 61–67.

Seney E. E. and J. A. Musick. 2006. Historical diet analysis of loggerhead sea turtles (*Caretta caretta*) in Virginia. *Copeia* 2007(2): 478–489.

Shamblin, B. M., K. A. Bjorndal, A. B. Bolten et al. 2012. Mitogenomic sequences better resolve stock structure of southern Greater Caribbean green turtle rookeries. *Molecular Ecology* 21: 2330–2340.

Shillinger, G. L., E. Di Lorenzo, H. Luo et al. 2012. On the dispersal of leatherback turtle hatchlings from Mesoamerican nesting beaches. *Proceedings of the Royal Society B* 279: 2391–2395.

Shoop, C. R. and R. D. Kenney. 1992. Seasonal distribution and abundances of loggerhead and leatherback sea turtles in northeastern United States waters. *Herpetological Monograph* 6: 43–67.

Smith, M. M. and M. Salmon. 2009. A comparison between habitat choices made by hatchling and juvenile green turtles (*Chelonia mydas*) and loggerheads (*Caretta caretta*). *Marine Turtle Newsletter* 126: 9–13.

Snover, M. L., A. A. Hohn, L. B. Crowder, and S. A. Macko. 2010. Combining stable isotopes and skeletal growth marks to detect habitat shifts in juvenile loggerhead sea turtles *Caretta caretta*. *Endangered Species Research* 13: 25–31.

Sponaugle, S., T. Lee, V. Kourafalou, and D. Pinkard. 2005. Florida current frontal eddies and the settlement of coral reef fishes. *Limnology and Oceanography* 50: 1033–1048.

Spotila, J. R., M. P. O'Connor, and F. Paladino. 1997. Thermal biology. In *The Biology of Sea Turtles*, eds. P. L. Lutz and J. A. Musick, pp. 297–314. Boca Raton, FL: CRC Press.

Spotila, J. R. and E. A. Standora. 1985. Environmental constraints on the thermal energetics of sea turtles. *Copeia* 3: 694–702.

Tomczak, M. and J. S. Godfrey. 1994. *Regional Oceanography: An Introduction*. New York: Elsevier Science Inc.

Turtle Expert Working Group. 2009. *An Assessment of the Loggerhead Turtle Population in the Western North Atlantic Ocean*. NOAA Technical Memorandum NMFS-SEFSC-575.

Wallace, B. P., S. S. Heppell, R. L. Lewison, S. Kelez, and L. B. Crowder. 2008. Reproductive values of loggerhead turtles in fisheries bycatch worldwide. *Journal of Applied Ecology* 45: 1076–1085.

Wang, J. H., L. C. Boles, B. Higgins, and K. J. Lohmann. 2007. Behavioral responses of sea turtles to lightsticks used in longline fisheries. *Animal Conservation* 10: 176–182.

Watson, J. W., S. P. Epperly, A. K. Shah, and D. G. Foster. 2005. Fishing methods to reduce sea turtle mortality associated with pelagic longlines. *Canadian Journal of Fisheries and Aquatic Science* 62: 965–981.

Witherington, B. E. 1991. Orientation of hatchling loggerhead turtles at sea off artificially lighted and dark beaches. *Journal of Experimental Marine Biology and Ecology* 149: 1–11.

Witherington, B. E. 2002. Ecology of neonate loggerhead turtles inhabiting lines of downwelling near a Gulf Stream front. *Marine Biology* 140: 843–853.

Witherington, B., R. Herren, and M. Bresette. 2006. *Caretta caretta*—Loggerhead sea turtle. In *Biology and Conservation of Florida Turtles*, ed. P. A. Meylan, Vol. 3, pp. 74–90. Lunenburg, MA: Chelonian Research Monographs.

Witherington, B., S. Hirama, and R. Hardy. 2012. Young sea turtles of the pelagic *Sargassum*-dominated drift community: Habitat use, population density and threats. *Marine Ecology Progress Series* 463: 1–22.

Witt, M. J., L. A. Hawkes, M. H. Godfrey, B. J. Godley, and A. C. Broderick. 2010. Predicting the impacts of climate change in globally distributed species: The case of the loggerhead turtle. *Journal of Experimental Biology* 213: 901–911.

Witzell, W. N. 1999. Distribution and relative abundance of sea turtles caught incidentally by the U.S. pelagic longline fleet in the western North Atlantic Ocean, 1992–1995. *Fishery Bulletin* 97: 200–211.

Witzell, W. N. 2002. Immature Atlantic loggerhead turtles (*Caretta caretta*): Suggested changes to the life history model. *Herpetological Review* 33(4): 226–269.

# 9 Feeding Biology
## *Advances from Field-Based Observations, Physiological Studies, and Molecular Techniques*

T. Todd Jones and Jeffrey A. Seminoff

## CONTENTS

## 9.1 INTRODUCTION

Since Archie Carr's seminal work in the 1960s and 1970s and efforts by Karen Bjorndal and others in the 1980s and 1990s, feeding biology has been a relatively well-studied facet of sea turtle biology. This is opportune for the science of sea turtles considering that nutrient acquisition strategies are among the most important components of a sea turtle's life history, influencing key demographic parameters such as somatic growth, age-at-maturity, and timing of reproductive migrations. Over the past two decades, however, the advent of new research fields such as physiological monitoring, biologging, and stable isotope analysis (SIA) have helped strengthen this understanding even further. These tools have provided insights that have in some cases confirmed earlier wisdom about how a sea turtle makes a living, and in other cases have redefined long-standing biological paradigms.

Considering the new information that has come available, it is clear that the ecological strategies of some species are much more diverse than originally considered. For example, green turtles (*Chelonia mydas*), long-considered obligate neritic herbivores instead eat large amounts of animal matter in many places (e.g., Heithaus et al., 2002; Seminoff et al., 2002a; Cardona et al., 2009; Carrion-Cortez et al., 2010), and at least in the Pacific are commonly high-seas dwellers, even as adults (Hatase et al., 2006; Kelez, 2011; Parker et al., 2011). Hawksbill turtles (*Eretmochelys imbricata*), the "coral reef dwelling" turtle, are turning up in the strangest of places. In the eastern Pacific, for example, adult hawksbills inhabit mangrove estuaries during non-breeding periods, a huge departure from our belief that the species was tied to coral reefs (Gaos et al., 2012). In the Caribbean and Indian Ocean, hawksbills are now known to depend on seagrass pastures for foraging and residence (Bjorndal and Bolten, 2010; J. Mortimer, unpubl. data). Leatherbacks (*Dermochelys coriacea*), historically defined as "high-seas inhabitants," are now seen in coastal habitats more than ever before (James et al., 2006; Benson et al., 2011; Fossette et al., 2011). These and other novel revelations about feeding biology are at least partly due to the globalization of sea turtle research and an everexpanding toolbox at the disposal of field and laboratory scientists. Indeed, more research with both traditional and novel tools is conducted in more parts of the world than ever before, and we are now gaining an appreciation of just how complex and adaptive sea turtles can be.

Much new biological information has emerged in the published literature since *The Biology of the Sea Turtles* (Volume 1; Lutz and Musick, 1997) was first published and a thorough update is warranted, particularly for aspects relating to feeding biology. In this chapter we present new information for all seven sea turtle species, building on Karen Bjorndal's chapter on Feeding Biology in Volume 1 that summarized what was known at that time. In Section 9.2 we present the latest information about sea turtle diet and feeding biology. Here we describe new diet items and novel foraging tactics that are reshaping our perceptions about the types of prey consumed and methods by which sea turtles access food resources. In Section 9.3 we focus on the feeding physiology of sea turtles (e.g., specific dynamic action (SDA), digestive efficiency, and passage rates of digesta), a still-understudied area of sea turtle feeding biology, but one that is expanding thanks to additional field and lab-based scientific research. Understanding how sea turtle energy acquisition is constrained by physiological and environmental factors is important as these data factor into growth rates, residency times, population demographics, bioenergetics and energy budgets, and reproductive output. In Sections 9.4 (stable isotopes) and 9.5 (fatty acids and trace elements) we explore the "molecular-based" techniques that are showing great promise for establishing diet, trophic status, and foraging movements of sea turtles. Clearly, the advent of these approaches allows us to learn much about the types of foods consumed by turtles based on the analysis of their own body tissues.

As described earlier, the feeding biology of sea turtles is a broad topic with many nuances. Together the established (e.g., stomach content analysis, esophageal lavage) and emerging (e.g., SIA, fatty acids) techniques give greater insight and understanding into the unique foraging strategies of sea turtles both intra- and interspecifically and through life-history stages. Studies of feeding physiology then begin to tie together what, when, and where sea turtles eat with why and how they eat to meet daily and yearly energy demands of maintenance, growth, and reproduction. In the end, our goal is to provide an update on the current knowledge of sea turtle feeding biology and share a perspective of how our understanding has evolved in the past decades.

## 9.2  DIET COMPOSITION

### 9.2.1  Leatherback Turtle, *Dermochelys coriacea*

Leatherback turtles feed primarily on gelatinous zooplankton from several days post-emergence (Salmon et al., 2004), through juvenile maturation (Iverson and Yoshida, 1956, authors personal observations), and adulthood (see Bjorndal, 1997). While observations of feeding during the pelagic juvenile years are rare, the authors had the opportunity to observe the stomach contents of several leatherback turtles (60–80 cm SCL) from fishery bycatch in the western Pacific Ocean, all of which contained remnants of gelatinous zooplankton. The majority of reports on leatherback diet come from direct observations of surface or water column feeding (Bacon, 1970; Duron, 1978; Eisenberg and Frazier, 1983; Grant and Ferrell, 1993; James and Herman, 2001; Salmon et al., 2004; Fossette et al., 2011; Heaslip et al., 2012), from alimentary tract contents from stranded or bycaught turtles (e.g., Bleakney, 1965; Brongersma, 1972; Den Hartog, 1980; Den Hartog and Van Nierop, 1984; Davenport and Balazs, 1991), and most recently from SIA (Dodge et al., 2011; Seminoff et al., 2012). The observed prey items consist mostly of the phylum Cnidaria, class Scyphozoa (i.e., true jellies) including *Aurelia* spp., *Catostylus* spp., *Chrysaora* spp., *Cyanea* spp., *Pelagia* spp., *Rhizostoma* spp., and *Stomolophus* spp. (Duguy, 1982; Den Hartog and Van Nierop, 1984; Duron-Dufrenne, 1987; Grant and Ferrell, 1993; Limpus and McLachlan, 1994; Davenport, 1998; James and Herman, 2001; Salmon et al., 2004). To date we only know of three reports indicating that leatherbacks forage on the class Hydrozoa, orders Leptomedusae (*Aequorea* spp.) and Siphonophorae (*Apolemia* spp., *Physalia* spp.) (Bacon, 1970; Den Hartog, 1980; Den Hartog and Van Nierop, 1984). Further it was suggested that the presence of the Leptomedusae (*Aequorea* spp.) in the leatherback alimentary tract may be a result of contamination, as Scyphomedusae (targeted leatherback prey) feed on *Aequorea* spp. (Den Hartog and Van Nierop, 1984). Leatherbacks have also been reported to feed on pyrosomes (Thaliacea) (Davenport and Balazs, 1991; pers. observ.) as well as ctenophores (Nuda) and gelatinous fish egg sacs (Actinopterygii eggs) (Salmon et al., 2004). Most recently, Dodge et al. (2011) have noted sea butterflies (Gastropoda) as a diet item of leatherbacks foraging in the eastern North Atlantic Ocean through SIA. This is the first documentation of a marine gastropod (family Cymbulioidea) in the leatherback diet. Squid, octopus, and fish have been noted in the alimentary tract of three leatherbacks caught in fishing gear (Brongersma, 1969, 1972; Limpus, 1984; Bello et al., 2011) as well as numerous crab species (e.g., *Libinia spinosa*) (Brongersma, 1969; Frazier et al., 1985). Frazier et al. (1985), however, clarified that the species of small crab and even the fish are probably jelly commensals and ingested incidentally.

Leatherbacks have long been thought to be oceanic-pelagic throughout their life-history (Bolten, 2003); however, recent evidence suggests that leatherback subadults and adults frequent coastal foraging areas (James et al., 2006; Benson et al., 2011) feeding on a vast array of dense gelatinous zooplankton. While these findings may be a departure from the strict oceanic paradigm what has not shifted is the epipelagic foraging strategy used by leatherbacks whether they are foraging in oceanic or neritic waters. The gelatinous diet of leatherbacks is varied across several phyla and classes (Table 9.1); however, the majority of data from feeding observations, gut content analysis, and use of SIA suggests that Scyphozoa (i.e., true jellies) remain the main diet component and that leatherbacks are obligate jelly (gelatinous zooplankton) consumers throughout their ontogeny.

### 9.2.2  Green Turtle, *Chelonia mydas*

Green turtles occur in tropical and temperate waters worldwide and have been shown to consume a wide variety of seagrass, marine algae, and invertebrates (see Bjorndal, 1997). They have an oceanic–neritic developmental pattern (Bolten, 2003) that comes with a concomitant diet shift from omnivory to primarily herbivory (Table 9.1) at neritic recruitment (20–35 cm carapace length [CL] in the Atlantic [Bjorndal and Bolten, 1988]; 35+ cm CL in the Pacific [Seminoff et al., 2002b; Arthur et al., 2008]). However, central North Pacific green turtles have recently been found to forage

**TABLE 9.1**

**Diet and Foraging Habitat of Sea Turtles throughout Development**

| Developmental Stage/Habitat | Leatherback | Green | Loggerhead | Hawksbill | Olive Ridley | Kemp's Ridley | Flatback |
|---|---|---|---|---|---|---|---|
| Post-hatchling diet | Scyphozoa, Nuda, Actinopterygii eggs | Seagrass, algae, Thaliacea, Actinopterygii eggs | Hydrozoa, Scyphozoa, Bryozoa, Gastropoda, Polychaeta, Maxillopoda, Malacostraca, Insecta, Actinopterygii, algae | Not observed | Not observed | Gastropoda, Malacostraca, algae | Not observed |
| Oceanic diet (juvenile) | Thaliacea (pyrosomes) | Gastropoda, Scyphozoa, Thaliacea, Malacostraca, Maxillopoda, algae, Insecta | Hydrozoa, Scyphozoa, Thaliacea, Actinopterygii eggs, Cephalopoda, Gastropoda, Malacostraca, Maxillopoda, Insecta, Polychaeta, algae | Algae, Maxillopoda, Malacostraca, Thaliacea, Actinopterygii eggs | Thaliacea, Gastropoda, Actinopterygii | Not observed | N/A |
| Neritic diet (juvenile) | N/A | N/A | N/A | N/A | N/A | N/A | Gastropoda, Hydrozoa, Scyphozoa, Anthozoa, Bryozoa |
| Neritic diet (post-recruitment) | N/A | Seagrass, algae, mangrove fruit/leaves, Gastropoda, Thaliacea, Porifera, Polychaeta, Anthozoa, Malacostraca, Cyanophyceae | Malacostraca, Merostomata, Actinopterygii (fisheries), Maxillopoda, Gastropoda, Bivalvia, Echinoidea, Anthozoa, Holothuroidea, Chondrichthyes, Scyphozoa, Cephalopoda, Bryozoa, Polychaeta, Porifera, Foraminifera, Porifera, Ophiuroidea, Brachiopoda, Demospongiae, Turbellaria, Sipunculida, Ascidiacea, seagrass, algae | Porifera, Anthozoa, Gastropoda, Polychaeta, Hydrozoa, Bryozoa, Scyphozoa, Malacostraca, Echinoidea, Actinopterygii, algae, seagrass | Gastropoda, Malacostraca | Malacostraca, Bivalvia, Gastropoda, Ascidiacea, seagrass, algae, Merostomata, Ctenophora, Actinopterygii | N/A |

| | | | | | | |
|---|---|---|---|---|---|---|
| Oceanic diet (adult) | Scyphozoa, Hydrozoa, Thaliacea, Gastropoda | N/A | N/A | N/A | Thaliacea, Gastropoda, Actinopterygii | N/A | Gastropoda, Hydrozoa, Scyphozoa, Anthozoa, Bryozoa |
| Neritic diet (adult) | See Oceanic diet (adult)[b] | See Neritic diet (post-recruitment) | See Neritic diet (post-recruitment) | See Neritic diet (post-recruitment) | Gastropoda, Malacostraca, Bryozoa, Sipunculidea, Ascidiacea, Actinopterygii, algae | Malacostraca, see Neritic diet (post-recruitment) | N/A |

[a] See Section 9.2 for complete details and cited references. Diet items listed in order of importance, abundance, or frequency when possible (Benson et al., 2011; Fossette et al., 2011; Heaslip et al., 2012).

[b] Recent observations suggest that adult leatherbacks feed in epipelagic–neritic waters.

in oceanic waters up to 70 cm CL (Parker et al., 2011) suggesting alternate life-history patterns. Little is known of the green turtle diet during oceanic development with only a few observations of surface or shallow depth foraging (Frick, 1976; Salmon et al., 2004), gut content analysis (Hughes, 1974; Boyle and Limpus, 2008), and more recently SIA (Reich et al., 2007; Arthur et al., 2008). Salmon et al. (2004) observed post-hatchling green turtles (<10 weeks of age) in the Gulf Stream (off the coast of eastern Florida, United States) feeding at the surface on floating bits of seagrass and algae (*Thalassia* sp. and *Sargassum* sp.) and at shallow depths on ctenophores (Nuda) and gelatinous egg sacs (unknown species) but they avoided larger Scyphozoa such as *Aurelia* sp. This corroborates a sighting of a green turtle hatchling eating a ctenophore 1 m below the surface off the shore of Bermuda (Frick, 1976). Gut content analysis revealed marine gastropod mollusks (Gastropoda) in green turtles off South Africa and eastern Australia (Hughes, 1974; Boyle and Limpus, 2008), as well as Scyphozoa, crustaceans (Malacostraca), plant material, and terrestrial insects (Insecta) (eastern Australia, Boyle and Limpus, 2008). North Atlantic green turtles have stable isotope signatures consistent with carnivory for their first 3–5 years of life while occupying the oceanic zone (Reich et al., 2007). Similar signatures were found for green turtles <44 cm CL of the southwestern Pacific (Arthur et al., 2008). Parker et al. (2011) described the gut contents of 10 oceanic green turtles (30–70 cm CL) from the central North Pacific (commercial fishery bycatch). The diet items included pyrosomes (Thaliacea), goose barnacles (Maxillopoda), amphipods (Malacostraca), sea snails and butterflies (Gastropoda), Ctenophora (Nuda), and jellyfish (Scyphozoa) (Parker et al., 2011); however, the carapace lengths are indicative of turtles recruiting to neritic zones (Bjorndal, 1997; Bolten, 2003). Therefore, plasticity may exist in recruitment length or turtles may shuttle between neritic and oceanic habitats.

Once green turtles recruit to neritic waters they primarily forage on benthic organisms (see Table 9.1). As reviewed by Bjorndal (1997), the diet mostly consists of seagrass and algae. Lopez-Mendilaharsu et al. (2005) recorded (through esophageal lavage) that eastern Pacific green turtles on the Pacific Coast of the Baja Peninsula fed primarily on red algae in the inner bays but that larger turtles fed on seagrass along the outer coastline. On the other side of the Baja California Peninsula (Gulf of California), also using esophageal lavage, Seminoff et al. (2002a) documented that neritic green turtles fed primarily on the red algae *Gracilariopsis lemaneiformis*; however, they also noted high animal matter consumption, including sponges (Porifera), tube worms (Polychaeta), sea pens (Anthozoa), and sea hares (Gastropoda). This was later corroborated by Crittercam analysis (Seminoff et al., 2006a) observing the consumption of algae as well as previously undocumented species of Cnidaria (yellow-polyp black coral, *Antipathes galapagensis*) and Annelida (fanworm, *Bispiria* sp.). Farther south, off of the South American coast, Carrion-Cortez et al. (2010) recorded mostly algae species and red mangrove in esophageal lavage samples of eastern Pacific green turtles of the Galapagos Islands, whereas Amorocho and Reina (2007) found (in order of abundance) tunicates (Thaliacea), red mangrove fruits, algae (Rhodophyta, Chlorophyta, Cyanophyta), small crustaceans (Malacostraca), and plant leaves as well as coral/shell fragments in esophageal contents of juveniles and adults (average 58 cm CL) in Gorgona National Park, Colombia. The gut content analysis of green turtles from the Sultanate of Oman also revealed a high rate of animal matter consumption (Ferreira et al., 2006), with algae and animal matter constituting 49% and 26% of the gut contents dry weight, respectively.

In southeast Queensland, Australia, the dominant diet item of green turtles is seagrass with algae a close second (Brand-Gardner et al., 1999; Limpus et al., 2005). Limpus et al. (2005) reported that 96.6% of lavage samples of green turtles from Shoalwater Bay had seagrass present (*Halodule* sp., *Zostera* sp., and *Halophila* sp.) and when present it constituted 50% or more of the sample. Mangrove leaves, fruits, and seedlings have also been reported in the diet of Australian green turtles (Pendoley and Fitzpatrick, 1999; Limpus and Limpus, 2000; Limpus et al., 2005) to the extent that Limpus (1998) described the herbivorous diet of green turtles in Australia as "seagrasses, algae, and mangroves." Three studies employing the use of animal-borne video cameras have documented the previously unreported consumption of gelatinous zooplankton by green turtles off Western

Australia (Heithaus et al., 2002) and eastern Australia (Arthur et al., 2007), and in the Gulf of California, Mexico (Seminoff et al., 2006a). Arthur et al. (2006) documented the consumption of the toxic filamentous cyanobacterium *Lyngbya majuscula* in the gut contents of green turtles in Shoalwater Bay, Queensland, Australia. Fifty-one percent of green turtles sampled had the cyanobacteria present (during a 2002 bloom), however, it only constituted 2% of stomach contents (Arthur et al., 2006) and did not relate to lessened body condition.

Non-native (alien) algae have been observed in green turtle diets in Hawaii, such as *Acanthophora spicifera* and *Hypnea musciformis* (Russell and Balazs, 1994), and with increasing abundance are becoming the dominant diet item (Russell and Balazs, 1994; Arthur and Balazs 2008). Green turtles in Australia were found to selectively consume algae that were low in fiber and high in nitrogen (Brand-Gardner et al., 1999) typical of invasive species. This preference for high nitrogen-invasive species may be of consequence for green turtle populations as Van Houtan et al. (2010) recently linked the consumption of invasive algae (high in nitrogen and arginine) to fibropapilloma tumors.

### 9.2.3 LOGGERHEAD TURTLE, *CARETTA CARETTA*

Loggerhead turtles follow an oceanic–neritic developmental pattern (Bolten, 2003) departing nesting beaches and entraining in prevailing oceanic currents as float-and-wait foragers (Witherington, 2002) until recruiting to neritic foraging grounds at >25 cm CL (Bjorndal et al., 2000; Gardner and Nichols, 2001; Lazar et al., 2008). Although recent evidence suggests a prolonged oceanic stage or perhaps a plasticity in shuttling between habitats (Parker et al., 2005) persisting even through the commencement of breeding (Hatase et al., 2002, 2006, 2010). During the oceanic stage (generally lasting 7–10 years; Bjorndal et al., 2000) loggerhead post-hatchlings and juveniles are primarily carnivorous (Bjorndal, 1997). Post-hatchlings (4–8 cm CL) leaving the beaches of southeast Florida swim offshore to the Gulf Stream where they feed on a diet of Hydrozoa, Scyphozoa (jellyfish), Bryozoa, Gastropoda, Polychaeta, Maxillopoda (copepods), Malacostraca (crabs, shrimp), Insecta, Actinopterygii (fish), as well as feeding on bits of floating seagrasses, algae, and filamentous cyanobacteria (Witherington, 2002). Loggerhead post-hatchlings (5–11 cm CL) of the southwest Pacific feed on similar items (unidentified Cnidaria, Gastropoda, Malacostraca, and seagrasses; Boyle and Limpus, 2008). Frick et al. (2009) examined the gut content of larger oceanic loggerheads (9–56 cm CL) near the Azores; the turtles had remnants of Hydrozoa, Scyphozoa, Thaliacea, Actinopterygii eggs, Cephalopoda, Gastropoda, Malacostraca, Maxillopoda, and Insecta. Oceanic loggerheads (14–74 cm CL) of the central North Pacific feed solely on epipelagic items largely including Gastropoda (sea snails), Hydrozoa, Maxillopoda (goose barnacles), Malacostraca (hitchhiker crab, *Planes* spp.), and Thaliacea (tunicates) as well as bits of Cephalopoda, Actinopterygii, Actinopterygii eggs, Polychaeta, and algae (Parker et al., 2005). The diet of oceanic stage loggerheads is purely pelagic/epipelagic unlike the mixed diet of neritic stage loggerheads (Bolten, 2003). Recruitment to the neritic, however, may not come with a concomitant change in diet. There is a transitional stage where new recruits continue to feed on epipelagic organisms while gradually including benthic organisms in their diet (Bolten, 2003; Casale et al., 2008). The length that loggerheads recruit to neritic areas also varies geographically 25 to >60 cm CL (Bjorndal et al., 2000; Seminoff et al., 2004; Boyle and Limpus, 2008; Casale et al., 2008; Wallace et al., 2009).

Peckham et al. (2011) studied the movements, habitat selection, and diet of neritic loggerheads (Baja California Peninsula) of the same length (50–90 cm CL) as oceanic loggerheads of the central North Pacific (Parker et al., 2005). The diets differed considerably between the study groups with neritic loggerheads feeding mostly on fish (Actinopterygii) and crabs (Malacostraca), mostly the pelagic red crab (*Pleuroncodes planipes*). The average energy density of diet items was 60% greater (on dry mass basis) for the neritic loggerheads (Peckham et al., 2011) than for the diet of oceanic loggerheads (Parker et al., 2005). Peckham et al. (2011) suggested that loggerheads that recruit to neritic waters benefit from greater productivity and energy-dense diet items; however, neritic foraging comes with a cost of greater mortality rates. The energy densities were compared on a dry mass basis, however, and this may underestimate the handling and volume of food needed to meet energy demands. Wet mass energy densities and handling costs associated with capture and/ or dealing with exoskeletons may offset lessening the energy density disparity found between the habitats or increase them. The increased fish in the diet was surmised to be from fishery discard of the local gillnet fishery (Peckham et al., 2011). The consumption of fish from commercial fisheries discard and gear is perhaps a regionally learned response observed in Baja California (Peckham et al., 2011), the Mediterranean (Tomas et al., 2001), and in the western North Atlantic (Seney and Musick, 2007). Across the Pacific to the nesting beaches and associated foraging grounds off Japan, Hatase et al. (2002) observed a relationship between adult body length and diet (through SIA of deposited eggs). Larger loggerheads were associated with a neritic-benthic diet from the East China Sea whereas smaller loggerheads had diet signatures associated with oceanic-pelagic organisms of the Kuroshio Current; the researchers suggested the relationship reflects recruitment of immature loggerheads where early recruitment to neritic environments is rewarded by faster growth rates due to nutrient-rich prey vs. the loggerheads that remain in the oceanic habitats (Hatase et al., 2002). This study corroborates the observations of Peckham et al. (2011) that neritic recruitment exposes the loggerheads to a greater abundance of energy-dense food.

The diet records of immature and adult loggerheads of eastern and western Australia include over 100 taxa, but the principle diet consists of Gastropoda, Bivalvia, and Malacostraca (portunid and hermit crabs). Of lesser abundance are Scyphozoa, Echinoidea (sea urchins), Anthozoa (anemones), Holothuroidea (sea cucumbers), and Actinopterygii (Limpus, 2008a).

In the western North Atlantic along the U.S. east coast loggerhead diet is dominated by crabs (Malacostraca). Seney and Musick (2007) analyzed the gut contents of 297 loggerheads (33–99 cm CL) from 1983 to 2002, with the gut contents consisting of Merostomata (horseshoe crabs), Malacostraca (e.g., blue crab, rock crab, spider crab), Maxillopoda (e.g., Acorn barnacle, crab barnacle), Gastropoda (e.g., whelks, mud snail), Bivalvia (e.g., razor clam, blue mussel), Actinopterygii, Chondrichthyes (e.g., clearnose skate, Atlantic sharpnose shark) as well as seagrass, algae, Cephalopoda, Bryozoa, Polychaeta, Porifera (sponges), and other unidentified gelatinous organisms. Diets were dominated by horseshoe crab and blue crabs in the 1980s and early1990s, but after 1993 diets were dominated by fish species, a trend perhaps the result of declining crab populations causing loggerheads to forage in nets or on discarded fishery bycatch (Seney and Musick, 2007). Farther south (North Carolina coast) SIA of 42–102 cm CL neritic loggerheads suggested a preference for Malacostraca (blue crabs), Gastropoda (whelks), and Scyphozoa (cannonball jellyfish) and that the loggerheads rarely supplemented their diet with fish discards or fish caught in gear of commercial fisheries. McClellan et al. (2010) revealed through SIA that loggerheads (53–82 cm CL) shuttle between foraging habitats; loggerheads that had recruited to neritic environments maintained oceanic diet signatures during overwintering periods and juvenile loggerheads first recruiting to neritic habitats continued to feed on oceanic-pelagic organisms.

Interestingly, in the western South Atlantic Carranza et al. (2011) reported the diet of neritic loggerheads (51–112 cm CL) in the Rio de la Plata Estuary (Uruguay) to consist solely of the invasive whelk (*Rapana venosa*, Gastropoda). The five loggerheads examined had on average 136 opercula in their stomach.

In the past decade studies have revealed a range of strategies by loggerheads that settle in the Mediterranean, including strict oceanic vs. neritic strategies as well as shuttling between strategies. Lazar et al. (2008) examined the gut contents of early recruitment loggerheads (25–39 cm CL) of the Adriatic Sea and found that the diet was composed of benthic and epipelagic organisms. The benthic organisms consisted of Foraminifera, Porifera, Echinoidea, Ophiuroidea, Malacostraca, Brachiopoda, Anthozoa, Bryozoa, Gastropoda, Polychaeta, seagrasses, and algae while the epipelagic organisms consisted of Cephalopoda, Actinopterygii, and Insecta (Lazar et al., 2008). The most abundant diet items by rank were Anthozoa (anemones), Malacostraca, and Gastropoda (Lazar et al., 2008). Analysis of the gut content and feces of larger neritic loggerheads (25–80 cm CL) in the central Mediterranean Sea indicate an abundance of benthic organisms (Demospongiae, Turbellaria, Sipunculida, Bivalvia, Gastropoda, Cephalopoda, Polychaeta, Malacostraca, Bryozoa, Echinoidea, Holothuroidea, Ophiuroidea, Ascidiacea, Chondrichthyes, Osteichthyes, seagrasses, and algae) with the most abundant organisms encompassing Malacostraca, Gastropoda, and Echinoidea (Casale et al., 2008). Neritic loggerheads (34–69 cm CL) in the western Mediterranean feed on Actinopterygii, Ascidiacea, Thaliacea, Malacostraca, Polychaeta, Echinoidea, Ophiuroidea, Anthozoa, Porifera, Cephalopoda, Gastropoda, Insecta, seagrasses, and algae (Tomas et al., 2001). The prevalence of Actinopterygii (fish) in the gut contents of the western Mediterranean study, however, is probably due to consumption while captured or to bycatch waste from the fisheries (Tomas et al., 2001). In a study of similar length loggerheads of the Balearic Archipelago (Revelles et al., 2007a) SIA suggested the diet composition was primarily epipelagic organisms (Cephalopoda [squid] and Scyphozoa [jellyfish]); furthermore, the stable isotope values did not differ from shelf and pelagic (assumed oceanic) loggerheads. Gut content analysis revealed the loggerheads also fed on Thaliacea, Malacostraca, Maxillopoda, and Actinopterygii—all of which were epipelagic. Revelles et al. (2007b) suggested that longline bait was a likely source of some of the prey species. In contrast, loggerheads of northwestern Mediterranean (off mainland Spain) had distinct isotopic values that differentiated between oceanic and neritic cohorts (CL data not provided), suggesting that gelatinous zooplankton is a primary diet item for oceanic loggerheads but not neritic turtles (Cardona et al. 2012).

Studies on the diet of immature and adult loggerheads have broadened the classic oceanic–neritic developmental pattern, expressing more plasticity in the use of habitats (Hatase et al., 2002; Revelles et al., 2007a; McClellan et al., 2010) and the ability to adapt to changing prey landscapes (Seney and Musick, 2007), which at times includes invasive species (Carranza et al., 2011). However, in all studies loggerheads were carnivorous from post-hatchling through adulthood with the main differences between populations coming from the proportion of benthic or pelagic fauna in the diet (see Table 9.1).

## 9.2.4   Hawksbill Turtle, *Eretmochelys imbricata*

Hawksbill turtles are thought to follow the oceanic–neritic developmental pattern as observed in loggerheads and green turtles, although the evidence is not as strong (Bolten, 2003). Early reports of post-hatchling hawksbills suggest they associate with sargassum rafts in oceanic habitats (Carr, 1987). Bjorndal (1997) summarized the gray literature noting that gut contents from stranded juvenile hawksbills (14–21 cm CL) in Florida (United States) had species of algae, Maxillopoda, Malacostraca, Thaliacea, and Actinopterygii eggs. This finding is similar to what Salmon et al. (2004) observed in post-hatchling green turtles feeding on in the Gulf Stream off the Florida coast. Limpus (2009) reports that the oceanic stage lasts for nearly 5 years for Australian hawksbills and that the post-hatchlings/juveniles feed on macroplankton.

Hawksbills recruit to neritic habitats at >20 cm CL in the Atlantic (Meylan, 1988) and at >30 cm CL in the Pacific (Limpus, 1992, 2009). Once recruiting to neritic habitats hawksbills have been observed foraging over coral reefs and rocky substrate, seagrass pastures, and in mangrove-fringed bays (Bjorndal and Bolten, 1988; Bjorndal, 1997). The main diet of Caribbean hawksbills after recruitment to neritic habitats is sponges (Porifera) with *Chondrilla nucula* the most abundant (Meylan, 1988). The diet of the hawksbill in the Caribbean and wider Atlantic region has been thoroughly reviewed (Meylan, 1988; Bjorndal, 1997), and Bjorndal (2003) further provides a case study on the impact of hawksbill spongivory in the Caribbean. Few accounts of the hawksbill diet have been given since. Other noted diet items (Table 9.1) include Actinopterygii, Gastropoda, Polychaeta, Hydrozoa, Bryozoa, Anthozoa, Malacostraca, Echinoidea (Carr and Stancyk, 1975; Den Hartog, 1980; Bjorndal, 1997). Blumenthal et al. (2009) observed recruitment to neritic habitats of the Cayman Islands at greater than 20 cm CL with diet components comprising benthic sponges (Porifera) and epipelagic jellyfish (Scyphozoa). A study of neritic hawksbills (19–50 cm CL) of the Dominican Republic revealed that Cnidarian corals (Corallimorpharian) dominate the diet of hawksbills in certain habitats (81% of diet volume) while in other habitats the sponges were still the mainstay (59% by volume) and, to a much lesser extent, these hawksbills fed on algae (Leon and Bjorndal, 2002). The preference or selectivity for Corallimorpharian (*Ricordea florida*) was also noted by Rincon-Diaz et al. (2011) where hawksbills of Puerto Rico selected *R. florida* when it was of low abundance in the environment. Hawksbill turtles (20–50 cm CL) in the Caribbean waters off the coast of Honduras fed primarily on sponges (e.g., *Melophlus ruber*) comprising 59% of ingesta, smaller hawksbills, however, consumed mostly algae and this was suggested to reflect new recruits that were feeding mostly in the epipelagic waters (Berube et al., 2012). Bjorndal and Bolten (2010) postulate that seagrass pastures may become important habitat for hawksbills as coral reef ecosystem health declines. A 30-year study of green and hawksbill turtles suggests that seagrass pastures can support healthy productive populations of hawksbills and that a sixfold increase in green turtle abundance did not affect the health or productivity of the associated hawksbills (Bjorndal and Bolten, 2010).

Neritic hawksbills (26–83 cm CL) of northern and western Australia (as well as in the Indian Ocean, Cocos Islands) feed in tidal and subtidal coral and rocky reef habitats (Limpus, 1992, 2009). The diet of Australian neritic hawksbills consists of algae, seagrass, and sponges (Porifera). Western Australian hawksbills rely less on spongivory than their Caribbean counterparts. Their diet is primarily algae (76%) with only 20% sponge consumption by volume (Limpus, 2009). Parker et al. (2009) reported the diet of Hawaiian hawksbills to consist of Porifera (sponge) and algae. In the eastern Pacific, hawksbills have also been reported to consume sponges (*Haliclona* sp.; Seminoff et al., 2003). A recent study by Gaos et al. (2012), although not reporting diet per se, indicated that hawksbills forage in mangrove estuaries, where sponges are presumably less frequent than in coral reef ecosystems.

Obura et al. (2010) observed that adult neritic hawksbills of the Aldabra Atoll feed on hard corals (Anthozoa) such as the bubble coral (*Physogyra lichtensteinii*). Reports of the same phenomena came from the Seychelles, Thailand, Madagascar, and the Red Sea. Obura et al. (2010) suggested that consumption of hard corals is widespread throughout the Indian Ocean and takes place in the Caribbean and Pacific but to a much lesser extent. Sponge health and abundance are closely related to reef health, and as coral bleaching and coral health continues to decline we may see a paradigm shift from hawksbills having a diet of sponge and corals (Leon and Bjorndal, 2002; Obura et al., 2010; Rincon-Diaz et al., 2011), to one including alternatives such as algae (Limpus, 2009), and reliance on seagrass pastures (Bjorndal and Bolten, 2010; J. Mortimer, unpublished data).

### 9.2.5  Olive Ridley Turtle, *Lepidochelys olivacea*

Olive ridley turtles develop in oceanic waters and recruitment to neritic waters as juveniles/adults may be regional, with at least some individuals maintaining an oceanic existence (Bolten, 2003). Australian and western Atlantic populations have been widely reported to recruit to neritic habitats (Pritchard, 1976; Limpus, 2008b) whereas the eastern Pacific olive ridley may maintain an oceanic life-history pattern.

To our knowledge, there are no data on post-hatchling diet of olive ridley turtles. Oceanic olive ridleys in the central North Pacific prefer warmer water (23–28 C SST) and spend 40% of their time diving below 40 m (Polovina et al., 2004). Gut content analysis from olive ridleys of the central North Pacific suggests they feed on Thaliacea (pyrosomes and salps), Gastropoda (sea snails), and Actinopterygii (e.g., cowfish) (unpublished data, Pacific Islands Fisheries Science Center, NMFS).

In Australia immature and adult olive ridleys recruit to shallow benthic–neritic habitats, feeding mainly on Gastropoda and Malacostraca (Conway, 1994; Limpus, 2008b). Reports of stomach contents (Thaliacea, Gastropoda, Malacostraca, Bryozoa, Sipunculidea, Ascidiacea [sea squirts], Actinopterygii, and algae) of adult olive ridleys along the Mexican coast are of females near nesting beaches and not reflective of strict neritic foraging habitat (see Bjorndal, 1997, for review).

There remains a paucity of data on olive ridley life-history patterns, foraging habitat, and diet; however, all reports so far suggest an omnivorous diet (Table 9.1) throughout ontogeny and plasticity in use of oceanic and neritic habitats.

### 9.2.6 KEMP'S RIDLEY TURTLE, *LEPIDOCHELYS KEMPII*

Kemp's ridley turtles have an oceanic–neritic developmental pattern (Collard and Ogren, 1990; Bolten, 2003) recruiting to benthic–neritic habitats at 20+ cm CL (Snover et al., 2007; Seney and Landry, 2011). Post-hatchling Kemp's ridleys (~10 cm CL) that stranded in the Gulf of Mexico had Gastropoda (sea snails), Malacostraca (crabs), and algae (e.g., *Sargassum* sp.) in their gut (Shaver, 1991). No other reports of oceanic post-hatchlings/juveniles are available. Immature Kemp's ridleys recruiting to benthic habitats can be found from the western North Atlantic (e.g., New York state waters) to the Gulf of Mexico (Shaver, 1991; Burke et al., 1994). Benthic–neritic Kemp's ridleys (23–69 cm CL) of the western Atlantic and Gulf of Mexico continental shelf feed primarily on Malacostraca (e.g., blue crabs), Bivalvia (e.g., blue mussels), Gastropoda (e.g., mud snails, whelks), seagrass, and algae, and to a lesser extent Merostomata (i.e., horseshoe crab), Ctenophora (Nuda), and Actinopterygii (Shaver, 1991; Burke et al., 1993, 1994; Creech and Allman, 1998; Seney and Musick, 2005; Witzell and Schmid, 2005). Witzell and Schmid (2005) observed that immature Kemp's ridleys that recruited to benthic–neritic waters of southwestern Florida fed primarily on Ascidiacea (e.g., tunicates, *Molgula occidentalis*). Adult Kemp's ridleys are found primarily in continental shelf waters of the Gulf of Mexico and feed on Malacostraca (i.e., crabs) (Frick and Mason, 1998; Morreale et al., 2007). The aforementioned diet analyses and movement data suggest that Kemp's ridleys feed on epipelagic organisms during their oceanic period and once recruiting to benthic–neritic habitats, as juveniles, their diet is dominated by bottom-dwelling fauna throughout maturation and adulthood (see Table 9.1).

### 9.2.7 FLATBACK TURTLE, *NATATOR DEPRESSUS*

The flatback turtle is believed to have a completely neritic life history (Walker and Parmenter, 1990; Walker, 1991; Bolten, 2003), feeding within the waters of the Australian continental shelf (Limpus, 2007). Bolten (2003) suggested that juvenile sea turtles with a neritic developmental pattern probably feed at the surface and within the upper water column until buoyancy control is developed and dive depth-duration increased. The stomach contents of juvenile and adult flatbacks had remnants of soft-bodied invertebrates (e.g., Gastropoda [sea snails], Hydrozoa [siphonophore], Scyphozoa [jellyfish], Bryozoa, Anthozoa [sea pens and soft corals]) (Walker and Parmenter, 1990; Walker, 1991). The foraging ecology of flatbacks remains poorly studied (Table 9.1), a biological review by Limpus (2007) summarized that post-hatchling flatbacks feed on macrozooplankton and that immature and adult flatbacks are carnivorous feeding primarily on soft-bodied invertebrates, adding holothurians (e.g., sea cucumbers) to the list of flatback diet items.

## 9.3 FEEDING PHYSIOLOGY

Paramount to sustaining life is the attainment, storage, and use of energy. Sea turtles obtain chemical energy through ingesting prey, that is, eating gelatinous zooplankton, crabs, mollusks, algae, plants, etc. (see Section 9.2). Once consumed, chemical energy is either used by the animal to fuel production and respiration or lost as heat and metabolic waste (Speakman, 1997). The terms used to describe this process (e.g., digestion, absorption, and assimilation) are often used broadly or incorrectly. For purposes of this chapter we define their use. In multicellular organisms with well-developed digestive tracts (e.g., the canal from mouth to cloaca opening of sea turtles), extracellular digestion takes place. Digestion is the process of breaking down large complex molecules with the aid of digestive enzymes secreted into the gut (Schmidt-Knielsen, 1997), making the chemical energy available to the animal. The energy is then absorbed through the digestive tract lining (absorption). The non-absorbed energy travels through the digestive tract and is lost as feces or gas production (e.g., methane). Of the energy absorbed into the body, some will be lost through excretion as urine (e.g., urea, uric acid, and ammonia—nitrogenous wastes), the absorbed energy minus excretion is the assimilated energy. Assimilated energy is stored by the animal (e.g., glycogen and lipid stores), used for production (e.g., growth and biosynthesis), used to perform work external to the animal (e.g., locomotion), or lost to the environment as heat. How well the animal performs each of these processes (i.e., digestion, absorption, and assimilation) is termed efficiency and defined as the amount of ingested chemical energy actually made available to the organism. All of these processes combined lead to an increase in postprandial metabolic rate (MR), also known as SDA. While there have been many field based studies of the diet (see Section 9.2) of sea turtles, few studies have been undertaken to understand energy acquisition, and the costs associated with freeing chemical energy from diet items for use in energy consuming activities (e.g., maintenance, growth, locomotion, homeostasis, and reproduction). Nevertheless, knowledge about how physiological and environmental factors influence energy acquisition is critical, as these are among the most fundamental aspects of sea turtle ecology.

When describing the concepts for each section below, we also provide examples of the seminal studies that have paved the way for our understanding of sea turtle feeding physiology. We start with describing the current knowledge on intake rate as handling and consumption are the initial steps in the attainment of chemical energy. We then discuss the digestion and assimilation of food energy and the trade-offs between intake, passage rate, and efficiency. We finally discuss the big picture of the metabolic costs associated with the processes of feeding physiology (i.e., SDA).

### 9.3.1 INTAKE RATE

Of all the aspects of feeding physiology the study of food consumption rates (i.e., intake rates) have expanded the most in the past decade. Since the landmark studies of daily intake rates of 0.24%–0.33% of body mass for green turtles grazing on seagrass (Bjorndal, 1980), observation of a leatherback feeding in the North Atlantic Ocean consuming an estimated 200 kg day$^{-1}$ (Duron, 1978), and post-hatchling leatherbacks needing to consume 100% of body mass per day of jellyfish to meet energy demands (Lutcavage and Lutz, 1986), new tools, and analyses have expanded our understanding. Several researchers have used energy budgets to estimate daily consumption rates, for instance, Jones et al. (2004) used a 36/64 resting vs. active diel activity model with measurements of oxygen consumption at rest and while active to estimate intake rates of 3.3% of body mass daily for eastern Pacific green turtles. Wallace et al. (2006a) estimated reproductive energy budgets for nesting female leatherbacks, suggesting that eastern Pacific and North Atlantic leatherbacks require 70–90 kg of jellyfish per day and up to 87–113 kg depending on nesting remigration intervals. Jones et al. (2012) measured gross food conversion of juvenile leatherbacks and estimated that adults need to consume prey quantities equivalent to 20%–30% of their body mass (60–108 kg of jellyfish) to meet daily energy requirements.

Biologging devices have allowed indirect and direct measurements of feeding bouts. Southwood et al. (2005) and Casey et al. (2010) measured possible feeding through stomach temperature pills that signal intake of cold prey or water. Casey et al. (2010) recorded limited daytime foraging events for internesting leatherbacks off St. Croix suggesting leatherbacks fed opportunistically, thus energy reserves required prior to nesting migrations are critical for success. Hochscheid et al. (2005) used Hall effect transducers and magnets attached to the beaks of loggerheads (intermandibular angle sensor [IMASEN]) to record mouth openings and general buccal movements. Validation of the sensor with video of foraging allowed for differentiation of consumption of hard or soft prey and moving and sessile prey (Hochscheid et al., 2005). Fossette et al. (2008) used IMASENs on gravid female leatherbacks nesting on French Guiana, the leatherbacks expressed various mouth-opening patterns for different dive patterns, and the authors interpreted that the turtles forage opportunistically on occasional prey, with possibly 17% of the dives hosting successful foraging bouts.

The use of turtle-borne video cameras in recent years has allowed direct observation of sea turtles foraging, many times negating previous assumptions of diet and foraging selectivity (Heithaus et al., 2002; Seminoff et al., 2006a; Arthur et al., 2007; see Section 9.2). Heaslip et al. (2012) used turtle-borne cameras on adult leatherbacks of the temperate North Atlantic recording ingestion rates of 330–840 kg day$^{-1}$ of large Lion's mane jellyfish (*Cyanea* sp.) and moon jellies (*Aurelia aurita*) extrapolated from video clips of approximately 3.6 h or less. Fossette et al. (2011) used a hand-held submersible camera while diving to collect footage of leatherbacks feeding, analysis of the 39 s clip suggested that leatherbacks could consume up to 150 kg day$^{-1}$ of small *Linuche unguiculata* jellyfish provided they ate continuously at the recorded rate of one jelly every 2.3 s. Even with the expansion in new technologies there still remains a fundamental lack of information on consumption rates across the turtle species. The handling and consumption of food items is the first and crucial step in the daily acquisition of energy requirements.

## 9.3.2 DIGESTIBILITY AND ASSIMILATION

Bjorndal (1997) thoroughly reviewed the studies of digestibility and assimilation efficiency of green turtles feeding on natural diets of seagrasses and sponges (in the wild), and artificial diets of trout chow high in protein (in captivity). Many authors use assimilation efficiency and digestibility interchangeably, however, as urine is generally not accounted for in the studies then digestibility (the proportion of a feed or diet which can be digested by the normal animal of the subject species) or digestive efficiency is more appropriate (and what we will use here). The energy lost as nitrogenous wastes (urine) from a meal may be as high as that lost in feces (Merker and Nagy, 1984); therefore, the energy available to the turtles may be considerably less than reported in digestibility studies. For instance, Merker and Nagy (1984) found striped plateau lizards (*Sceloporus virgatus*) fed crickets and mealworms have digestive efficiencies of $0.82 \pm 0.01$ and assimilation efficiencies of $0.63 \pm 0.01$, thus only 63% of the energy in the meal was available to the lizards for respiration and production. Wild green turtles (8–66 kg) were found to have energy digestibilities of 22%–71% and protein digestibilities of 14%–57% for a diet of seagrass (Bjorndal, 1980). For green turtles (7–68 kg) feeding on the sponge *C. nucula* energy and protein digestibility ranged from $40\% \pm 15\%$ to $43\% \pm 9\%$ and $52\% \pm 11\%$ to $55\% \pm 7\%$, respectively (Bjorndal, 1990). These studies on wild turtles feeding on natural diets are in stark contrast with captive turtles fed pelleted (trout chow) diets. In a study of captive green turtles (4–22 kg) fed a pelleted diet to satiation (1.2% of body mass), Wood and Wood (1981) recorded protein digestive efficiencies of 82%–90% at 30°C. For juvenile green turtles (<0.9 kg) fed a high protein (40%–50%) diet of trout chow, at 25°C, the digestive efficiencies were $76\% \pm 6\%$ for energy and $86\% \pm 6\%$ for protein (Hadjichristophorou and Grove, 1983). These data corroborate with Davenport and Scott (1993) who recorded digestive efficiencies of juvenile green turtles (<0.8 kg), fed a similar diet of trout chow at 25°C, of $68.0\% \pm 5.9\%$, $89.3\% \pm 1.9\%$, and $60.9\% \pm 7.6\%$ for energy, protein, and lipids, respectively. Furthermore, Davenport and Scott (1993) observed a positive correlation between digestive efficiency and growth rate but not with MR, suggesting that the efficiency of physiological processes of digestion, absorption, and assimilation plays a greater role in growth rate, development, and stage-specific residency times than do consumption rate and MR. Turtles with low digestive efficiencies were associated with increased consumption rates (Davenport and Scott, 1993), presumably to offset poor digestion/assimilation to meet daily energy requirements.

In the only study since Bjorndal's (1997) review, Amorocho and Reina (2008) measured digestibilities of eastern Pacific green turtles ($32.3 \pm 6.7$ kg) fed protein (fish), plant (mixed leaves), and mixed diets at 28°C in an in-water enclosure. The digestibilities for the protein, plant, and mixed diets were 85%–91%, 67%, and 77%, respectively (Amorocho and Reina, 2008). While the plant digestibility was similar to that found by Bjorndal (1980), the protein digestibility corroborated that of captive turtles fed trout chow. The lack of digestion and assimilation studies of the other six species of carnivorous sea turtle is troubling. Data on digestive and assimilation efficiency are important for studies of physiological ecology and energetic budgeting that seek to understand energy acquisition and subsequent investment in various life-history demands (e.g., reproduction, growth, physiological regulation, etc.) (Wallace and Jones, 2008). Several authors have assumed assimilation efficiencies of 80% for studies of reproductive budgets and individual and population-level consumption rates of leatherbacks (Wallace et al., 2006a; Jones et al., 2012), and remigration intervals of loggerheads based on energy budgets (Hatase and Tsukamoto, 2008). The assumption of 80% assimilation efficiency was based on measurements of freshwater turtles (*Trachemys scripta*) fed a diet of similar protein content to that of jellyfish and the rapid rate of digestion of gelatinous organisms by fishes (Avery et al., 1993; Malej et al., 1993; Arai et al., 2003).

### 9.3.3 PASSAGE RATE

The passage rate of food along the digestive tract and digestive efficiency are competing interests for the assimilation of energy. Animals need to move food along the digestive tract to make room for continued consumption, although the efficiency of digestive and absorptive processes is reduced at rapid passage rates (Karasov and Levy, 1990). Therefore, we might assume passage rates may vary significantly among sea turtle species that specialize in carnivorous (high digestive efficiency) and herbivorous (low digestive efficiency) diets. However, like the other processes of feeding physiology, these processes are poorly studied in sea turtles. The time a given meal spends in the alimentary canal of an animal has been called by many names, e.g., total gut clearance time, gastric emptying time (Hadjichristophorou and Grove, 1983; Davenport and Oxford, 1984), passage rate of digesta (Bjorndal, 1997), digesta retention time (Brand et al., 1999), total digestive transit time (Di Bello et al., 2006), ingesta passage time (Valente et al., 2008), and intake passage time (Amorocho and Reina, 2008). Furthermore, the names have slightly different meanings, for example, gastric emptying time is simply clearance of the stomach, whereas total digestive transit time is clearance of the entire digestive tract. The passage of food along the gut in 500–900 g green turtles, fed a satiation diet (2.5% of body mass) of trout chow, took $4.6 \pm 0.5$ days (gastric emptying) and $7.3 \pm 0.7$ days (total gut clearance) at 25°C (Hadjichristophorou and Grove, 1983) whereas 60 g post-hatchling green turtles fed a similar diet at 25°C had passage rates of 16 days for turtles continuously fed a daily satiation meal (2.6% of body mass) and 23 days for turtles that were fasted after feeding (Davenport and Oxford, 1984). Davenport and Oxford (1984) reported that feces production was halted after 16 days of starvation; however, in larger 22 kg green turtles feces production continued through 25 days of fasting in 24°C water temperature (T.T. Jones, personal observation; study Jones et al., 2009). Brand et al. (1999) observed that plastic markers were still in the gut of juvenile green turtles (50–55 cm CCL) 5–9 days after release into the wild; based on the passage of markers within the gut at time of recapture they estimated that the passage rate was 6.5–13.5 days (for 50% of markers to pass; water temperatures upward of 28°C). Using biodegradable colored ribbon, McDermott et al. (2006) measured mean initial passage time of $15 \pm 1$ day (range 5 to >35 days) for $42 \pm 12$ kg eastern Pacific green turtles, the turtles were allowed to feed ad libitum consuming 0.7%–1.75% of their body mass of red algae (*G. lemaneiformis*) daily (water temperature 27°C ± 2°C). In an experiment using an in-water enclosure to keep green turtles near shore, Amorocho and Reina (2008) observed passage rates of $23.3 \pm 6.6$ days (73% of markers had passed) for $32.3 \pm 6.7$ kg juveniles fed plastic markers along with various diets (water temperature 28°C). Birse and Davenport (1987) measured passage rates of 4–18 days for juvenile loggerheads (1–2 kg) fed a daily satiation ration of 3.7% of body mass of trout chow. Valente et al. (2008) observed the passage of markers in loggerheads (4–22 kg) fed a daily fish ration of 1.5%–2.5% of body mass. The time to first elimination was $9 \pm 3$ days and the time until 50% and 85% of the markers had passed was $12 \pm 5$ days and $13 \pm 5$ days, respectively. The Valente et al. (2008) study took place in water temperature ranging from 16°C to 24°C; however, there was no significant difference in passage rates with temperature whereas Birse and Davenport (1987) observed a 60% increase in passage rate from 20°C to 30°C. Di Bello et al. (2006) measured gastric emptying times of 1.4–11.0 days and total clearance times of 8–40 days;

however, the study was of post-surgery loggerheads. The data on intake passage rates are limited and the myriad experimental designs do not allow for clear determinations of diet types, species, and temperature effects on passage rate through the gut. As with all aspects of feeding physiology, there remain large gaps in our understanding of interspecific differences, the role of diet habituation, seasonal temperature fluctuations, and body mass or age on passage rate of food items along the gut or how this affects the fitness of the turtles.

### 9.3.4 Specific Dynamic Action

When animals consume food there is an accompanying increase in their MR. The postprandial increase in MR has been referred to as SDA and more recently as heat increment of feeding (HIF). Regardless of the name, the process refers to the increase in MR associated with consumption, digestion, absorption, and assimilation of food items (Secor, 2009). Only one study (Davenport et al., 1982) has directly measured SDA in sea turtles (i.e., the accumulated energy expended above standard metabolic rate [SMR]). Davenport et al. (1982) report SDA of 75–153 kJ for juvenile green turtles (0.8–2.2 kg) fed satiation meals (1.8%–2.0% of body mass) of trout chow (43.6% protein and 4.5% fiber). The SDA coefficient (SDA divided by the energy in the meal) ranged from 15% to 25% (Davenport et al., 1982). The SDA coefficient for juvenile green turtles is similar to that of freshwater turtles and tortoises (17.9% ± 1.3%) fed mealworms, kale, and beef (see Secor, 2009, for review). Several studies report the factorial scope of SDA (i.e., the peak postprandial MR divided by the SMR of the turtle); these data can also be gleaned from studies of MR of fasted and fed turtles. Davenport et al. (1982) reported a doubling of MR (factorial scope of 2) for post-absorptive (fasted) turtles fed a satiation meal; the peak in postprandial MR was 2 h post-consumption and the heightened MR lasted for 5 days. Kowalski (2006) measured a postprandial scope of 1.6–1.8 for green and loggerhead post-hatchlings. Jones et al. (2009) measured a scope of 2.1 for juvenile green turtles (22.4 ± 3.1 kg) fed a diet (trout chow, gelatin, vitamins [41% protein, 4% fiber]) 1%–2% of their body mass after 25 days of fasting. Postprandial MR scope for freshwater turtles and tortoises averages 1.9 ± 0.1 and the increased MR typically lasts for 3 days (Secor, 2009). While a doubling of MR is considerable, boa constrictors have postprandial scopes up to 19 times SMR (Secor and Diamond, 2000) and 100% of their scope of activity. The scope of activity for sea turtles is up to four times SMR (Wallace and Jones, 2008), therefore SDA can cause increases in MR of 50% of the turtles maximum MR. SDA accounts for all the energetic costs associated with consumption to assimilation including numerous pre-absorptive, absorptive, and post-absorptive physiological processes (McCue, 2006). If the post-absorptive processes associated with SDA do not include biosynthesis (somatic growth), then the higher the SDA the less energy available to the organism for growth and reproduction (Secor, 2009). Therefore, the more efficient turtles are at consumption, digestion, absorption, and assimilation the lower the cost of SDA and the greater amount of energy available to the turtle for production (e.g., growth, biosynthesis), storage (e.g., fat), and work (e.g., locomotion, migration).

As with other emerging areas of scientific inquiry, our knowledge about these and other aspects of feeding physiology in sea turtles will benefit from continued research. So far, however, the majority of studies have focused on green turtles and we know very little about the feeding physiology of carnivorous sea turtles. We encourage both field studies and controlled laboratory experiments on all species, particularly species other than green turtles. These multiple lines of inquiry will hopefully shed light on the biological and ecological drivers that result in the varying behaviors and physiologies of each species. Basic research on intake rates, energy assimilation, and associated costs elucidate how sea turtles meet the energy demands of their various developmental patterns and ever changing environments. These data also allow inferences on distribution and demographic patterns of sea turtles based on current and future resource availability. Research on diet components and acquisition and use of chemical energy can be greatly expanded upon by the use of emerging technologies such as stable isotope, fatty acid, and trace element analysis.

These technologies may reveal differential use of stored energy in diet items (e.g., fat, proteins), as well as the spatial and temporal timing of feeding events and diet composition. Thus by melding feeding physiology with emerging technologies, we can better understand the dynamic and often complex nature of the foraging ecology of sea turtles.

## 9.4   STABLE ISOTOPE ANALYSIS IN SEA TURTLES

Among the various new tools that are available to sea turtle researchers, perhaps no application has blossomed more than SIA (Table 9.2). Stable isotopes are various forms of the same element (e.g., carbon, nitrogen, oxygen, sulfur) that differ in the numbers of neutrons in their atomic nucleus and therefore have minute differences in atomic mass (recall that the mass of an atom is constituted by total mass of protons, neutrons, and electrons in a single atom). The slight differences in atomic mass ultimately translate into variability in the way isotope ratios change in the environment, body tissues, and metabolic pathways (e.g., digestion, excretion) of animals (Miniwaga and Wada, 1984; Peterson and Fry, 1987) or in nutrient flow within ecosystems (Rubenstein and Hobson, 2004; Table 9.2). The predictable ways that isotope ratios change in body tissues as a result of these processes allow researchers to gain insights about a variety of biological aspects such as feeding biology, ontogeny, and habitat use (Miniwaga and Wada, 1984; Peterson and Fry, 1987). Because digestion and excretion processes are major influences on stable isotope ratios—especially of stable nitrogen and carbon—insights gained from SIA are often associated with consumer trophic status and diet. Indeed, SIA for sea turtle research has focused almost exclusively on stable carbon ($^{13}C/^{12}C = \delta^{13}C$) and stable nitrogen ($^{15}N/^{14}N = \delta^{15}N$), which are the most instructive elements for feeding studies, and are our primary focus in this chapter. It should be noted, however, that stable sulfur ($^{34}S/^{32}S = \delta^{34}S$) and stable oxygen ($^{18}O:^{16}O = \delta^{18}O$) analyses are becoming more widespread, and where appropriate we describe these studies later.

Stable isotope analysis in the context of animal ecology—including sea turtles—is based on the tenet that "you are what you eat." That is, dietary inferences based on stable isotopic profiles in animal tissues are possible because the isotope compositions of a consumer's body tissues are ultimately derived from those in its diet. However, rather than reflecting the exact stable isotope values of the foods they eat, predators such as sea turtles exhibit predictable differences (i.e., discriminate) from their prey. Because $\delta^{13}C$ signatures undergo only slight trophic enrichment, they are not necessarily good indicators of trophic levels, but more effectively describe different carbon sources and flow pathways (DeNiro and Epstein, 1978; Peterson and Fry, 1987; Table 9.2, Figure 9.1). On the other hand, due to their large enrichment factor, $\delta^{15}N$ signatures are useful for the identification of trophic level of the organism or trophic structure of the system of interest (DeNiro and Epstein, 1981; Miniwaga and Wada, 1984; Table 9.2). Stable sulfur and oxygen are also of value for studying wildlife, although in these cases there are spatial gradients in isotope concentrations, which make their study more appropriate for elucidating animal movements and habitat use rather than trophic status. For example, whereas $\delta^{34}S$ can discern the relative importance of estuarine vs. marine habitats and/or benthic vs. pelagic habitats for sea turtles (Thode, 1991), $\delta^{18}O$ elucidates latitudinal movements due to decreasing $\delta^{18}O$ in surface waters of higher latitudes (Lajtha and Marshall, 1994; Table 9.2)

Assuming discrimination values are known, the stable isotope technique is particularly useful when an organism's diet is difficult to establish with conventional approaches such as esophageal lavage (i.e., stomach flushing) and fecal analysis. Because isotope analysis provides information on nutrients assimilated over extended periods, the approach is much less affected by short-term temporal change in diet than other methods, which only provide dietary "snapshots" of recently consumed food items (Peterson and Fry, 1987; Hobson et al., 1996). When examined for multiple individuals within a population, or multiple organisms within an ecosystem, stable isotope analyses provide unique insights to trophic variability and niche width, or specificity with which an individual or population forages (Gu et al., 1997; Bearhop et al., 2004; Araújo et al., 2007), and nutrient transport within an ecosystem (e.g., Vander Zanden et al., 2012a).

## TABLE 9.2

## Summary of Stable Isotopes Used in Sea Turtle Research

| Stable Isotope Notation of Isotope Ratios | Natural Processes that Influence Isotope Abundance | Natural Environmental Patterns (i.e., without Human Influence) | | | |
|---|---|---|---|---|---|
| | Biological and/or Biochemical | Terrestrial | Marine | Terrestrial vs. Marine | Reference(s) |
| Carbon $^{13}C:^{12}C$ $\delta^{13}C$ | Vary in plant tissue with Isotopic fractionation during photosynthesis[a] Ambient conditions that limit enzymatic reactions during photosynthesis or alter stomatal opening | Decrease with increasing latitude[b,c] Increase with increasing altitude[b] Mesic (i.e., dry) habitats more enriched compared to xeric (i.e., wet) habitats[d] | Decrease with increasing latitude[e] Northern oceans more enriched compared to southern oceans[e] Benthic more enriched compared to pelagic[e] | Marine more enriched compared to terrestrial[f] | Kelly (2000) Lajta and Marshall (1994) Hobson (1999) |
| Oxygen $^{18}O:^{16}O$ $\delta^{18}O$ | Vary in meteoric waters with Precipitation patterns Temperature Elevation Relative humidity | Decrease with increasing latitude[g] Decrease with increasing altitude[g] Highest in summer and lowest in winter above 30° latitude[g] Decrease moving inland[g] | No patterns | Marine more enriched compared to terrestrial[h] | Lajta and Marshall (1994) Poage and Chamberlain (2001) |

*(continued)*

## TABLE 9.2 (continued)
## Summary of Stable Isotopes Used in Sea Turtle Research

| Stable Isotope Notation of Isotope Ratios | Natural Processes that Influence Isotope Abundance — Biological and/or Biochemical | Natural Environmental Patterns (i.e., without Human Influence) | | | Reference(s) |
|---|---|---|---|---|---|
| | | Terrestrial | Marine | Terrestrial vs. Marine | |
| Nitrogen $^{15}N:^{14}N$ $\delta^{15}N$ | Vary in plant tissue according to how plants fix N; Symbiotic fixation; Direct conversion of atmospheric N; Increase with trophic level[i] | Xeric (i.e., wet) habitats more enriched compared to mesic (i.e., dry) habitats[j] | Northern oceans more enriched compared to southern oceans[k] | Marine more enriched compared to terrestrial[l] | Kelly (2000); Hobson (1999); Hobson (2003) |
| Sulfur $^{34}S:^{32}S$ $\delta^{34}S$ | Vary in nature with; Distribution of light and heavy sulfides in bedrock; Quality of plant growing conditions (i.e., aerobic vs. anaerobic); Atmospheric deposition from natural sources | No patterns | Estuarine and marsh more enriched compared to marine[m]; Benthic (i.e., inshore) more enriched compared to pelagic (i.e., offshore)[m] | Marine more enriched compared to terrestrial[n] | Thode (1991); Hobson (2003) |

*Source:*	Modified from Rubenstein, D.R. and Hobson, K.A., *Trends Ecol. Evol.*, 19, 256, 2004.

a Plants use one of three different types of photosynthetic pathway (C3, C4 or crassulacean acid metabolism [CAM]) that utilize different $CO_2$-fixing enzymes and result in varying ranges of $\delta^{13}C$ values (Latja and Marshall, 1994; Kelly, 2000).

b Due to temperature differences (Latja and Marshall, 1994; Kelly, 2000).

c Due to differences in abundance of C4 plants (Kelly, 2000).

d In C3-based systems 0 wing to difference in water use efficiency (Latja and Marshall, 1994).

e Due to temperature differences, surface-water $CO_2$ concentrations and differences in plankton biosynthesis or metabolism (Kelly, 2000).

f Due to bicarbonate as a carbon source and slower diffusion of $CO_2$ in marine environment (Kelly, 2000).

g Due to rainfall patterns and temperature differences (Latja and Marshall, 1994; Poage and Chamberlain, 2001).

h Due to evaporation of seawater and subsequent condensation of cloud moisture over land (Latja and Marshall, 1994).

i Due to discrimination and fractionation during prey nutrient assimilation by predators (DeNiro and Epstein, 1981).

j Reasons are unclear (Kelly, 2000).

k Due to inorganic nitrogen being an important contributing factor in marine environment (Hobson, 2003).

l Due to anoxic conditions (Thode, 1991).

m Due to bacterial sulfate reduction in marine environment (Thode, 1991).

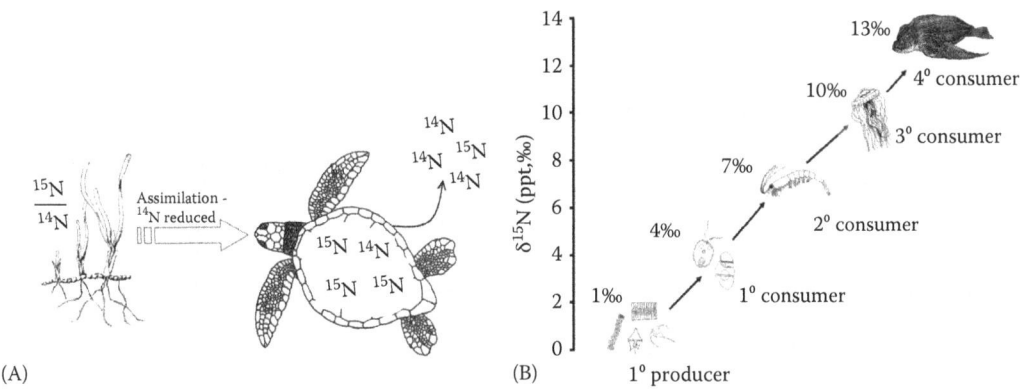

FIGURE 9.1    Schematic of $^{15}$N trophic enrichment in sea turtles. (A) Conceptual presentation of retention of heavier $^{15}$N isotopes and higher rate of purging of lighter $^{14}$N isotopes during digestion. (B) Rendition of progressive enrichment in $^{15}$N at 3 ppt per trophic step.

The initial sea turtle stable isotope study was conducted by Killingley and Lutcavage (1983) who measured stable oxygen profiles in commensal barnacle shells to reconstruct loggerhead movements, but it was not until Godley et al.'s (1998) efforts that isotopic measurements focused on sea turtle body tissues—in their case to explore trophic status of sea turtles in the Mediterranean. Since then, the application of this technique has rapidly expanded, and has resulted in deeper insights about sea turtle diet and trophic status, habitat use, movements, and demography (see later).

### 9.4.1  LIMITATIONS OF STABLE ISOTOPE ANALYSIS

The recent and rapid expansion of stable isotope applications to address questions about wildlife ecology has garnered panacea status for SIA. However, there are several limitations that must be considered prior to applying this research technique. Chief among these are for dietary investigations, the power of stable isotopes to determine the trophic level and/or specific prey of a consumer requires that isotopic discrimination and residence time be well understood. Yet for sea turtles these key aspects have not been determined for most species, diets, and life stages (see Table 9.3). A second challenge is that the ability of stable isotope analyses to decipher dietary complexity in a predator is only possible if the putative diet items differ appreciably in their isotope values (Boecklen et al., 2011). In other words, if all the potential foods in a foraging habitat have the same or similar stable isotope values, it is difficult to distinguish diet signals of each food type. Low isotopic diversity among species within a foraging habitat can thus limit the predictive power of stable isotope mixing models, and often requires that isotopically similar species are lumped into prey groups (Phillips and Gregg, 2003; Phillips et al., 2005), which thereby undermines the primary reason why mixing models are used. This problem can be ameliorated through incorporating elemental concentration dependency into modeling efforts (Phillips and Koch, 2002) and using a greater number of isotopes in such models (e.g., adding $\delta^{34}$S to ongoing $\delta^{13}$C and $\delta^{15}$N analyses). However, expanding the palette of isotopes in a study is costly and may be prohibitive of maximizing sample sizes.

An additional challenge to the broad application of stable isotope analysis, particularly as it relates to studies of habitat use and animal movements, is that patterns of isotopic abundances (i.e., isoscapes) are poorly resolved for most ocean regions. While spatial gradients in stable isotope values at the base of food webs have been described on very large scales (Somes et al., 2010; Deutsch et al., 2011), few regional maps are available and little is known about temporal shifts in isotope abundances in natural systems (Faure, 1986; Boecklen et al., 2011). For high-order consumers, the

**TABLE 9.3**

**Summary of Stable Isotope Discrimination and Isotope Retention Time in Turtles**

| Species | Lifestage | Size (SCL) cm | Diet Type | Tissue Type | $\Delta_{dt}$ Carbon | $\Delta_{dt}$ Nitrogen | C Turnover (d) | N Turnover (d) | References |
|---|---|---|---|---|---|---|---|---|---|
| *Caretta caretta* | Hatchling | 4.3–4.9 | Soy protein pellet | Skin | 2.62±0.34 | 1.65±0.12 | 83.0±7.02 | 66.7±7.36 | Reich et al. (2008) |
| | | | | Scute | −0.86±0.57 | 0.61±0.16 | 62.5±7.31 | 45.5±5.48 | |
| | | | | RBC | −0.64±0.73 | −0.25±0.30 | 76.9±11.34 | 71.4±10.66 | |
| | | | | Plasma | 0.29±0.20 | 0.32±0.09 | 20.0±6.34 | 18.5±4.25 | |
| | | | | Whole blood | 0.19±0.08 | 0.92±0.34 | 43.5±2.34 | 35.7±2.73 | |
| *Caretta caretta* | Juvenile | 9.0–13.1 | Animal protein pellet | Skin | 1.11±0.17 | 1.60±0.07 | 46.1±7.02 | 44.9±3.1 | Reich et al. (2008) |
| | | | | Scute | 1.77±0.58 | −0.64±0.09 | 50.9±13.14 | 16.2±2.3 | |
| | | | | RBC | 1.53±0.17 | 0.16±0.08 | 40.1±3.4 | 36.3±3.4 | |
| | | | | Plasma | −0.38±0.21 | 1.50±0.17 | 39.6±9.1 | 22.5±5.1 | |
| | | | | Whole blood | 1.11±0.18 | 0.14±0.06 | 46.1±8.9 | 27.7±3.5 | |
| *Chelonia mydas* | Juvenile | 43.0–47.5 | Soy and fish-meal protein pellet | Skin | 0.17±0.03 | 2.80±0.11 | | | Seminoff et al. (2006b) |
| | | | | RBC | −1.11±0.05 | 0.22±0.03 | | | |
| | | | | Plasma | −0.12±0.03 | 2.92±0.03 | | | |
| | | | | Whole blood | −0.92±0.06 | 0.57±0.09 | | | |
| *Chelonia mydas* | Juvenile | 28.5–35.8 | Soy protein pellet | Plasma | 2.62±0.66 | 3.18±0.66 | | | Seminoff et al., unpublished data |
| *Dermochelys coriacea* | Juvenile | 25.8±1.0a | Fortified squid gelatin | Skin | 2.26±0.61a | 1.85±0.50a | | | Seminoff et al. (2009) |
| | | | | RBC | 0.46±0.35a | 1.49±0.76a | | | |
| | | | | Plasma | −0.58±0.53a | 2.86±0.82a | | | |
| | | | | Whole blood | 0.35±0.33a | 1.98±1.14a | | | |

| Species | Life stage | | Diet | Tissue | | | Reference |
|---|---|---|---|---|---|---|---|
| *Trachemys scripta* | Adult male | 15.5–24.0 | Soy pellet | RBC | 1.9±0.2b | | Seminoff et al. (2007) |
| | | | | Plasma | 3.8±0.1b | | |
| | | | | Whole blood | 2.2±0.2b | | |
| | | | | Liver | 3.0±0.3b | | |
| | | | | Brain | 2.9±0.3b | | |
| | | | | PM muscle | 2.7±0.3b | | |
| | | | | PI muscle | 3.4±0.4b | | |
| *Trachemys scripta* | Adult male | 15.5–24.0 | Fish-meal pellet | Plasma | −0.8±0.8b | 38.7c | Seminoff et al. (2007) |
| | | | | Whole blood | 2.5±0.8b | 35.6c | |
| | | | | Liver | 0.4±0.5b | 52.5c | |
| *Gopherus agassizii* | Adults | | Soy-plant based | Red Blood Cells | 0.2±0.3 | 126.7±40.3 | Murray and Wolf (2012) |
| | | | | Plasma | 1.0±0.2 | 32.9±14.5 | |
| | | | | Scute | 0.8±0.1 | | |

a = SD, b = SE, c = halflife.

elaboration of foraging habitats is further obscured by the substantial difference in stable isotope values between their tissues and those of basal producers (Post, 2002; Phillips et al., 2009; Boecklen et al., 2011). These differences are due to stepwise isotope enrichment occurring at each trophic level (DeNiro and Epstein, 1981) along with poorly understood extrinsic and intrinsic factors affecting isotope enrichment along the food chain (Post, 2002; Phillips et al., 2009; Boecklen et al., 2011).

### 9.4.2 ISOTOPIC DISCRIMINATION

Fundamental to the interpretation of stable isotope data is the need to understand the patterns by which diet isotopic values are reflected in consumer tissues (i.e., isotope discrimination, or $\Delta_{dt}$; Cerling and Harris, 1999). For example, $\delta^{13}C$ and $\delta^{15}N$ of consumer and prey tissues are usually not identical, and instead exhibit predictable differences due to selectivity for lighter isotopes during a consumer's metabolic processes (DeNiro and Epstein, 1978, 1981). Although there is variation among species, most endothermic species show an isotopic increase of 0‰–1‰ for $\delta^{13}C$ and 3‰–5‰ for $\delta^{15}N$ per trophic level (DeNiro and Epstein, 1978, 1981; Miniwaga and Wada, 1984; Post, 2002; Figure 9.1). However, stable isotopic discrimination may vary widely, with some species $^{13}C$-depleted by ~1‰ and $^{13}C$-enriched by ~3‰ relative to their prey, and $^{15}N$-enriched by less than 3 ‰ (Hesslein et al., 1993; Pinnegar and Polunin, 1999; McCutchan et al., 2003; Reich et al., 2008).

The mechanisms for this enrichment are not thoroughly understood, but may result from a variety of biochemical pathways. For example, there is relatively greater differential excretion of $^{14}N$ in urine; resulting in a net gain of $^{15}N$ in consumer tissues (Miniwaga and Wada, 1984; Peterson and Fry, 1987) (Figure 9.1). Fractionation during amino acid (AA) amination and transamination further contributes to isotopic enrichment (Abelson and Hoering, 1961; Macko et al., 1986; Tieszen and Boutton, 1989; Hobson and Clark, 1992; Tieszen and Fagre, 1993). Variations in stable isotope ratios between consumer tissues and mean diet may also be influenced by consumer age (Roth and Hobson, 2000), nutrition status (Sealy et al., 1987; Hobson and Clark, 1992; McCue and Pollock, 2008), diet (Caut et al., 2009), trophic level (Tieszen et al., 1983; Hobson and Clark, 1992; Pearson et al., 2003), body temperature (Pinnegar and Polunin, 1999), digestive strategy (Macrae and Reeds, 1980), and (or) environmental conditions (Michener and Lajtha, 2007).

Within a given organism, stable isotope ratios may differ substantially among different tissue types. Such differences have been attributed to variations in the protein and AA compositions of different tissues because these components can differ in their nitrogen isotope content (Peterson and Fry, 1987). Selective routing of exogenous nutrients during tissue maintenance and construction, and differential mobilization of endogenous resources into tissues or tissue components, may also contribute to these tissue-level differences (Macrae and Reeds, 1980; Peterson and Fry, 1987; Gannes et al., 1997).

To date, discrimination in sea turtles has been established only for post-hatchling and small juvenile loggerheads (Reich et al., 2008), juvenile and adult green turtles (Seminoff et al., 2006b; Vander Zanden et al., 2012a), and juvenile leatherbacks (Seminoff et al., 2009) (Table 9.3). As found in other taxa, there is evidence that discrimination varies with size class, somatic growth rates, and diet (Seminoff et al., 2006b; Reich et al., 2008; Vander Zanden et al., 2012b). Clearly there are numerous influences on predator–prey isotope discrimination, many of which remain poorly understood. Additional controlled sea turtle studies that focus on novel taxa, different life stages, and varied diets will be vital for providing discrimination values that will enable effective and accurate determinations of prey isotope values based on sea turtle bulk tissue data.

### 9.4.3 ISOTOPIC RESIDENCE TIME

Another key aspect is the determination of the isotopic residence time (i.e., turnover rate) in turtle body tissues. That is, when a turtle adopts a new diet, how long does it take for the old dietary signal to be completely turned over to the new diet signal. Incorporation of diet-derived stable isotopic signatures into consumer body tissues occurs at varying rates based primarily on tissue-specific metabolism (Gannes et al., 1997). As a result, different tissues may provide dietary information that is integrated over different time scales. Tissues with higher metabolic activity (liver, whole blood) will reflect more recent diet history, and those with lower metabolism (integument, bone) will represent an integration of diet over a substantially longer period, perhaps approaching the full life of the consumer (Tieszen et al., 1983; Hobson and Clark, 1992).

Isotopic residence times have so far been established for relatively few sea turtle species and life stages (Table 9.3). Within sea turtles, isotopic residence times have only been established for juvenile loggerhead turtles (Reich et al., 2008) and green turtles (Vander Zanden et al., 2012a), although there have also been studies of freshwater turtles (Seminoff et al., 2007) and tortoises (Murray and Wolf, 2012) that are instructive. In all cases it is evident that the turnover rates in turtles are much slower than for mammals and birds (Hobson and Clark, 1992; B. Wolf unpublished data). There is also a general trend within the reptilia that stable carbon turnover rates are much slower than those for stable nitrogen (Seminoff et al., 2007; Reich et al., 2008; Murray and Wolf, 2012).

### 9.4.4 Diet and Trophic Status

Applications of stable isotope analytical techniques have provided many new insights about the ecology of sea turtles and other marine consumers. Among the most fundamental questions that can be addressed are those relating to trophic status and/or diet of a species. That is, what is the trophic level of a turtle or general population, and what are the specific prey species or prey groups consumed. This assumes knowledge is available regarding the turtle life stage and diet-specific isotope discrimination as well as the stable isotope baseline for primary producers within the marine system in which study animals forage. If information is available on the stable isotope values of individual putative prey species, stable isotope mixing models can be used to determine the relative contribution of various potential prey species to the overall diet composition of a consumer (Phillips, 2001; Semmens and Moore, 2008; Inger et al., 2010).

Stable isotope analyses have complemented conventional diet techniques to decipher dietary contributions for sea turtles in many localities worldwide. Stable carbon and nitrogen have been used to study the diet of loggerheads in the western North Atlantic (Wallace et al., 2009) and Mediterranean (Revelles et al., 2007a,b); green turtles in the eastern Pacific (Lemons et al., 2011), Australia (Arthur et al., 2009), and Japan (Hatase et al., 2006), hawksbills in Hawaii (Graham, 2009) and the Bahamas (Bjorndal and Bolten, 2010); and leatherbacks in the western North Atlantic (Dodge et al., 2011). In several cases, isotopic inquiry has also revealed novel insights otherwise unknown via conventional techniques. For example, green turtles are not obligate neritic herbivores throughout their range as once thought (Hatase et al., 2006; Arthur et al., 2009; Kelez, 2011), whereas hawksbill turtles appear to have a greater dependency on seagrass habitats than previously believed (Bjorndal and Bolten, 2010).

Stable isotopes have also been used to illustrate ontogenic shifts in diet and habitat use, as well as differences in foraging behavior of individuals within a population. For instance, stable isotope analysis revealed polymodal foraging in adult female loggerheads in Florida (Reich et al., 2009) and individual dietary specialization in loggerheads from the same region (Vander Zanden et al., 2010). A study by Reich et al. (2007) was the first to confirm the oceanic foraging by "lost year" turtles based on isotope analysis of sequential scute layers. Likewise, Arthur et al. (2008) used $\delta^{13}C$ and $\delta^{15}N$ to reveal ontogenic changes in foraging behavior in Australian green turtles, and Cardona et al. (2009, 2010) used $\delta^{13}C$, $\delta^{15}N$, and $\delta^{34}S$ to reveal ontogenic diet shifts in green turtles from West Africa and the Mediterranean. Size-related differences in feeding habitat use and foraging dichotomies in adult loggerhead turtles around Japan have also been revealed via SIA (Hatase et al., 2002, 2010). In one of the first isotope studies to shed light on reproductive output parameters, Caut et al. (2008) used $\delta^{13}C$ and $\delta^{15}N$ to show that French Guianan nesting groups with differing remigration intervals had significantly different $\delta^{13}C$ and $\delta^{15}N$ values, likely owing to isoscape differences between each groups' unique foraging areas.

### 9.4.5  Isoscapes and Foraging Movements

In addition to the value of isotopes for determining the trophic status and diet of sea turtles, SIA can decipher the key foraging habitats used by sea turtles. While not as precise as satellite telemetry, SIA is much lower in cost, and a viable tool for tracking animal movements because the isotopic compositions of consumer tissues integrate isoscape information from foraging environments (DeNiro and Epstein, 1981). Thus, when a sea turtle moves among spatially discrete food webs that are isotopically distinct (i.e., isoscapes), stable isotope values of its tissues can provide unambiguous information about its previous location (Hobson, 1999; Rubenstein and Hobson, 2004; Hobson and Wassenaar, 2008; Newsome et al., 2010).

There have been several studies that have shed light on marine isoscapes, and this is an area of rapid advancement. The first such example was a study of leatherback turtles by Wallace et al. (2006b) that revealed significantly higher $\delta^{15}N$ in body tissues of leatherbacks nesting in the eastern Pacific (Playa Grande, Costa Rica) compared to those nesting in the western North Atlantic (St. Croix, USVI). The $\delta^{15}N$ disparity was likely due to differences in the prevailing nitrogen cycling regimes between the two ocean regions, with high levels of denitrification driving down $\delta^{15}N$ values in the eastern Pacific and greater $N_2$ fixation reducing the $\delta^{15}N$ values in the Atlantic (Montoya, 2007). Pajuelo et al. (2010) later found the same pattern in loggerhead turtles from Peru vs. those found near the Azores Islands, and also attributed this dichotomy to effects from baseline differences in nitrogen cycling between the two regions. However, while these two studies were paramount in showing that the influence of baseline stable isotope signatures is conserved up the food chain within a particular region, they lacked the ability to unequivocally determine the specific whereabouts of turtles when they incorporated observed stable isotope values.

Elucidating the whereabouts of a turtle during the assimilation of specific isotope signatures would appear to be a key need in isotopic research. An ideal, if not costly tool, to provide this spatial information is satellite telemetry, and studies blending isotope and satellite tracking are increasing, the result of which is a firmer understanding of marine isoscape patterns (Hobson and Wassenaar, 2008). A study by McClellan et al. (2010) of loggerhead turtles in the eastern United States was the first to our knowledge that used satellite telemetry to establish the spatial patterns of marine isoscapes; in their case, showing an isotopic dichotomy between juvenile loggerhead sea turtles that foraged coastally vs. those that foraged in offshore waters. Zbinden et al. (2011) similarly showed a migratory dichotomy and associated phenotypic variation in Mediterranean loggerhead turtles, while Ceriani et al. (2012) and Pajuelo et al. (2012) found that variation in $\delta^{13}C$ and $\delta^{15}N$ in post-nesting Florida loggerheads can be explained by differences in food-web baseline isotopic signatures rather than differences in loggerhead trophic levels. Seminoff et al. (2012) provided similar insights for Indonesian leatherbacks that used eastern and western Pacific foraging areas.

### 9.4.6  Compound-Specific Isotope Analysis of Amino Acids

As a complimentary tool to bulk tissue stable isotope analysis, compound-specific isotope analysis of individual AAs (CSIA) can provide additional information that can separate source and metabolic changes in the bulk isotopic signature. The CSIA technique has been utilized in the fields of geology and paleoecology for over four decades (Evershed et al., 2007), but only recently has it been applied to studies of marine vertebrates and turtles specifically. CSIA is the measurement of the isotopic composition of specific AAs in body tissues and it can substantially enhance the value of bulk tissue stable isotope measurements for understanding the trophic status of marine species. This is possible because some AAs, such as phenylalanine, retain the isotopic composition of source nitrogen at the base of the food web (i.e., "source" AAs), whereas other AAs such as glutamic acid are significantly enriched in $^{15}N$ as they move through the food web (i.e., "trophic" AAs; McClelland and Montoya, 2002). Baseline and trophic information can thus be obtained from a consumer's tissue without need for analyses of prey items or of basal food web samples (McClelland and Montoya, 2002; Popp et al., 2007). It is believed that these "trophic" AAs are either synthesized by the consumer or undergo significant transamination and deamination reactions that preferentially favor the retention of the heavier isotope (McClelland and Montoya, 2002), and it is assumed that this is also the mechanism underlying $^{15}$N enrichment observations in bulk tissues (Figure 9.1).

Compound-specific isotope analysis of AAs has been conducted in a small number of ecological studies, being applied to differentiate metabolic and trophic-level relationships in a food web from changes in isotopic composition at the base of the food web (Uhle et al., 1997; Fantle et al., 1999; McClelland and Montoya, 2002; Popp et al., 2007). To our knowledge there have only been two studies that have applied CSIA technique to sea turtles. Vander Zanden et al. (in review) investigated the $\delta^{15}$N of AAs and bulk tissues in green turtles nesting at Tortuguero, Costa Rica, in an effort to discern the foraging destinations of individual nesters. Similarly, Seminoff et al. (2012) used CSIA coupled with satellite telemetry and bulk tissue SIA to track the foraging migrations of leatherback turtles nesting at Jamursba Medi, Indonesia. Both studies show great promise for the CSIA technique, and it is clear that this approach is a powerful complement to bulk tissue SIA.

## 9.5   FATTY ACID AND TRACE ELEMENT ANALYSES

### 9.5.1   Fatty Acid Analysis

Traditional methods for determining diet have relied on visually identifying the contents of the gastro-intestinal (GI) tract and the degree to which the components in the diet have been digested often complicate this assessment. Establishing dietary composition by monitoring fatty acid profiles for organisms has become increasingly common as a means for overcoming these difficulties and determining the relative importance of a dietary constituent for wildlife species (Iverson et al., 1997).

Fatty acids (FAs), the primary components of most lipids, are composed of a hydrocarbon chain with a methyl group ($CH_3$) at one end and a carboxyl group (COOH) at the other. Fatty acids with no double bonds are considered saturated because they contain the maximum number of hydrogen atoms. Those with one or more double bonds are referred to as monounsaturated or polyunsaturated, respectively, and the number and position of double bonds, as well as the length of the hydrocarbon chain, give a FA its particular biochemical properties. FAs are most commonly referred to by the following shorthand notation: A:Bn-C where A is the number of carbon atoms in the hydrocarbon chain, B is the number of double bonds, and C is the position of the first double bond relative to the methyl end (IUPAC-IUB Commission on Biochemical Nomenclature, 1967).

Unlike proteins and carbohydrates that are degraded during digestive processes, dietary FAs are often deposited in the fat stores of a consumer with little or no modification (Ackman, 1989). Thus, predators have biochemical limitations on the types of FA they can synthesize, and FAs longer than 14:0 are predictably incorporated into consumer tissue (Cook, 1991; Galli and Rise, 2006). As a result, the FA composition of the diet is thought to be reflected in the stored fat of consumers (Holland et al., 1990). The FA composition of a consumer can therefore provide insight into the types of foods that were previously consumed.

There have been relatively few studies on sea turtles that employ fatty acid analysis. The technique was first applied in a study of leatherback diet by Holland et al. (1990) that linked the FAs in sea jellies to those in the fat tissue of leatherback turtles. Guitart et al. (1999) studied loggerhead turtles to similarly determine their prior diet. Green turtles have been the sea turtle species for which FA analysis has been most common, with Ackman et al. (1992) exploring FAs in green turtles from the Hawaiian Islands and Johnston Atoll, Seaborn et al. (2005) measuring FAs in green turtles from Hawaii, and Craven et al. (2008) exploring FAs in the yolks of captive green turtles.

Despite some rapid advancement over the past decade, FA signature analysis is still in development. Many important physiological and statistical issues need to be resolved to maximize its applicability to studies of sea turtle ecology. For example, much as with stable isotopes, if potential food sources all have very similar FA compositions, consumers will likely be relatively uniform in their FA signatures and the FA approach will have low power for discriminating diet source. Conversely, if a consumer group ingests a wide variety of different foods that each has highly variable FA compositions, predator signatures may be too variable to reflect meaningful ecological patterns. Another challenge to the FA approach is that the exact temporal scale of FA information is still unclear in most cases. Evidence from captive feeding studies indicates that the FA signatures of seals reflect diet composition integrated over weeks to months (Kirsch et al., 2000), but more information on the FA temporal window for other species is necessary. For sea turtles, calibration of this timing, as well as greater understanding about the effects of dietary FA variability on the power of FA analysis, could be determined via captive feeding studies.

### 9.5.2 Trace Element Analysis

The value of trace elemental analysis (TEA) is less resolved than for SIA and FA analysis. The theory behind the TEA is similar to that of stable isotopes and FA in that it is expected that prey species' and spatially explicit background chemical signatures in the water column and/or food source are incorporated into body tissues of consumers. For diet investigations, it is believed that the minute trace levels of a host of heavy metals and pollutants (including but not limited to arsenic, antimony, cadmium, cobalt, copper, lead, mercury, selenium, and zinc) in foods ingested by consumers will bioaccumulate in the consumer tissues. Thus, when both consumer tissues and prey are analyzed for trace element concentrations, the elemental "signatures" of these groups can be linked to determine which potential prey species were featured in the diet. Likewise, for studies of animal movements, trace element concentrations in consumer tissues are presumed to be an integration of trace elements present in foraging environments, and thus, when an animal moves among spatially discrete food webs that have different trace element occurrences, TEA can provide unambiguous information about an animal's previous location (Ramos et al., 2009).

Few TEA studies have been conducted on sea turtles to our knowledge, although there are many studies underway (e.g., Lopez Castro, unpublished data). Bergeron et al. (2007) studied diet composition in four species of freshwater turtles using trace elements, Ikonomopoulou et al. (2011) studied trace element concentrations in nesting flatback turtles in Australia, and van de Merwe et al. (2010) studied long-distance migrations of green turtles from Southeast Asia. Although TEA is in its infancy, this approach shows promise as another tool to examine foraging ecology and animal movements across broad spatial scales.

## 9.6 CONCLUSIONS

Since publication of *The Biology of the Sea Turtles* (Volume 1; Lutz and Musick, 1997), there have been many advances in our understanding of sea turtle biology. These insights have been gained via continued application of many tried and true, and traditionally "low tech" research tools and benefited more recently from the development of new research fields such as biologging, fine-scale physiological measurements, stable isotope analysis, and fatty acid analysis. However, despite their value in sea turtle research, there are still improvements that can be achieved.

More than a decade ago in Volume 1, Karen Bjorndal (Chapter 7, Foraging ecology and nutrition of sea turtles) stressed the importance of studies on all aspects of feeding physiology and in quantifying nutritional requirements, since that time only a handful of studies have emerged (see Section 9.3). The reproductive success of turtles, and inevitably their rebound from past declines, requires the successful attainment of chemical energy and the conversion of that energy into useable components to fuel somatic and reproductive growth, trans-oceanic migrations, temperate water foraging, and all the physiological processes of life (i.e., homeostasis). With impending changes in climate and temperature, prey landscapes are changing (Mills, 2001); human perturbations such as commercial fishing reduce target species (e.g., crabs) but offer free lunch in the form of fisheries discard (Seney and Musick, 2007). How well turtles adapt to these changes both behaviorally and physiologically and how they maintain energy intake and absorption despite ever changing conditions such as dietary dilution from plastic ingestion (McCauley and Bjorndal, 1999), is yet to be determined. Therefore, a decade and a half later, we end with a call for more studies into the diet selection and feeding physiology of sea turtles and an incorporation of emerging techniques and technologies to elucidate aspects of feeding biology.

In the realm of molecular analyses, we need greater emphasis on technical aspects such as tissue preservation for sea turtle stable isotope studies (e.g., Barrow et al., 2008; Lemons et al., 2012). Knowledge like this, coupled with efforts to standardize tissue collection and analytical protocols will benefit molecular approaches. For analysis of consumer and prey isotope values, the disparities in discrimination factors and isotope retention time found among taxa underscore the need for additional feeding trials under controlled conditions for the technique to be reliably used in unstudied groups. Likewise, we need a more solid understanding of the spatiotemporal isotopic patterns in marine fauna at basal and higher-order trophic levels. Clarification about the baseline influence on isotopic compositions of tissues of higher-order marine consumers is also required for maximizing the value of isotopic analyses (Graham et al., 2010). In the coming years we encourage studies to address these issues and believe that through greater emphasis on controlled experimental studies and additional data collected in situ, we will be able to maximize the value of SIA for addressing questions about sea turtle ecology (see Gannes et al., 1997; Newsome et al., 2007; Martínez del Rio et al., 2009). We also urge additional studies that combine techniques (e.g., biologging with SIA, SIA with FA analysis). Research efforts will yield their greatest insights when tools are used in combination. It is our hope that another 15 years will see the increase in numbers of sea turtles and that novel and traditional tools will result in even larger data streams, with a need to provide another update on the biology and ecology of feeding.

## ACKNOWLEDGMENTS

We thank the editors of this book for inviting us to contribute this chapter. It is an honor to build off of the foundation created in Volume 1 by Karen Bjorndal. We gratefully acknowledge Garrett Lemons, Brad MacDonald, Elizabeth Whitman for assistance with the drafting of this chapter. An anonymous reviewer provided valuable comments that improved earlier versions of this chapter. JAS thanks Karen Bjorndal and Alan Bolten of the Archie Carr Center for Sea Turtle Research, where he initially got involved with stable isotope analysis. TTJ is indebted to Michael Salmon, Peter Lutz, and David Jones for nurturing his interest and giving him the tools to study "how" sea turtles fuel their behaviors.

## REFERENCES

Abelson, P. H. and T. C. Hoering. 1961. Carbon isotope fractionation in formation of amino acids by photosynthetic organisms. *Proc. Natl Acad. Sci. U.S.A.* 47:623–632.

Ackman, R. G. 1989. Fatty acids. In *Marine Biogenic Lipids, Fats, and Oils*, ed. R. G. Ackman, pp. 103–137. Boca Raton, FL: CRC Press.

Ackman, R. G., T. Takeuchi, and G. H. Balazs. 1992. Fatty-acids in depot fats of green turtles *Chelonia mydas* from the Hawaiian-islands and Johnston Atoll. *Comp. Biochem. Physiol. B* 102(4):813–819.

Amorocho, D. F. and R. D. Reina. 2007. Feeding ecology of the East Pacific green sea turtle *Chelonia mydas agassizii* at Gorgona National Park, Colombia. *Endanger. Species Res.* 3:43–51.

Amorocho, D. F. and R. D. Reina. 2008. Intake passage time, digesta composition and digestibility in East Pacific green turtles (*Chelonia mydas agassizii*) at Gorgona National Park, Colombian Pacific. *J. Exp. Mar. Biol. Ecol.* 360:117–124.

Arai, M. N., D. W. Welch, A. L. Dunsmuir, M. C. Jacobs, and A. R. Ladouceur. 2003. Digestion of pelagic Ctenophora and Cnidaria by fish. *Can. J. Fish. Aquat. Sci.* 60:825–829.

Araújo, M. S., D. I. Bolnick, G. Machado, A. A. Giaretta, and S. F. dos Reis. 2007. Using d¹³C stable isotopes to quantify individual-level diet variation. *Oecologia* 152:643–654.

Arthur, K. E. and G. H. Balazs. 2008. A comparison of immature green turtle (*Chelonia mydas*) diets among seven sites in the main Hawaiian Islands. *Pac. Sci.* 62(2):205–217.

Arthur, K. E., M. C. Boyle, and C. J. Limpus. 2008. Ontogenetic changes in green sea turtle (*Chelonia mydas*) foraging behaviour as demonstrated by $\delta^{13}C$ and $\delta^{15}N$ stable isotope analysis. *Mar. Ecol. Prog. Ser.* 362:303–311.

Arthur, K. E., C. J. Limpus, C. M. Roelfsema, J. W. Udy, and G. R. Shaw. 2006. A bloom of *Lyngbya majuscula* in Shoalwater Bay, Queensland, Australia: An important feeding ground for the green turtle (*Chelonia mydas*). *Harmful Algae* 5(3):251–265.

Arthur, K. E., K. M. McMahon, C. J. Limpus, and W. C. Dennison. 2009. Feeding ecology of green turtles (*Chelonia mydas*) from Shoalwater Bay, Australia. *Mar. Turtle Newslett.* 123:6–11.

Arthur, K. E., J. M. O'Neil, and C. J. Limpus. 2007. Using animal-borne imaging to assess green turtle (*Chelonia mydas*) foraging ecology in Moreton Bay, Australia. *Mar. Technol. Soc. J.* 41(4):9–13.

Avery, H. W., J. R. Spotila, J. D. Congdon, R. U. Fischer Jr., E. A. Standora, and S. B. Avery. 1993. Roles of diet protein and temperature in the growth and nutritional energetics of juvenile slider turtles, *Trachemys scripta*. *Physiol. Zool.* 66:902–925.

Bacon, P. R. 1970. Studies on the leatherback turtle, *Dermochelys coriacea* (L.), in Trinidad, West Indies. *Biol. Conserv.* 2:213–217.

Barrow, L. M., K. A. Bjorndal, and K. Reich. 2008. Effects of preservation method on stable carbon and nitrogen isotope values. *Physiol. Biochem. Zool.* 81:688–693.

Bearhop, S., C. E. Adams, S. Waldron, R. A. Fuller, and H. Macleod. 2004. Determining trophic niche width: A novel approach using stable isotope analysis. *J. Anim. Ecol.* 73:1007–1012.

Bello, G., A. Travaglini, and F. Bentivegna. 2011. *Histioteuthis bonnellii* (Cephalopoda: Histioteuthidae): A new prey item of the leatherback turtle *Dermochelys coriacea* (Reptilia: Dermochelidae). *Mar. Biol. Res.* 7(3):314–316.

Benson, S. R., T. Eguchi, D. G. Foley et al. 2011. Large-scale movements and high-use areas of western Pacific leatherback turtles, *Dermochelys coriacea*. *Ecosphere* 2(7):art84.

Bergeron, C. M., J. F. Husak, J. M. Unrine, C. S. Romanek, and W. A. Hopkins. 2007. Influence of feeding ecology on blood mercury concentrations in four species of turtles. *Environ. Toxicol. Chem.* 26(8):1733–1741.

Berube, M. D., S. G. Dunbar, K. Rutzler, and W. K. Hayes. 2012. Home range and foraging ecology of juvenile hawksbill sea turtles (*Eretmochelys imbricata*) on inshore reefs of Honduras. *Chelonian Conserv. Biol.* 11(1):33–43.

Birse, R. F. and J. Davenport. 1987. A study of gut function in young loggerhead sea turtles, *Caretta caretta* L. at various temperatures. *Herpetol. J.* 1:170–175.

Bjorndal, K. A. 1980. Nutrition and grazing behavior of the green turtle, *Chelonia mydas*. *Mar. Biol.* 56:147–154.

Bjorndal, K. A. 1990. Digestibility of the sponge *Chondrilla nucula* in the green turtle, *Chelonia mydas*. *Bull. Mar. Sci.* 47(2):567–570.

Bjorndal, K. A. 1997. Foraging ecology and nutrition of sea turtles. In *The Biology of Sea Turtles*, eds. P. Lutz and J. Musick, pp. 199–232. Boca Raton, FL: CRC Press.

Bjorndal, K. A. and A. B. Bolten. 1988. Growth rates of immature green turtles, *Chelonia mydas*, on feeding grounds in the southern Bahamas. *Copeia* 1988:555–564.

Bjorndal, K. A. and A. B. Bolten. 2010. Hawksbill sea turtles in seagrass pastures: Success in a peripheral habitat. *Mar. Biol.* 157:135–145.

Bjorndal, K. A., A. B. Bolten, and H. R. Matins. 2000. Somatic growth model of juvenile loggerhead sea turtles *Caretta caretta*: Duration of pelagic stage. *Mar. Ecol. Prog. Ser.* 202:265–272.

Bleakney, J. S. 1965. Reports of marine turtles from New England and eastern Canada. *Can. Field Nat.* 79(2):120–128.

Blumenthal, J. M., T. J. Austin, C. D. L. Bell et al. 2009. Ecology of hawksbill turtles, *Eretmochelys imbricata*, on a western Caribbean foraging ground. *Chelonian Conserv. Biol.* 8(1):1–10.

Boecklen, W. J., C. T. Yarnes, B. A. Cook, and A. C. James. 2011. On the use of stable isotopes in trophic ecology. *Annu. Rev. Ecol. Evol. Syst.* 42:411–440.

Bolten, A. B. 2003. Variation in sea turtle life history patterns: Neritic vs. oceanic developmental stages. In *The Biology of Sea Turtles*, eds. P. Lutz, J. Musick, and J. Wyneken, Vol. II, pp. 243–258. Boca Raton, FL: CRC Press.

Borjndal, K. A. and J. B. C. Jackson. 2003. Roles of sea turtles in marine ecosystems: Reconstructing the past. In *The Biology of Sea Turtles Volume II.*, eds. P. Lutz, J. Musick, and J. Wyneken, 259–274. Boca Raton: CRC Press.

Boyle, M. C. and C. J. Limpus. 2008. The stomach contents of post-hatchling green and loggerhead sea turtles in the southwest Pacific: An insight into habitat association. *Mar. Biol.* 155(2):233–241.

Brand, S. J., J. M. Lanyon, and C. J. Limpus. 1999. Digesta composition and retention times in wild immature green turtles, *Chelonia mydas*: A preliminary investigation. *Mar. Freshwater Res.* 50:145–147.

Brand-Gardner, S. J., J. M. Lanyon, and C. J. Limpus. 1999. Diet selection by immature green turtles, *Chelonia mydas*, in subtropical Moreton Bay, south-east Queensland. *Aust. J. Zool.* 47(2):181–191.

Brongersma, L. D. 1969. Miscellaneous notes on turtles. IIA, IIB. *Proc. K. Ned. Akad. Wet. C* 72:76–102.

Brongersma, L. D. 1972. European Atlantic turtles. *Leiden Zool. Verhandel.* 121:1–318.

Burke, V. J., S. J. Morreale, and E. A. Standora. 1994. Diet of the Kemp's ridley sea turtle, *Lepidochelys kempii*, in New York waters. *Fish. Bull.* 92:26–32.

Burke, V. J., E. A. Standora, and S. J. Morreale. 1993. Diet of juvenile Kemp's ridley and loggerhead sea turtles from Long Island, New York. *Copeia* 1993:1176–1180.

Cardona, L., I. Alvarez de Quevedo, A. Borrell, and A. Aguilar. 2012. Massive consumption of gelatinous zooplankton by Mediterranean apex predators. *PLoS One* 7(3):e31329.

Cardona, L., P. Campos, Y. Levy, A. Demetropoulos, and D. Margaritoulis. 2010. Asynchrony between dietary and nutritional shifts during the ontogeny of green turtles (*Chelonia mydas*) in the Mediterranean. *J. Exp. Mar. Bio. Ecol.* 393(1–2):83–89.

Cardona, L., L. Pazos, and L. Aguilar. 2009. Delayed ontogenic dietary shift and high levels of omnivory in green turtles (*Chelonia mydas*) from the NW coast of Africa. *Mar. Biol.* 156:1487–1495.

Carr, A. 1987. New perspectives on the pelagic stage of sea turtle development. *Conserv. Biol.* 1:103–121.

Carr, A. and S. Stancyk. 1975. Observations on the ecology and survival outlook of the hawksbill turtle. *Biol. Conserv.* 8:161–172.

Carranza, A., A. Estrades, F. Scarabino, and A. Segura. 2011. Loggerhead turtles *Caretta caretta* (Linnaeus) preying on the invading gastropod *Rapana venosa* (Valenciennes) in the Rio de la plata estuary. *Mar. Ecol.* 32(2):142–147.

Carrion-Cortez, J. A., P. Zarate, and J. A. Seminoff. 2010. Feeding ecology of the green sea turtle (*Chelonia mydas*) in the Galapagos Islands. *J. Mar. Biol. Assoc. U.K.* 90(5):1005–1013.

Casale, P., G. Abbate, D. Freggi, N. Conte, M. Oliverio, and R. Arganto. 2008. Foraging ecology of loggerhead sea turtles: *Caretta caretta* in the central Mediterranean Sea: Evidence for a relaxed life history model. *Mar. Ecol. Prog. Ser.* 372:265–276.

Casey, J., J. Garner, S. Garner, and A. S. Williard. 2010. Diel foraging behavior of gravid leatherback sea turtles in deep waters of the Caribbean Sea. *J. Exp. Biol.* 213:3961–3971.

Caut, S., E. Angulo, and F. Courchamp. 2009. Variation in discrimination factors ($\delta^{15}$N and $\delta^{13}$C): The effect of diet isotopic values and applications for diet reconstruction. *J. Appl. Ecol.* 46:443–453.

Caut, S., E. Guirlet, E. Angula, K. Das, and M. Girondot. 2008. Isotope analysis reveals foraging area dichotomy for Atlantic leatherback turtles. *PLoS One* 3:1845–1853.

Ceriani, S. A., J. D. Roth, D. R. Evans, J. F. Weishampel, and L. M. Ehrhart. 2012. Inferring foraging areas of nesting loggerhead turtles using satellite telemetry and stable isotopes. *PLoS ONE* 7(9): e45335. doi:10.1371/journal.pone.0045335.

Cerling, T. E. and J. M. Harris. 1999. Carbon isotope fractionation between diet and bioapatite in ungulate mammals and implications for ecological and paleontological studies. *Oecologia* 120:347–363.

Collard, S. B. and L. H. Ogren. 1990. Dispersal scenarios for pelagic post-hatchling sea turtles. *Bull. Mar. Sci.* 47:233–243.

Conway, S. P. 1994. Diets and feeding biology of adult olive ridley (*Lepidochelys olivacea*) and loggerhead (*Caretta caretta*) sea turtles in Fog Bay (Northern Territory). Unpublished Dip. Sci. Diss., Northern Territory University, Darwin, Australia.

Cook, H. W. 1991. Fatty acid desaturation and chain elongation in Eukaryotes. In *Biochemistry of Lipids and Membranes*, eds. D. E. Vance and J. E. Vance, pp. 181–212. Menlo Park, CA: Benjamin/Cummings.

Craven, K. S., J. Parsons, S. A. Taylor, C. N. Belcher, and D. W. Owens. 2008. The influence of diet on fatty acids in the egg yolk of green sea turtles, *Chelonia mydas*. *J. Comp. Physiol. B* 178(4):495–500.

Creech, L. and P. E. Allman. 1998. Stomach and gastrointestinal contents of stranded Kemp's ridley (*Lepidochelys kempii*) sea turtles in Georgia. In *Proceedings of the Seventeenth Annual Sea Turtle Symposium*, Orlando, FL, compilers S. P. Epperly and J. Braun, NOAA Tech Memo NMFS-SEFSC-415, March 4–8, 1997, p. 167.

Davenport, J. 1998. Sustaining endothermy on a diet of cold jelly: Energetics of the leatherback turtle *Dermochelys coriacea*. *Br. Herpetol. Soc. Bull.* 62:4–8.

Davenport, J. and G. H. Balazs. 1991. "Fiery pyrosomas"—Are pyrosomas an important item in the diet of leatherback turtles? *Br. Herpetol. Soc. Bull.* 37:33–38.

Davenport, J., G. Inagle, and A. K. Hughes. 1982. Oxygen uptake and heart rate in young green turtles (*Chelonia mydas*). *J. Zool.* 198(3):399–412.

Davenport, J. and P. J. Oxford. 1984. Feeding, gut dynamics, digestion and oxygen consumption in hatchling green turtles (*Chelonia mydas*). *Br. J. Herpetol.* 6:351–358.

Davenport, J. and C. R. Scott. 1993. Individuality of growth, appetite, metabolic rate and assimilation of nutrients in young green turtles (*Chelonia mydas* L.). *Herpetol. J.* 3:26–31.

Den Hartog, J. C. 1980. Notes on the food of sea turtles: *Eretmochelys imbricata* (Linnaeus) and *Dermochelys coriacea* (Linnaeus). *Neth. J. Zool.* 30(4):595–610.

Den Hartog, J. C. and M. M. Van Nierop. 1984. A study on the gut contents of 6 leathery turtles *Dermochelys coriacea* (Linnaeus) (Reptilia: Testudines: Dermochelyidae) from British waters and The Netherlands. *Zool. Verhandel.* 209:1–31.

DeNiro, M. J. and S. Epstein. 1978. Influence of diet on the distribution of carbon isotopes in animals. *Geochim. Cosmochim. Acta* 42:495–506.

DeNiro, M. J. and S. Epstein. 1981. Influence of diet on the distribution of nitrogen isotopes in animals. *Geochim. Cosmochim. Acta* 45:341–351.

Deutsch, C. A., N. P. Gruber, R. M. Key, J. L. Sarmiento, and A. Ganachaud. 2011. Denitrification and $N_2$ fixation in the Pacific Ocean. *Global Biogeochem. Cycles* 15:483–506.

Di Bello, A., C. Valastro, F. Staffieri, and A. Crovace. 2006. Contrast radiography of the gastrointestinal tract in sea turtles. *Vet. Radiol. Ultrasound* 47:351–354.

Dodge, K. L., J. M. Logan, and M. E. Lutcavage. 2011. Foraging ecology of leatherback sea turtles in the western North Atlantic determined through multi-tissue stable isotope analyses. *Mar. Biol.* 158:2813–2824.

Duguy, R. 1982. Note sur les méduses des Pertuis Charentais. *Annales de la Société des Sciences Naturelles de la Charente-Maritime* 6:1029–1034.

Duron, M. 1978. Contribution à l'étude de la biologie de *Dermochelys coriacea* (Linné) dans les Pertuis Charentais. Thèse Universite Bordeaux, France.

Duron-Dufrenne, M. 1987. Premier suivi par satellite en Atlantique d'une tortue luth *Dermochelys coriacea*. *Comptes Rendus de l'Academie des Sciences Paris* 304:339–402.

Eisenberg, J. F. and J. Frazier. 1983. A leatherback turtle (*Dermochelys coriacea*) feeding in the wild. *J. Herpetol.* 17(1):81–82.

Evershed, R. P., I. D. Bull, L. T. Corr et al. 2007. Compound-specific stable isotope analysis in ecology and paleoecology. In *Stable Isotopes in Ecology and Environmental Science*, eds. R. H. Michener and K. Lajtha, pp. 480–540. Malden, MA: Blackwell Publishing Ltd.

Fantle, M. S., A. I. Dittel, S. M. Schwalm, C. E. Epifanio, and M. L. Fogel. 1999. A food web analysis of the juvenile blue crab, *Callinectes sapidus*, using stable isotopes in whole animals and individual amino acids. *Oecologia* 120:416–426.

Faure, G. 1986. *Principles of Isotope Geology*, 2nd edn. New York: John Wiley & Sons Ltd.

Ferreira, B., M. Garcia, B. P. Jupp, and A. Al-Kiumi. 2006. Diet of the green turtle (*Chelonia mydas*) at Ra's Al Hadd, Sultanate of Oman. *Chelonian Conserv. Biol.* 5:141–146.

Fossette, S., P. Gaspar, Y. Handrich, Y. Le Maho, and J.-Y. Georges. 2008. Dive and beak movement patterns in leatherback turtles *Dermochelys coriacea* during internesting intervals in French Guiana. *J. Anim. Ecol.* 77(2):236–246.

Fossette, S., A. C. Gleiss, J. P. Casey, A. R. Lewis, and G. C. Hays. 2011. Does prey size matter? Novel observations of feeding in the leatherback turtles (*Dermochelys coriacea*) allow a test of predator-prey size relationships. *Biol. Lett.* 8(3):351–354.

Frazier, J., M. D. Meneghel, and F. A. Achaval. 1985. A clarification on the feeding habits of *Dermochelys coriacea*. *J. Herpetol.* 19(1):159–160.

Frick, J. 1976. Orientation and behavior of hatchling green turtles (*Chelonia mydas*) in the sea. *Anim. Behav.* 24:849–857.

Frick, M. G. and P. A. Mason. 1998. *Lepidochelys kempii* (Kemp's ridley sea turtle) diet. *Herpetol. Rev.* 29:166–168.

Frick, M. G., K. L. Williams, A. B. Bolten, K. A. Bjorndal, and H. R. Martins. 2009. Foraging ecology of oceanic-stage loggerhead turtles *Caretta caretta*. *Endanger. Species Res.* 9:91–97.

Galli, C. and P. Rise. 2006. Origin of fatty acids in the body: Endogenous synthesis versus dietary intakes. *Eur. J. Lipid Sci. Technol.* 108:521–525.

Gannes, L. Z., C. Martinez del Rio, and P. Koch. 1997. Stable isotopes in animal ecology: Assumptions, caveats, and a call for more laboratory experiments. *Ecology* 78:1271–1276.

Gaos, A. R., R. R. Lewison, I. L. Yañez et al. 2012. Shifting the life-history paradigm: Discovery of novel habitat use by hawksbill turtles. *Biol. Lett.* 8:54–56.

Gardner S. C. and W. J. Nichols. 2001. Assessment of sea turtle mortality rates in the Bahia Magdalena region, Baja California Sur, Mexico. *Chelonian Conserv. Biol.* 4:197–199.

Godley, B. J., D. R. Thompson, S. Waldron, and R. W. Furness. 1998. The trophic status of marine turtles as determined by stable isotope analysis. *Mar. Ecol. Prog. Ser.* 166:277–284.

Graham, S. C. 2009. Analysis of the foraging ecology of hawksbill turtles (*Eretmochelys imbricata*) on Hawaìi Island: An investigation utilizing satellite tracking and stable isotopes. MS thesis, University of Hawaìi at Hilo, Hawaìi.

Graham, B. S., P. L. Koch, S. D. Newsome, K. W. McMahon, and D. Aurioles. 2010. Using isoscapes to trace the movements and foraging behavior of top predators in oceanic ecosystems. In *Understanding Movement, Pattern, and Process on Earth through Isotope Mapping*, eds. J. B. West, G. J. Bowen, T. E. Dawson, and K. P. Tu, pp. 299–318. New York: Springer.

Grant, G. S. and D. Ferrell. 1993. Leatherback turtle, *Dermochelys coriacea* (Reptilia, Dermochelidae)—Notes on near-shore feeding behavior and association with Cobia. *Brimleyana* 19:77–81.

Gu, B., C. L. Schelske, and M. V. Hoyer. 1997. Intrapopulation feeding diversity in blue tilapia: Evidence from stable-isotope analysis. *Ecology* 78:2263–2266.

Guitart, R., A. M. Silvestre, X. Guerrero, and R. Mateo. 1999. Comparative study on the fatty acid composition of two marine vertebrates: Striped dolphins and loggerhead turtles. *Comp. Biochem. Physiol. B Biochem. Mol. Biol.* 124(4):439–443.

Hadjichristophorou, M. and D. J. Grove. 1983. A study of appetite, digestion and growth in juvenile green turtle (*Chelonia mydas L.*) fed on artificial diets. *Aquaculture* 30:191–201.

Hatase, H., K. Omuta, and K. Tsukamoto. 2010. Oceanic residents, neritic migrants: A possible mechanism underlying foraging dichotomy in adult female loggerhead turtles (*Caretta caretta*). *Mar. Biol.* 157(6):1337–1342.

Hatase, H., K. Sato, M. Yamaguchi, K. Takahashi, and K. Tsukamoto. 2006. Individual variation in feeding habitat use by adult female green sea turtles (*Chelonia mydas*): Are they obligately neritic herbivores? *Oecologia* 149:52–64.

Hatase, H., N. Takai, Y. Matsuzawa et al. 2002. Size-related differences in feeding habitat use of adult female loggerhead turtles *Caretta caretta* around Japan determined by stable isotope analyses and satellite telemetry. *Mar. Ecol. Prog. Ser.* 233:273–281.

Hatase, H. and K. Tsukamoto. 2008. Smaller longer, larger shorter: Energy budget calculations explain intra-population variation in remigration intervals for loggerhead sea turtles (Caretta caretta). *Canadian Journal of Zoology* 86(7):595–600.

Heaslip, S. G., S. J. Iverson, W. D. Bowen, and M. C. James. 2012. Jellyfish support high energy intake of leatherback sea turtles (*Dermochelys coriacea*): Video evidence from animal-borne cameras. *PLoS One* 7(3): e33259. doi:10.1371/journal.pone.0033259.

Heithaus, M. R., J. L. McLash, A. Frid, L. M. Dill, and G. J. Marshall. 2002. Novel insights into green sea turtle behavior using animal-borne video cameras. *J. Mar. Biol. Assoc. U.K.* 82(6):1049–1050.

Hesslein, R. H., K. A. Hallard, and P. Ramlal. 1993. Replacement of sulfur, carbon, and nitrogen in tissue of growing broad whitefish (*Coregonus nasus*) in response to a change in diet traced by d34S, d13C, and d15N. *Can. J. Fish. Aquat. Sci.* 50:2071–2076.

Hobson, K. A. 1999. Tracing origins and migration of wildlife using stable isotopes: A review. *Oecologia* 120:314–326.

Hobson, K. A. 2003. Making migratory connections with stable isotopes. In *Avian Migration* (Berthold, P. et al., eds), pp. 379–391, Springer-Verlag.

Hobson, K. A. and R. G. Clark. 1992. Assessing avian diets using stable isotopes II: Factors influencing diet-tissue fractionation. *Condor* 94:189–197.

Hobson, K. A., D. M. Schell, D. Renoug, and E. Noseworthy. 1996. Stable carbon and nitrogen isotopic fractionation between diet and tissues of captive seals: Implications for dietary reconstructions involving marine mammals. *Can. J. Fish. Aquat. Sci.* 53:528–533.

Hobson, K. A. and L. I. Wassenaar. 2008. *Tracking Animal Migration with Stable Isotopes*. London, U.K.: Academic Press.

Hochscheid, S., F. Maffucci, F. Bentivegna, and R. P. Wilson. 2005. Gulps, wheezes, and sniffs: How measurement of beak movement in sea turtles can elucidate their behaviour and ecology. *J. Exp. Mar. Biol. Ecol.* 316:45–53.

Holland, D. L., J. Davenport, and J. East. 1990. The fatty acid composition of the leatherback turtle *Dermochelys coriacea* and its jellyfish prey. *J. Mar. Biol. Assoc. U.K.* 70:761–770.

Hughes, G. R. 1974. The sea turtles of south-east Africa II. The biology of the Tongaland loggerhead turtle *Caretta caretta* L. with comments on the leatherback turtle *Dermochelys coriacea* L. and the green turtle *Chelonia mydas* L. in the study region. Oceanography Research Institute, Durban, South Africa, Investigative Report No. 36, pp. 1–96.

Ikonomopoulou, M. P., H. Olszowy, C. Limpus, R. Francis, and J. Whittier. 2011. Trace element concentrations in nesting flatback turtles (*Natator depressus*) from Curtis Island, Queensland, Australia. *Mar. Environ. Res.* 71(1):10–16.

Inger, R., A. Jackson, A. Parnell, and S. Bearhop. 2010. SIAR V4: Stable isotope analysis in R. An ecologist's guide.

IUPAC-IUB Commission on Biochemical Nomenclature. 1967. The nomenclature of lipids. *Journal of Biological Chemistry* 212:4845–4849.

Iverson, S. J., Frost, K. J., and Lowry, L. F. 1997. Fattyacid signatures reveal fine scale structure of foraging distribution of harbor seals and their prey in Prince William Sound, Alaska. *Mar. Ecol. Prog. Ser.* 151:255–271.

Iverson, E. S. and H. O. Yoshida. 1956. Longline fishing for tuna in the central Equatorial Pacific, 1954. Washington, DC: USFWS, Special Scientific Report Fisheries No. 184.

James, M. C. and T. B. Herman. 2001. Feeding of *Dermochelys coriacea* on medusae in the northwest Atlantic. *Chelonian Conserv. Biol.* 4:202–205.

James, M. C., S. A. Sherill-Mix, K. Martin, and R. A. Myers. 2006. Canadian waters provide critical foraging habitat for leatherback sea turtles. *Biol. Conserv.* 133:347–357.

Jones, T. T., B. L. Bostrom, M. D. Hastings, K. S. Van Houtan, D. P. Pauly, and D. R. Jones. 2012. Resource requirements of the Pacific leatherback turtle population. *PLoS One* 7(10): e45447. doi:10.1371/journal.pone.0045447.

Jones, T. T., M. D. Hastings, B. L. Bostrom, R. D. Andrews, and D. R. Jones. 2009. Validation of the use of doubly labeled water for estimating metabolic rate in the green turtle (*Chelonia mydas* L.): A word of caution. *J. Exp. Biol.* 212:2635–2644.

Jones, T. T., J. A. Seminoff, A. Resendiz, and P. L. Lutz. 2004. Energetics of the East Pacific green turtle at a Gulf of California foraging habitat. In Coyne, M. S. and R. D. Clark (compilers). *Proceedings of the Twenty-First Annual Symposium on Sea Turtle Biology and Conservation*. NOAA Technical Memorandum NMFS-SEFSC-528.

Karasov W. H. and D. J. Levey. 1990. Digestive system trade-offs and adaptations of frugivorous passerine birds. *Physiol. Zool.* 63(6):1248–1270.

Kelez, S. 2011. Bycatch and foraging ecology of sea turtles in the eastern Pacific. PhD dissertation, Duke University, Durham, NC.

Kelly, J. F. 2000. Stable isotopes of carbon and nitrogen in the study of avian and mammalian trophic ecology. *Can. J. Zool.* 78:1–27.

Killingley, J. S. and M. Lutcavage. 1983. Loggerhead turtle movements reconstructed from $^{18}O$ and $^{16}O$ profiles from commensal barnacle shells. *Estuar. Coast. Shelf Sci.* 16:345–349.

Kirsch, P. E., S. J. Iverson, and W. D. Bowen. 2000. Effect of a low-fat diet on body composition and blubber fatty acids of captive juvenile harp seals (*Phoca groenlandica*). *Physiol. Biochem. Zool.* 73:45–59.

Kowalski, A. 2006. Specific dynamic action in hatchling and posthatchling green (*Chelonia mydas*) and loggerhead (*Caretta caretta*) sea turtles. MS thesis, Florida Atlantic University, Boca Raton, FL.

Lajtha, K. and J. D. Marshall. 1994. Sources of variation in the stable isotopic composition of plants. In *Stable Isotopes in Ecology and Environmental Science*, eds. K. Lajtha and R. H. Michener, pp. 1–21. Oxford, U.K.: Blackwell Scientific Publications.

Lazar, B., R. Gracan, D. Zavodnik, and N. Tvrtkovic. 2008. Feeding ecology of "pelagic" loggerhead turtles, *Caretta caretta*, in the northern Adriatic Sea: Proof of an early ontogenetic habitat shift. In *Proceedings of the Twenty-Fifth Annual Symposium on Sea Turtle Biology and Conservation*, Savannah GA, December 2008, compilers H. Kalb, A. S. Rohde, K. Gayheart, and K. Shanker. NOAA Technical Memorandum NMFS-SEFSC-582.

Lemons, G., T. Eguchi, B. D. Lyon, R. LeRoux, and J. A. Seminoff. 2012. Effects of blood preservatives on stable isotope values ($\delta^{13}C$ and $\delta^{15}N$) of sea turtle blood tissue: Implications for studies at remote field sites. *Aquat. Biol.* 14:201–206.

Lemons, G., R. Lewison, L. Komoroske et al. 2011. Trophic ecology of green sea turtles in a highly urbanized bay: Insights from stable isotopes and mixing models, *J. Exp. Mar. Biol. Ecol.* 405:25–32.

Leon, Y. M. and K. A. Bjorndal. 2002. Selective feeding in the hawksbill turtle, an important predator in coral reef ecosystems. *Mar. Ecol. Prog. Ser.* 245:249–258.

Limpus, C. J. 1984. A benthic feeding record from neritic waters for the leathery turtle (*Dermochelys coriacea*). *Copeia* 1984:522–523.

Limpus, C. J. 1992. The hawksbill turtle, *Eretmochelys imbricata*, in Queensland: Population structure within a southern Great Barrier Reef feeding ground. *Wildlife Res.* 19:489–505.

Limpus, C. J. 1998. Overview of marine turtle conservation and management in Australia. In *Marine Turtle Conservation and Management in Northern Australia*, eds. R. Kennett, A. Webb, G. Duff, M. Guinea, and G. Hill, pp. 1–8. Darwin, Australia: Northern Territory University.

Limpus, C. J. 2007. *A Biological Review of Australian Marine Turtle Species. 5. Flatback Turtle, Natator depressus* Garman. Brisbane, Queensland, Australia: Queensland Environmental Protection Agency.

Limpus, C. J. 2008a. *A Biological Review of Australian Marine Turtle Species. 1. Loggerhead Turtle, Caretta caretta* Linnaeus. Brisbane, Queensland, Australia: Queensland Environmental Protection Agency.

Limpus, C. J. 2008b. *A Biological Review of Australian Marine Turtle Species. 4. Olive Ridley Turtle Lepidochelys olivacea* (Eschscholtz). Brisbane, Queensland, Australia: Queensland Environmental Protection Agency.

Limpus, C. J. 2009. *A Biological Review of Australian Marine Turtle Species. 3. Hawksbill Turtle, Eretmochelys* imbricata Linnaeus. Brisbane, Queensland, Australia: Queensland Environmental Protection Agency.

Limpus, C. J. and D. J. Limpus. 2000. Mangroves in the diet of *Chelonia mydas* in Queensland, Australia. *Mar. Turtle Newslett.* 89:13–15.

Limpus, C. J., D. J. Limpus, K. E. Arthur, and C. J. Parmenter. 2005. Monitoring green turtle population dynamics in Shoalwater Bay: 2000–2004. Australian Government, Great Barrier Reef Marine Park Authority, Research Publication No. 83.

Limpus, C. J. and N. C. McLachlan. 1994. The conservation status of the leatherback turtle, *Dermochelys coriacea*, in Australia. In *Proceedings of the Australian Marine Turtle Conservation Workshop*, Queensland, Australia, 1994, p. 68.

Lopez-Mendilaharsu, M., Gardner, S. C., Seminoff, J. A., and Riosmena-Rodriguez, R. 2005. Identifying critical foraging habitats of the green turtle (*Chelonia mydas*) along the Pacific coast of the Baja California peninsula, Mexico. *Aquat. Conserv.* 15(3):259–269.

Lutcavage, M. E. and P. L. Lutz. 1986. Metabolic rate and food energy requirements of the leatherback sea turtle *Dermochelys coriacea. Copeia* 1986:796–798.

Lutz, P. L. and J. Musick. 1997. *The Biology of the Sea Turtles*, Vol. 1. Boca Raton, FL: CRC Press.

Macko, S. A., M. L. Fogel-Estep, M. H. Engel, and P. E. Hare. 1986. Kinetic fractionation of nitrogen isotopes during amino acid transamination. *Geochim. Cosmochim. Acta* 50:2143–2146.

Macrae, J. C. and P. J. Reeds. 1980. Prediction of protein deposition in ruminants. In *Protein Deposition in Animals*, eds. P. J. Buttery and D. B. Lindsay, pp. 225–249. London, U.K.: Butterworths.

Malej, A., J. Faganeli, and J. Pezdič. 1993. Stable isotope and biochemical fractionation in the marine pelagic food chain: The jellyfish *Pelagia noctiluca* and net zooplankton. *Mar. Biol.* 116:565–570.

Martínez del Rio, C., N. Wolf, S. A. Carleton, and L. Z. Gannes. 2009. Isotopic ecology ten years after a call for more laboratory experiments. *Biol. Rev.* 84:91–111.

McCauley, S. J. and K. A. Bjorndal. 1999. Conservation implications of dietary dilution from debris ingestion: Sublethal effects in post-hatchling loggerhead sea turtles. *Conserv. Biol.* 13:925–929.

McClellan, C. M., J. Braun-McNeill, L. Avens, B. P. Wallace, and A. J. Read. 2010. Stable isotopes confirm a foraging dichotomy in juvenile loggerhead sea turtle. *J. Exp. Mar. Biol. Ecol.* 387:44–51.

McClelland, J. W. and J. P. Montoya. 2002. Trophic relationships and the nitrogen isotopic composition of amino acids in plankton. *Ecology* 83:2173–2180.

McCue, M. D. 2006. Specific dynamic action: A century of investigation. *Comp. Biochem. Physiol. A* 144:381–394.

McCue, M. D. and E. D. Pollock. 2008. Stable isotopes may provide evidence for starvation in reptiles. *Rapid Commun. Mass Spectrom.* 22:2307–2314.

McCutchan Jr., J. M., W. M. Lewis Jr., C. Kendall, and C. C. McGrath. 2003. Variation in trophic shift for stable isotope ratios of carbon, nitrogen, and sulfur. *Oikos* 102:378–390.

McDermott, A. J., J. A. Seminoff, T. T. Jones, and A. Resendiz. 2006. Food intake and retention time in green turtles (*Chelonia mydas*) from the Gulf of California: Preliminary development of a digestive model. In *Proceedings of the Twenty-Third Annual Symposium on Sea Turtle Biology and Conservation*, compilers N. J. Pilcher. Kuala Lumpur, Malaysia. NOAA Technical Memorandum NMFS-SEFSC-536, March 17–21, 2003.

Merker, G. P. and K. A. Nagy. 1984. Energy utilization by free-ranging *Sceloporus virgatus* lizards. *Ecology* 65(2):575–581.

van de Merwe, J. P., M. Hodge, J. M. Whittier, K. Ibrahim, and S. Y. Lee. 2010. Persistent organic pollutants in the green sea turtle *Chelonia mydas*: Nesting population variation, maternal transfer, and effects on development. *Mar. Ecol. Prog. Ser.* 403:269–278.

Meylan, A. 1988. Spongivory in hawksbill turtles: A diet of glass. *Science* 239:393–395.

Michener, R. and K. Lajtha. 2007. *Stable Isotopes in Ecology and Environmental Science*, 2nd edn. Malten, MA: Blackwell Publishing, 566 pp.

Mills, C. E. 2001. Jellyfish blooms: Are populations increasing globally in response to changing ocean conditions? *Hydrobiologia* 451:55–68.

Miniwaga, M. and W. Wada. 1984. Stepwise enrichment of $d^{15}N$ along food chains: Further evidence of the relation between $d^{15}N$ and animal age. *Geochim. Cosmochim. Acta* 48:1135–1140.

Montoya, J. P. 2007. Natural abundance of $^{15}N$ in marine planktonic ecosystems. In *Stable Isotopes in Ecology and Environmental Science*, eds. R. Michener and K. Lajtha, pp. 176–201. Malden, MA: Blackwell Publishing Ltd.

Morreale, S. J., P. T. Plotkin, D. J. Shaver, and K. J. Kalb. 2007. Adult migration and habitat utilization: Ridley turtles in their element. In *Biology and Conservation of Ridley Sea Turtles*, ed. P. T. Plotjin, pp. 213–229. Baltimore, MD: John Hopkins University Press.

Murray, I. W. and B. O. Wolf. 2012. Tissue carbon incorporation rates and diet-to-tissue discrimination in ectotherms: Tortoises are really slow. *Physiol. Biochem. Zool.* 85:96–105.

Newsome, S. D., M. R. Clementz, and P. L. Koch. 2010. Using stable isotope biogeochemistry to study marine mammal ecology. *Mar. Mamm. Sci.* 26:509–572.

Newsome, S. D., C. M. del Rio, S. Bearhop, and D. L. Phillips. 2007. A niche for isotopic ecology. *Front. Ecol. Environ.* 5(8):429–436.

Obura, D. O., A. Harvey, T. Young, M. M. Eltayeb, and R. von Brandis. 2010. Hawksbill turtles as significant predators on hard coral. *Coral Reefs* 29:759.

Pajuelo, M., K. A. Bjorndal, J. Alfaro-Shigueto, J. A. Seminoff, J. Mangel, and A. B. Bolten. 2010. Stable isotope variation in loggerhead turtles reveals Pacific-Atlantic oceanographic differences. *Mar. Ecol. Prog. Ser.* 417:277–285.

Pajuelo, M., K. A. Bjorndal, K. J. Reich, M. D. Arendt, and A. B. Bolten. 2012. Foraging habitat use and movement patterns of male loggerhead turtles (*Caretta caretta*) as revealed by stable isotopes and satellite telemetry. *Mar. Biol.* 159(6):1255–1267.

Parker, D. M., G. H. Balazs, C. S. King, L. Katahira, and W. Gilmartin. 2009. Short-range movements of hawksbill turtles (*Eretmochelys imbricata*) from nesting to foraging areas within the Hawaiian Islands. *Pac. Sci.* 63(3):371–382.

Parker, D. M., W. J. Cooke, and G. H. Balazs. 2005. Diet of oceanic loggerhead sea turtles (*Caretta caretta*) in the central North Pacific. *Fish. Bull.* 103:142–152.

Parker, D., P. H. Dutton, and G. H. Balazs. 2011. Oceanic diet and distribution of haplotypes for the green turtle, *Chelonia mydas*, in the central North Pacific. *Pac. Sci.* 65(4):419–431.

Pearson, S. F., D. J. Levey, C. H. Greenberg, and C. Martinez del Rio. 2003. Effects of elemental composition on the incorporation of dietary nitrogen and carbon isotopic signatures in an omnivorous songbird. *Oecologia* 135(4):516–523.

Peckham, S. H., D. Maldonado-Diaz, Y. Tremnlay et al. 2011. Demographic implications of alternative foraging strategies in juvenile loggerhead turtles *Caretta caretta* of the North Pacific Ocean. *Mar. Ecol. Prog. Ser.* 425:269–280.

Pendoley, K. and J. Fitzpatrick. 1999. Browsing on mangroves by green turtles in Western Australia. *Mar. Turtle Newslett.* 84:10.

Peterson, B. J. and B. Fry. 1987. Stable isotopes in ecosystem studies. *Annu. Rev. Ecol. Syst.* 18:293–320.

Phillips, D. L. 2001. Mixing models in analyses of diet using multiple stable isotopes: A critique. *Oecologia* 127:166–170.

Phillips, R. A., S. Bearhop, R. A. R. McGill, and D. A. Dawson. 2009. Stable isotopes reveal individual variation in migration strategies and habitat preferences in a suite of seabirds during the breeding period. *Oecologia* 160:795–806.

Phillips, D. L. and J. W. Gregg. 2003. Source partitioning using stable isotopes: Coping with too many sources. *Oecologia* 136:261–269.

Phillips, D. L. and P. L. Koch. 2002. Incorporating concentration dependence in stable isotope mixing models. *Oecologia* 130:114–125.

Phillips, D. L., S. D. Newsome, and J. W. Gregg. 2005. Combining sources in stable isotope mixing models: Alternative methods. *Oecologia* 144:520–524.

Pinnegar, J. K. and N. V. C. Polunin. 1999. Differential fractionation of $\delta^{13}$C and $\delta^{15}$N among fish tissues: Implications for the study of trophic interactions. *Funct. Ecol.* 13:225–231.

Poage, M. A. and C. P. Chamberlain. 2001. Empirical relationships between elevation and the stable isotope composition of precipitation and surface waters: Considerations for studies of paleoelevation change. *Am. J. Sci.* 301:1–15.

Polovina, J. J., G. H. Balazs, E. A. Howell, D. M. Parker, M. P. Seki, and P. H. Dutton. 2004. Forage and migration habitat of loggerhead (*Caretta caretta*) and olive ridley (*Lepidochelys olivacea*) sea turtles in the central North Pacific Ocean. *Fish. Oceanogr.* 13(1):1–16.

Popp, B. N., B. S. Graham, R. J. Olson et al. 2007. Insight into the trophic ecology of yellowfin tuna, *Thunnus albacares*, from compound-specific nitrogen isotope analysis of proteinaceous amino acids. In *Stable Isotopes as Indicators of Ecological Change*, eds. T. Dawson and R. Siegwolf, pp. 173–190. Amsterdam, the Netherlands: Elsevier Academic Press.

Post, D. M. 2002. Using stable isotopes to estimate trophic position: Models, methods, and assumptions. *Ecology* 83:703–718.

Pritchard, P. C. H. 1976. Post-nesting movements of marine turtles (Cheloniidae and Dermochelyidae) tagged in the Guianas. *Copeia* 1976:749–754.

Ramos, R., J. Gonzalez-Solis, J. P. Croxall, D. Oro, and X. Ruiz. 2009. Understanding oceanic migrations with intrinsic biogeochemical markers. *PLoS One* 4(7):e6236.

Reich, K. J., K. A. Bjorndal, and A. B. Bolten. 2007. The 'lost years' of green turtles: Using stable isotopes to study cryptic life stages. *Biol. Lett.* 3:712–714.

Reich, K. J., K. A. Bjorndal, M. G. Frick, B. E. Witherington, C. Johnson, and A. B. Bolten. 2009. Polymodal foraging in adult female loggerheads (*Caretta caretta*). *Mar. Biol.* 157:113–121.

Reich, K. J., K. A. Bjorndal, and C. Martínez del Rio. 2008. Effects of growth and tissue type on the kinetics of $^{13}$C and $^{15}$N incorporation in a rapidly growing ectotherm. *Oecologia* 155:651–663.

Revelles, M., L. Cardona, A. Aguilar, A. Borrell, G. Fernández, and M. San Félix. 2007a. Concentration of stable C and N isotopes in several tissues of the loggerhead sea turtle *Caretta caretta* from the western Mediterranean and dietary implications. *Sci. Mar.* 71:87–93.

Revelles, M., L. Cardona, A. Aguilar, and G. Fernández. 2007b. The diet of pelagic loggerhead sea turtles *Caretta caretta* off the Balearic archipelago (western Mediterranean): Relevance of long-line baits. *J. Mar. Biol. Assoc. U.K.* 87:805–813.

Rincon-Diaz, M. P., C. E. Diez, R. P. Van Dam, and A. M. Sabat. 2011. Foraging selectivity of the hawksbill sea turtle (*Eretmochelys imbricata*) in the Culebra Archipelago, Puerto Rico. *J. Herpetol.* 45(3):277–282.

Roth, J. D. and K. A. Hobson. 2000. Stable carbon and nitrogen isotopic fractionation between diet and tissue of captive red fox: Implications for dietary reconstruction. *Can. J. Zool.* 78:848–852.

Rubenstein, D. R. and K. A. Hobson. 2004. From birds to butterflies: Animal movement patterns and stable isotopes. *Trends Ecol. Evol.* 19:256–263.

Russell, D. J. and G. H. Balazs. 1994. Colonization by the alien marine alga *Hypnea musciformis* (Wulfen) J. Ag. (Rhodophyta: Gigartinales) in the Hawaiian Islands and its utilization by the green turtle, *Chelonia mydas* L. *Aquat. Bot.* 47:53–60.

Salmon, M., T. T. Jones, and K. Horch. 2004. Ontogeny of diving and feeding behavior in juvenile sea turtles: A comparison study of green turtles (*Chelonia mydas* L.) and leatherbacks (*Dermochelys coriacea* L.) in the Florida current. *J. Herpetol.* 38:36–43.

Schmidt-Nielsen, K. 1997. Animal physiology: Adaptation and environment. Cambridge University Press. 612 p.

Seaborn, G. T., M. K. Moore, and G. H. Balazs. 2005. Depot fatty acid composition in immature green turtles (*Chelonia mydas*) residing at two near-shore foraging areas in the Hawaiian Islands. *Comp. Biochem. Physiol. B* 140(2):183–195.

Sealy, J. C., N. J. Van Der Merwe, J. A. L. Thorp, and J. A. Lanham. 1987. Nitrogen isotopic ecology in southern Africa: Implications for environmental and dietary tracing. *Geochim. Cosmochim. Acta* 51:2707–2717.

Secor, S. M. 2009. Specific dynamic action: A review of the postprandial metabolic response. *J. Comp. Physiol. B* 179:1–56.

Secor, S. M. and J. Diamond. 2000. Evolution of regulatory response to feeding in snakes. *Physiol. Biochem. Zool.* 73:123–141.

Seminoff, J. A., S. R. Benson, K. E. Arthur, P. H. Dutton, T. Eguchi, R. Tapilatu, and B. N. Popp. 2012. Stable isotope tracking of endangered sea turtles: Validation with satellite telemetry and $\delta^{15}$N analysis of amino acids. *PLoS One* 7(5): e37403. doi:10.1371/journal.pone.0037403.

Seminoff, J. A., K. A. Bjorndal, and A. B. Bolten. 2007. Stable carbon and nitrogen isotope discrimination and turnover in pondsliders *Trachemys scripta*: Insights for trophic study of freshwater turtles. *Copeia* 2007:534–542.

Seminoff, J. A., T. T. Jones, T. Eguchi, D. R. Jones, and P. H. Dutton. 2006b. Stable isotope discrimination ($\delta^{13}$C and $\delta^{15}$N) between soft tissues of the green sea turtle *Chelonia mydas* and its diet. *Mar. Ecol. Prog. Ser.* 308:271–278.

Seminoff, J. A., T. T. Jones, T. Eguchi, M. Hastings, and D. R. Jones. 2009. Stable carbon and nitrogen isotope discriminations in soft tissues of the leatherback turtle (*Dermochelys coriacea*): Insights for trophic studies of marine turtles. *J. Exp. Mar. Biol. Ecol.* 381(1):33–41.

Seminoff, J. A., T. T. Jones, and G. J. Marshall. 2006a. Underwater behavior of green turtles monitored with video-time-depth recorders: What's missing from dive profiles? *Mar. Ecol. Prog. Ser.* 322:269–280.

Seminoff, J. A., A. Resendiz, and W. J. Nichols. 2002a. Diet of east Pacific green turtles (*Chelonia mydas*) in the central Gulf of California, México. *J. Herpetol.* 36(3):447–453.

Seminoff, J. A., A. Resendiz, W. J. Nichols, and L. Brooks. 2003. Occurrence of hawksbill turtles, *Eremochelys imbricata*, near Baja California. *Pac. Sci.* 57:9–16.

Seminoff, J. A., A. Resendiz, W. J. Nichols, and T. T. Jones. 2002b. Growth rates of wild green turtles (*Chelonia mydas*) at a temperate foraging area in the Gulf of California, Mexico. *Copeia* 2002(3):610–617.

Seminoff, J. A., A. Resendiz, B. Resendiz, and W. J. Nichols. 2004. Occurrence of loggerhead sea turtles (*Caretta caretta*) in the Gulf of California, Mexico: Evidence of life-history variation in the Pacific Ocean. *Herpetol. Rev.* 35(1):24–27.

Semmens, B. X. and J. W. Moore. 2008. MixSIR: A Bayesian stable isotope mixing model, Version 1.04. http://www.ecologybox.org

Seney, E. E. and A. M. Landry Jr. 2011. Movement patterns of immature and adult female Kemp's ridley sea turtles in the northwestern Gulf of Mexico. *Mar. Ecol. Prog. Ser.* 440:241–254.

Seney, E. E. and J. A. Musick. 2005. Diet analysis of Kemp's ridley sea turtles (*Lepidochelys kempii*) in Virginia. *Chelonian Conserv. Biol.* 4:864–871.

Seney, E. E. and J. A. Musick. 2007. Historical diet analysis of loggerhead sea turtles (*Caretta Caretta*) in Virginia. *Copeia* 2007(2):478–489.

Shaver, D. J. 1991. Feeding ecology of wild and head-started Kemp's ridley sea turtles in South Texas waters. *J. Herpetol.* 25:327–334.

Snover, M. L., A. A. Hohn, L. B. Crowder, and S. S. Heppell. 2007. Age and growth in Kemp's ridley sea turtles: Evidence from mark-recapture skeletochronology. In *Biology and Conservation of Ridley Sea Turtles*, ed. P. T. Plotjin, pp. 89–105. Baltimore, MD: John Hopkins University Press.

Somes, C. J., A. Schmittner, E. D. Galbraith et al. 2010. Simulating the global distribution of nitrogen isotopes in the ocean. *Global Biogeochem. Cycles* 24:GB4019.

Southwood, A. L., R. D. Andrews, F. V. Paladino, and D. R. Jones. 2005. Effects of swimming and diving behavior on body temperatures of Pacific leatherbacks in tropical seas. *Physiol. Biochem. Zool.* 78:285–297.

Speakman, J. R. 1997. Doubly labelled water: Theory and practice. London: Chapman & Hall. p. 399.

Thode, H. G. 1991. Sulfur isotopes in nature and the environment: An overview. In *Stable Isotopes: Natural and Anthropogenic Sulfur in the Environment*, eds. H. R. Krouse and V. A. Grinenko, pp. 1–26. New York: John Wiley & Sons.

Tieszen, L. L. and T. W. Boutton. 1989. Stable carbon isotopes in terrestrial ecosystem research. In *Stable Isotopes in Ecological Research*, eds. P. W. Rundel, J. R. Ehleringer, and K. A. Nagy, pp. 167–195. New York: Springer-Verlag.

Tieszen, L. L., T. W. Boutton, K. G. Tesdahl, and N. A. Slade. 1983. Fractionation and turnover of stable carbon isotopes in animal tissues: Implications for $\delta^{13}$C analysis of diet. *Oecologia* 57:32–37.

Tieszen, L. L. and T. Fagre. 1993. Effect of diet quality and composition on the isotopic composition of respiratory $CO_2$, bone collagen, bioapatite, and soft tissues. In *Molecular Archeology of Prehistoric Human Bone*, eds. J. Lambert and G. Grupe, pp. 123–135. Berlin, Germany: Springer-Verlag.

Tomas, J., F. J. Aznar, and J. A. Raga. 2001. Feeding ecology of the loggerhead turtle *Caretta caretta* in the western Mediterranean. *J. Zool.* 255(4):525–532.

Uhle, M. E., S. A. Macko, H. J. Spero, M. H. Engel, and D. W. Lea. 1997. Sources of carbon and nitrogen in modern planktonic foraminifera: The role of algal symbionts as determined by bulk and compound specific stable isotopic analyses. *Org. Geochem.* 27:103–113.

Valente, A. L., I. Marco, M. L. Parga, S. Lavin, F. Alegre, and R. Cuenca. 2008. Ingesta passage and gastric emptying times in loggerhead sea turtles (*Caretta caretta*). *Res. Vet. Sci.* 84:132–139.

Van Houtan, K. S., S. K. Hargrove, and G. H. Balazs. 2010. Land use, macroalgae, and a tumor-forming disease in marine turtles. *PLoS One* 5(9):e12900.

Vander Zanden, H. B., K. E. Arthur, A. B. Bolten et al. In review. Interpreting the isotopic niche of a breeding population to understand the trophic ecology of a highly migratory species. *Mar. Ecol. Prog. Ser.*

Vander Zanden, H. B., K. A. Bjorndal, P. W. Inglett, and A.B. Bolten. 2012a. Marine derived nutrients from green turtle nests subsidize terrestrial beach ecosystems. *Biotropica* 44:294–301.

Vander Zanden, H. B., K. A. Bjorndal, W. Mustin, J. M. Ponciano, and A. B. Bolten. 2012b. Inherent variation in stable isotope values and discrimination factors in two age-classes of green turtles. *Physiol. Biochem. Zool.* 85(5):431–441.

Vander Zanden, H. B., K. A. Bjorndal, K. J. Reich, and A. B. Bolten. 2010. Individual specialists in a generalist population: Results from a long-term stable isotope series. *Biol. Lett.* 6:711–714. doi:10.1098/rsbl.2010.0124.

Walker, T. A. 1991. Juvenile flatback turtles in proximity to coastal nesting islands in the Great Barrier Reef Province. *J. Herpetol.* 25:246–248.

Walker, T. A. and C. J. Parmenter. 1990. Absence of a pelagic phase in the life cycle of the flatback turtle, *Natator depressa* (Garman). *J. Biogeogr.* 17:275–278.

Wallace, B. P., L. Avens, J. Braun-McNeill, and C. M. McClellan. 2009. The diet composition of immature loggerheads: Insights on trophic niche, growth rates, and fisheries interactions. *J. Exp. Mar. Biol. Ecol.* 373:50–57.

Wallace, B. P. and T. T. Jones. 2008. What makes marine turtles go: A review of metabolic rates and their consequences. *J. Exp. Mar. Bio. Ecol.* 356:8–24.

Wallace, B. P., S. S. Kilham, F. V. Paladino, and J. R. Spotila. 2006a. Energy budget calculations indicate resource limitation in Eastern Pacific leatherback turtles. *Mar. Ecol. Prog. Ser.* 318:263–270.

Wallace, B. P., J. A. Seminoff, S. S. Kilham, J. R. Spotila, and P. H. Dutton. 2006b. Leatherback turtles as oceanographic indicators: Stable isotope analyses reveal a trophic dichotomy between ocean basins. *Mar. Biol.* 149:953–960.

Witherington, B. E. 2002. Ecology of neonate loggerhead turtles inhabiting lines of downwelling near a Gulf Stream front. *Mar. Biol.* 140:843–853.

Witt, M. J., A. C. Broderick, D. J. Johns et al. 2007. Prey landscapes help identify potential foraging habitats for leatherback turtles in the northeast Atlantic. *Mar. Ecol. Prog. Ser.* 337:231–244.

Wood, J. R. and F. E. Wood. 1981. Growth and digestibility for the green turtle (*Chelonia mydas*) fed diets containing varying protein levels. *Aquaculture* 25(2-3):269–274.

Zbinden, J. A., S. Bearhop, P. Bradshaw et al. 2011. Migratory dichotomy and associated phenotypic variation in marine turtles revealed by satellite tracking and stable isotope analysis. *Mar. Ecol. Prog. Ser.* 421:291–302.

# 10 Predators, Prey, and the Ecological Roles of Sea Turtles

*Michael R. Heithaus*

## CONTENTS

## 10.1 INTRODUCTION

Predator–prey interactions are critical in shaping behavior, ecology, population dynamics, and life histories and are central to community and ecosystem dynamics. Sea turtles undergo dramatic shifts in body size and often in habitats (e.g., Musick and Limpus 1997, Plotkin 2003). Not surprisingly, their interactions with both predators and prey vary considerably through ontogeny. In addition, the need to come to the surface to replenish oxygen stores has important implications for both foraging behavior and predator avoidance. In this chapter, I place predator–prey (including primary producers) interactions of sea turtles in a theoretical framework from both a behavioral and a trophic perspective and conclude with a consideration of how these interactions might shape the ecological roles of sea turtles. I do not investigate the diets of individual species or food selection, which are covered in Chapter 9 (see also Bjorndal 1997 for a review). I hope that this chapter will stimulate further studies of sea turtles as both prey and consumers that will ultimately increase our understanding of the roles, and importance, of these reptiles in their ecosystems.

## 10.2 SEA TURTLES AS PREY

Because of changes in body size, the risk that sea turtles face from predators drops considerably as they grow. Large juveniles and adults are characterized by high survival rates (e.g., Heppell et al. 2003). Predators on large juvenile and adults, however, may still be important in shaping turtle behavior, populations (e.g., Heithaus et al. 2008b; Section 10.2.1.2), and life histories. For example, van Buskirk and Crowder (1994) hypothesized that the rate of predation on adults relative to earlier life history stages could be responsible for similar geographic patterns of variation in body sizes across four turtle species (green turtles *Chelonia mydas*, hawksbills *Eretmochelys imbricata*, loggerheads *Caretta caretta*, leatherbacks *Dermochelys coriacea*). Although data necessary to test this hypothesis are not yet available, the authors suggested that relatively small body sizes in the Indian and Pacific Ocean compared to the Caribbean might be explained by heavy predation pressure on adults and lower predation pressure on eggs and hatchlings in the Indian and Pacific Oceans and the opposite pattern of predation pressure in the Caribbean.

In this section, I begin by considering the predators of sea turtles at various life history stages then describe the behavioral responses of turtles to these predators and the potential for predators to influence turtle population sizes.

### 10.2.1 PREDATORS OF SEA TURTLES AND PREDATION RATES

#### 10.2.1.1 Eggs, Hatchlings, and Small Juveniles

Sea turtle eggs, hatchlings, and small juveniles experience considerable risks from predators. There are a variety of predators on sea turtle nests including insects (e.g., Marcos et al. 2003, Donlan et al. 2004, Caut et al. 2006), birds (e.g., Tomillo et al. 2010), native and introduced mammals (e.g., Engeman et al. 2006, 2010, Ficetola 2008), large lizards (Blamires 2004), and crocodiles (Whiting and Whiting 2011). Rates of nest loss vary considerably depending on predator populations and the presence of predator control programs, which generally are highly effective. Overall, it appears that turtle populations are relatively resilient to high rates of predation on nests and hatchlings (e.g., Heppell et al. 2003), but in some areas egg predators have been linked to recruitment failure (e.g., Chaloupka and Limpus 2001). Also, protecting nests from predators may have aided population recovery in some cases (e.g., Dutton et al. 2005).

In 2002 and 2003, beetle larvae were identified as the most destructive egg predator of loggerhead nests on a southeast Florida beach (Donlan et al. 2004), and in French Guiana mole cricket (*Scapteriscus didactylus*) nymphs were found to prey upon 3%–40% of yolked eggs in leatherback nests (Marcos et al. 2003). In the Southeast United States, invasive fire ants (*Solenopsis invicta*) are present on many sea turtle nesting beaches and can kill sea turtle hatchlings while they are emerging from eggs or just after their emergence from eggs while hatchlings are still belowground (Moulis 1997, Allen et al. 2001, Parris et al. 2002).

Sea turtle eggs were the major prey for goannas (*Varanus panoptes*) in Fog Bay, Australia, during the dry season, but it did not appear that goannas changed their use of beaches in response to turtle nesting. It was unclear whether goanna predation impacted turtle population sizes (Blamires 2004).

Mammalian predators can have heavy impacts on sea turtle nesting success. In many cases, mammals may destroy more than 80% of nests not protected by humans (e.g., Engeman et al. 2006, 2010, Ficetola 2008). For example, 80% of hawksbill nests in Qatar were completely destroyed by feral cats and foxes. All nests were disturbed to some degree (Ficetola 2008). In northeastern Australia, more than 90% of nests were depredated by foxes before fox baiting began (Limpus and Limpus 2003a). Along the Gulf of Mexico coast of Florida, 60%–84% of nests were depredated (Engeman et al. 2010). Raccoons (*Procyon lotor*) and armadillos (*Dasypus novemcinctus*) historically destroyed up to 95% of green turtle, leatherback, and loggerhead nests at Hobe Sound, Florida (Engeman et al. 2006). In Turkey, foxes and golden jackals preyed upon up to 75% of green turtle nests (Brown and MacDonald 1995).

Reducing nest predation requires an understanding of how risk to nests varies in space and time as well as how predators detect nests. Some predators, like coatis (*Nasua*) likely rely on olfactory clues to find nests while others, like small Asian mongooses (*Herpestes javanicus*) foraging on hawksbill nests, use sand disturbance (Leighton et al. 2008). Leighton et al. (2011) used survival analysis, a statistical method to estimate how daily probabilities of nest survival vary with nest age and the factors that influence daily survival probabilities at a given age, to determine that Asian mongoose predation on hawksbill nests in the Caribbean could be predicted with a high degree of accuracy. Predation—which ranged from 17.9% to 38.9% per year—was highest for new nests, declined rapidly, and then increased again before hatching. There was, however, an increase in overall predation rates through the nesting season. Predation rates were highest in and near the vegetation that mongooses preferred—more than 50% of nests were preyed upon—compared to open sand. The avoidance of open sand by mongooses, however, likely was driven by avoidance of humans (Leighton et al. 2010) so predation rates of nests in open sand habitats may be higher for less disturbed nesting beaches. In addition, despite the tendency for mongooses to avoid open sand habitats, predation rates in this habitat increased as turtle nest densities increased.

Spatial variation in nest predation across open and vegetated habitats varies with predator type. For example, mongooses (Leighton et al. 2011) and mole crickets (Caut et al. 2006) are much more likely to attack nests near vegetation while other predators—like raccoons and canids—move across habitat types and may prey upon turtle nests in open and vegetated areas equally (e.g., Brown and MacDonald 1995, Whiting et al. 2007). Therefore, predicting nest predation rates requires an understanding of the identity and behavior of potential nest predators at a beach.

Species interactions among predators also can influence predation rates on nests. For example, raccoons and ghost crabs are major predators of loggerhead turtle eggs, but raccoons consume ghost crabs. This "intraguild predation" has important implications for predation rates on turtle nests (Barton and Roth 2008). On beaches in Florida, there generally are more crabs at beaches with fewer raccoons, and the highest rates of nest predation by *both* are found at beaches with the highest crab densities and lowest raccoon densities. This could be because high numbers of crab burrows facilitate olfactory discovery of nests by raccoons (Barton and Roth 2008). These findings suggest that attempts to maintain raccoons at very low abundances may not be the optimal conservation strategy for minimizing egg predation (Barton and Roth 2008).

Predation rates on hatchlings that are crossing beaches vary among locations and species. At Playa Grande, Costa Rica, predators—including ghost crabs (*Ocypode occidentalis*; 48% of observed predation), great blue herons (*Ardea herodias*), yellow-crowned night herons (*Nycticorax violaceus*), and crested caracaras (*Caracara plancus*)—killed 12% of hatchling leatherback turtles on the beach (Tomillo et al. 2010). Nocturnal birds and diurnal predators had the largest impact per nest that they attacked, but attacked far fewer nests than ghost crabs. At Playa Ostional, Costa Rica, less than 4% of olive ridley (*Lepidochelys olivacea*) hatchlings were consumed by predators—primarily large hermit crabs and ghost crabs (Madden et al. 2008). On southern Great Barrier Reef loggerhead rookeries, less than 2% of the hatchlings were consumed by predators on the beach (Limpus and Limpus 2003a). Predation rates on flatback turtle (*Natator depressus*) hatchlings vary considerably—between 3% and 38% were captured by predators while crossing an Australian beach (Limpus et al. 1983). The probability of predation also can vary with hatchling size for at least some predators. For example, crabs (*Ocypode cursor*) tended to kill smaller green turtle hatchlings at Bijagós Archipelago, Guinea-Bissau (Rebelo et al. 2012).

Several studies have investigated predation rates on hatchlings once they enter the ocean, where hatchlings can be consumed by a variety of teleosts, sharks, birds, and even squid (e.g., Walker and Parmenter 1990, Gyuris 1994, Wyneken et al. 2000, Stewart and Wyneken 2004). In eastern Australia, fishes consumed an average of around 30% of green turtle hatchlings (range 0%–85%) followed from the beach to the reef crest (Gyuris 1994). Predation rates were highest during low tide. It appears that for this population the majority of first-year mortality occurs within the first hour in the ocean and crossing the reef crest (Gyuris 1994). Predation rates appear to be higher on green turtle hatchlings departing a Malaysian beach (Pilcher et al. 2000). In the first 2 h after reaching the ocean, an average of 46% of released hatchlings was taken by predators. Hatchlings that entered deeper waters experienced lower predation rates (21%) than those that remained in shallower waters (77%). Overall predation rates also were higher when hatchlings were released in larger groups, and extremely high predation rates may reflect management practices where hatchlings are released predictably in space and time (Pilcher et al. 2000, Stewart and Wyneken 2004).

Predation rates appear to be lower along the Atlantic coast of Florida where only 5% of 217 loggerhead turtle hatchlings were taken in the first 15 min in the ocean (corresponding to the estimated amount of time needed to cross the nearshore reef line) (Stewart and Wyneken 2004). Even lower predation rates were found for loggerhead hatchlings at one site on the west coast of Florida (1%; Whelan and Wyneken 2007). Predation rates from 15 min to 1 h after entering the water appear to be even lower (Witherington and Salmon 1992). Off Florida's east coast, tarpon (*Megalops atlanticus*) were most frequently observed consuming hatchlings with a carcharhinid shark and a catfish also recorded as predators (Stewart and Wyneken 2004). Lutjanid and carangid fish were the most common potential predators captured off nesting beaches. Survival was lower at a nearby hatchery site, where 28% of 125 hatchlings were taken in their first 15 min at sea (Wyneken and Salmon 1997). It is possible that, at hatchery sites, the unnaturally high densities of hatchlings that enter the water more predictably than at natural beaches may attract more predators, which results in abnormally high predation rates (Stewart and Wyneken 2004). Also, predation rates on hatchlings may vary throughout the hatching season. For example, there are increases in predation rates for loggerhead hatchlings late in the hatching season along Florida's Atlantic coast (Stewart and Wyneken 2004, Whelan and Wyneken 2007).

Comparatively little is known about predation on juvenile turtles once they have entered pelagic waters. Presumably, they would be at risk from birds, large teleosts and sharks, but remaining in algal mats likely minimizes this risk for at least some species (see Section 10.2.2.1). Young juvenile flatback turtles, which inhabit the same shallow coastal waters as adults, fall prey to sea eagles (*Haliaeetus leucogaster*), and flatback juveniles may be important prey for eagles (Walker and Parmenter 1990). Presumably juvenile flatbacks would also face the same suite of potential predators as sympatric adults.

## 10.2.1.2 Adults and Large Juveniles

In general, predation on large juvenile and adult sea turtles has received relatively little attention because these age/size classes typically experience low predation rates (Bjorndal et al. 2003; see Heithaus et al. 2008b for an earlier detailed review of lethal and nonlethal effects of predators on adult turtles). Also, large magnitude and widespread declines in populations of potential predators (e.g., Ferretti et al. 2010) occurred before many of the current extensive research programs on adult sea turtles began, making it difficult to infer historic levels of predation. Recent studies from diverse systems, however, have shown that predation rates need not be high for predators to play an important role in shaping the dynamics of populations and communities (see Heithaus et al. 2008a, Creel and Christianson 2008, Creel 2011 for recent reviews).

For the past several hundred years, humans certainly have been the most significant predator of adult turtles (e.g., Jackson 1997, Spotila et al. 2000), and human-inflicted mortality rates are likely orders of magnitude greater than those inflicted by natural predators for many of sea turtle populations. Here, however, I focus on nonhuman predators. Unfortunately, very few studies have been conducted on predators of adult sea turtles. Most insights come from stomach contents analysis of potential predators, observations of predator-inflicted injuries, and observed predation events during restricted periods. These accounts show that, although predation is infrequent, adults of all sea turtle species are susceptible to predators.

### 10.2.1.2.1   Terrestrial Mammals

Feral dogs are a threat to nesting turtles (e.g., Caut et al. 2006), and throughout the Americas, jaguars (*Panthera onca*) have been recorded killing nesting green turtles, olive ridleys, hawksbills, and leatherbacks. In Suriname, an individual jaguar killed up to 13 green turtles over the period of a few days, and jaguars killed at least 82 green turtles over 11 years (Aurtar 1994). Predation by jaguars may be even more common at Tortuguero, Costa Rica (Troëng 2000, Harrison et al. 2005, Veríssimo et al. 2012). Over a 5 year period (2005–2010), at least 676 turtles (n = 672 green turtles, n = 3 hawksbills, n = 1 leatherback) were killed by jaguars on the nesting beach (Veríssimo et al. 2011, 2012). The number of turtles killed per survey increased steadily from 2005/2006 to 2009/2010, when at least 189 turtles were killed. Given the relatively high, and increasing, rate of jaguar predation on nesting turtles at Tortuguero, further studies are warranted to assess jaguar impacts on green turtle populations and the importance of green turtles to jaguar diets (Veríssimo et al. 2012).

### 10.2.1.2.2   Crocodilians

Several species of crocodiles and alligators (*Alligator mississippiensis*) may kill large juvenile and adult turtles on land or in the water. In Costa Rica, American crocodiles (*Crocodylus acutus*) killed nine olive ridley turtles before and after *arribada*s (Ortiz et al. 1997). In the Indo-Pacific, leatherbacks, olive ridleys, and flatbacks are taken by saltwater crocodiles (*Crocodylus porosus*) on or near nesting beaches (e.g., Hirth et al. 1993, Sutherland and Sutherland 2003, Whiting and Whiting 2011). At Crab Island, Australia, saltwater crocodiles took at least one flatback a week during the nesting season (Sutherland and Sutherland 2003). In the Northern Territory and northern Western Australia, saltwater crocodile predation on nesting olive ridley and flatback turtles occurs regularly on the beach and at the water's edge (Whiting and Whiting 2011). About 5% of nesting olive ridleys suffer damage from crocodile attacks; ca. 1% of the nesting population may be taken in areas where observations can be made (in-water predation rates are unknown) during the peak of nesting (Whiting and Whiting 2011). Increasing crocodile populations could be an important source of mortality for sea turtles in northern Australia (Whiting and Whiting 2011). Although unlikely to be a major mortality source, recent evidence suggests that American alligators may take sea turtles where their distributions overlap (Nifong et al. 2011).

### 10.2.1.2.3   Marine Mammals

Mediterranean monk seals (*Monachus monachus*) have been recorded killing loggerhead turtles, including adults, but it is unlikely that monk seals are common predators of sea turtles (see Fertl and Fulling, 2007). Killer whales (*Orcinus orca*) take leatherback turtles and may prey upon olive ridley and green turtles (e.g., Pitman and Dutton 2004, Fertl and Fulling 2007). It is unclear how frequently killer whales might prey upon turtles, but they could be one of the major threats to adult sea turtles in some areas.

*10.2.1.2.4  Sharks*

At sea, sharks—particularly tiger sharks (*Galeocerdo cuvier*)—are the most common predators of large juvenile and adult turtles (e.g., Marquez 1990, see Heithaus et al. 2008b for a review). In coastal waters, juvenile turtles and adults of smaller species may also be at risk from bull sharks (*Carcharhinus leucas*) (Cliff and Dudley 1991). White shark (*Carcharodon carcharias*) attacks have been recorded on leatherback turtles (Fergusson et al. 2000).

Tiger sharks are found in coastal and offshore habitats of tropical and warm-temperate waters where many sea turtle species are abundant. In Shark Bay, Australia, tiger sharks prefer to forage over shallow seagrass habitats where turtles forage (Heithaus et al. 2002a, 2006). Tiger sharks are adapted for consuming large prey, like sea turtles. They reach over 4.5 m in length and have a broad head and large serrated teeth in both upper and lower jaws that can cut in both directions (Randall 1992, Motta and Wilga 2001). In many locations, large juvenile and adult turtles, including green turtles, loggerheads, hawksbills, and olive ridleys, are regularly found in the diets of tiger sharks (e.g., Witzell 1987, Simpfendorfer 1992, Lowe et al. 1996; see Heithaus et al. 2008b for more details). A tiger shark attack has been recorded on a leatherback (Keinath and Musick, 1993). Tiger sharks generally increase the proportion of turtles in their diets at larger body sizes (e.g., >3 m TL; e.g., Lowe et al. 1996, Simpfendorfer et al. 2001). Although some authors (e.g., Witzell 1987) have speculated that tiger sharks are selective predators on sea turtles, their dietary importance to tiger sharks varies spatially. Also, the contribution of sea turtles to shark diets could be somewhat overestimated because of slow digestion rates of turtle shells and beaks relative to other prey types.

Injury rates have been used to investigate inter- and intraspecific variation in the risk of predation when predation events are rarely observed (e.g., Heithaus 2001a,b). There are numerous caveats to using this approach because injury probabilities vary with body sizes of predators and prey, escape abilities of the prey, efficiency of predators (Heithaus 2001a), and difficulties in ascertaining the origin and timing of injuries (Witzell 2007). In the absence of other data, however, injury rates can provide initial insights into predation risk to sea turtles and aid in the development of testable hypotheses.

Shark-inflicted injuries appear to be relatively rare in ridley turtles (Shaver 1998, Witzell 2007). Because of their small body sizes, however, ridley turtles that are attacked would likely be consumed whole or killed so the lack of injuries does not necessarily reflect low predation risk. In contrast, much larger leatherback turtles would be expected to be consumed whole less frequently and injuries would be more likely. Two of 34 (6%) nesting leatherback turtles observed in Papua New Guinea had portions of their carapace removed by sharks (Hirth et al. 1993). Off the Cayman Islands, 15% of injuries to hawksbills may have been due to shark attacks (Blumenthal et al. 2009a), but the overall rate of potential shark-inflicted injuries was low at ca. 3% of turtles.

Along the northeast coast of Australia, shark predation may be relatively infrequent on adult green and loggerhead turtles at their feeding grounds. No fresh shark-inflicted injuries were recorded on 862 green turtles and only one was found among 320 loggerhead turtles (Limpus et al. 1994a,b). New juvenile recruits, however, appear to face a much greater threat; 24% of 84 loggerheads recently recruited to neritic feeding habitats had scars from recent shark bites (Limpus and Limpus 2003a). In contrast, on foraging habitats of Shark Bay, Australia, tiger sharks attack adults and large juveniles of both species. Shark-inflicted injuries are relatively infrequent on green turtles (<10%), but around 50% of male and 25% of female loggerheads have shark-inflicted injuries; many are fresh and feature tooth marks from tiger sharks (Heithaus et al. 2002b, 2005). Interspecific differences in injury rates may reflect higher rates of successful attack on loggerheads since green turtles should be more likely to escape from the grasp of a tiger shark (Heithaus et al. 2002b). The reason for sex-differences in injury rates of loggerhead turtles is unknown, but is not due to differences in escape ability (Wirsing et al. 2008). Plausible hypotheses include sex-differences in risk-taking or males being in positions more vulnerable to attack during mating.

## 10.2.2  AVOIDING PREDATORS

Turtles can employ an array of tactics to avoid predators. These range from responses to an imminent threat such as flight, to longer-term tactics like selecting habitats that have low predator encounter rates. Relatively little work has focused on how sea turtles might respond to predation risk, which remains a fertile area for future work.

### 10.2.2.1  Habitat Use

Prey can greatly reduce their probability of being killed by choosing appropriate habitats. In some cases, this involves selecting areas where predator numbers are lowest. Ontogenetic shifts in habitat use by most turtle species likely occur so that small juveniles avoid higher predator abundances in neritic habitats until they reach a size where risk is relatively low (e.g., Bolten 2003). Still, newly settled loggerhead turtles in northeast Australia appear to experience higher predation rates than larger turtles (Limpus and Limpus 2003a,b). While in oceanic habitats, young juvenile loggerhead selection for algal mats may serve the dual functions of providing foraging and resting substrates while also reducing risk from potential predators through crypsis (e.g., Witherington 2002). Small juvenile loggerheads, due to their coloration, appear to benefit from reduced detection by predators when they inhabit mats despite likely higher densities of potential predators attracted to these structures (Witherington 2002, Smith and Salmon 2009). Although juvenile green turtles also will associate with algal mats under laboratory and field conditions, they are not cryptic. They may, however, benefit from the physical shelter of the mats and flee from approaching predators before they are detected (Smith and Salmon 2009).

Although rarely studied, in some situations predation risk appears to be important in shaping turtle habitat use decisions in neritic habitats. For example, green turtle selection of shallow lava beaches in Hawaii could be driven by avoidance of tiger sharks (Wabnitz et al. 2010), although these low-risk areas may also be nutritionally most profitable. In the Marquesas Keys, green turtles <65 cm SCL select shallow waters that likely serve as a refuge from predation. When tides drop, turtles seek refuge in depressions that would not allow shark access and avoid channels of similar depth where sharks could occur (Bresette et al. 2010). Juvenile green in nearby coastal Everglades waters select shallow waters where there is abundant forage, but also dead mangrove trees that could serve as refuges from predation (Hart and Fujisaki 2010). In Shark Bay, Australia, newly settled green turtles select shallow mangrove habitats and nearshore waters that limit predator accessibility (Heithaus et al. 2005).

In some cases, overall predation risk is not correlated with predator densities or encounter rates (see Heithaus et al. 2009). Indeed, the probability that prey are killed by a predator in a habitat is determined by the product of the probability of encountering a predator and the probability that prey is killed in an encounter situation (see Hugie and Dill 1994, Heithaus 2001c). The latter probability can be influenced by the presence of cover (habitat complexity), substrate color, light level, water depth, and water turbidity (e.g., Werner and Hall, 1988, Lima and Dill 1990, Hugie and Dill 1994, Heithaus et al. 2009). Prey should be occupy or even prefer, habitats with higher predator encounter rates if features of a habitat sufficiently reduced the probability of death once a predator is encountered (Heithaus and Dill 2006, Heithaus et al. 2009, Wirsing et al. 2010). Green turtles in Shark Bay appear to select the edges of banks in order to reduce the risk of tiger shark predation (Heithaus et al. 2007a). The risk to turtles at edges, however, is determined more by the ability of turtles to escape sharks once they are encountered rather than lower shark encounter rates (Heithaus et al. 2007a, 2009). Kemp's ridley turtles (*Lepidochelys kempii*) off west-central Florida preferentially spent time near rocky outcroppings within their foraging ranges (Schmid et al. 2003). This could function to reduce predation risk or be a response to spatial variation in food resources. Similarly, the distribution of juvenile hawksbills in Puerto Rico did not correlate with that of their prey; instead turtles selected habitats with high structural complexity that would provide shelter from predators (Rincon-Diaz et al. 2011).

Female turtles may be able to reduce the risk of nest predation or predation on hatchlings by modifying nesting locations in response to the abundance and type of predators, but few studies have investigated this possibility. In Oman, green turtles appear to be more likely to nest in areas that are less disturbed by humans and/or natural predators (Meddonça et al. 2010). In highly disturbed sites, turtles tended to nest closer to the ocean, possibly to reduce predation on hatchlings. Burial depth of eggs also might be considered a form of "habitat use" by nesting females. Leighton et al. (2009) found that mongooses identify nests based on surface cues, but will modify their digging intensity based on olfactory cues while excavating. Hawksbill eggs buried at shallower depths were much more likely to be preyed upon. Future studies investigating burial depth decisions and associated trade-offs, however, are needed.

### 10.2.2.2 Diving Behavior

Recent theoretical (Heithaus and Frid 2003, Frid et al. 2007b) and empirical studies (Frid et al. 2007a, Dunphy-Daly et al. 2010, Wirsing et al. 2011) demonstrate that diving vertebrates—including large-bodied species—should, and sometimes do, modify their diving and surfacing patterns in response to the risk of predation. In general, the surface portion of a dive cycle is predicted to be the most dangerous because of the hunting tactics of predators and vigilance/detection disadvantages of an animal at the surface (e.g., Heithaus and Frid 2003). Therefore, divers are predicted to shorten their time spent vulnerable at the surface. How they arrange dive cycles to achieve this may vary based on the relationship between instantaneous risk of predation and time at the surface during a single surfacing about as well as the relationship between bottom time and energy intake rates (Heithaus and Frid 2003). Although specific empirical tests of predator impacts on sea turtle diving behavior have not yet been undertaken, several studies have suggested that predators influence turtle diving behavior. For example, U-shaped dives of green turtles at Ascension Island might function as both resting dives and predator avoidance (Hays et al. 2000), and the location of these dives might be influenced by the distribution of predators (Martin 2003). Also, many turtles approach the surface slowly with a passive ascent that may allow them to scan the habitat for predators before surfacing (Glen et al. 2001). Indeed, loggerhead turtles in Shark Bay, Australia, have been observed turning in circles before taking breaths at the surface (unpublished data) and juvenile hawksbills appear more wary during surfacing events and react more strongly to divers (potentially perceived as predators) (von Brandis et al. 2010). Also, deep dives by migrating green turtles may minimize the risk of predation from sharks by reducing silhouetting against the surface (Hays et al. 2001). Small juvenile flatback turtles are capable of engaging in long dives with relatively short surface intervals and can engage in faster burst swimming compared to other sea turtle species (Salmon et al. 2010). Together, these behaviors might allow young turtles to minimize exposure to, and enhance escape probability from, predatory white-bellied sea eagles. Short surface intervals, however, may also allow turtles to quickly relocate foraging sites in the turbid waters they inhabit (Salmon et al. 2010).

### 10.2.2.3 Activity Levels and Patterns

Choosing an appropriate time of day to be active can reduce predation risk, and changes in light level often cause prey to modify their activities because of increased susceptibility to predators (Lima and Dill 1990). Some sea turtle populations show diel variation in behavior (Section 10.3.1.3). These patterns may be in response to predation risk, but often are likely to be associated with foraging or other behavior. For example, in eastern Australia, green turtle hatchling emergence at night appears to be driven by avoiding heat and dehydration rather than predator avoidance (Gyuris 1994), although more work is required to understand the complex interplay of thermal conditions and predation risk in driving emergence patterns (e.g., Glen et al. 2005).

Prey also can reduce predation risk by reducing their activity level (i.e., movement speed or duration) (e.g., Taylor 1984, Werner and Anholt 1993, Anholt and Werner 1995). Lower risk can be achieved through reduced encounter rates with predators as well as reduced probability of being detected by a predator. However, in some cases, increased activity can be beneficial to rapidly transit high-risk areas. For example, hatchling turtles could reduce their risk of predation by reducing the time spent transiting the dangerous beach and nearshore habitats. The evolution of interspecific differences in activity levels of hatchlings upon reaching the ocean likely has been shaped by predation risk (e.g., Salmon et al. 2009). Loggerhead, leatherback, and green turtle hatchlings exhibit a period of frenzy swimming (which varies in duration among locations) that likely helps them rapidly traverse areas where predators are concentrated (Wyneken and Salmon 1987, Wyneken and Salmon 1992, Pereira et al. 2011). In contrast, hawksbill hatchlings do not display frenzy swimming, are less active than green turtles, and adopt slower and less conspicuous swimming gaits (Chung et al. 2009a,b). These behavioral patterns likely represent an antipredator tactic based on hiding in flotsam or minimizing detection probabilities by fish predators. Such a tactic may be necessary for hawksbill turtles, which are smaller and more vulnerable to predators than green turtles (Chung et al. 2009a,b). Although larger than green and loggerhead hatchlings, flatback hatchlings do not expend as much energy in swimming once they reach the ocean. After an initial ca. 2 h burst of swimming, activity drops dramatically compared to greens and loggerheads (Pereira et al. 2011), but nighttime activity remains relatively high after this period when compared to greens and loggerheads (Salmon et al. 2009). The activity patterns of flatbacks, therefore, appear to reflect energy conservation while remaining in coastal habitats after moving through the high-risk waters adjacent to the nesting beach (Pereira et al. 2011). Flatback hatchlings and posthatchlings maintain activity throughout the day and night in response to generally higher risk than hatchlings that move into oceanic waters (Salmon et al. 2009).

### 10.2.2.4 Hiding

Hiding behavior and cryptic coloration can reduce the probability of being killed by a predator, and examples of both are found in sea turtles. Many sea turtles are cryptically colored (e.g., Witherington 2002, Salmon et al. 2009, Smith and Salmon 2009), which may reduce detection rates by predators. Juvenile green turtles in reef habitats often rest under rocks or coral ledges, primarily at night, possibly to minimize the risk of predation (e.g., Makowski et al. 2006). Also, the swimming style of hatchling hawksbill turtles (as mentioned earlier) likely represents hiding from predators (Chung et al. 2009a,b).

### 10.2.2.5 Group Formation

Grouping can reduce the risk of predation, which has been suggested as the selective force leading to sociality in many taxa (Bertram 1978, Pulliam and Caraco 1984). Groups can reduce the risk of predation through (1) diluting the probability that a particular individual is captured by a predator that detects a group, (2) increasing the probability that predators are detected in time for effective escape responses, (3) confusing a predator when a group takes flight, (4) reducing per capita encounter rates with predators, and (5) actively defending against predators. At least the first three benefits may accrue to sea turtles in some situations. It is, however, often hard to separate these benefits and they are not mutually exclusive (e.g., Bednekoff and Lima 1998). Dilution is likely a benefit that promotes synchronous hatching by sea turtles, and possibly synchronous nesting in some species. For example, during the peak in nesting of olive ridleys, densities of egg predators peak but the probability of a particular nest being attacked is lowest (Madden et al. 2008). Similar dilution effects during emergence events likely outweigh any increase in predator abundances. On Playa Grande, however, there was a positive relationship between the number of leatherback hatchlings emerging and the number killed by crabs (but not other predators), suggesting that leaving as a group increases detection probability (Tomillo et al. 2010). In this case, it is unclear if dilution counterbalances this increased encounter rate with predators. The *arribada* of nesting olive ridleys may serve an antipredator function both for eggs—arribada nests experienced lower predation rates (7.6%) than first-night solitary nests (51%; Eckrich and Owens 1995)—as well as possibly reducing the risk of crocodile predation on adults during nesting (e.g., Ortiz et al. 1997).

Grouping behavior on foraging grounds has not been studied extensively. Satellite tracking of loggerheads has suggested that, at least in some areas, individuals may spread out on foraging grounds (e.g., Hart et al. 2012). In other situations, foraging turtles may be found in close proximity. For example, large juvenile and adult green turtles in the Florida Keys are found in groups, sometimes foraging within 1–3 m of each other (Bresette et al. 2010). Such groups, which also are found for green turtles in Shark Bay (personal observation), likely serve an antipredator function. Although there are no measures of potential antipredator benefits of this association, turtles in Shark Bay often initiate flight behavior in response to movements of other nearby turtles (personal observation) and, therefore, may gain both detection and dilution benefits.

### 10.2.2.6  Vigilance

Vigilance, where prey stop to watch for predators, is commonly used to reduce predation risk. Optimal vigilance level depends on both the risk of attack and the size of the group an animal is in, with higher levels when predation risk is high and when group size is low (e.g., Lima and Dill 1990, Brown, et al. 1999). The surfacing behaviors described in Section 10.2.2.3 may be a form of vigilance, but tests of these hypotheses have yet to be conducted.

### 10.2.2.7  Deterrence

Pursuit-deterrence signals, where prey signal to predators that they have been observed, can result in a reduced probability of predatory attack when signals are honest (Caro 1995). There are currently no studies of such signaling in sea turtles, but such behavior could be effective. For example, tiger sharks that have encountered vigilant turtles have not attacked, despite orienting to inspect them (Heithaus et al. 2002a). Thus, forms of turtle behavior that signal their readiness to flee might reduce attack probability.

### 10.2.2.8  Flight

Fleeing involves decisions of escape direction, at what distance from the predator to initiate flight (flight initiation distance) and escape velocity. Because flight is costly (in terms of both energetic expenditure and lost foraging or mating opportunities), animals should not necessarily flee as soon as a predator is detected; there should be an optimal flight response (Ydenberg and Dill 1986, Dill 1990). The distance to safety, angle, and speed of a predator's approach and the cost of flight influence escape responses (e.g., Ydenberg and Dill 1986, Dill 1990, Lima and Dill 1990, Bonenfant and Kramer 1994). Often, distance or time to safety is equivalent to the distance to a physical refuge. While this may be the case for turtles living in reef environments, in many marine systems time to safety might be the amount of time it takes a prey animal to reach a critical velocity or maneuvering ability such that they are inaccessible to a predator (Heithaus et al. 2002a). Sea turtles confronted by imminent shark threat in relatively deep water have been observed, and filmed, swimming in circles while keeping their shell angled toward the shark. This keeps the shark from gaining an adequate angle of attack. I am aware of no studies of flight initiation distances in sea turtles. However, green turtles and loggerheads in Shark Bay, Australia, flee toward deep water when disturbed by boats, which would allow greater maneuverability relative to their tiger shark predators (Heithaus et al. 2007a).

### 10.2.3  CAN PREDATORS INFLUENCE SEA TURTLE POPULATION SIZES?

The importance of predation on eggs, hatchlings, and young juveniles for sea turtle population recovery has been explored through models (Mazaris et al. 2006) and has been supported by empirical studies (Dutton et al. 2005). Although extreme levels of predation on nests and hatchlings could have major impacts on population viability, population models suggest that survivorship of larger juveniles and adults usually is more important (e.g., Crouse et al. 1987, Heppell et al. 1996, Heppell 1998).

Could predators of adult and large juvenile turtles influence, or have historically influenced, turtle population sizes? This question has received relatively little attention (but see Heithaus et al. 2008b). For example, green turtle populations have been assumed to be limited by the availability of food resources (e.g., Jackson 1997, Bjorndal et al. 2000). This assumption is based largely on the notion that large-bodied terrestrial herbivores are able to escape regulation by predators, but population sizes of many terrestrial herbivores, including large-bodied ones, are limited by a combination of direct predation, predation risk, and their interaction (e.g., white tailed deer *Odocoileus virginianus*, elk *Cervus elaphus*; Binkley et al. 2006, Creel et al. 2007). Furthermore, terrestrial herbivores cited as being immune to population-level effects of predators are characterized by much larger body sizes relative to their predators—and are more dangerous—than are sea turtles compared to their predators (see Heithaus et al. 2008b). Thus, drawing inferences about marine turtle susceptibility to predators from terrestrial herbivores must be done cautiously.

Suggestions that predators are unimportant to green turtle populations in the Caribbean are not surprising in light of likely large-scale declines in tiger shark populations in the region. Although fisheries' catches suggest that tiger shark populations in the Gulf of Mexico and in the Atlantic Ocean off the southeastern coast of the United States have remained stable, or possibly increased marginally, since the early 1990s (Carlson et al. 2012), several lines of evidence suggest that population sizes likely are substantially less than they were historically (e.g., Baum et al. 2005, Heithaus et al. 2007b, Ward-Paige et al. 2010). Results from Shark Bay, Australia—where tiger shark abundances are substantially higher than other locations where comparable data are available (Heithaus et al. 2012)—suggest that predation risk can be important to adult turtles. Indeed, green turtles respond to tiger sharks in a manner analogous to elk responses to wolves (Heithaus et al. 2007a). Because elk populations are influenced by wolves (Creel et al. 2007), this raises the possibility that turtle populations in Shark Bay, and in other locations historically, could be regulated by bottom-up forces (largely within safer areas) *and* top-down forces (by restricting the use of resources in dangerous areas; see Heithaus et al. 2007a). Such dynamics, known as the "Foraging Arena," have been identified as important in population dynamics of other marine organisms (Walters and Juanes 1993, Walters and Martel 2004). Therefore, assuming that access to food at the adult life history stage sets the carrying capacity of green turtles in Shark Bay, the presence of predators could impact green turtle populations by restricting their access to foraging locations (Heithaus et al. 2008b). Indeed, population estimates for green turtles are far below what would be predicted based on energetic assumptions alone despite minimal anthropogenic impacts in Shark Bay (Heithaus et al. 2008b). Whether sharks or other large predators historically could have influenced turtle population sizes remains an important area of research for setting management goals and predicting ecosystem dynamics (Heithaus et al. 2008a,b).

Our current understanding of the possible effects of predators on sea turtles, both behaviorally and at the population level, is still in its infancy. It is likely that the importance of predation varies both within and among regions and with species of sea turtle. Also, the impacts of predators could occur over restricted spatial and temporal scales. For example, tiger sharks congregate at seasonally available resources (e.g., Simpfendorfer et al. 2001, Wirsing et al. 2007) and thus they might concentrate off of nesting beaches or in mating areas and take advantage of unaware or exhausted turtles. Studies of nesting green turtles suggest that depressed endocrine responses could be an adaptation to realize reproductive success in the face of predator-inflicted injuries (e.g., Jessop et al. 2004). Thus, studies of predator–prey interactions off nesting beaches are difficult but would be valuable.

## 10.3   SEA TURTLES AS CONSUMERS

Sea turtles feed on an array of species from seagrasses and macroalgae to large jellyfish, benthic invertebrates, and teleosts. A review of sea turtle diets (see Bjorndal 1997) and forage/prey selection is beyond the scope of this chapter (see Chapter 9). Instead, I focus on the foraging tactics of sea turtles and only explore turtle diets from the perspective of individual variation in foraging tactics.

**10.3.1**  FINDING PREY

**10.3.1.1  Habitat Use**

Selection of the proper forging habitat can greatly influence energy intake rates of consumers. A diverse array of theory has been developed to explore optimal habitat use strategies and tactics. In the absence of other individuals, the optimal strategy may entail selecting the habitat with the highest food abundance. However, for herbivores, nutrient acquisition often is more important than energy intake and for mobile prey, prey density or abundance may not correlate with its availability. For example, prey in complex environments like reefs or seagrass beds may escape easily and become unavailable. Also, when other factors such as the presence of competitors or a third trophic level (the prey's food) are considered, then there may not be a good relationship between the abundance of prey and the distribution of predators even if energy intake is the only factor determining the distribution of the predator (e.g., Hugie and Dill 1994). Finally, if a predator is at risk of predation itself, then its distribution may be unrelated to prey distributions (see Section 10.3.3).

Habitat use of many species of turtles in both oceanic and neritic habitats appears to be driven by the abundance of food, but studies often are constrained by the difficulty of collecting rigorous measurements of food densities or availability. Many leatherback turtles in the northeast Atlantic appear to congregate at large, consistent, aggregations of jellyfish prey in coastal waters (Houghton et al. 2006) while those nesting in the Caribbean appear to remain within offshore oceanographic features that concentrate prey (Eckert 2006).

Green turtles in shallow waters of the Lakshadweep Islands, Indian Ocean, were found primarily over seagrass habitats (Lal et al. 2010), and Ballorain et al. (2010) found that juvenile and adult green turtles varied in their use of shallow multispecies seagrass habitats. Smaller individuals foraged in the shallowest locations, dominated by *Halodule uninervis*, where the largest individuals were not found. Size-based variation in habitat use may be due to differences in food requirements and physiology (Ballorain et al. 2010), or related to smaller turtles selecting shallower habitats to reduce the risk of predation. In Baja California Sur, Mexico, juvenile green turtles traverse multiple habitat types over relatively short time frames but select for seagrass habitats and moderate depths and avoid waters >10 m depth (Senko et al. 2010). Based on stable isotopic analysis, green turtles in San Diego Bay, California appear to use eelgrass habitats for foraging, but likely in response to the availability of invertebrate prey in these habitats rather than just to access seagrasses (Lemons et al. 2011). Off Palm Beach, Florida, juvenile green turtles inhabit small ranges of 0.7–5 km$^2$ that feature abundant algal resources as well as shelters (Makowski et al. 2006). Similarly, small home ranges are found in Hawaii where resources are abundant (Brill et al. 1995), but much larger home ranges occur in areas where resources are widely distributed (Seminoff et al. 2002).

In the absence of other factors, foragers should begin using a patch only when the energy intake rate available in that habitat is above that, or equal to, the energy intake rate available in other areas within the environment. In some cases, this results in threshold foraging responses. Although threshold responses have not been specifically identified, they likely occur for turtles in oceanic environments, which congregate in areas where prey should be abundant. Density-dependent effects on growth (e.g., reduced prey availability) may lead to large juvenile hawksbill turtles leaving foraging grounds near Barbados (Krueger et al. 2011).

Green turtle foraging, at least at moderate levels, can improve the nutritional quality of seagrass blades (see Section 10.4.3).In the Caribbean, individual turtles take advantage of these changes and continually return to restricted foraging patches (Bjorndal 1980, Ogden et al. 1983). Grazing in this manner results in turtles consuming newer seagrass leaves with higher nitrogen content and lower lignin content, which is energetically beneficial (Bjorndal 1980, Moran and Bjorndal 2005, 2007). Such continued copping of particular areas by the same green turtles, however, is not universal.

The intake rate of an individual predator often is influenced by the number of other predators in a habitat. Numerous models have been developed to describe such situations. The most basic is the ideal free distribution (IFD; Fretwell and Lucas 1970). This model assumes that (1) food distribution is fixed, (2) animals forage to maximize energy intake rate, (3) resources are split evenly among consumers, and (4) energy intake of each consumer is reduced with the addition of another consumer. Under this model, consumers are expected to be distributed across habitats proportional to availability or productivity of food resources. This results in the basic prediction that intake rates of individuals residing in patches differing in food availability will be identical because more foragers will exploit the more productive patches. Although there is empirical support for this model in some non-sea turtle systems (e.g., Tregenza 1995), many factors may cause deviations from this distribution, including differences in competitive ability among foragers (see Tregenza 1995, Hamilton 2010) and predation risk (see Section 10.2.2.1). Few studies of sea turtles allow testing of IFD-based models.

One prediction of the IFD is that, all else being equal, fitness should be equal among patches through differential reproductive success through time or individuals moving to maximize residual reproductive value. There are no studies that explicitly test this prediction in sea turtles. However, a 30 year study of hawksbill turtles in the Caribbean demonstrated similar size distributions, residence times, and body condition index values for turtles in reef habitats and those in seagrass habitats, which are thought to be peripheral (Bjorndal and Bolten 2010). These results are consistent with predications of the IFD. In other locations, however, hawksbills show variation in growth rates among habitats that appears to be linked to differences in prey availability (Diez and van Dam 2002), a result that contrasts with IFD predictions.

Growth rates may vary considerably among foraging habitats at relatively large spatial scales, which likely is due to variation in food resources available in these habitats. For example, there is significant variation in growth rates across four widely separated green turtle foraging grounds of the Great Barrier Reef genetic stock that may be driven by differences in food resources (Chaloupka et al. 2004). Similar spatial variation in growth rates among juvenile green turtle foraging locations have been observed in Florida, where there also appeared to be density-dependent reductions in growth at one location (Kubis et al. 2009). Also, Eastern Pacific leatherback turtles are smaller and have around half the reproductive output of leatherbacks in the Atlantic and western Indian Oceans, likely because of lower food resources and higher temporal variability in these resources in the eastern Pacific (Saba et al. 2008b). Within the eastern Pacific populations, nesting peaks are associated with productivity peaks during La Nina events (Saba et al. 2008a). It remains to be determined why turtles that could, theoretically, move among potential foraging grounds to maximize growth rates do not. Spatial variation in predation risk (see Section 10.2.2.1) or early experience and strong site fidelity could lead to these patterns.

A key aspect of sea turtle habitat use for many populations is an ontogenetic shift in the types of habitats occupied and the foraging behavior adopted within these habitats. For example, both green and loggerhead turtles generally shift from being oceanic drifting predators after hatching to neritic benthic consumers once they attain sufficient body size. Recent studies, however, have shown that these habitat shifts do not necessarily occur for all individuals and may be more complex than previously thought (see Section 10.3.4). For example, green turtles generally were thought to transition to coastal habitats at body sizes at or less than 40 cm CCL. In the Pacific Ocean, however, individuals up to 70 cm CCL are found in pelagic waters consuming a carnivorous diet (Parker et al. 2011). This suggests that some individuals may delay their recruitment to neritic habitats or facultatively shift back into pelagic waters (Parker et al. 2011). Similar results have been found for loggerhead turtles, where juvenile shifts from pelagic waters into neritic habitats are not universal and are reversible (e.g., McClellan and Read 2007, Mansfield et al. 2009).

Seasonal migrations, in which individuals move great distances, can be considered an extreme case of habitat selection with preferred habitats widely separated. For many populations, migrations between foraging and nesting areas appear to occur through relatively consistent "corridors" and individuals often show fidelity to the same foraging areas across multiple migrations (e.g., Broderick et al. 2007). However, for some populations considerable foraging may occur during migration. For example, leatherback turtles in the Atlantic Ocean appear to travel relatively continuously but modify their foraging behavior in response to local conditions (Hays et al. 2006).

Remigration intervals to nesting beaches can vary within and among populations and may be linked to foraging success. In loggerheads, remigration intervals are shorter for individuals foraging in neritic habitats than oceanic ones, likely reflecting the higher prey availability in neritic habitats (Hatase and Tsukamoto 2008). At Tortuguero, Costa Rica, green turtles show shorter remigration intervals than other populations, which could reflect greater foraging opportunities, shorter distances between nesting and foraging grounds, or more favorable environmental conditions (Troëng and Chaloupka 2007).

### 10.3.1.2  Search Patterns

The probability of prey encounter and capture can be increased within a habitat by adopting an optimal searching strategy. This may involve variation in the range over which turtles search for food in broad habitats. For example, in the Mediterranean, neritic loggerheads have ranges on the order of 10 s of km$^2$ while those in oceanic habitats have ranges of ca. 1000 km$^2$ (Schofield et al. 2010). Such variation likely reflects variation in the density and distribution of prey, with more widely distributed and sparse prey in oceanic habits.

Much recent work has focused on elucidating optimal search strategies in oceanic habitats where prey availability and predictability could vary substantially but there are not obvious habitat differences that could be used to reliably locate prey. Optimal searching behavior should change with variation in the distribution, abundance, and predictability of resources. Random movements (Brownian motion) are optimal when prey are abundant, while different movement tactics are required in situations where prey are less abundant or unpredictable. Levy flights are a type of random movement, typified by many short moves with less frequent long-distance displacements (Sims et al. 2008). Such movement can maximize search efficiency, and minimize energetic costs, when predators are confronted with an uncertain prey environment (see Reynolds and Rhodes 2009). Recent studies suggest that leatherback turtles, like many other large predators in open oceans, adopt such search behavior (Sims et al. 2008, Humphries et al. 2010).

McCarthy et al. (2010) used a straightness index based on satellite telemetry to gain insights into habitat preferences and movement tactics of juvenile loggerhead turtles in pelagic waters of the North Atlantic. Turtles tended to move in straight lines, suggestive of lower quality habitats, in the warmest waters and in less powerful currents. Turtles adopted highly sinuous tracks, indicative of preferred foraging habitats, where chlorophyll concentrations were high and in shallower waters. This led to turtles spending more time near seamounts and in upwelling locations where prey likely are concentrated.

Adult leatherback turtles leaving rookeries around the world disperse widely. For example, in the South Atlantic females tend to move into areas of high surface chlorophyll concentration but may move into temperate habitats off South America or Africa or the equatorial Atlantic (Witt et al. 2011). Leatherbacks departing Costa Rican nesting beaches move relatively quickly through warmer waters until reaching cooler waters further south where travel rates slowed, indicative of foraging behavior (Shillinger et al. 2011). Slower travel speeds were associated with higher productivity (chlorophyll a concentrations and upwelling). Interestingly, mesoscale features were not areas where turtles spent more time (i.e., slower travel rates), even though leatherback foraging often is associated with these features in several populations (e.g., Doyle et al. 2008, Lambardi et al. 2008). Rapid transit times through areas of relatively high productivity may occur to reach convergence zones and frontal systems that aggregate their gelatinous zooplankton prey (e.g., Lambardi et al. 2008). Bailey et al. (2008) found that although parameters of leatherback tracks (e.g., mean turning angle and autocorrelation in speed and direction) are similar between turtles in the eastern Pacific and Atlantic Oceans, leatherbacks in the Pacific spent more time in foraging phases inferred from tracks and these phases were more dispersed, suggesting that food patches were less predictable.

## 10.3.1.3 Activity Levels and Patterns

Increased activity levels can result in increased access to food through increased encounter rates with potential food items. However, sea turtles may be capable of meeting energetic requirements in relatively short time periods (e.g., Fossette et al. 2011).

In some situations, sea turtles could also increase the probability of both encountering and capturing prey by selecting an appropriate time of day to feed. Many turtles exhibit diel patterns of habitat use and diving behavior. For example, green turtles in the Cayman Islands (Caribbean), Mayotte Island (Indian Ocean), and off northeast Australia appear to forage during the day and rest at night (Bjorndal 1980, Taquet et al. 2006, Blumental et al. 2009b, 2010, Hazel et al. 2009). Similar results have been obtained for hawksbills in the Caribbean, which move into shallower waters at night to rest (vanDam and Dietz 1996, Blumenthal et al. 2009b). However, green turtles in the Indian Ocean foraged at night when light levels were relatively high (Taquet et al. 2006). Diel patterns of habitat use appear to be variable in green turtles. In northeast Australia and the Caribbean resting occurred in deeper waters than foraging (Bjorndal 1980, Hazel et al. 2009), but greens in Bahía de los Angeles, Gulf of California exhibited an opposite pattern. Turtles were found in deeper waters during the day and moved into the shallows at night, perhaps to avoid heavy boat traffic in shallow areas during the day (Seminoff et al. 2002).

Leatherback turtles in many areas appear to forage predominantly at night when vertically migrating prey become more easily accessible (e.g., Eckert 2006, Myers and Hays 2006, Shillinger et al. 2011; see Section 10.3.1.4). These patterns also may be driven by the need for turtles to thermoregulate during the day, when they can bask at the surface to warm their bodies (James et al. 2005). The need to bask, however, may be somewhat reduced in leatherbacks by fat deposits that help isolate cold prey from the large blood vessels and the respiratory tract (Davenport et al. 2009). In other situations, leatherback turtles may switch between diurnal and nocturnal foraging, likely in response to changes in the vertical distribution of prey (Hays et al. 2006). Off St. Croix, leatherback turtles feeding during the interesting period appear to concentrate their foraging during the day, but overall prey intake rates likely are relatively unimportant to overall energy budgets (Casey et al. 2010).

Seasonal patterns of activity are also common in sea turtles inhabiting waters near their thermal tolerance levels. For example, green and loggerhead turtles in some locations will move to warmer overwintering habitats and/or make very long resting dives through much of the winter and likely cease foraging (e.g., Broderick et al. 2007, Mansfield et al. 2009). Loggerhead turtles, however, appear to resume activity to forage during these "hibernation" periods (Hochscheid et al. 2007).

## 10.3.1.4 Diving Behavior

As a major component of turtles' lives at sea, diving behavior is a critical component of foraging and energy budgets. Theory predicts that there should be a general increase in both dive and surface times as the depth of prey increases (Kramer 1988), and sea turtles conform to this prediction in many cases (e.g., vanDam and Diez 1996, Hays et al. 2004, Hazel et al. 2009). However, sea turtles do not always follow this pattern, probably because of variation in activity levels during dives and the fact that the majority of dives may not involve foraging. Also, the use of lungs as buoyancy during diving can complicate relationships between dive times and depth (e.g., Hays et al. 2004). Leatherback turtles are the most accomplished sea turtle divers and can engage in exceptionally deep (>1000 m) and long dives (>80 min) (Houghton et al. 2008, López-Mendilaharsu et al. 2009). These long and deep dives, however, are rare and appear to be related to prey detection and active foraging rather than thermoregulation or predator avoidance (Houghton et al. 2008). Turtles that detect sufficient prey during deep dives may then wait in an area to access these prey when they ascend at night. Leatherbacks in the Eastern Pacific engaged in deeper but shorter dives during the day, which may have been exploratory dives, and then dove shallower but longer at night, likely to forage (Houghton et al. 2008, Shillinger et al. 2011). Such behavior, which likely is in response to increased accessibility of vertically migrating prey, appears to be common in leather-backs (e.g., Eckert 2006, Myers and Hays 2006). Hawksbill turtles in the Seychelles tend to engage in shallow, relatively short dives that are well within aerobic limits, but diving depths likely are influenced by food resources (von Brandis et al. 2010).

There is considerable variation in diving behavior among even populations of the same species. For example, in the Eastern Pacific, leatherback dives were deeper and longer on high-latitude foraging grounds than during the transit period (Shillinger et al. 2011), while in the North Atlantic dives are shallower at high-latitude foraging grounds (Hays et al. 2006, James et al. 2006, Jonsen et al. 2007).

## 10.3.2 Food Capture and Consumption

Because of their morphology and prey types, sea turtles exhibit a relatively restricted array of tactics for capturing and consuming food.

### 10.3.2.1 Herbivory

Green turtles are the only species of sea turtle that is primarily herbivorous and they may consume an array of seagrasses and macroalgae (e.g., Bjorndal 1997). A review of their foraging behavior is beyond the scope of this chapter (see Chapter 9). However, it is now apparent that green turtles in some locations are more omnivorous than previously thought (e.g., Heithaus et al. 2002c, Hatase et al. 2006, Seminoff et al. 2006, Cardona et al. 2009, Burkholder et al. 2011, Lemons et al. 2011). Even in seagrass habitats, animal tissue may form an important component of green turtle diets. For example, mobile invertebrates appear to be one of the most important diet items for green turtles in San Diego Bay followed by seagrasses and sessile invertebrates (Lemons et al. 2011). Juvenile green turtles in the Mediterranean may shift to a diet largely made of seagrasses when they recruit to coastal foraging areas but still rely largely on animal-derived nutrients for some time after recruit-ment (Cardona et al. 2010).

### 10.3.2.2 Benthic Foraging Tactics

Loggerheads forage primarily on benthic prey when in neritic habitats. While many of these prey occur above the sediment, others are buried and need to be excavated. Loggerheads have been observed mining such prey from the substrate (Limpus et al. 1994b, Preen 1996). Sometimes this behavior involves using the foreflippers to dig for prey, resulting in considerable displacement of sediment, while in other instances the turtles use just the beak to pull bivalves from the sediment (Preen 1996, Schofield et al. 2006, Thomson et al. 2011, 2012).

### 10.3.2.3 Scavenging

Scavenging does not appear to be a primary foraging tactic of turtles, but loggerheads will scavenge fisheries discards and from fishing gear (e.g., Houghton et al. 2000, Tomás et al. 2001). In Virginia, declines in the availability of preferred prey appear to have led loggerheads to scavenge teleosts from nets (Seney and Musick 2007). A green turtle was observed scavenging a large jellyfish in Australia (Heithaus et al. 2002c), but the relative importance of this foraging tactic remains unknown. The decreasing cost and increasing availability of animal-borne video technologies (e.g., Seminoff et al. 2006, Thomson et al. 2011, 2012) may help provide further insights into specific foraging tactics like scavenging.

## 10.3.3 Trade-Offs

Trade-offs are a fundamental aspect of ecological dynamics because time and resources are limited and demands are often in conflict. One of the most common trade-offs is that between foraging and avoiding predators. This is because behavior that increases foraging efficiency (e.g., increased

activity levels) often increases the risk of being killed by a predator and habitats that are most energetically productive often are the most dangerous (Lima and Dill 1990, Lima 1998). In general,
animals are willing to forego foraging opportunities to enhance safety (see Lima and Dill 1990,
Lima 1998, Brown and Kotler 2004 for reviews).

Although predator-inflicted mortality rates are low for most age classes of sea turtles, energy
intake-safety trade-offs are likely to be important for many populations. Indeed, the risk of predation can be an important determinant of prey behavior even in situations where predation rates are
low (e.g., Heithaus and Dill 2002, 2006, Creel 2011). Species with relatively "slow" life history
characteristics, like sea turtles, should invest more heavily in antipredator behavior because of the
high fitness costs of an early death (Warner 1998, Heithaus et al. 2008a, Frid et al. 2012).

There have been very few studies that investigate food-risk trade-offs in sea turtles. However,
juvenile hawksbills in Puerto Rico appear to trade foraging opportunities for enhanced safety.
Turtles select habitats based on the presence of refuges despite lower availability of prey in these
areas (Rincon-Diaz et al. 2011). Also, the neritic-oceanic foraging dichotomy found in loggerheads (see Section 10.3.4.2) may be the result of a foraging-safety trade-off. Demographic modeling suggests that only a slightly higher juvenile survivorship in oceanic habitats than neritic ones
could counterbalance substantially greater prey availabilities and growth rates in neritic habitats
(Peckham et al. 2011). Such differences in survivorship are plausible. Although tiger sharks, the
primary predator of large juvenile and adult loggerheads, are found in oceanic habitats, they are far
more common in coastal waters. Thus, the two habitat use tactics of juvenile loggerheads may result
in similar fitness, with oceanic habitats representing a low-risk slow-growth habitat, and neritic ones,
a riskier, high-growth habitat (Peckham et al. 2011). Similar food-safety trade-offs could potentially
explain the persistence of the foraging dichotomy found in adult loggerheads and green turtles.
Although not considered here, in rare cases turtles may face other mortality risks when foraging.
For example, a recent study documented two green turtles killed by the toxin of blue-lined octopus
(*Hapalochlaena fasciata*) ingested while turtles were feeding on seagrass (Townsend et al. 2012).

One important factor that can influence an individual's willingness to accept greater risks to
obtain additional foraging rewards is their "state." An individual's state can include energy reserves
or expected future reproduction. Individuals in lower condition should, in general, be willing to
accept higher risk in order to obtain more energy (e.g., Clark 1994, Sinclair and Arcese 1995,
Heithaus et al. 2007a, 2008a). Green turtles threatened by tiger sharks conform to this pattern.
Adult and large juvenile green turtles modify their use of shallow seagrass banks based on the risk
posed by tiger sharks as well as their foraging needs (Heithaus et al. 2007a). Turtles that are in poor
energetic condition forage in the middle of banks where food quality is greater, but the risk from
tiger sharks is higher. In contrast, turtles in good condition, forage along bank edges where foraging
options are of lower quality, but risk from tiger sharks is less. The selection of edges by turtles in
good condition is reduced when tiger shark abundance drops.

Age-sex classes that realize the greatest fitness benefits of enhanced energy intake are often willing to incur higher predation risk (e.g., Cresswell 1994, Corti and Shackleton 2002). Although there
are no specific tests of this in sea turtles, the higher shark-inflicted injury rates of male loggerhead
turtles, compared to females, might be an indication of greater risk-taking (Heithaus et al. 2002b).

Some turtles also appear to make other risk-related trade-offs. For example, flatback turtles lay
many fewer but considerably larger eggs that develop into larger hatchlings than other hard-shelled
turtles, but nest at similar frequencies (Walker and Parmenter 1990, Van Buskirk and Crowder 1994).
This pattern likely reflects the need for flatbacks produce larger hatchlings that are more likely to
survive in the higher-risk neritic habitats where they reside. Other risk-reproduction trade-offs that
turtles face include those of nest site selection and burial depth. Laying nests near vegetation may
provide a more favorable thermal environment for egg development, but at the cost of higher nest
predation risk or nest destruction by roots (Conrad et al. 2011). Burying nests deeper may reduce
the probability of nest predation (Leighton et al. 2009) but likely at a cost of decreased emergence
success. Whether, and how, turtles respond to these trade-offs remains largely unexplored.

### 10.3.4   Variation in Feeding Strategies and Tactics

Sea turtles show considerable variation in foraging tactics. This variation may occur within an individual, when individuals vary their foraging tactics in response to internal or external changes in conditions. For example, seasonal or interannual changes in prey abundance or variation in an individual's condition may result in changes in foraging strategies and tactics. Changes may also occur over longer time scales; ontogenetic shifts in diets, foraging strategies, and tactics are common (see Chapter 9 for a review of ontogenetic shifts in diets). Consistent differences among individuals of a species or a population may also occur. Although some of these differences can be attributed to sex- or size-differences in foraging or regional variation within a species, there is increasing recognition that individuals within an age-sex class foraging in the same area can exhibit consistent differences in foraging patterns (individual specialization; Bolnick et al. 2003). All of these sources of variation in foraging strategies and tactics can play an important role in population, community, and ecosystem dynamics as well as optimal conservation strategies.

### 10.3.4.1   Intraindividual Variation

Variation in foraging behavior within individuals appears to be common in some species of sea turtles. This variation may occur over multiple temporal scales from ontogenetic shifts changing foraging tactics or movements based on diel or short-term variation in prey availability or prey types. For example, loggerheads appear to switch their foraging tactics and foraging habitats to take advantage of abundant prey types (e.g., Seney and Musick 2007, Mansfield et al. 2009, Carranza et al. 2011) and leatherback turtles off of Florida shift foraging habitats seasonally from the continental shelf in spring through fall to off the shelf in winter (Eckert et al. 2006).

#### 10.3.4.1.1   Ontogenetic Variation

Ontogenetic variation in habitat use and foraging characterize most sea turtles. Size-related changes in diving behavior are important for shaping foraging behavior. Green turtles and leatherbacks overlap in their diets and in the depths at which they forage shortly after hatching, but over the first weeks of life leatherback turtles develop deeper-diving capabilities than green turtles, leading to a rapid divergence in feeding niches (Salmon et al. 2004). Also, size-related variation in diving abilities of hawksbills may result in partitioning of vertical reef habitat by size (Blumenthal et al. 2009b).

Ontogenetic shifts in diet and habitat use do not necessarily end once major habitat shifts (e.g., oceanic to neritic) occur. The shift from pelagic omnivory to coastal herbivory in green turtles may not be as abrupt as previously thought in some areas. For example, off NW Africa newly settled green turtles sometimes resumed consuming animal matter for extended periods even after they had started consuming seagrasses (Cardona et al. 2009). Off southern Florida, green turtles on seagrass foraging grounds segregated by size over small (3–5 km) spatial scales into turtles <65 cm SCL and those >65 cm SCL (Bresette et al. 2010). Small turtles were found in shallower habitats that likely reduced their risk of predation (Bresette et al. 2010). Similarly, in Shark Bay, Australia, the smallest size classes of green turtles (<45 cm CCL) are generally found nearshore along beaches or in mangrove habitats while larger individuals are found on offshore seagrass beds and associated deeper habitats (Heithaus et al. 2005). In Mexico, smaller individual green turtles are found in sheltered bays whereas larger individuals tend to be found in higher energy coastal habitats. In Shark Bay, size-related changes in susceptibility to predation is the likely driver of this pattern (Heithaus et al. 2005), while in Mexico size-related changes in energetics (i.e., smaller individuals cannot afford the energy expended in high-energy zones outside the bay) appear to create this pattern (Seminoff et al. 2003, López-Mendilaharsu et al. 2005).

For leatherback turtles, the highest latitude foraging sites appear to be inaccessible to smaller animals due to thermal constraints (Witt et al. 2007), and off Massachusetts juveniles appear to spend more time in oceanic habitats, where prey are less abundant, compared to adults that spend more time closer to shore (Dodge et al. 2011). In the Mediterranean, there are size-related differences in large-scale loggerhead distribution (Eckert et al. 2008). In addition, larger logger-heads foraging in oceanic waters were more likely to be associated with environmental features that concentrate prey, which may be the result of better detection capabilities or greater energy requirements (Eckert et al. 2008).

### 10.3.4.2 Interindividual Variation

Variation among individuals in foraging habitats is widespread in sea turtles. For example, although leatherbacks are generally considered to be pelagic foragers, turtles from nesting beaches in Florida used coastal areas relatively frequently (Eckert et al. 2006) and while some leatherbacks tracked in the North Atlantic spend considerable time in "hotspots" (e.g., near oceanographic features) others move nearly continuously (Hays et al. 2006). Green turtles from different populations show different patterns of diving and habitat use during the internesting period based on the availability of food resources near nesting beaches (Hays et al. 2002). There may be interpopulation variation in the extent to which olive ridleys make use of oceanic currents for foraging (Polovina et al. 2004). Also, adult loggerhead turtles nesting on Japanese beaches exhibit size-related variation in foraging habitats (pelagic or neritic) that leads to differences in prey types (planktonic vs. benthic) (Hatase et al. 2002). Similarly, displayed consistent inter-individual differences in feeding habits loggerhead turtles found on a nesting beach in Florida, but whether these differences were driven by consistent use of different foraging areas or specialization in feeding habitats within foraging sites remains unclear (Vander Zanden et al. 2010). Stable isotopic and animal-borne video data collected from a foraging area in Shark Bay, Australia, suggest that it is likely that loggerheads are site specialists but generalists once on foraging grounds (Thomson et al. 2011, 2012). Also, loggerhead females departing Greek nesting grounds settled in one of two main foraging areas. Turtles from the northern foraging area tended to be larger than those of the southern foraging area and laid more eggs (in a single clutch) for a given length (Zbinden et al. 2011).

Green and loggerhead turtles, which were once thought to transition from oceanic habitats as young juveniles to neritic habitats as larger juveniles and adults, are now known to display habitat polymorphisms as adults. Some individuals recruit to neritic habitats while others continue to show fidelity to oceanic habitats even after reaching maturity (e.g., Hatase et al. 2006, Hawkes et al. 2006, Reich et al. 2010). It is now apparent that the oceanic versus pelagic foraging dichotomy is widespread in loggerhead turtles, with evidence from Japan, the Cape Verde Islands (Hawkes et al. 2006), the Arabian Sea (Rees et al. 2010), and the Southeast United States (e.g., Reich et al. 2010).

On Japanese nesting beaches there are no consistent genetic differences between adult female loggerhead turtles with divergent foraging habitats (oceanic vs. neritic) (Watanabe et al. 2011). This lack of population substructuring suggests that individual differences in foraging habitats likely are the result of phenotypic plasticity with smaller individuals adopting, and then maintaining, an oceanic foraging tactic while larger individuals forage in neritic habitats. Although oceanic turtles are smaller than neritic turtles, they appear to nest for the first time at similar ages (Hatase et al. 2010). Therefore, it appears juvenile turtles that experience higher growth rates early in life in oceanic waters may remain in these habitats throughout their lives while individuals that experience poor growth early in development shift to neritic waters where growth rates are faster and they are able to attain a greater length at first reproduction than oceanic individuals (Hatase et al. 2010). An interesting question is how this foraging dichotomy is maintained since neritic adults should have higher fecundity due to larger body sizes and shorter average remigration intervals due to higher prey availability and predictability (Hatase et al. 2010). However, fitness between these two foraging strategies could be similar if survivorship is higher for turtles in oceanic habitats (e.g., Peckham et al. 2011).

Interestingly, the clear size-based differences shown by loggerhead turtles in some locations are not universally evident. For example, off the SE United States (Reich et al. 2010) and the Arabian Sea (Rees et al. 2010) there are no clear size-differences between neritic and oceanic individuals and even some large turtles use pelagic habitats. Green turtles display a foraging dichotomy similar to loggerheads, but the sizes of turtles adopting the two basic tactics do not vary (Hatase et al. 2006).

Immature loggerheads also show dichotomies in their use of foraging habitat. For example, off the Atlantic coast of the United States, some turtles remain over the shelf and make north–south seasonal migrations while others remain in pelagic waters (Mansfield et al. 2009). In the Mediterranean, foraging appears to be more variable than previously thought with very small loggerheads adopting benthic foraging habitats and large individuals foraging in pelagic waters (Casale et al. 2008).

### 10.3.4.2.1  Individual Specialization within Habitats

Ecologists are increasingly recognizing that there can be consistent differences in behavior among individuals of the same age-sex class and in the same basic habitats (see Bolnick et al. 2003). This "individual specialization" results in individuals consuming, or using, a subset of the resources of a population as a whole and has important implications for evolutionary, ecological, and population dynamics as well as conservation (e.g., Baird et al. 1992, Bolnick et al. 2003). In some cases, populations that appear to be generalists are actually made up of groups of specialists (e.g., Bolnick et al. 2003, Quevedo et al. 2009, Matich et al. 2011).

Although there has been considerable recent work on individual variation in basic habitat use tactics by individual turtles leaving a common nesting ground or using different foraging habitats (e.g., Schmid et al. 2003, Section 10.3.4), relatively little work has been done that documents individual specialization within foraging areas. Loggerhead turtles appear to be generalists at both the foraging ground and individual level (e.g., Thomson et al. 2012), while several studies of green turtles suggest that individuals may specialize within foraging grounds. For example, individual green turtles in Shark Bay, Australia, may specialize on different mixes of seagrass, algae, and animal matter (primarily jellyfish and ctenophores). Stomach contents analysis from other locations, where some individuals consume mostly seagrass and others mostly algae, support this possibility (e.g., Bjorndal 1997, Fuentes et al. 2006). The potential drivers of specialization in green turtles are not well known but could include variation in the microflora needed to digest macroalgae versus seagrasses (e.g., Bjorndal et al. 1991, Bjorndal 1997), different microhabitat use patterns and exposure to different relative abundances of food, or relative scarcity of resources and enhanced efficiency due to specialization on a subset of resources (e.g., Bolnick et al. 2003, Tinker et al. 2008, 2012, Burkholder et al. 2011). Future studies of individual specialization within foraging areas likely will provide important insights into the ecological roles of turtles and factors influencing foraging behavior.

## 10.4  ROLES OF SEA TURTLES IN MARINE ECOSYSTEMS

Gaining a greater understanding of the diverse ecological roles of sea turtles, including as consumers, ecosystem engineers, nutrient transporters, prey, and facilitators, is considered a critical research priority (Hamann et al. 2010). Indeed, given large changes in turtle abundances, often negative, but sometimes positive, understanding their roles and how these relate to their ecological importance (i.e., the consequences of changes in their abundances to ecosystem structure and function; e.g., Heithaus et al. 2008b, 2010) is necessary for implementing appropriate conservation and management measures for turtles and their ecosystems. Elucidating ecological roles of sea turtles is complicated by massive reductions in turtle populations before most studies began (e.g., Jackson 1997, Bjorndal and Jackson 2003; see Bjorndal 1997, 2003 for previous reviews).

### 10.4.1  Positive Species Interactions

Sea turtles may have positive impacts, both direct and indirect, on numerous species in their communities. For example, turtles, especially loggerheads, provide habitat for more than a hundred of species of epibionts (see Bjorndal 2003, Bjorndal and Jackson 2003; Chapter 15). Sea turtles also provide a resource for a variety of cleaning organisms, especially fishes that may consume dead skin, parasites, or algae growing on turtles' shells (e.g., Losey et al. 1994, Grossman et al. 2006, Scofield et al. 2006, Sazima et al. 2007, 2010).

Turtles may also facilitate fish foraging and safety. A variety of fishes follow green turtles and feed on invertebrates displaced during grazing (Sazima et al. 2004). Similarly, in Australia, loggerhead turtles are frequently attended by pilot fish and several species of fish are attracted to feeding turtles and consume pieces of prey that have been extracted from bivalve shells (personal observation). Hawksbills facilitate angelfish (Pomacanthidae) foraging by biting through the outer wall of sponges, which allows fish to feed on the sponge's interior or small particles that are dislodged (Blumenthal et al. 2009a). In offshore waters, birds (e.g., boobies) use olive ridley turtles as roosting platforms and feed on fish that aggregate under the turtles (Pitman 1993). Hawksbills likely promote coral growth by reducing abundances of space competitors like sponges and cnidarians (León and Bjorndal 2002). Finally, there may be indirect facilitation among sea turtle species. Seagrass pastures that are heavily grazed by green turtles may enhance their value as hawksbill foraging habitat (Bjorndal and Bolten 2010).

## 10.4.2 Nutrient Transport and Dynamics

Sea turtles can play important roles in moving nutrients across ecosystem boundaries as well as in nutrient dynamics within foraging habitats. The best known example is green turtles short-circuiting detrital cycles in seagrass ecosystems (e.g., Thayer et al. 1982, Bjorndal 1997). Green turtle grazing greatly reduces the decomposition time of seagrass leaves by quickly reducing particle sizes to ones that are easily consumed by detritivores and recycling N more quickly (Thayer et al. 1982).

Loggerhead turtles can play a role in nutrient dynamics of sand-bottom communities. Loggerheads crush the shells of their prey and quickly reduce particle sizes, likely reducing their decomposition times and impacting nutrient dynamics (Bjorndal 2003). Also, bioturbation can be critical to maintaining high biological activity in marine sediments of coastal zones (see Kogure and Wada 2005 for a review). Even small-scale resuspension events can release buried nutrients into the water column, where they become available to microorganisms and primary producers (Yahel et al. 2002, 2008). Resulting increases in microbial activity can significantly reduce the amount of organic carbon sequestered in sediments (Yahel et al. 2008). Bioturbation by rays, which often forage by extracting prey from sediments, can increase rates of organic matter accumulation and modify invertebrate population and community dynamics (e.g., VanBlaricom 1982, Thrush et al. 1991; see Heithaus et al. 2010 for a review). Benthic excavation by loggerhead turtles can result in considerable turnover of sediment that could impact benthic communities and nutrient dynamics. For example, in the Adriatic Sea, loggerheads may remove up to 33 tons of mollusks per year, resulting in considerable bioturbation (Lazar et al. 2011). Studies that quantify the amounts of sediment turned over by loggerheads and their impacts on nutrient dynamics currently are lacking.

During migrations, sea turtles will translocate nutrients from foraging habitats to waters off of nesting beaches (Bjorndal 2003, Bjorndal and Jackson 2003), but nutrient transport into beach ecosystems likely has a much larger impact on ecosystem dynamics. For example, loggerhead turtles nesting on Atlantic beaches of Florida move large quantities of marine-derived nutrients and energy into beach habitats that may be important for supporting predator populations and dune vegetation (Bouchard and Bjorndal 2000). Estimates from a 21 km stretch of beach in 2006 suggest that only 25% of organic matter, 27% of energy, 34% of lipids, 29% of nitrogen, and 39% of phosphorus from nests reentered marine habitats as hatchlings (Bouchard and Bjorndal 2000). The considerable nutrients that remained in terrestrial habitats included eggs that did not hatch and eggs and hatchlings that were consumed by predators. Nutrients from eggs that do not hatch promote the growth of plants when roots grow into nests. Excretion by egg and hatchling predators (Bouchard and Bjorndal 2000) and caching (i.e., burying) of eggs further inland by predators (e.g., red foxes, Macdonald et al. 1994) also promote plant growth. As nesting loggerhead density increases, so does the amount of marine-derived nitrogen in nearby plants (Hannan et al. 2007). On Playa Ostional, Costa Rica, olive ridleys were estimated to lay more than 8 million eggs in a single month, but more than 7 million of these did not hatch, representing a large nutrient influx from marine to beach habitats (Madden et al. 2008). These nutrients facilitate insect populations on the beach (Madden et al. 2008) and may be further transported inland by insect and egg predators that leave the beach area. At Tortuguera, Costa Rica, green turtle nesting appears to influence plant communities. A substantial amount of marine-derived nitrogen—likely from sea turtle eggs—is found in beach vegetation and the dominant plant species varies between locations with high and low nesting density (Vander Zanden et al. 2012). A very different possible impact on beach ecosystems by nesting turtles is in digging up seedlings near the edges of dunes, thereby preventing the encroachment of vegetation down beaches (Rogers 1989, Bjorndal 2003). Taken together, these studies suggest that sea turtle impacts on beach and surrounding ecosystems are likely substantial in many nesting areas. These turtle-coastal ecosystem interactions are intriguing and invite future studies that quantify the overall importance of sea turtle-transported nutrients to dune and associated ecosystems across a range of turtle nesting densities and terrestrial community contexts.

### 10.4.3 TOP-DOWN IMPACTS

Despite their ability to consume considerable quantities of food in marine communities, especially before human-caused population declines, the impacts of turtle foraging on populations of their prey are largely unknown. Two exceptions are green turtles, and to some extent, hawksbills (e.g., Bjorndal and Jackson 2003).

Hawksbill turtle abundance in the Caribbean is greatly reduced from historical levels, which likely has disrupted their role in reef ecosystems (Jackson 1997, Bjorndal and Jackson 2003). A major prey item of hawksbills, the sponge *Chondrilla nucula*, is an aggressive competitor with corals. In the Caribbean, exclusion of sponge predators (including hawksbills) results in increased overgrowth of corals and greater loss of coral cover (Hill 1998). Because hawksbills preferentially forage on a number of aggressive space competitors on reefs, they likely promote biodiversity (León and Bjorndal 2002). Therefore, the loss of the majority of the Caribbean hawksbill population, which may have numbered 540,000 and consumed 200–420 million kg wet mass of sponges a year (Bjorndal and Jackson 2003), likely has led to shifts in reef communities and/or changes in the relative importance of other spongivores (e.g., angelfish) in minimizing sponge overgrowth of corals (Bjorndal and Jackson 2003). Not all sea turtle impacts on coral reefs are positive, however. Green turtles in Hawaii have been observed rubbing against coral heads, causing heavy damage (Bennett et al. 2002).

The impacts of green turtle foraging on the biomass, community structure, and nutrient dynamics of seagrass habitats has received considerable attention (see Bjorndal 1997, Bjorndal and Jackson 2003 for earlier reviews). Bjorndal and Jackson (2003) estimated that at an estimated historical population of 100 million (based on the assumption of bottom-up regulation of populations) green turtles in the Caribbean would have consumed $6.2 \times 10^9$ kg dry mass of the seagrass *Thalassia testudinum*, approximately half of the estimated annual production. Although this estimate is necessarily rough, it provides evidence of the central and historically important role that green turtles likely played in the dynamics of seagrass ecosystems (Bjorndal and Jackson 2003, Heck and Valentine 2006).

Simulated grazing has been used to investigate the impacts of green turtle gazing in multiple locations. Off northeastern Australia, one bout of simulated green turtle grazing can cause shifts in seagrass species composition from *Zostera capricorni* dominance to *Halophila ovalis* (Aragones and Marsh 2000). Simulated cropping by turtles also resulted in increased N content and decreased fiber content relative to controls, but turtles that are grazing plots that had previously been cropped would only realize benefits if seagrasses were left ungrazed for more than a month (Aragones et al. 2006). Another experiment revealed variation among three species of seagrasses in their responses to 3.5 months of simulated grazing over the summer (Kuiper-Linley et al. 2007). *H. ovalis*, the preferred forage species, showed increases in leaf regrowth rate and water-soluble carbohydrates in clipped plots and no change in total plant material. In contrast, two less preferred seagrass species (*Z. capricorni* and *Cymodocea serrulata*) showed more than a 40% reduction in whole plant biomass.

In general, responses to simulated grazing in this area may be driven partially by variation among species in seasonal growth patterns and in timing of experiments (Aragones and Marsh 2000).

In the Caribbean, 16 months of simulated green turtle grazing in a *T. testudinum* seagrass pasture resulted in thinner detrital layers and seagrass blades that were shorter and narrower. But, shoot density, number of blades per shoot, and rhizome biomass did not change (Moran and Bjorndal 2005). Therefore, structural complexity was reduced, but productivity did not change. However, clipping resulted in a higher productivity to biomass ratio (Moran and Bjorndal 2005). In addition, in clipped plots, seagrass blades had higher energy and nutrient content, but also tannins (a secondary compound), than blades in unclipped plots (Moran and Bjorndal 2007). Nitrogen and organic matter in rhizomes, however, declined in clipped plots. Thus, continued simulated grazing may have eventually led to reduced nutrient content in blades (Moran and Bjorndal 2007).

Other methods that have been used to investigate impacts of green turtle foraging on seagrasses include correlating variation in seagrass communities with densities of grazing turtles and excluding turtles from grazing in particular locations. In the Indian Ocean, increasing populations of green turtles at the Lakshadweep Islands have had large impacts on seagrass communities. Areas with high densities of turtles had lower seagrass shoot densities, lower aboveground biomass, and lower flowering intensity (Lal et al. 2010). Turtle grazing also appears to have reduced seagrass diversity, with a fast-growing species becoming dominant. Similarly, in Bermuda, apparent increases in turtle populations may be driving declines in seagrasses (Murdoch et al. 2007, Fourqurean et al. 2010). When turtles were experimentally excluded from grazing for a year, the length and width of seagrass blades increased compared to unprotected seagrasses. This led to greater seagrass biomass and canopy structural complexity in the absence of grazing (Fourqurean et al. 2010). In addition, areas where turtles were excluded had higher rates of productivity. The loss of seagrasses under heavy grazing appears to be driven by the reduction in photosynthetic capacity with the removal of leaves and resulting reduction in carbon fixation (Fourqurean et al. 2010). In contrast, heavy grazing by turtles in tropical seagrass ecosystems may help to alleviate the negative effects of eutrophication through the stimulation of seagrass production and increases in nutrient uptake (Christianen et al. 2012). Impacts of green turtles on seagrass beds are likely to cascade through ecosystems. Seagrasses provide important habitat for many species and changes in the structure of seagrass habitats will modify their value as shelter and foraging habitat (Heck et al. 2003).

Results from an ecological model of a coral reef ecosystem within a Hawaiian marine park suggest that green turtles have reached carrying capacity and help maintain lower algal cover (Wabnitz et al. 2010). Although the large biomass of sea urchins likely have a greater impact on algae, turtles appear to be important in enhancing coral reef resilience by helping maintain low algal cover. By foraging on introduced species of algae, green turtles may further enhance the resilience of reef ecosystems (Wabnitz et al. 2010).

## 10.4.4  SEA TURTLES AS PREY AND TRANSMITTERS OF INDIRECT EFFECTS

Sea turtle eggs and hatchlings may be important food sources for a number of predators (see Section 10.2.1.1) and may make up the majority of predator diets during the nesting season. Although the implications for predator populations are unknown in the majority of cases, sometimes they may be maintained at higher levels than would be possible without consuming turtles. For example, rats in New Caledonia compensate for shortages in other food sources by switching to prey heavily on green turtle hatchlings, which may promote rat population persistence (Caut et al. 2008). Although a relatively large proportion of the diets of tiger sharks in some locations may be made up of sea turtles (Heithaus et al. 2008b; Section 10.2.1.2.4), the importance of turtle prey for maintaining population sizes of sharks remains unknown.

Because of their potential impacts on their ecosystems, turtles may be important players in marine trophic cascades, not only as initiators (e.g., green turtle-induced changes to seagrass habitats and resulting impacts on communities), but also in transmitting indirect effects of top predators (see Heithaus et al. 2008a,b). Such indirect effects could be transmitted through predators consuming turtles ("direct predation" or "consumptive effects"), predator-induced modifications of turtle behavior ("risk" or "trait-mediated" effects), or the interaction of these two (see Werner and Peacor 2003, Creel and Christianson 2008, Heithaus et al. 2008a, Creel 2011). While the majority of ecological literature has been focused on the effects of direct predation, or assumed direct predation as the primary mechanism through which predators impact prey (Peckarsky et al. 2008), it is now clear that in some cases risk effects (which can interact with effects of direct predation) may equal or exceed the impacts of direct predation (e.g., Schmitz et al. 1997, 2004, Dill et al. 2003, Werner and Peacor 2003, Preisser et al. 2005, Heithaus et al. 2008b). Often, risk effects are most important because direct mortality usually removes a limited number of individuals from a population, which may result in decreased intraspecific feeding or reproductive competition. This, in turn, can result in increased reproduction or growth for remaining prey individuals (compensatory reproduction or growth) that has an end result of no reduction in population size. In contrast, antipredator behavior, which may include leaving high-risk but high-productivity habitats or reduced foraging rates, is generally performed by all (or most) individuals in a population and can result in lower access to food and a reduction in the population's reproductive potential. Because of their high adult survival rates and long lifespans, risk effects are likely to be the primary mechanism through which predators of large juvenile and adult sea turtles affect turtle populations and their impacts on ecosystems and communities (see Warner 1998, Heithaus et al. 2008b, Frid et al. 2012).

Behaviorally mediated indirect interactions (BMII) occur when the presence or absence of one "initiator" species affects the behavior of a second ("transmitter") species, that in turn impacts a third species (Dill et al. 2003). BMII may create, enhance, ameliorate, or even reverse the sign (e.g., a species actually has a positive effect on its competitor) of interactions between species (see Dill et al. 2003, Heithaus et al. 2008a). For example, the presence of people, even at low densities, on nesting beaches has a positive indirect effect on hawksbill nests by inducing mongooses, a nest predator, to largely abandon portions of nesting beaches (Leighton et al. 2010). In Shark Bay, green turtles may be an important transmitter of a BMII between tiger sharks and primary producer communities. In this interaction, tiger sharks cause spatial shifts in green turtle foraging locations, which, in turn, may indirectly modify seagrass community composition and nutrient content (Heithaus et al. 2007a, 2008a,b). Experimental studies to test the second step of this BMII are ongoing.

## 10.5 SUMMARY

Our understanding of sea turtle predator–prey interactions is growing rapidly from the perspective of sea turtles as consumers, but there remains much to be done to gain a more general understanding of the foraging tactics of sea turtles and their potential to impact their communities and ecosystems. The emerging view that large juvenile and adult sea turtles within the same population may display large differences in their foraging habits and behavior (even within the same habitats) suggests that sea turtles may play more varied roles in marine communities than previously thought.

Our understanding of the potential for predators to affect sea turtle populations and behavior— or to have done so historically—is still in its infancy. Few sea turtle species have been studied from this perspective, and none have been studied in depth. Yet, such studies are important given large-scale changes in population sizes of both turtles and their potential predators. Furthermore, because human disturbance often is perceived in a manner similar to predation risk (Frid and Dill 2002, Frid and Heithaus 2010), studies of turtle–predator interactions may provide insights into impacts of anthropogenic disturbance on turtles. Ultimately, additional studies are needed to provide a functional understanding of sea turtle behavior relative to their resources and potential predators. Such an understanding should provide the ability to make predictions about how changes in ecological conditions will affect sea turtles, and how changes in sea turtle populations and behavior are likely to influence marine and coastal ecosystems.

## ACKNOWLEDGMENTS

I thank the editors for inviting me to contribute to this work and to the editors and an outside reviewer for helpful comments on this chapter. Financial support was provided by the National Science Foundation grant OCE0745606. This is publication 60 of the Shark Bay Ecosystem Research Project (www.sberp.org).

## REFERENCES

Allen, C. R., E. A. Forys, K. G. Rice, and D. P. Wojcik. 2001. Effects of fire ants (Hymenoptera: Formicidae) on hatching turtles and prevalence of fire ants on sea turtle nesting beaches in Florida. *Florida Entomologist* 84: 250–253.

Anholt, B. R. and E. E. Werner. 1995. Interaction between food availability and predation mortality mediated by adaptive behavior. *Ecology* 76: 2230–2234.

Aragones, L. V., I. R. Lawler, W. J. Foley, and H. Marsh. 2006. Dugong grazing and turtle cropping: Grazing optimization in tropical seagrass systems? *Oecologia* 149: 635–647.

Aragones, L. V. and H. Marsh. 2000. Impact of dugong grazing and turtle cropping on tropical seagrass communities. *Pacific Science* 5: 277–288.

Aurtar, L. 1994. Sea turtles attacked and killed by jaguars in Suriname. *Marine Turtle Newsletter* 67: 11–12.

Bailey, H., G. Shillinger, D. Palacius et al. 2008. Identifying and comparing phases of movement by leatherback turtles using state-space models. *Journal of Experimental Marine Biology and Ecology* 356: 128–135.

Baird, R. W., P. A. Abrams, and L. M. Dill. 1992. Possible indirect interactions between transient and resident killer whales: Implications for the evolution of foraging specializations in the genus *Orcinus*. *Oecologia* 89: 125–132.

Ballorain, K., S. Ciccione, J. Bourjea, H. Grizel, M. Enstipp, and J. Georges. 2010. Habitat use of a multispecific seagrass meadow by green turtles *Chelonia mydas* at Mayotte Island. *Marine Biology* 157: 2581–2590.

Barton, B. T. and J. D. Roth. 2008. Implications of intraguild predation for sea turtle nest protection. *Biological Conservation* 141: 2139–2145.

Baum, J. K., D. Kehler, and R. A. Myers. 2005. Robust estimates of decline for pelagic shark populations in the northwest Atlantic and Gulf of Mexico. *Fisheries* 30: 27–29.

Bednekoff, P. A. and S. L. Lima. 1998. Re-examining safety in numbers: Interactions between risk dilution and collective detection depend upon predator targeting behaviour. *Proceedings of the Royal Society B* 265: 2021–2026.

Bennett, P., U. Keuper-Bennett, and G. H. Balazs. 2002. Changing the landscape: Evidence for detrimental impacts to coral reefs by Hawaiian marine turtles. In A. Mosier, A. Foley, and B. Brost (eds.). *Proceedings of the 20th Annual Symposium on Sea Turtle Biology and Conservation*, February 29–March 4, 2000, Orlando, FL, NOAA Technical Memorandom NMFS-SEFSC-477. pp. 287–288.

Bertram, B. C. R. 1978. Living in groups: Predators and prey. In *Behavioural Ecology: An Evolutionary Approach*, eds. J. R. Krebs and N. B. Davies, pp. 64–96. Oxford, U.K.: Blackwell Press.

Binkley, D., M. M. Moore, W. H. Romme, and P. M. Brown. 2006. Was Aldo Leopold right about the Kaibab deer herd? *Ecosystems* 9: 227–241.

Bjorndal, K. A. 1980. Nutrition and grazing behavior of the green turtle, *Chelonia mydas*. *Marine Biology* 56: 147–154.

Bjorndal, K. A. 1997. Foraging ecology and nutrition of sea turtles. In *The Biology of Sea Turtles*, eds. P. L. Lutz and J. A. Musick, pp. 199–232. Boca Raton, FL: CRC Press.

Bjorndal, K. A. 2003. Roles of loggerhead sea turtles in marine ecosystems. In *Loggerhead Sea Turtles*, eds. A. B. Bolten and B. E. Witherington, pp. 235–254. Washington, DC: Smithsonian Books.

Bjorndal, K. A. and A. B. Bolten. 2010. Hawksbill sea turtles in seagrass pastures: Success in a peripheral habitat. *Marine Biology* 157: 135–145.

Bjorndal, K. A., A. B. Bolten, and M. Y. Chaloupka. 2000. Green turtle somatic growth model: Evidence for density dependence. *Ecological Applications* 10: 269–282.

Bjorndal, K. A., A. B. Bolten, and M. Y. Chaloupka. 2003. Survival probability estimates for immature green turtles *Chelonia mydas* in the Bahamas. *Marine Ecology Progress Series* 252: 273–281.

Bjorndal, K. A. and J. B. C. Jackson. 2003. Roles of sea turtles in marine ecosystems: Reconstructing the past. In *The Biology of Sea Turtles*. Vol. II, eds. P. L. Lutz, J. A. Musick, and J. Wyneken, pp. 259–274. Boca Raton, FL: CRC Press.

Bjorndal, K. A., H. Suganuma, and A. B. Bolten. 1991. Digestive fermentation in green turtles, *Chelonia mydas*, feeding on algae. *Bulletin of Marine Science* 48: 166–171.

Blamires, S. J. 2004. Habitat preferences of coastal goannas (*Varanus panoptes*): Are they exploiters of sea turtle nests at Fog Bay, Australia? *Copeia* 2004: 370–377.

Blumenthal, J. M., T. J. Austin, C. D. L. Bell et al. 2009a. Ecology of hawksbill turtles, *Eretmochelys imbricata*, on a western Caribbean foraging ground. *Chelonian Conservation and Biology* 8: 1–10.

Blumenthal, J. M., T. J. Austin, J. B. Bothwell et al. 2009b. Diving behavior and movements of juvenile hawksbill turtles *Eretmochelys imbricata* on a Caribbean coral reef. *Coral Reefs* 28: 55–65.

Blumenthal, J. M., T. J. Austin, J. B. Bothwell et al. 2010. Life in (and out) of the lagoon: Fine-scale movements of green turtles tracked using time-depth recorders. *Aquatic Biology* 9: 113–121.

Bolnick D. I., R. Svanbäck, J. A. Fordyce, L. H. Yang, J. M. David, C. D. Hulsey, and M. L. Forister. 2003. The ecology of individuals: Incidence and implications of individual specialization. *American Naturalist* 161: 1–28.

Bolten, A. B. 2003. Active swimmers—Passive drifters: The oceanic juvenile stage of loggerheads in the Atlantic system. In *Loggerhead Sea Turtles*, eds. A. B. Bolten and B. E. Witherington, pp. 63–78. Washington, DC: Smithsonian Books.

Bonenfant, M. and D. L. Kramer. 1994. The influence of distance to burrow on flight initiation distance in the woodchuck, *Marmota monax*. *Behavioral Ecology* 7: 299–303.

Bouchard, S. S. and K. A. Bjorndal. 2000. Sea turtles as biological transporters of nutrients and energy from marine to terrestrial ecosystems. *Ecology* 81: 2305–2313.

Bresette, M. J., B. E. Witherington, R. M. Herren et al. 2010. Size-class partitioning and herding in a foraging group of green turtles *Chelonia mydas*. *Endangered Species Research* 9: 105–116.

Brill, R. W., G. H. Balazs, K. N. Holland, R. K. C. Chang, S. Sullivan, and J. C. George. 1995. Daily movements, habitat use, and submergence intervals of normal and tumor-bearing juvenile green turtles (*Chelonia mydas* L.) within a foraging area in the Hawaiian Islands. *Journal of Experimental Marine Biology and Ecology* 185: 203–218.

Broderick, A. C., M. S. Coyne, W. J. Fuller, G. Glen, and B. J. Godley. 2007. Fidelity and over-wintering of sea turtles. *Proceedings of the Royal Society B* 274: 1533–1538.

Brown, J. S. and B. Kotler. 2004. Hazardous duty pay and the foraging cost of predation. *Ecology Letters* 10: 999–1014.

Brown J. S., J. W. Laundré, and M. Gurung. 1999. The ecology of fear: Optimal foraging, game theory, and trophic interactions. *Journal of Mammalogy* 80: 385–399.

Brown, L. and D. W. MacDonald. 1995. Predation on green turtle *Chelonia mydas* nests by wild canids at Akyatan Beach, Turkey. *Biological Conservation* 71: 55–60.

Burkholder, D. A., M. R. Heithaus, J. A. Thomson, and J. W. Fourqurean. 2011. Diversity in trophic interactions of green sea turtles *Chelonia mydas* on a relatively pristine coastal foraging ground. *Marine Ecology Progress Series* 439: 277–293.

Cardona, L., A. Aguilar, and L. Pazos. 2009. Delayed ontogenetic dietary shift and high levels of omnivory in green turtles (*Chelonia mydas*) from the NW coast of Africa. *Marine Biology* 156: 1487–1495.

Cardona, L., P. Campos, Y. Levy, A. Demetropoulos, and D. Margaritoulis. 2010. Asynchrony between dietary and nutritional shifts during the ontogeny of green turtles (*Chelonia mydas*) in the Mediterranean. *Journal of Marine Experimental Biology and Ecology* 393: 83–89.

Carlson, J. K., L. F. Hale, A. Morgan, and G. Burgess. 2012. Relative abundance and size of coastal sharks derived from commercial shark longline catch and effort data. *Journal of Fish Biology* 80: 1749–1764.

Caro, T. M. 1995. Pursuit-deterrence revisited. *Trends in Ecology and Evolution* 10: 500–503.

Carranza, A., A. Estrades, F. Scarabino, and A. Segura. 2011. Loggerhead turtles *Caretta caretta* (Linnaeus) preying on the invading gastropod *Rapana venosa* (Valenciennes) in the Rio de la Plata estuary. *Marine Ecology: An Evolutionary Perspective* 32: 142–147.

Casale, P., G. Abbate, D. Freggi, N. Conte, M. Oliverio, and R. Argano. 2008. Foraging ecology of loggerhead sea turtles *Caretta caretta* in the central Mediterranean Sea: Evidence for a relaxed life-history model. *Marine Ecology Progress Series* 372: 265–276.

Casey, J., J. Garner, S. Garner, and A. Willard. 2010. Diel foraging behavior of gravid leatherback sea turtles in deep waters of the Caribbean Sea. *Journal of Experimental Biology* 213: 2961–3971.

Caut, S., E. Angulo, and F. Courchamp. 2008. Dietary shift of an invasive predator: Rats, seabirds and sea turtles. *Journal of Applied Ecology* 45: 428–437.

Caut, S., E. Guirlet, P. Jouquet, and M. Girondot. 2006. Influence of nest location and yolkless eggs on the hatching success of leatherback turtle clutches in French Guiana. *Canadian Journal of Zoology* 84: 908–915.

Chaloupka, M. and C. J. Limpus. 2001. Trends in the abundance of sea turtles resident in southern Great Barrier Reef waters. *Biological Conservation* 102: 235–249.

Chaloupka, M., C. Limpus, and J. Miller. 2004. Green turtle somatic growth dynamics in a spatially disjunct Great Barrier Reef metapopulation. *Coral Reefs* 23: 325–335.

Christianen, M. J. A., L. L. Grovers, T. J. Bouma et al. 2012. Marine megaherbivore grazing may increase seagrass tolerance to high nutrient loads. *Journal of Ecology* 100: 546–560.

Chung, F. C., N. J. Pilcher, M. Salmon, and J. Wyneken. 2009a. Offshore migratory activity of hawksbill turtle (*Eretmochelys imbricata*) hatchlings I. Quantitative analysis of activity with comparisons to green turtles (*Chelonia mydas*). *Chelonian Conservation and Biology* 8: 28–34.

Chung, F. C., N. J. Pilcher, M. Salmon, and J. Wyneken. 2009b. Offshore migratory activity of hawksbill turtle (*Eretmochelys imbricata*) hatchlings II. Swimming gaits, swimming speed, and morphological comparisons. *Chelonian Conservation and Biology* 8: 35–42.

Clark, C. W. 1994. Antipredator behaviour and the asset–protection principle. *Behavioral Ecology* 5: 159–170.

Cliff, G. and S. F. J. Dudley. 1991. Sharks caught in the protective gill nets off Natal, South Africa. 4. The bull shark *Carcharhinus leucas* Valenciennes. *South African Journal of Marine Science* 10: 253–270.

Conrad, J. R., J. Wyneken, J. A. Garner, and S. Garner. 2011. Experimental study of dune vegetation impact and control on leatherback sea turtle *Dermochelys coriacea* nests. *Endangered Species Research* 15: 13–27.

Corti, P. and D. M. Shackleton. 2002. Relationship between predation-risk factors and sexual segregation in Dall's sheep (*Ovis dalli dalli*). *Canadian Journal of Zoology* 80: 2108–2117.

Creel, S. 2011. Toward a predictive theory of risk effects: Hypotheses for prey attributes and compensatory mortality. *Ecology* 92: 2190–2195.

Creel, S. and D. Christianson. 2008. Relationships between direct predation and risk effects. *Trends in Ecology and Evolution* 23: 194–201.

Creel, S., D. Christianson, S. Liley, and J. A. Winnie. 2007. Predation risk affects reproductive physiology and demography of elk. *Science* 315: 960.

Cresswell, W. 1994. Age-dependent choice of redshank (*Tringa tetanus*) feeding location: Profitability or risk? *Journal of Animal Ecology* 63: 589–600.

Crouse, D. T., L. B. Crowder, and H. Caswell. 1987. A stage-based population model for loggerhead sea turtles and implications for conservation. *Ecology* 68: 1412–1423.

Davenport, J., J. Fraher, E. Fitzherald et al. 2009. Fat head: An analysis of head and neck insulation in the leatherback turtle (*Dermochelys coriacea*). *Journal of Experimental Biology* 212: 2753–2759.

Diez, C. E. and R. P. van Dam. 2002. Habitat effect on hawksbill turtle growth rates on feeding grounds at Mona and Monito Islands, Puerto Rico. *Marine Ecology Progress Series* 234: 301–309.

Dill, L. M. 1990. Distance-to-cover and the escape decisions of an Africa cichlid fish, *Melanochromis chipokae*. *Environmental Biology of Fishes* 27: 147–152.

Dill, L. M., M. R. Heithaus, and C. J. Walters. 2003. Behaviorally mediated indirect interactions in marine communities and their conservation implications. *Ecology* 84: 1151–1157.

Dodge, K. L., J. M. Logan, and M. E. Lutcavage. 2011. Foraging ecology of leatherback sea turtles in the western North Atlantic determined through multi-tissue stable isotope analysis. *Marine Biology* 158: 2813–2324.

Donlan, E. M., J. H. Townsend, and E. A. Golden. 2004. Predation of *Caretta caretta* (Testudines: Cheloniidae) eggs by larvae of *Lanelater sallei* on Key Biscayne, Florida. *Caribbean Journal of Science* 40: 415–420.

Doyle, T. K., J. D. R. Houghton, P. F. O'Súilleabháin et al. 2008. Leatherback turtles satellite-tagged in European waters. *Endangered Species Research* 4: 23–31.

Dunphy-Daly, M. M., M. R. Heithaus, A. J. Wirsing, J. S. F. Mardon, and D. A. Burkholder. 2010. Predation risk influences the diving behavior of a marine mesopredator. *Open Ecology Journal* 3: 8–15.

Dutton, D. L., P. H. Dutton, M. Chaloupka, and R. H. Boulon. 2005. Increase of a Caribbean leatherback turtle *Dermochelys coriacea* nesting population linked to long-term nest protection. *Biological Conservation* 126: 186–194.

Eckert, S. A. 2006. High-use oceanic areas for Atlantic leatherback sea turtles (*Dermochelys coriacea*) as identified using satellite telemetered location and dive information. *Marine Biology* 149: 1257–1267.

Eckert, S. A., D. Bagley, S. Kubis et al. 2006. Internesting and postnesting movements and foraging habitats of leatherback sea turtles (*Dermochelys coriacea*) nesting in Florida. *Chelonian Conservation and Biology* 5: 239–248.

Eckert, S. A., J. E. Moore, D. C. Dunn, R. S. van Buiten, K. L. Eckert, and P. N. Halpin. 2008. Modeling loggerhead turtle movement in the Mediterranean: Importance of body size and oceanography. *Ecological Applications* 18: 290–308.

Eckrich, C. E. and D. W. Owens. 1995. Solitary versus arribada nesting in the olive ridley sea turtles (*Lepidochelys olivacea*): A test of the predator-satiation hypothesis. *Herptologica* 51: 349–354.

Engeman, R., A. Duffiney, S. Braem et al. 2010. Dramatic and immediate improvements in insular nesting success for threatened sea turtles and shorebirds following predator management. *Journal of Experimental Marine Biology and Ecology* 395: 147–152.

Engeman, R. M., R. E. Martin, H. T. Smith et al. 2006. Impact on predation of sea turtle nests when predator control was removed midway through the nesting season. *Wildlife Research* 33: 187–192.

Fergusson, I. K., L. J. V. Conpagno, and M. A. Marks. 2000. Predation by white sharks *Carcharodon carcharias* (Chondrichthyes: Lamnidae) upon chelonians, with new records from the Mediterranean Sea and a first record of the ocean sunfish *Mola mola* (Osteichthyes: Molidae) as stomach contents. *Environmental Biology of Fishes* 58: 447–453.

Ferretti, F., B. Worm, B. G. L. Britten, M. R. Heithaus, and H. K. Lotze. 2010. Patterns and ecosystem consequences of shark declines in the ocean. *Ecology Letters* 13: 1055–1071.

Fertl, D. and G. L. Fulling. 2007. Interactions between marine mammals and turtles. *Marine Turtle Newsletter* 115: 4–8.

Ficetola, G. F. 2008. Impacts of human activities and predators on the nest success of the hawksbill turtle, *Eretmochelys imbricata*, in the Arabia Gulf. *Chelonian Conservation and Biology* 7: 255–257.

Fossette, S., A. C. Gleiss, J. P. Casey, A. R. Lewis, and G. C. Hays. 2011. Does prey size matter? Novel observations of feeding in the leatherback turtle (*Dermochelys coriacea*) allow a test of predator-prey size relationships. *Biology Letters* 8: 351–354. doi: 10.1098/rsbl.2011.0965.

Fourqurean, J. W., S. Manuel, K. A. Coates, W. J. Kenworthy, and S. R. Smith. 2010. Effects of excluding sea turtle herbivores from a seagrass bed: Overgrazing may have led to loss of seagrass meadows in Bermuda. *Marine Ecology Progress Series* 419: 223–232.

Frid, A. and Dill, L. M. 2002. Human caused disturbance stimuli as a form of predation risk. *Conservation Ecology* 6: 11. http://www.consecol.org/vol6/iss1/art11

Frid, A., L. M. Dill, R. E. Thorne, and G. M. Blundell. 2007a. Inferring prey perception of relative danger in large-scale marine systems. *Evolutionary Ecology Research* 9: 635–649.

Frid, A. and M. R. Heithaus. 2010. Conservation and anti-predator behavior. In *Encyclopedia of Animal Behavior*, eds. M. D. Breed and J. Moore, pp. 366–376. Amsterdam, The Netherlands: Elsevier.

Frid, A., M. R. Heithaus, and L. M. Dill. 2007b. Dangerous dive cycles and the proverbial ostrich. *Oikos* 116: 893–902.

Frid, A., J. Marglive, and M. R. Heithaus. 2012. Interspecific variation in life history relates to antipredator decisions by marine mesopredators on temperate reefs. *PLoS One* 7(6): e40083. doi:10.1371/journal. pone.0040083.

Fretwell, S. D. and H. L. Lucas. 1970. On territorial behavior and other factors influencing habitat distribution in birds. *Acta Biotheoretica* 19: 16–36.

Fuentes, M. M. P. B., I. R. Lawler, and E. Gyuris. 2006. Dietary preferences of juvenile green turtles (*Chelonia mydas*) on a tropical reef flat. *Wildlife Research* 33: 671–678.

Glen, F., A. C. Broderick, B. J. Godley, and G. C. Hays. 2005. Patterns in the emergence of green (*Chelonia mydas*) and loggerhead (*Caretta caretta*) turtle hatchlings from their nests. *Marine Biology* 146: 1039–1049.

Glen, F., A. C. Broderick, B. J. Godley, J. D. Metcalfe, and G. C. Hays. 2001. Dive angles for a green turtle (*Chelonia mydas*). *Journal of the Marine Biological Association of the United Kingdom* 81: 683–686.

Grossman, A., C. Sazima, C. Bellini, and I. Sazima. 2006. Cleaning symbiosis between turtles and reef fishes at Fernando de Noronha Archipelago, off northeast Brazil. *Chelonian Conservation and Biology* 5: 284–288.

Gyuris, E. 1994. The rate of predation by fishes on hatchlings of the green sea turtle (*Chelonia mydas*). *Coral Reefs* 13: 137–144.

Hamann, M., M. H. Dogfrey, J. A. Seminoff et al. 2010. Global research priorities for sea turtles: Informing management and conservation in the 21st century. *Endangered Species Research* 11: 245–269.

Hamilton, I. M. 2010. Habitat selection. In *Encyclopedia of Animal Behavior*, eds. M. D. Breed and J. Moore, pp. 38–43. Amsterdam, The Netherlands: Elsevier.

Hannan, L. B., J. D. Roth, L. M. Ehrhart, and J. F. Weishampel. 2007. Dune vegetation fertilization by nesting sea turtles. *Ecology* 88: 1053–1058.

Harrison, E., S. Troëng, and M. Fletcher. 2005. Jaguar predation of green turtles (*Chelonia mydas*) at Tortuguero, Costa Rica—Current trends and conservation implications. Abstracts of paper presented at *25th International Sea Turtle Symposium*, Savannah, GA, 2005. Available online at: http://www. conserveturtles.org/symposiumpresentations.php?page=harrison-2005

Hart, K. M. and I. Fujisaki. 2010. Satellite tracking reveals habitat use of juvenile green turtles *Chelonia mydas* in the Everglades, Florida, USA. *Endangered Species Research* 11: 221–232.

Hart, K. M., M. M. Lamont, I. Fujisaki, A. D. Tucker, and R. R. Carthy. 2012. Common coastal foraging areas for loggerheads in the Gulf of Mexico: Opportunities for marine conservation. *Biological Conservation* 145: 185–194.

Hatase, H., K. Omuta, and K. Tsukamoto. 2010. Oceanic residents, neritic migrants: A possible mechanism underlying foraging dichotomy in adult female loggerhead turtles (*Caretta caretta*). *Marine Biology* 157: 1337–1342.

Hatase, H., K. Sato, M. Yamaguchi, K. Takahashi, and K. Tsukamoto. 2006. Individual variation in feeding habitat use by adult female green sea turtles (*Chelonia mydas*): Are they obligate neritic herbivores? *Oecologia* 149: 52–64.

Hatase, H., N. Takai, Y. Matsuzawa et al. 2002. Size-related differences in feeding habitat use of adult female loggerhead turtles *Caretta caretta* around Japan determined by stable isotope analysis and satellite telemetry. *Marine Ecology Progress Series* 233: 273–281.

Hatase, H. and K. Tsukamoto. 2008. Smaller longer, larger shorter: Energy budget calculations explain intrapopulation variation in remigration intervals for loggerhead turtles. *Canadian Journal of Zoology* 86: 595–600.

Hawkes, A., A. C. Broderick, M. S. Coyne et al. 2006. Phenotypically linked dichotomy in sea turtle foraging requires multiple conservation approaches. *Current Biology* 16: 990–995.

Hays, G. C., C. Adams, A. C. Broderick et al. 2000. The diving behavior of green turtles at Ascension Island, south Atlantic. *Journal of the Marine Biological Association of the United Kingdom* 75: 405–411.

Hays, G. C., S. Akesson, A. C. Broderick et al. 2001. The diving behavior of green turtles undertaking oceanic migrations to and from Ascension Island: Dive durations, dive profiles, and depth distribution. *Journal of Experimental Biology* 204: 4093–4098.

Hays, G. C., F. Glen, A. C. Broderick, B. J. Godley, and J. D. Metcalfe. 2002. Behavioral plasticity in a large marine herbivore: Contrasting patterns of depth utilisation between two green turtle (*Chelonia mydas*) populations. *Marine Biology* 141: 985–990.

Hays, G. C., V. J. Hobson, J. D. Metcalfe, D. Righton, and D. W. Sims. 2006. Flexible foraging movements of leatherback turtles across the northern Atlantic Ocean. *Ecology* 87: 2647–2656.

Hays, G. C., J. D. Metcalfe, and A. W. Walne. 2004. The implications of lung-regulated buoyancy control for dive depth and duration. *Ecology* 85: 1137–1145.

Hazel, J., I. R. Lawler, and M. Hamann. 2009. Diving at the shallow end: Green turtle behaviour in near-shore foraging habitat. *Journal of Experimental Marine Biology and Ecology* 371: 84–92.

Heck Jr., K. L, C. G. Hays, and R. J. Orth. 2003. Critical evaluation of the nursery hypothesis for seagrass meadows. *Marine Ecology Progress Series* 253: 123–136.

Heck Jr., K. L. and J. F. Valentine. 2006. Plant-herbivore interactions in seagrass meadows. *Journal of Experimental Marine Biology and Ecology* 330: 420–436.

Heithaus, M. R. 2001a. Predator–prey and competitive interactions between sharks (order Selachii) and dolphins (suborder Odontoceti): A review. *Journal of Zoology (London)* 253: 53–68.

Heithaus, M. R. 2001b. Shark attacks on bottlenose dolphins (*Tursiops aduncus*) in Shark Bay, Western Australia: Attack rate, bite scar frequencies, and attack seasonality. *Marine Mammal Science* 17: 526–539.

Heithaus, M. R. 2001c. Habitat selection by predators and prey in communities with asymmetrical intraguild predation. *Oikos* 92: 542–554.

Heithaus, M. R., D. Burkholder, R. E. Hueter, L. I. Heithaus, H. W. Pratt Jr., and J. C. Carrier. 2007b. Spatial and temporal variation in shark communities of the lower Florida Keys and evidence for historical population declines. *Canadian Journal of Fisheries and Aquatic Sciences* 64: 1302–1313.

Heithaus, M. R. and L. M. Dill 2002. Food availability and tiger shark predation risk influence bottlenose dolphin habitat use. *Ecology* 83: 480–491.

Heithaus M. R. and L. M. Dill. 2006. Does tiger shark predation risk influence foraging habitat use by bottlenose dolphins at multiple spatial scales? *Oikos* 114: 257–264.

Heithaus, M. R., L. M. Dill, G. J. Marshall, and B. Buhleier. 2002a. Habitat use and foraging behavior of tiger sharks (*Galeocerdo cuvier*) in a seagrass ecosystem. *Marine Biology* 140: 237–248.

Heithaus, M. R., and A. Frid. 2003. Optimal diving under the risk of predation. *Journal of Theoretical Biology* 223: 79–93.

Heithaus, M. R., A. Frid, and L. M. Dill. 2002b. Shark-inflicted injury frequencies, escape ability, and habitat use of green and loggerhead turtles. *Marine Biology* 140: 229–236.

Heithaus, M. R., A. Frid, J. J. Vaudo, B. Worm, and A. J. Wirsing. 2010. Unraveling the ecological importance of elasmobranchs. In *Sharks and Their Relatives II*, eds. J. C. Carrier, J. A. Musick, and M. R. Heithaus, pp. 611–637. Boca Raton, FL: CRC Press.

Heithaus, M. R., A. Frid, A. J. Wirsing, L. Bejder, and L. M. Dill. 2005. Biology of sea turtles under risk from tiger sharks at a foraging ground. *Marine Ecology Progress Series* 288: 285–294.

Heithaus, M. R., A. Frid, A. J. Wirsing et al. 2007a. State-dependent risk-taking by green turtles mediates top-down effects of tiger sharks. *Journal of Animal Ecology* 76: 837–844.

Heithaus, M. R., A. Frid, A. J. Wirsing, and B. Worm. 2008a. Predicting ecological consequences of marine top predator declines. *Trends in Ecology and Evolution* 23: 202–210.

Heithaus, M. R., I. M. Hamilton, A. J. Wirsing, and L. M. Dill. 2006. Validation of a randomization procedure to assess animal habitat preferences: Microhabitat use of tiger sharks in a seagrass ecosystem. *Journal of Animal Ecology* 75: 666–676.

Heithaus, M. R., J. M. McLash, A. Frid, L. M. Dill, and G. J. Marshall. 2002c. Novel insights into the behavior of sea turtles from animal-borne cameras. *Journal of the Marine Biological Association of the United Kingdom* 82: 1049–1050.

Heithaus, M. R., A. J. Wirsing, D. Burkholder, J. A. Thomson, and L. M. Dill. 2009. Towards a predictive framework for predator risk effects: The interaction of landscape features and prey escape tactics. *Journal of Animal Ecology* 78: 556–562.

Heithaus, M. R., A. J. Wirsing, and L. M. Dill. 2012. The ecological importance of intact top predator populations: A synthesis of fifteen years of research in a seagrass ecosystem. *Marine and Freshwater Research* 63(11): 1039–1050.

Heithaus, M. R., A. J. Wirsing, J. A. Thomson, and D. Burkholder. 2008b. A review of lethal and non-lethal effects of predators on adult marine turtles. *Journal of Experimental Marine Biology and Ecology* 356: 43–51.

Heppell, S. S. 1998. Application of life-history theory and population model analysis to turtle conservation. *Copeia* 1998: 367–375.

Heppell, S. S., C. J. Limpus, D. T. Crouse, N. B. Frazer, and L. B. Crowder. 1996. Population model analysis for the loggerhead sea turtle, *Caretta caretta*, in Queensland. *Wildlife Research* 23: 143–159.

Heppell, S. S., M. L. Snover, and L. B. Crowder. 2003. Sea turtle population ecology. In *The Biology of Sea Turtles*. Vol. II, eds. P. L. Lutz, J. A. Musick, and J. Wyneken, pp. 275–306. Boca Raton, FL: CRC Press.

Hill, M. S. 1998. Spongivory on Caribbean reefs releases corals from competition with sponges. *Oecologia* 117: 143–150.

Hirth, H. F., J. Kasu, and T. Mala. 1993. Observations on a leatherback turtle *Dermochelys coriacea* nesting population near Piguwa, Papua New Guinea. *Biological Conservation* 65: 77–82.

Hochscheid, S., F. Bentivegna, M. N. Bradai, and G. C. Hays. 2007. Overwintering behavior in sea turtles: Dormancy is optimal. *Marine Ecology Progress Series* 340: 287–298.

Houghton, J. D. R., T. K. Doyle, J. Davenport, R. P. Wilson, and G. C. Hays. 2008. The role of infrequent and extraordinary deep dives in leatherback turtles (*Dermochelys coriacea*). *Journal of Experimental Marine Biology and Ecology* 211: 2566–2575.

Houghton, J. D. R., T. K. Doyle, R. P. Wilson, J. Davenport, and G. C. Hays. 2006. Jellyfish aggregations and leatherback turtle foraging patterns in a temperate coastal environment. *Ecology* 87: 1967–1972.

Houghton, J. D. R., A. Woolmer, and G. C. Hays. 2000. Sea turtle diving behaviour around the Greek Island of Kefalonia. *Journal of Experimental Marine Biology and Ecology* 80: 761–762.

Hugie, D. M. and L. M. Dill. 1994. Fish and game: A game theoretic approach to habitat selection by predators and prey. *Journal of Fish Biology* 45(Suppl. A): 151–169.

Humphries, N. E., N. Queiroz, J. R. M. Dyer et al. 2010. Environmental context explains Lévy and Brownian movement patterns of marine predators. *Nature* 465: 1066–1069.

Jackson, J. B. C. 1997. Reefs since Columbus. *Coral Reefs* 16: S23–S32.

James, M. C., R. A. Myers, and C. A. Ottensmeyer. 2005. Behaviour of leatherback sea turtles, *Dermochelys coriacea*, during the migratory cycle. *Proceedings of the Royal Society B* 272: 1547–1555.

James, M. C., C. A. Ottensmeyer, S. A. Eckert, and R. A. Myers. 2006. Changes in diel diving patterns accompany shifts between northern foraging and southern migration in leatherback turtles. *Canadian Journal of Zoology* 84: 754–765.

Jessop, T., J. Sumner, V. Lance, and C. Limpus. 2004. Reproduction in shark-attacked sea turtles is supported by stress-reduction mechanisms. *Proceedings of the Royal Society B* 271: S91–S94.

Jonsen, I. D., R. A. Myers, and M. C. James. 2007. Identifying leatherback turtle foraging behavior from satellite telemetry using a switching state-space model. *Marine Ecology Progress Series* 337: 255–264.

Keinath, J. A. and J. A. Musick. 1993. Movements and diving behavior of a leatherback turtle, *Dermochelys coriacea*. *Copeia* 1993: 1010–1017.

Kogure, K. and M. Wada. 2005. Impacts of macrobenthic bioturbation in marine sediment on bacterial metabolic activity. *Microbes and Environments* 20: 191–199.

Kramer, D. L. 1988. The behavioral ecology of air breathing by aquatic animals. *Canadian Journal of Zoology* 66: 89–94.

Krueger, B. H., M. Y. Chaloupka, P. A. Leighton, J. A. Dunn, and J. A. Horrocks. 2011. Somatic growth rates for a hawksbill turtle population in coral reef habitat around Barbados. *Marine Ecology Progress Series* 432: 269–276.

Kubis, S., M. Chaloupka, L. Ehrhart, and M. Bresette. 2009. Growth rates of juvenile green turtles *Chelonia mydas* from three ecologically distinct foraging habitats along the east central coast of Florida, USA. *Marine Ecology Progress Series* 389: 257–269.

Kuiper-Linley, M., C. R. Johnson, and J. M. Lanyon. 2007. Effects of simulated green turtle regrazing on seagrass abundance, growth, and nutritional status in Moreton Bay, south-east Queensland, Australia. *Marine and Freshwater Research* 58: 492–503.

Lal, A., R. Arthur, N. Marba, A. Lill, W. T. Adrian, and T. Alcoverro. 2010. Implications of conserving an ecosystem modifier: Increasing green turtle (*Chelonia mydas*) densities substantially alters seagrass meadows. *Biological Conservation* 143: 2730–2738.

Lambardi, P., J. R. E. Lutjeharms, R. Mencacci, C. G. Hays, and P. Luschi. 2008. Influence of ocean currents on long-distance movement of leatherback sea turtles in the Southwest Indian Ocean. *Marine Ecology Progress Series* 353: 289–301.

Lazar, B., R. Gracan, J. Gatic, D. Zavodnik, A. Jaklin, and N. Tvrtkovic. 2011. Loggerhead sea turtles (*Caretta caretta*) as bioturbators in neritic habitats: An insight through the analysis of benthic mollusks in the diet. *Marine Ecology: An Evolutionary Perspective* 32: 65–74.

Leighton, P. A., J. A. Horrocks, and D. L. Kramer. 2009. How depth alters detection and capture of buried prey: Exploitation of sea turtle eggs by mongooses. *Behavioral Ecology* 20: 1299–1306.

Leighton, P. A., J. A. Horrocks, and D. L. Kramer. 2010. Conservation and the scarecrow effect: Can human activity benefit threatened species by displacing predators? *Biological Conservation* 143: 2156–2163.

Leighton, P. A., J. A. Horrocks, and D. L. Kramer. 2011. Predicting nest survival in sea turtles: When and where are eggs most vulnerable to predation. *Animal Conservation* 14: 186–195.

Leighton, P. A., J. A. Horrocks, B. H. Krueger, J. A. Beggs, and D. L. Kramer. 2008. Predicting species interactions from edge responses: Mongoose predation on hawksbill sea turtle nests in fragmented beach habitat. *Proceedings of the Royal Society B* 275: 2465–2472.

Lemons, G., R. Lewison, L. Komoroske, A. Gaos, C. Lai, P. Dutton, T. Eguchi, R. LeRoux, and J. A. Seminoff. 2011. Trophic ecology of green sea turtles in a highly urbanized bay: Insights from stable isotopes and mixing models. *Journal of Experimental Marine Biology and Ecology* 405: 25–32.

León, Y. M. and K. A. Bjorndal. 2002. Selective feeding in the hawksbill turtle, an important predator in coral reef ecosystems. *Marine Ecology Progress Series* 245: 249–258.

Lima, S. L. 1998. Stress and decision making under the risk of predation: Recent developments from behavioral, reproductive, and ecological perspectives. *Advances in the Study of Behavior* 27: 215–290.

Lima, S. L. and Dill, L. M. 1990. Behavioural decisions made under the risk of predation: A review and prospectus. *Canadian Journal of Zoology* 68: 619–640.

Limpus, C. J., P. J. Couper, and M. A. Read. 1994a. The loggerhead turtle, *Caretta caretta*, in Queensland: Population structure in a warm temperate feeding area. *Memoirs of the Queensland Museum* 37: 195–204.

Limpus, C. J., P. J. Couper, and M. A. Read. 1994b. The green turtle, *Chelonia mydas*, in Queensland: Population structure in a warm temperate feeding area. *Memoirs of the Queensland Museum* 35: 139–154.

Limpus, C. J. and D. J. Limpus. 2003a. Loggerhead turtles in the equatorial and southern Pacific Ocean: A species in decline. In *Loggerhead Sea Turtles*, eds. A. B. Bolten and B. E. Witherington, pp. 199–209. Washington, DC: Smithsonian Books.

Limpus, C. J. and D. J. Limpus. 2003b. Biology of the loggerhead turtle in western South Pacific foraging areas. In *Loggerhead Sea Turtles*, eds. A. B. Bolten and B. E. Witherington, pp. 93–113. Washington, DC: Smithsonian Books.

Limpus, C. J., C. J. Parmenter, V. Baker, and A. Fleay. 1983. The Crab Island sea turtle rookery in the northeastern Gulf of Carpentaria. *Australian Wildlife Research* 10: 173–184.

López-Mendilaharsu, M., S. C. Gardner, J. A. Seminoff, and R. Riosmena-Rodriguez. 2005. Identifying critical foraging habitats of the green turtle (*Chelonia mydas*) along the Pacific coast of the Baja California peninsula, Mexico. *Aquatic Conservation: Marine and Freshwater Ecosystems* 15: 259–269.

López-Mendilaharsu, M., C. F. D. Rocha, P. Miller, A. Domingo, and L. Prosdocimi. 2009. Insights on leatherback turtle movements and high use areas in the southwest Atlantic Ocean. *Journal of Experimental Marine Biology and Ecology* 378: 31–39.

Losey, G. S., G. H. Balazs, and L. A. Privittera. 1994. Cleaning symbiosis between the wrasse, *Thalassoma duperrey*, and the green turtle *Chelonia mydas*. *Copeia* 1994: 684–690.

Lowe, C. G., B. M. Wetherbee, G. L. Crow, and A. L. Tester. 1996. Ontogenetic dietary shifts and feeding behavior of the tiger shark, *Galeocerdo cuvier*, in Hawaiian waters. *Environmental Biology of Fishes* 47: 203–211.

Macdonald, D. W., L. Brown, S. Yerli, and A. Canbolat. 1994. Behavior of red foxes, *Vulpes vulpes*, caching eggs of loggerhead turtles, *Caretta caretta*. *Journal of Mammalogy* 75: 985–988.

Madden, D., J. Ballestero, C. Calvo, R. Carlson, E. Christians, and E. Madden. 2008. Sea turtle nesting as a process influencing a sandy beach ecosystem. *Biotropica* 40: 758–765.

Makowski, C., J. A. Seminoff, and M. Salmon. 2006. Home range and habitat use of juvenile Atlantic green turtles (*Chelonia mydas* L.) on shallow reef habitats in Palm Beach, Florida, USA. *Marine Biology* 148: 1167–1179.

Mansfield, K. L., V. S. Saba, J. A. Keinath, and J. A. Musick. 2009. Satellite tracking reveals a dichotomy in migration strategies among juvenile loggerhead turtles in the northwest Atlantic. *Marine Biology* 156: 2555–2570.

Marcos, A., A. Louveaux, M. H. Godfrey, and M. Girondot. 2003. *Scapteriscus didactylus* (Orthoptera, Gryllotalpidae), predator of leatherback turtle eggs in French Guiana. *Marine Ecology Progress Series* 249: 289–296.

Marquez, R. 1990. FAO species catalogue. Volume 11: Sea turtles of the world. An annotated and illustrated catalogue of sea turtle species known to date. *FAO Fisheries Synopsis* 125: 1–81.

Martin, C. S. 2003. The behaviour of free-living marine turtles: Underwater activities, migrations, and seasonal occurrences. PhD thesis, University of Wales Swansea, Swansea, U.K.

Matich, P., M. R. Heithaus, and C. R. Layman. 2011. Contrasting patterns of individual specialization and trophic coupling in two marine apex predators. *Journal of Animal Ecology* 80: 294–305.

Mazaris, A. D., B. Broder, and Y. G. Matsinos. 2006. An individual based model of a sea turtle population to analyze effects of age dependent mortality. *Ecological Modeling* 198: 174–182.

McCarthy, A. L., S. Heppell, F. Royer, C. Freitas, and T. Dellinger. 2010. Identification of likely for aging habitat of pelagic loggerhead sea turtles (*Caretta caretta*) In the North Atlantic through analysis of telemetry track sinuosity. *Progress in Oceanography* 86: 224–231.

McClellan, C. M. and A. J. Read. 2007. Complexity and variation in loggerhead turtle life history. *Biology Letters* 3: 592–594.

Meddonça, S. V. M., S. Al Saady, A. Al Kiyumi, and K. Erzini. 2010. Interactions between green turtles (*Chelonia mydas*) and foxes (*Vulpes vulpes arabica*, *V. rueppellii sabaea*, and *V. cana*) on turtle nesting grounds in the northwestern Indian Ocean: Impacts of the fox community on the behavior of nesting sea turtles at the Ras Al Hadd Turtle Reserve, Oman. *Zoological Studies* 49: 437–452.

Moran, K. L. and K. A. Bjorndal. 2005. Simulated green turtle grazing affects structure and productivity of seagrass pastures. *Marine Ecology Progress Series* 305: 235–247.

Moran, K. L. and K. A. Bjorndal. 2007. Simulated green turtle grazing affects nutrient composition of the seagrass *Thalassia testudinum*. *Marine Biology* 150: 1083–1092.

Motta, P. J. and C. D. Wilga. 2001. Advances in the study of feeding behaviors, mechanisms, and mechanics of sharks. *Environmental Biology of Fishes* 60: 131–156.

Moulis, R. A. 1997. Predation by the imported fire ant (*Solenopsis invicta*) on loggerhead sea turtle (*Caretta caretta*) nests on Wassaw National Wildlife Refuge, Georgia. *Chelonian Conservation and Biology* 2: 433–436.

Murdoch, T. J. T., A. F. Glasspool, M. Outerbridge et al. 2007. Large-scale decline of offshore seagrass meadows In Bermuda. *Marine Ecology Progress Series* 339: 123–130.

Musick, J. A. and C. J. Limpus. 1997. Habitat utilization and migration in juvenile sea turtles. In *The Biology of Sea Turtles*, eds. P. L. Lutz and J. A. Musick, pp. 137–164. Boca Raton, FL: CRC Press.

Myers, A. E. and C. G. Hays. 2006. Do leatherback turtles *Dermochelys coriacea* forage during the breeding season? A combination of data-logging devices provide new insights. *Marine Ecology Progress Series* 322: 259–267.

Nifong, J. C., M. G. Frick, and S. F. Eastman. 2011. Putative predation and scavenging of two sea turtle species by the American alligator, *Alligator mississippiensis*, in coastal southeastern United State. *Herpetological Review* 42: 511–513.

Ogden, J. C., L. Robinson, K. Whitlock, H. Daganhardt, and R. Cebula. 1983. Diel foraging patterns in juvenile green turtles (*Chelonia mydas* L.) in St. Croix, U.S. Virgin Islands. *Journal of Experimental Marine Biology and Ecology* 66: 199–205.

Ortiz, R. M., P. T. Plotkin, and D. W. Owens. 1997. Predation upon olive ridley sea turtles (*Lepidochelys olivacea*) at Playa Nancite, Costa Rica. *Chelonian Conservation and Biology* 2: 585–587.

Parker, D. M., P. H. Dutton, and G. H. Balazs. 2011. Oceanic diet and distribution of haplotypes for the green turtle, *Chelonia mydas*, in the central North Pacific. *Pacific Science* 65: 419–431.

Parris, L. B., M. M. Lamont, and R. R. Carthy. 2002. Increased incidence of red imported fire ant (Hymenoptera: Formicidae) presence in loggerhead sea turtle (Testudines: Cheloniidae) nests and observations of hatchling mortality. *Florida Entomologist* 85: 514–517.

Peckarsky, B. L., P. A. Abrams, D. I. Bolnick et al. 2008. Revisiting the classics: Considering nonconsumptive effects in textbook examples of predator-prey interactions. *Ecology* 89: 2416–2425.

Peckham, S. H., D. Maldonado-Diaz, Y. Tremblay et al. 2011. Demographic implications of alternative foraging strategies in juvenile loggerhead turtles *Caretta caretta* of the North Pacific Ocean. *Marine Ecology Progress Series* 425: 269–280.

Pereira, C. M., D. T. Booth, and C. J. Limpus. 2011. Locomotor activity during the frenzy swim: Analyzing early swimming behaviour of hatchling sea turtles. *Journal of Experimental Biology* 214: 3972–3976.

Pilcher, N. J., S. Enderby, T. Stringell, and L. Bateman. 2000. Nearshore turtle hatchling distribution and predation. In *Sea turtles of the Indo Pacific: Research, Management, and Conservation*, eds. N. J. Pilcher and G. Ismail, pp. 151–166. Malaysia: Asean Academic Press.

Pitman, R. L. 1993. Seabird associations with marine turtles in the eastern Pacific Ocean. *Colonial Waterbirds* 16: 194–201.

Pitman, R. L. and P. H. Dutton. 2004. Killer whale predation on a leatherback turtle in the northeast Pacific. *Pacific Science* 58: 497–498.

Plotkin, P. 2003. Adult migrations and habitat use. In *The Biology of Sea Turtles*. Vol. II, eds. P. L. Lutz and J. A. Musick, pp. 225–242. Boca Raton, FL: CRC Press.

Polovina J. J., G. H. Balazs, E. A. Howell, D. M. Parker, M. P. Seki, and P. H. Dutton. 2004. Forage and migration habitat of loggerhead (*Caretta caretta*) and olive ridley (*Lepidochelys olivacea*) sea turtles in the central North Pacific Ocean. *Fisheries Oceanography* 13: 36–51.

Preen, A. R. 1996. Infaunal mining: A novel foraging method of loggerhead turtles. *Journal of Herpetology* 30: 94–96.

Preisser, E. L., D. I. Bolnick, and M. F. Benard. 2005. Scared to death? The effects of intimidation and consumption in predator-prey interactions. *Ecology* 86: 501–509.

Pulliam, H. R. and T. Caraco. 1984. Living in groups: Is there an optimal group size? In *Behavioural Ecology: An Evolutionary Approach*, eds. J. R. Krebs and N. B. Davies, pp. 122–147. Oxford, U.K.: Blackwell Press.

Quevedo, M., R. Svanbäck, and P. Eklöv. 2009. Intrapopulation niche partitioning in a generalist predator limits food web connectivity. *Ecology* 90: 2263–2274.

Randall, J. E. 1992. A review of the biology of the tiger shark (*Galeocerdo cuvier*). *Australian Journal of Marine and Freshwater Research* 43: 21–31.

Rebelo, R., C. Barbosa, J. P. Granadeiro et al. 2012. Can leftovers from predators be reliably used to monitor marine turtle hatchling sex-ratios? The implications of prey selection by ghost crabs. *Marine Biology* 159: 613–620.

Rees, A. F., S. Al Saady, A. C. Broderick, M. S. Coyne, N. Papathanasopoulou, and B. J. Godley. 2010. Behavioural polymorphism in one of the world's largest populations of loggerhead sea turtles *Caretta caretta*. *Marine Ecology Progress Series* 418: 201–212.

Reich, K. J., K. A. Bjorndal, M. G Frick, B. E. Witherington, C. Johnson, and A. B. Bolten. 2010. Polymodal foraging in adult female loggerheads (*Caretta caretta*). *Marine Biology* 157: 113–121.

Reynolds, A. M. and C. J. Rhodes. 2009. The Levy flight paradigm: Random search patterns and mechanisms. *Ecology* 90: 877–887.

Rincon-Diaz, M. R., C. E. Diez, R. P. van Dam, and A. M. Sabat. 2011. Effect of food availability on the abundance of juvenile hawksbill sea turtles (*Eretmochelys imbricata*) in inshore aggregation areas of the Culebra Archipelago, Puerto Rico. *Chelonian Conservation and Biology* 10: 213–221.

Rogers, R. W. 1989. The influence of sea turtles on the terrestrial vegetation of Heron Island, Great Barrier Reef. *Proceedings of the Royal Society of Queensland* 100: 67–70.

Saba, V. S., G. L. Shillinger, A. M. Swithenbank et al. 2008a. An oceanographic context for the foraging ecology of eastern Pacific leatherback turtles: Consequences of ENSO. *Deep Sea Research Part I* 55: 646–660.

Saba, V. S., J. R. Spotila, F. P. Chavez, and J. A. Musick. 2008b. Bottom-up and climatic forcing on the worldwide population of leatherback turtles. *Ecology* 89: 1414–1427.

Salmon, M., M. Hamann, and J. Wyneken. 2010. The development of early diving behavior by juvenile flatback sea turtles (*Natator depressus*). *Chelonian Conservation and Biology*. 9: 8–17.

Salmon, M., M. Hamann, J. Wyneken, and C. Schauble. 2009. Early swimming activity of hatchling flatback turtles *Natator depressus*: A test of the 'predation risk' hypothesis. *Endangered Species Research* 9: 41–47.

Salmon, M., T. T. Jones, and K. W. Horch. 2004. Ontogeny of diving and feeding behavior in juvenile sea turtles: Leatherback sea turtles (*Dermochelys coriacea* L) and green sea turtles (*Chelonia mydas* L) in the Florida Current. *Journal of Herpetology* 38: 36–43.

Sazima, C., A. Grossman, and I. Sazima. 2010. Turtle cleaners: Reef fishes foraging on epibionts of sea turtles in the tropical southwestern Atlantic, with a summary of this association type. *Neotropical Ichthyology* 8: 187–192.

Sazima, C., J. Krajewski, R. M. Bonaldo, and I. Sazima. 2007. Nuclear-follower associations of reef fishes and other animals at an oceanic archipelago. *Environmental Biology of Fishes* 80: 351–361.

Sazmina, C., A. Grossman, C. Bellini, and I. Sazima. 2004. The moving gardens: Reef fishes grazing, cleaning and following green turtles in the SW Atlantic. *Cybium* 28: 47–53.

Schmid, J. R., A. B. Bolten, K. A. Bjorndal, W. J. Lindberg, H. F. Percival, and P. D. Zwick. 2003. Home range and habitat use by Kemp's ridley turtles in west-central Florida. *Journal of Wildlife Management* 67: 196–206.

Schmitz, O. J., A. P. Beckerman, K. M. O'Brien. 1997. Behaviorally mediated trophic cascades: Effects of predation risk on food web interactions. *Ecology* 78: 1388–139.

Schmitz, O. J., V. Krivan, and O. Ovadia. 2004. Trophic cascades: The primacy of trait-mediated indirect interactions. *Ecology Letters* 7: 153–163.

Schofield, G., V. J. Hobson, S. Fossette, M. K. S. Lilley, K. A. Katselidis, and G. C. Hays. 2010. Fidelity to foraging sites, consistency of migration routes and habitat modulation of home range by sea turtles. *Diversity and Distributions* 16: 840–953.

Schofield, G., K. A. Katselidis, P. Dimopoulos, J. D. Pantis, and G. C. Hays. 2006. Behaviour analysis of the loggerhead sea turtle *Caretta caretta* from direct in-water observation. *Endangered Species Research* 2: 71–79.

Seminoff, J. A., T. T. Jones, and G. J. Marshall. 2006. Underwater behavior of green turtles monitored with video-time-depth recorders: What's missing from dive profiles? *Marine Ecology Progress Series* 322: 269–280.

Seminoff, J. A., T. T. Jones, A. Resendiz, W. J. Nichols, and M. Y. Chaloupka. 2003. Monitoring green turtles (*Chelonia mydas*) at a coastal foraging area in Baja California, Mexico: Multiple indices describe population status. *Journal of the Marine Biological Association of the United Kingdom* 83: 1355–1362.

Seminoff, J. A., A. Resendiz, and W. J. Nichols. 2002. Home range of green turtle *Chelonia mydas* at a coastal foraging area in the Gulf of California, Mexico. *Marine Ecology Progress Series* 242: 253–265.

Senko, J., V. Koch, W. M. Megill et al. 2010. Fine scale daily movements and habitat use of East Pacific green turtles at a shallow coastal lagoon in Baja California Sur, Mexico. *Journal of Experimental Marine Biology and Ecology* 391: 92–100.

Seney, E. E. and J. A. Musick. 2007. Historical diet analysis of loggerhead sea turtles (*Caretta caretta*) in Virginia. *Copeia* 2007: 459–461.

Shaver, D. J. 1998. Sea turtle strandings along the Texas coast, 1980–94. NOAA Technical Report NMFS 143: 57–72.

Shillinger, G. L., A. M. Swithenbank, H. Bailey et al. 2011. Vertical and horizontal habitat preferences of post-nesting leatherback turtles in the South Pacific Ocean. *Marine Ecology Progress Series* 422: 275–289.

Simpfendorfer, C. A. 1992. Biology of tiger sharks (*Galeocerdo cuvier*) caught by the Queensland shark meshing program off Townsville, Australia. *Australian Journal of Marine and Freshwater Research* 43: 33–43.

Simpfendorfer, C. A., A. B. Goodreid, and R. B. McAuley. 2001. Size, sex and geographic variation in the diet of the tiger shark, *Galeocerdo cuvier*, from Western Australian waters. *Environmental Biology of Fishes* 61: 37–46.

Sims, D. W., E. J. Southall, N. E. Humphries et al. 2008. Scaling laws of marine predator search behavior. *Nature* 451: 1098–1102.

Smith, M. M. and M. Salmon. 2009. A Comparison between the habitat choices made by hatchling and juvenile green turtles (*Chelonia mydas*) and loggerheads (*Caretta caretta*). *Marine Turtle Newsletter* 126: 9–13.

Sinclair, A. R. E. and P. Arcese. 1995. Population consequences of predation-sensitive foraging: The Serengeti wildebeest. *Ecology* 76: 882–891.

Spotila, J. R., R. D. Reina, A. C. Steyermark, P. T. Plotkin, and F. V. Paladino. 2000. Pacific leatherback turtles face extinction. *Nature* 405: 529–530.

Stewart, K. R. and J. Wyneken. 2004. Predation risk to loggerhead hatchlings at a high-density nesting beach in Southeast Florida. *Bulletin of Marine Science* 74: 325–335.

Sutherland, R. W. and E. G. Sutherland. 2003. Status of the flatback sea turtle (*Natator depressus*) rookery on Crab Island, Australia, with notes on predation by crocodiles. *Chelonian Conservation and Biology* 4: 612–619.

Taquet, C., M. Taquet, T. Depster et al. 2006. Foraging of the green sea turtle *Chelonia mydas* on seagrass beds at Mayotte Island (Indian Ocean), determined by acoustic transmitters. *Marine Ecology Progress Series* 306: 295–302.

Taylor, R. J. 1984. *Predation*, 166pp. London, U.K.: Chapman & Hall.

Thayer, G. W., D. W. Engel, and K. A. Bjorndal. 1982. Evidence for short-circuiting of the detrital cycle of seagrass beds by the green turtle, *Chelonia mydas* L. *Journal of Experimental Marine Biology and Ecology* 62: 173–183.

Thomson, J. A., M. R. Heithaus, D. A. Burkholder, J. J. Vaudo, A. J. Wirsing, and L. M. Dill. 2012. Site specialists, diet generalists? Isotopic variation, site fidelity and foraging by loggerhead turtles in Shark Bay, Western Australia. *Marine Ecology Progress Series* 453: 213–226.

Thomson, J. A., M. R. Heithaus, and L. M. Dill. 2011. Informing the interpretation of dive profiles using animal-borne video: A marine turtle case study. *Journal of Experimental Marine Biology and Ecology* 410: 12–20.

Thrush, S. F., R. D. Pridmore, J. E. Hewitt, and V. J. Cummings. 1991. Impact of ray feeding disturbances on sandflat macrobenthos: Do communities dominated by polychaetes or shellfish respond differently? *Marine Ecology Progress Series* 69: 245–252.

Tinker, M. T., G. Benthall, and J. A. Estes. 2008. Food limitation leads to behavioral diversification and dietary specialization in sea otters. *Proceedings of the National Academy of Sciences* 105: 560–565.

Tinker, M. T., P. R. Guimarães, M. Novak et al. 2012. Structure and mechanism of diet specialization: Testing models of individual variation in resource use with sea otters. *Ecology Letters* 15: 475–483. doi: 10.1111/j.1461-0248.2012.01760.x.

Tomás, J., F. J. Aznar, and J. A. Raga. 2001. Feeding ecology of the loggerhead turtle *Caretta caretta* in the western Mediterranean. *Journal of Zoology* 255: 525–532.

Tomillo, P. S., F. V. Paladino, J. S. Suss, and J. R. Spotila. 2010. Predation of leatherback turtle hatchlings during the crawl to the water. *Chelonian Conservation and Biology* 9: 18–25.

Townsend, K. A., J. Altvater, M. C. Thomas, Q. A. Schuyler, and G. W. Nette. 2012. Death in the octopus' garden: Fatal blue-lined octopus envenomations of adult green sea turtles. *Marine Biology* 159: 689–695.

Tregenza, T. 1995. Building on the ideal free distribution. *Advances in Ecological Research* 26: 253–307.

Troëng, S. 2000. Predation of green (*Chelonia mydas*) and leatherback (*Dermochelys coriacea*) turtles by jaguars (*Panthera onca*) at Tortuguero National Park, Costa Rica. *Chelonian Conservation and Biology* 3: 751–753.

Troëng, S. and M. Chaloupka. 2008. Variation in adult annual survival probability and remigration intervals of sea turtles. *Marine Biology* 151: 1721–1730.

Van Buskirk, J. and L. B. Crowder. 1994. Life-history variation in marine turtles. *Copeia* 1994: 66–81.

VanBlaricom, G. R. 1982. Experimental analyses of structural regulation in a marine sand community exposed to oceanic swell. *Ecological Monographs* 52: 283–305.

vanDam, R. P. and C. E. Diez. 1996. Diving behavior of immature hawksbills (*Eretmochelys imbricata*) in a Caribbean cliff wall habitat. *Marine Biology* 127: 171–178.

Vander Zanden, H. B., K. A. Bjorndal, P. W. Inglett, and A. B. Bolten. 2012. Marine-derived nutrients from green turtle nests subsidize terrestrial beach ecosystems. *Biotropica* 44: 294–301. doi: 10.1111/j.1744-7429.2011.00827.x.

Vander Zanden, H. B., K. A. Bjorndal, K. J. Reich, and A. B. Bolten. 2010. Individual specialists in a generalist population: Results from a long-term stable isotope series. *Biology Letters* 6: 711–714.

Veríssimo, D., D. A. Jones, R. Chaverri, and S. R. Meyer. 2012. Jaguar *Panthera onca* predation of marine turtles: Conflict between flagship species in Tortuguero, Costa Rica. *Oryx*, http:/dx.doi.org/10.1017/S0030605311001487.

von Brandis, R. G., J. A. Mortimer, and B. K. Reilly. 2010. In-water observations of the diving behavior of immature hawksbill turtles, *Eretmochelys imbricata*, on a coral reef at D'Arros Island, Republic of Seychelles. *Chelonian Conservation and Biology* 9: 26–32.

Wabnitz, C. C. C., G. Balazs. S. Neavers et al. 2010. Ecosystem structure and processes at Kaloko Honokōhai, focusing on the role of herbivores, including the green sea turtle *Chelonia mydas*, in reef resilience. *Marine Ecology Progress Series* 420: 27–44.

Walker, T. A. and C. J. Parmenter. 1990. Absence of a pelagic phase in the life cycle of the flatback turtle, *Natator depressa* (Garman). *Journal of Biogeography* 17: 275–278.

Walters, C. and F. Juanes. 1993. Recruitment limitation as a consequence of natural selection for use of restricted feeding habitats and predation risk taking by juvenile fishes. *Canadian Journal of Fisheries and Aquatic Sciences* 50: 2058–2070.

Walters, C. and S. J. D. Martel. 2004. *Fisheries Management and Ecology*, 448pp. Princeton, NJ: Princeton University Press.

Ward-Paige, C. A., C. Mora, H. K. Lotze et al. 2010. Large-scale absence of sharks on reefs in the greater-Caribbean: A footprint of human pressures. *PLoS One* 5: e11968. doi: 10.1371/journal.pone.0011968.

Warner, R. R. 1998. The role of extreme iteroparity and risk avoidance in the evolution of mating systems. *Journal of Fish Biology* 53: 82–93.

Watanabe, K. K., H. Hatase, M. Kinoshita et al. 2011. Population structure of the loggerhead turtle *Caretta caretta*, a large marine carnivore that exhibits alternative foraging behaviors. *Marine Ecology Progress Series* 424: 273–287.

Werner, E. E. and B. R. Anholt. 1993. Ecological consequences of the trade-off between growth and mortality rates mediated by foraging activity. *American Naturalist* 142: 242–272.

Werner, E. E. and D. J. Hall. 1988. Ontogenetic habitat shifts in bluegill: The foraging rate-predation risk trade-off. *Ecology* 69: 1352–1366.

Werner E. and S. Peacor. 2003. A review of trait-mediated indirect interactions in ecological communities. *Ecology* 84: 1083–1100.

Whelan, C. L. and J. Wyneken. 2007. Estimating predation levels and site-specific survival of hatchling loggerhead sea turtles (*Caretta caretta*) from South Florida beaches. *Copeia* 2007: 745–754.

Whiting, S. D., J. L. Long, K. M. Hadden, A. D. K. Lauder, and A. U. Koch. 2007. Insights into size, seasonality and biology of a nesting population of the olive ridley turtle in northern Australia. *Wildlife Research* 34: 200–210.

Whiting, S. D. and A. U. Whiting. 2011. Predation by the saltwater crocodile (*Crocodylus porosus*) on sea turtle adults, eggs, and hatchlings. *Chelonian Conservation and Biology* 10: 198–205.

Wirsing, A. J., R. Abernethy, and M. R. Heithaus. 2008. Speed and maneuverability of adult loggerhead turtles (*Caretta caretta*) under simulated predatory attack: Do the sexes differ? *Journal of Herpetology* 42: 411–413.

Wirsing, A. J., K. E. Cameron, and M. R. Heithaus. 2010. Spatial responses to predators vary with prey escape mode. *Animal Behaviour* 79: 531–537.

Wirsing, A. J., M. R. Heithaus, and L. M. Dill. 2007. Can measures of prey availability improve our ability to predict the abundance of large marine predators? *Oecologia* 153: 563–568.

Wirsing, A. J., M. R. Heithaus, and L. M. Dill. 2011. Predator induced modifications to diving behavior vary with foraging mode. *Oikos* 120: 1005–1012.

Witherington, B. E. 2002. Ecology of neonate loggerhead turtles inhabiting lines of downwelling near a Gulf Stream front. *Marine Biology* 140: 843–853.

Witherington, B. E. and M. Salmon. 1992. Predation on loggerhead turtle hatchlings after entering the sea. *Journal of Herpetology* 26: 226–228.

Witt, M. J., E. A. Bonguno, A. C. Broderick et al. 2011. Tracking leatherback turtles form the world's largest rookery: Assessing threats across the South Atlantic. *Proceedings of the Royal Society B: Biological Sciences* 278: 2338–2347.

Witt, M. J., A. C. Broderick, J. D. Johns et al. 2007. Prey landscapes help identify potential foraging habitats for leatherback turtles in the NE Atlantic. *Marine Ecology Progress Series* 337: 231–243.

Witzell, W. N. 1987. Selective predation on large cheloniid sea turtles by tiger sharks (*Galeocerdo cuvier*). *Japanese Journal of Herpetology* 12: 22–29.

Witzell, W. N. 2007. Kemp's Ridley (*Lepidochelys kempi*) shell damage. *Marine Turtle Newsletter* 115: 16–17.

Wyneken, J. and M. Salmon. 1992. Frenzy and postfrenzy swimming activity in loggerhead, leatherback, and green sea turtles. *Copeia* 1992: 478–484.

Wyneken, J. and M. Salmon. 1997. Assessment of reduced density open beach hatcheries and "spread-the-risk strategies" in managing sea turtles on Hillsboro Beach, Florida. Technical Report 97–04, Broward County Board of Commissioners, Ft. Lauderdale, FL. 37p.

Wyneken, J., M. Salmon, L. Fisher, and S. Weege. 2000. Managing relocated sea turtle nests in open-beach hatcheries. Lessons in hatchery design and implementation in Hillsboro Beach, Broward, County, Florida, USA. pp. 193–194. In H. Kalb and T. Wibbels, compilers. *Proceedings of the 19th Annual Sea Turtle Symposium*. U.S. Department of Commerce. NOAA Technical Memorandum NMFS-SEFSC-443. NMFS/SEFSC Miami Laboratory, Miami, FL.

Yahel, R., G. Yahel, and A. Genin. 2002. Daily cycles of suspended sand at coral reefs: A biological control. *Limnology and Oceanography* 47: 1071–1083.

Yahel, G., R. Yahel, T. Katz, B. Lazar, B. Herut, and V. Tunnicliffe. 2008. Fish activity: A major mechanism for sediment resuspension and organic matter remineralization in coastal marine sediments. *Marine Ecology Progress Series* 372: 195–209.

Ydenberg, R. C. and L. M. Dill. 1986. The economics of fleeing from predators. *Advances in the Study of Behaviour* 16: 229–249.

Zbinden, J. A., S. Bearhop, R. Bradsaw et al. 2011. Migratory dichotomy and associated phenotypic variation in marine turtles revealed by satellite tracking and stable isotope analysis. *Marine Ecology Progress Series* 421: 291–302.

# 11 Exposure to and Effects of Persistent Organic Pollutants

*Jennifer M. Keller*

## CONTENTS

## 11.1 BACKGROUND ON PERSISTENT ORGANIC POLLUTANTS

Man-made chemicals have enhanced our quality of life for centuries through improvements in healthcare, industrial efficiency, food production, and consequently increased economic profits. Other benefits include improved fire safety and simplification of daily activities (e.g., stain and stick repellant chemicals make cleaning household spills easier). However, some chemicals have consequences, especially those that persist for long periods as contaminants in the environment, preferentially accumulate in animal tissues (termed bioaccumulative), and have known toxicities. Chemical contaminants with these characteristics that are also organic in structure (consist of a carbon backbone) have been termed persistent organic pollutants (POPs).

Two historical events in the last century heightened awareness of the negative effects of POPs, which subsequently lead to environmental legislation. In 1962, Rachel Carson described the harmful effects of environment pollutants in wildlife in her book, "Silent Spring" (Carson, 1962). Her vivid imagery of a world devoid of bird song because of highly toxic organochlorine insecticides, like dichlorodiphenyltrichloroethane (DDT), prompted the United States and some other countries to ban certain POPs in the 1970s and 1980s. The second event took place in 2001, when an international treaty known as the United Nations Stockholm Convention on Persistent Organic Pollutants was signed. The Stockholm Convention originally named twelve chemicals (or chemical classes) as POPs, which are considered to be too persistent, too bioaccumulative,

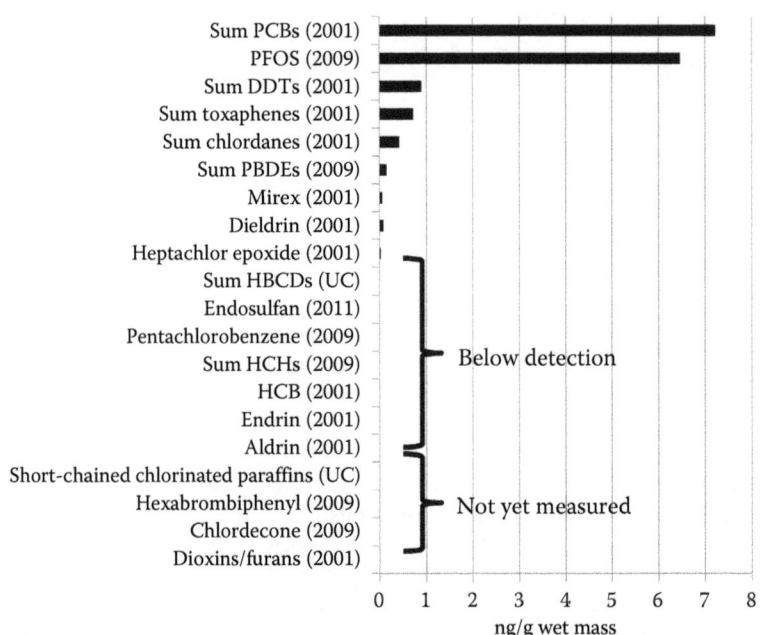

**FIGURE 11.1** Relative concentrations of persistent organic pollutants (POPs) measured in loggerhead sea turtle blood components along the east coast of the United States. Arithmetic means were converted from Keller et al. (2004a, 2005a) and taken from Ragland et al. (2011) and O'Connell et al. (2010). The year the chemical class was listed on the Stockholm Convention is shown in parentheses along the y-axis; UC = under consideration for the Stockholm Convention. Of the chemicals indicated as "not yet measured," none have been measured in sea turtles except for dioxins and furans in green sea turtle blood from Australia (Hermanussen et al., 2006). The high ranking of toxaphenes may be misleading, because toxaphene data are available only from adult males that likely have higher concentrations relative to juveniles that were used for most other POP classes shown here. Chemical abbreviations are PCBs = polychlorinated biphenyls; PFOS = perfluorooctane sulfonate; DDTs = 4,4'-dichlorodiphenyltrichloroethane-related compounds; PBDEs = polybrominated diphenyl ethers; HBCDs = hexabromocyclododecanes; HCHs = hexachlorohexanes; and HCB = hexachlorobenzene.

and too toxic for continued widespread use; many of these chemicals were named by Carson four decades earlier. Figure 11.1 shows the so-called Dirty Dozen compounds listed in 2001 as well as recent additions to the Stockholm Convention's list of chemicals to eliminate or restrict from production and use or to reduce unintentional releases. Abbreviations for these compound classes are also provided in Figure 11.1.

More than 22,000 chemicals are listed on selected United States and Canadian chemical registries, which do not account for all chemicals in use or production. A recent assessment of their chemical structure and properties demonstrated that about 610 of them might be persistent and bioaccumulative, two of the three characteristics of POPs (Howard and Muir, 2010). The majority of these have not yet been discussed by the Stockholm Convention. Furthermore, only 47 of these chemicals are routinely monitored in the environment (Howard and Muir, 2010) and only 16 have been reported in the sea turtle literature (Figure 11.1).

## 11.2 INTRODUCTION TO EXPOSURE AND EFFECTS

Understanding the exposure and effects of POPs on sea turtles is important because societies care about conserving their populations. Moreover, POPs have contributed to population declines of several wildlife species (Fox, 2001; Guillette et al., 1994). For example, an American alligator

(*Alligator mississippiensis*) population inhabiting Lake Apopka in Florida declined by 85% within 3 years of a spill of a pesticide called Dicofol containing DDT (Guillette et al., 1994), and egg shell thinning caused by widespread DDT use contributed to sharp population declines of birds in the 1950s and 1960s (Fox, 2001). Adding to evidence of cause and effect, the populations began to recover after DDT was banned. While no studies have directly investigated the effects of environmental pollutants on sea turtles at the population level, many scientists suspect that contaminants contribute to health problems, disease prevalence, altered embryonic growth, mortality or reduced reproductive success of sea turtles (Aguirre et al., 1994; Herbst and Klein, 1995; Keller et al., 2004c; van de Merwe et al., 2010b). While a handful of studies have provided correlative evidence that chemical pollutants may affect the health, survival, or reproduction of sea turtles, much more research and a weight of evidence approach is needed to better understand the toxic effects and more importantly to provide resource managers information to determine mortality risk due to this threat. In fact, a committee of 35 sea turtle researchers named chemical pollution as a top global priority for future sea turtle research (Hamann et al., 2010). Furthermore, the National Oceanic and Atmospheric Administration convened a *"Sea Turtles and Contaminants Workshop"* in May 2010 to address the uncertainty surrounding this threat; the workshop report has been delayed in part because of the Deepwater Horizon Oil incident, which coincidentally occurred 2 weeks before the workshop.

While POPs are the topic of this chapter, it is important to note that POPs are only one class of environmental chemical pollutants. To put POPs into perspective, other contaminant classes include metals like mercury, lead, copper, cadmium, and zinc; organometallics, such as methylmercury and tributyltin; petroleum products of several types; oil dispersants; polycyclic aromatic hydrocarbons (PAHs); plastics; plasticizers; surfactants; current-use pesticides like organophosphates and carbamates; excess nitrogen loading from fertilizer; sewage and urban runoff; nanoparticles like fullerenes and quantum dots; pharmaceuticals including antibiotics, hormones, and antidepressants to name only a few; and even naturally produced toxic chemicals from harmful algal blooms. A few of these chemical classes have been assessed in sea turtles, particularly metals (see Pugh and Becker [2001] for a comprehensive bibliography prior to 2001 or more recent tissue-specific or species-specific reviews [Aguirre et al., 2006; D'Ilio et al., 2011; Guirlet et al., 2008; Perrault et al., 2011]). Additionally, antibiotic resistance has been detected in bacterial swabs from sea turtles, suggesting that environmental exposure to antibiotics is changing the microbial communities that sea turtles confront (Al-Bahry et al., 2009; Foti et al., 2009). Several studies have also investigated harmful algal toxins in sea turtles (Arthur et al., 2008; Harris et al., 2011; Pierce and Henry, 2008; Takahashi et al., 2008; Walsh et al., 2010). Many more studies are needed to understand the effects of not only POPs but all chemical classes in sea turtles.

## 11.3   SEA TURTLE EXPOSURE TO POPS

### 11.3.1   GLOBAL ACCOUNT OF STUDIES

Currently, the number of peer-reviewed papers attempting to measure POP concentrations in sea turtles is only 51 (Table 11.1; Figure 11.2). This is a meager number considering that there are seven species inhabiting wide ranges of habitat types globally from remote oceanic realms to highly contaminated urban harbors. In addition, most of these studies focus on one or two POP classes of chemicals and are spread over more than four decades (Table 11.1). For perspective, this number represents only 6% of the POP literature covering three categories of marine megafauna (sea turtles, seabirds, and marine mammals). Fortunately, the interest in this field has been increasing exponentially over four decades (Figure 11.2).

The first two published measurements of POP concentrations in sea turtle samples occurred in 1974. Interestingly, one came from an unlikely remote location, Ascension Island in the central South Atlantic Ocean, where Thompson (Thompson et al., 1974) sampled green turtle (*Chelonia mydas*)

**TABLE 11.1**

**List of Published Studies on Persistent Organic Pollutant Concentrations in Sea Turtle Tissues**

| Published Study | Species | Tissue | Location | Years Sampled | Compound Class Measured |
|---|---|---|---|---|---|
| Hillestad et al. (1974) | Cc | Egg | South Carolina-Georgia | NR | OCPs, metals, radionuclide |
| Thompson et al. (1974) | Cm | Egg yolk | Ascension Island | 1972 | PCBs, OCPs |
| Clark and Krynitsky (1980) | Cc, Cm | Egg contents | Merritt Island, Florida | 1976 | PCBs, OCPs |
| McKim and Johnson (1983) | Cc, Cm | Liver, muscle | East Florida, USA | NR | PCBs, 4,4'-DDE |
| Clark and Krynitsky (1985) | Cc | Egg contents | Merritt Island, Florida | 1979 | OCPs |
| Davenport et al. (1990) | Dc | Blubber | Wales, United Kingdom | 1988 | PCBs, OCPs, metals |
| Aguirre et al. (1994) | Cm, Cc | Liver, adipose, kidney, egg shells, hatchling tissues | Hawaii | NR | PCBs, OCPs, OPs, carbamates, metals |
| Lake et al. (1994) | Cc, Lk | Body fat, liver | Long Island, New York | 1980–1989 | PCBs, OCPs |
| Rybitiski et al. (1995) | Cc, Lk | Fat, liver, muscle, kidney | Virginia-North Carolina | 1991–1992 | PCBs, OCPs |
| Cobb and Wood (1997) | Cc | Egg contents, CAM | Cape Island, South Carolina | 1993 | PCBs |
| Godley et al. (1998) | Dc | Adipose | United Kingdom | 1993–1996 | PCBs, OCPs, PAHs, metals |
| Podreka et al. (1998) | Cm | Egg contents except shell membranes | Heron Island, Queensland, Australia | 1995 | 4,4'-DDE |
| Mckenzie et al. (1999) | Cc, Dc, Cm | Egg, liver, adipose, hatchling | Cyprus, Greece, and Scotland | 1993–1995 | PCBs, OCPs |
| Alam and Brim (2000) | Cc | Egg contents, all stages | Northwest Florida | 1992 | PCBs, OCPs, PAHs, metals |
| Storelli and Marcotrigiano (2000) | Cc | Liver, kidney, muscle, heart, lung | Southern Adriatic Sea and Ionian Sea | 1990–1991 | PCBs, OCPs |
| Corsolini et al. (2000) | Cc | Adipose, liver, muscle | Northeast Italy, Adriatic Sea | 1993 | PCBs |
| Vetter et al. (2001) | Cm | Fat | Australia | 1998 | PCBs, OCPs, unknown Br compounds |
| Miao et al. (2001) | Cm | Liver, adipose | Oahu, Hawaii | 1992–1993 | PCBs |
| Gardner et al. (2003) | Cm, Lo, Cc | Adipose, liver, muscle, kidney | Baja California Peninsula | NR | PCBs, OCPs |

| Reference | Species | Tissue | Location | Years | Pollutants |
|---|---|---|---|---|---|
| Keller et al. (2004b) | Cc | Plasma, whole blood, RBC | Core Sound, North Carolina | 1998–2001 | PCBs, OCPs |
| Keller et al. (2004a) | Cc, Lk, Dc, Cm | Adipose, blood | North Carolina and Massachusetts | 1999–2001 | PCBs, OCPs |
| Keller et al. (2004c) | Cc | Adipose, blood | Core Sound, North Carolina | 2000–2001 | PCBs, OCPs |
| Keller et al. (2005b) | Cc, Lk | Plasma | North Carolina-Florida | 2003 | PFCs |
| Keller et al. (2005a) | Cc | Plasma | South Carolina-Florida | 2003 | PCBs, PBDEs |
| Deem et al. (2006) | Dc | Plasma | Gabon, Africa | 2001–2002 | PCBs, OCPs, metals |
| Perugini et al. (2006) | Cc | Liver, muscle, fat | Southern and central Adriatic Sea | 2003–2004 | PCBs, OCPs |
| Hermanussen et al. (2006) | Cm | Whole blood | Moreton Bay, Australia | 2000 | Dioxins, furans |
| Alava et al. (2006) | Cc | Egg yolk | Boca Raton and Sarasota, Florida | 2002 | PCBs, OCPs |
| Storelli et al. (2007) | Cc | Liver, kidney, lung, muscle | Southern Adriatic Sea and Ionian Sea | 1999–2001 | PCBs, DDTs |
| Innis et al. (2008) | Lk | Plasma, liver, kidney, fat, brain | Cape Cod, Massachusetts | 2005 | OCPs, metals |
| Hermanussen et al. (2008) | Cm, Ei, Nd | Muscle, liver, adipose, blood, plasma | Queensland, Australia | 2004–2006 | PBDEs |
| Monagas et al. (2008) | Cc | Fat, liver | Canary Islands | 2003–2004 | 2,4'-DDTs |
| Orós et al. (2009) | Cc, Cm, Dc | Liver, fat | Canary Islands | 2002–2005 | PCBs |
| van de Merwe et al. (2009b) | Cm | Egg yolk + albumen | Heron Island, Queensland, Australia | 1998 | PCBs, OCPs, PBDEs |
| van de Merwe et al. (2009a) | Cm | Egg yolk + albumen plasma | Malaysian markets | 2006 | PCBs, OCPs, PBDEs |
| Deem et al. (2009) | Cc | Liver | Georgia-Florida | 2000–2004 | PCBs, OCPs, metals |
| Richardson et al. (2010) | Cc, Cm, Lo | Liver | Baja California Peninsula | 2001–2003 | PCBs |
| Swarthout et al. (2010) | Lk, Cm | Whole blood | Gulf of Mexico and South Carolina-Florida | 2001–2002 | PCBs, OCPs, PBDEs |
| van de Merwe et al. (2010a) | Cm | Liver, muscle, kidney, whole blood | Queensland, Australia (rehab) | 2006–2007 | PCBs, OCPs, PBDEs, metals |
| van de Merwe et al. (2010b) | Cm | Maternal blood, egg contents, hatchling blood | Peninsular Malaysia | 2004 | PCBs, OCPs, PBDEs |

(continued)

**TABLE 11.1 (continued)**
**List of Published Studies on Persistent Organic Pollutant Concentrations in Sea Turtle Tissues**

| Published Study | Species | Tissue | Location | Years Sampled | Compound Class Measured |
|---|---|---|---|---|---|
| Guirlet et al. (2010) | Dc | Maternal blood, egg contents | French Guiana | 2006 | PCBs, OCPs |
| O'Connell et al. (2010) | Cc | Plasma or serum | Maryland-Florida | 2000–2008 | PFCs |
| Lazar et al. (2011) | Cc | Yellow fat | Eastern Adriatic Sea | 2001–2002 | PCBs, OCPs |
| Ragland et al. (2011) | Cc | Plasma | Cape Canaveral, Florida | 2006–2007 | PCBs, OCPs, PBDEs, HBCDs, pentachlorobenzene |
| Stewart (2011) | Dc | Maternal blood, egg contents, fat, blubber | North Carolina, South Carolina, Florida | 1999–2003 | PCBs, OCPs, PBDEs |
| Alava et al. (2011) | Cc | Egg yolk | North Carolina and Florida | 2002 | PCBs, OCPs, PBDEs |
| Malarvannan et al. (2011) | Cm, Ei, Cc | Liver | Japan | 1998–2006 | PCBs, OCPs, PBDEs |
| Labrada-Martagon et al. (2011) | Cm | Plasma | Baja California Peninsula | 2005 and 2007 | OCPs, metals |
| Harris et al. (2011) | Dc | Serum | Central California and St. Croix, USVI | 2005–2007 | PCBs, OCPs, metals |
| Komoroske et al. (2011) | Cm | Plasma | San Diego Bay | 2007–2009 | OCPs, PBDEs, metals |
| Keller et al. (2012) | Cm, Ei, Dc, Cc, Lk | Plasma | North Carolina and eastern Florida | 2006–2007 | PFCs |

Loggerhead (Cc), green (Cm), leatherback (Dc), Kemp's ridley (Lk), olive ridley (Lo), hawksbill (Ei), flatback (Nd) sea turtles, chorioallantoic membrane (CAM), red blood cells (RBCs), not reported (NR), organochlorine pesticides (OCPs), polychlorinated biphenyls (PCBs), organophosphate pesticides (OPs), polycyclic aromatic hydrocarbons (PAHs), perfluorinated contaminants (PFCs), polybrominated diphenylethers (PBDEs), hexabromocyclododecanes (HBCDs).

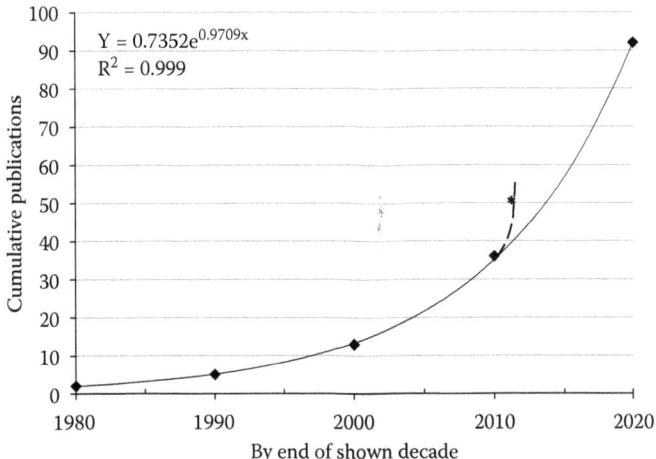

FIGURE 11.2 Cumulative number of peer-reviewed publications on persistent organic pollutant concentrations in sea turtles by decade. The exponential curve was fit through the data from the 1970s to 2010 to predict the number of publications by 2020. The asterisk is the current number of publications and the dotted line shows how the field's growth may exceed the predicted exponential trend.

eggs for PCBs and DDTs. This site has not been sampled since. The other study analyzed loggerhead sea turtle (*Caretta caretta*) eggs from South Carolina and Georgia (Hillestad et al., 1974), which has become one of the top sampled regions and species for POPs exposure assessment (Figure 11.3). This focus is not because the Southeastern United States is the most contaminated region or that the loggerhead is the species of most concern, but because the highest density of sea turtle biologists inhabits this region where loggerhead turtles are abundant.

Spatially, the 51 available studies span the globe providing POP baseline concentrations in sea turtles from many regions, but major data gaps still exist (Figure 11.3). While the list of regions sampled suggests wide coverage (e.g., from Hawaii to the Mediterranean Sea to Australia), the distribution shown in Figure 11.3 can be misleading. The compiled studies represent different tissue types sampled from different age classes during different decades, and they report on different POP classes or used different methods, which cannot be compared. For example, the loggerheads sampled from Florida Bay represent one study on juvenile turtle plasma for a newer contaminant class, the perfluorinated contaminants (PFCs) (O'Connell et al., 2010). Our knowledge is devoid of other POP classes, such as PCBs or pesticide concentrations, in any sea turtle species or age from this location (but a study is underway). Worst yet, the data from several of the pie charts shown in Figure 11.3 have limited usefulness because of low sample sizes or use of analytical methods that were not sensitive enough for the species or tissue chosen. For example, 18 green turtles have been analyzed for POPs from Hawaii (Aguirre et al., 1994; Miao et al., 2001), but only one of these studies used methods that could detect the target compounds; so, known baseline concentrations are available on only three turtles from this region. Additionally, plasma or serum samples from 53 loggerheads from South Carolina and Georgia, 18 Kemp's ridleys (*Lepidochelys kempii*) from Massachusetts, and 33 leatherbacks (*Dermochelys coriacea*) from California, St. Croix, and Gabon were wasted because of high detection limits of the chosen method (Deem et al., 2006, 2009; Harris et al., 2011; Innis et al., 2008). These examples teach a lesson; that it is important to choose analytical methods that will detect the contaminant range expected from the sampled species and tissue type to avoid wasting time, money, and valuable samples. Additionally, it is very important for analytical laboratories to use quality assurance and quality control practices, including analyzing field and procedural blanks and certified reference materials, using high quality authentic chemical standards plus internal standards, and participating in inter-laboratory comparison exercises.

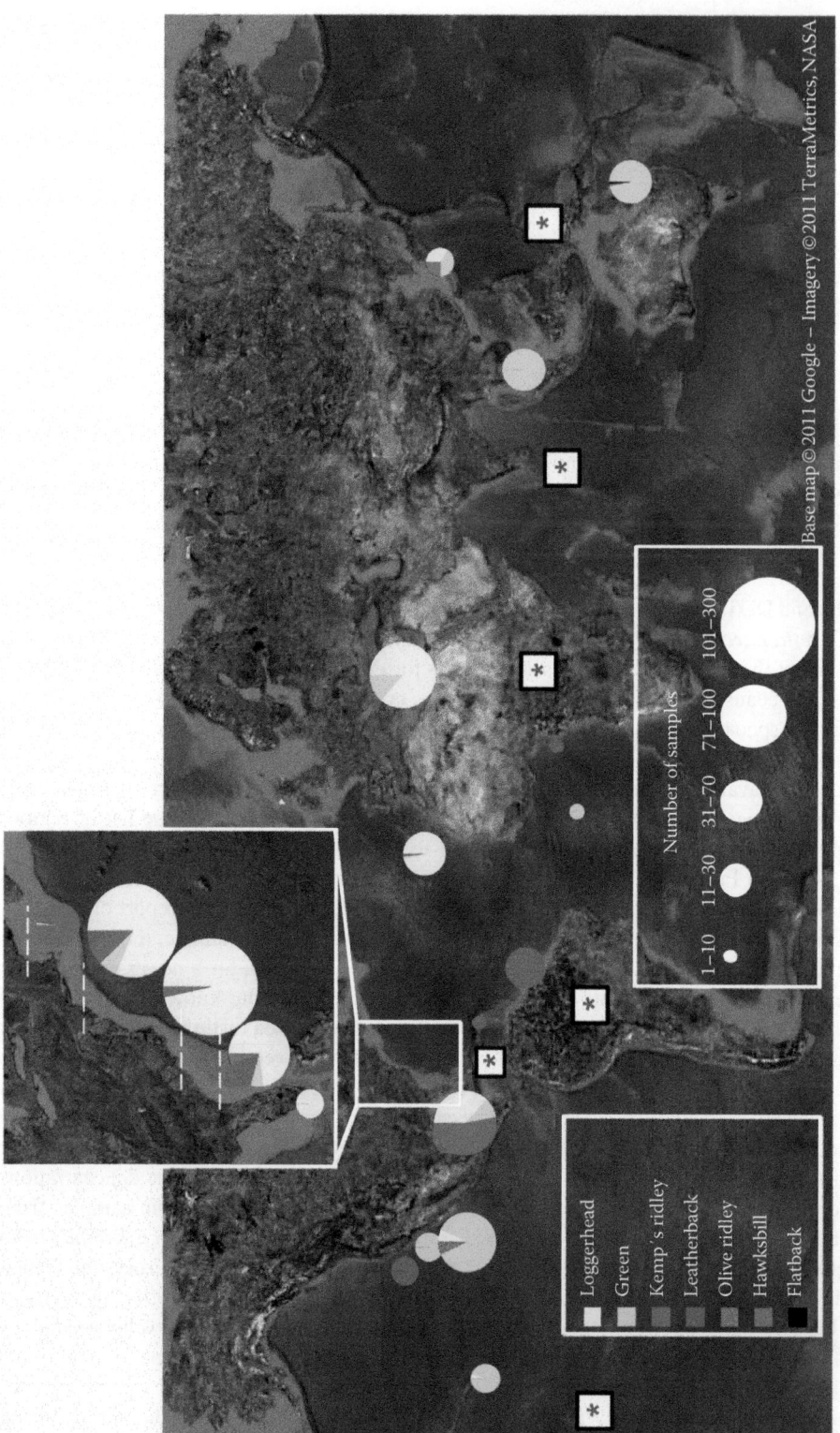

FIGURE 11.3 Global map of studies examining persistent organic pollutant (POP) concentrations in sea turtle tissues from 1974 to 2012. The number of samples included in studies assessing each species by region is shown by different sized pie-charts. The proportion of samples by species is shown inside each pie-chart. Red asterisks signify regions that have major data gaps. Dashed horizontal lines indicate latitudinal delineations of regions along the east coast of North America.

Low sample size is an issue easily deciphered from Figure 11.3. This problem is especially important for the less studied species, like the hawksbill sea turtle (*Eretmochelys imbricata*). Only three studies have reported POPs in hawksbills, all using small sample sizes: only three liver samples were analyzed from Japanese waters, a single blood sample from Australia, and five plasma samples from Florida (Hermanussen et al., 2008; Keller et al., 2012; Malarvannan et al., 2011). Worst yet is the flatback turtle (*Natator depressus*) with only a single turtle ever sampled for any POP (Hermanussen et al., 2008). Small sample sizes are likely the result of limited funding, opportunistic field sampling, and/or difficulty in accessing these endangered or threatened species, but studies with small sample sizes like these cannot provide an accurate baseline concentration for the species or location.

Despite these problems, Figure 11.3 accurately portrays that the loggerhead sea turtle, followed by the green turtle, has been the most frequently analyzed species. The east coast of the United States greatly outnumbers other regions, but the recent additions of locations, such as French Guiana, Malaysia, Baja California, and Japan (Gardner, 2003; Guirlet et al., 2010; Labrada-Martagon et al., 2011; Malarvannan et al., 2011; van de Merwe et al., 2009a, 2010b; Richardson et al., 2010) are beginning to provide a more descriptive understanding of baseline POP concentrations in sea turtles across the globe. However, very large spatial data gaps still exist, notably the South Pacific Ocean, Caribbean Sea, both coasts of South America, Africa, the Indian Ocean, and Southeast Asia (Figure 11.3).

Because of the different confounding factors, detection problems, and sample size limitations, a robust meta-analysis for assessing geographic hotspots or temporal trends is not possible even after four decades of work. More research is needed to gain a better picture of temporal and spatial trends and to simply understand baseline concentrations in vast regions globally and in species of certain age classes not yet assessed. With the predicted exponential growth of this field, surely more species and locations will be better assessed with time.

## 11.3.2 REPORTED SEA TURTLE POP CONCENTRATIONS

Since the last review of chemical pollutants in all sea turtle tissues in 2001 (Pugh and Becker, 2001), a larger diversity of tissues, species, and locations has been added to the literature. Tables 11.2 through 11.5 show measured POP concentrations from selected sea turtle studies that analyzed blood components, adipose, liver, and egg contents, respectively. The white space within each table is another indication of data gaps.

It is important to acknowledge that not all studies or all concentrations are shown in Tables 11.2 through 11.5. Selection criteria excluded studies or portions of data from studies that could not detect POPs (Aguirre et al., 1994; Deem et al., 2006; Harris et al., 2011; Innis et al., 2008), chose compounds that were not typically found in sea turtles or other wildlife (Monagas et al., 2008), reported their data in another publication with a larger sample size (Alava et al., 2006; Cobb and Wood, 1997; Keller et al., 2004a,b, 2006b), or presented data in units that could not be converted into the units shown (Davenport et al., 1990; Miao et al., 2001; Vetter et al., 2001). Also excluded are portions of data reported for other less studied tissues, like muscle, heart, lung, or kidney.

The units used for reporting POP concentrations also caused difficulty in creating comprehensive tables. POP tissue concentrations are most commonly reported as either wet mass, wet volume, or lipid normalized (e.g., ng/g wet tissue mass, ng/mL plasma, or ng/g lipid in tissue, respectively), and rarely in dry mass (e.g., ng/g dry tissue mass). Conversions are required for direct comparison of POP concentrations reported with different units, but often conversions must be done with estimates of water content, density, or lipid content, which can lead to inaccuracies in the POP concentrations. Normalizing POP concentrations to the actual lipid content of each sample is often argued to be the best choice for most POPs (not PFCs) in most tissues (not in blood) for comparisons. Unfortunately, the most commonly reported unit for sea turtle fat and liver is on a wet mass basis. Reporting both wet mass and lipid-normalized concentrations is suggested so that future comparisons can select which values to use.

TABLE 11.2

**Persistent Organic Pollutant Levels in Blood Matrices of Sea Turtles from Selected Studies**

| Species | Stage/Sex | Location | Year | Tissue | N | ΣPCBs | 4,4'-DDE | ΣChlordanes | Dieldrin | Mirex | Σ2,3,7,8-PCDD/F | ΣPBDEs | PFOS | Reference |
|---|---|---|---|---|---|---|---|---|---|---|---|---|---|---|
| Lk | JMF | South Carolina-Florida | 2001–2002 | WB | 3 | 6.610 (12.200)[a]<br>NA<br>(2.400–24.700) | 0.773 (1.800)[a]<br>NA<br>(0.188–3.530) | 1.140 (1.850)[a]<br>NA<br>(0.353–3.870) | 0.500 (0.409)[a]<br>NA<br>(0.206–1.020) | 0.00917 (NA)[a]<br>NA<br>(<0.0006–0.00952) | | 0.105 (0.141)[a]<br>NA<br>(0.038–0.307) | | Swarthout et al. (2010) |
| Lk | JMF | Cape Cod, Massachusetts | 1999 | WB | 8 | 4.33 (4.52)[b]<br>2.42<br>(0.304–11.4)[b] | 0.894 (0.731)[b]<br>0.783<br>(0.186–2.38)[b] | 0.417 (0.406)[b]<br>0.333<br>(0.057–1.29)[b] | 0.114 (0.0834)[b]<br>0.0741 | 0.0460 (0.0646)[b]<br>0.0195 | | | | Keller et al. (2004a) |
| Lk | JMF | Louisiana-Texas, Gulf of Mexico | 2001–2002 | WB | 46 | 3.190 (3.620)[a]<br>NA<br>(0.227–21.590) | 0.472 (0.633)[a]<br>NA<br>(0.0439–4.110) | 0.163 (0.249)[a]<br>NA<br>(0.0109–1.190) | 0.199 (0.119)[a]<br>NA<br>(<0.0748–0.609) | 0.0100 (0.0042)[b]<br>NA<br>(<0.0006–0.019) | | 0.146 (0.273)[a]<br>NA<br>(0.0195–1.450) | | Swarthout et al. (2010) |
| Lk | JMF | South Carolina-Florida | 2003 | P | 6 | | | | | | | | 39.4 (17.1)<br>41.9<br>(13.8–60.2) | Keller et al. (2005b) |
| Lk | JMF | Core Sound, North Carolina | 2006 | P | 10 | | | | | | | | 15.7 (9.86)<br>10.8<br>(6.85–35.0) | Keller et al. (2012) |
| Cc | AM | Transients; migrated north to SC-NJ | 2006–2007 | P | 9 | 13.1 (17.2)<br>7.27 (NA) | 1.557 (2.656)<br>0.701 (NA) | 0.335 (NA)[c]<br>0.241 (NA)[c] | | 0.0783 (0.0625)<br>0.0528 (NA) | | 0.0876 (0.0965)<br>0.043 (NA) | | Ragland et al. (2011) |
| Cc | JMF | Core Sound, North Carolina | 2000–2001 | WB | 48 | 7.23 (6.86)[b]<br>5.56<br>(0.157–31.1)[b] | 0.830 (0.887)[b]<br>0.584<br>(<0.030–4.94)[b] | 0.293 (0.260)[b]<br>0.207<br>(<0.013–1.28)[b] | 0.0874 (0.181)[b]<br>0.477<br>(<0.013–1.24)[b] | 0.0633 (0.0893)[b]<br>0.0202<br>(<0.013–0.384)[b] | | | | Keller et al. (2004c) |
| Cc | AM | Residents; stayed at Cape Canaveral, Florida | 2006–2007 | P | 11 | 5.38 (2.25)<br>5.03 (NA) | 0.0917<br>(0.0323)<br>0.106 (NA) | 0.0615 (NA)[c] | | 0.0230 (0.0161)<br>0.0144 (NA) | | NA<br>0.0221 (NA) | | Ragland et al. (2011) |

| Species | Class | Location | Year | Tissue | n | | | | | | Reference |
|---|---|---|---|---|---|---|---|---|---|---|---|
| Cc | JMF | South Carolina-Florida | 2003 | P | 29 | 3.38 (3.43)[d] 2.34 (0.210–13.4)[d] | 0.467 (0.644)[d] 0.242 (<0.0286–3.05)[d] | | 0.0277 (0.0260)[d] 0.0161 (<0.006–0.0937)[d] | 0.152 (0.182)[d] 0.0780 (<0.0177–0.698)[d] | Keller et al. (2005a) |
| Cc | AF, JMF | Georgia-Florida | 2000–2004 | P | 53 | <20 | <20 | <20 | | | Deem et al. (2009) |
| Cc | JMF, AM | North Carolina-Florida | 2003 | P | 73 | | | | | 11.0 (17.2) 5.45 (1.40–96.8) | Keller et al. (2005b) |
| Cc | JMF | Chesapeake Bay, Maryland | 2005–2006 | P or S | 14 | | | | | 9.34 (11.4) 3.54 (NA) | O'Connell et al. (2010) |
| Cc | JMF | Core Sound, North Carolina | 2006 | P | 15 | | | | | 6.47 (7.52) 3.13 (0.305–26.5) | Keller et al. (2012); O'Connell et al. (2010) |
| Cc | JMF | Charleston, South Carolina | 2005–2006 | P | 9 | | | | | 3.86 (2.92) 2.87 (NA) | O'Connell et al. (2010) |
| Cc | JMF | Florida Bay | 2005–2006 | P | 11 | | | | | 3.67 (1.03) 3.801 (NA) | O'Connell et al. (2010) |
| Cc | JMF | Cape Canaveral, Florida | 2005–2006 | P | 10 | | | | | 1.44 (0.94) 1.24 (NA) | O'Connell et al. (2010) |
| Dc | AF | Juno Beach, Florida | 2003 | WB | 6 | 2.50 (2.27) 1.62 (0.162–6.54) | 0.424 (0.235) 0.317 (0.211–0.865) | 0.140 (0.100) 0.097 (0.040–0.328) | <LOD <LOD (<0.009–0.062) | 0.198 (0.190) 0.155 (<0.050–0.510) | Stewart et al. (2011) |
| Dc | AF | French Guiana | 2006 | WB | ≤44 | 1.26 (0.71) NA (0.86–4.04) | | 0.328 (0.158) 0.281 (0.081–0.507) | | | Guirlet et al. (2010) |
| Dc | JMF | Juno Beach, Florida | 2007 | P | 7 | | | | | 3.95 (2.51) 4.42 (0.884–7.83) | Keller et al. (2012) |
| Cm | JAMF | San Diego Bay, California | 2007–2009 | P | 20 | | 0.736 (0.097)[e] 0.750 (<LOD–1.56) | 0.775 (NA)[e] 0.846 (NA)[e] | | <0.200 (<0.200–0.760)[f] | Komoroske et al. (2011) |
| Cm | JAMF | Punta Abreojos, Baja California | 2005 and 2007 | P | 39 | <LOD[g] | 0.230227 (0.001–1.31)[g] | <LOD | | | Labrada-Martagon et al. (2011) |

*(continued)*

## TABLE 11.2 (continued)
## Persistent Organic Pollutant Levels in Blood Matrices of Sea Turtles from Selected Studies

| Species | Stage/Sex | Location | Year | Tissue | N | ΣPCBs | 4,4'-DDE | ΣChlordanes | Dieldrin | Mirex | Σ2,3,7,8-PCDD/F | ΣPBDEs | PFOS | Reference |
|---|---|---|---|---|---|---|---|---|---|---|---|---|---|---|
| Cm | JAMF | Bahía Magdalena, Baja California | 2005 and 2007 | | 13 | | <LODᵉ | 0.313725 (0.001–0.84)ᵉ | <LOD | | | | | Labrada-Martagon et al. (2011) |
| Cm | HMF | Peninsular Malaysia | 2004 | WB | 11 pools | 0.8508 (0.1052)ᵉ NA (0.5594–1.4566) | | | | 0.1324 (0.0340)ᵉ NA (<LOD–0.3406) | | 0.0830 (0.0144)ᵉ NA (0.0238–0.1732) | | van de Merwe et al. (2010b) |
| Cm | JMF | Queensland, Australia (rehab) | 2006–2007 | WB | 16 | 0.6839 (0.1528)ᵉ NA (0.1983–1.6948) | <0.035 | | <0.035 | 0.0428 (0.0162)ᵉ NA (<0.035–0.1157) | | 0.0793 (0.0108)ᵉ NA (0.0424–0.1208) | | van de Merwe et al. (2010a) |
| Cm | AF | Peninsular Malaysia | 2004 | WB | 11 | 0.5789 (0.0856)ᵉ NA (0.3164–1.2065) | | | | 0.161 (0.0430)ᵉ NA (<LOD–0.4763) | | 0.1208 (0.0141)ᵉ NA (0.0575–0.2243) | | van de Merwe et al. (2010b) |
| Cm | JMF | Texas, Gulf of Mexico | 2001–2002 | WB | 9 | 0.331 (0.701)ᵃ NA (0.117–2.330) | 0.0664 (0.110)ᵃ NA (<0.0101–0.348) | 0.0164 (0.0257)ᵃ NA (<0.0022–0.0643) | 0.0960 (NA)ᵃ NA (<0.0748–0.096) | 0.0111 (0.0114)ᵃ NA (<0.0006–0.0377) | | 0.0806 (0.217)ᵃ NA (0.0244–0.623) | | Swarthout et al. (2010) |
| Cm | JAMF | Queensland, Australia | 2004–2006 | P | 1 pool of 7 | | | | | | | 0.00900 | | Hermanussen et al. (2008) |
| Cm | JAMF | Queensland, Australia | 2004–2006 | WB | 1 pool of 7 | | | | | | | 0.00444 | | Hermanussen et al. (2008) |
| Cm | MF | Polluted site, Moreton Bay, Australia | NR | WB | 7 | | | | | | 0.910 (NA)ᵇ NA (0.510–1.400)ᵇ | | | Hermanussen et al. (2006) |
| Cm | MF | Variable site, Moreton Bay, Australia | NR | WB | 6 | | | | | | 0.580 (NA)ᵇ NA (0.160–1.900)ᵇ | | | Hermanussen et al. (2006) |

| | | | | | | | | |
|---|---|---|---|---|---|---|---|---|
| Cm | MF | Background site, Moreton Bay, Australia | NR | WB | 16 | 0.090 (NA)[h] NA (0.030–0.200)[h] | | Hermanussen et al. (2006) |
| Cm | JMF | Core Sound, North Carolina | 2006 | P | 10 | | 2.41 (1.12) 2.37 (0.871–3.87) | Keller et al. (2012) |
| Ei | JF | Queensland, Australia | 2004–2006 | WB | 1 | 0.01300 | | Hermanussen et al. (2008) |
| Ei | JMF | Juno Beach, Florida | 2006 | P | 5 | | 11.9 (6.27) 11.9 (5.45–21.2) | Keller et al. (2012) |
| Nd | AF | Queensland, Australia | 2004–2006 | WB | 1 | 0.00609 | | Hermanussen et al. (2008) |

Mean (SD) and/or median (range) in ng/g wet mass, unless otherwise stated. Ranked generally from highest to lowest by species and location.

Kemp's ridley (Lk), loggerhead (Cc), leatherback (Dc), green (Cm), hawksbill (Ei), flatback (Nd) sea turtles, juvenile (J), adult (A), male (M), female (F), hatchling (H), whole blood (WB), plasma (P), serum (S), polychlorinated biphenyls (PCBs), polychlorinated dibenzo-p-dioxin/furan (PCDD/F), perfluorooctane sulfonate (PFOS), not available (NA), not reported (NR), below detection (<LOD).

a   Geometric mean (standard deviation).

b   Original data corrected by multiplying values in ng/g wet mass by 1.3 because of immiscible solvent used for internal standards (Keller et al., 2009); summary statistics recalculated using Helsel (2005) recommendations.

c   Sum of means or medians of detected compounds.

d   Summary statistics recalculated using original data in ng/g wet mass and Helsel (2005) recommendations.

e   Arithmetic mean (standard error).

f   Values include only PBDE 47.

g   4,4'-DDE was not detected, but 2,4'-DDD was. Total chlordane values represent only trans-chlordane and cis-chlordane; other chlordanes were <LOD.

h   Units are ng/g lipid, no lipid content was provided for a conversion.

## TABLE 11.3
## Persistent Organic Pollutant Levels in Fat of Sea Turtles from Selected Studies

| Species | Stage/Sex | Location | Years | Tissue | N | ΣPCBs | 4,4'-DDE | ΣChlordanes | ΣHCHs | HCB | Dieldrin | Mirex | Heptachlor Epoxide | ΣPBDEs | % Lipid | Reference |
|---|---|---|---|---|---|---|---|---|---|---|---|---|---|---|---|---|
| Lk | JMF | Long Island, New York | 1985 | Body fat | 7 | 1250 (985) | 386 (250) | | | | | | | | | Lake et al. (1994) |
| Lk | J | Virginia-North Carolina | 1991 | Fat | 3 | 660 (333) 794 (281–904) | 194 (98.2) 194 (95.7–292) | | | | | | | | | Rybitski et al. (1995) |
| Lk | JMF | Cape Cod, Massachusetts | 1999 | Yellow fat | 9 | 701 (893)[a] 368 (49.0–2920)[a] | 99.6 (76.4)[a] 75.2 (19.2–238)[a] | 103 (95.3)[a] 58.5 (12.7–305)[a] | 52.5 (73.3)[a] 10.9 (<1–205)[a] | 13.6 (9.8)[a] 12.2 (<1–30.7)[a] | 21.3 (13.7)[a] 22.5 (5.32–43.8)[a] | 3.83 (3.09)[a] 2.20 (<1–11.1)[a] | 16.7 (23.0)[a] 7.52 (0.998–75.3)[a] | | 65.8 (10.6) 70.7 (57.7–74.8) | Keller et al. (2004a) |
| Lk | JMF | Cape Cod, Massachusetts | 1999 | Brown fat | 10 | 525 (545)[a] 317 (16.6–1680)[a] | 91.0 (72.0)[a] 75.7 (6.00–219)[a] | 84.1 (70.7)[a] 52.0 (5.92–206)[a] | 60.8 (62.9)[a] 47.9 (<1–170)[a] | 10.5 (9.09)[a] 4.95 (<1–25.5)[a] | 18.3 (13.3)[a] 14.3 (1.20–40.4)[a] | 3.96 (2.78)[a] 2.54 (<1–10.9)[a] | 22.1 (28.4)[a] 13.3 (<1–91.5)[a] | | 62.0 (23.9) 72.8 (0.521–80.3) | Keller et al. (2004a) |
| Lk | JMF | Long Island, New York | 1989 | Body fat | 6 | 476 (273) | 232 (157) | | | | | | | | | Lake et al. (1994) |
| Lk | J | Cape Cod, Massachusetts | 2005 | Fat | 3 | | 54 (51–209) | <10[b] | <10 | | <10 (<10–10) | | <10 | | | Innis et al. (2008) |
| Cc | JAMF | Cyprus and Greece | 1994–1995 | Adipose | 3 | 840 (60.0) 853 (775–893) | 509 (173) 446 (376–705) | 19.7 (11.6) 14 (12–33) | | | 5.4 (4.2) 6.2 (<1.8–9.2) | | | | | Mckenzie et al. (1999) |
| Cc | NR | NR (probably U.S. East coast) | 1986 | Body fat | NR (maybe 1) | 647 | 300 | | | | | | | | | Lake et al. (1994) |
| Cc | JAMF | Virginia-North Carolina | 1991–1992 | Fat | 20 | 551 (473) 365 (55.4–1730) | 195 (266) 99.0 (2.86–1210) | | | | | | | | | Rybitski et al. (1995) |

| Species | Code | Location | Years | Tissue | n | | | | | | | | | | Reference |
|---|---|---|---|---|---|---|---|---|---|---|---|---|---|---|---|
| Cc | JAMF | Eastern Adriatic Sea | 2001–2002 | Yellow fat | 27 | 474 (547) 312 (177–2934) | 92.4 (53.0) 81.0 (19.1–282) | | 1.9 (1.3) 1.8 (<LOD–5.9) | | | | | 47.9 (13.6) 51.1 (25.9–68.2) | Lazar et al. (2011) |
| Cc | JMF | Southern and central Adriatic Sea | 2003–2004 | Fat | 9 | 459.7 (NR) NA (2.9–1472.1) | 280.9 (NR) NA (1.5–621) | | | | | | | | Perugini et al. (2006) |
| Cc | JMF | Canary Islands | 2002–2005 | Fat | 30 | 450 (1700) NA (<LOD–9800) | | | | | | | | | Orós et al. (2009) |
| Cc | JMF | Northeast Italy, Adriatic Sea | 1993 | Adipose | 4 | 334 (179) 136–563 | | | | | | | | | Corsolini et al. (2000) |
| Cc | JMF | Core Sound, North Carolina | 2000–2001 | Fat | 44 | 256 (269)[a] 171 (7.99–1360)[a] | 64.9 (64.3)[a] 43.2 (<0.09–273)[a] | 2.15 (7.15)[a] 0.0655 (<1–43.5)[a] | 1.13 (2.38)[a] 0.233 (<1–12.6)[a] | 5.04 (3.90)[a] 3.54 (<0.09–16.7)[a] | 4.52 (4.06)[a] 3.34 (<0.09–18.8)[a] | 3.11 (2.23)[a] 2.25 (<1–10.6)[a] | | 26.3 (20.6) 26.1 (4.68–42.6) | Keller et al. (2004a) |
| Cc | NR | Baja California | NR | Adipose | 1 | <3 | | | <3 | <3 | | | | | Gardner et al. (2003) |
| Dc | AMF, JF | South Carolina–North Carolina | 1999–2003 | Blubber | 7 | 193 (384) 66.9 (1.52–1061) | 41.5 (64.4) 22.7 (4.80–185) | <LOD | 0.700 (0.335) 0.657 (0.323–1.11) | 8.39 (10.3) 4.67 (2.92–31.6) | <LOD (<0.352–7.60) | 1.73 (2.79) 0.170 (<0.349–7.80) | 25.7 (41.1) 9.99 (<2.94–116) | 52.7 (16.9) 58.5 (14.9–63.2) | Stewart et al. (2011) |
| Dc | AM | Scotland and Wales | 1993–1996 | Adipose | 3 | 178 (47–230) | 57 (10–68) | | 2 (<1–3) | 13 (<1–13) | | | | 50 (41–74) | Godley et al. (1998) |
| Dc | AM | Scotland | 1993–1995 | Adipose | 2 | NA (47–178) | NA (10–57) | | NA (12–22) | NA (13–19) | | | | | Mckenzie et al. (1999) |
| Dc | AMF, JF | South Carolina–North Carolina | 1999–2003 | Fat | 7 | 90.1 (65.9) 75.1 (4.87–188) | 19.7 (10.9) 20.0 (5.14–35.7) | <LOD | 0.628 (0.363) 0.743 (0.121–1.18) | 4.41 (1.92) 4.71 (2.16–7.92) | 0.379 (0.404) 0.164 (<0.046–0.944) | 0.929 (0.716) 0.430 (<0.225–2.29) | 15.4 (6.67) 13.2 (<1.67–26.0) | 48.1 (23.9) 60.6 (0.697–67.9) | Stewart et al. (2011) |
| Dc | AF | Canary Islands | 2002–2005 | Fat | 1 | 77 | | | | | | | | | Orós et al. (2009) |
| Cm | JF | Canary Islands | 2002–2005 | Fat | 1 | 144 | | | | | | | | | Orós et al. (2009) |
| Cm | JMF | Cyprus | 1995 | Adipose | 3 | 109 (39–261) | 6 (2.4–19) | | | 1.5 (<1.9–3.5) | | | | | Mckenzie et al. (1999) |

*(continued)*

## TABLE 11.3 (continued)
## Persistent Organic Pollutant Levels in Fat of Sea Turtles from Selected Studies

| Species | Stage/ Sex | Location | Years | Tissue | N | ΣPCBs | 4,4'-DDE | ΣChlordanes | ΣHCHs | HCB | Dieldrin | Mirex | Heptachlor Epoxide | ΣPBDEs | % Lipid | Reference |
|---------|-----------|----------|-------|--------|---|-------|----------|-------------|-------|-----|----------|-------|--------------------|--------|---------|-----------|
| Cm | JMF | Baja California | NR | Adipose | 7 | NA (<3–49.5) | | NA (<3–65.1) | | <3 | <3 | | | | | Gardner et al. (2003) |
| Cm | AF | Queensland, Australia | 2004–2006 | Adipose | 1 | | | | | | | | | 0.25740 | 78.00 | Hermanussen et al. (2008) |
| Lo | NR | Baja California | NR | Adipose | 1 | 18.4 | | 8.1 | | <3 | <3 | | | | | Gardner et al. (2003) |

Mean (SD) and/or median (range) in ng/g wet mass, unless otherwise stated. Ranked generally from highest to lowest by species and location.

Kemp's ridley (Lk), loggerhead (Cc), leatherback (Dc), green (Cm), olive ridley (Lo) sea turtles, juvenile (J), adult (A), male (M), female (F), polychlorinated biphenyls (PCBs), hexachlorocyclohexanes (HCHs), hexachlorobenzene (HCB), polybrominated diphenylethers (PBDEs), not available (NA), not reported (NR), below detection (<LOD).

a   Summary statistics recalculated using original data in ng/g wet mass and Helsel (2005) recommendations.

b   *Trans*- and *cis*-chlordane were the only chlordanes measured.

**TABLE 11.4**

**Persistent Organic Pollutant Levels in Liver of Sea Turtles from Selected Studies**

| Species | Stage/Sex | Location | Years | N | ΣPCBs | 4,4'-DDE | ΣChlordanes | ΣHCHs | HCB | Dieldrin | Mirex | Heptachlor Epoxide | ΣPBDEs | % Lipid | Reference |
|---|---|---|---|---|---|---|---|---|---|---|---|---|---|---|---|
| Lk | JMF | Long Island, New York | 1985 | 8 | 738 (737) | 253 (162) | | | | | | | | | Lake et al. (1994) |
| Lk | J | Virginia-North Carolina | 1991 | 3 | 360 (158–608) | 55.5 (54.2–56.8) | | | | | | | | | Rybitski et al. (1995) |
| Lk | JMF | Long Island, New York | 1989 | 6 | 272 (126) | 137 (85.3) | | | | | | | | | Lake et al. (1994) |
| Lk | J | Cape Cod, Massachusetts | 2005 | 3 | | 14 (<10–15) | <10[a] | <10 | | <10 | | <10 | | | Innis et al. (2008) |
| Cc | JMF | Canary Islands | 2002–2005 | 30 | 1980 (5320) NA (<LOD–34900) | | | | | | | | | | Orós et al. (2009) |
| Cc | NR | NR (likely east coast U.S.) | 1986 | 1 | 360[b] | 110[b] | | | | | | | | | Lake et al. (1994) |
| Cc | J | Scotland | 1995 | 1 | 159 | 149 | NR | | | NR | | | | | Mckenzie et al. (1999) |
| Cc | JAMF | Virginia-North Carolina | 1991–1992 | 18 | 145 (158) 99.7 (7.46–514) | 47.5 (104) 18.9 (<2–458) | | | | | | | | | Rybitski et al. (1995) |
| Cc | JMF | Northeast Italy, Adriatic Sea | 1993 | 4 | 119 (60) 69–205 | | | | | | | | | | Corsolini et al. (2000) |
| Cc | NR | NR (likely east coast U.S.) | 1988 | 1 | 110[b] | 50[b] | | | | | | | | | Lake et al. (1994) |
| Cc | JMF AM | Cyprus and Greece | 1994–1995 | 4 | 84.0 (24.3) 92.0 (50–102) | 60.5 (13.7) 58.5 (49–76) | 3.0 (1.8) 2.1 (1.8–5) | | | 1.8 (1.3) 2.5 (0.3–2.7) | | | | | Mckenzie et al. (1999) |
| Cc | JMF | Southern and central Adriatic Sea | 2003–2004 | 11 | 83 (NR) NA (7.6–247.3) | 40.6 (NR) NA (3.6–217.3) | | | | | | | | | Perugini et al. (2006) |
| Cc | JMF | Southern Adriatic Sea and Ionian Sea | 1990–1991 | 5 | 77.2[c] | 6.31[c] | | | 0.616[c] | | | | | 1.54 (0.59) NA (0.90–2.42) | Storelli and Marcotrigiano (2000) |

*(continued)*

## TABLE 11.4 (continued)
## Persistent Organic Pollutant Levels in Liver of Sea Turtles from Selected Studies

| Species | Stage/Sex | Location | Years | N | ΣPCBs | 4,4'-DDE | ΣChlordanes | ΣHCHs | HCB | Dieldrin | Mirex | Heptachlor Epoxide | ΣPBDEs | % Lipid | Reference |
|---|---|---|---|---|---|---|---|---|---|---|---|---|---|---|---|
| Cc | JMF | East Florida | NR | 8 | 64.4 (61.1)<br>40.5 (8.0–182) | 21.5 (32.8)<br>8.5 (2–100) | | | | | | | | | McKim and Johnson (1983) |
| Cc | AF | Southern Adriatic Sea and Ionian Sea | 1990–1991 | 6 | 55.7[c] | 65.2[c] | | | 6.36[c] | | | | | 15.91 (9.64) | Storelli and Marcotrigiano (2000) |
| Cc | JMF | Southern Adriatic Sea and Ionian Sea | 1999–2001 | 19 | 52.32 (74.99)<br>NA (6.85–297.49) | | | | | | | | | 6.5 (5.7)<br>NA (0.5–23.3) | Storelli et al. (2007) |
| Cc | NR | Baja California | NR | 1 | 41.0 | | <3 | | <3 | <3 | | | | | Gardner et al. (2003) |
| Cc | JMF | Baja California | 2001–2003 | 4 | 19.3 (5.80) | | | | | | | | | | Richardson et al. (2010) |
| Cc | JMF | Ishigaki Island and Kochi, Japan | 1998–2006 | 4 | 2.10 (1.21)<br>1.71 (1.185–3.8) | 2.46 (0.94)[d]<br>2.16 (1.7–2.37)[d] | 0.711 (0.506)<br>0.553 (0.292–1.444) | 0.257 (0.172)<br>0.183 (0.150–0.513) | 0.0901 (0.0383)<br>0.0869 (0.05688–0.1159) | | | | 0.269 (0.391)<br>0.081 (0.060–0.855) | 10.5 (5.7)<br>8.2 (6.5–19) | Malarvannan et al. (2011) |
| Lo | NR | Baja California | NR | 1 | 58.1 | | 45.3 | | 3.5 | <3 | | | | | Gardner et al. (2003) |
| Lo | JMF | Baja California | 2001–2003 | 4 | 15.2 (6.17) | | | | | | | | | | Richardson et al. (2010) |
| Dc | AF | Canary Islands | 2002–2005 | 1 | 445 | | | | | | | | | | Orós et al. (2009) |
| Dc | AM | Scotland | 1993–1995 | 2 | NA (3.1–3.7) | NA (1.7–6.5) | NA (2.3–2.3) | | | NA (2.5–3.1) | | | | | Mckenzie et al. (1999) |
| Cm | JF | Canary Islands | 2002–2005 | 1 | 116 | | | | | | | | | | Orós et al. (2009) |
| Cm | JMF | East Florida | NR | 4 | 65.3 (15.7)<br>69 (43–80) | 3.4 (4.5)<br>1.5 (<1–10) | | | | | | | | | McKim and Johnson (1983) |
| Cm | JMF | Cyprus | 1995–1996 | 9 | 33.6 (22.2)<br>34 (<1.1–7.7) | 4.89 (6.40)<br>2.7 (<1.0–2.1) | 1.07 (1.35)<br>0.45 (<0.4–3.7) | | | 1.43 (1.07)<br>1.5 (<0.4–3) | | | | | Mckenzie et al. (1999) |
| Cm | JMF | Baja California | NR | 7 | NA (<3–44.7) | NA (<3–10.4) | | | NA (<3–18.6) | <3 | | | | | Gardner et al. (2003) |

| Species | Sex | Location | Years | n | | | | | | | | | | Reference |
|---|---|---|---|---|---|---|---|---|---|---|---|---|---|---|
| Cm | JMF | Baja California | 2001–2003 | 3 | 10.5 (4.92) | | 0.138[f] | 0.131[g] | | | | | | Richardson et al. (2010) |
| Cm | JMF | Queensland, Australia (rehab) | 2006–2007 | 16 | 1.1527 (0.3092)[c]; NA; (0.4111–3.0823) | 0.0355 (0.0137)[c]; NA; (<0.010–0.1144) | | 0.0404 (0.0120)[c]; NA; (<0.010–0.1200) | 0.3871 (0.0716)[c]; NA; (0.0766–0.7770) | 0.07351 (0.03528)[c]; NA; (0.00592–0.35739) | 0.1692 (0.0403)[c]; NA; (0.0218–0.4109) | 0.1201 (0.0213)[c]; NA; (0.0544–0.2360) | 2.37 (0.44)[c]; NA; (1.15–4.84) | van de Merwe et al. (2010a) |
| Cm | AF | Queensland, Australia | 2004–2006 | 1 | | | | | | | | 0.08080 | 5.05 | Hermanussen et al. (2008) |
| Cm | JMF | Ishigaki Island, Japan | 2003–2005 | 5 | 0.257 (0.375); 0.0975; (0.0403–0.924) | 0.0809 (0.113)[d]; 0.0225; (<0.003–0.104)[d] | 0.123 (0.094); 0.0840; (0.045–0.26) | 0.0439 (0.0152); 0.0468; (0.03–0.0645) | | | | 0.150 (0.260); 0.029; (0.0084–0.613) | 11.8 (3.5); 13.0; (7.7–15.0) | Malarvannan et al. (2011) |
| Ei | JMF | Ishigaki Island, Japan | 2002–2004 | 3 | 7.34 (5.59); 6.12; (2.46–13.44) | 3.43 (4.84)[d]; 0.72 (0.56–9.02)[d] | 1.12 (0.41); 1.312; (0.648–1.4) | 0.0245 (0.0267); 0.0134; (0.00504–0.0549) | | | | 1.85 (2.77); 0.386; (0.115–5.04) | 5.8 (2.3); 5.6 (3.6–8.2) | Malarvannan et al. (2011) |

Mean (SD) and/or median (range) in ng/g wet mass, unless otherwise stated. Ranked generally from highest to lowest by species and location. Kemp's ridley (Lk), loggerhead (Cc), leatherback (Dc), green (Cm), olive ridley (Lo) sea turtles, juvenile (J), adult (A), male (M), female (F), polychlorinated biphenyls (PCBs), hexachlorocyclohexanes (HCHs), hexachlorobenzene (HCB), polybrominated diphenylethers (PBDEs), not available (NA), not reported (NR), below detection (<LOD).

[a] Summary statistics recalculated using original data in ng/g wet mass and Helsel (2005) recommendations.

[b] *Trans-* and *cis-*chlordane were the only chlordanes measured.

**TABLE 11.5**

## Persistent Organic Pollutant Levels in Eggs and Hatchlings of Sea Turtles from Selected Studies

| Species | Location | Years | Tissue | N Nests (Eggs/Nest) | Live Eggs Sacrificed? | ΣPCBs | ΣDDTs | ΣChlordanes | ΣHCHs | HCB | Dieldrin | Mirex | Heptachlor Epoxide | ΣToxaphenes | ΣPBDEs | % Lipid | Reference |
|---|---|---|---|---|---|---|---|---|---|---|---|---|---|---|---|---|---|
| Cc | Northwest Florida, Gulf of Mexico | 1992 | Egg contents | 20 (4–12) | No | 240–3720 | <LOD–178[a] | <LOD | <LOD | <LOD | <LOD | <LOD | <LOD | | | | Alam and Brim (2000) |
| Cc | Cape Lookout, North Carolina | 2002 | Egg yolk | 9 (2–7) | No | 117 (118)[b] 70.8 (2.52–273) | 53.2 (57.1)[b] 61.9 (0.369–164) | 28.6 (33.1)[b] 27.8 (0.285–97.2) | 0.239 (0.314)[b] 0.0707 (<0.049–0.989) | | 2.35 (2.73)[b] 0.623 (<0.180–6.94) | 0.789 (0.679)[b] 0.708 (0.0334–2.24) | 4.37 (5.34)[b] 2.76 (<0.048–16.2) | 0.240 (0.244)[b] 0.162 (0.0177–0.677) | 1.04 (1.09)[b] 0.522 (0.0392–2.77) | 7.68 (0.33)[b] 7.41 (6.51–9.51) | Alava et al. (2011) |
| Cc | South Carolina-Georgia | NR | Egg | >3 (NR) | NR | | NA (58–305) | | | | NA (<LOD–56.4) | | | | | | Hillestad et al. (1974) |
| Cc | Cyprus | 1995 | Whole egg | 1 (1) | No | 89 | 155 | 1.8 | | | 0.6 | | | | | | Mckenzie et al. (1999) |
| Cc | Merritt Island, eastern Florida | 1979 | Egg contents | 1 (56) | Yes | | 99 (56–150)[c] | | | | | | | | | | Clark and Krynitsky (1985) |
| Cc | Merritt Island, eastern Florida | 1976 | Egg contents | 9 (20+) | Yes and no | 78 (32–201)[d] | 47 (18–200)[c,d] | NA (<LOD–26[c]) | | | NA (<LOD–28) | NA (<LOD–5) | NA (<LOD–6) | | | | Clark and Krynitsky (1980) |
| Cc | Cyprus | 1995 | Hatchling | ≤4 (1+) | No | 34.0 (22–71) | 36.5 (5.3–113) | 3.0 (0.9–7.9) | | | 1.3 (<0.2–9.2) | | | | | 7.7 (4.8–8.6) | Mckenzie et al. (1999) |

| Species | Location | Year | Tissue | n (range) | | | | | | | | | | | Reference |
|---|---|---|---|---|---|---|---|---|---|---|---|---|---|---|---|
| Cc | eastern Florida | 2002 | Egg yolk | 24 (1–10) | No | 18.4 (29.0)[b] 10.1 (0.449–113) | 8.46 (12.8)[b] 3.13 (0.046–55.5) | 0.0956 (0.165)[b] 0.0423 (<0.049–0.620) | 0.734 (0.797)[b] 0.447 (<0.180–3.28) | 0.505 (1.31)[b] 0.124 (<0.049–6.21) | 1.45 (1.96)[b] 0.705 (<0.048–8.84) | 0.131 (0.148)[b] 0.0499 (0.006–0.546) | 0.220 (0.195)[b] 0.175 (<0.019–0.657) | 7.68 (0.54)[b] 7.40 (4.53–13.1) | Alava et al. (2011) |
| Cc | Sarasota, Florida, Gulf of Mexico | 2002 | Egg yolk | 11 (2–8) | No | 3.14 (5.20)[b] 0.842 (0.158–16.5) | 1.94 (3.38)[b] 0.618 (<0.049–11.4) | 0.0282 (0.0159)[b] 0.0241 (<0.049–0.0624) | 0.457 (0.490)[b] 0.321 (0.204–1.88) | 0.0926 (0.185)[b] 0.00939 (<0.049–0.603) | 0.477 (0.575)[b] 0.256 (<0.048–1.81) | 0.0291 (0.0227)[b] 0.0172 (<0.005–0.0678) | 0.0936 (0.0851)[b] 0.0586 (<0.017–0.245) | 8.65 (0.82)[b] 8.42 (2.6–12.7) | Alava et al. (2011) |
| Dc | Juno Beach, Florida | 2003 | Yolk + albumen | 8 (1–6) | No | 8.45 (7.59) 4.15 (0.441–19.9) | 2.28 (1.71) 1.36 (0.562–5.39) | <LOD; 0.225 (0.076) 0.207 (0.150–0.368) | 0.535 (0.347) 0.450 (0.132–1.16) | <0.084 (<0.083 to <0.084) | 0.219 (0.091) 0.183 (0.096–0.362) | 0.074 (0.029) 0.061 (0.048–0.121) | 0.845 (0.630) 0.689 (0.121–1.64) | 5.00 (0.380) 4.89 (4.67–5.69) | Stewart et al. (2011) |
| Dc | French Guiana | 2006 | Yolk + albumen | 46 (1) | Yes | 6.98 (5.02) NA (1.18–23.62) | 1.44 (1.26) NA (0.08–5.82) | 0.41 (0.26) NA (0.08–1.0) | | | | | | 12.9 (5.1) NA (3.8–25.5) | Guirlet et al. (2010) |
| Cm | Ascension Island | 1972 | Egg yolk | 4 (10) | Yes | 76 (64) 45 (20–220) | | | | | | | | | Thompson et al. (1974) |
| Cm | Cyprus | 1995 | whole egg | 1 (1) | No | 6.1 | | <0.3 | <0.3 | | | | | | Mckenzie et al. (1999) |
| Cm | Merritt Island, eastern Florida | 1976 | Egg contents | 2 (5+) | Yes and no | 4.3 | | NA (<LOD–47)[f] | | | | | | | Clark and Krynitsky (1980) |
| Cm | Heron Island, Queensland, Australia | 1995 | Egg contents | 4 (1–5) | Yes | 1.7 (0.2)[g] 1.6 (1.5–2.0)[g] | | | | | | | | | Podreka et al. (1998) |
| Cm | Cyprus | 1995 | Hatchling | 3 (1) | No | 1.1 (<0.4–13) | 0.2 (<0.4–5.8) | <LOD | <LOD | | | | | 7.53 (0.55) | Mckenzie et al. (1999) |
| Cm | Peninsular Malaysia | 2004 | Yolk + albumen | 11 (3) | Yes | 0.5536 (0.0546)[b] NA (0.3928–0.8394) | 0.0183 (0.0009)[b,h] NA (0.0138–0.0223)[b] | 0.1723 (0.0074)[b,i] NA (0.1378–0.2078)[i] | | 0.0094 (0.0011)[b] NA (<LOD–0.0128) | | | 0.1293 (0.0081)[b] NA (0.0617–0.1638) | 8.9 (0.2)[b] NA (6.8–10.9) | van de Merwe et al. (2010b) |

*(continued)*

## TABLE 11.5 (continued)
## Persistent Organic Pollutant Levels in Eggs and Hatchlings of Sea Turtles from Selected Studies

| Species | Location | Years | Tissue | N Nests (Eggs/Nest) | Live Eggs Sacrificed? | ΣPCBs | ΣDDTs | ΣChlordanes | ΣHCHs | HCB | Dieldrin | Mirex | Heptachlor Epoxide | ΣToxaphenes | ΣPBDEs | % Lipid | Reference |
|---|---|---|---|---|---|---|---|---|---|---|---|---|---|---|---|---|---|
| Cm | Peninsular Malaysian markets | 2006 | Yolk + albumen | Perhaps 55 (1) | Yes | 0.4705 (0.0833)[b] NA (0.1466– 3.6915) | 0.0835 (0.0183)[b] NA (<LOD– 0.7019) | 0.0575 (0.0094)[b] NA (0.0247– 0.5142) | 0.0688 (0.0087)[b] NA (0.0132– 0.2301) | | | | | | 0.0214 (0.0066)[b] (<LOD– 0.3527) | 9.33 (0.14)[b] NA (7.3–13.54) | van de Merwe et al. (2009a) |
| Cm | Heron Island, Australia | 1998 | Yolk + albumen | 10 reps of 1 pool of 15 eggs from 3–4 nests | Yes | 0.25616[f] | | 0.04319[a] | | 0.01294 (0.00073)[b] | 0.4225 (0.0039)[b] | | 0.01974 (0.00026)[b] | | 0.30956[f] | | van de Merwe et al. (2009b) |

Mean (SD) and/or median (range) in ng/g wet mass, unless otherwise stated. Ranked generally from highest to lowest by species and location.

Loggerhead (Cc), leatherback (Dc), green (Cm) sea turtles, polychlorinated biphenyls (PCBs), sum of dichlorodiphenyltrichloroethane-related compounds (ΣDDTs), hexachlorocyclohexanes (HCHs), hexachlorobenzene (HCB), polybrominated diphenylethers (PBDEs), not available (NA), not reported (NR), below detection (<LOD).

a Value represents only 4,4'-DDD and was estimated from ng/g dry weight using a conversion factor calculated from the PCB concentrations that were reported as both wet mass and dry mass (0.223 g dry mass/g wet mass).
a Arithmetic mean (standard error).
c Geometric mean (range) of only 4,4'-DDE.
d Geometric mean (range).
e Maximum value estimated by summing trans-nonachlor and oxychlordane maximums.
f Maximum value estimated by summing 4,4'-DDE and 4,4'-DDT maximums.
g Values represent only 4,4'-DDE.
h Values represent only trans-chlordane, the only chlordane detected.
i Values represent only γ-HCH, the only HCH isomer detected.
j Value represents sum of means of 16 individual PCB congeners reported.
k Value represents sum of means of cis- and trans-chlordane and cis-nonachlor.
l Value represents sum of means of seven individual PBDE congeners reported.

Another complication is that most POP data are not normally distributed, so the best estimates of central tendency are medians or geometric means, rather than arithmetic means, but few studies provide these. Thus, Tables 11.2 through 11.5 show concentrations reported only in wet mass units (or estimated using conversions), and the tables show both the arithmetic mean as well as a preferred estimate of central tendency (median or geometric mean) when available.

### 11.3.2.1 Comparison of Different POP Classes

The predominant POP classes are shown in Tables 11.2 through 11.5. Chemicals were excluded if they were of very low concentrations, seldom measured, or could be represented in totals of a class of POP (represented here with the symbol Σ). Regardless of the tissue, the different classes of POPs generally ranked from highest to lowest concentration in sea turtles in the following order: ΣPCBs, ΣDDTs, other organochlorine pesticides (Σchlordanes, Σtoxaphenes, mirex, dieldrin), and ΣPBDEs (Figure 11.1; Tables 11.2 through 11.5). HCB and ΣHCHs are often very low in concentrations, when detected. In blood, however, PFOS can rival or exceed ΣPCB concentrations. No study has measured every single POP listed on the Stockholm Convention in the same turtle, but Figure 11.1 is the best across-study comparison for the largest number of POP classes. In loggerhead blood from the southeastern region of the United States, ΣPCBs are usually the predominant POP class in concentration, followed closely by the single compound of PFOS and then three classes of OCPs, ΣPBDEs, and a few more OCPs. Several compound classes were below detection in loggerhead blood from the southeastern coast of the United States, but this pattern may be different in other species, tissues, or regions. Four POP classes have not been measured in this population of loggerheads (dioxins/furans, chlordecone, the paraffins, and hexabromobiphenyls) nor in any other sea turtle, except for dioxins/furans, which have been detected in green turtles from Australia (Hermanussen et al., 2006).

Many of these POP concentrations are reported as totals or sums, which create additional confounding complexity when comparing results from different studies. For example, there are 209 possible PCB congeners (or configurations of chlorine atoms around the biphenyl structure), and usually the highly recalcitrant bioaccumulative congener, PCB 153, dominates the PCB pattern. Because each laboratory measures and sums a different suite and number of PCB congeners, it is not always accurate to compare summed or total PCB concentrations. Thus, comparing only one PCB congener, like PCB 153, may make a better comparison. However, individual congener concentrations have not been reported in most sea turtle studies, leaving only ΣPCB concentrations to be compared here. The same problem exists for the six DDT-related compounds (of which 4,4′-DDE, the final and very persistent metabolite of the DDT pesticide formulation, nearly always predominates), the several chlordane compounds (trans-nonachlor and oxychlordane predominate but detections of cis- and trans-chlordane, and cis-nonachlor are also common), the large number of toxaphenes (usually Parlars 26 and 50 are the highest), and the 209 possible PBDE congeners (PBDE 47 most always predominates in wildlife tissues).

The POP patterns measured in sea turtles generally match those measured in other wildlife with a few exceptions in localized areas. Surprisingly, 4,4′-DDE was not the predominant DDT in sea turtles from Baja California, where 2,4′-DDD and 4,4′-DDD were predominant (Gardner, 2003; Labrada-Martagon et al., 2011) or in loggerhead eggs from northwestern Florida, where 4,4′-DDD was predominant (Alam and Brim, 2000). These patterns are difficult to explain and should be validated. Additionally, an interesting trend in PBDE patterns is emerging in the sea turtle literature. Globally, PBDE 47 is the PBDE congener in highest concentration in most wildlife (Hites, 2004). However, some sea turtles have higher proportions of PBDEs 99, 100, 153, and/or 154 instead. This deviation from the expected pattern is found exclusively in turtles (sea turtles, diamondback terrapins, and freshwater turtles) inhabiting the eastern United States between 35°N and 45°N latitude (Basile et al., 2011; Carlson, 2006; Moss et al., 2009; Ragland et al., 2011). Other wildlife or fish within this region (Chen et al., 2010; Hale et al., 2006), and freshwater and sea turtles just outside of this region accumulate the expected pattern (Keller et al., 2005a; Ragland et al., 2011; de Solla et al., 2007; Stewart et al., 2011; Swarthout et al., 2010). Thus, these findings are not specific to one sea turtle species, rather to turtles in general within this region. Some sea turtles in regions far from the United States also have PBDE patterns that deviate from the expected pattern (Hermanussen et al., 2008; van de Merwe et al., 2010b). This is interesting because it suggests that sea turtles are good bioindicator species of release of unusual environmental contaminants at least at the regional scale.

### 11.3.2.2 Biological Factors Influencing POP Concentrations

It is crucial to examine Tables 11.2 through 11.5 with the understanding that POP concentrations are driven not only by localized uses and releases of the chemicals within certain watersheds but also by environmental factors on scales ranging from local to global (e.g., water currents, air movements, temperature, precipitation, salinity, and organic matter content) and biological factors. Here, biological factors (lipid content, body condition, trophic status, age, and sex) influencing POP concentrations and accumulation in sea turtles will be discussed.

#### *11.3.2.2.1 Lipid Content Influences Tissue Differences*

Lipid content of various organs drives the distribution of POPs throughout the body. Most POPs, except for PFCs, distribute into tissues with the highest lipid content; therefore, fat, adipose, or blubber tissues (30%–80% lipid) followed by liver (5%–15% lipid) have higher lipid and thus higher POP concentrations than other bodily tissues. These are often the tissues chosen for POP assessment, but they can only be obtained from a dead animal or with invasive surgical procedures. Eggs, containing 5%–10% lipid, are also a choice sample. Earlier studies focused on these tissues, while more recent studies have switched to blood sampling (Table 11.1).

Because blood has very low lipid content (<1%), POP levels in blood are much lower than concentrations in fat, liver, or eggs (Tables 11.2 through 11.5). Blood analysis, therefore, requires highly sensitive methods for detection. This issue has been a major disadvantage for several unsuccessful studies whose detection limits for individual compounds were 10 or 20 ng/g wet mass (Deem et al., 2006; Harris et al., 2011; Innis et al., 2008), which would not allow detection of the maximum concentration in sea turtle blood analyzed to date (Table 11.2). Because of the low lipid level in blood, POPs likely also bind to plasma proteins, which is one argument against normalizing blood POP concentrations to solely lipid content. Additionally, blood concentrations expressed more simply on a per wet mass or volume basis are more relevant when assessing toxicity as it is easier to compare sea turtle levels to those in laboratory-exposed animals. However, lipid normalization is recommended when comparing concentrations in blood to other more lipid-rich tissues from the same animal. This was done in studies demonstrating that POP concentrations in blood were significantly correlated to those in fat or other internal tissues from the same loggerhead, Kemp's ridley, or green sea turtles (Keller et al., 2004a; van de Merwe et al., 2010a). These results indicated that blood, a less invasive sample, reasonably represents the concentration stored in more routinely analyzed fatty tissues. POPs have been measured in several blood components of sea turtles, including plasma/serum, whole blood, and red blood cells (RBCs). Keller et al. (2004b) and Carlson (2006) showed that POPs preferentially distribute into the plasma rather than the RBC fraction, and linear regressions were provided to convert concentrations measured in one matrix to another.

Early studies analyzed internal organs from stranded animals or eggs for POPs; however, because blood has been proven to be a reasonable tissue for POP measurements, many recent projects are choosing this nonlethal, simple tissue sampling methodology. Additional advantages of using blood over tissues from stranded animals are that a less biased population of live animals can be sampled and that more simultaneous health and biological data can be collected from the same animals. For these reasons, six of the seven sea turtle species have been monitored for blood POP concentrations (Table 11.2); one more species than liver and fat. Blood studies also contain more POP classes than the traditional fat and liver studies, adding PBDEs, PFOS, toxaphenes, and dioxins/furans to the sea turtle literature.

PFOS has been measured exclusively in blood samples from sea turtles, and it holds the record for the highest concentration of a single compound among all POPs measured in sea turtle blood

(Keller et al., 2005b, 2012; O'Connell et al., 2010). This compound, along with other PFCs, differs from the other POP classes because it associates with proteins instead of lipids. Thus, PFCs preferentially distribute into plasma and liver of animals rather than fat, which makes these tissues the target for analysis. An important point to emphasize regarding blood is that POP concentrations in this tissue are being circulated and perfusing internal organs that could be targets for toxicological effects, rather than being stored away in fatty depots.

Eggs are one of the best sampling compartments for measuring POP exposure and have been used extensively to monitor temporal and geographical trends of POPs in birds from Antarctica to the Arctic (Braune et al., 2007; Corsolini et al., 2011). POP concentrations in eggs represent the contamination of the maternal foraging grounds (Alava et al., 2011), because the compounds are transferred into the egg during yolk production, which occurs months before females migrate to nesting beaches often far distances from their foraging grounds. Evidence of maternal transfer comes from the significant correlations between maternal blood POP concentrations and those in her eggs for green and leatherback sea turtles (Guirlet et al., 2010; van de Merwe et al., 2010b; Stewart et al., 2011). Guirlet et al. (2010) also showed that POP concentrations declined with subsequent clutches within a season from the same mother, and the decline was related to a coincident decline in lipid content of the eggs. Leatherback turtles that took 3 years off between nesting also appeared to make eggs with higher POP concentrations than those that only took 2 years (Guirlet et al., 2010). Often people are curious if POPs can transfer across the shell during incubation from contaminated sand. Basile (2010) demonstrated that negligible amounts of POPs transfer from highly contaminated sediments into diamondback terrapin (*Malaclemys terrapin*) eggs and that indeed >98% of the POP concentration within an egg was maternally transferred.

Eggs from the same sea turtle nest have very similar concentrations (Alava et al., 2006; van de Merwe et al., 2010b), whereas clutches from different females show great variability (Table 11.5). Several studies have used unhatched sea turtle eggs after live hatchlings emerge, which makes this egg sampling technique nonlethal (Table 11.5) (Alam and Brim, 2000; Alava et al., 2006, 2011; Mckenzie et al., 1999; Stewart et al., 2011). When using unhatched eggs, it is important to consider the effects of embryonic development on egg POP concentrations. Alava et al. (2006) showed that if only the yolk is measured for POPs, the concentrations within the yolk increase through development mainly because the yolk sac becomes smaller as water is taken up and lipids are concentrated during embryonic growth. Thus, the yolk sac taken from a late stage embryo will not have the same concentration as the yolk taken from an egg without development. Standardized sampling protocols using unhatched eggs should either homogenize the entire egg contents or collect subsamples (e.g., yolk) from only one stage of development. A recent study has shown that unhatched loggerhead eggs have the same concentration as freshly laid eggs from the same nest (Keller, unpublished data).

Since egg sampling is so simple, it is surprising how few studies have used this meaningful packet from sea turtles. Only loggerheads, leatherbacks, and greens have been examined for egg POP concentrations (Table 11.5). Moreover, loggerhead eggs have been sampled from only the southeastern United States aside from one egg from the Mediterranean Sea. Exposure of other nesting loggerhead subpopulations is completely unknown. Ironically, no green sea turtle egg has been analyzed from the southeastern United States since 1976, and the majority of the available egg concentration data for this species now come from Australia and Malaysia.

### *11.3.2.2.2 Body Condition Influences POP Concentrations*

POP levels are known to fluctuate in all tissues during significant weight changes, but most drastically in blood. POPs (except PFCs) are normally found in fatty tissues because they preferentially bind to lipids. During weight loss, these POPs along with associated lipids will move out of fat depots into blood (Chevrier et al., 2000; Debier et al., 2006; Hall et al., 2008). The body utilizes the lipids for energy, but the POPs remain circulating in the blood stream, because these compounds are difficult to metabolize or eliminate even in healthy animals. Thus, the higher circulating POP levels are more available to enter and cause toxic effects to vulnerable target organs (liver, kidney, brain, gonads, etc.) until foraging resumes and fat stores are replenished. Because POP concentrations can be dramatically affected by body condition, this factor must be considered when interpreting POP concentrations (Ross, 2004). In apparently healthy loggerhead sea turtles and bottlenose dolphins (*Tursiops truncatus*), blood POP concentrations increase as fatty tissue lipid content decreases (Keller et al., 2004a; Yordy et al., 2010b). Likewise, loggerhead turtles from North Carolina with poorer body condition had significantly higher dieldrin blood concentrations (Keller et al., 2004c). Similar correlations were seen between body condition and ΣPCBs in livers from loggerhead turtles that stranded in the Canary Islands (Orós et al., 2009) and plasma HCHs in green turtles from Baja California (Labrada-Martagon et al., 2011). Monagas et al. (2008) grouped dead stranded loggerhead turtles into emaciated and normal body condition to look for differences in fat and liver concentrations of 2,4'-DDTs. Within the emaciated group, liver concentrations were higher than fat concentrations (both on wet mass basis), suggesting that the pesticides had mobilized from fat into other less lipid-rich and more protein-rich tissues. Furthermore, several extremely emaciated and sick loggerhead sea turtles (debilitated syndrome) were found to have nine times higher blood POP concentrations and much lower lipid content in their fat depots compared to their healthy counterparts along the southeastern coast of the United States (Keller et al., 2006a).

### 11.3.2.2.3   Trophic Status Influences Species Differences

Species differences are apparent and consistent across all tissues (Tables 11.2 through 11.5) with blood and liver samples allowing the most comparisons among species (Tables 11.2 and 11.4). Broad comparisons across all studies support the general conclusion that has been stated in many past sea turtle studies that POP concentrations are highest in Kemp's ridley sea turtles, followed by loggerhead, leatherback, and finally green sea turtles. This ranking follows closely with their trophic status, which is an expected relationship because POPs are known to biomagnify up food webs. Thus, the turtle species with the highest trophic status will accumulate the highest POP concentrations. The diets of the different sea turtle species were reviewed by Bjorndal (1997), and Chapter 9 summarizes dietary analyses including C and N isotopic signatures that characterize trophic and regional feeding. Kemp's ridley sea turtles feed mostly on crabs, while loggerheads are omnivorous carnivores choosing a variety of marine invertebrates, including crustaceans and mollusks. Leatherbacks eat almost exclusively gelatinous zooplankton, and green turtles are primarily herbivores. Future studies should pair contaminant and stable isotope analysis to validate that these species differences are primarily caused by differences in trophic status. Additionally, dietary shifts are known in sea turtles, especially for the green turtle where younger pelagic turtles are omnivores and older neritic stages are herbivores (Bjorndal, 1997). These shifts could influence POP concentrations.

Tissues from olive ridley (*Lepidochelys olivacea*), hawksbill, and flatback sea turtles have been measured so infrequently that it is unwise to make any strong conclusions as to their POP exposure level relative to the other species. However, one olive ridley along Baja California had POP fat and liver concentrations higher than a loggerhead from this region, potentially placing it on a similar exposure level as the related Kemp's ridley sea turtle (Tables 11.3 and 11.4) (Gardner, 2003). In contrast, a more recent paper on levels in liver places the olive ridley nearly equivalent or slightly below the loggerhead sea turtle (Table 11.4) (Richardson et al., 2010). Surprisingly the spongivorous hawksbill sea turtle may turn out to be the most exposed species, as demonstrated by higher PCB concentrations in liver than loggerhead and green sea turtles within Japanese waters (Table 11.4) (Malarvannan et al., 2011). Similarly, one hawksbill blood sample had higher PBDE concentrations than green sea turtles and one flatback turtle from Australia (Hermanussen et al., 2008). These findings from very small sample sizes were supported by recent PFC measurements, in which hawksbill plasma was found to have the highest concentrations of PFOS compared to four other species along the southeastern U.S. coast (Table 11.2) (Keller et al., 2012). With only one sample from flatback sea turtles, too few values are available to determine their placement within the sea turtle ranks.

Along the lines of prey choices, sea turtles are known to ingest plastic trash, probably mistaking it for food. While this causes a mortality risk to sea turtles from entanglement or gastrointestinal blockage, a new concern with this interaction has emerged (Moore, 2008). Plastic polymers are indeed persistent pollutants containing several organic and toxic chemicals, but they also adsorb and concentrate POPs and other hydrophobic contaminants from the water. Ingestion of these plastics could be an additional source of exposure to POPs and other pollutants for sea turtles. This new hypothesis deserves some research attention.

### 11.3.2.2.4  Sex and Age Class Influences on POP Concentrations

POPs are known to accumulate through the life of animals, increasing with age for males and increasing until reproductive maturity for females, which is especially well documented in marine mammals (Yordy et al., 2010a). Females are known to offload a significant portion of the POP burden into their offspring and this offloading, or maternal transfer, is known for both oviparous and live-bearing reproductive strategies. For sea turtles, all classes of POPs are expected to transfer from mother to egg during yolk deposition into ovarian follicles. This occurs while the female is acquiring and storing energy resources on her foraging grounds and developing her next season's follicles. Thus, the concentrations in eggs represent the contamination of her distant foraging grounds, not from the nesting beach where she ultimately lays her eggs. Thus as a female ages, she is able to offload her POP burden to her eggs, thereby decreasing her tissue concentrations. Males do not have a similar offloading mechanism, so they continually accumulate POPs. It is expected that adult female sea turtles would have lower POP concentrations than adult males, but only one study has ever measured a large enough sample size of adult males to make any comparison (Ragland et al., 2011). Ragland et al. (2011) found that adult male loggerhead plasma had the highest POP concentrations compared to any other age class, sex, or species of sea turtle studied (Table 11.2). During the juvenile and subadult stages, males and females are not expected to have different concentrations as supported by studies on PCBs, OCPs, and PFCs (Keller et al., 2004a, 2005b; Komoroske et al., 2011; Lazar et al., 2011; Malarvannan et al., 2011).

Some studies have examined whether POP concentrations change through age by analyzing sea turtles that span a range of carapace lengths. Unfortunately, live sea turtles cannot be aged, and turtle length, the only proxy for age, has its flaws because the relationship between age and length is quite variable. Nevertheless, observed relationships between length and POP concentrations are inconsistent. Some studies show expected increases with loggerhead turtle length for concentrations of PFCs (Keller et al., 2005b), PCB congeners 52 and 114 (Lazar et al., 2011), and several OCPs and ΣPCBs (Ragland et al., 2011). Other studies show no relationship between turtle length and POP concentrations (Komoroske et al., 2011; Labrada-Martagon et al., 2011; O'Connell et al., 2010; Swarthout et al., 2010), while others show negative correlations for loggerheads (Keller et al., 2004a) but mostly for green turtles (Malarvannan et al., 2011; Mckenzie et al., 1999; Richardson et al., 2010). The latter studies suggest growth dilution instead of accumulation with age, and this is highly likely in green turtles after they shift from being omnivorous as young juveniles to a more herbivorous diet later in the juvenile stage (Bjorndal, 1997). These ontogenetic shifts and the fact that tissue POP concentrations are a culmination of POPs accumulated through all past life stages make interpretation of POPs through age difficult. No study has strategically examined the POP burden of each life history stage of a subpopulation at major locations along its migratory routes.

Few studies have measured POPs in hatchling sea turtles (Aguirre et al., 1994; Mckenzie et al., 1999; van de Merwe et al., 2010b). One large and potentially crucial data gap is the concentrations of POPs in the blood of hatchlings during and after the frenzy period when they have metabolically utilized all the yolk resources. That flush of lipids and its associated POPs during this critical developmental stage in such a small animal could plausibly result in acute toxicity or nonlethal effects that lead to mortality indirectly. Van de Merwe et al. (2010b) measured blood POP concentrations in hatchlings and also in eggs and maternal blood. The hatchling blood samples were collected quickly after emergence from the nest, just before the frenzy period, and had 1.5 times higher mean concentrations of PCBs than the adult females (Table 11.2). During the rapid use of the absorbed yolk sac in the following few days, the blood concentration of these hatchlings would likely increase dramatically. Thus, it remains to be shown if hatchlings struggling to swim offshore may also face some of the highest POP concentrations among the sea turtle age classes.

### 11.3.2.3 Nonbiological Factors Influencing POP Concentrations

POPs are known to change across space and time. These nonbiological factors could be examined with a meta-analysis approach if enough quality data were available without major confounding factors mentioned previously. In the next sections an attempt is made to examine spatial and time trends across available studies in Tables 11.2 through 11.5 as well as to highlight a few recent studies that have examined these nonbiological factors directly within a controlled sampling regimen with statistical hypothesis testing.

#### 11.3.2.3.1 Spatial Trends

A global spatial delineation is becoming apparent with newer studies on fat, liver, and eggs (Tables 11.3 through 11.5) showing that loggerhead and green sea turtles in the North Atlantic and Mediterranean Sea have higher exposure to PCBs and OCPs than those sampled along Baja California, Japan, Malaysia, and Australia. More sampling is needed in these and other regions with larger sample sizes along with standardized collection years and analytical methods to make global comparisons more robust.

Several recent studies have directly sought to assess spatial differences in sea turtle POP exposure. These studies have determined that even highly migratory sea turtles can be used as indicator species of marine contamination on a regional scale (Alava et al., 2011; Hermanussen et al., 2006; Keller et al., 2005b; O'Connell et al., 2010; Ragland et al., 2011; Swarthout et al., 2010). These studies selected samples from the same species, age class, and years of collection and analyzed samples in a single laboratory with standardized methods to eliminate confounding factors in order to compare only across locations. The results show significant differences among sampling locations. Hermanussen et al. (2006) linked dioxin concentrations in green sea turtles from Australia to measured concentrations in the abiotic environment (sediments) that the turtles were inhabiting. Most other studies focused on loggerhead sea turtles sampled within the United States, and taken together, they conclude that higher POP concentrations are detected in loggerheads inhabiting regions further north along the U.S. Atlantic coastline (Figure 11.4). These findings suggest that subpopulations utilizing areas of higher contamination (e.g., along the mid-Atlantic and northeastern regions of the United States) accumulate higher concentrations and could be at higher risk for toxic effects. One specific location that deserves more attention is near Brunswick, Georgia. This coastal city hosts three U.S. EPA Superfund sites and its estuary and coastal habitats are contaminated with a unique and highly chlorinated PCB mixture, mercury, and toxaphenes (Balmer et al., 2011; Maruya and Lee, 1998). Preliminary data show that loggerhead turtles captured offshore of Brunswick have a pattern of PCBs primarily composed of higher chlorination groups indicative of exposure from this Superfund site (Keller et al., unpublished data). This is one known hot spot for a mixture of POPs that should be examined for sea turtle exposure and toxic effects.

Among the spatial studies, two U.S. studies have attempted to relate POP concentrations to known migratory pathways of sea turtles. Loggerhead turtles nesting in North Carolina had higher concentrations of PCBs, OCPs, and PBDEs, represented in their eggs, than loggerheads nesting on the east or west coast of Florida (Alava et al., 2011). Using previously published satellite tracks of females nesting in these three regions, Alava et al. (2011) showed that North Carolina nesters forage in areas further north than those from Florida. Likewise, Ragland et al. (2011) measured POP concentrations in plasma of adult male loggerheads that were satellite tagged during the mating season off of Cape Canaveral, Florida. After the mating season, two major migratory pathways were evident. The residents, those males that stayed near Cape Canaveral into and through the

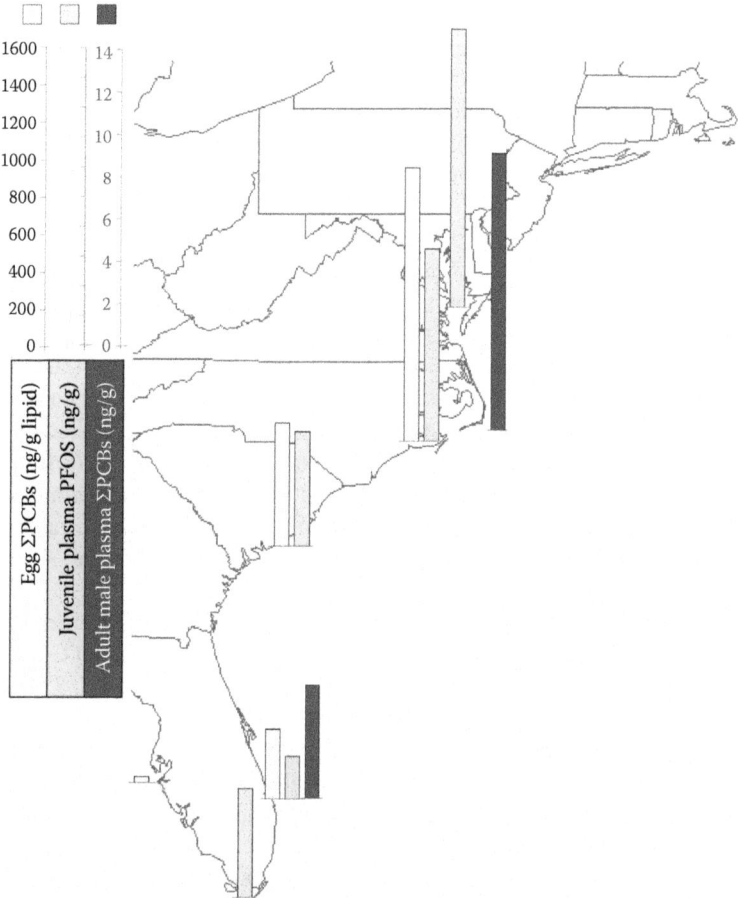

FIGURE 11.4   Spatial comparison of average persistent organic pollutant (POP) concentrations in logger-head sea turtle tissues along the coast of the United States. Data were taken from Alava et al. (2011) and Keller (unpublished data) (yellow), O'Connell et al. (2010) (blue), and Ragland et al. (2011) (red).

winter, had lower POP concentrations than the transients that chose to forage in northern regions off of South Carolina to New Jersey. Their choice of foraging habitat influenced their POP exposure, and these satellite tracking studies provide critical information about previously unknown foraging habitat selection.

*11.3.2.3.2   Temporal Trends*

Temporal trend studies are very important for monitoring changes in POP concentrations after the onset of manufacture or restriction in use of a compound. POP concentrations have been shown to change through time in the environment, but trends are often species and location dependent (Tuerk et al., 2005). Using Tables 11.2 through 11.5, it is very difficult to extract time trends in sea turtle tissues. Blood sampling has only occurred in the last decade, so no across-study comparisons are available (Table 11.2). Kemp's ridley fat and liver collected from stranded animals in the north-eastern United States generally show a decline in PCBs and OCPs from 1985 to 2005 (Tables 11.3 and 11.4) (Innis et al., 2008; Keller et al., 2004a; Lake et al., 1994; Rybitski et al., 1995). Likewise, PCB and 4,4′-DDE concentrations seem to have declined in loggerhead fat from Virginia-North Carolina from 1991–1992 to 2000–2001 (Table 11.3) (Keller et al., 2004a; Rybitski et al., 1995) and in loggerhead eggs from eastern Florida between 1976 and 2002 (Table 11.5) (Alava et al., 2011; Clark and Krynitsky, 1980).

Aside from these general comparisons, specimen banks, or well-maintained archives of samples collected over several years and stored appropriately, are excellent resources of samples for assessing temporal trends of environmental pollutants. Two studies have made use of informal banks of sea turtle plasma to directly test temporal trends in POP concentrations (Carlson, 2006; O'Connell et al., 2010). Concentrations of PCBs, OCPs, and PBDEs were assessed in five juvenile loggerhead plasma samples per year from Core Sound, North Carolina, from 1998 to 2006 (Carlson, 2006). Respective mean concentrations by year in pg/g wet mass were: 2340, 6130, 2950, 2730, 1440, 8710, 3030, 1150, and 9580 for ΣPCBs; 251, 159, 182, 262, 117, 663, 171, 117, and 312 for 4,4′-DDE; and 156, 42.0, 65.7, 63.7, 10.5, 139, 90.6, 14.8, and 62.4 for ΣPBDE (data not shown in Table 11.2). No significant temporal trends were observed for any of these POPs. While declining trends might be optimistically anticipated for PCBs and DDTs, which were banned from use in the United States in the 1970s, the lack of trend is not too surprising. In other U.S. locations, like the Great Lakes, PCB concentrations in fish exponentially declined from the 1970s until 1990, but then the decline slowed (Stow et al., 2004). Thus, this nine-year sampling design for sea turtles centered around 2000 may be too late or too short to detect a trend. During this study's sampling of sea turtles, PBDE use was ongoing in the United States until certain states began to restrict certain formulations in the late 2000s; thus, the stable PBDE concentrations indicate no significant increase or decrease into this sea turtle habitat during this time period. Future samples would be expected to show lower concentrations of all three POP classes.

Another study measured PFC concentrations in 10 or more juvenile loggerhead plasma samples per year from offshore of Charleston, South Carolina, from 2000 to 2008 (O'Connell et al., 2010). Respective median PFOS concentrations by year in ng/g wet mass were 9.43, 7.11, 7.24, 6.35, 2.10, 1.26, 2.99, 1.91, and 3.07 (data not shown in Table 11.2). These data indicated a significant decline in PFOS over this time period, and more importantly, a successful and positive environmental response within sea turtle habitat quickly after PFOS use was discontinued around 2001. These temporal trend studies are good examples of using a specimen bank, and sea turtle contaminant and health research will benefit from a longer-term, formal specimen banking program, similar to the Marine Turtle Molecular Research Collection at the Southwest Fisheries Science Center or the Marine Environmental Specimen Bank at the National Institute of Standards and Technology (Pugh et al., 2008; Serra-Valente et al., 2010). The latter of which is developing a sea turtle specimen bank project called the Biological and Environmental Monitoring and Archival of Sea Turtle tissues (BEMAST).

### 11.3.3 SEA TURTLE POP EXPOSURE COMPARED TO OTHER WILDLIFE AND AS FOOD FOR HUMANS

In comparison to other marine/coastal wildlife, sea turtle POP concentrations are often orders of magnitude lower than those in sharks, alligators, seabirds, and marine mammals from similar locations (Bazan, 2011; Blus et al., 1974; Harper et al., 1999; Heinz et al., 1991; Houde et al., 2006; Yordy et al., 2010b). This ranking is seen in blood as well as eggs for PCBs and pesticides (Figures 11.5 and 11.6). For PFCs, few data are available from plasma of marine wildlife from the southeast U.S. coast for comparison, except for bottlenose dolphins (Houde et al., 2005). Dolphins from Charleston, South Carolina, in 2003 had a geometric mean PFOS concentration of 1171 ng/g wet mass in their plasma (Houde et al., 2005), compared to 6.35 ng/g wet mass, the median in loggerhead turtle plasma from the same location and year (O'Connell et al., 2010). These differences are due mainly to different trophic levels. Most of these other species are fish eaters, except alligators, which eat a variety of vertebrates, placing their trophic status higher than any sea turtle species. One exception to this common ranking is that dioxin/furan toxic equivalencies (TEQs or summed concentrations of dioxins and furans that are weighted based on the toxicity of the individual compounds) in green turtles in more polluted sites within Moreton Bay, Australia were found to be higher than TEQs measured in some marine mammals from locations often considered to be

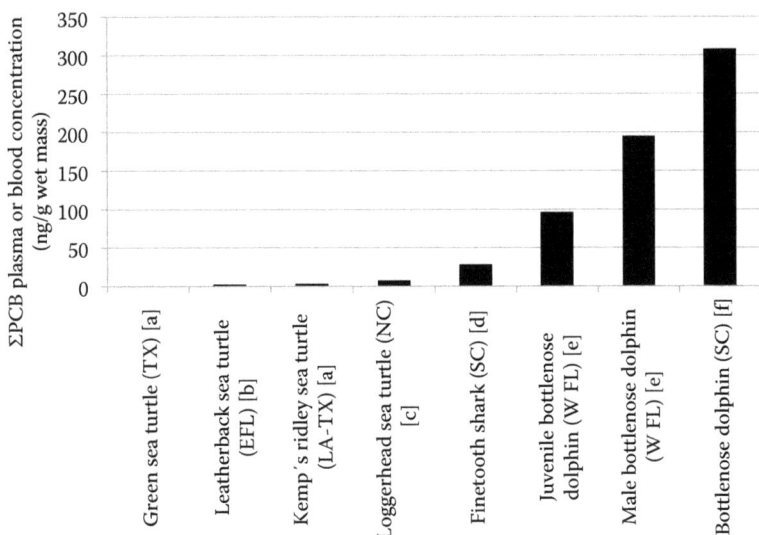

**FIGURE 11.5** Comparison of average blood or plasma concentrations of total PCBs among sea turtle species and other coastal marine organisms. Location abbreviations are TX=Texas, E FL=eastern Florida, LA=Louisiana, NC=North Carolina, SC=South Carolina, W FL=western Florida. Data from [a] Swarthout et al. (2010), [b] Stewart et al. (2011), [c] Keller et al. (2004a), [d] Bazan (2011), [e] Yordy et al. (2010b), and [f] Houde et al. (2006).

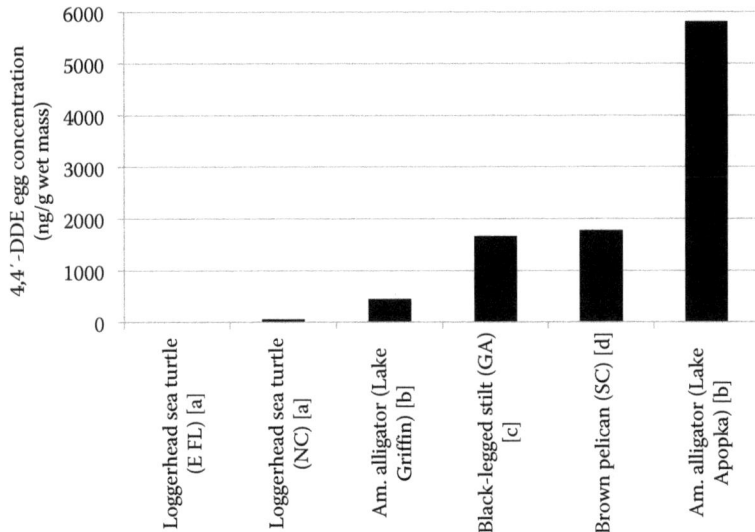

**FIGURE 11.6** Comparison of average egg concentrations of 4,4′-DDE among sea turtle species and other organisms from the Southeastern United States. Location abbreviations are E FL=eastern Florida, NC=North Carolina, GA=Georgia, SC=South Carolina. Data from [a] Alava et al. (2011), [b] Heinz et al. (1991), [c] Harper et al. (1999), and [d] Blus et al. (1974).

much more polluted, like the Mediterranean Sea (Hermanussen et al., 2006). This study emphasizes the need to examine sea turtle POP exposure in more urbanized bays and harbors.

Two papers have addressed POP concentrations in sea turtle tissues as an issue for human consumption (Aguirre et al., 2006; van de Merwe et al., 2009a). Green turtle eggs from Malaysian markets had concentrations of coplanar PCBs that exceed the acceptable daily intake limits established by the World Health Organization (WHO) by threefold or more (van de Merwe et al., 2009a). The coplanar PCBs (PCBs 77, 126, and 169) are the PCB congeners most like dioxin and contribute the most to TEQs, but they are difficult to measure with accuracy. Often other PCB congeners interfere and lead to overestimation of the concentrations, but these authors showed in a separate paper that their analytical methods only overestimated these congeners by approximately 35% in a fish tissue Standard Reference Material (van de Merwe et al., 2009b). Therefore, even after correcting for this overestimation, the minimum egg would still have concentrations higher than the WHO limit. Aguirre et al. (2006) reviewed the literature on environmental contaminants in sea turtle meat (muscle) and eggs and found that these sea turtle tissues have OCPs and PCBs within the range of or exceed safe exposure limits set by the WHO and the U.S. Agency for Toxic Substances and Disease Registry. They note that sea turtle products contain lower levels than some other seafood choices, but developmental effects of POPs are a concern even at these levels, especially for nursing mothers and children. These studies emphasize that POP accumulation in sea turtles is a concern, despite the fact that sea turtles generally have lower POP concentrations than other higher trophic level marine species.

## 11.4 TESTING THE TOXICITY OF POPS IN SEA TURTLES

While other marine species have higher POP concentrations than sea turtles, and many of their populations are doing very well despite their exposure, the sensitivity of sea turtles to toxic effects of POPs is not well understood. Testing for the toxicity, or determining the adverse biological effects of exposure to a chemical pollutant, can be done using numerous experimental designs that range in their ability to definitively document toxic effects. The most definitive tests use direct laboratory exposure of the species in question to a known and administered dose of a chemical. These are rare in the sea turtle literature because of their conservation status and the ethical/legal issues of sacrificing or intentionally harming endangered sea turtles. Only one study has used this approach for testing sea turtles' sensitivity to POP effects. Podreka et al. (1998) applied 4,4′-DDE to green sea turtle eggshells, incubated the eggs at temperatures that are known to produce males, and found that it did not cause embryonic sex reversal. They applied 4,4′-DDE concentrations of 3,300–66,500 ng to the shell, resulting in egg concentrations up to 543 ng/g, which exceeded the maximum ever measured in any sea turtle egg (see Table 11.5). Thus, they concluded that lower concentrations should not present a problem for sea turtle sexual determination, as well as other measured endpoints, including incubation time, hatching success, incidence of body deformities, and hatchling length and weight. This conclusion may have been premature because Willingham (2004) showed that lower doses of 4,4′-DDE (7 ng applied) caused more sex reversal in red-eared slider turtles (*Trachemys scripta*) than higher doses (28 ng). This suggests that this endocrine disruptive effect follows a hormetic dose–response curve (upside down U shape), which is very common in toxicology (Calabrese and Blain, 2011) and that Podreka et al. (1998) may have used a dose too high to observe an effect in green turtles. Podreka et al. (1998) also did not assess other toxic endpoints that might be more sensitive. Future studies should focus on the developing immune system, other parts of the endocrine system like thyroid hormones, or the nervous system, all of which disrupted at such an early developmental stage could cause lasting and profound effects on future survival or reproduction (Willingham and Crews, 2000).

Another study used the direct laboratory exposure approach to assess the effects of crude oil, which is not a POP but is worth mentioning here. Lutcavage et al. (1995) exposed juvenile loggerhead sea turtles to a layer of crude oil on the surface of the water in their tanks and demonstrated that the oil caused skin necrosis, dermal hemorrhage, and extensive inflammatory cells infiltrating the epidermis, as well as salt gland dysfunction and hematological changes. Aside from these direct exposures, other studies use less definitive approaches.

Another possibly more acceptable approach is to expose captive sea turtles to wild-caught prey items that come from a more contaminated area. This approach has not been used for sea turtles, but was modeled by a chronic study on POPs in which captive harbor seals (*Phoca vitulina*) were fed fish caught for human consumption from two different ocean basins, one more heavily contaminated than the other (Ross et al., 1996). The seals fed the more contaminated fish exhibited suppression of several immune functions, including natural killer cell activity, lymphocyte proliferation, delayed type hypersensitivity, and antibody responses. Differences in nutritional content and other non-POP contaminant concentrations in the two batches of fish created difficulties in linking immunosuppression with solely POP exposure.

### 11.4.1 Surrogate Species Toxicity Testing

Examining the effects of POPs on surrogate species is commonly used to test the toxicity of chemicals to set exposure limits for humans, who like sea turtles cannot usually be directly exposed to known harmful chemicals. Laboratory mammals, like mice or rats, are exposed to the chemical using the route of exposure expected for humans and then examined for toxic endpoints ranging from the most severe/least sensitive (death) to the more subtle/most sensitive (e.g., neurological, immune, endocrine, or developmental toxicity). Because each species, as well as different sexes or races, have different sensitivities to a chemical, a safety factor of 100 or more is commonly used to divide the concentration just below the lowest concentration that causes the toxic effect in a surrogate species (the no adverse effect level or NOAEL) to calculate a safe margin of exposure, or a safe exposure limit, for the species of concern (human or sea turtle). This safety factor is quite arbitrary and some may argue is not safe enough. One review provides reasons against a safety factor of 100 (McConnell, 1985); the toxic concentration of dioxin on six commonly used laboratory mammal species ranged four orders of magnitude; which would require a safety factor of 1000 to account for the species difference between the guinea pig and the hamster (McConnell, 1985), and that is among only mammals. It is important to mention that turtles are considered to be highly sensitive to the effects of POPs (Bishop and Gendron, 1998; Sheehan et al., 1999). Surrogate species proposed for predicting toxic effects in sea turtles include the diamondback terrapin because they are aquatic turtles that use the same unique organ, a salt gland, for tolerating ingestion of saltwater and can be reared in captivity with success (Keller et al., 2006c). The red-eared slider, although a freshwater turtle, is another good option because of their commercial availability and their extensive use in reptilian toxicology research (Willingham and Crews, 2000). The surrogate species approach has been used intentionally only once to determine POP toxicity to sea turtles. This study exposed mice to PFOS at concentrations known to be in plasma of three species of interest (sea turtles, dolphins, and humans) (Peden-Adams et al., 2008). Male mice exhibited immune suppression (T-cell dependent IgM antibody response) at concentrations found in all three species. This lack of a margin of safety suggests that if sea turtles are as sensitive as mice, then they too might have suppressed immune systems because of their environmental exposure to PFOS. Margins of safety were assessed for five sea turtle species to investigate their risk of toxic effects based on their PFOS exposure and known effects in laboratory species on the liver, neurodevelopment, endocrine, and immune systems (Keller et al., 2012). Several margins of safety were <100, including all species for immune suppression, indicating possible risk. Aside from these studies, a wealth of literature exists on the toxic effects of POPs on red-eared sliders and other reptiles that could be used for a risk assessment for sea turtles. The collective conclusions from these studies are that reptiles are indeed quite sensitive to the effects of POPs and a multitude of toxic effects have been documented ranging from altered growth rates, sex reversal, and altered hormone concentrations (Sheehan et al., 1999; Willingham and Crews, 2000).

### 11.4.2 In Vitro Toxicity Testing

In vitro study designs are yet another laboratory approach that allow for direct exposure of the target organism's cells to a pollutant, but without using the whole animal. Because only one or a few cell or tissue types are present in these experiments, it is questionable whether the results are representative of what would occur in the whole animal. Aside from this disadvantage, in vitro approaches provide the ability to answer mechanistic questions at the tissue, cellular, or subcellular level. At least three studies have used an in vitro approach to assess POP effects on sea turtles (Ikonomopoulou et al., 2009; Keller and McClellan-Green, 2004; Keller et al., 2006b). Keller et al. (2006b) exposed loggerhead peripheral blood leukocytes in culture to PCBs and 4,4'-DDE at concentrations that included known blood concentrations in this species. The lymphocytes in that cell mixture were tested for their ability to be stimulated by mitogens and begin proliferation, which is an early step in immune reactions. They found that both POPs enhanced lymphocyte proliferation relative to the nonexposed control cells at approximately 10 ng/mL total PCBs and 1 ng/mL 4,4'-DDE as the lowest observed adverse effect level (LOAEL). These doses were strikingly similar to the blood concentrations in wild-caught chronically exposed loggerhead turtles with the maximal lymphocyte proliferation response and supported correlative findings also reported in that study. Keller and McClellan-Green (2004) assessed the effects of OCPs on cytochrome P450 aromatase (an enzyme that converts testosterone to estradiol and is thus important for reproductive development and function) in an in vitro system using a green turtle testis cell line. Atrazine is a pesticide not categorized as a POP but has endocrine-disrupting activity like many POPs. It significantly induced aromatase activity in the cells. 4,4'-DDE inhibited activity but only at the highest tested concentration (31,000 ng/mL), which was cytotoxic to the cells and more than two orders of magnitude higher than concentrations found in the fat of sea turtles. Ikonomopoulou et al. (2009) assessed the ability of 4,4'-DDT, 4,4'-DDE, and dieldrin to affect the binding affinity of sex-steroid binding proteins in the plasma of nesting green turtles to testosterone and estradiol. In the two different assays performed, dieldrin and 4,4'-DDT affected the ability of binding proteins to associate with the steroids, but sometimes in unexpected ways. These effects occurred at concentrations of 1000 ng/mL or more, which is three orders of magnitude higher than maximum concentrations measured in sea turtle blood samples (Table 11.2).

### 11.4.3  COMPARATIVE OR CORRELATIVE TOXICITY FIELD STUDIES

Alternative approaches not requiring laboratory exposures, including comparative or correlative field studies, are the most common toxicity studies for sea turtles. These approaches take advantage of exposures in the wild and either (1) compare the exposure level of two or more groups that have different health statuses, (2) compare the health of two or more groups that have different levels of exposure, or (3) use correlative relationships to compare individuals within the same population that have varying levels of exposure or health status. The number of sea turtle studies using this type of approach is still too few to make concrete conclusions about whether the health of sea turtles in the wild are being affected by POP exposure, but they provide early evidence to assess whether a cause-and-effect relationship exists (as part of a weight of evidence approach) (Fox, 1991). The first study to attempt this approach with sea turtles was performed by Aguirre et al. (1994) in which tissues from 10 green turtles with fibropapillomatosis from Hawaii were analyzed for POPs with the objective to compare them to 2 green turtles without tumors. Unfortunately, the limits of detection were too high to detect any POPs and no comparison could be made. Sixteen years later, no study has yet answered the question of whether POPs or other environmental contaminants contribute to the incidence or severity of this debilitating disease. This speculation is plausible, especially since POPs affect the immune system and could make it more difficult for turtle immune cells to detect and destroy viruses or tumor cells.

A decade later, Keller et al. (2004c, 2006b) examined approximately 48 wild-caught juvenile loggerhead sea turtles from Core Sound, North Carolina, for PCB and OCP concentrations in blood and fat biopsies, while also measuring a suite of health parameters. Most turtles appeared healthy based on external exams, plasma chemistry panels, and hematology. However, concentrations of several POPs, depending upon the compound and tissue, significantly correlated with many health parameters. Significant positive correlations were observed for total white blood cell counts, the heterophil:lymphocyte ratio (Keller et al., 2004c), and the ability of B- and T-lymphocytes to proliferate after mitogen stimulation (Keller et al., 2006b); negative correlations were observed for plasma lysozyme activity (Keller et al., 2006c). These findings suggest that POPs could be altering the number of immune cells in the peripheral blood and affecting the way those cells respond to non-self antigens. The positive correlation with the heterophil:lymphocyte ratio indicates that turtles with higher POP exposure may respond physiologically as if they are experiencing more stress, and the negative correlations with lysozyme suggests that more highly exposed turtles are producing less antibacterial enzymes, possibly leaving them more susceptible to infections. Aside from immunology, significant positive correlations were seen between POP concentrations and blood urea nitrogen (BUN), aspartate aminotransferase (AST), and plasma sodium concentrations (Keller et al., 2004c). Significant negative correlations were observed for body condition, plasma glucose, albumin, albumin:globulin ratio, alkaline phosphatase (ALP) activity, gamma glutamyl transferase (GGT) activity, magnesium, and markers of anemia (hemoglobin, hematocrit and red blood cell counts) (Keller et al., 2004c). The authors concluded from these correlations that exposure to environmental PCB and OCP concentrations may affect a wide variety of biological functions in loggerhead sea turtles, including immunity, organ health, and homeostasis of proteins, carbohydrates, and ions. What is most striking about these correlations is that they were seen in turtles that appeared mostly healthy and were foraging in a fairly pristine location, supporting prior conclusions that turtles may be exquisitely sensitive to POPs (Bishop and Gendron, 1998; Sheehan et al., 1999).

Three recent studies have repeated this type of study while focusing on sea turtle populations that are more highly exposed. Correlative studies have been performed for juvenile Kemp's ridley sea turtles from the Gulf of Mexico (Swarthout et al., 2010), adult male loggerheads captured near Cape Canaveral, Florida (Ragland et al., 2011), and green turtles from San Diego Bay, California (Komoroske et al., 2011). In all three, blood POP concentrations were similar or higher than the loggerhead study from North Carolina (Table 11.2). In approximately 20 Kemp's ridley turtles, significant negative correlations were seen between ΣDDT blood concentrations and T-lymphocyte proliferation, suggesting immunosuppression in the more exposed turtles (Swarthout et al., 2010). This correlation is in the opposite direction as was seen in loggerhead sea turtles, indicating that more research is needed to understand how POPs modulate the sea turtle immune system. Significant positive correlations were noted between BUN and Σchlordane concentrations (similar to the finding in loggerhead turtles by Keller et al. [2004c]) and between ΣDDTs and plasma potassium concentrations. Significant negative correlations were observed between ΣPCBs and creatine phosphokinase (CPK) activity, which is an unexpected direction similar to the correlations seen for ALP and GGT in loggerheads, and between dieldrin and testosterone concentrations in female turtles. Some of these correlations were similar to the previous loggerhead study, while others differed. Differences could be due to the smaller sample size of this Kemp's ridley study or caused by species or location differences. In the 19 adult male loggerheads assessed by Ragland et al. (2011), no significant correlations were observed between plasma ΣPCBs, ΣDDTs, Σchlordanes, or ΣPBDEs and any hematological or plasma chemistry value. It is difficult to determine why no correlations were observed as these turtles were the most exposed group of sea turtle ever monitored by blood sampling, but the sample size could have limited the power to detect relationships. In addition, nothing is known about normal ranges of these health indicators in adult male sea turtles, let alone how those values might change during the mating season, which is when these males were sampled. Understanding the toxic effects of POPs on male reproductive success is important especially if climate change results in warmer nest temperatures, skewing the sex ratio toward fewer males. Temperature is the most important factor for sex determination in developing sea turtles, but hormones and hormone disruptors like POPs also play a role. In 20 green turtles from San Diego Bay, γ-HCH, *trans*-chlordane, and 4,4′-DDE plasma concentrations correlated with a few health parameters (γ-HCH negatively with eosinophils and total protein and positively with ALP activity and albumin:globulin ratio; *trans*-chlordane negatively with heterophils; and 4,4'-DDE positively with hematocrit, albumin and uric acid). None of these correlations are similar in endpoint or direction as compared to the findings of Keller et al. (2004c), and could indicate differences in sensitivity among sea turtle species. Again, it is important to note that correlative relationships cannot prove cause and effect, and the lack of consistency among studies could be due to confounding factors that affect health endpoints or POP concentrations. Alternatively, some relationships could be due to statistical chance; most significant correlation coefficients were above 0.4, but the full range (0.18–0.79) included some weak correlations.

Peden-Adams et al. (2005) presented correlations between loggerhead health parameters and concentrations of plasma PFOS and other PFCs. According to them PFOS significantly and positively correlated with T-lymphocyte proliferation, AST activity, and concentrations of plasma globulin, glucose, potassium, total protein, and BUN. Likewise, PFOS significantly and negatively correlated with plasma lysozyme activity. These correlations suggest PFOS may be modulating the immune system, damaging liver or other tissues, and altering other plasma chemistries. Since these specific effects have been documented in laboratory animals exposed to PFOS, it is plausible that environmental exposure to PFOS may be causing these effects in loggerhead turtles. Future studies should focus on other sea turtles that have even higher PFOS exposure, like Kemp's ridleys and hawksbills (Table 11.2).

Another correlative study measured PCB concentrations in liver and fat of stranded loggerhead turtles in relation to body condition and bacterial infections (Orós et al., 2009). Liver concentrations were significantly higher in turtles with poorer body condition, as discussed in a previous section regarding lipid mobilization. The PCB concentrations were not significantly correlated to septicemia (defined by the authors as multiple bacterial infections found throughout the body).

Similar to the study on green turtle FP and POPs, a study addressed whether POP exposure could contribute to another sea turtle disease or syndrome. Loggerhead debilitated syndrome, identified as turtles with extreme emaciation, lethargy, and a heavy barnacle cover on skin, was on the rise in the early 2000s along the southeast coast of the United States. These debilitated turtles suffer from a range of secondary bacterial and parasitic infections (Norton et al., 2005). Turtles in this condition were measured for fat and blood POP concentrations and compared to apparently health loggerhead turtles (Keller et al., 2006a, 2007). PCB and OCP concentrations on a lipid-normalized basis in fat and on a wet mass basis in blood were approximately seven to nine times higher in these turtles than the healthy turtle group. The suggestion that POPs contributed to the onset of this syndrome was discounted after considering lipid mobilization of POPs. As debilitated turtles regained fat stores during rehabilitation, their plasma POP concentrations decreased to concentrations typically observed in healthy turtles, likely sequestering the POPs back into fat stores (Keller et al., 2007). This finding suggests that these turtles had average POP concentrations before they became ill. During recovery, plasma chemistries and hematology coincidentally improved as plasma POP concentrations decreased, so it is possible that higher POPs in blood due to mobilization from fat during weight loss contributed to the progression of this syndrome. Robust statistical analyses of these data are ongoing.

One correlative study has examined POP exposure and possible effects on hatchling green sea turtles (van de Merwe et al., 2010b). The authors found that among seven nest success parameters, only hatchling body condition (mass/SCL) was negatively related to ΣPOP egg concentrations. Hatching success, emergence success, percentage of abnormal hatchlings, as well as hatchling mass, length, and abnormality index were not correlated to POP concentrations. If the observed correlation is causative, meaning exposure to POPs in the egg are causing hatchlings to have lower body condition, then the more highly exposed hatchlings could be less fit during the swim out to sea and more susceptible to predation because of their thinner body condition. Much more research like this study is needed to address very possible lethal and sublethal effects of POPs on the sensitive early life stages of sea turtles.

## 11.4.4 Toxicological Biomarker Studies

Within only the last 5 years, studies have begun to look at traditional toxicological biomarkers in sea turtles, primarily enzymes that combat oxidative stress and a protein that indicates endocrine disruption (Casini et al., 2010; Labrada-Martagon et al., 2011; Richardson et al., 2009, 2010; Valdivia et al., 2007; Zaccaroni et al., 2010). Richardson et al. (2009) confirmed the presence and measured the activity of glutathione *S*-transferase (GST) enzyme in livers of four species (loggerhead, green, olive ridley, and hawksbill sea turtles). GST is an enzyme that is commonly upregulated in animal tissues when antioxidant defenses are needed, such as during oxidative stress when an animal is attempting to detoxify or biotransform a chemical pollutant. Hawksbill turtles had the highest GST activity, which is intriguing since this species feeds on sponges that create toxic natural chemicals and has been shown preliminarily to be the highest exposed sea turtle species to POPs (Hermanussen et al., 2008; Keller et al., 2012; Malarvannan et al., 2011). A follow-up study measured PCB concentrations in these animals and looked for correlations between the PCBs and activity of GST as well as expression of cytochrome P450 enzymes (Richardson et al., 2010). Many cytochrome P450 enzymes are commonly upregulated when an organism is exposed to certain POPs, like dioxins and PCBs, in order to attempt the first phase of metabolism and ultimately to excrete the compounds. While Richardson et al. (2010) was able to confirm the expression of CYP2K and CYP3A in three species of sea turtles, they could not detect CYP1A, which is the most commonly upregulated cytochrome P450 enzyme following exposure to PCBs and dioxins. GST activity and cytochrome P450 enzyme expression did not correlate significantly with PCB concentrations. It is important to note that these studies were meant primarily to characterize and develop the methods to measure these biomarkers, and sample sizes were small so statistical power was probably low to detect relationships.

A second set of studies to address traditional toxicological biomarkers focused on oxidative stress in green turtles from Baja California (Labrada-Martagon et al., 2011; Valdivia et al., 2007). Valdivia et al. (2007) tested methods to measure oxidative stress in livers, particularly superoxide radical production, lipid peroxidation, and activities of several antioxidant enzymes. Labrada-Martagon et al. (2011) expanded this work by comparing OCP concentrations in plasma of green turtles from two sites as well as lipid peroxidation and antioxidant enzyme activity levels in the red blood cells from the same animals. Turtles from the Punta Abreojos site, where a larger sample size of 35 was collected, showed several significant correlations. Elevated lipid peroxidation is an indicator of oxidative stress, in which reactive oxygen species are being created at destructive levels. Exposure to chemicals, as well as other environmental stressors like UV or heat, can cause oxidative stress and increase lipid peroxidation. Lipid peroxidation was negatively correlated with β-HCH concentrations, which is the opposite from the expected direction and difficult to interpret, but positively correlated with aldrin and ΣDDT concentrations. GST activity levels in the red blood cells were positively correlated to all three HCH isomers, ΣHCHs, and Σheptachlor concentrations measured in the plasma. Likewise, catalase (CAT), glutathione peroxidase (GPx), and superoxide dismutase (SOD) enzymes are commonly upregulated in tissues during oxidative stress. Few of these enzymatic activities correlated with POP levels in the green turtles, except CAT correlated positively with endrin sulfate. At the other site (Bahía Magdalena), different correlations were seen with a smaller sample size of 11 turtles. *Trans*-chlordane was significantly and positively correlated with lipid peroxidation, GST, CAT, and GPx. ΣHeptachlors were positively correlated with lipid peroxidation, and ΣDDTs were correlated with CAT. Most of these correlations are positive and suggest that the more highly exposed turtles to OCPs have greater oxidative stress and are expressing a stronger antioxidant enzyme response.

Several sea turtle studies have measured vitellogenin (VTG), a yolk protein precursor produced in the liver normally only by adult oviparous females in response to estradiol, which then circulates in the blood and is deposited into egg yolk. VTG expression is possible in juveniles and males when they are exposed to estrogen or an estrogen-mimicking compound, and this has been documented in juvenile sea turtles after exposure to estradiol (Heck et al., 1997). In red-eared sliders, 2,4'-DDT and PCBs can mimic estrogen and turn on the expression of VTG in males (Palmer and Palmer, 1995; Smelker and Valverde, 2012), verifying that VTG can be a biomarker of estrogenic exposure. Zaccaroni et al. (2010) measured this protein biomarker in plasma of 61 juvenile to adult loggerhead turtles of both sexes from Italy with the goal of using it as a biomarker of exposure to endocrine-disrupting contaminants. A percentage of loggerhead turtles considered juveniles were expressing this protein, concluding that they were either precocious or had been exposed to an estrogenic compound. However, no contaminants were measured in these animals for comparison. Similarly, a study that screened over 400 loggerheads from North Carolina also discovered precocious females as well as a male (confirmed by laparoscopy) expressing VTG (Keller et al., 2003). POP concentrations measured in some of these animals were on the higher end of loggerhead exposure (Keller, unpublished data), but a statistical analysis of the data has not yet been performed.

Casini et al. (2010) have tested a suite of nonlethal biomarkers using blood and skin biopsies from live loggerhead turtles from the Mediterranean Sea. They have shown cytochrome P450 1A induction in blood lymphocytes and skin biopsies exposed in vitro to POPs. Additionally, they are investigating correlations between POPs and plasma lipid peroxidation, genotoxic effects in whole blood using the comet assay and erythrocytic nuclear abnormalities, plasma liver enzyme activities, VTG, and butyrylcholinesterase activity.

## 11.5 CONCLUSIONS

This review leaves no doubt that sea turtles are exposed to POPs, some species and locations more so than others. While this field has been exponentially growing since 1974, large data gaps still exist in understanding POP exposure to sea turtles globally, and future studies need to improve sample sizes and use analytical approaches with appropriate sensitivity and that assure quality data. POP concentrations in sea turtles are driven by the same factors that influence POP accumulation in other species, such as lipid content, trophic status, age, and sex. Interesting spatial and temporal trends in POP concentrations are emerging in sea turtle studies, and tissues stored in formal specimen bank programs would allow for more of these types of studies.

For decades, sea turtle researchers and conservationists have suspected that environmental contaminants cause or contribute to several sea turtle health issues and diseases, ranging from green turtle fibropapillomatosis (Aguirre and Lutz, 2004; Foley et al., 2005; Herbst and Klein, 1995) to low hatch success in leatherback sea turtles (Bell et al., 2003). However, too few studies have directly examined these questions, and preliminary results are weak, inconsistent, or just correlative, emphasizing that much more research is warranted. One important and consistent confounding factor shown in at least three sea turtle studies is that body condition affects POP concentrations in ways that might confuse conclusions regarding causes of health issues (Keller et al., 2004a, 2007; Orós et al., 2009). Future studies should always consider this factor from initial sampling design to final data interpretation. Finally, the preliminary studies showing correlations between POP concentrations and health parameters or toxicological biomarkers are the first line of evidence suggesting that chronic environmental exposure to POPs could be causing sublethal effects on sea turtle populations. These studies pave the way for future more directed studies that should strive to include endpoints that are important to conservation managers, such as the vital rates of survival at each stage, growth at each stage, and reproductive output by both sexes. These kinds of data would allow managers to understand the relative risk of POP effects to other better known threats, like fisheries bycatch and harvest for meat or eggs. As a top priority, toxicity testing should be focused on sensitive life stages as well as subpopulations that are more exposed. For example, a study could assess POP effects on mortality and fitness of loggerhead hatchlings from North Carolina or the Mediterranean Sea, where they are more exposed. This is just one example of the vast research gaps available to scientists eager to approach this field.

## DISCLAIMER

Certain commercial equipment or instruments are identified in the chapter to specify adequately the experimental procedures. Such identification does not imply recommendations or endorsement by the NIST nor does it imply that the equipment or instruments are the best available for the purpose.

## REFERENCES

Aguirre, A. A., Balazs, G. H., Zimmerman, B., and Galey, F. D. 1994. Organic contaminants and trace metals in the tissues of green turtles (*Chelonia mydas*) afflicted with fibropapillomas in the Hawaiian islands. *Mar Pollut Bull* 28: 109–114.

Aguirre, A. A., Gardner, S. C., Marsh, J. C. et al. 2006. Hazards associated with the consumption of sea turtle meat and eggs: A review for health care workers and the general public. *EcoHealth* 3: 141–153.

Aguirre, A. A. and Lutz, P. L. 2004. Marine turtles as sentinels of ecosystem health: Is fibropapillomatosis an indicator? *EcoHealth* 1: 275–283.

Al-Bahry, S., Mahmoud, I., Elshafie, A. et al. 2009. Bacterial flora and antibiotic resistance from eggs of green turtles *Chelonia mydas*: An indication of polluted effluents. *Mar Pollut Bull* 58: 720–725.

Alam, S. K. and Brim, M. S. 2000. Organochlorine, PCB, PAH, and metal concentrations in eggs of loggerhead sea turtles (*Caretta caretta*) from Northwest Florida, USA. *J Environ Sci Health* B35: 705–724.

Alava, J. J., Keller, J. M., Kucklick, J. R. et al. 2006. Loggerhead sea turtle (*Caretta caretta*) egg yolk concentrations of persistent organic pollutants and lipid increase during the last stage of embryonic development. *Sci Total Environ* 367: 170–181.

Alava, J. J., Keller, J. M., Wyneken, J. et al. 2011. Geographical variation of persistent organic pollutants in eggs of threatened loggerhead sea turtles (*Caretta caretta*) from southeastern United States. *Environ Toxicol Chem* 30: 1677–1688.

Arthur, K., Limpus, C., Balazs, G. et al. 2008. The exposure of green turtles (*Chelonia mydas*) to tumour promoting compounds produced by the cyanobacterium *Lyngbya majuscula* and their potential role in the aetiology of fibropapillomatosis. *Harmful Algae* 7: 114–125.

Balmer, B. C., Schwacke, L. H., Wells, R. S. et al. 2011. Relationship between persistent organic pollutants (POPs) and ranging patterns in common bottlenose dolphins (*Tursiops truncatus*) from coastal Georgia, USA. *Sci Total Environ* 409: 2094–2101.

Basile, E. R. 2010. Persistent organic pollutants in diamondback terrapin (*Malaclemys terrapin*) tissues and eggs, and sediments in Barnegat Bay, New Jersey. PhD dissertation, Drexel University, Philadelphia, PA.

Basile, E. R., Avery, H. W., Keller, J. M., Bien, W. F., and Spotila, J. R. 2011. Diamondback terrapins as indicator species of persistent organic pollutants: Using Barnegat Bay, New Jersey as a case study. *Chemosphere* 82: 137–144.

Bazan, K. L. 2011. Persistent organic pollutants in shark blood plasma from estuaries along the southeast U.S. coast. MS thesis, College of Charleston, Charleston, SC.

Bell, B. A., Spotila, J. R., Paladino, F. V., and Reina, R. D. 2003. Low reproductive success of leatherback turtles, *Dermochelys coriacea*, is due to high embryonic mortality. *Biol Conserv* 115: 131–138.

Bishop, C. and Gendron, A. 1998. Reptiles and amphibians: Shy and sensitive vertebrates of the Great Lakes basin and St. Lawrence River basin. *Environ Monit Assess* 53: 225–244.

Bjorndal, K. 1997. Foraging ecology and nutrition of sea turtles. In *The Biology of Sea Turtles*, eds. P. Lutz and J. Musick, pp. 199–231. Boca Raton, FL: CRC Press.

Blus, L. J., Neely, B. S. J., Belisle, A. A., and Prouty, R. M. 1974. Organochlorine residues in brown pelican eggs: Relation to reproductive success. *Environ Pollut* 7: 81–91.

Braune, B. M., Mallory, M. L., Gilchrist, H. G., Letcher, R. J., and Drouillard, K. G. 2007. Levels and trends of organochlorines and brominated flame retardants in ivory gull eggs from the Canadian Arctic, 1976 to 2004. *Sci Total Environ* 378: 403–417.

Calabrese, E. J. and Blain, R. B. 2011. The hormesis database: The occurrence of hormetic dose responses in the toxicological literature. *Regul Toxicol Pharmacol* 61: 73–81.

Carlson, B. K. R. 2006. Assessment of organohalogen contaminants in benthic juvenile loggerhead sea turtles, *Caretta caretta*, from coastal North Carolina, including method development, blood compartment partitioning, and temporal trend analysis with emphasis on polybrominated diphenyl ethers. MS thesis, College of Charleston, Chaleston, SC.

Carson, R. 1962. *Silent Spring*. Boston, MA: Houghton Mifflin Co.

Casini, S., Caliani, I., Marsili, L. et al. 2010. A non-lethal multi-biomarker approach to investigate the ecotoxicological status of Mediterranean loggerhead sea turtle (*Caretta caretta*, Linneo,1758). *Comp Biochem Physiol A Mol Integr Physiol* 157: S23–S24.

Chen, D., Hale, R. C., Watts, B. D. et al. 2010. Species-specific accumulation of polybrominated diphenyl ether flame retardants in birds of prey from the Chesapeake Bay region, USA. *Environ Pollut* 158: 1883–1889.

Chevrier, J., Dewailly, E., Ayotte, P. et al. 2000. Body weight loss increases plasma and adipose concentrations of potentially toxic pollutants in obese individuals. *Int J Obes* 24: 1272–1278.

Clark, D. R., Jr. and Krynitsky, A. J. 1980. Organochlorine residues in eggs of loggerhead and green sea turtles nesting at Merritt Island, Florida—July and August 1976. *Pest Monit J* 14: 7–10.

Clark, D. R., Jr. and Krynitsky, A. J. 1985. DDE residues and artificial incubation of loggerhead sea turtle eggs. *Bull Environ Contam Toxicol* 34: 121–125.

Cobb, G. P. and Wood, P. D. 1997. PCB concentrations in eggs and chorioallantoic membranes of loggerhead sea turtles (*Caretta caretta*) from the Cape Romain National Wildlife Refuge. *Chemosphere* 34: 539–549.

Corsolini, S., Aurigi, S., and Focardi, S. 2000. Presence of polychlorobiphenyls (PCBs) and coplanar congeners in the tissues of the Mediterranean loggerhead turtle *Caretta caretta*. *Mar Pollut Bull* 40: 952–960.

Corsolini, S., Borghesi, N., Ademollo, N., and Focardi, S. 2011. Chlorinated biphenyls and pesticides in migrating and resident seabirds from East and West Antarctica. *Environ Int* 37: 1329–1335.

D'Ilio, S., Mattei, D., Blasi, M. F., Alimonti, A., and Bogialli, S. 2011. The occurrence of chemical elements and POPs in loggerhead turtles (*Caretta caretta*): An overview. *Mar Pollut Bull* 62: 1606–1615.

Davenport, J., Wrench, J., McEvoy, J., and Camacho-Ibar, V. 1990. Metal and PCB concentrations in the "Harlech" leatherback. *Mar Turtle Newslett* 48: 1–6.

Debier, C., Chalon, C., Le Boeuf, B. J. et al. 2006. Mobilization of PCBs from blubber to blood in northern elephant seals (*Mirounga angustirostris*) during the post-weaning fast. *Aquat Toxicol* 80: 149–157.

Deem, S. L., Dierenfeld, E. S., Sounguet, G. P. et al. 2006. Blood values in free-ranging nesting leatherback sea turtles (*Dermochelys coriacea*) on the coast of the Republic of Gabon. *J Zoo Wildlife Med* 37: 464–471.

Deem, S. L., Norton, T. M., Mitchell, M. et al. 2009. Comparison of blood values in foraging, nesting, and stranded loggerhead turtles (*Caretta caretta*) along the coast of Georgia, USA. *J Wildl Dis* 45: 41–56.

Foley, A. M., Schroeder, B. A., Redlow, A. E., Fick-Child, K. J., and Teas, W. G. 2005. Fibropapillomatosis in stranded green turtles (*Chelonia mydas*) from the eastern United States (1980–98): Trends and associations with environmental factors. *J Wildl Dis* 41: 29–41.

Foti, M., Giacopello, C., Bottari, T. et al. 2009. Antibiotic resistance of gram negative isolates from loggerhead sea turtles (*Caretta caretta*) in the central Mediterranean Sea. *Mar Pollut Bull* 58: 1363–1366.

Fox, G. A. 1991. Practical causal inference for ecoepidemiologists. *J Toxicol Environ Health* 33: 359–373.

Fox, G. A. 2001. Wildlife as sentinels of human health effects in the Great Lakes-St. Lawrence Basin. *Environ Health Perspect* 109: 853–861.

Gardner, S., Pier, M. D., Wesselman, R., and Juárez, J. A. 2003. Organochlorine contaminants in sea turtles from the Eastern Pacific. *Mar Pollut Bull* 46: 1082–1089.

Godley, B. J., Gaywood, M. J., Law, R. J. et al. 1998. Patterns of marine turtle mortality in British waters (1992–1996) with reference to tissue contaminant levels. *J Mar Biol Assoc UK* 78: 973–984.

Guillette, L. J., Jr., Gross, T. S., Masson, G. R. et al. 1994. Developmental abnormalities of the gonad and abnormal sex hormone concentrations in juvenile alligators from contaminated and control lakes in Florida. *Environ Health Perspect* 102: 680–688.

Guirlet, E., Das, K., and Girondot, M. 2008. Maternal transfer of trace elements in leatherback turtles (*Dermochelys coriacea*) of French Guiana. *Aquat Toxicol* 88: 267–276.

Guirlet, E., Das, K., Thomé, J.-P., and Girondot, M. 2010. Maternal transfer of chlorinated contaminants in the leatherback turtles, *Dermochelys coriacea*, nesting in French Guiana. *Chemosphere* 79: 720–726.

Hale, R. C., La Guardia, M. J., Harvey, E., Gaylor, M. O., and Mainor, T. M. 2006. Brominated flame retardant concentrations and trends in abiotic media. *Chemosphere* 64: 181–186.

Hall, A. J., Gulland, F. M., Ylitalo, G. M., Greig, D. J., and Lowenstine, L. 2008. Changes in blubber contaminant concentrations in California sea lions (*Zalophus californianus*) associated with weight loss and gain during rehabilitation. *Environ Sci Technol* 42: 4181–4187.

Hamann, M., Godfrey, M. H., Seminoff, J. A. et al. 2010. Global research priorities for sea turtles: Informing management and conservation in the 21st century. *Endanger Species Res* 11: 245–269.

Harper, F. D., Waldrop, V. C., Jeffers, R. D., Duncan, C. D., and Cobb, G. P. 1999. Organochlorine and poly-chlorinated biphenyl contamination in black neck stilt, *Himantopus mexicanus*, eggs from the Savannah and Tybee National Wildlife Refuges. *Chemosphere* 39: 151–163.

Harris, H. S., Benson, S. R., Gilardi, K. V. et al. 2011. Comparative health assessment of western Pacific leatherback turtles (*Dermochelys coriacea*) foraging off the coast of California, 2005–2007. *J Wildl Dis* 47: 321–337.

Heck, J., MacKenzie, D. S., Rostal, D., Medler, K., and Owens, D. 1997. Estrogen induction of plasma vitel-logenin in the Kemp's ridley sea turtle (*Lepidochelys kempi*). *Gen Comp Endocr* 107: 280–288.

Heinz, G. H., Percival, H. F., and Jennings, M. L. 1991. Contaminants in American alligator eggs from Lake Apopka, Lake Griffin, and Lake Okeechobee, Florida. *Environ Monit Assess* 16: 277–285.

Helsel, D. R. 2005. *Nondetects and Data Analysis: Statistics for Censored Environmental Data*. Hoboken, NJ: John Wiley & Sons.

Herbst, L. H. and Klein, P. A. 1995. Green turtle fibropapillomatosis: Challenges to assessing the role of environmental cofactors. *Environ Health Perspect* 103 (Suppl 4): 27–30.

Hermanussen, S., Limpus, C. J., Päpke, O., Connell, D. W., and Gaus, C. 2006. Foraging habitat contamination influences green sea turtle PCDD/F exposure. *Organohalogen Compounds* 68: 592–595.

Hermanussen, S., Matthews, V., Päpke, O., Limpus, C. J., and Gaus, C. 2008. Flame retardants (PBDEs) in marine turtles, dugongs and seafood from Queensland, Australia. *Mar Pollut Bull* 57: 409–418.

Hillestad, H. O., Reimold, R. J., Stickney, R. R., Windom, H. L., and Jenkins, J. 1974. Pesticides, heavy metals and radionuclide uptake in loggerhead sea turtles from South Carolina and Georgia. *Herpetol Rev* 5: 75.

Hites, R. A. 2004. Polybrominated diphenyl ethers in the environment and in people: A meta-analysis of con-centrations. *Environ Sci Technol* 38: 945–956.

Houde, M., Pacepavicius, G., Wells, R. S. et al. 2006. Polychlorinated biphenyls and hydroxylated polychlori-nated biphenyls in plasma of bottlenose dolphins (*Tursiops truncatus*) from the western Atlantic and the Gulf of Mexico. *Environ Sci Technol* 40: 5860–5866.

Houde, M., Wells, R. S., Fair, P. A. et al. 2005. Polyfluoroalkyl compounds in free-ranging bottlenose dol-phins (*Tursiops truncatus*) from the Gulf of Mexico and the Atlantic Ocean. *Environ Sci Technol* 39: 6591–6598.

Howard, P. H. and Muir, D. C. G. 2010. Identifying new persistent and bioaccumulative organics among chemicals in commerce. *Environ Sci Technol* 44: 2277–2285.

Ikonomopoulou, M. P., Olszowy, H., Hodge, M., and Bradley, A. J. 2009. The effect of organochlorines and heavy metals on sex steroid-binding proteins in vitro in the plasma of nesting green turtles, *Chelonia mydas*. *J Comp Physiol B* 179: 653–662.

Innis, C., Tlusty, M., Perkins, C. et al. 2008. Trace metal and organochlorine pesticide concentrations in cold-stunned juvenile Kemp's ridley turtles (*Lepidochelys kempii*) from Cape Cod, Massachusetts. *Chelonian Conserv Biol* 7: 230–239.

Keller, J. M., Alava, J. J., Aleksa, K., Young, B., and Kucklick, J. R. 2005a. Spatial trends of polybromi-nated diphenyl ethers (PBDEs) in loggerhead sea turtle eggs and plasma. *Organohalogen Compd* 67: 610–611.

Keller, J. M., Kannan, K., Taniyasu, S. et al. 2005b. Perfluorinated compounds in the plasma of loggerhead and Kemp's ridley sea turtles from the southeastern coast of the United States. *Environ Sci Technol* 39: 9101–9108.

Keller, J. M., Kucklick, J. R., Harms, C. A., and McClellan-Green, P. D. 2004a. Organochlorine contaminants in sea turtles: Correlations between whole blood and fat. *Environ Toxicol Chem* 23: 726–738.

Keller, J. M., Kucklick, J. R., Harms, C. A. et al. 2006a. Organic contaminant concentrations are higher in debilitated loggerhead turtles compared to apparently healthy turtles. Paper presented at the *26th Annual Symposium on Sea Turtle Biology and Conservation*, Crete, Greece, pp. 63.

Keller, J. M., Kucklick, J. R., and McClellan-Green, P. D. 2004b. Organochlorine contaminants in loggerhead sea turtle blood: Extraction techniques and distribution among plasma and red blood cells. *Arch Environ Contam Toxicol* 46: 254–264.

Keller, J. M., Kucklick, J. R., Stamper, M. A., Harms, C. A., and McClellan-Green, P. D. 2004c. Associations between organochlorine contaminant concentrations and clinical health parameters in loggerhead sea turtles from North Carolina, USA. *Environ Health Perspect* 112: 1074–1079.

Keller, J. M. and McClellan-Green, P. 2004. Effects of organochlorine compounds on cytochrome P450 aroma-tase activity in an immortal sea turtle cell line. *Mar Environ Res* 58: 347–351.

Keller, J. M., McClellan-Green, P. D., Kucklick, J. R., Keil, D. E., and Peden-Adams, M. M. 2006b. Effects of organochlorine contaminants on loggerhead sea turtle immunity: Comparison of a correlative field study and in vitro exposure experiments. *Environ Health Perspect* 114: 70–76.

Keller, J. M., Ngai, L., Braun McNeill, J. et al. 2012. Perfluoroalkyl contaminants in plasma of five sea turtle species: Comparisons in concentration and potential health risks. *Environ Toxicol Chem* 31: 1223–1230.

Keller, J. M., Owens, D. W., Kucklick, J. R. et al. 2003. Abnormal vitellogenin production in vivo and alterations of aromatase activity in vitro due to organochlorine contaminants in sea turtles. Paper presented at the *23rd Annual Symposium on Sea Turtle Biology and Conservation*, Kuala Lumpur, Malaysia, pp. 253–254.

Keller, J. M., Peden-Adams, M. M., and Aguirre, A. A. 2006c. Immunotoxicology and implications for reptilian health. In *Toxicology of Reptiles*, eds. S. C. Gardner and E. Oberdorster, pp. 199–240. Boca Raton, FL: CRC Press.

Keller, J. M., Swarthout, R. F., Carlson, B. K. et al. 2009. Comparison of five extraction methods for measuring PCBs, PBDEs, organochlorine pesticides, and lipid content in serum. *Anal Bioanal Chem* 393: 747–760.

Keller, J. M., Thorvalson, K., Sheridan, T. et al. 2007. Organic contaminant concentrations change in debilitated loggerhead turtle plasma during recovery in rehabilitation. Paper presented at the *27th Annual Symposium on Sea Turtle Biology and Conservation*, Myrtle Beach, SC, pp. 20–21.

Komoroske, L. M., Lewison, R. L., Seminoff, J. A., Deheyn, D. D., and Dutton, P. H. 2011. Pollutants and the health of green sea turtles resident to an urbanized estuary in San Diego, CA. *Chemosphere* 84: 544–552.

Labrada-Martagon, V., Rodriguez, P. A. T., Mendez-Rodriguez, L. C., and Zenteno-Savin, T. 2011. Oxidative stress indicators and chemical contaminants in East Pacific green turtles (*Chelonia mydas*) inhabiting two foraging coastal lagoons in the Baja California peninsula. *Comp Biochem Phys C* 154: 65–75.

Lake, J. L., Haebler, R., McKinney, R., Lake, C. A., and Sadove, S. S. 1994. PCBs and other chlorinated organic contaminants in tissues of juvenile Kemp's ridley turtles (*Lepidochelys kempi*). *Mar Environ Res* 38: 313–327.

Lazar, B., Maslov, L., Romanić, S. H. et al. 2011. Accumulation of organochlorine contaminants in loggerhead sea turtles, *Caretta caretta*, from the eastern Adriatic Sea. *Chemosphere* 82: 121–129.

Lutcavage, M. E., Lutz, P. L., Bossart, G. D., and Hudson, D. M. 1995. Physiologic and clinicopathologic effects of crude oil on loggerhead sea turtles. *Arch Environ Contam Toxicol* 28: 417–422.

Malarvannan, G., Takahashi, S., Isobe, T. et al. 2011. Levels and distribution of polybrominated diphenyl ethers and organochlorine compounds in sea turtles from Japan. *Mar Pollut Bull* 63: 541–547.

Maruya, K. A. and Lee, R. F. 1998. Aroclor 1268 and toxaphene in fish from a southeastern U.S. estuary. *Environ Sci Technol* 32: 1069–1075.

McConnell. 1985. Comparative toxicity of PCBs and related compounds in various species of animals. *Environ Health Perspect* 60: 29–33.

Mckenzie, C., Godley, B. J., Furness, R. W., and Wells, D. E. 1999. Concentrations and patterns of organochlorine contaminants in marine turtles from Mediterranean and Atlantic waters. *Mar Environ Res* 47: 117–135.

McKim, J. M., Jr. and Johnson, K. L. 1983. Polychlorinated biphenyls and p,p′-DDE in loggerhead and green postyearling Atlantic sea turtles. *Bull Environ Contam Toxicol* 31: 53–60.

van de Merwe, J. P., Hodge, M., Olszowy, H. A. et al. 2009a. Chemical contamination of green turtle (*Chelonia mydas*) eggs in peninsular Malaysia: Implications for conservation and public health. *Environ Health Perspect* 117: 1397–1401.

van de Merwe, J. P., Hodge, M., Olszowy, H. A., Whittier, J. M., and Lee, S. Y. 2010a. Using blood samples to estimate persistent organic pollutants and metals in green sea turtles (*Chelonia mydas*). *Mar Pollut Bull* 60: 579–588.

van de Merwe, J. P., Hodge, M., Whittier, J. M., Ibrahim, K., and Lee, S. Y. 2010b. Persistent organic pollutants in the green sea turtle *Chelonia mydas*: Nesting population variation, maternal transfer, and effects on development. *Mar Ecol Progress Ser* 403: 269–278.

van de Merwe, J. P., Hodge, M., Whittier, J. M., and Lee, S. Y. 2009b. Analysing persistent organic pollutants in eggs, blood and tissue of the green sea turtle (*Chelonia mydas*) using gas chromatography with tandem mass spectrometry (GC-MS/MS). *Anal Bioanal Chem* 393: 1719–1731.

Miao, X.-S., Balazs, G. H., Murakawa, S. K. K., and Li, Q. X. 2001. Congener-specific profile and toxicity assessment of PCBs in green turtles (*Chelonia mydas*) from the Hawaiian Islands. *Sci Total Environ* 281: 247–253.

Monagas, P., Orós, J., Araña, J., and González-Díaz, O. M. 2008. Organochlorine pesticide levels in loggerhead turtles (*Caretta caretta*) stranded in the Canary Islands, Spain. *Mar Pollut Bull* 56: 1949–1956.

Moore, C. J. 2008. Synthetic polymers in the marine environment: A rapidly increasing, long-term threat. *Environ Res* 108: 131–139.

Moss, S., Keller, J. M., Richards, S., and Wilson, T. P. 2009. Concentrations of persistent organic pollutants in plasma from two species of turtle from the Tennessee River Gorge. *Chemosphere* 76: 194–204.

Norton, T. M., Keller, J. M., Peden-Adams, M. M. et al. 2005. Debilitated loggerhead turtle (*Caretta caretta*) syndrome along the southeastern U.S. coast: Incidence, pathogenesis and monitoring. Paper presented at the *25th Symposium on Sea Turtle Biology and Conservation*, Savannah, GA, pp. 36.

O'Connell, S. G., Arendt, M., Segars, A. et al. 2010. Temporal and spatial trends of perfluorinated compounds in juvenile loggerhead sea turtles (*Caretta caretta*) along the east coast of the United States. *Environ Sci Technol* 44: 5202–5209.

Orós, J., González-Díaz, O. M., and Monagas, P. 2009. High levels of polychlorinated biphenyls in tissues of Atlantic turtles stranded in the Canary Islands, Spain. *Chemosphere* 74: 473–478.

Palmer, B. C. and Palmer, S. K. 1995. Vitellogenin induction by xenobiotic estrogens in the red-eared turtle and African clawed frog. *Environ Health Perspect* 103 (Suppl 4): 19–25.

Peden-Adams, M. M., Kannan, K., Lee, A. M. et al. 2005. Perfluorinated contaminants measured in sea turtle blood correlate to modulations in plasma chemistry values and immune function measurements. Paper presented at the *25th Symposium on Sea Turtle Biology and Conservation*, Savannah, GA, pp. 37.

Peden-Adams, M. M., Keller, J. M., EuDaly, J. G. et al. 2008. Suppression of humoral immunity in mice following exposure to perfluorooctane sulfonate. *Toxicol Sci* 104: 144–154.

Perrault, J., Wyneken, J., Thompson, L. J., Johnson, C., and Miller, D. L. 2011. Why are hatching and emergence success low? Mercury and selenium concentrations in nesting leatherback sea turtles (*Dermochelys coriacea*) and their young in Florida. *Mar Pollut Bull* 62: 1671–1682.

Perugini, M., Giammarino, A., Olivieri, V. et al. 2006. Polychlorinated biphenyls and organochlorine pesticide levels in tissues of *Caretta caretta* from the Adriatic Sea. *Dis Aquat Organ* 71: 155–161.

Pierce, R. H. and Henry, M. S. 2008. Harmful algal toxins of the Florida red tide (*Karenia brevis*): Natural chemical stressors in South Florida coastal ecosystems. *Ecotoxicology* 17: 623–631.

Podreka, S., Georges, A., Maher, B., and Limpus, C. J. 1998. The environmental contaminant DDE fails to influence the outcome of sexual differentiation in the marine turtle *Chelonia mydas*. *Environ Health Perspect* 106: 185–188.

Pugh, R. S. and Becker, P. R. 2001. Sea turtle contaminants: A review with annotated bibliography. NISTIR 6700. National Institute of Standards and Technology, Charleston, SC.

Pugh, R. S., Becker, P. R., Porter, B. J. et al. 2008. Design and applications of the National Institute of Standards and Technology's (NIST's) environmental specimen banking programs. *Cell Preserv Technol* 6: 59–72.

Ragland, J. M., Arendt, M. D., Kucklick, J. R., and Keller, J. M. 2011. Persistent organic pollutants in blood plasma of satellite-tracked adult male loggerhead sea turtles (*Caretta caretta*). *Environ Toxicol Chem* 30: 1549–1556.

Richardson, K. L., Gold-Bouchot, G., and Schlenk, D. 2009. The characterization of cytosolic glutathione transferase from four species of sea turtles: Loggerhead (*Caretta caretta*), green (*Chelonia mydas*), olive ridley (*Lepidochelys olivacea*), and hawksbill (*Eretmochelys imbricata*). *Comp Biochem Physiol C Toxicol Pharmacol* 150: 279–284.

Richardson, K. L., Lopez Castro, M., Gardner, S. C., and Schlenk, D. 2010. Polychlorinated biphenyls and biotransformation enzymes in three species of sea turtles from the Baja California Peninsula of Mexico. *Arch Environ Contam Toxicol* 58: 183–193.

Ross, P. 2004. Response to Beckman et al. *Mar Pollut Bull* 48: 806–807.

Ross, P. S., De Swart, R. L., Addison, R. et al. 1996. Contaminant-induced immunotoxicity in harbour seals: Wildlife at risk? *Toxicology* 112: 157–169.

Rybitski, M. J., Hale, R. C., and Musick, J. A. 1995. Distribution of organochlorine pollutants in Atlantic sea turtles. *Copeia* 1995: 379–390.

Serra-Valente, G. N., Robertson, K. M., LeRoux, R. A. et al. 2010. Got samples? Introducing the Southwest Fisheries Science Center Marine Turtle Molecular Research Collection. Paper presented at the *30th Annual Symposium on Sea Turtle Biology and Conservation*, San Diego, CA.

Sheehan, D. M., Willingham, E., Gaylor, D., Bergernon, J. M., and Crews, D. 1999. No threshold dose for estradiol-induced sex reversal of turtle embryos: How little is too much? *Environ Health Perspect* 107: 155–159.

Smelker, K. S. and Valverde, R. A. 2012. Vitellogenin induction by PCBs in the turtle *Trachemys scripta*. Paper presented at the *Annual Meeting of the Society for Integrative and Comparative Biology*, Charleston, SC, pp. 335.

de Solla, S. R., Fernie, K. J., Letcher, R. J. et al. 2007. Snapping turtles (*Chelydra serpentina*) as bioindicators in Canadian areas of concern in the Great Lakes Basin. 1. Polybrominated diphenyl ethers, polychlorinated biphenyls, and organochlorine pesticides in eggs. *Environ Sci Technol* 41: 7252–7259.

Stewart, K. R., Keller, J. M., Templeton, R., Kucklick, J. R., and Johnson, C. 2011. Monitoring persistent organic pollutants in leatherback turtles (*Dermochelys coriacea*) confirms maternal transfer. *Mar Pollut Bull* 62: 1396–1409.

Storelli, M. M., Barone, G., and Marcotrigiano, G. O. 2007. Polychlorinated biphenyls and other chlorinated organic contaminants in the tissues of Mediterranean loggerhead turtle *Caretta caretta*. *Sci Total Environ* 373: 456–463.

Storelli, M. M. and Marcotrigiano, G. O. 2000. Chlorobiphenyls, HCB, and organochlorine pesticides in some tissues of *Caretta caretta* (Linnaeus) specimens beached along the Adriatic Sea, Italy. *Bull Environ Contam Toxicol* 64: 481–488.

Stow, C. A., Lamon, E. C., Qian, S. S., and Schrank, C. S. 2004. Will Lake Michigan lake trout meet the Great Lakes strategy 2002 PCB reduction goal? *Environ Sci Technol* 38: 359–363.

Swarthout, R. F., Keller, J. M., Peden-Adams, M. et al. 2010. Organohalogen contaminants in blood of Kemp's ridley (*Lepidochelys kempii*) and green sea turtles (*Chelonia mydas*) from the Gulf of Mexico. *Chemosphere* 78: 731–741.

Takahashi, E., Arthur, K., and Shaw, G. 2008. Occurrence of okadaic acid in the feeding grounds of dugongs (*Dugong dugon*) and green turtles (*Chelonia mydas*) in Moreton Bay, Australia. *Harmful Algae* 7: 430–437.

Thompson, N. P., Rankin, P. W., and Johnston, D. W. 1974. Polychlorinated biphenyls and p,p' DDE in green turtle eggs from Ascension Island, South Atlantic Ocean. *Bull Environ Contam Toxicol* 11: 399–406.

Tuerk, K. J. S., Kucklick, J. R., Becker, P. R., Stapleton, H. M., and Baker, J. E. 2005. Persistent organic pollutants in two dolphin species with focus on toxaphene and polybrominated diphenyl ethers. *Environ Sci Technol* 39: 692–698.

Valdivia, P. A., Zenteno-Savín, T., Gardner, S. C., and Alonso Aguirre, A. 2007. Basic oxidative stress metabolites in eastern Pacific green turtles (*Chelonia mydas agassizii*). *Comp Biochem Physiol C Toxicol Pharmacol* 146: 111–117.

Vetter, W., Scholz, E., Gaus, C., Müller, J., and Haynes, D. 2001. Anthropogenic and natural organohalogen compounds in blubber of dolphins and dugongs (*Dugong dugon*) from northeastern Australia. *Arch Environ Contam Toxicol* 41: 221–231.

Walsh, C. J., Leggett, S. R., Carter, B. J., and Colle, C. 2010. Effects of brevetoxin exposure on the immune system of loggerhead sea turtles. *Aquat Toxicol* 97: 293–303.

Willingham, E. 2004. Endocrine-disrupting compounds and mixtures: Unexpected dose-response. *Arch Environ Contam Toxicol* 46: 265–269.

Willingham, E. and Crews, D. 2000. The red-eared slider turtle: An animal model for the study of low doses and mixtures. *Am Zool* 40: 421–428.

Yordy, J. E., Wells, R. S., Balmer, B. C. et al. 2010a. Life history as a source of variation for persistent organic pollutant (POP) patterns in a community of common bottlenose dolphins (*Tursiops truncatus*) resident to Sarasota Bay, FL. *Sci Total Environ* 408: 2163–2172.

Yordy, J. E., Wells, R. S., Balmer, B. C. et al. 2010b. Partitioning of persistent organic pollutants between blubber and blood of wild bottlenose dolphins: Implications for biomonitoring and health. *Environ Sci Technol* 44: 4789–4795.

Zaccaroni, A., Zucchini, M., Segatta, L. et al. 2010. Vitellogenin (VTG) conservation in sea turtles: Anti-VTG antibody in *Chelonia mydas* versus *Caretta caretta*. *Physiol Biochem Zool* 83: 191–195.

# 12 Fisheries Bycatch of Marine Turtles

## Lessons Learned from Decades of Research and Conservation

*Rebecca Lewison, Bryan Wallace, Joanna Alfaro-Shigueto,
Jeffrey C. Mangel, Sara M. Maxwell, and Elliott L. Hazen*

## CONTENTS

## 12.1 BIOLOGY AND SIGNIFICANCE OF SEA TURTLE BYCATCH

Sea turtles spend the majority of their lives in coastal or pelagic waters, making in-water sources of mortality critical to population viability. Sea turtles have been negatively impacted by a number of human-mediated factors including oil spills (Antonio et al., 2011), contaminants (van de Merwe et al., 2010; Swarthout et al., 2010; Komoroske et al., 2011; Stewart et al., 2011), and other types of marine pollution, namely debris ingestion and entanglement (Lazar and Gracan, 2011; do Sul et al., 2011). Coastal and in-water shoreline development also have been shown to degrade ocean habitat, which can negatively affect resident turtles (Harewood and Horrocks, 2008; Pike, 2008). While all of these factors likely have some negative effect on sea turtle populations, the human activity that has the largest impact on sea turtles is fisheries bycatch (Lewison et al., 2004a; Wallace et al., 2011). Although directed take of turtles is one form of fisheries impact, and in some regions opportunistic take of captured turtles is still prevalent (Alfaro-Shigueto et al., 2011), turtles are generally an unwanted and unwelcome byproduct of fishing activities. Because fishing is an important source of protein and livelihood for millions of people worldwide, incidental capture, or bycatch, of sea turtles continues to be the most pressing human impact on sea turtle populations globally. In this chapter, we review the current state of knowledge about global marine turtle bycatch, including how characteristics of sea turtle biology and fishing practices interact to result in bycatch, assessments of population-level impacts of turtle bycatch, descriptions of where and how turtle bycatch occurs across distinct fisheries sectors, a summary of techniques and approaches to bycatch reduction, and new ways forward for bycatch research and management.

## 12.2   UNDERSTANDING HOW SEA TURTLE BYCATCH HAPPENS

Fisheries bycatch occurs at the intersection of sea turtle ecology, behavior, distribution, and fisheries activity (Figure 12.1). Bycatch is a result of an individual's *vulnerability* to capture, which is influenced by behavior, ecological, and intrinsic life history attributes, as well as the *susceptibility* of an individual due to spatial or temporal overlap with fishing gear. The likelihood of capture, or overall sea turtle catchability, is a function of a combination of these two elements.

Vulnerability reflects a combination of ecological characteristics including foraging behavior (e.g., likelihood to chase baited hooks), migratory routes, and distributions at depth (e.g., proportion of shallow vs. deep dives), as well as aggregations of individuals in time and space for breeding and/or feeding. As air-breathers, turtles must return to the surface periodically to replenish oxygen stores. This physiological constraint on diving behavior exposes them to particular bycatch threats, in terms of the depth of the fishing gear as well as in the amount of time that fishing gear is left

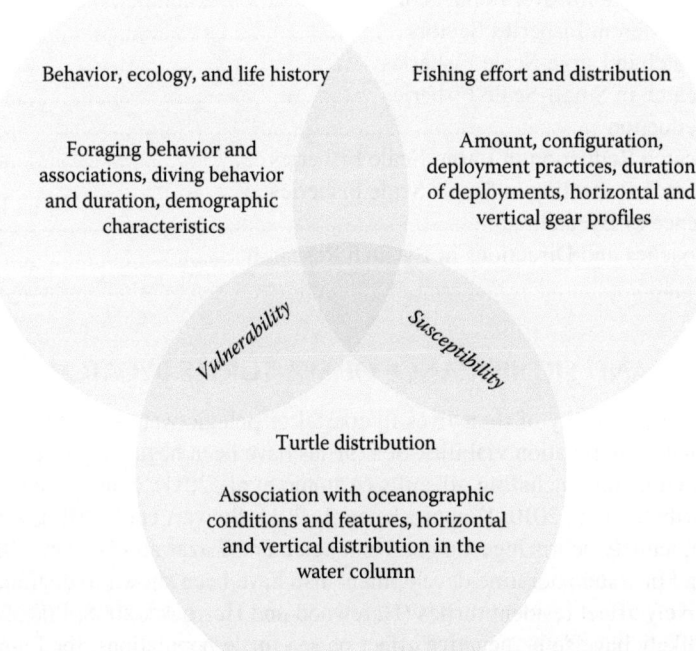

FIGURE 12.1   A conceptual model of the different factors that drive sea turtle bycatch. Vulnerability is primarily driven by ecological and life history attributes and characteristics that govern behavior and distribution. Susceptibility, in contrast, is driven largely by the horizontal and vertical overlap of fishing vessels and sea turtles, and represents the elements in the system that can be managed.

in the water (Poiner and Harris, 1996). The different ecological functions of diving (e.g., foraging, thermoregulation, predator evasion) as well as diel dive patterns, i.e., daily and seasonal patterns of dive frequency and duration, also influence the likelihood of encountering fishing gear (Howell et al., 2010). Satellite telemetry research has demonstrated that sea turtles occupy particular ranges of water temperatures to optimize the efficiency of physiological processes and/or to take advantage of resource availability related to these temperatures (Polovina et al., 2000, 2003, 2004; Wingfield et al., 2011). Foraging sea turtles, including loggerheads and leatherbacks, also exhibit a close association to more productive waters, where they aggregate and forage in thermohaline fronts, convergence zones, upwellings, or mesoscale eddies, where primary productivity is high and turtle prey tend to be aggregated (Kobayashi et al., 2008; McCarthy et al., 2010; Benson et al., 2011; Ferreira et al., 2011; Shillinger et al., 2011; Bailey et al., 2012).

Another dimension of bycatch vulnerability is the demographic sensitivity of sea turtles to bycatch mortality. Because sea turtles have delayed sexual maturity (9–30 years, see Chapter 5), sea turtle populations are most sensitive to impacts that kill individuals from older age classes (Crouse et al., 1987; Heppell, 1998). These individuals have higher per capita reproductive values, where reproductive value (RV) is the number of offspring a member of a given age group can produce between any specific age and their death; RV tends to be highest at the onset of reproductive maturity (Fisher, 1930). Elasticity analyses provide additional insight into the relative contribution of individual age classes to overall population growth rate, or lambda, taking into account the duration of those age classes (Heppell et al., 2000a,b). Elasticity analyses across turtle species have demonstrated that population growth rates depend strongly on the survival of turtles nearing and reaching sexual maturity (i.e., large benthic juveniles, subadults, and adults; Heppell, 1998; Heppell et al., 2000a; NMFS, 2001), which are age classes commonly caught as bycatch (Lewison and Crowder, 2007).

Susceptibility refers to the overlap in space and time of fishing effort with turtle habitats and is an essential element of the bycatch equation. In contrast to the ecological and life history traits that drive vulnerability, susceptibility is driven by factors that can be managed. The level of overlap is largely because fishing fleets, like turtles, favor areas of high productivity. Distributions of many target species, e.g., swordfish, have been shown to closely associate with convergence areas (Hazin and Erzini, 2008). Transition zones and fronts in the Azores, North Pacific, the Costa Rica Dome, off the western coast of Baja California Peninsula, and along the Gulf Stream in the Western Atlantic are examples of areas where sea turtles aggregate, where fishing pressure is intense, and consequently are areas where the probability of bycatch is likely to be high (Polovina et al., 2000, 2003, 2004; Hawkes et al., 2007; Howell et al., 2008; Shillinger et al., 2008; Wingfield et al., 2011; Ferreira et al., 2011). Similar aggregations can be found in continental shelf zones (Casale et al., 2012). As with turtle distribution, fishing activity shifts considerably within and among years, often in response to the same oceanographic features that attract turtles. Because sea turtle movements can be described in three dimensions—i.e., horizontal distribution as well as vertical dive-depth—bycatch susceptibility is also driven by spatial location, depth and vertical profile of gear. The species-specific seasonal and regional dive behaviors that sea turtles exhibit may also account for differences in susceptibility among sea turtle species within and among regions (Godley et al., 2008). Fishing activity can overlap foraging grounds, migration corridors, or areas adjacent to nesting grounds, each with different population-level ramifications.

## 12.2.1 Differences among Fishing Gears

Sea turtle bycatch occurs in a diversity of fishing gears throughout turtles' broad geographic ranges in the ocean. Vessels from large-scale and small-scale fisheries (SSFs) using trawls (Lewison et al., 2003), gillnets (Murray, 2009), seine nets, pound nets (Gilman et al., 2010), longlines (Witzell, 1999; Watson et al., 2005; Casale, 2010), and many other gears all incur sea turtle bycatch. Sea turtle bycatch has been most widely documented in four broad categories of fishing gear, although

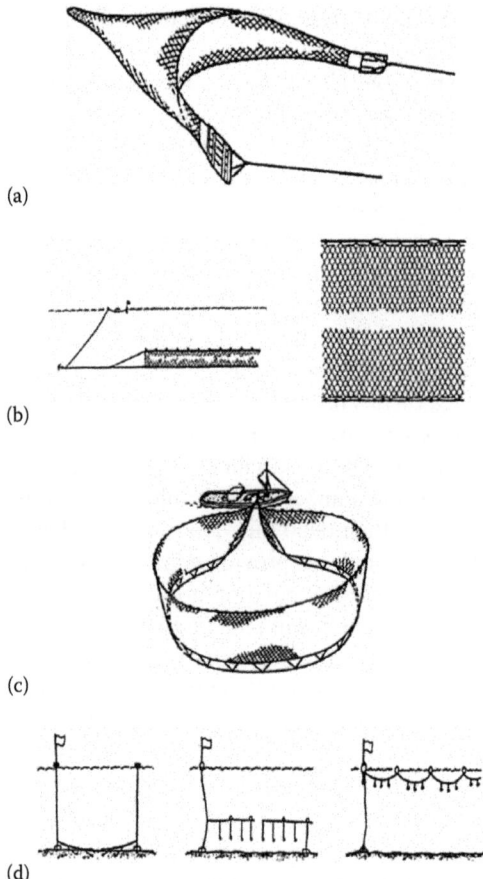

**FIGURE 12.2** Illustrations of the general fishing gear in which sea turtle bycatch has been most widely documented. (From FAO Fisheries Technical Paper. No. 222. Revision 1, ftp:/ftp.fao.org/docrep/fao/008/t0367t/t0367t04.pdf). The four gear categories that are shown (a) trawl, (b) nets, (c) purse, and (d) longline represent very broad categories of fishing gear and within each category is a wide diversity of fishing gear types. For more information on fishing gears, go to http:/www.fao.org/fishery/topic/1617/en.

within each of these categories there is a wide diversity of gear types (see http:/www.fao.org/fishery/topic/1617/en for more detailed gear descriptions, see Figure 12.2 for illustrations). These include the following:

*Trawls*: Trawl vessels typically pull one or more large funnel-shaped nets through the water where the target species are captured in a bag at the end of the net, termed a cod-end. Trawls can be deployed at different depths depending on the target species. For sea turtles, coastal or shallow bottom trawls used to capture shrimp and other coastal flatfish can result in high bycatch (Finkbeiner et al., 2011). Once sea turtles enter the cod-end, they are unable to escape and will die if the duration of a trawl operation exceeds the physiological capacity for a sea turtle to remain submerged without surfacing to breathe. There also may be sublethal effects to sea turtles from trawl capture and recapture (Caillouet et al., 1996).

*Nets*: Nets are another broad gear category of fishing gears that are vertically oriented in the water column either tethered to the substrate or left to drift. Gillnets are one common type of net gear, comprising panels of nets that are used to form walls of nets of varying lengths. They catch a wide assortment of species based on the mesh sizes. The primary threat to sea turtles is entanglement in the net mesh, which can result in injury or death from drowning. Another type of net gear, pound nets, are stationary nets usually supported by poles pounded into the substrate. Pound nets corral migrating fish through a series of funnels into a holding pen. Turtles may become entangled in a leader net, which is set perpendicular to shore to divert fish to the mouth of the pound net. In some regions, the holding pen is open (i.e., has no roof) and in others, it is enclosed. If turtles enter a pen that is enclosed, they are unable to reach the surface to breathe and drown.

*Purse seines*: Purse seines consist of a wall of netting that is set in a circle around a school of targeted fish. The bottom of the net is pulled shut, or pursed, to form a bag, and the catch is hauled on board the ship. Purse seiners often set nets around natural floating debris and fish-aggregating devices (FADs, Fonteneau et al., 2000) because fish species aggregate at these objects. Smaller size classes of turtles can become entangled in the FADs' tethered ropes, buoys, or floats. Existing data suggest that FAD setting has resulted in an increased bycatch of sea turtles (Gilman and Lundin, 2009).

*Longlines*: Longlines are a series of hundreds or thousands of hooks that hang off a mainline of variable length set at discrete depths to target fish species, often tuna and swordfish. Much of the bycatch of sea turtles occurs when the lines are set shallowly (0–100 m), a depth range where all sea turtle species dive extensively. Sea turtles can be hooked while trying to ingest bait from baited hooks or become entangled when their flippers encounter the hooked branch or mainlines. Bottom set longlines can also lead to bycatch (Jribi et al., 2008).

### 12.2.2 BYCATCH RATES AND MORTALITY AMONG GEARS

Bycatch rates vary widely within and among gears, fleets, and fishing areas. Bycatch rates vary substantially, in part, because of different gear configurations and fishing practices but also because of turtle and fishing vessel movement. Lewison and Crowder (2007) compared published bycatch rates for a single gear type, pelagic longlines, and found that even among four different longline fleets deploying tuna (deep) sets in the Pacific, maximum bycatch rates of leatherbacks for each fleet ranged from 30% to 60% of the highest overall rate see references in Lewison and Crowder 2007. In a more detailed comparison, Wallace et al., 2010a synthesized reported sea turtle bycatch records and fishing effort in gillnets, longlines, and trawls by major fishing regions (see Table 2 in Wallace et al. [2010a]). This comprehensive data compilation confirmed what previous studies have asserted, i.e., bycatch rates are highly variable within and among gears and regions.

Sea turtle mortality is not synonymous with bycatch across gear types. Whereas in some gear, turtles die as a result of becoming captured or entangled, in other gear types, a turtle can be released within little or no injury depending on the type of gear and the type of interaction. If mortality is not directly observed during gear retrieval, it may occur after the turtle is released. Although post-capture mortality estimates are essential to understanding the impact bycatch may be having on sea turtle populations, it is a major knowledge gap. While it is difficult to estimate and compare post-capture mortality rates across gears and among sea turtles species, the existing estimates suggest that post-capture mortality varies substantially among gear types and sea turtle species, reflecting likely variation among sea turtle populations, oceanographic conditions in which bycatch occurs, and gear-related differences.

In general, existing mortality estimates suggest that sea turtle mortality is higher in net and trawl gear than in longlines. Henwood and Stuntz (1987) published some of the earliest estimates of mortality in shrimp trawl vessels in the southeastern U.S. waters, estimating overall mortality rate for the Gulf of Mexico is 29% (34%, 22%, 38% for the eastern, central, western Gulf, respectively). For the U.S. Atlantic coast, these authors estimated a mortality rate of 21% reflecting the shorter average duration of trawl tows on this coast. Sea turtle mortality in trawls in the Eastern Tropical Pacific Ocean was estimated at 37% without the use of bycatch reduction devices (Arauz et al., 1998). A study examining artisanal drift gillnets in the Caribbean found that 27% of leatherbacks caught were hauled on board dead (Lum, 2006), similar to estimates for gillnets in the Mediterranean of 20%–30% of loggerheads caught (Gerosa and Casale, 1999). Forty per cent of turtles caught in drift gillnets in the northeast Atlantic were recorded as dead, and this was considered to be largely a function of soak time (Murray, 2009). Direct mortality estimates from longlines vary from 8% to over 30%, and are related to factors such as hook type, set depth, and whether the turtle was hooked in the mouth, stomach, or externally (Chaloupka et al., 2004; Casale et al., 2008). Tagging studies of turtles caught in longlines indicated that mortality was greater for individuals that swallowed or ingested the hook (referred to as deep-hooked individuals), and some data suggest that, with proper hook removal, lightly hooked turtles may not experience a reduction in annual survival (Sasso and Epperly, 2007).

## 12.3 INTERPRETING THE BYCATCH LANDSCAPE

### 12.3.1 Characterizing Bycatch

There are two basic data types needed to characterize sea turtle bycatch. The first essential data type is direct reports of observed bycatch. Bycatch data are typically collected in two ways: (1) by data recorded by trained observers on fishing vessels (termed observer data) by resource agencies and independent scientists, or (2) data collected during dockside interviews and surveys. Information on bycatch is usually reported in the form of a bycatch rate, or bycatch per unit effort (BPUE). Bycatch rates are generally calculated as the number of turtles captured relative to the associated amount of fishing effort observed. Comparisons among bycatch records are hindered substantially by the diversity in fishing effort metrics used to report bycatch (see Table 1 in Wallace et al. [2010a]). This lack of conformity can be overcome (see Wallace et al., 2010a), but it presents a substantial challenge to comparing or assessing bycatch effects among fisheries, gear types, or ocean regions.

Observer bycatch data have been shown to provide high-resolution information by providing a more accurate and precise estimate of the number of turtles caught as well as locations where bycatch occurred. Although observer data are an essential ingredient to characterizing and quantifying bycatch, the precision of the data is influenced by the amount of fishing effort upon which the data are based (Tuck, 2011). Sims et al. (2008) found that high or low bycatch rates of sea turtles in gillnets in the northwest Atlantic Ocean tended to occur where relatively low fishing effort were observed, illustrating potential biases in bycatch rates based on relatively low levels of observed fishing effort. This finding was confirmed with a similar assessment using global-scale bycatch data across geographic regions and different gear categories (Wallace et al., 2010a).

Observer data are collected primarily in large-scale fisheries; however, observers typically monitor small proportions of a fishing fleet's total effort (typically <5% with some exceptions; Finkbeiner et al., 2011). Fisher interviews have been used effectively in many SSFs, which are typically data-poor, to capture bycatch occurrences and spatial extents of fishing activities (Moore et al., 2010; Alfaro-Shigueto et al., 2007, 2011). The costs of implementing observer programs in developing countries are often prohibitive, especially given that SSFs consist of large numbers of boats distributed diffusely (as opposed to in centralized ports) along the coasts (see Section 12.4.2). In the absence of empirical datasets, researchers have increasingly relied on the knowledge of local fishermen to characterize bycatch in this fishing sector (Moore et al., 2010). Despite the limitations of social survey data (Kennelly, 1999; Huntington, 2000; Gilchrist et al., 2005), structured interviews have provided useful information about marine mammal and sea turtle bycatch in both small- and large-scale fisheries when observer data were limited or not feasible to collect (Moore et al., 2010; Alfaro-Shigueto et al., 2011).

The second kind of information that is needed to evaluate fisheries impacts is the amount of fishing gear deployed, or fishing effort. Data on the intensity and spatial locations of fishing effort are needed to quantify and monitor bycatch risk for sea turtles and other nontarget species (Bellman et al., 2005). The most commonly reported measure of fisheries production is the amount of catch (Maunder and Punt, 2004). This is due in part to relative ease of data collection; catch data can be collected at ports or landing sites. While catch data provides important information on the quantity (i.e., number or biomass) of target species harvested, it does not necessarily provide information on the expended effort, which is likely to be a better indicator of bycatch of nontarget species like sea turtles (Caillouet et al., 1996).

Although these two data types form the foundation of bycatch assessments for sea turtles, the vulnerability/susceptibility framework in Figure 12.1 demonstrates the complex suite of factors that impact bycatch occurrence for sea turtle species. Collecting data on these other axes of influence continues to be a priority to provide an accurate characterization of species and location-specific bycatch likelihoods.

### 12.3.2  MAPPING THE BYCATCH LANDSCAPE

Maps provide a visual representation of processes and patterns, highlighting relationships between map objects, themes, and regions (Nelson and Boots, 2008). Creating maps of the bycatch landscape is challenging given data gaps and nonstandard data reporting. However, mapping bycatch data as well as fishing effort distribution provides an important tool to analyze spatial patterns, reveal areas where bycatch data are lacking, and to identify fishing areas where multinational efforts or regional oversight are needed (*sensu* Small, 2005).

Maps of fishing effort have been hindered by lack of data reporting; many agencies report catch, not effort (see Section 12.2.1). There have been a number of attempts to directly map fishing effort in the context of bycatch in large ocean regions (Tuck et al., 2003; Lewison et al., 2004b; Stewart et al., 2010; Waugh et al., 2011). Even with inherent imprecision in these mapping exercises, these studies can serve as the foundation for spatially explicit bycatch risk assessments (Waugh et al., 2011). Fishing effort mapping exercises also provide gross estimates of total fishing effort (Lewison et al., 2004b; Stewart et al., 2010), which by itself can help frame the potential risk bycatch poses to sea turtle populations in particular fishing areas.

Given the strong effect that spatial distribution of fishing vessels and sea turtles has on bycatch, spatial analyses are an important part of bycatch characterization. Analyzing spatial patterning and extent of sea turtle bycatch at the scale of a fleet can be used to inform management and develop strategies designed to reduce bycatch (Gardner et al., 2008; Sims et al., 2008; Lewison et al., 2009; Kot et al., 2010). However, sea turtle bycatch also occurs at spatial scales far larger than that of a single fleet. At larger scales, mapping bycatch becomes hampered by the nonstandard bycatch metrics used within and among fishing areas and fleets. One way to overcome the obstacle of bycatch metric variability is to use expert opinion to create relative ranks for bycatch records. Using a comprehensive sea turtle bycatch database from 1990 to 2008 (see Wallace et al., 2010 for citations), Lewison et al. (in review) used independent bycatch experts to rank bycatch records on a standard scale (from low to high severity). The resultant map demonstrates the diverse bycatch landscape (Figure 12.3). Although some of the variability among data records may reflect effective bycatch mitigation strategies employed by some but not all fleets, the differences among records demonstrate that even within gear categories, sea turtle bycatch is highly variable. Furthermore, spatial variation in bycatch in different gear types could reflect regional differences in fishing gears used, distributions of observer coverage, or reporting biases.

Some of the newer and more innovative bycatch reduction measures acknowledge this variability and are based on more complex maps that capture the dynamic nature of sea turtle catchability. Howell et al. (2008) developed a software product, TurtleWatch (http:/www.pifsc.noaa.gov/eod/turtlewatch.php), which is based on extensive research that has identified an association between the geographic distributions of loggerheads and sea surface temperature (SST) isotherms, and data showed that seasonal habitat use tends to track these temperature boundaries (Polovina et al., 2001, 2004; Kobayashi et al., 2008). TurtleWatch maps changing conditions in SSTs and the associated likelihood of loggerhead turtle presence in fishing areas, and provides this information to help fishermen avoid turtle bycatch. When its recommendations are heeded, TurtleWatch has been effective at reducing loggerhead bycatch in the Hawaii-based pelagic longline fishery (Howell et al., 2008).

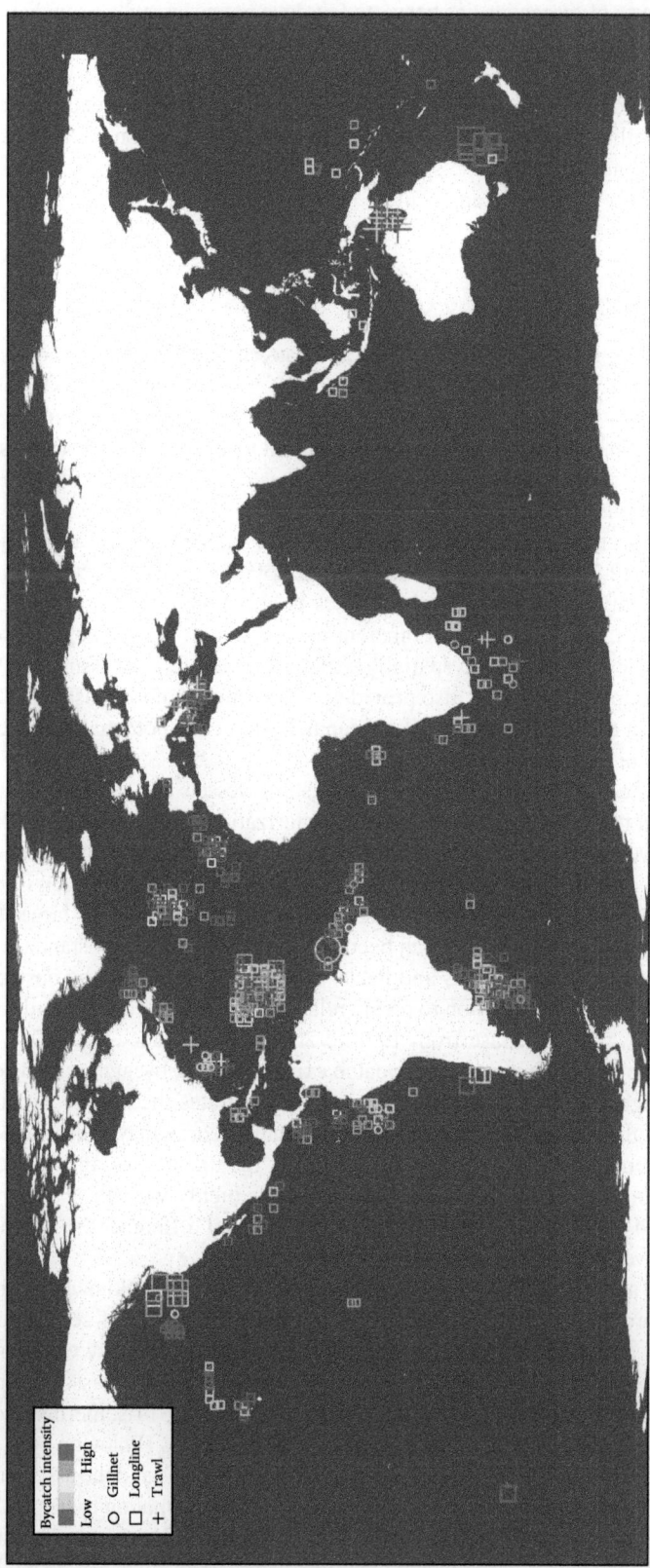

**FIGURE 12.3** Map of global documented bycatch records of sea turtles from 1990 to 2008, across gillnets (o), longlines (□), and trawls (+) where bycatch intensity of each record is ranked from high (red) to low (blue). Symbol size corresponds to the amount of fishing effort for each data record. (From Lewison et al., in review.)

## 12.4    ASSESSING POPULATION-LEVEL IMPACTS OF BYCATCH

To assess population-level impacts of bycatch within and across fishing gears, several constituent pieces of information are necessary, and must be considered together. As discussed in Section 12.2.1, some measure of the frequency or magnitude of bycatch is needed, which is usually in the form of BPUE. However, the fact that bycatch occurred does not *de facto* demonstrate mortality has occurred, i.e., bycatch rates only indicate the number of turtles caught. Specific information on mortality rates (i.e., the proportion of turtles caught that die as a result of bycatch interactions, see Section 12.1) is required to estimate the number of turtle deaths due to bycatch, which is more directly useful to population projections. Another critical element needed to assess the population-level impacts of bycatch is the relative "importance" of turtles taken as bycatch, namely the RV of the turtles caught. Using RVs as a scalar for absolute bycatch numbers can allow for comparisons of relative impacts of different fisheries on sea turtle populations (Wallace et al., 2008). Finally, but perhaps most importantly, information on population viability is necessary as a foundation for interpreting the aforementioned variables in a population context. Specifically, estimates of population abundance and trends, as well as other characteristics that might make a population more or less vulnerable to bycatch (and other threats)—e.g., geographic distributions, feeding ecology, life history traits—provide a "common denominator" for comparisons of different bycatch impacts across sea turtle populations. Effective assessments of population-level impacts of bycatch for purposes of identifying conservation priorities in different gears, regions, or for different populations require a combination of all of these pieces of information—bycatch rates, mortality rates, RVs, and population characteristics.

A far more detailed understanding of the affected populations is required to identify the drivers of observed population trends, create conservation targets, and to prioritize limited conservation resources to reduce bycatch and leverage the greatest recovery outcomes (Wallace et al., 2010a). This type of threat assessment must be conducted at biologically appropriate scales to permit population-relevant evaluations and subsequent management responses. To this end, Wallace et al. (2010b) established regional management units (RMUs) for sea turtles worldwide to provide an appropriate biogeographic and population framework for such assessments. Within this RMU context, expert evaluation of available data was used to assess the conservation status of all marine turtle RMUs by evaluating population viability and relative impacts of various threats (Wallace et al., 2011). In this assessment, bycatch was identified as the highest threat for sea turtles globally (Table 1 in Wallace et al. [2011]) and was determined to be a moderate or high threat for more than three-fourths of all sea turtle RMUs globally. Furthermore, this evaluation demonstrated that different gear types were driving the RMU-specific bycatch threats across regions and species (Wallace et al., 2011).

A more detailed analysis of the Wallace et al. (2011) results reveals differences in the relative impacts of bycatch among species and gear types (Table 12.1). Loggerheads, olive ridleys, and leatherbacks had the highest average bycatch scores, with 80% of loggerhead RMUs, 75% of olive ridley RMUs, and 50% of leatherback RMUs scored as high bycatch RMUs. The average bycatch scores for other species were moderate, and no other species had more than 30% of its RMUs scored as high bycatch. Gillnets were identified as a gear of primary concern most frequently for leatherbacks, green turtles, and hawksbills, while longlines were identified for loggerheads, and trawls for olive ridleys. These interspecific differences in which gears have highest impacts might be explained by variations in life histories and habitat use, as well as in different fishing gears operating in individual RMUs (see Section 12.1). Looking across all RMUs for all species, gillnets were identified as the primary bycatch gear for 18 RMUs, followed by trawls (13 RMUs), longlines (10 RMUs), and others (2 RMUs), suggesting that nets may be the gear category of highest conservation concern for sea turtles globally. Although these results point to some general patterns in sea turtle bycatch at broad species-level and global scales, bycatch reduction strategies are not "one size fits all." The strategies must take into account biological (e.g., RMU, nesting stock), geographical factors (e.g., proximity to nesting beaches, high-density feeding areas), and fisheries sectors (large vs. small scale) to ensure long-term population recoveries (see Section 12.5).

TABLE 12.1

**Bycatch Scores by Species Based on RMU-Specific Assessments by Wallace et al. (2011)**

| Species | No. RMUs (No. Scored)[a] | No. RMUs Scored "High Bycatch" | Average Bycatch Score (Low to High, 1–3) | Gear Types Identified for "High Bycatch" RMUs (No. RMUs in Which Each Gear was Mentioned) |
|---|---|---|---|---|
| Loggerheads | 10 (10) | 8 | 2.80 | Longlines (7), trawls (5), gillnets (5), IUU[b] (1) |
| Olive ridleys | 8 (8) | 6 | 2.63 | Trawls (6), gillnets (3), longlines (2) |
| Leatherbacks | 7 (6) | 3 | 2.50 | Gillnets (3), longlines (1) |
| Kemp's ridleys | 1 (1) | 0 | 2.00 | NA |
| Flatbacks | 2 (2) | 0 | 2.00 | NA |
| Green turtles | 17 (16) | 5 | 1.97 | Gillnets (5), trawls (2) |
| Hawksbills | 13 (13) | 2 | 1.69 | Gillnets (2), bomb fishing (1) |

Relative impacts of bycatch (and other threats) were assessed for each RMU as "low," "medium," or "high" (scores of 1, 2, or 3) based on expert evaluation of available data and the likelihood of an RMU going extinct in the future if current levels of bycatch continue unabated (see Wallace et al. [2011] for details).

[a] Two RMUs received scores of "data deficient" for bycatch impacts.

[b] Illegal, unreported, and unregulated fisheries.

## 12.5 BYCATCH IN DIFFERENT FISHERIES SECTORS

Like fishing gear types, variability among fishing sectors also can underlie differences in sea turtle bycatch rates and associated mortality. Although differentiation between large-scale and small-scale fishing sectors can be imperfect and imprecise (Ruttan et al., 2000), the generalizable characteristics of the two sectors correspond to recognizable patterns in sea turtle bycatch.

### 12.5.1 BYCATCH IN LARGE-SCALE FISHERIES

Large-scale fisheries are commercial operations commonly involving at-sea processing or extensive storage, enabling fishing activities to continue without the need to offload landings frequently at port. Information on bycatch from large-scale fisheries varies greatly from region to region, as well as fishery to fishery, with some fisheries within a jurisdiction collecting high-resolution bycatch data and others collecting virtually none. In the most data-rich fisheries, dedicated observers record information such as gear configuration, catch and bycatch coordinates, species composition, gear set or soak times, as well as date and volume of catch. Large-scale, industrial fisheries are a recognized source of bycatch and mortality for sea turtles, as well as other marine megafauna, including seabirds, sharks, and marine mammals (Brothers, 1991; Baum et al., 2003; Lewison et al., 2004b). Indeed, a number of these fisheries have been implicated in contributing to dramatic declines in sea turtle populations (Chan and Liew, 1996; Spotila et al., 2000; Fujiwara and Caswell, 2001). Because of the high amounts of fishing effort that large-scale fleets exert, even relatively low bycatch rates from vessels in this sector can have high cumulative effects on sea turtle populations due to the sheer magnitude of total interactions across all fishing operations, e.g., in a single year, pelagic longline fleets from 40 nations set an estimated 1.4 billion hooks in the water, which is equivalent to ca. 3.8 million hooks every day (Lewison et al., 2004b). The cumulative nature of the effects from large-scale fisheries as well as the management infrastructure and oversight has led to a high level of scrutiny and action to reduce sea turtle bycatch in many regions in this sector (Gilman et al., 2011).

## 12.5.2   BYCATCH IN SMALL-SCALE FISHERIES

In recent years, attention has shifted to bycatch in SSFs, which has been identified as an equally important source of sea turtle mortality (Lewison and Crowder, 2007; Soykan et al., 2008; Wallace et al., 2010a). SSFs, also often called "artisanal" fisheries, use a wide range of fishing methods including set and drift nets, pound nets, trawls and seines, surface, midwater or demersal gear, longlines, and traps. Most attempts to define SSF focus on fleet characteristics such as their general reliance upon manual labor, relatively small vessel or engine size and storage capacities, dispersed vessel ownership, and relatively coastal fishing locations. Despite restricted local scales of individual SSFs, the aggregated SSF sector has economic importance globally, and serves as a source of food and employment for *ca.* 1 billion people (Béné, 2006). Small-scale fleets are particularly common in developing countries where they often form the mainstay of the fisheries sector (Béné, 2006). What distinguishes SSFs from the industrial fisheries described earlier is the low degree of capital investment, smaller vessel size, limited mechanization, and the decentralization of effort and resources.

Despite being defined as small-scale, SSF fleet sizes can be vast, with many thousands of vessels operating in a country or region (Alfaro-Shigueto et al., 2010; Stewart et al., 2010). These fleets are often spread along long stretches of coastline, operating out of remote coastal communities. The fleets themselves are often dynamic, switching between gear types throughout the year to target seasonally abundant species. These communities are often economically and politically marginalized, which typically means that few bycatch reduction measures and limited enforcement of existing bycatch mitigation measures exist in SSFs. Furthermore, bycatch monitoring and management are often hard to assess due to the nature of SSFs themselves, i.e., diffuse effort, remote landing sites, and political and economic marginalization (Chuenpagdee et al., 2006).

Research in recent years has shown that SSF fleets can have high, possibly unsustainable, levels of sea turtle bycatch (Godley et al., 1998; Lewison and Crowder, 2007; Peckham et al., 2007; Gilman et al., 2009; Alfaro-Shigueto et al., 2011; Casale, 2011). Sea turtle bycatch by SSFs has been reported for many nations and regions around the globe, including Trinidad and Tobago (Lum, 2006), Brazil (Gallo et al., 2006), Tunisia (Echwikhi et al., 2010), the Mediterranean (Godley et al., 1998; Casale, 2011), Peru (Alfaro-Shigueto et al., 2011), and parts of Africa and Asia (Chaloupka et al., 2004; Moore et al., 2010), and likely includes all species of sea turtles (Chaloupka et al., 2004; Limpus, 2007; Gilman et al., 2009; Wallace et al., 2010a; Casale, 2011). It is also clear that bycatch occurs in many of the different gear types employed by SSFs, including longlines, demersal gillnets, driftnets, pound nets, and trawls (Arauz et al., 1998; Peckham et al., 2007; Gilman et al., 2009; Alfaro-Shigueto et al., 2011; Casale, 2011, also see refs in Lewison and Crowder, 2007; Wallace et al., 2010a). Small-scale gillnet fisheries in particular are a source of growing concern, given their high observed bycatch and mortality rates (Peckham et al., 2007; Gilman et al., 2009; Alfaro-Shigueto et al., 2011), and a number of studies have highlighted assessments of sea turtle bycatch in SSF as an urgent research priority (Salas et al.; 2007; Gilman et al., 2009; Casale, 2011; Wallace et al., 2011).

Estimates from SSFs suggest that the amount of sea turtle bycatch in SSF may be comparable to bycatch levels in industrial fleets (Lewison and Crowder, 2007). In a study of sea turtle bycatch by SSFs operating in Baja California, Mexico, Peckham et al. (2007) estimated an annual bycatch of ca. 1000 loggerheads and suggested that this value is similar in magnitude to the Pacific-wide industrial longline fleet. Similarly, Alfaro-Shigueto et al. (2011) estimated that ca. 5900 sea turtles are taken annually in Peruvian SSFs operating out of just three ports, but suggested that the true total likely numbers in the tens of thousands of sea turtles caught each year if cumulative impacts of the numerous and widespread Peruvian SSFs is considered. However, many bycatch studies in SSFs are based on a relatively low amount of observed effort, which typically correspond to low-confidence bycatch estimates, and/or could reflect a reporting bias, wherein researchers are more likely to report high bycatch rates than low or absent bycatch rates (see Sims et al., 2008; Wallace et al., 2010a). Nonetheless, even in cases where bycatch rates may be low, the vast number of boats that operate in SSFs can lead to large numbers of total interactions (Peckham et al., 2007; Alfaro-Shigueto et al., 2011; Casale, 2011). Moreover, SSFs sometimes have high observed mortality rates (Peckham et al., 2008; Echwikhi et al., 2010) or some of the incidentally caught turtles, while captured alive, may be used for human consumption (Peckham et al., 2008; Alfaro-Shigueto et al., 2011). For all of these reasons, we echo previous calls for enhanced and urgent efforts directed toward observation, monitoring, management, and reduction of sea turtle bycatch in SSF.

## 12.6  BYCATCH REDUCTION

Despite many remaining challenges, there have been major improvements and developments in sea turtle bycatch reduction in the past decade. The best strategies to reduce bycatch integrate sea turtle ecology with fishing patterns or practices to minimize overlap and entanglement risk with fishing gear, with minimal impact on target species yield (Gilman et al., 2011). Modifications to gear, bait types, set locations, and timing and duration of sets have all been explored as possible bycatch reduction measures (Gilman et al., 2007). Some bycatch reduction measures have been shown to be relevant and effective across both large-scale and SSFs. However, given fundamental differences in the management framework and infrastructure between the two fishery sectors, reduction efforts vary according to fishery-specific circumstances.

### 12.6.1  Bycatch Reduction in Large-Scale Fisheries

Direct gear and fishery modifications such as changes to bait type, modifying gear to make it less visible or attractive to sea turtles, making gear less likely to cause direct mortality, or changing the way that gear is deployed are all examples of bycatch mitigation techniques that have been employed to reduce sea turtle bycatch in trawl, passive net, and longline large-scale fisheries. Here, we outline the range of techniques, highlighting the bycatch reduction achievements within each gear. However, considerable work remains to be done to further reduce bycatch across gears, and bycatch reduction strategies that have been successful in one fishery or one region may not work well in a similar fishery in a different part of the world. Mitigation techniques need to be tested and tailored to the specific fishery in which they are being utilized (Cox et al., 2007; Read, 2007).

Trawls became the focus of sea turtle bycatch reduction efforts in the 1980s, a focus that continues today. Shrimp fisheries in the Gulf of Mexico have been historically one of the largest sources of sea turtle bycatch in U.S. waters, as bycaught turtles would be held underwater and drowned over the duration of a multi-hour tow (Finkbeiner et al., 2011). Bycatch of large juvenile and adult logger-heads, in particular, has been identified as the greatest source of mortality for the southeastern U.S. loggerhead turtle population (Finkbeiner et al., 2011), and stage-based population models showed that reduction of capture in trawl nets was necessary for population recovery (Crouse et al., 1987). To decrease bycatch of loggerheads and other species, particularly Kemp's ridley and leatherback sea turtles, turtle excluder devices (TEDs) were developed to allow turtles to escape from trawl nets. TEDs usually consist of metal bars inserted into the neck of a trawl; when a turtle encounters the bars, it is forced out of an opening in the bottom of the net while shrimp continue through the bars into the bag end of the trawl. In 1991, year-round TED regulations were put into effect in U.S. waters. Subsequent to that action, there was a significant decrease in stranding rates of both species, particularly Kemp's ridleys (Crowder et al., 1995, Lewison et al., 2003, Heppell et al., 2005). The escape opening in TEDs was first increased in some areas and fishing seasons in 1994 to accommodate for the larger-size turtles like leatherbacks and adult loggerheads (Federal Register, 1994). A second increase was mandated in 2002 when predicted bycatch reductions were not realized (Epperly and Teas, 2002). A number of studies have shown that the effectiveness of TEDs is more complex than simply mandating their use in key areas and during key times of the year; shifting effort of trawlers, proper use of installed TEDs, limited requirement of TEDs with enlarged escape openings, and, particularly, compliance of TED use are critical for the recovery of turtle populations (Epperly and Teas, 2002; Lewison et al., 2003; Cox et al., 2007; Finkbeiner et al., 2011). However, with full compliance and proper implementation, TEDs can dramatically decrease sea turtle bycatch and mortality, as shown in a multidecade synthesis of sea turtle bycatch in U.S. fisheries (Finkbeiner et al., 2011). Indeed, TEDs are used effectively in other fisheries, most notably Australia's northern prawn fishery and Queensland's east coast trawl fishery (Brewer et al., 2006). Following a World Trade Organization ruling that TED requirements were a permissible requirement for shrimp imported into the United States, TEDs have been implemented in a number of countries, although compliance may be poorly enforced (Alio et al., 2010; Sala et al., 2011).

Sea turtle bycatch in longline fisheries has received substantial scrutiny in several regions, and as a result, a number of effective sea turtle bycatch reduction strategies have been implemented in longlines. Gear depth, set and soak time, and hook type have all shown to be important elements of gear configuration that affect bycatch rates. For example, shallow longlines set less than 50 m deep have higher bycatch rates than deeper sets (Gilman et al., 2006; Beverly et al., 2009); sea turtles are caught more often on hooks closer than 30 m from floats than those further away (Seco Pon et al., 2007); leatherback turtles are caught more often during nighttime longline sets compared to the day; and increased soak times result in higher catches of loggerhead turtles in the U.S. Atlantic longline fishery (Gilman et al., 2006). These differences in bycatch rates among gear deployment practices and gear configurations have driven many of the effective bycatch reduction strategies in longline vessels, which include changing the time of day of sets or setting at depths in the water column less frequently used by sea turtles, changing to bait types less likely to be consumed by turtles, changing hook type, size, and shape to decrease ingestion of the hook, and spatial and temporal management of fishing effort (Polovina et al., 2003; Gilman et al., 2006; Howell et al., 2008; Lucchetti and Sala, 2010; Piovano et al., 2012). Switching from J to circle hooks that tend to decrease the severity of hooking, as well as switching to larger hooks that are more difficult for turtles to ingest, have shown promise in several fisheries, particularly because these fixes have resulted in little impact on catch rates (Watson et al., 2005; Gilman et al., 2006; Read, 2007; Pacheco et al., 2011, Swimmer et al., 2011). Changes in hook and bait type have been successfully regulated or applied voluntarily in a number of fisheries around the world including the Mediterranean, U.S. Atlantic, Pacific and Gulf of Mexico longline fisheries, and in the Western and Central Pacific longline fisheries (Gilman et al., 2006, 2010; Lucchetti and Sala, 2010; Curran and Bigelow, 2011). In one of the most successful examples, the Hawaii-based longline swordfish fishery switched from J hooks with squid bait to large circle hooks with fish bait, which resulted in a significant decline in loggerhead (83%) and leatherback bycatch (90%), and a concomitant increase in swordfish catch (16%) (Gilman et al., 2007).

While fewer direct gear modifications have been made to large-scale gillnet fisheries, set modifications, as well as spatial and temporal restrictions, have been employed to reduce interactions between turtles and gillnets (Murray et al., 2009). In Japan and in the U.S., there also has been some attention focused on developing pound net escape devices (PEDs) to reduce sea turtle bycatch and mortality (Ishihara et al., 2011). For gillnet fisheries in the U.S. Atlantic, latitude, temperature, and net mesh size all were significant predictors of bycatch rates where larger mesh, southern latitudes, and warmer temperatures result in higher catches of loggerhead turtles (Murray, 2009). In addition, increasing the depth of gillnet sets from the surface would decrease the likelihood of capture as the fishing gear would reside outside of the typical range of turtle vertical habitat (Lucchetti and Sala, 2010). Reduction in mesh size and increased rigidity of the net leaders has helped reduce the impact of pound nets to turtles (Gilman et al., 2010).

Time-area closures have been another successful technique for reducing bycatch (Dunn et al., 2011). These may be seasonal or permanent closures based on known areas of high bycatch, or can be more precautionary and dynamic in nature, based on the probability of turtles being present. For example, since 2000, the area off South Padre Island, Texas, has been closed to shrimp trawling from July 15 to December 1 in order to protect nesting Kemp's ridleys (Lewison et al., 2003). A large-scale, annual 3 month closure of the drift gillnet fishery in California and Oregon has resulted in zero leatherback bycatch in this fishery (Moore et al., 2009). Furthermore, along the U.S. west coast, the drift gillnet fishery may be closed during El Niño events in order to reduce bycatch of loggerhead turtles that move further north on the warm El Niño currents from Mexico into U.S. waters (Federal Register, 2007). Off the coast of central West Africa, Mayumba National Park in Gabon, and the adjacent Conkouati National Park in the Republic of Congo were created as permanent no-take areas to protect leatherback and olive ridley sea turtles from bycatch, with additional seasonal closures in adjacent areas during peak nesting seasons (Witt et al., 2008; Maxwell et al., 2011).

Time-area closures may also be enacted for other purposes, such as when a fishery hits a "bycatch quota" (i.e., an area is closed to fishing when a certain level of bycatch has been reached), or to protect target catch during key times of the year, but these closures may simultaneously protect bycatch species. An example of a bycatch quota forcing a fishery closure is the Hawaii-based shallow-set longline fishery, which is limited to 16 interactions with leatherbacks and 17 interactions with loggerheads in a calendar year; if more interactions occur, the fishery is closed, as it was in 2011 for reaching the leatherback take limit (Federal Register, 2011). Lewison et al. (2003) described closures in Texas to protect shrimp stocks that also resulted in a reduction in sea turtle strandings, even if that was not the intention of the closure. Closures, while effective in many areas, may have unintended negative consequences on bycatch, however, by shifting fishing effort to new areas, potentially ones with higher concentrations of turtles or other species vulnerable to bycatch (Abbott and Haynie, 2012). Being able to anticipate and adapt to fisher's responses to closures is key for successful, long-term bycatch reduction.

### 12.6.2  Bycatch Reduction in Small-Scale Fisheries

The characteristics that define SSF (e.g., large, dispersed fisheries, economically marginalized, little regulation) present significant challenges to implementation of bycatch mitigation measures. As many of these fisheries operate in impoverished communities, the costs associated with new technologies can be prohibitive. Moreover, mitigation products used in large-scale may not be regularly available to fishers in this sector, requiring the creation of new markets. Given the geographic dispersion of SSF fleets, proper implementation of mitigation and monitoring to ensure compliance can also be problematic. Initiatives such as changes to fishing methods can address some of these challenges (Eckert et al., 2008; Peckham et al., 2009). Fishery certifications or eco-labeling could also provide incentives for SSF to implement bycatch mitigation measures, if obstacles to compliance monitoring could be overcome. Small-scale fishers can directly benefit from sea turtle bycatch reduction; fewer turtles can mean less gear damage, bait loss, and time savings for fishers. There is a clear need to find the opportunities and mutual benefits for fishers to engage in potential bycatch solutions.

Mitigation measures that have been tested in small-scale longline fisheries include the use of circle hooks to reduce hooking rates and severity coupled with dehookers to facilitate hook removal (see references in Read, 2007). Mitigation measures tested in gillnet fisheries include a number of gear changes that are designed to reduce turtle attraction and incidence of entanglement (Gilman et al., 2010). These include net illumination (Wang et al., 2010), eliminating floats from main lines (Peckham et al., 2009; Gilman et al., 2010), alteration to net tie-downs and net height, and removal or reduction of floats (Gilman, 2009). The use of at-sea advisory programs, in which fishers share bycatch information with land-based biologists via radio to facilitate bycatch avoidance as well as safe handling and release of bycaught turtles, have also been used as a way to help fishermen select their fishing areas and minimize the likelihood of bycatch (Alfaro-Shigueto et al., 2012). Developing alternative food sources or conservation incentives has also been proposed as a means to reduce bycatch in SSF (Peckham et al, 2007; Ferraro and Gjertsen, 2009). Switching from higher to lower bycatch gear capable of targeting the same target species is another promising mitigation technique that has been shown to be an effective bycatch reduction strategy in SSFs (Chuenpagdee et al., 2003; Peckham et al., 2009).

## 12.7   SOCIAL SCIENCE OF BYCATCH

The ability to address the global issue of sea turtle bycatch has been challenged by a number of different factors, some of which relate more directly to facets of social science than biological science, e.g., social capital, the level of ecological awareness, governance structure of management and fisher communities, existence of policies to regulate and mitigate bycatch (Lewison et al., 2011). The multidisciplinary nature of these challenges, coupled with the need to work across local to ocean-wide scales, provides support for the assertion that effective bycatch reduction requires an integrated approach involving researchers from multiple disciplines working with partners from local communities up through international governance regimes (Figure 12.4). Although this level of cross-disciplinary integration has not been achieved, ongoing efforts within these various fields are redefining the ability to effectively address the issue of bycatch in small and large fisheries. In some developing Central and Latin American countries, community involvement, coordination, and collaborations have been established to address bycatch in SSFs (Hall et al., 2007; Peckham et al., 2007; Peckham and Maldonado-Diaz, 2012), yielding promising results. Combining education, outreach, and cooperative fisheries management, these efforts provide a clear model of participatory bycatch assessments and ultimately bycatch mitigation (Hall et al., 2007).

Engaging fishermen, fishing cooperatives, and the communities in which they live may be essential to reducing sea turtle bycatch (Gutierrez et al., 2011). Work by Jenkins (2010) clearly demonstrates that for two large-scale fisheries, U.S. trawl and purse seine, the most effective and successful bycatch reduction technology and strategies were invented and designed by fishers. This evaluation of successful bycatch reduction accounts for both sea turtle bycatch reduction achieved as well as fisher adoption and compliance, two essential elements for meaningful and long-term sea

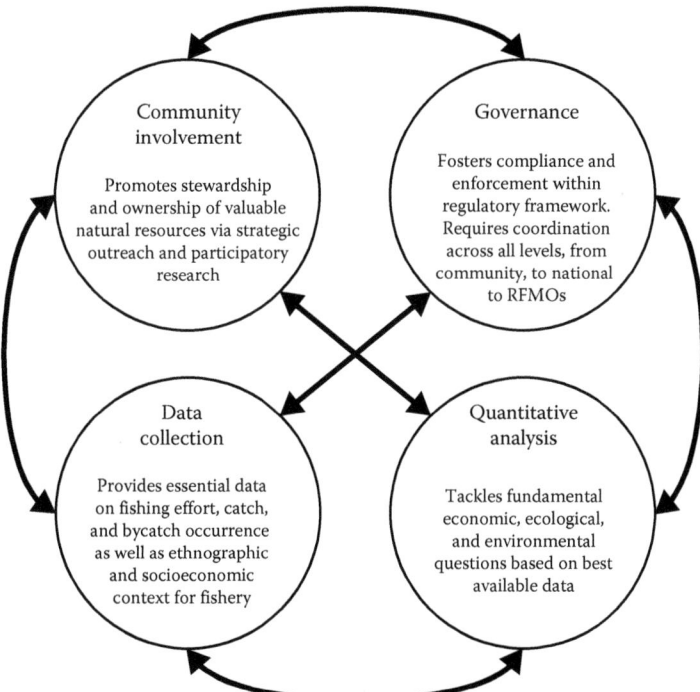

FIGURE 12.4   Cross-sectoral and multidisciplinary framework for bycatch reduction. (From Lewison, R.L., Soykan, C.U., Cox, T., Peckham, H., Pilcher, N., LeBoeuf, N., McDonald, S. Moore, J.E., Safina, C., Crowder, L.B. 2011. Ingredients for addressing the challenges of fisheries bycatch. *Bulletin of Marine Science* 87(2): 235–250.)

turtle bycatch reduction (Jenkins, 2010). For SSFs, in particular, command-and-control approaches, such as fisheries closures and mandated technological fixes, are often impractical and may only provide short-term solutions (Berkes et al., 2001; Hilborn et al., 2005; McClanahan et al., 2006). Numerous studies have shown that engaging fishermen from the outset of bycatch research and reduction initiatives can augment the development and adoption of long-term solutions (Hall et al., 2000; Kennelly, 2007; Campbell and Cornwell, 2008; Jenkins, 2010), in part because investment in the conservation process may increase fishers' subsequent adoption of conservation strategies (Cox et al., 2007; Jenkins et al., 2008). In the context of SSFs, which predominantly occur in developing nations where management and enforcement are limited, engaging fishers and their communities can be particularly important because bycatch mitigation programs are essentially voluntary (McClanahan et al., 2006; Jackson, 2007).

## 12.8   NEW APPROACHES AND DIRECTIONS IN BYCATCH RESEARCH

There have been substantial advances in the recognition and assessment of the significant threat that fisheries bycatch poses to sea turtle population worldwide. At the same time, the development of bycatch reduction measures also has yielded some promising and effective approaches. The problem of sea turtle bycatch is still largely one of scale; while some fleets require and enforce bycatch reduction measures, the vast majority do not. Although large-scale and SSFs are faced with different challenges in terms of bycatch reduction, both sectors are likely exerting population-level effects on sea turtle populations that, in many cases, are already in decline (Wallace et al., 2011). Innovations, such as at-sea advisory programs that provide real-time information to small-scale fishers on observed bycatch (Alfaro-Shigueto et al., 2012), and tri-national programs that connect small-scale fishers on opposite sides of the ocean to gain a clearer understanding of sea turtle status (Tri-National Fisherman's Exchange, Grupo Tortuguero, Peckham and Maldonado-Diaz, 2012), are creating new possibilities to tackle the daunting issue of sea turtle bycatch in SSFs. Likewise, rapid bycatch assessments, which are interview-based surveys that characterize gear use, fishing effort and obtain semi-quantitative estimates of bycatch, are proving to be a powerful approach to gathering general bycatch information from widely distributed and difficult-to-monitor SSFs (*sensu* Moore et al., 2010).

For large-scale fisheries, technological advances are paving the way to more effective bycatch reduction. A number of new promising approaches serve to integrate multiple factors that drive sea turtle bycatch vulnerability, i.e., insights from sea turtle ecology, life history, and physiology, gained from sea turtle telemetry and tracking. Integration of these data in the context of a dynamic ocean environment will yield a new generation of innovative and effective bycatch reduction strategies. One of the best contemporary examples of this, TurtleWatch (Howell et al., 2008), provides management recommendations to the Hawaii longline fishery that are based on documented seasonal relationships between SST and turtle distribution, with the overall aim of reducing loggerhead bycatch. In the U.S. mid-Atlantic coast gillnet fisheries, managers also use SSTs to enact rolling closures based on the probability that turtles aggregate in predictable temperatures zones (Murray et al., 2009). Comparable studies that have also shown relationships between water temperatures or movement patterns and seasonal distributions of other sea turtle species (McMahon and Hays, 2006; Hawkes et al., 2007; Sherrill-Mix et al., 2007; Gardner et al, 2008; Benson et al., 2011; Shillinger et al., 2011) provide the foundation on which to develop similar bycatch management strategies in other ocean regions.

As the field of bycatch research has developed, new perspectives and definitions of bycatch have emerged. Defining bycatch as "any catch that is unwanted and unmanaged" (Davies et al., 2009), we can consider sea turtle bycatch in an integrated multi-species catch management context. This type of integration of bycatch and catch patterns has been employed by a small number of fleets. In the Eastern Australian longline fishery, managers use a combination of satellite tracking and remote sensing to create forecasting models of where sensitive bycatch species will occur, creating tiered fishing zones based on predicted distribution of multiple target catch species and bycatch species that also takes into account the potential economic yield given real-time quota levels (Hobday et al., 2011). Because these kinds of approaches require large amounts of high-resolution data on biological and physical oceanography, fleet-specific behavior, and economic parameters, they are difficult to develop and apply widely. However, the synoptic nature of these tools provides a template for how it might be possible to simultaneously reduce bycatch of protected species like sea turtles, while maintaining sustainable catch levels. Given that a third of all fish stocks are overexploited or depleted (Worm et al., 2009) and the ongoing concerns about the viability of many sea turtle populations, creating assessment tools that can consider sea turtle bycatch reduction within the broader context of fisheries sustainability is an essential next step.

Effective bycatch research and mitigation will rely on the continued integration of sea turtle ecology, fisheries management, and social science. As demonstrated in this volume, research on sea turtle ecology over the past decade has transformed our understanding of these species. Likewise, over the past 10 years, quantitative analyses of bycatch data have developed substantially and played an important role in refining our understanding of the population-level effects of bycatch, and the oceanographic variables associated with sea turtle bycatch. More recent programs on education, outreach, and cooperative fisheries management approaches have also provided powerful models of the importance of participatory bycatch assessment and bycatch mitigation. Maximizing the integration of ecological data within an oceanographic, fisheries and social context will be essential in balancing the survival of sea turtles and sustainable fisheries.

## REFERENCES

Abbott, J. K. and A. C. Haynie. 2012. What are we protecting? Fisher behavior and the unintended consequences of spatial closures as a fishery management tool. *Ecological Applications* 22:762–777.

Alfaro-Shigueto J., P. H. Dutton, M. F. Van Bressem, J. C. Mangel. 2007. Interactions between leatherback turtles and Peruvian artisanal fisheries. *Chelonian Conservation and Biology* 6:129–134.

Alfaro-Shigueto, J., J. C. Mangel, F. Bernedo et al. 2011. Small-scale fisheries of Peru: A major sink for marine turtles in the Pacific. *Journal of Applied Ecology* 48:1432–1440.

Alfaro-Shigueto, J., J. C. Mangel, P. H. Dutton, J. A. Seminoff, and B. J. Godley. 2012. Trading information for conservation: A novel use of radio broadcasting to reduce turtle bycatch. *Oryx* 46(3):332–339.

Alio, J. J., L. A. Marcano, and D. E. Altuve, D. E. 2010. Incidental capture and mortality of sea turtles in the industrial shrimp trawling fishery of northeastern Venezuela. *Ciencias Marinas* 36(2):161–178.

Antonio, F. J., R. S. Mendes, and S. M. Thoma. 2011. Identifying and modeling patterns of tetrapod vertebrate mortality rates in the Gulf of Mexico oil spill. *Aquatic Toxicology* 105(1–2):177–179.

Arauz, R., R. Vargas, I. Naranjo, and C. Gamboa. 1998. Analysis of the incidental capture and mortality of sea turtles in the shrimp fleet of Pacific Costa Rica. In: S. P. Epperly and J. Braun (eds.), *Proceedings of the 17th Annual Sea Turtle Symposium*, Orlando, FL, pp. 1–5. NOAA Technical Memorandum NMFS-SEFSC-415.

Baum, J. K., M. A. Myers, D. G. Kehler et al. 2003. Collapse and conservation of shark populations in the northwest Atlantic. *Science* 299:389–392.

Bailey, H., S. R. Benson, SR, G. L. Shillinger et al. 2012. Identification of distinct movement patterns in Pacific leatherback turtle populations influenced by ocean conditions. *Ecological Applications* 22:735–747.

Bellman, M. A., S. A. Heppell, C. Goldfinger. 2005. Evaluation of a US west coast groundfish habitat conservation regulation via analysis of spatial and temporal patterns of trawl fishing effort. *Canadian Journal of Fisheries and Aquatic Sciences* 62:2886–2900.

Béné, C. 2006. Small-scale fisheries: Assessing their contribution to rural livelihoods in developing countries. FAO Fisheries Circular No. 1008, 46 p.

Benson, S. R., T. Eguchi, D. G. Foley et al. 2011. Large-scale movements and high-use areas of western Pacific leatherback turtles, *Dermochelys coriacea. Ecosphere* 2(7):84.

Berkes F., R. Mahon, P. McConney, et al. 2001. Managing small-scale fisheries: Alternative directions and methods. International Development Research Centre, Ottawa.

Beverly S., D. Curran, M. Musyl, and B. Molony. 2009. Effects of eliminating shallow hooks from tuna longline sets on target and non-target species in the Hawaii-based pelagic tuna fishery. *Fisheries Research* 96:281–288.

Brewer D., D. Heales, D. Milton et al. 2006. The impact of turtle excluder devices and bycatch reduction devices on diverse tropical marine communities in Australia's northern prawn trawl fishery. *Fisheries Research* 81:176–188.

Brothers, N. 1991. Albatross mortality and associated bait loss in the Japanese longline fishery in the Southern Ocean. *Biological Conservation* 55:255–268.

Caillouet, C. W., Shaver, D. J., Teas, W. G., Nance, J. M., Revera, D. B., and Cannon, A. C.1996. Relationship between sea turtle stranding rates and shrimp fishing intensities in the northwestern Gulf of Mexico: 1986–1989 versus 1990–1993. *Fishery Bulletin* 94:237–249.

Campbell, L. M. and M. L. Cornwell. 2008. Human dimensions of bycatch reduction technology: Current assumptions and directions for future research. *Endang Species Res.* 5:325–334.

Casale, P. 2010. Incidental catch of marine turtles in the Mediterranean Sea: Captures, mortality, priorities. *Fish and Fisheries* 12:299–316.

Casale, P. 2011. Sea turtle by-catch in the Mediterranean. *Fish and Fisheries* 12:299–316.

Casale, P., A. C. Broderick, D. Freggi et al. 2012. Long-term residence of juvenile loggerhead turtles to foraging grounds: A potential conservation hotspot in the Mediterranean. *Aquatic Conservation-Marine and Freshwater Ecosystems* 22(2):144–154.

Casale, P., D. Freggi, and M. Rocco. 2008. Mortality induced by drifting longline hooks and branchlines in loggerhead sea turtles, estimated through observation in captivity *Aquatic Conservation- Marine and Freshwater Ecosystems* 18(6):945–954.

Chaloupka, M., P. Dutton, and H. Nakano. 2004. Status of sea turtle stocks in the Pacific. In *Expert Consultation on Interactions between Sea Turtles and Fisheries within an Ecosystem Context.* FAO Fisheries Report No. 738, pp. 135–164, Supplement.

Chan, E.-H. and H.-C. Liew. 1996. Decline of the leatherback population in Terengganu, Malaysia, 1956–1995. *Chelonian Conservation and Biology* 2:196–203.

Chuenpagdee, R., L. Liguori, M. L. D. Palomares, and D. Pauly. 2006. Bottom-up, global estimates of small-scale marine fisheries catches. Fisheries Centre Research Reports, Vancouver, British Columbia, Canada.

Chuenpagdee, R., L. E. Morgan, S. M. Maxwell, E. A. Norse, and D. Pauly. 2003. Shifting gears: Assessing collateral impacts of fishing methods in us waters. *Frontiers in Ecology and the Environment* 1:517–524.

Crouse, D., L. Crowder, and H. Caswell. 1987. A stage-based population model for loggerhead sea turtles and implications for conservation. *Ecology* 68:1412–1423.

Crowder, L. B., D. T. Crouse, S. S. Heppell, and T. H. Martin. 1994. Predicting the impact of turtle excluder devices on loggerhead sea-turtle populations. *Ecological Applications* 4:437–445.

Crowder, L. B., S. R. Hopkins-Murphy, and J. A. Royle. 1995. Effects of turtle excluder devices (TEDs) on loggerhead sea turtle strandings with implications for conservation. *Copeia* 1994(4):773–779.

Curran, D. and K. Bigelow. 2011. Effects of circle hooks on pelagic catches in the Hawaii-based tuna longline fishery. *Fisheries Research* 109:265–275.

Cox, T. M., R. L. Lewison, R. Zydelis, L. B. Crowder, C. Safina, and A. J. Read. 2007. Comparing effectiveness of experimental and implemented bycatch reduction measures: The ideal and the real. *Conservation Biology* 21:1155–1164.

Davies, R. W. D., Cripps, S. J., Nicksona, A., and G. Porter. 2009. Defining and estimating global marine fisheries bycatch. *Marine Policy* 33(4):661–672. doi:10.1016/j.marpol.2009.01.003.

Dunn, D. C., A. M. Boustany, and P. N. Halpin PN. 2011. Spatio-temporal management of fisheries to reduce by-catch and increase fishing selectivity. *Fish and Fisheries* 12:110–119.

Echwikhi, K., I. Jribi, M. N. Bradai, and A. Bouain. 2010. Gillnet fishery—Loggerhead turtle interactions in the Gulf of Gabes, Tunisia. *Herpetological Journal* 20:25–30.

Eckert, S. A., J. Gearhart, C. Bergmann, and K. L. Eckert. 2008. Reducing leatherback sea turtle bycatch in the surface drift-gillnet fishery in Trinidad. *Bycatch Communication Newsletter* 8:2–6.

Epperly, S. P. and W. Teas. 2002. Turtle excluder devices—Are the escape openings large enough? *Fishery Bulletin* 100(3):466–474.

Federal Register. 1994. Sea turtle conservation; approved turtle excluder devices. *US Federal Register* 25827–25831.

Federal Register. 2007. Fisheries off west coast states; highly migratory species fisheries. *US Federal Register* 31756–31757.

Federal Register. 2011. Western Pacific pelagic fisheries; closure of the Hawaii shallow-set pelagic longline fishery due to reaching the annual limit on sea turtle interactions. *US Federal Register* 72643–72644.

Ferraro, P. J. and H. Gjertsen. 2009. A global review of incentive payments for sea turtle conservation. *Chelonian Conservation and Biology* 8:48–56.

Ferreira, R. L., H. R. Martins, A. B. Bolten, M. A. Santos, and K. Erzini. 2011. Influence of environmental and fishery parameters on loggerhead sea turtle by-catch in the longline fishery in the Azores archipelago and implications for conservation. *Journal of the Marine Biological Association of the United Kingdom* 91(8):1697–1705.

Finkbeiner, E. M., B. P. Wallace, J. E. Moore, R. L. Lewison, L. B. Crowder, and A. J. Read. 2011. Cumulative estimates of sea turtle bycatch and mortality in USA fisheries between 1990 and 2007. *Biological Conservation* 144:2719–2727.

Fisher, R. A. 1930. The genetical theory of natural selection. Oxford University Press, London, UK.

Fonteneau, A., J. Ariz, D. Gaertner et al. 2000. Observed changes in the species composition of tuna schools in the Gulf of Guinea between 1981 and 1999, in relation with the Fish Aggregating Device fishery. *Aquatic Living Resources* 13(4):253–257.

Fujiwara M., H. Caswell. 2001. Demography of the endangered North Atlantic right whale. *Nature,* 414: 537–541.

Gallo, B. M. G., S. Macedo, B. D. B. Giffoni, J. H. Becker, and P. C. R. Barata. 2006. Sea turtle conservation in Ubatuba, Southeastern Brazil, a feeding area with incidental capture in coastal fisheries. *Chelonian Conservation and Biology* 5:93–101.

Gardner, B., P. J. Sullivan, S. J. Morreale et al. 2008. Spatial and temporal statistical analysis of bycatch data: Patterns of sea turtle bycatch in the North Atlantic. *Canadian Journal of Fisheries and Aquatic Sciences* 65(11):2461–2470.

Gerosa, G. and P. Casale. 1999. Interaction of marine turtles with fisheries in the Mediterranean. UNEP/MAP, RAC/SPA: Tunis, Tunisia. 59pp.

Gilman, E. (ed.). 2009. Proceedings of the *Technical Workshop on Mitigating Sea Turtle Bycatch in Coastal Net Fisheries,* January 20–22, 2009, Honolulu, HI. Western Pacific Regional Fishery Management Council, IUCN, Southeast Asian Fisheries Development Center, Indian Ocean South-east Asian Marine Turtle MoU, US National Marine Fisheries Service, Southeast Fisheries Science Center.

Gilman, E., J. Gearhart, B. Price S. Eckert, H. Milliken, J. Wang, Y. Swimmer, D. Shiode, O. Abe, S. H. Peckham, M. Chaloupka, M. Hall et al. 2010. Mitigating sea turtle by-catch in coastal passive net fisheries. *Fish and Fisheries* 11(1):57–88.

Gilman, E., D. Kobayashi, T. Swenarton, N. Brothers, P. Dalzell, and I. Kinan-Kelly. 2007. Reducing sea turtle interactions in the Hawaii-based longline swordfish fishery. *Biological Conservation* 139:19–28.

Gilman, E. and C. Lundin. 2009. Minimizing bycatch of sensitive species groups in marine capture fisheries: Lessons from commercial Tuna fisheries. In: Handbook of Marine Fisheries Conservation and Management (eds Q. Grafton, R. Hillborn, D. Squires, M. Tait and M. Williams). Oxford University Press, Oxford, 150–164.

Gilman, E., E. Zollett, S. Beverly et al. 2006. Reducing sea turtle by-catch in pelagic longline fisheries. *Fish and Fisheries* 7:2–23.

Godley, B. J., J.M. Blumenthal, A.C. Broderick, M.S. Coyne, et al. 2008. Satellite tracking of sea turtles: Where have we been and where do we go next? *Endangered Species Research* 4:3–22.

Godley, B. J., A. C. Gucii, A. C. Broderick, R. W. Furness, and S. E. Solomon. 1998. Interaction between marine turtles and artisanal fisheries in the eastern Mediterranean: A probable cause for concern? *Zoology in the Middle East* 16:49–64.

Gutierrez, N. L., R. Hilborn, and O. Defeo. 2011. Leadership, social capital and incentives promote successful fisheries. *Nature* 470:386–389.

Hall, M. A., D. L. Alverson, and K. I. Metuzals. 2000. By-catch: Problems and solutions. *Marine Pollution Bulletin* 41:1–6.

Hall M. A., H. Nakano, S. Clarke, S. Thomas et al. 2007. Working with fishers to reduce bycatches. In: *Bycatch Reduction in the World's Fisheries,* Kennelly SJ, editor. Springer-Verlag. doi:10.1007/978-1-4020-6078-6_8.

Hart, K. M., P. Mooreside, and L. B. Crowder. 2006. Interpreting the spatio-temporal patterns of sea turtle strandings: Going with the flow. *Biological Conservation* 129:283–290.

Harewood, A. and J. Horrocks. 2008. Impacts of coastal development on hawksbill hatchling survival and swimming success during the initial offshore migration. *Biological Conservation* 141(2):394–401.

Hawkes, L. A., A.C. Broderick, M.S. Coyne, M. H. Godfrey et al. 2007. Only some like it hot--quantifying the environmental niche of the loggerhead sea turtle. *Diversity and distributions* 13:447–457.

Hazin, H. and K. Erzini. 2008. Assessing swordfish distribution in the South Atlantic from spatial predictions. *Fisheries Research* 90(1–3):45–55.

Henwood, T. A. and W. E. Stuntz. 1987. Analysis of sea turtle captures and mortalities during commercial trawling. *Fishery Bulletin* 85(4):813–817.

Heppell, S. S. 1998. An application of life history theory and population model analysis to turtle conservation. *Copeia* 1998(2):367–375.

Heppell, S. S., H. Caswell, and L. B. Crowder. 2000a. Life histories and elasticity patterns: Perturbation analysis for species with minimal demographic data. *Ecology* 81(3):654–665.

Heppell, S. S., D. T. Crouse, L. B. Crowder et al. 2005. A population model to estimate recovery time, population size, and management impacts on Kemp's ridley sea turtles. *Chelonian Conservation and Biology* 4:767–773.

Heppell, S., C. Pfister, and H. de Kroon. 2000b. Elasticity analysis in population biology: Methods and application. *Ecology* 81(3):605–606.

Hilborn R., J. M. Orensanz, and A. M. Parma. 2005. Institutions, incentives and the future of fisheries. *Philos Trans R Soc London Biol.* 360:47–57.

Hobday, A. J., J. R. Hartog, C. M. Spillman, and O. Alves. 2011. Seasonal forecasting of tuna habitat for dynamic spatial management. *Canadian Journal of Fisheries and Aquatic Sciences* 68:898–911.

Howell, E. A., P. H. Dutton, J. J. Polovina, H. Bailey, D. M. Parker, and G. H. Balazs, G. H. 2010. Oceanographic influences on the dive behavior of juvenile loggerhead turtles (*Caretta caretta*) in the North Pacific Ocean. *Marine Biology* 157(5):1011–1026.

Howell, E., D. Kobayashi, D. Parker, and G. Balazs. 2008. Turtlewatch: A tool to aid in the bycatch reduction of loggerhead turtles *Caretta caretta* in the Hawaii-based pelagic longline fishery. *Endangered Species Research* 5:267–278.

Huntington, H. P. 2000. Using traditional ecological knowledge in science: Methods and applications. *Ecological Applications* 10:1270–1274.

Ishihara, T., Y. Matsuzawa, J. Wang, and H. Peckham. 2011. 2nd International Workshop to Mitigate Bycatch of Sea Turtles in Japanese Pound Nets. *Marine Turtle Newsletter* 130:27–28.

Jackson, J. B. C. 2007. Economic incentives, social norms and the crises of fisheries. *Ecological Research* 22:16–18.

Jenkins, L. D. 2010. Profile and influence of the successful fisher-inventor of marine conservation technology. *Conservation and Society* 8(1):44–54.

Jenkins L. D., R. B. Mast, B. J. Hutchinson, and A. H. Hutchinson. 2008. Key factors in the invention and diffusion of marine conservation technology: A case study of TEDs. NOAA Technical Memorandum NMFS SEFSC: 73.

Jribi, I., K. Echwikhi, M. N. Bradai, and A. Bouain. 2008. Incidental capture of sea turtles by longlines in the Gulf of Gabes (South Tunisia): A comparative study between bottom and surface longlines. *Scientia Marina* 72(2):337–342.

Kennelly, S. J. 1999. The role of fisheries monitoring programmes in identifying and reducing problematic bycatches. In: Nolan, C.P. (Ed.), *Proceedings of the International Conference on Integrated Fisheries Monitoring*, Sydney, Australia, 1–5 February 1999, pp. 75–82.

Kennelly, S. J. 2007. *Bycatch Reduction in the World's Fisheries*. Springer Verlag.

Kobayashi, D. R., J. J. Polovina, D. M. Parker et al. 2008. Pelagic habitat characterization of loggerhead sea turtles, *Caretta caretta*, in the North Pacific Ocean (1997–2006): Insights from satellite tag tracking and remotely sensed data. *Journal of Experimental Marine Biology and Ecology* 356(1–2):96–114.

Komoroske, L. M., R. L. Lewison, J. A. Seminoff, D. D. Deheyn, and P. H. Dutton. 2011. Pollutants and the health of green sea turtles resident to an urbanized estuary in San Diego, CA. *Chemosphere* 84(5):544–552.

Kot, C. Y., A. M. Boustany, and P. N. Hallpin. 2010. Temporal patterns of target catch and sea turtle bycatch in the US Atlantic pelagic longline fishing fleet. *Canadian Journal of Fisheries and Aquatic Science* 67:42–57.

Lazar, B. and R. Gracan. 2011. Ingestion of marine debris by loggerhead sea turtles, *Caretta caretta*, in the Adriatic Sea. *Marine Pollution Bulletin* 62(1):43–47.

Lewison, R. L. and L. B. Crowder. 2007. Putting longline bycatch of sea turtles into perspective. *Conservation Biology* 21:79–86.

Lewison, R. L., L. B. Crowder, A. J. Read, and S. A. Freeman. 2004a. Understanding impacts of fisheries bycatch on marine megafauna. *Trends in Ecology and Evolution* 19:598–604.

Lewison, R. L., L. B. Crowder, and D. J. Shaver. 2003. The impact of turtle excluder devices and fisheries closures on loggerhead and Kemp's ridley strandings in the western Gulf of Mexico. *Conservation Biology* 17:1089–1097.

Lewison, R. L., L. B. Crowder, B. Wallace et al. Global patterns of marine megafauna bycatch, In review.

Lewison, R. L., S. A. Freeman, and L. B. Crowder. 2004b. Quantifying the effects of fisheries on threatened species: The impact of pelagic longlines on loggerhead and leatherback sea turtles. *Ecology Letters* 7:221–231.

Lewison, R. L., C. Soykan, and J. Franklin. 2009. Mapping the bycatch seascape: Multispecies and multi-scale spatial patterns of fisheries bycatch. *Ecological Applications* 19(4):920–930.

Lewison, R. L., C. U. Soykan, T. Cox, H. Peckham, N. Pilcher, N. LeBoeuf, S. McDonald, J. E. Moore, C. Safina, and L.B. Crowder. 2011. Ingredients for addressing the challenges of fisheries bycatch. *Bulletin of Marine Science* 87(2):235–250. doi:10.5343/bms.2010.1062.

Limpus, C. 2007. A biological review of Australian marine turtles: The flatback turtle *Natator depressus* (Garman). Queensland Environmental Protection Agency, Brisbane, Queensland, Australia. http:/www.derm.qld.gov.au/register/p02340aa.pdf

Lucchetti, A. and A. Sala. 2010. An overview of loggerhead sea turtle (*Caretta caretta*) bycatch and technical mitigation measures in the Mediterranean Sea. *Reviews in Fish Biology and Fisheries* 20:141–161.

Lum, L. L. 2006. Assessment of incidental sea turtle catch in the artisanal gillnet fishery in Trinidad and Tobago, West Indies. *Applied Herpetology* 3:357–368.

Maunder, M. N. and A. E. Punt. 2004. Standardizing catch and effort data: A review of recent approaches. *Fisheries Research* 70:141–159.

Maxwell, S. M., G. A. Breed, B. A. Nickel et al. 2011. Using satellite tracking to optimize protection of long-lived marine species: Olive ridley sea turtle conservation in central Africa. *PLoS One* 6:e19905.

McCarthy, A. L., S. S. Heppell, F. Royer, C. Freitas, and T. Dellinger. 2010. Identification of likely foraging habitat of pelagic loggerhead sea turtles (*Caretta caretta*) in the North Atlantic through analysis of telemetry track sinuosity. *Progress in Oceanography* 86(1–2):224–231.

McClanahan T. R., M. J. Marnane, J. E. Cinner, and W. E. Kiene. 2006. A comparison of marine protected areas and alternative approaches to coral-reef management. *Current Biology* 16:1408–1413.

McDaniel, C. J., L. B. Crowder, and J. A. Priddy. 2000. Spatial dynamics of sea turtle abundance and shrimping intensity in the US Gulf of Mexico. *Conservation Ecology* 4(1):15.

McMahon, C. R. and G. C. Hays. 2006. Thermal niche, largescale movements and implications of climate change for a critically endangered marine vertebrate. *Global Change Biology* 12:1330–1338.

van de Merwe, J. P., M. Hodge, H. A. Olszowy et al. 2010. Using blood samples to estimate persistent organic pollutants and metals in green sea turtles (*Chelonia mydas*). *Marine Pollution Bulletin* 60(4):579–588.

Moore, J. E., T. M. Cox, R. L. Lewison et al. 2010. An interview-based approach to assess marine mammal and sea turtle captures in artisanal fisheries. *Biological Conservation* 143:795–805.

Moore, J. E., B. P. Wallace, R. L. Lewison, R. Žydelis, T. M. Cox, and L. B. Crowder. 2009. A review of marine mammal, sea turtle and seabird bycatch in USA fisheries and the role of policy in shaping management. *Marine Policy* 33:435–451.

Murray, K. T. 2009. Characteristics and magnitude of sea turtle bycatch in US mid-Atlantic gillnet gear. *Endangered Species Research* 8:211–224.

National Marine Fisheries Service (NMFS). 2001. Stock assessment of loggerhead and leatherback sea turtles and an assessment of the impact of the pelagic longline fishery on the loggerhead and leatherback sea turtles of the Western North Atlantic. U.S. Department of Commerce, NOAA Technical Memorandum NFMS SEFSC-455, 343 pp.

Nelson, T. A. and B. Boots. 2008. Detecting spatially explicit hot spots in landscape-scale ecology. *Ecography* 31(5):556–566.

Pacheco, J. C., D. W. Kerstetter, F. H. Hazin et al. 2011. A comparison of circle hook and J hook performance in a western equatorial Atlantic Ocean pelagic longline fishery. *Fisheries Research* 107:39–45.

Peckham, S. H., D. M. Diaz, A. Walli, G. Ruiz, L. B. Crowder, and W. J. Nichols. 2007. Small-scale fisheries bycatch jeopardizes endangered Pacific loggerhead turtles. *PLoS One* 2:e1041.

Peckham, S. H. and D. Maldonado-Diaz. 2012 Empowering small-scale fishermen to be conservation heroes: A trinational fishermen's exchange to protect loggerhead turtles. In: J. A. Seminoff and B. P. Wallace (eds.), *Sea Turtles of the Eastern Pacific: Advances in Research and Conservation*. University of Arizona Press, Tucson, AZ, pp. 279–301.

Peckham, S. H., D. Maldonado-Diaz, V. Koch et al. 2008. High mortality of loggerhead turtles due to bycatch, human consumption and strandings at Baja California Sur, Mexico, 2003 to 2007. *Endangered Species Research* 5:171–183.

Peckham, S. H., D. Maldonado-Diaz, J. Lucero, A. Fuentes-Montalvo and A. Gaos. 2009. Loggerhead bycatch and reduction off the Pacific coast of Baja California Sur, Mexico. In: E. Gilman (ed.), *Proceedings of the Technical Workshop on Mitigating Sea Turtle Bycatch in Coastal Net Fisheries*, pp. 58–60. Pascagoula, MS: IUCN, Western Pacific Regional Fishery Management Council, Southeast Asian Fisheries Development Center, Indian Ocean—South-East Asian Marine Turtle MoU, U.S. National Marine Fisheries Service, Southeast Fisheries Science Center.

Pike, D. A. 2008. Natural beaches confer fitness benefits to nesting marine turtles. *Biology Letters* 4(6):704–706.

Piovano, S., G. Basciano, Y. Swimmer, and C. Giacoma. 2012. Evaluation of a bycatch reduction technology by fishermen: A case study from Sicily. *Marine Policy* 36:272–277.

Poiner, I. R. and A. N. M. Harris. 1996. Incidental capture, direct mortality and delayed mortality of sea turtles in Australia's Northern Prawn Fishery. *Marine Biology* 125(4):813–825.

Polovina, J. J., G. H. Balazs, E. A. Howell et al. 2004. Forage and migration habitat of loggerhead (*Caretta caretta*) and olive ridley (*Lepidochelys olivacea*) sea turtles in the central North Pacific Ocean. *Fisheries Oceanography* 13(1): 36–51.

Polovina, J. J., E. Howell, D. M. Parker, and G. H. Balazs. 2003. Dive-depth distribution of loggerhead (*Caretta caretta*) and olive ridley (*Lepidochelys olivacea*) sea turtles in the central North Pacific: Might deep long-line sets catch fewer turtles? *Fishery Bulletin* 101:189–193.

Polovina, J. J., D. R. Kobayashi, D. M. Parker, M. P. Seki et al. 2000. Turtles on the edge: Movement of loggerhead turtles (*Caretta caretta*) along oceanic fronts, spanning longline fishing grounds in the central North Pacific, 1997-1998. *Fisheries Oceanography* 9:71–82.

Read, A. J. 2007. Do circle hooks reduce the mortality of sea turtles in pelagic longlines? A review of recent experiments. *Biological Conservation* 135(2):155–169.

Ruttan, L. M., F. C. Gayanilo, U. R. Sumaila, and D. Pauly. 2000. Small versus large-scale fisheries: A multi-species, multi-fleet model for evaluation their interactions and potential benefits. In: D. Pauly and T. J. Pitcher (eds.), *Methods for Evaluating the Impacts of Fisheries on North Atlantic Ecosystems*. Fisheries Center Research Reports, Vancouver, British Columbia, Canada, pp. 64–78.

Sala, A., A. Lucchetti, and A. Marco. 2011. Effects of turtle excluder devices on bycatch and discard reduction in the demersal fisheries of Mediterranean Sea. *Aquatic Living Resources* 24(2):183–192.

Salas, S., R. Chuenpagdee, J. C. Seijo, and A. Charles. 2007. Challenges in the assessment and management of small-scale fisheries in Latin America and the Caribbean. *Fisheries Research* 87:5–16.

Sasso, C. R. and S. P. Epperly. 2007. Survival of pelagic juvenile loggerhead turtles in the open ocean. *Journal of Wildlife Management* 71:1830–1835.

Seco Pon, J. P., P. A. Gandini, et al. 2007. Effect of longline configuration on seabird mortality in the Argentine semi-pelagic kingclip *Genypterus blacodes* fishery. *Fisheries Research* 85(1-2):101–105.

Sherrill-Mix, S. A., M. C. James, and R. A. Myers. 2007. Migration cues and timing in leatherback sea turtles. *Behavioral Ecology* 19:231–236.

Shillinger G. L., D. M. Palacios, H. Bailey , S. J. Bograd et al. 2008. Persistent leatherback turtle migrations present opportunities for conservation. *PLoS Biology* 6(7): e171.doi:10.1371/journal.pbio.0060171.

Shillinger, G. L., A.M. Swithenbank, H. Bailey et al. 2011. Characterization of leatherback turtle post-nesting habitats and its application to marine spatial planning. *Marine Ecology Progress Series* 422:275–289.

Sims, M., T.C. Cox, R. L. Lewison. 2008. Modeling spatial patterns in fisheries bycatch: Improving bycatch maps to aid fisheries management. *Ecological Applications* 18(3), 649–661.

Small, C. J. 2005. Regional Fisheries Management Organisations: Their duties and performance in reducing bycatch of albatrosses and other species. Cambridge, U.K.: BirdLife International.

Soykan, C. U., J. E. Moore, R. Zydelis, L. B. Crowder, C. Safina, and R. L. Lewison. 2008. Why study bycatch? An introduction to the theme section on fisheries bycatch. *Endangered Species Research* 5:91–102.

Spotila, J. R., R. D. Reina, A. C. Steyermark, P. T. Plotkin, and F. V. Paladino. 2000. Pacific leatherback turtles face extinction. *Nature* 405:529–530.

Stewart, K. R., J. M. Keller, R. Templeton et al. 2011. Monitoring persistent organic pollutants in leatherback turtles (*Dermochelys coriacea*) confirms maternal transfer. *Marine Pollution Bulletin* 62(7):1396–1409.

Stewart, K. R., R. L. Lewison, D. C. Dunn et al. 2010. Characterizing fishing effort and spatial extent of coastal fisheries. *PLoS ONE* 5:e14451.

do Sul, J. A. I., J. Assuncao, I. R. Santos, A. C. Friedrich, A. Matthiensen, and G. Fillman. 2011. Plastic pollution at a sea turtle conservation area in NE Brazil: Contrasting developed and undeveloped beaches. *Estuaries and Coasts* 34(4):814–823.

Swarthout, R. F., J. M. Keller, M. Peden-Adams et al. 2010. Organohalogen contaminants in blood of Kemp's ridley (*Lepidochelys kempii*) and green sea turtles (*Chelonia mydas*) from the Gulf of Mexico. *Chemosphere* 78(6):731–741.

Swimmer, Y., J. Suter, R. Arauz, K. Bigelow, A. Lopez, I. Zanela, A. Bolanos, J. Ballestero, R. Suarez, J. Wang, and C. Boggs. 2011. Sustainable fishing gear: The case of modified circle hooks in a Costa Rican longline fishery. *Marine Biology* 158(4):757–767. doi: 10.1007/s00227-010-1604-4.

Tuck, G. N. 2011. Are bycatch rates sufficient as the principal fishery performance measure and method of assessment for seabirds? *Aquatic Conservation: Marine and Freshwater Ecosystems* 21:412–422.

Tuck, G. N., T. Polacheck, and C. M. Bulman. 2003. Spatio-temporal trends of longline fishing effort in the Southern Ocean and implications for seabird bycatch. *Biological Conservation* 114(1):1–27.

Wallace, B. P., A. D. DiMatteo, A. B. Bolten M. Y. Chaloupka, B. J. Hutchinson, F. A. Abreu-Grobois, J. A. Mortimer, J. A. Seminoff, D. Amorocho, K. A. Bjorndal, J. Bourjea, et al. 2011. Global conservation priorities for marine turtles. *PLoS ONE* 6:e24510.

Wallace, B.P., A. D. DiMatteo, B. J. Hurley, E. M. Finkbeiner, A. B. Bolten, M. Y. Chaloupka, B. J. Hutchinson, F. A. Abreu-Grobois, D. Amorocho, K. A. Bjorndal, J. Bourjea et al. 2010b. Regional Management Units for marine turtles: A novel framework for prioritizing conservation and research across multiple scales. *PLoS ONE* 5(12): e15465.doi:10.1371/journal.pone.0015465.

Wallace, B. P., S. S. Heppell, R. L. Lewison, S. Kelez, and L. B. Crowder. 2008. Impacts of fisheries bycatch on loggerhead turtles worldwide inferred from reproductive value analyses. *Journal of Applied Ecology* 45:1076–1085.

Wallace, B. P., R. L. Lewison, S. McDonald et al. 2010a. Global patterns of marine turtle bycatch in fisheries. *Conservation Letters* 3:131–142.

Wang, J. H., S. Fisler, and Y. Swimmer. 2010. Developing visual deterrents to reduce sea turtle bycatch in gillnet fisheries. *Marine Ecology Progress Series* 408:241–250.

Ward, P., S. Epe, D. Kreutzb, E. Lawrencea, C. Robinsc, and A. Sandsa. 2009. The effects of circle hooks on bycatch and target catches in Australia's pelagic longline fishery. *Fisheries Research* 97:253–262.

Watson, J. W., S. P. Epperly, A. K. Shah, and D. G. Foster. 2005. Fishing methods to reduce sea turtle mortality associated with pelagic longlines. *Canadian Journal of Fisheries and Aquatic Sciences* 62:965–981.

Wingfield, D. K., S. H. Peckham, D. G. Foley et al. 2011. The making of a productivity hotspot in the coastal ocean. PLoS ONE 6(11): e27874. doi:10.1371/journal.pone.0027874.

Witt, M. J., A. C. Broderick, M. S. Coyne et al. 2008. Satellite tracking highlights difficulties in the design of effective protected areas for Critically Endangered leatherback turtles *Dermochelys coriacea* during the inter-nesting period. *Oryx* 42:296–300.

Witzell, W. N. 1999. Distribution and relative abundance of sea turtles caught incidentally by the U.S. pelagic longline fleet in the western North Atlantic Ocean, 1992–1995. *Fisheries Bulletin* 97:200–211.

Worm, B., R. Hilborn., Baum, J. K., et al. 2009. Rebuilding global fisheries. *Science* 325(5940):578–585.

# 13 Climate Change and Marine Turtles

*Mark Hamann, Mariana M.P.B. Fuentes, Natalie C. Ban, and Véronique J.L. Mocellin*

## CONTENTS

## 13.1 INTRODUCTION

Climate change is increasingly being seen as a ubiquitous issue across the globe. The current and potential impacts of climate change on society and ecology, positive, negative, or neutral, are of utmost and increasing importance. As such, climate change spans all disciplines of science and policy. From a science perspective, climate change research has made substantial progress over the past two to three decades, with compelling evidence of a warming planet and changes to ecological and social systems (Harley et al., 2006; Hoegh-Guldberg and Bruno, 2010; Walther et al., 2002a,b).

The Fourth Assessment by the Intergovernmental Panel on Climate Change (IPCC) concluded that warming of the climate system is unequivocal (IPCC, 2007). Compelling evidence indicates that the world's oceans are warming, likely driving thermal expansion, increased stratification, changes to wind and ocean circulation, storm activity, and the hydrological cycle. Furthermore, increased sea surface temperatures will have numerous impacts on primary productivity, biogeochemistry, fisheries, and human populations dependent on marine resources (Behrenfeld et al., 2006; Cheung et al., 2009). While the volume and scope of science has been increasing, considerable uncertainty remains in the direction and magnitude of impacts. The uncertainty stems from the inherent challenge of predicting the magnitude of change for each climate process, lack of knowledge around cumulative and/or competing risks, synergetic impacts, and a lack of knowledge about species' biology, thresholds of concern, and local versus global-scale impacts.

Understanding the patterns, process, and impacts in the marine environment lags behind that of the terrestrial environment, but there has been substantial improvement in our knowledge (see Harley et al., 2006; Hoegh-Guldberg and Bruno, 2010). Traditionally, most investigations on climate change in the marine environment were focused on either temperature or sea level rise and their effect on coral reefs and coastal species, with fewer studies investigating ecosystem impacts (Hoegh-Guldberg and Bruno, 2010). Over the past decade, a shift occurred toward research on other marine habitats, such as the pelagic zone, subtidal benthic habitats, sea grass, and mangrove habitats, and also other abiotic factors such as pH and ocean circulation. In addition, increasing attention is being paid to the potential impacts of multiple and synergetic stressors in marine systems, in part made possible by improvement of computing power and increasing availability of remotely sensed data. Irrespective of the gaps, the issue of climate change and vulnerability of marine ecosystems, and people who rely on them, is gaining substantial momentum.

Marine turtles are a useful case in point. Key habitats for marine turtles—sandy beaches, sea grass meadows, coral reefs, subtidal and deep water benthic communities, and the pelagic zone—are projected to alter as the climate changes (Hawkes et al., 2009; Poloczanska et al., 2009). While current marine turtle species have survived historical climate events, marine turtles are sensitive to changes in their environment, and many species/populations have been negatively impacted by human disturbances to their habitats. One major concern is that climate change will impart additional stress onto already diminished marine turtle populations, likely reducing resilience and in some cases causing further declines.

Climate change was first identified as a potential issue for marine turtles in the 1980s. Several authors now regard it as a pervasive threat, not only to species and their critical habitats but also for human populations that reply on marine turtles and coastal ecosystems. The past decade has seen the topic of climate change and marine turtles emerge as important issue in marine turtle conservation. In 2003 it was listed among the 12 Burning Issues for marine turtle conservation by the IUCN Marine Turtle Specialist Group (Mast et al., 2004, 2005). Further, since the seminal work of Mrosovsky et al. (1984) and Davenport (1989), the impact of climate change on marine turtles has seen increased empirical research (Hamann et al., 2007; Hawkes et al., 2009; Poloczanska et al., 2009), and was highlighted as a priority research area (Hamann et al., 2010; Wallace et al., 2011). In this chapter, we synthesize the latest research on climate change and marine turtles, highlight case studies and approaches for studying climate change impacts on marine turtles, and outline the main issues and topics for future research and management.

## 13.2 WHAT IS CLIMATE CHANGE?

In this chapter, we use the IPCC definition of climate change: "a change in the state of the climate that can be identified (e.g., using statistical tests) by changes in the mean and/or the variability of its properties, and that persists for an extended period, typically decades or longer." This definition refers to any change in climate over time, whether due to natural variability or as a result of human activity.

## 13.3 CLIMATE CHANGE AND ANCIENT POPULATIONS

Extant species of marine turtles arose from ancient species that existed millions of years ago in the Jurassic (Brochu, 2001; Heatwole, 1999; Pritchard, 1997). They have persisted through several large-scale climatic and sea level changes, including periods of ocean warming similar in magnitude to patterns predicted for the next 50–100 years (Hamann et al., 2007; Poloczanska et al., 2009).

Even though no precise historical data exist to indicate how populations of extant species may have coped with historical climate changes, they likely adapted by redistributing their nesting sites and developing new migratory routes (Poloczanska et al., 2009), as well as possible changes to behavior, phenotypic plasticity, and adaptation to new conditions. The world's coastal regions, and thus marine turtle nesting and foraging areas, were vastly different 10,000 years ago, and today's turtles have new nesting and foraging distributions, and migratory routes. We can therefore speculate that marine turtles have evolved to—and will likely continue to—cope with climate change. However, the pertinent contemporary question is how individual populations or species will cope with the current fast rate of climate change while being simultaneously impacted by an array of anthropogenic activities. The resilience of marine turtles to additional threats is likely compromised as a result of widespread increases in the type and scale of anthropogenic impacts over the past century, which have depleted several populations and threatened others.

## 13.4  KEY CLIMATE CHANGE CONCERNS FOR MARINE TURTLES

Marine turtles will be affected by changes to multiple climatic processes (e.g., increased air and sea temperature, sea level rise, precipitation, and storm activity) at all life stages and at different temporal and geographical scales (Fuentes et al., 2011b; Hawkes et al., 2009). Marine turtles will likely be affected indirectly through changes to other biochemical factors such as increased atmospheric $CO_2$ concentrations and declines in ocean pH. Arguably, the more detectable impacts of climate change to marine turtles will occur during their terrestrial reproductive phase (egg laying, egg incubation, and hatchling success) because there are clear, and relatively straightforward, effects of increased temperature, sea level rise, and cyclonic activity on nesting sites and reproductive output (Fuentes and Abbs, 2010; Fuentes et al., 2009a,c; Hawkes et al., 2007b; Witt et al., 2010). Later, we describe the trends of existing knowledge and each of the main climate change–related threats to marine turtles. For each threat, we focus on relevant aspects of their biology.

### 13.4.1  Overview of Research Effort

To gain an understanding of the research effort to date on marine turtles and climate change, we searched ISI Web of Science for scientific literature that mentioned marine or sea turtles and climate change or global warming in the title or abstract (as per Hawkes et al., 2009). Between 1984 and 2011 (Nov), there were 98 research papers. We then categorized each paper into species (each species singularly or multi), geographic scope (nesting beach, foraging location, population level [either foraging or nesting], national level, regional level [i.e., ocean basin or multicountry] or global), and finally the climate process the study addresses (air temperature, sea temperature, sea level rise, extreme weather events, ocean scale climate drivers, rainfall, or multiple). We found a clear species, geographic region, and process trend. Nearly 66% of studies involved either loggerhead (29%) or green (37%) turtles, with 17% of studies investigating two or more species. In terms of geographic location, 44% of studies were conducted on nesting beaches, either on nesting turtles, eggs, or hatchlings, and over 85% of these focused on a single beach/island and did not address population-scale change. Finally, 46% of studies involved air temperature and 15% investigated sea temperature. We did not examine whether a similar species and geographic bias exists in more general marine turtle literature, but it is possible. Clear gaps exist in our knowledge of aspects of climate change for most species, geographic areas, and processes.

### 13.4.2  Temperature: Nesting Beaches and Nesting Females

Marine turtles are vulnerable to changes in global temperatures because they are seasonal breeding ectotherms with temperature-dependent sex determination. The timing and length of the breeding season varies considerably within and among species and the season's length is determined, at least to some degree, by whether the beach is warm enough to incubate eggs and whether off-shore currents favor the dispersal of hatchlings (Hamann et al., 2002). Then, once eggs are laid in sandy beaches, embryo development, sex, and phenotype for each marine turtle species are highly influenced by the temperature of the sand surrounding the nest (Miller, 1985; Spotila and Standora, 1985; Standora and Spotila, 1985).

Given the strong links between temperature and embryo development, and the relative ease of conducting research on marine turtle nesting beaches, it is not surprising that much of the collective research output on potential climate change impacts have focused on understanding the relationships between rising air and sand temperatures on sex ratios, incubation success, and hatchling phenotype (see reviews by Ackerman, 1997; Hawkes et al., 2009; Miller, 1985; Rhen and Lang, 1995). Determination of hatchling sex depends on sand temperatures during the middle third of the incubation period, with cooler and warmer sand producing more males and females, respectively (Georges et al., 1994; Mrosovsky et al., 1984). Pivotal temperatures are generally between 27°C and 30°C, successful embryo development is constrained between average incubation temperatures of between 25°C and 34°C, and the transitional range from 100% male to 100% female is typically ~2°C (Wibbels, 2003). Hence, because air, sea, and nest depth temperatures are related (see Fuentes et al., 2009b,d; Hays et al., 2003), increased global warming, even at conservative levels predicted by the (IPCC, 2007) of 1.1°C–2.9°C by 2099, could lead to large shifts in female biased sex ratios.

In general, most empirical studies have demonstrated that beaches produce female biased sex ratios of hatchlings (see Hawkes et al., 2009; Witt et al., 2010), and the relationship between air and sand temperatures indicates that some beaches have produced female bias for several decades (Hays et al., 1999, 2003). However, for most marine turtle populations, the pivotal temperature and/or nesting beach temperatures have not been measured; and while expert judgment can inform pivotal temperature based upon the same species from other locations, beach temperature data need to be collected for nesting sites for all populations (Fuentes et al., 2011b; Hamann et al., 2010). Aside from the logistical and cost challenges, determining natural sex ratios for populations or rookeries is difficult because sand temperatures are not constant throughout the incubation period and vary by proximate environmental or geographic features (see Ackerman, 1997). For example, incubation temperature and sex ratios can be influenced by features such as natural beach shading, beach slope, sand color, sand type, and nest depth and, at a larger scale, the variation of beaches used by the population (Booth and Freeman, 2006; Fuentes et al., 2009a; Hays et al., 2001a, 2003). Therefore, along with temperature, nesting beach fidelity and nest-site selection can play important roles in the determination of within and across season sex ratios.

Nesting beaches typically offer a variety of microhabitats, such as vegetated and nonvegetated zones, variation in dune height and width, moisture content, sand color, and texture. This variation, together with marine turtles generally laying multiple clutches of eggs in a season, gives rise to significant within-beach variation in the sex ratio of clutches (Fuentes et al., 2009d). Nest site selection has received increased research attention in the past decade, and significant inter- and intraspecies variation exist due to fidelity to a single beach, and to specific sections of a beach (Hays et al., 1995; Kamel, 2006; Kamel and Mrosovsky, 2004, 2005; Pfaller et al., 2009). While some data indicate that nest site selection is genetically determined, this conclusion is controversial (Mrosovsky, 2006, 2008; Pike, 2008a,b). Therefore, understanding the role of nest beach and nesting site fidelity was identified as a research priority (Hamann et al., 2010) to enable adaptive management to boost or maintain resilience in marine turtle populations in response to increased temperatures.

Most research on sand temperature impacts used restricted spatial and/or temporal scales, and less is known about population-scale impacts and longer-term decadal relationships. The latter will be more relevant to understanding the impacts of climate change to long-lived species that utilize broad range of nesting sites, and/or nesting sites spread over a latitudinal gradient (Fuentes et al., 2009d, 2011b; Hawkes et al., 2007a,b). Other key knowledge gaps are the degree of female bias a species/population can sustain, and whether bias in sex ratios of hatchlings equates to similar bias in juvenile turtles when they recruit back to the coastal zones (Hamann et al., 2010). Furthermore, most adult sex ratios now or in the past remain elusive, and little information exists on how hatchling sex ratios relate to operational sex ratios, and whether observed female bias in hatchlings is a recent phenomenon. Key research needs include understanding operational sex ratios and number of males needed to maintain fitness.

In addition to effects on sex determination, increased sand temperatures can also influence embryology and shorten the incubation duration of eggs (Miller, 1985). Warmer nests typically incubate quicker, and hatchlings arising from nests with shorter incubation periods have lower residual yolk mass, which consequently influences hatchling phenotype and performance (Booth and Astill, 2001a). Furthermore, although studies are limited, eggs incubated at the upper thermal limits (i.e., ~35°C) likely produce hatchlings with higher rates of scale and morphological abnormalities (Mast and Carr, 1989; Miller, 1985). Essentially, warmer sands can influence the sex and condition of hatchlings, and these in turn influence performance (Booth and Astill, 2001b; Booth and Evans, 2011) and may have implications for neonate survivorship (Booth and Evans, 2011).

In general, marine turtle breeding seasons appear to be limited by mean air temperatures of around 20°C in the northern hemisphere and 25°C in the southern hemisphere (Poloczanska et al., 2009; Witt et al., 2010), and breeding seasons in equatorial and tropical zones tend to be longer or more diffuse while those in subtropical environments are more temporally constrained (Hamann et al., 2002; Miller, 1997). One response to increasing temperatures could be shifts in timing of reproductive events such as migrations and breeding (Hoegh-Guldberg and Bruno, 2010; Parmesan, 2007; Parmesan and Yohe, 2003). To date six studies have used long-term datasets to investigate the relationship between temperature, timing, and duration of marine turtle nesting seasons. Three studies from the south-east United States and one from Greece indicate that loggerhead turtles start nesting earlier in years with warmer sea surface temperatures (Hawkes et al., 2007a; Pike et al., 2006; Weishampel et al., 2004), although (Weishampel et al., 2010) subsequently found a more complex relationship, at least in Florida, than first thought. A sixth study (Pike, 2009) then used a long-term dataset on green turtles (comparable years to Pike et al., 2006) from Florida to demonstrate that, unlike loggerhead turtles, green turtle breeding phenology is not related to sea temperatures. These studies indicate likely inter- and intraspecies variation in links between sea temperatures and timing of breeding events. This variation in reproductive cycles and phenology could be driven by other environmental factors at the foraging areas, the location and geographic spread of foraging areas, ecological variation or the timing and location of other reproductive events such as vitellogenesis, migration, or courtship (see review by Hamann et al., 2002). There is a clear need for more research in this area.

### 13.4.3 TEMPERATURE CHANGES AT SEA

Marine turtles are long-lived ectotherms and spend the majority of their lives in the marine environment. Among and within species considerable variation exists in habitat preference and diet, which influences growth rates, reproduction rates, age, and size at maturity. One of the central proximate influences on marine turtles in the marine environment is the direct effect of temperature on physiological processes, behavior, dietary ecology (food metabolism), and reproduction, and the indirect effect of temperature through changes to primary productivity, plankton community structure, and ocean circulation (Jennings et al., 2008). In terms of likely exposure, the IPCC predict various levels of increased global warming depending on the emissions scenario. Projections range from very conservative (1.1°C–2.9°C by 2099 (B1) to least conservative (2.4°C–6.4°C by 2099) (IPCC, 2007), with increasing sea surface temperatures throughout the world's oceans (IPCC, 2007). Indeed, recent research used by the IPCC indicates that, on average, the temperature of the upper layer of the world's oceans has increased by 0.6°C over the past 100 years (IPCC, 2007). Therefore, although local and regional variation in the magnitude and rates of sea temperature change is likely, especially along coastal zones, marine turtles will be exposed to increased sea temperatures at all age classes and across much of their range.

Marine turtle distributions are strongly influenced by sea temperatures. For example, Hawkes et al. (2007a) and McMahon and Hays (2006) combined satellite telemetry data, observation data, and oceanographic data to demonstrate thermal distribution constraints of ~15°C for loggerhead and leatherback turtles. Furthermore, the 15°C isotherm has shifted pole-wards by 330 km over the past 17 years, effectively increasing the available habitat for leatherback turtles in the northeast Atlantic by around 400 km in two decades (McMahon and Hays, 2006). More recently, Witt et al. (2010) used oceanographic data to model the past, current, and future distribution range of logger-head turtles and demonstrated a pole-ward shift by up to 100 km in habitat suitability within the Atlantic Ocean, and in the Mediterranean Sea the western basin is predicted to become increasingly favorable to year-round use by loggerhead turtles.

In addition to distributional shifts, ocean temperature increases influence primary productiv-ity (Jennings et al., 2008), recruitment, and population dynamics (van Houtan and Halley, 2011). Temperature affects annual breeding rates of loggerhead turtles, with higher temperatures related to lower ocean productivity and lower rates of breeding (Chaloupka et al., 2008), and influence breed-ing rates in green turtles and leatherback turtles (positive relationships with El Niño—Limpus and Nicholls, 1988); (Saba et al., 2007). While studies such as these provide extremely valuable insight into some impacts of warming oceans on marine turtles, many important gaps in our knowledge remain at large scales (e.g., distribution range models for other species/populations; understand-ing the influence of large scale ocean and atmosphere patterns on distribution and productivity), regional/local scales (localized shifts in habitat use in coastal and near-shore waters, Hawkes et al., 2007a), about the influence of temperature on seasonal and/or developmental migrations, and across scales (e.g., the combined impacts of changing proximate environmental conditions, changes to ecosystem processes and food availability, and the influence of biological/ecological variation such as growth rates, body size, diet, density effects).

While the impacts of temperature and other climate change processes on coral reef systems are well studied (Hoegh-Guldberg and Bruno, 2010), a rapidly growing body of literature investigates temperature impacts on habitats important for sea turtles such as sea grass (Orth et al., 2006; Short et al., 2011; Waycott et al., 2005), mangrove (Lovelock and Ellison, 2007), and inter/subtidal benthic habitats (Hutchings et al., 2007; Laurance et al., 2011; Schiel et al., 2004). While results generally suggest that marine flora and invertebrate fauna are sensitive to factors such as increased water temperatures, other factors such as ultra-violet radiation, pH, $CO_2$, nutrient load, suspended sedi-ments, and other biogeochemical factors are also driving change. Aside from the likely synergistic effects of biochemical changes on food webs and habitat quality, an aspect worthy of attention is the relationship(s) between biochemical and biological change on disease etiology. Essentially, knowl-edge of how interactions between stressors may influence biological processes is largely unknown, and is a noted knowledge gap (see Harley et al. [2006] for a discussion on nonlinear response and dependent effects). Thus, despite the increase in research attention, providing causal links between climate change drivers and change to biological and/or ecological attributes in marine taxa and habitats remains challenging.

## 13.4.4 SEA LEVEL RISE

Marine turtles will be exposed to changes in sea level through nesting beach availability and/or sta-bility, especially for sites in low-lying nonvegetated islands, mangrove communities, or between the tidal zone and coastal development (Fish et al., 2005; Mazaris et al., 2009; Woodroffe et al., 1999). The Fourth IPCC assessment concluded that global sea level rise has been consistent with warming of the climate, and average sea levels have risen 1.8 mm/year from 1961 to 3.1 mm/year since 1993. Additionally, under a conservative scenario (B1), the IPCC predict that by 2090–2099 sea levels will have risen 0.18–0.38 m. Clearly, because these are global averages, substantial regional and local variation is likely in the magnitude and potential impact of sea level rise due to local oceanographic processes and land movements. Downscaling global predictions to suit more appropriate management scales, while challenging, is ongoing in many locations (Nicholls and Cazenave, 2010). Aiding the development of downscaled predictions of potential sea level rise impact are advances in GIS-based tools, GPS technology, computing power, LIDAR imagery, and remote-sensing capabilities. Indeed, standard GIS-based flooding models derived from digital elevation modeled (DEM) data are now commonly used by coastal urban planners and managers and have proven useful for developing generalized predictions of sea level rise vulnerability to natural ecosystems (Baker et al., 2006; Fuentes et al., 2009c).

Assessing the vulnerability of marine turtle species to sea level rise, even at smaller scales such as populations, is complicated because, in addition to variation in abiotic beach factors described earlier, considerable intra- and interspecies variation is evident in physical characteristics of types and locations of beaches used. Given those challenges, studies at the population level, rather than beach-specific studies, are rare, and are essentially flooding-based models that use beach elevation profiles and sea level rise predictions to quantify the proportion of habitat likely to be impacted by rising sea levels. Two studies were conducted at a population scale—Baker et al. (2006) investigated the impact on the Hawaiian green turtle population and Fuentes et al. (2009c) investigated the vulnerability of green turtles from the northern Great Barrier Reef. Both studies assumed that beach height would remain constant and then used different predictions of sea level rise out to 2100 to demonstrate considerable variation in percent inundation among islands and across flooding scenarios. Although informative for management, these studies do not consider island stability or short- and long-term island change (patterns of erosion and accretion). Two recent geomorphological studies highlight the need to examine reef/island systems either at longer time scales or in a more holistic setting. Webb and Kench (2010) used a multidecade model of shoreline change to demonstrate that islands are dynamic landforms and sea level rise is one of several influencing factors on island shape and stability. Similarly, Dawson and Smithers (2010) used four decades of survey data to highlight that changes to beach structure on Raine Island, an important green turtle rookery, were generated by a complex series of patterns such as seasonal winds, wave action, reef productivity, sediment type, and the island's shape. Essentially, Dawson and Smithers (2010) found that Raine Island did not have a net loss of sand, rather the island had both erosion and accretion zones and was changing shape. Thus, when examining the vulnerability of sea level rise on marine turtle nesting sites, the scope studies should be expanded to include more complex models that predicted vulnerability (Cowell et al., 2006; Webb and Kench, 2010).

Many of the world's marine turtle populations utilize developed or partially developed coastlines for breeding, where understanding vulnerability of marine turtles to sea level rise is closely aligned with broader coastal management issues, and other anthropogenic pressures such as consumptive and nonconsumptive use, tourism, fishing, and light pollution. In areas with development pressures, vulnerability of marine turtles is linked to vulnerability of coastal communities, including social and economic structures, livelihoods, and coastal infrastructure. While these aspects are well investigated in their own right, studies that combine coastal development with marine turtle, or ecosystem, vulnerability are few. One study, Fish et al. (2008), used predictive models of sea level rise in the Caribbean and demonstrated that the landward shift of beaches was constrained by local Government development setback regulations (<90 m). Hence, the risk to beach ecosystems and infrastructure can be minimized by maintaining or expanding buffer zones. However, many small islands or highly developed coastal zones are not large enough to allow such buffers to be legislated, or sufficient land does not often exist to allow a managed retreat or other mitigation.

In addition to the inundation of nesting sites, sea level rise can increase exposure of clutches to salt water inundation, over-topping of beaches by storm surges or wave energy, and beach erosion (Caut et al., 2009; Fish et al., 2008; Fuentes et al., 2009c; Klein and Nichols, 1999). Gaining detailed site-specific knowledge of impacts is challenging and costly (time and money) because

(1) beach-specific impacts will vary according to beach vegetation assemblage, geomorphic profile, development profiles, coastal buffer zones (i.e., distance between tidal marks and coastal development), shoreline structure (i.e., rock walls, jetties, groynes), wave energy, storm frequency, shoreline type and shape, physical and the biological structure of the intertidal zone; and (2) understanding vulnerability will normally require gaining site-specific information from numerous beaches and/or along an irregular coastline. Most research has been focused on vulnerable areas—such as low-lying populated island nations, or coastal cities (Church et al., 2006; Cowell et al., 2006; Webb and Kench, 2010).

### 13.4.5 SEVERE WEATHER EVENTS

Many of the world's marine turtle populations are impacted by severe weather events such as tropical cyclones/hurricanes (Fuentes et al., 2011a; Limpus and Reed, 1985; Pike and Stiner, 2007a,b), tsunamis (Brodie et al., 2008; Hamann et al., 2006), and extreme flooding of coastal rivers. While severe weather events are short-term, aperiodic drivers of change over time, repeated and/or regular exposure could influence the distribution of nesting sites (Fuentes et al., 2011a; Pike and Stiner, 2007b). The IPCC (2007) concluded that the frequency of severe weather events such as tropical cyclones will likely increase. However, as with other climate processes, downscaling the global predictions of storm frequency to local scales is problematic and resultant regionalized models have varying degrees of certainty (Fuentes and Abbs, 2010). Analyses of long-term datasets from the south eastern United States and eastern Australia have demonstrated that cyclones can decrease annual hatchling production, cause slight differences in the timing of breeding seasons, or alter locations of nesting sites for different species (Fuentes and Abbs, 2010; Fuentes et al., 2011a; Pike and Stiner, 2007b). Interestingly, combining cyclone exposure data with data on reproductive timing could demonstrate that exposure risk varies temporally and could therefore be influenced by global warming (Pike, 2009; Pike and Stiner, 2007a; Weishampel et al., 2004). Analyses at longer multi-decadal time scales (Fuentes et al., 2011a) demonstrated that the frequency of coastline exposure to severe weather events can be an important factor in shaping the biophysical dynamics and geomorphological structure of beaches and could influence the distribution of nesting sites.

Aside from nesting beach impacts, severe weather events can also be aperiodic drivers of change in coastal foraging habitats. Severe weather and associated heavy rainfall, coastal flooding, storm surge, and increased wave action have each been implicated in the destruction of coastal intertidal and subtidal habitats such as sea grass. While species groups such as sea grass have evolved to cope with aperiodic events, repeated and/or more frequent events, or out of season events, have the potential to influence the ecosystem processes of the intertidal and subtidal habitats—which can impact on foraging marine turtles that rely on these habitats.

### 13.4.6 PRECIPITATION

Marine turtles will be exposed to precipitation both directly through nesting beach hydrology (i.e., cooling of sands, altered gas transport, and changes to sediment dynamics and beach hydrology) and indirectly through increased freshwater inputs into coastal ecosystems such as sea grass, inshore coral reefs, and mangroves. However, there is high variability and uncertainty in predicting how precipitation patterns will change in a changing climate. As with other processes, substantial regional and local variation is likely. Although few empirical data exist, some studies report links between biological characteristics and rainfall, such as seasonal shifts in sand temperatures and consequent changes in sex ratios (Houghton et al., 2007), changes to mortality rates (Foley et al., 2006; Kraemer and Bell, 1980), changes to nesting success of females, or changes to nesting activity (Limpus et al., 2001; Mortimer and Carr, 1987; Pike, 2008b). Given the importance of improving knowledge and accuracy of predictions for population level sex ratios, understanding the impact of seasonal and longer-term changes to precipitation patterns will be important.

### 13.4.7 OCEAN pH

A growing empirical research base indicates that greenhouse gas emissions (of which $CO_2$ is a main source) will continue to rise over coming decades, resulting in ocean acidification. The ocean has absorbed close to a 1/3 of the $CO_2$ produced by human activities, and one result of this has been a steady decrease in ocean pH (Hoegh-Guldberg et al., 2007). Studies on the effect of lowered ocean pH have mostly focused on coral reef systems, indicating that increased rates of change are predicted to influence the structure and resilience of coral reef habitat (Anthony et al., 2011). This could have indirect influence on marine turtle species that utilize coral reef systems, such as hawksbill, loggerhead, and green turtles (Hamann et al., 2007; Poloczanska et al., 2009). Further, changes to ocean pH will potentially influence reproductive success, growth, fertilization rates, etc. of other marine organisms such as jellyfish, salps, crustaceans, mollusks, sponges, etc., potentially indirectly impacting marine turtles. This is clearly an area for future research.

### 13.4.8 OCEAN CIRCULATION

Marine turtle dispersal and distributions are linked to some degree by ocean and coastal current circulation, especially during hatchling dispersal from nesting beaches, the dispersal of post hatchlings and young juveniles through oceanic waters, and the migration of adult turtles to and from breeding areas. Indeed, the role of currents and other environmental factors in marine turtle dispersal has received increased attention (Godley et al., 2010; Hamann et al., 2011; Santana Garcon et al., 2010). In general, studies found strong relationships between locations of favorable nesting areas and proximity to ocean currents that facilitate hatchling dispersal and shape future migrations and movements of individuals. Although there is high uncertainty about the magnitude and temporal and spatial variation, a likely consequence of a warming climate, especially disproportionate heating, is changes to the global wind patterns, large-scale ocean-atmosphere patterns and the strength, direction, and behavior of major current systems (Hoegh-Guldberg, 2011). Given uncertainty in both predicting changes to wind systems and oceanic currents, and understanding how marine turtles use current systems, predicting how turtles will be affected is not yet possible.

## 13.5 CLIMATE AND MANAGEMENT SCALES

Marine turtles utilize large geographic areas, are long-lived, have diverse diets, and there is substantial inter- and intraspecies variation in threat exposure, management need, available biological data, and population trajectory. In addition, few data exist on thresholds of concern or predicted exposure rates among and between species. However, challenges of predicting and managing impacts are not solely limited to climate change, but exist for other broad-scale pervasive threats such as bycatch, coastal development, pollution, and consumptive use. The variation within and among species gives rise to the question, what is the most adequate scale to monitor and manage marine turtle populations? Indeed, over the past 15 years, substantial debate occurred about appropriate management units to assess marine turtle status and condition (Godfrey and Godley, 2008; Mrosovsky, 2000; Seminoff and Shanker, 2008). In some regions, for some species, substantial data on population structure exist and hence management at a population level is possible (i.e., management units, stocks, genetic populations). For example, the Australian region uses management units or stocks (Dethmers et al., 2006; Moritz et al., 2001). Recently, Conant et al. (2009) and Wallace et al. (2010) defined distinct population segments and regional management units (RMUs), respectively, as an approach for delineating management boundaries. RMUs incorporate molecular information, nesting distribution data, and migration data. As such, multiscale RMUs should reflect population structure and variation in habitat use, and are likely to be relevant for monitoring and managing when genetic population structures, or the boundaries of the structure, are unknown (Wallace et al., 2010).

Global climate models that project temperature generally agree that, at large spatial scales such as ocean basins, significant variation in exposure is likely, making them relevant for species with large ecological scales like marine turtles. To illustrate the likely variation and potential influence on marine turtles, we examined how exposure to increased temperature varied across marine turtle RMUs. Fifty-eight RMUs were identified by Wallace et al. (2010), each was scored using combinations of empirical data and expert opinion on a 2D matrix for threats and risk (Wallace et al., 2011). One result was the identification of the 11 RMUs considered to be the world's most threatened. We use those RMUs (n = 11) to demonstrate the variability in vulnerability, a technique that can be broadly applied to other RMUs or management units. Our purpose is to illustrate how climate change data can be used with some examples of data and models. Our intention is not to provide definitive answers. Rather, we focus on exposure to increasing air and sea surface temperatures because marine turtles are known to be sensitive to changes therein, and because many climate models provide projections for temperature variables.

### 13.5.1 Climate Change Data Availability

Many research centers around the world have developed global projections of climate change through climate models that are called coupled ocean-atmosphere global circulation models—they link models of ocean and atmospheric change. Results of these models were used in the IPCC's fourth assessment report. As part of an effort to make the climate modeling outputs widely available and accessible to researchers, policy makers, NGOs, and other interested parties, a data portal was set up to provide access to the outputs of these global models (WCRP CMIP3 Multi-Model Data, https:// esg.llnl.gov:8443/index.jsp). This data portal provides access to climate data, giving researchers the opportunity to examine projections of relevance to marine turtles such as temperature.

Global climate models vary in their projections because each model of changes in magnitude and exposure for various climate change factors (e.g., air and sea temperature, sea level rise) uses different assumptions about how the earth might be changing over the next 100 years. While a review of the different models is beyond the scope of this chapter, much of the variation occurs because models can be developed using a range of scenarios and can therefore include a range of assumptions about greenhouse gas emissions, global human population growth, and development of new technologies. The IPCC Special Report on Emissions Scenarios (IPCC, 2000) document six families of scenarios based on different emission trajectories (referred to as A1F1, A1B, A1T, A2, B1, and B2). Within these families are several variants based on slight modifications to one or more elements about population growth, energy source, and technological change. Model outputs vary because each has their own way of implementing the scenarios, they can use different model structures and assumptions, and have various levels of complexity and resolution. Thus models might develop projections that differ from one another in global patterns of change for the same scenario.

To illustrate the degree of variation in how marine turtle management units, in this case RMUs, will be exposed to increased temperature, we analyzed three models for illustrative purposes. We selected models that had a relatively high resolution for both air and sea surface temperatures. The models we used were CSIRO Mk3.5 (Gordon et al., 2010) http://www.cawcr.gov.au/publications/technicalreports/CTR_021.pdf), MIROC (http://www.ccsr.u-tokyo.ac.jp/kyosei/hasumi/MIROC/tech-repo.pdf) (Hasumi and Emori, 2004), and NCAR CCSM3 (http://www.cgd.ucar.edu/cms/ccm3/). For each of the three models, we analyzed three of the scenarios from the IPCC Special Report on Emissions Scenarios (IPCC, 2000). These represent high (A2), medium (A1B), and low (B1) emissions trajectories (see IPCC, 2007, for detailed descriptions of each). In general, high-emissions scenario (A2) describes a heterogeneous world with high population growth, slow economic development, and slow technological change. The medium scenario (A1B) assumes a world of very rapid economic growth, a population that peaks mid-century and rapid introduction of new, efficient technologies that balance fossil intensive and non-fossil energy sources. The low-emission scenario (B1) assumes a world with more rapid changes in economic structures toward a service and information economy (IPCC, 2000, 2007).

## 13.5.2 ANALYSES OF AIR AND SEA SURFACE TEMPERATURE CLIMATE CHANGE PROJECTIONS FOR RMUS

The climate models produce outputs of projected temperature changes for the air and the sea over 100 years for each scenario. Because the raw temperature data for each month fluctuates with seasons and we were interested in examining long-term trends, we analyzed the data with season removed. Removing seasons involves calculating the anomalies for each model so that season does not bias the results but retains the long-term trends.

While the climate models provide outputs of projected temperatures over the next 100 years, the main interest for marine turtle researchers are the trends in those temperature changes, rather than the projected temperatures themselves. There are two ways one can think about these temperature trends. First, they can be thought of in terms of the consistency of change—i.e., in a region, are temperatures increasing, decreasing, or not changing much? We used a statistic called the Mann-Kendall monotonic trend to calculate this consistency of change (Eastman, 2009). A second way to think about and summarize temperature trends is in terms of the rate of change—i.e., in any region, how much are temperatures changing per month or decade? To calculate this trend, we used the ordinary least squares linear trend (Eastman, 2009). We then summarized the consistency and rate of change for each of the 11 RMUs.

The aforementioned approach demonstrates that, as expected, the RMUs are projected to be variably affected by increases to air and sea surface temperature (Tables 13.1 and 13.2). The spatial variation within and between RMUs means that some RMUs are likely to be more severely affected by temperature increases than others, which may affect approaches to management of particular RMUs (e.g., different management strategies may be needed in RMUs more severely affected). Within models, spatial patterns were generally similar between scenarios (i.e., for any model, the spatial patterns seen in B1, A1B, A2 were similar), but the magnitude of those changes varied (i.e., greater rates of change are projected for the higher greenhouse gas emissions scenarios).

Generally, it is informative to use outputs from global climate change projections to assess projected changes for marine turtle management units such as RMUs or genetic stocks, as it allows for comparison between and among species. The ecological scales of marine turtle populations, however, are large and contain considerable variation in temperature projections within them (Tables 13.1 and 13.2). Figure 13.1 shows examples of climate change trends for one climate model (CSIRO) and scenario (A2) over 100 years. Consistency of change can range from −1 to +1, with −1 indicating temperatures are consistently decreasing, +1 consistently increasing, and 0 stable or no pattern. The rate of change is measured in degrees Celsius change per month, with positive numbers indicating increases in temperature, and negative numbers indicating decreases. The polygons and numbers in Figure 13.1a indicate regional management units for marine turtles: 07 = Northeast Indian Ocean (arribada) *Lepidochelys olivacea*, 08 = West Indian Ocean *L. olivacea*, 11 = East Atlantic Ocean *Eretmochelys imbricata*, 13 = East Pacific Ocean *E. imbricata*, 16 = Northeast Indian Ocean *E. imbricata*, 21 = West Pacific Ocean *E. imbricata*, 22 = West Pacific Ocean *E. imbricata*, 23 = Northeast Atlantic Ocean (Cape Verde) *Caretta caretta*, 32 = Northeast Indian Ocean *Caretta caretta*, 31 = North Pacific Ocean *Caretta caretta*, 55 = East Pacific Ocean *Dermochelys coriacea*. Using the 11 most vulnerable RMUs (Wallace et al., 2011) as a case study we indicate that, despite the variability, some RMUs will be exposed to higher rates and more consistent changes to both air and sea surface temperatures than others. Indeed, Tables 13.1 and 13.2 demonstrate that RMUs 8 (western Indian Ocean olive ridley turtles), 11 (east Atlantic hawksbill turtles), and 21 (western Pacific Ocean hawksbill turtles) have relatively high consistency of change compared to the other RMUs, and RMUs 8 and 11 have relatively high rates of change; however, because of the large size of RMUs, rates of change show greater variability than consistency of change. Clearly though, this type of broad analysis is most relevant at global and large regional scales, and as improved downscaled climate models become available, they could be used for smaller regions. Also, an understanding of local scale variability in exposure, sensitivity, and adaptive capacity remains crucially important, and is necessary for site-specific management.

**TABLE 13.1**

**Summary of the Mean Consistency of Change for Air Temperature for RMUs from Three Climate Model Projections Over 100 Years**

| | | | Mean Consistency of Change ± Standard Deviation | | | | | | | | |
|---|---|---|---|---|---|---|---|---|---|---|---|
| | | | CCSM3 | | | CSIRO | | | Miroc | | |
| RMU | Species | Basin | A2 | A1B | B1 | A2 | A1B | B1 | A2 | A1B | B1 |
| 07 | *Lepidochelys olivacea* | Northeast Indian Ocean (arribadas) | 0.76 ± 0.06 | **0.68 ± 0.06** | 0.42 ± 0.08 | **0.67 ± 0.05** | 0.63 ± 0.05 | 0.49 ± 0.07 | **0.79 ± 0.03** | **0.79 ± 0.03** | **0.68 ± 0.05** |
| 08 | *L. olivacea* | West Indian Ocean | **0.77 ± 0.08** | **0.68 ± 0.09** | **0.45 ± 0.1** | **0.66 ± 0.06** | **0.68 ± 0.04** | **0.56 ± 0.04** | 0.76 ± 0.08 | **0.79 ± 0.05** | **0.70 ± 0.06** |
| 11 | *Eretmochelys imbricate* | East Atlantic Ocean | **0.79 ± 0.04** | **0.73 ± 0.04** | **0.52 ± 0.06** | **0.66 ± 0.47** | **0.68 ± 0.03** | **0.58 ± 0.04** | 0.73 ± 0.03 | **0.79 ± 0.03** | **0.68 ± 0.04** |
| 13 | *E. imbricate* | East Pacific Ocean | 0.64 ± 0.09 | 0.55 ± 0.09 | 0.34 ± 0.08 | 0.55 ± 0.07 | 0.56 ± 0.05 | 0.44 ± 0.06 | 0.67 ± 0.04 | 0.74 ± 0.03 | 0.66 ± 0.05 |
| 16 | *E. imbricate* | Northeast Indian Ocean | 0.72 ± 0.05 | 0.64 ± 0.05 | 0.37 ± 0.06 | 0.64 ± 0.05 | 0.60 ± 0.5 | 0.45 ± 0.06 | **0.77 ± 0.04** | **0.79 ± 0.03** | **0.68 ± 0.05** |
| 21 | *E. imbricate* | West Pacific Ocean | **0.82 ± 0.04** | **0.74 ± 0.04** | **0.49 ± 0.06** | **0.68 ± 0.08** | **0.69 ± 0.05** | **0.57 ± 0.06** | **0.77 ± 0.02** | **0.82 ± 0.02** | **0.71 ± 0.02** |
| 22 | *E. imbricate* | West Pacific Ocean | 0.75 ± 0.09 | 0.65 ± 0.10 | 0.41 ± 0.11 | 0.56 ± 0.10 | 0.56 ± 0.08 | 0.44 ± 0.09 | 0.75 ± 0.05 | 0.76 ± 0.05 | 0.65 ± 0.07 |
| 23 | *Caretta caretta* | Northeast Atlantic Ocean (Cape Verde) | 0.55 ± 0.09 | 0.49 ± 0.08 | 0.25 ± 0.08 | 0.54 ± 0.06 | 0.58 ± 0.07 | 0.47 ± 0.06 | 0.68 ± 0.04 | 0.66 ± 0.05 | 0.58 ± 0.05 |
| 32 | *Caretta caretta* | Northeast Indian Ocean | 0.75 ± 0.06 | 0.66 ± 0.06 | 0.41 ± 0.08 | 0.50 ± 0.08 | 0.55 ± 0.06 | 0.41 ± 0.07 | 0.68 ± 0.04 | 0.68 ± 0.04 | 0.56 ± 0.05 |
| 31 | *Caretta caretta* | North Pacific Ocean | 0.60 ± 0.08 | 0.49 ± 0.09 | 0.26 ± 0.07 | 0.65 ± 0.05 | 0.61 ± 0.05 | 0.47 ± 0.07 | **0.78 ± 0.03** | **0.79 ± 0.03** | **0.68 ± 0.05** |
| 55 | *Dermochelys coriacea* | East Pacific Ocean | 0.62 ± 0.09 | 0.51 ± 0.09 | 0.29 ± 0.09 | 0.53 ± 0.07 | 0.56 ± 0.05 | 0.45 ± 0.06 | 0.66 ± 0.05 | 0.71 ± 0.05 | 0.62 ± 0.07 |

*Note:* −1 = Trend is consistently decreasing; +1 = trend is consistently increasing; and 0 = stable. Bold results correspond to most affected RMUs for each scenario and model.

**TABLE 13.2**

## Summary of the Mean Rate of Change per RMU per Month from Three Climate Model Projections Over 100 Years

Mean Rate of Change ± Standard Deviation

| RMU | Species | CCSM3 | | | CSIRO | | | Miroc | | |
|---|---|---|---|---|---|---|---|---|---|---|
| | | A2 | A1B | B1 | A2 | A1B | B1 | A2 | A1B | B1 |
| 07 | *Lepidochelys olivacea* | **0.0025 ± 6.2×10⁻⁵** | 0.0016 ± 7.1×10⁻⁵ | 7.0×10⁻⁴ ± 3.8×10⁻⁵ | 0.0019 ± 6.7×10⁻⁵ | 0.0019 ± 5.3×10⁻⁵ | 0.0012 ± 7.5×10⁻⁵ | **0.0029 ± 1×10⁻⁴** | **0.0032 ± 7.6×10⁻⁵** | 0.0019 ± 7.8×10⁻⁵ |
| 08 | *L. olivacea* | **0.0026 ± 2.2×10⁻⁴** | 0.0017 ± 2.1×10⁻⁴ | **7.8×10⁻⁴ ± 9.4×10⁻⁵** | **0.0022 ± 2.2×10⁻⁴** | **0.0023 ± 2.8×10⁻⁴** | **0.0015 ± 1.7×10⁻⁴** | **0.0033 ± 3.1×10⁻⁴** | **0.0033 ± 1.5×10⁻⁴** | **0.002 ± 1.3×10⁻⁴** |
| 11 | *Eretmochelys imbricata* | **0.0025 ± 1.8×10⁻⁴** | **0.0018 ± 1.5×10⁻⁴** | **8.4×10⁻⁴ ± 4.9×10⁻⁵** | **0.0021 ± 1.5×10⁻⁴** | **0.0022 ± 1.5×10⁻⁴** | **0.0015 ± 1.5×10⁻⁴** | 0.0026 ± 1.5×10⁻⁴ | **0.0032 ± 2.1×10⁻⁴** | 0.0019 ± 9.4×10⁻⁵ |
| 13 | *E. imbricata* | 0.0022 ± 4.8×10⁻⁴ | 0.0015 ± 3.6×10⁻⁴ | 7.7×10⁻⁴ ± 1.8×10⁻⁴ | **0.002 ± 3×10⁻⁴** | 0.0022 ± 3×10⁻⁴ | **0.0015 ± 1.7×10⁻⁴** | 0.0024 ± 4.1×10⁻⁴ | 0.003 ± 3.6×10⁻⁴ | **0.002 ± 2.1×10⁻⁴** |
| 16 | *E. imbricata* | **0.0025 ± 9.1×10⁻⁵** | 0.0016 ± 0.1×10⁻³ | 6.9×10⁻⁴ ± 4.3×10⁻⁵ | 0.0019 ± 6.8×10⁻⁵ | 0.0020 ± 7.4×10⁻⁵ | 0.0012 ± 7.2×10⁻⁵ | **0.0029 ± 1.4×10⁻⁴** | **0.0032 ± 8.9×10⁻⁵** | 0.0019 ± 8.3×10⁻⁵ |
| 21 | *E. imbricata* | 0.0024 ± 7.5×10⁻⁵ | 0.0016 ± 6.6×10⁻⁵ | 7.0×10⁻⁴ ± 5.8×10⁻⁵ | **0.002 ± 1.5×10⁻⁴** | **0.0022 ± 2.6×10⁻⁴** | 0.0013 ± 1.6×10⁻⁴ | 0.0025 ± 1.4×10⁻⁴ | 0.0029 ± 1.8×10⁻⁴ | 0.0018 ± 1.1×10⁻⁴ |
| 22 | *E. imbricata* | 0.0024 ± 1.4×10⁻⁴ | 0.0016 ± 8.8×10⁻⁵ | 6.8×10⁻⁴ ± 7.3×10⁻⁵ | 0.0018 ± 1.9×10⁻⁴ | 0.0020 ± 2×10⁻⁴ | 0.0013 ± 1.7×10⁻⁴ | 0.0028 ± 2.2×10⁻⁴ | 0.003 ± 1.6×10⁻⁴ | 0.0018 ± 1.1×10⁻⁴ |
| 23 | *Caretta caretta* | 0.0022 ± 4.6×10⁻⁴ | **0.0018 ± 3.9×10⁻⁴** | 7.1×10⁻⁴ ± 1.9×10⁻⁴ | 0.0017 ± 3.5×10⁻⁴ | 0.0017 ± 2.4×10⁻⁴ | 0.0012 ± 1.2×10⁻⁴ | 0.0027 ± 1.8×10⁻⁴ | 0.0026 ± 2.8×10⁻⁴ | 0.0018 ± 1.4×10⁻⁴ |
| 32 | *Caretta caretta* | **0.0028 ± 7.3×10⁻⁴** | **0.0018 ± 4.3×10⁻⁴** | **8.8×10⁻⁴ ± 3.2×10⁻⁴** | 0.0018 ± 2.8×10⁻⁴ | 0.0021 ± 2.6×10⁻⁴ | 0.0014 ± 2.1×10⁻⁴ | **0.0033 ± 7.2×10⁻⁴** | **0.0033 ± 3.5×10⁻⁴** | **0.0021 ± 2.4×10⁻⁴** |
| 31 | *Caretta caretta* | **0.0025 ± 6.8×10⁻⁵** | 0.0016 ± 6.4×10⁻⁵ | 6.9×10⁻⁴ ± 3.6×10⁻⁵ | 0.0019 ± 6.3×10⁻⁵ | 0.0020 ± 6.3×10⁻⁵ | 0.0012 ± 7×10⁻⁵ | **0.0029 ± 1.2×10⁻⁴** | **0.0032 ± 8.6×10⁻⁵** | 0.0019 ± 7.6×10⁻⁵ |
| 55 | *Dermochelys coriacea* | 0.0020 ± 5.1×10⁻⁴ | 0.0014 ± 3.7×10⁻⁴ | 6.5×10⁻⁴ ± 2.2×10⁻⁴ | 0.0017 ± 4.6×10⁻⁴ | 0.0020 ± 3.7×10⁻⁴ | 0.0014 ± 2.1×10⁻⁴ | 0.0022 ± 4.1×10⁻⁴ | 0.0028 ± 4.8×10⁻⁴ | 0.0018 ± 2.9×10⁻⁴ |

Bold results correspond to most affected RMUs for each scenario and model.

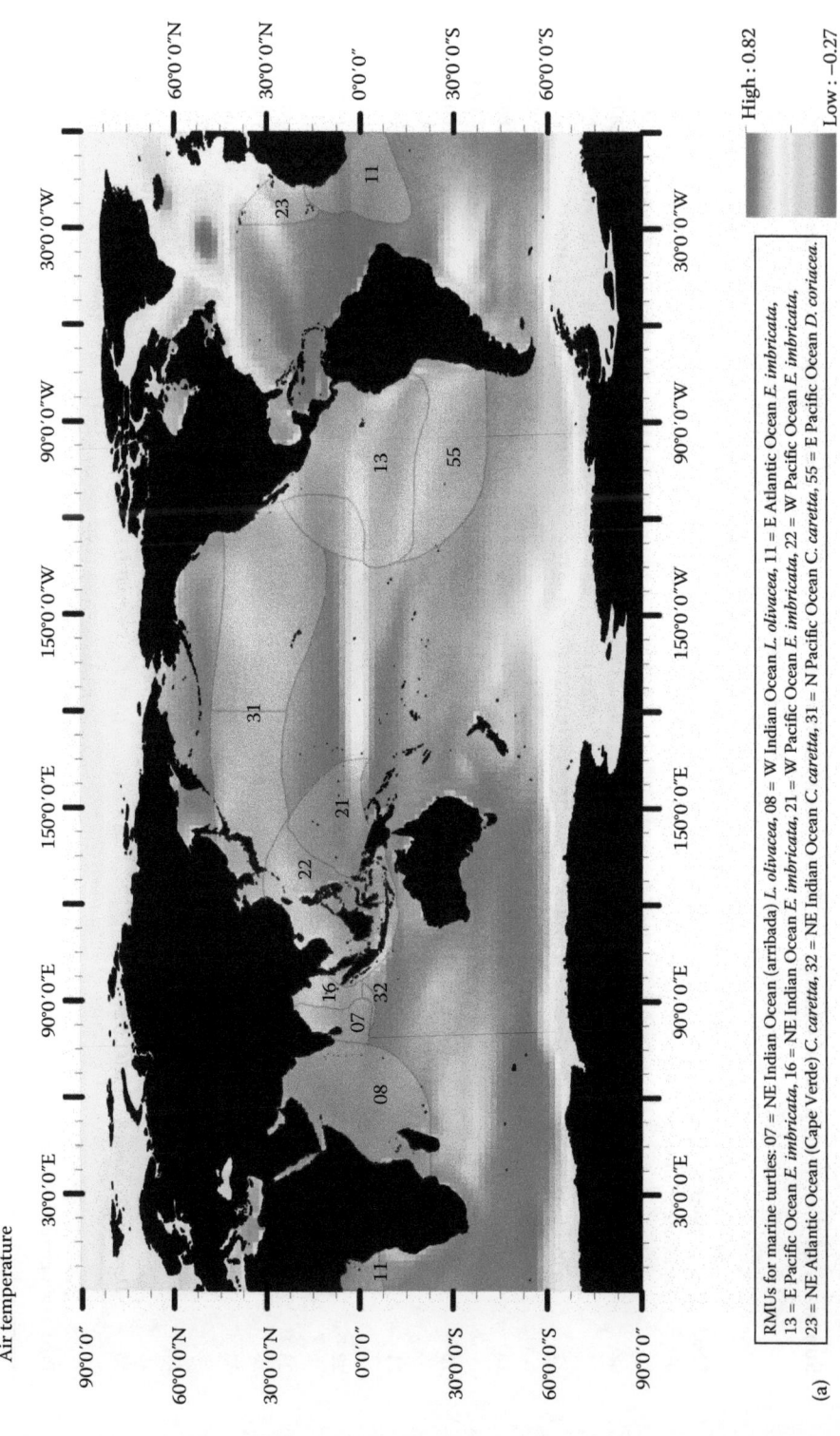

Air temperature

RMUs for marine turtles: 07 = NE Indian Ocean (arribada) *L. olivacea*, 08 = W Indian Ocean *L. olivacea*, 11 = E Atlantic Ocean *E. imbricata*, 13 = E Pacific Ocean *E. imbricata*, 16 = NE Indian Ocean *E. imbricata*, 21 = W Pacific Ocean *E. imbricata*, 22 = W Pacific Ocean *E. imbricata*, 23 = NE Atlantic Ocean (Cape Verde) *C. caretta*, 32 = NE Indian Ocean *C. caretta*, 31 = N Pacific Ocean *C. caretta*, 55 = E Pacific Ocean *D. coriacea*.

High : 0.82

Low : −0.27

(a)

**FIGURE 13.1** Parts (a)–(d) show examples of climate change trends for one climate model (CSIRO) and scenario (A2) over 100 years. (a) Trends in air temperature— Consistency of change. Values are −1 to +1, with −1 indicating temperatures are consistently decreasing, +1 consistently increasing, and 0 stable or no pattern.

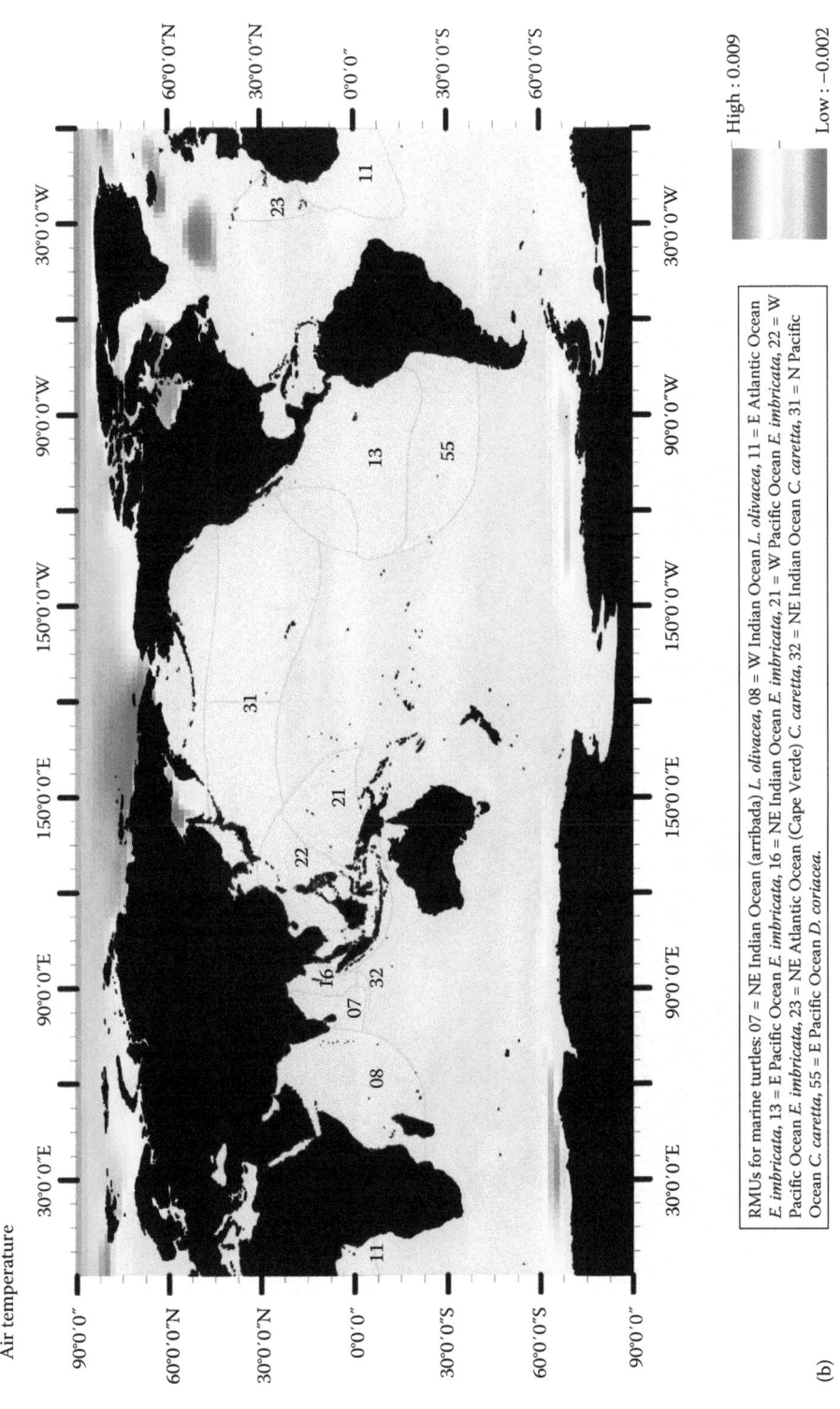

**FIGURE 13.1 (continued)** (b) Trends in air temperature—Rate of change, measured in degrees Celsius change per month, with positive numbers indicating increases in temperature, and negative numbers indicating decreases.

Air temperature

RMUs for marine turtles: 07 = NE Indian Ocean (arribada) *L. olivacea*, 08 = W Indian Ocean *L. olivacea*, 11 = E Atlantic Ocean *E. imbricata*, 13 = E Pacific Ocean *E. imbricata*, 16 = NE Indian Ocean *E. imbricata*, 21 = W Pacific Ocean *E. imbricata*, 22 = W Pacific Ocean *E. imbricata*, 23 = NE Atlantic Ocean (Cape Verde) *C. caretta*, 32 = NE Indian Ocean *C. caretta*, 31 = N Pacific Ocean *C. caretta*, 55 = E Pacific Ocean *D. coriacea*.

High : 0.009

Low : −0.002

(b)

Sea surface temperature

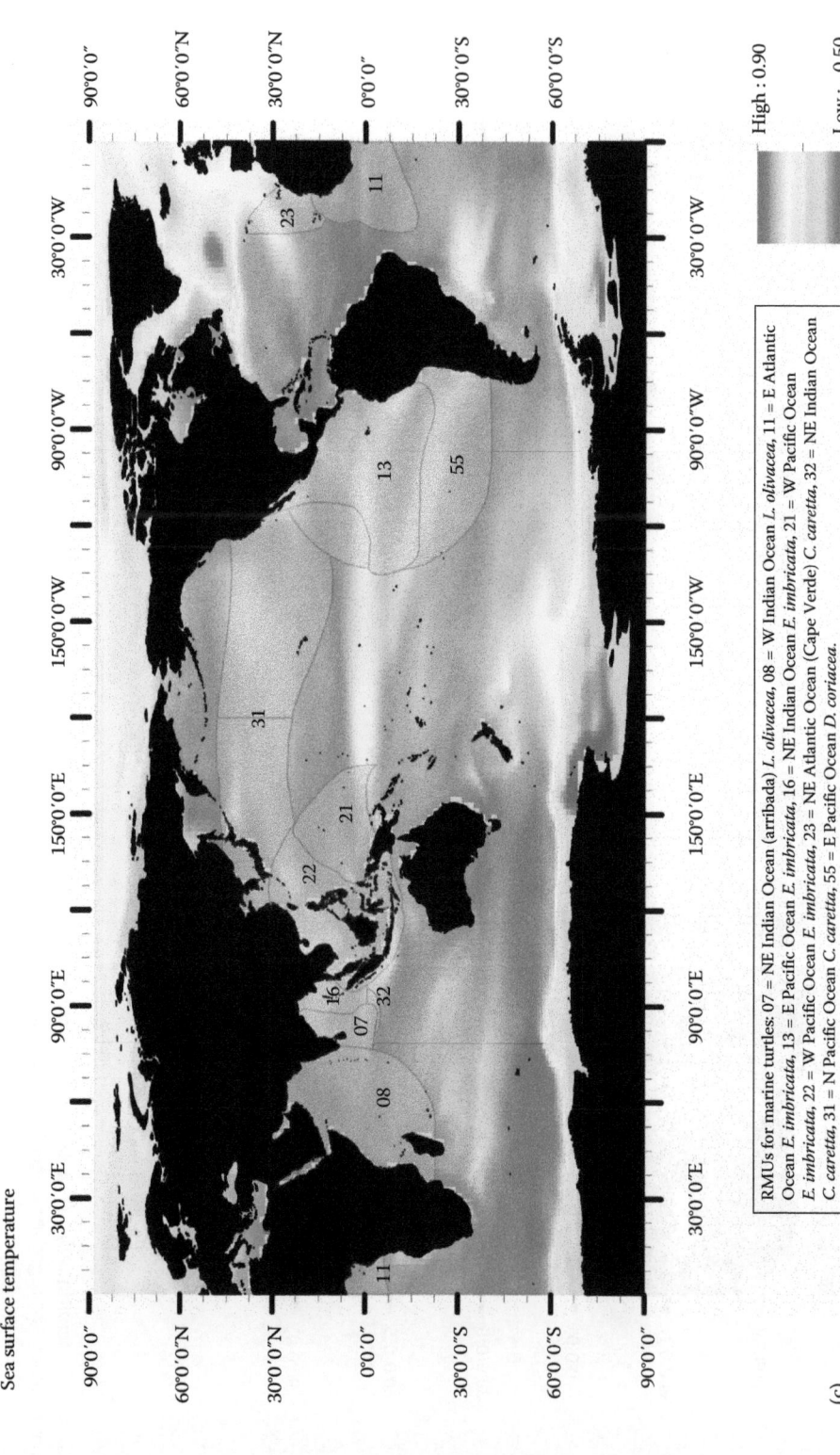

RMUs for marine turtles: 07 = NE Indian Ocean (arribada) *L. olivacea*, 08 = W Indian Ocean *L. olivacea*, 11 = E Atlantic Ocean *E. imbricata*, 13 = E Pacific Ocean *E. imbricata*, 16 = NE Indian Ocean *E. imbricata*, 21 = W Pacific Ocean *E. imbricata*, 22 = W Pacific Ocean *E. imbricata*, 23 = NE Atlantic Ocean (Cape Verde) *C. caretta*, 32 = NE Indian Ocean *C. caretta*, 31 = N Pacific Ocean *C. caretta*, 55 = E Pacific Ocean *D. coriacea*.

High : 0.90

Low : −0.50

(c)

**FIGURE 13.1 (continued)** (c) Trends in sea surface temperature—Consistency of change. Values are −1 to +1, with −1 indicating temperatures are consistently decreasing, +1 consistently increasing, and 0 stable or no pattern.

Sea surface temperature

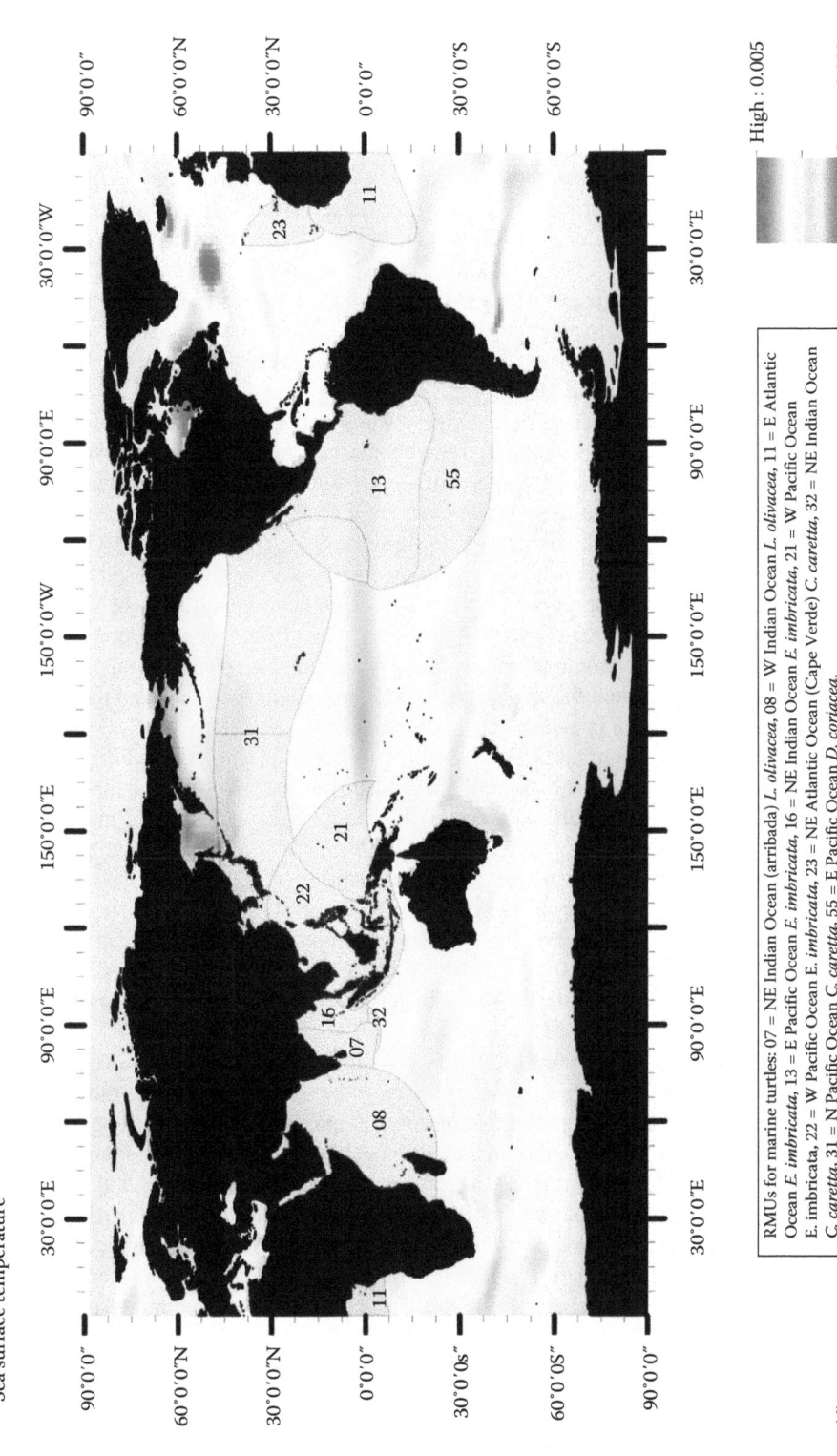

RMUs for marine turtles: 07 = NE Indian Ocean (arribada) *L. olivacea*, 08 = W Indian Ocean *L. olivacea*, 11 = E Atlantic Ocean *E. imbricata*, 13 = E Pacific Ocean *E. imbricata*, 16 = NE Indian Ocean *E. imbricata*, 21 = W Pacific Ocean *E. imbricata*, 22 = W Pacific Ocean *E. imbricata*, 23 = NE Atlantic Ocean (Cape Verde) *C. caretta*, 32 = NE Indian Ocean *C. caretta*, 31 = N Pacific Ocean *C. caretta*, 55 = E Pacific Ocean *D. coriacea.*

High : 0.005

Low : −0.003

(d)

**FIGURE 13.1 (continued)** (d) Trends in sea surface temperature—Rate of change, measured in degrees Celsius change per month, with positive numbers indicating increases in temperature, and negative numbers indicating decreases.

## 13.6   UNDERSTANDING THE VULNERABILITY OF MARINE TURTLES TO CLIMATE CHANGE

We need to understand how vulnerable particular species and habitat types are to the direct, indirect, and synergistic impacts of climate change. This will prove challenging for marine turtles given the expected variation in regional and local variation in climate change impacts, coupled with their large ecological (temporal and spatial) scale (see previous section), and incomplete understanding of life history traits and variation. Vulnerability of a species, management unit, or habitat depends on three general criteria: (1) exposure, (2) sensitivity, and (3) adaptive capacity (for a localized example, see Hamann et al., 2007) (Table 13.1). Exposure reflects the exposure for an area or species to a climatic process (e.g., Figure 13.1). According to Hamann et al. (2007), sensitivity refers to the level that climatic processes will impact each area or species. Adaptive capacity is the ability of the species or area to adapt to climatic impacts. Most studies to date have addressed vulnerability at species or habitat levels (see chapters contained in Johnson and Marshall [2007]), and single climate drivers (e.g., temperature).

The key challenge to understanding, predicting, and mitigating climate change impacts is to undertake studies that address multiple and synergistic impacts (Cheung et al., 2009; Walther et al., 2002a). Fuentes et al. (2011b) used a combination of empirical field-based data, climate modeling, and expert opinion to complete a vulnerability assessment for one life cycle stage (nesting and egg incubation) in a single genetic population of green turtle in Queensland (northern Great Barrier Reef), Australia. Aside from the paper's main conclusion that (1) nesting grounds used by the northern Great Barrier Reef green turtle population closer to the equator are the most vulnerable to climate change; and that (2) in the long term, reducing the threats from increased temperature may provide a greater return in conservation investment than mitigating the impacts from other climatic processes, their study demonstrated the exceptionally extensive data necessary, and the challenges in completing an assessment even in a data-rich environment.

One of the key challenges identified by several authors when assessing vulnerability is the high level of uncertainty related to exposure, sensitivity, and adaptive capacity of marine turtles (Fuentes et al., 2011b; Hamann et al., 2007, 2010; Hawkes et al., 2009). Some indication of marine turtles' ability to adapt to climate change may be provided by information on their current status, trend, the threats they face or being faced (e.g., predation, harvest, bycatch), the awareness and legislative compliance at several scales, their ability to shift habitats (such as nesting beaches or foraging habitats) or shift preferences for diet, and the stability and condition of their critical habitat sites (Fuentes et al., 2011b; Witt et al., 2010). Further insights may be obtained by understanding how sea turtles adapted to climate change in the past, which could possibly be obtained from advances in genetic studies.

In terms of understanding the mechanisms through which marine turtles may adapt to climate change, more is known about possible changes to the reproductive phase than the other life cycle stages. Among the literature, studies have used either/both empirical data and expert opinion to suggest that climate change will lead to changes in the timing of breeding (Pike, 2009; Pike et al., 2006; Weishampel et al., 2004, 2010), changes to nesting sites or foraging sites (Hamann et al., 2007; Hays et al., 2001b), and adapting in situ through adjusting pivotal temperatures (Davenport, 1989; Hawkes et al., 2007a). While adaptation is possible, the temporal scale over which they can occur (possibly multiple generations) may preclude them from occurring in a timeframe necessary for marine turtles to adapt to climate change. Furthermore, the ability of current and future monitoring programs to detect changes if they do occur might be limited by the long timeframes involved in some of these processes. Past sea levels and shoreline boundaries indicate that marine turtles currently use habitats that were either flooded, or many thousands of kilometers from the coast under historical sea level scenarios, which demonstrates they can redistribute nesting sites, migratory routes, and foraging areas (Poloczanska et al., 2009). However, anthropogenic pressure has likely impeded the adaptive capacity of most species and populations to climate change. Clearly, there is an important need for research and monitoring programs to address questions related to the adaptive capacity of marine turtles and their habitats to a changing climate.

## 13.7  MANAGING MARINE TURTLE THREATS IN A CHANGING WORLD

Concern over the potential impact of climate change on marine turtles has prompted discussion of management strategies to mitigate impacts. Strategies vary from boosting resilience or minimizing mortality through habitat protection and management to more active and direct manipulation of nests and the nesting habitat to reduce site-specific impacts. The majority of suggested strategies focus on the nesting environment, as this is where baseline knowledge on exposure rates and thresholds is strongest and where implementation and monitoring is logistically easiest (see Fuentes et al., 2012; Patino-Martinez et al., 2012). However, implementation of strategies (especially those that manipulate thermal profile of beaches, e.g., with shade and/or sprinklers), while certainly discussed, is constrained by limited information on the costs and benefits associated with implementation, and the effectiveness of strategies, at reducing impacts at relevant temporal and spatial scales. Interestingly, some strategies discussed in a climate change context have already been used with nonclimatic threats to sea turtles (e.g., hatcheries to prevent egg predation or consumption and replanting beach side vegetation for dune stability) and the conservation merit of these tools is debated (Morreale et al., 1982; Mrosovsky and Yntema, 1980) for examples of early discussion and (Kornaraki et al., 2006; Mrosovsky, 2006, 2008; van de Merwe et al., 2005, 2006) for recent discussion. Importantly, little discussion has occurred about strategies for mitigating in-water impacts from climate change or boosting resilience of foraging turtles, presumably because of data paucity.

Climate change impacts are diverse and will have considerable species and regional variation, and thus no universally best set of tools exists for managing climate-specific impacts to marine turtles. Rather, managing marine turtles in a changing climate will rely on a mix of approaches, including (1) reducing the rate and extent of climate change, (2) enhancing sea turtle resilience by mitigating other threats and protecting representative samples of important habitats, (3) understanding and managing at appropriate temporal and spatial scales, and (4) applying adaptive management techniques to implement and test adaptation strategies. From an ecological perspective, management strategies come with some degree of risk, such as impacts to other species, negative impacts due to inadvertent selection of bad genes (e.g., Mrosovsky, 2006), and changes to beach-dune ecosystems. Understanding and mitigating risk requires careful case-by-case consideration.

A growing body of literature indicates that climate change is more than an environmental issue because it will likely necessitate broad-scale social and economic changes. In addition to their ecosystem value, marine turtles are culturally valuable (e.g., in indigenous societies), economically valuable (Tisdell and Wilson, 2001), and have been used as a flagship species group for marine conservation (Eckert and Hemphill, 2005; Frazier, 2005). Thus planning for and potentially mitigating the negative impacts, as well as potentially benefiting from any positive impacts, of climate change on marine turtles will require social change at various scales—from community level up to broader national and international political scales. As iconic species, marine turtles could be used to promote an understanding of the synergistic impacts of climate change on marine and coastal biodiversity, which could provide incentives for effective management and support for research and conservation.

Marine turtles nest along and forage adjacent to the coastline of dozens of countries throughout the world. In many of these countries, turtles and humans utilize the same habitats; and in addition to climate change, coastal-based impacts such as light pollution, physical structures, consumptive use, inshore fisheries, and boat-related impacts have been implicated as major threats for many species and populations. Thus, additional future social/environmental challenges will likely need to be overcome if anthropogenic threats to marine turtles are to be managed in the future. Such additional challenges could be (1) human development expanding into existing or future marine turtle habitats—such as coastal systems, (2) rising sea levels narrowing the gap between the dune systems and development, (3) nesting turtles or foraging turtles adapting to increasing temperatures, or sea level rises, by shifting the location of nesting or foraging such that they encroach on existing human development/use, (4) human migration patterns changing (such as climate refugees) in relation to sea level rises or increased exposure to other climate processes, (5) changing value or use of marine turtles by human society, or communities, and (6) the structure and functioning of coastal fisheries impacted by a changing climate (Allison et al., 2009; Cheung et al., 2009; Rashid Sumaila et al., 2011). These are certainly not trivial issues; and although it is a generalized list, there is evidence of each of the challenges emerging in local/regional areas of the world. Moreover, other sources of future human/environment conflict or cooperation will likely emerge as both the climate and the way humans use the environment change (Allison et al., 2009; Harris and Tyrrell, 2001).

The aforementioned challenges, while related to marine turtles, extend beyond direct species-based management planning to managing synergistic impacts through approaches such as coastal planning, management of social livelihoods and social resilience, management of resource extraction and natural resource governance. Indeed, much of the research surrounding the social–environmental links with regard to understanding and mitigating climate change impacts have a fishery and/or coastal livelihood basis. More specifically, there has been recent attention toward developing techniques and improving knowledge toward implementing adaptive management (Cinner et al., 2011; Lovecraft and Meek, 2011), improving social and economic resilience of coastal communities in developing, or at risk, nations (Fisher, 2011; Tompkins, 2005) improving resilience in management and government networks (Perry et al., 2010a,b; Weiss et al., 2012), and understanding how use or value of ecosystem services (including marine turtles) will change with a changing climate. Given the high importance and status of marine turtles in many human societies, there is clear need to improve the understanding of the human–turtle interactions and the synergistic impacts from multiple stressors and how these may change.

## 13.8  CONCLUSIONS

The collective body of empirical and anecdotal science over the past 30 years indicate that the structure and the function of several marine and coastal ecosystems are changing. Indeed, some marine turtle populations/RMUs are already threatened and thus will be particularly vulnerable to climate change (Hawkes et al., 2007b; Poloczanska et al., 2009; Wallace et al., 2011; Witt et al., 2010). However, substantial gaps remain in our knowledge of climate change as a threat to marine turtles. Indeed, when assessing global threats to marine turtles, Wallace et al. (2011) found that the there was a lack of relevant data for 38 of 58 RMUs to assess the risk to them from climate change. Further, as we discuss earlier, substantial uncertainty exists in our understanding of the magnitude, direction, and spatial-temporal variation in the predicted change for most climate processes. Hence, it is inherently difficult to assess how climate change will impact marine turtles, their habitats, and the people that interact with them, especially when considering other competing, cumulative, and synergistic risks. Rather than attempting to disentangle the impact, or potential impact, of climate change from other pervasive pressures such as coastal development, fisheries bycatch, and consumptive uses, we advocate incorporating climate change scenarios, and implementing climate change adaptation strategies into all tiers of environmental and marine turtle management and relevant marine and coastal legislation, from local communities, local-scale government up to national level government. This will undoubtedly require, and initiate, social change.

From the perspective of marine turtle biology, the main current constraints to management are a lack of knowledge on (1) population-scale data on thresholds of concern, especially for foraging turtles; (2) the degree to which populations, rather than specific sites, will be exposed to changing climate processes (i.e., exposure at more relevant ecological or management scales); (3) potential for populations to adapt to exposure to various abiotic and biotic changes; (4) how important ecosystem processes will be exposed and potentially adapt; and (5) the species, multispecies, or ecological costs and benefits of management action. Filling these biological gaps, alongside improved understanding of synergistic impacts (from climate change and/or climate change management), and improved understanding of the human dimensions will help strengthen the management of threats to marine turtles and their habitats into the future.

## REFERENCES

Ackerman, R.A. 1997. The nest environment and the embryonic development of sea turtles. In: Lutz, P.L., Musick, J.A. (Eds.), *The Biology of Sea Turtles*. CRC Publishing, Boca Raton, FL, pp. 83–107.

Allison, E.H., Perry, A.L., Badjeck, M.-C., Adger, W.N., Brown, K., Conway, D., Halls, A.S., Pilling, G.M., Reynolds, J.D., Andrew, N.L., Dulvy, N.K. 2009. Vulnerability of national economies to the impacts of climate change on fisheries. *Fish and Fisheries* 10, 173–196.

Anthony, K.R.N., Maynard, J.A., Diaz-Pulido, G., Mumby, P.J., Marshall, P.A., Cao, L., Hoegh-Guldberg, O. 2011. Ocean acidification and warming will lower coral reef resilience. *Global Change Biology* 17, 1798–1808.

Baker, J., Littnan, C., Johnston, D. 2006. Potential effects of sea level rise on the terrestrial habitats of endangered and endemic megafauna in the northwestern Hawaiian Islands. *Endangered Species Research* 4, 1–10.

Behrenfeld, M.J., O'Malley, R.T., Siegel, D.A., McClain, C.R., Sarmiento, J.L., Feldman, G.C., Milligan, A.J., Falkowski, P.G., Letelier, R.M., Boss, E.S. 2006. Climate-driven trends in contemporary ocean productivity. *Nature* 444, 752–755.

Booth, D.T., Astill, K. 2001a. Incubation temperature, energy expenditure and hatchling size in the green turtle (*Chelonia mydas*), a species with temperature-sensitive sex determination. *Australian Journal of Zoology* 49, 389–396.

Booth, D.T., Astill, K. 2001b. Temperature variation within and between nests of the green sea turtle, *Chelonia mydas* (*Chelonia*: Cheloniidae) on Heron Island, Great Barrier Reef. *Australian Journal of Zoology* 49, 71–84.

Booth, D.T., Evans, A. 2011. Warm water and cool nests are best. How global warming might influence hatchling green turtle swimming performance. *PLoS One* 6, e23162.

Booth, D.T., Freeman, C. 2006. Sand and nest temperatures and an estimate of hatchling sex ratio from the Heron Island green turtle (*Chelonia mydas*) rookery, southern Great Barrier Reef. *Coral Reefs* 25, 629–633.

Brochu, C. 2001. Congruence between physiology, phylogenetics and the fossil record on crocodilian historical biogeography. In: Grigg, G., Seebacher, F., Franklin, C. (Eds.), *Crocodilian Biology and Evolution*. Surrey Beatty & Sons, Chipping Norton, Oxfordshire, U.K.

Brodie, J., Sanjayan, M., Corea, R., Helmy, O., Amarasiri, C. 2008. Effects of the 2004 Indian Ocean Tsunami on sea turtle populations in Sri Lanka. *Chelonian Conservation and Biology* 7, 249–251.

Caut, S., Guirlet, E., Girondot, M. 2009. Effect of tidal overwash on the embryonic development of leatherback turtles in French Guiana. *Marine Environmental Research* 69, 254–261.

Chaloupka, M., Kamezaki, N., Limpus, C. 2008. Is climate change affecting the population dynamics of the endangered Pacific loggerhead sea turtle? *Journal of Experimental Marine Biology and Ecology* 356, 136–143.

Cheung, W.W.L., Lam, V.W.Y., Sarmiento, J.L., Kearney, K., Watson, R., Zeller, D., Pauly, D. 2009. Large-scale redistribution of maximum fisheries catch potential in the global ocean under climate change. *Global Change Biology* 16, 24–35.

Church, J.A., White, N.J., Hunter, J.R. 2006. Sea-level rise at tropical Pacific and Indian Ocean islands. *Global and Planetary Change* 53, 155–168.

Cinner, J.E., McClanahan, T.R., Graham, N.A.J., Daw, T.M., Maina, J., Stead, S.M., Wamukota, A., Brown, K., Bodin, O. 2011. Vulnerability of coastal communities to key impacts of climate change on coral reef fisheries. *Global Environmental Change* 22: 12–20.

Conant, T.A., Dutton, P., Eguchi, T., Epperly, S., Fahy, C., Godfrey, M., MacPherson, S., Possardt, E., Schroeder, B., Seminoff, J., Snover, M., Upite, C., Witherington, B. 2009. Loggerhead sea turtle (*Caretta caretta*), 2009 status review under the U.S. Endangered Species Act. Report of the Loggerhead Biological Review Team to the National Marine Fisheries Service, p. 222.

Cowell, P.J., Thom, B.G., Jones, R.A., Everts, C.H., Simanovic, D. 2006. Management of uncertainty in predicting climate-change impacts on beaches. *Journal of Coastal Research* 22, 232–245.

Davenport, J. 1989. Sea turtles and the greenhouse effect. *British Herpetological Society Bulletin* 29, 11–15.

Dawson, J.L., Smithers, S.G. 2010. Shoreline and beach volume change between 1967 and 2007 at Raine Island, Great Barrier Reef, Australia. *Global and Planetary Change* 72, 141–154.

Dethmers, K.E.M., Broderick, D., Moritz, C., Fitzsimmons, N., Limpus, C.J., Lavery, S., Whiting, S., Guinea, M., Prince, R., Kennett, R. 2006. The genetic structure of Australasian green turtles (*Chelonia mydas*): Exploring the geographical scale of genetic exchange. *Molecular Ecology* 15, 3931–3946.

Eastman, J. 2009. *DRISI Taiga Guide to GIS and Image Processing*. Clark Labs, Clark University, Worcester, MA.

Eckert, K., Hemphill, A. 2005. Sea turtles as flagships for protection of the wider Caribbean region. *Maritime Studies* 3, 119–145.

Fish, M.R., Cote, I.M., Gill, J.A., Jones, A.P., Renshoff, S., Watkinson, A.R. 2005. Predicting the impact of sea-level rise on Caribbean sea turtle nesting habitat. *Conservation Biology* 19, 482–491.

Fish, M.R., Cote, I.M., Horrocks, J.A., Mulligan, B., Watkinson, A.R., Jones, A.P. 2008. Construction setback regulations and sea-level rise: Mitigating sea turtle nesting beach loss. *Ocean & Coastal Management* 51, 330–341.

Fisher, P.B. 2011. Climate change and human security in Tuvalu. *Global Change, Peace and Security* 23, 293–313.

Foley, A.M., Peck, S.A., Harman, G.R. 2006. Effects of sand characteristics and inundation on the hatching success of loggerhead sea turtle (*Caretta caretta*) clutches on low-relief mangrove islands in southwest Florida. *Chelonian Conservation and Biology* 5, 32–41.

Frazier, J. 2005. The role of flagship species in interactions between people and the sea. *Maritime Studies* 3, 5–38.

Fuentes, M.M.P.B., Abbs, D. 2010. Effects of projected changes in tropical cyclone frequency on sea turtles. *Marine Ecology Progress Series* 412, 283–292.

Fuentes, M.M.P.B., Bateman, B.L., Hamann, M. 2011a. Relationship between tropical cyclones and the distribution of sea turtle nesting grounds. *Journal of Biogeography* 38, 1886–1896.

Fuentes, M., Dawson, J.L., Smithers, S.G., Hamann, M., Limpus, C.J. 2009a. Sedimentological characteristics of key sea turtle rookeries: Potential implications under projected climate change. *Marine and Freshwater Research* 61, 464–473.

Fuentes, M.M.P.B., Fish, M.R., Maynard, J.A. 2012. Management strategies to mitigate the impacts of climate change on sea turtle's terrestrial reproductive phase. *Mitigation and Adaptation Strategies for Global Change* 17, 51–63.

Fuentes, M., Hamann, M., Limpus, C.J. 2009b. Past, current and future thermal profiles of green turtle nesting grounds: Implications from climate change. *Journal of Experimental Marine Biology and Ecology* 383, 56–64.

Fuentes, M., Limpus, C.J., Hamann, M. 2011b. Vulnerability of sea turtle nesting grounds to climate change. *Global Change Biology* 17, 140–153.

Fuentes, M., Limpus, C.J., Hamann, M., Dawson, J. 2009c. Potential impacts of projected sea-level rise on sea turtle rookeries. *Aquatic Conservation-Marine and Freshwater Ecosystems* 20, 132–139.

Fuentes, M.M.P.B., Maynard, J.A., Guinea, M., Bell, I.P., Werdell, P.J., Hamann, M. 2009d. Proxy indicators of sand temperature help project impacts of global warming on sea turtles in northern Australia. *Endangered Species Research* 9, 33–40.

Georges, A., Limpus, C., Stoutjesdijk, R. 1994. Hatchling sex in the marine turtle *Caretta caretta* is determined by proportion of development at a temperature, not daily duration of exposure. *Journal of Experimental Zoology* 270, 432–444.

Godfrey, M.H., Godley, B.J. 2008. Seeing past the red: Flawed IUCN global listings for sea turtles. *Endangered Species Research* 6, 155–159.

Godley, B.J., Barbosa, C., Bruford, M., Broderick, A.C., Catry, P., Coyne, M.S., Formia, A., Hays, G.C., Witt, M.J. 2010. Unravelling migratory connectivity in marine turtles using multiple methods. *Journal of Applied Ecology* 47, 769–778.

Gordon, H., OFarrell, S., Collier, M., Dix, M., Rotstayn, L., Kowalczyk, E., Hirst, T., Watterson, I. 2010. The CSIRO Mk3.5 Climate Model. Technical Report; 21. Centre for Australian Weather and Climate Research, Australia, p. 53.

Hamann, M., Godfrey, M.H., Seminoff, J.A., Arthur, K., Barata, P.C.R., Bjorndal, K.A., Bolten, A.B. et al. 2010. Global research priorities for sea turtles: Informing management and conservation in the 21st century. *Endangered Species Research* 11, 245–269.

Hamann, M., Grech, A., Wolanski, E., Lambrechts, J. 2011. Modelling the fate of marine turtle hatchlings. *Ecological Modelling* 222, 1515–1521.

Hamann, M., Limpus, C., Hughes, G., Mortimer, J., Pilcher, N. 2006. Assessment of the impact of the December 2004 tsunami on marine turtles and their habitats in the Indian Ocean and South East Asia. IOSEA Marine Turtle MoU Secretariat, Bangkok, Thailand.

Hamann, M., Limpus, C., Read, M. 2007. Vulnerability of marine reptiles to climate change in the Great Barrier Reef. In: Johnson, J., Marshal, P. (Eds.), *Climate Change and the Great Barrier Reef*. Great Barrier Reef Marine Park Authority and The Australian Greenhouse Office, Queensland, Australia.

Hamann, M., Owens, D., Limpus, C.J. 2002. Reproductive cycles in male and female sea turtles. In: Lutz, P.L., Musick, J.A., Wyneken, J. (Eds.), *The Biology of Sea Turtles*, Vol. 2, CRC Press, Boca Raton, FL.

Harley, C.D.G., Hughes, A.R., Hultgren, K.M., Miner, B.G., Sorte, C.J.B., Thornber, C.S., Rodriguez, L.F., Tomanek, L., Williams, S.L. 2006. The impacts of climate change in coastal marine systems. *Ecology Letters* 9, 228–241.

Harris, L.G., Tyrrell, M.C. 2001. Changing community states in the Gulf of Maine: Synergism between invaders, overfishing and climate change. *Biological Invasions* 3, 9–21.

Hasumi, H., Emori, S. 2004. K-1 Coupled GCM (MIROC) Description, K-1 Technical Report No. 1. Center for Climate System Research, University of Tokyo, Tokyo, Japan, p. 28.

Hawkes, L.A., Broderick, A.C., Coyne, M.S., Godfrey, M.H., Godley, B.J. 2007a. Only some like it hot—Quantifying the environmental niche of the loggerhead sea turtle. *Diversity and Distributions* 13, 447–457.

Hawkes, L.A., Broderick, A.C., Godfrey, M.H., Godley, B.J. 2007b. Investigating the potential impacts of climate change on a marine turtle population. *Global Change Biology* 13, 923–932.

Hawkes, L.A., Broderick, A.C., Godfrey, M.H., Godley, B.J. 2009. Climate change and marine turtles. *Endangered Species Research* 7, 137–154.

Hays, G.C., Ashworth, J.S., Barnsley, M.J., Broderick, A.C., Emery, D.R., Godley, B.J., Henwood, A., Jones, E.L. 2001a. The importance of sand albedo for the thermal conditions on sea turtle nesting beaches. *Oikos* 93, 87–94.

Hays, G.C., Broderick, A.C., Glen, F., Godley, B.J. 2003. Climate change and sea turtles: A 150-year reconstruction of incubation temperatures at a major marine turtle rookery. *Global Change Biology* 9, 642–646.

Hays, G.C., Dray, M., Quaife, T., Smyth, T.J., Mironnet, N.C., Luschi, P., Papi, F., Barnsley, M.J. 2001b. Movements of migrating green turtles in relation to AVHRR derived sea surface temperature. *International Journal of Remote Sensing* 22, 1403–1411.

Hays, G.C., Godley, B.J., Broderick, A.C. 1999. Long-term thermal conditions on the nesting beaches of green turtles on Ascension Island. *Marine Ecology Progress Series* 20, 297–299.

Hays, G.C., Mackay, A., Adams, C., Mortimer, J., Speakman, J., Boerema, M. 1995. Nest site selection by sea turtles. *Journal of the Marine Biological Association of the UK* 75, 667–674.

Heatwole, H. 1999. *Sea Snakes*. University of New South Wales Press Ltd, Sydney, New South Wales, Australia.

Hoegh-Guldberg, O. 2011. Coral reef ecosystems and anthropogenic climate change. *Regional Environmental Change* 11, S215–S227.

Hoegh-Guldberg, O., Bruno, J.F. 2010. The impact of climate change on the world's marine ecosystems. *Science* 328, 1523–1528.

Hoegh-Guldberg, O., Mumby, P.J., Hooten, A.J., Steneck, R.S., Greenfield, P., Gomez, E., Harvell, C.D. et al. 2007. Coral reefs under rapid climate change and ocean acidification. *Science* 318, 1737–1742.

Houghton, J.D.R., Myers, A.E., Lloyd, C., King, R.S., Isaacs, C., Hays, G.C. 2007. Protracted rainfall decreases temperature within leatherback turtle (*Dermochelys coriacea*) clutches in Grenada, West Indies: Ecological implications for a species displaying temperature dependent sex determination. *Journal of Experimental Marine Biology and Ecology* 345, 71–77.

van Houtan, K.S., Halley, J.M. 2011. Long-term climate forcing in loggerhead sea turtle nesting. *PLoS One* 6, e19043.

Hutchings, P., Ahyong, S., Byrne, M., Przeslawski, R., Woerheide, G. 2007. Vulnerability of benthic invertebrates of the Great Barrier Reef to climate change. In: Johnson, J., Marshall, P. (Eds.), *Climate Change and the Great Barrier Reef: A Vulnerability Assessment*. Great Barrier Reef Marine Park Authority and Australian Greenhouse Office, Queensland, Australia.

IPCC. 2000. *IPCC, 2000*. Emissions Scenarios. A Special Report of Working Group III of the Intergovernmental Panel on Climate Change. Cambridge University Press, Cambridge, U.K.

IPCC. 2007. Climate change 2007: The physical science basis. Contribution of working group 1 to the fourth assessment report of the Intergovernmental Panel on Climate Change. Cambridge University Press, Cambridge, U.K.

Jennings, S., Mélin, F., Blanchard, J.L., Forster, R.M., Dulvy, N.K., Wilson, R.W. 2008. Global-scale predictions of community and ecosystem properties from simple ecological theory. *Proceedings of the Royal Society B: Biological Sciences* 275, 1375–1383.

Johnson, J., Marshall, P. 2007. *Climate Change and the Great Barrier Reef.* Great Barrier Reef Marine park Authority and the Australian Greenhouse Office, Queensland, Australia.

Kamel, S.J. 2006. Inter-seasonal maintenance of individual nest site preferences in hawksbill sea turtles. *Ecology* 87, 2947–2952.

Kamel, S., Mrosovsky, N. 2004. Nest site selection in leatherbacks, *Dermochelys coriacea*: Individual patterns and their consequences. *Animal Behaviour* 68, 357–366.

Kamel, S., Mrosovsky, N. 2005. Repeatability of nesting preferences in the hawksbill sea turtle, *Eretmochelys imbricata*, and their fitness consequences. *Animal Behaviour* 70, 819–828.

Klein, R., Nichols, R. 1999. Assessment of coastal vulnerability to sea level rise. *Ambio* 28, 182–187.

Kornaraki, E., Matossian, D.A., Mazaris, A.D., Matsinos, Y.G., Margaritoulis, D. 2006. Effectiveness of different conservation measures for loggerhead sea turtle (*Caretta caretta*) nests at Zakynthos Island, Greece. *Biological Conservation* 130, 324–330.

Kraemer, J.E., Bell, R. 1980. Rain-induced mortality of eggs and hatchlings of loggerhead sea turtles (*Carette caretta*) on the Georgia coast. *Herpetologica* 36, 72–77.

Laurance, W.F., Dell, B., Turton, S.M., Lawes, M.J., Hutley, L.B., McCallum, H., Dale, P. et al. 2011. The 10 Australian ecosystems most vulnerable to tipping points. *Biological Conservation* 144, 1472–1480.

Limpus, C.J., Carter, D., Hamann, M. 2001. The green turtle, *Chelonia mydas*, in Queensland, Australia: The Bramble Cay rookery in the 1979–1980 breeding season. *Chelonian Conservation and Biology* 4, 34–46.

Limpus, C.J., Nicholls, N. 1988. The southern oscillation regulates the annual numbers of green turtles (*Chelonia mydas*) breeding around northern Australia. *Australian Wildlife Research* 15, 157–162.

Limpus, C.J., Reed, P.C. 1985. Green sea turtles (*Chelonia mydas*) stranded by Cyclone Kathy on the SouthWestern coast of the Gulf of Carpentaria (Australia). *Australian Wildlife Research* 12, 523–534.

Lovecraft, A.L., Meek, C.L. 2011. The human dimensions of marine mammal management in a time of rapid change: Comparing policies in Canada, Finland and the United States. *Marine Policy* 35, 427–429.

Lovelock, C., Ellison, J. 2007. Vulnerability of mangroves and tidal wetlands of the Great Barrier Reef to climate change. In: Johnson, J., Marshall, P. (Eds.), *Climate Change and the Great Barrier Reef: A Vulnerability Assessment.* Great Barrier Reef Marine Park Authority and Australian Greenhouse Office, Queensland, Australia, pp. 241–269.

Mast, R.B., Carr, J.L. 1989. Carapacial scute variation in Kemp's ridley sea turtle (*Lepidochelys kempi*) hatchlings and juveniles. In: Caillouet, C.W., Landry, A.M. (Eds.), *Proceedings of the First International Symposium on Kemp's Ridley Sea Turtle Biology, Conservation and Management*, A & M University, Sea Grant College, TX, TAMU-SG-89-105, pp. 202–219.

Mast, R., Hutchinson, B., Howgate, E., Pilcher, N. 2005. IUCN/SSC Marine Turtle Specialist Group hosts the second burning issues assessment workshop. *Marine Turtle Newsletter* 110, 13.

Mast, R., Hutchinson, B., Pilcher, N. 2004. IUCN/Species Survival Commission Marine Turtle Specialist Group news first quarter 2004. *Marine Turtle Newsletter* 104, 21–22.

Mazaris, A.D., Matsinos, G., Pantis, J.D. 2009. Evaluating the impacts of coastal squeeze on sea turtle nesting. *Ocean & Coastal Management* 52, 139–145.

McMahon, C.R., Hays, G.C. 2006. Thermal niche, large-scale movements and implications of climate change for a critically endangered marine vertebrate. *Global Change Biology* 12, 1330–1338.

van de Merwe, J., Ibrahim, K., Whittier, J. 2005. Effects of hatchery shading and nest depth on the development and quality of *Chelonia mydas* hatchlings: Implications for hatchery management in Peninsular, Malaysia. *Australian Journal of Zoology* 53, 205–211.

van de Merwe, J., Ibrahim, K., Whittier, J. 2006. Effects of nest depth, shading, and metabolic heating on nest temperatures in sea turtle hatcheries. *Chelonian Conservation and Biology* 5, 210–215.

Miller, J.D. 1985. Embryology of marine turtles. In: Gans, C., Billett, F., Maderson, P.F.A. (Eds.), *Biology of the Reptilia*. John Wiley & Sons, New York, pp. 269–328.

Miller, J.D., 1997. Reproduction in sea turtles. In: Lutz, P.L., Musick, J.A. (Eds.), *The Biology of Sea Turtles*. CRC Press, Boca Raton, FL, pp. 51–83.

Moritz, C., Broderick, D., Dethmers, K., Fitzsimmons, N., Limpus, C.J. 2001. Migration and genetics of Indo-Pacific marine turtles. Final Report to UNEP/CMS.

Morreale, S.J., Ruiz, G.J., Spotila, J.R., Standora, E.A. 1982. Temperature-dependent sex determination— Current practices threaten conservation of sea turtles. *Science* 216, 1245–1247.

Mortimer, J.A., Carr, A. 1987. Reproduction and migrations of the Ascension Island green turtle (*Chelonia mydas*). *Copeia* 1987, 103–113.

Mrosovsky, N. 2000. Sustainable use of sea turtles. *Proceedings of the 15th Working Meeting of the Crocodile Specialist Group*, IUCN-The world conservation Union, Gland, Switzerland.

Mrosovsky, N. 2006. Distorting gene pools by conservation: Assessing the case of doomed turtle eggs. *Environmental Management* 38, 523–531.

Mrosovsky, N. 2008. Against oversimplifying the issues on relocating turtle eggs. *Environmental Management* 41, 465–467.

Mrosovsky, N., Dutton, P.H., Whitmore, C.P. 1984. Sex ratios of 2 species of sea turtle nesting in Surinam. *Canadian Journal of Zoology* 62, 2227–2239.

Mrosovsky, N., Yntema, C.L. 1980. Temperature-dependence of sexual differentiation in sea turtles— Implications for conservation practices. *Biological Conservation* 18, 271–280.

Nicholls, R.J., Cazenave, A. 2010. Sea-level rise and its impact on coastal zones (June, pg 1517, 2007). *Science* 329, 628–628.

Orth, R.J., Carruthers, T., Dennison, W., Duarte, C., Fourqurean, J., Heck, K., Hughes, A. et al. 2006. A global crisis for seagrass ecosystems. *Bioscience* 5612, 987–996.

Parmesan, C. 2007. Influences of species, lattitudes and methodologies on estimates of phenological response to global warming. *Global Change Biology* 13, 1860–1872.

Parmesan, C., Yohe, G. 2003. A globally coherent fingerprint of climate change impacts across natural systems. *Nature* 421, 37–42.

Patino-Martinez, J., Marco, A., Quinones, L., Hawkes, L. 2012. A potential tool to mitigate the impacts of climate change to the Caribbean leatherback sea turtle. *Global Change Biology* 18, 401–411.

Perry, R.I., Barange, M., Ommer, R.E. 2010a. Global changes in marine systems: A social-ecological approach. *Progress in Oceanography* 87, 331–337.

Perry, R.I., Ommer, R.E., Barange, M., Werner, F. 2010b. The challenge of adapting marine social-ecological systems to the additional stress of climate change. *Current Opinion in Environmental Sustainability* 2, 356–363.

Pfaller, J.B., Limpus, C.J., Bjorndal, K.A. 2009. Nest-site selection in individual loggerhead turtles and consequences for doomed-egg relocation. *Conservation Biology* 23, 72–80.

Pike, D. 2008a. The benefits of nest relocation extend far beyond recruitment: A rejoinder to Mrosovsky. *Environmental Management* 41, 461–464.

Pike, D.A. 2008b. Environmental correlates of nesting in loggerhead turtles, *Caretta caretta. Animal Behaviour* 76, 603–610.

Pike, D.A. 2009. Do green turtles modify their nesting seasons in response to environmental temperatures? *Chelonian Conservation and Biology* 8, 43–47.

Pike, D.A., Antworth, R.L., Stiner, J.C. 2006. Earlier nesting contributes to shorter nesting seasons for the Loggerhead Seaturtle, *Caretta caretta. Journal of Herpetology* 40, 91–94.

Pike, D.A., Stiner, J.C. 2007a. Fluctuating reproductive output and environmental stochasticity: Do years with more reproducing females result in more offspring? *Canadian Journal of Zoology* 85, 737–742.

Pike, D.A., Stiner, J.C. 2007b. Sea turtle species vary in their susceptibility to tropical cyclones. *Oecologia* 153, 471–478.

Poloczanska, E.S., Limpus, C.J., Hays, G.C. 2009. Vulnerability of marine turtles to climate change. *Advances in Marine Biology* 56, 151–211.

Pritchard, P. 1997. Evolution, phylogeny and current status. In: Lutz, P.L., Musick, J.A. (Eds.), *The Biology of Sea Turtles*. CRC Press, Boca Raton, FL.

Rashid Sumaila, U., Cheung, W., Lam, V., Pauly, D., Herrick, S. 2011. Climate change impacts on the biophysics and economics of world fisheries. *Nature Climate Change* 1, 449–456.

Rhen, T., Lang, J.W. 1995. Phenotypic plasticity for growth in the common snapping turtle: Effects of incubation temperature, clutch, and their interaction. *American Naturalist* 146, 726–747.

Saba, V.S., Santidrian-Tomillo, P., Reina, R.D., Spotila, J.R., Musick, J.A., Evans, D.A., Paladino, F.V. 2007. The effect of the El Nino Southern Oscillation on the reproductive frequency of eastern Pacific leatherback turtles. *Journal of Applied Ecology* 44, 395–404.

Santana-Garcon, J.S., Grech, A., Moloney, J., Hamann, M. 2010. Relative Exposure Index: An important factor in sea turtle nesting distribution. *Aquatic Conservation: Marine and Freshwater Ecosystems* 20, 140–149.

Schiel, D.R., Steinbeck, J.R., Foster, M.S. 2004. Ten years of induced ocean warming causes comprehensive changes in marine benthic communities. *Ecology* 85, 1833–1839.

Seminoff, J.A., Shanker, K. 2008. Marine turtles and IUCN Red Listing: A review of the process, the pitfalls, and novel assessment approaches. *Journal of Experimental Marine Biology and Ecology* 356, 52–68.

Short, F.T., Polidoro, B., Livingstone, S.R., Carpenter, K.E., Bandeira, S., Bujang, J.S., Calumpong, H.P. et al. 2011. Extinction risk assessment of the world's seagrass species. *Biological Conservation* 144, 1961–1971.

Spotila, J.R., Standora, E.A. 1985. Environmental constraints on the thermal energetics of sea turtles. *Copeia* 1985, 694–702.

Standora, E.A., Spotila, J.R. 1985. Temperature dependent sex determination in sea turtles. *Copeia* 1985, 711–722.

Tisdell, C., Wilson, C. 2001. Wildlife-based tourism and increased support for nature conservation financially and otherwise: Evidence from sea turtle ecotourism at Mon Repos. *Tourism Economics* 7, 233–249.

Tompkins, E.L. 2005. Planning for climate change in small islands: Insights from national hurricane preparedness in the Cayman Islands. *Global Environmental Change* 15, 139–149.

Wallace, B.P., DiMatteo, A.D., Bolten, A.B., Chaloupka, M.Y., Hutchinson, B.J., Abreu-Grobois, F.A., Mortimer, J.A. et al. 2011. Global conservation priorities for marine turtles. *PLoS One* 6(9), 1–10, e24510.

Wallace, B.P., DiMatteo, A.D., Hurley, B.J., Finkbeiner, E.M., Bolten, A.B., Chaloupka, M.Y., Hutchinson, B.J. et al. 2010. Regional management units for marine turtles: A novel framework for prioritizing conservation and research across multiple scales. *PLoS One* 5(12), 1–11, e15465.

Walther, G.R., Post, E., Convey, P., Menzel, A., Parmesan, C., Beebee, T.J.C., Fromentin, J.M., Hoegh-Guldberg, O., Bairlein, F. 2002a. Ecological responses to recent climate change. *Nature* 416, 389–395.

Walther, G.R., Prost, E., Convey, P., Menzels, A., Parmesan, C., Beebee, T.J.C., Fromentin, J.M., Hoegh-Guldberg, O., Bairlein, F. 2002b. Ecological responses to recent climate change. *Nature* 416, 389–395.

Waycott, M., Longstaff, B.J., Mellors, J. 2005. Seagrass population dynamics and water quality in the Great Barrier Reef region: A review and future research directions. *Marine Pollution Bulletin* 51, 343–350.

Webb, A.P., Kench, P.S. 2010. The dynamic response of reef islands to sea-level rise: Evidence from multi-decadal analysis of island change in the central Pacific. *Global and Planetary Change* 72, 234–246.

Weishampel, J.F., Bagley, D.A., Ehrhart, L.M. 2004. Earlier nesting by loggerhead sea turtles following sea surface warming. *Global Change Biology* 10, 1424–1427.

Weishampel, J.F., Bagley, D.A., Ehrhart, L.M., Weishampel, A.C. 2010. Nesting phenologies of two sympatric sea turtle species related to sea surface temperatures. *Endangered Species Research* 12, 41–47.

Weiss, K., Hamann, M., Kinney, M., Marsh, H. 2012. Knowledge exchange and policy influence in a marine resource governance network. *Global Environmental Change-Human and Policy Dimensions* 22, 178–188.

Wibbels, T. 2003. Critical approaches to sex determination in sea turtles. In: Lutz, P.L., Musick, J.A., Wyneken, J. (Eds.), *The Biology of Sea Turtles*, Vol. 11, CRC Press, Boca Raton, FL, pp. 103–134.

Witt, M.J., Hawkes, L.A., Godfrey, M.H., Godley, B.J., Broderick, A.C. 2010. Predicting the impacts of climate change on a globally distributed species: The case of the loggerhead turtle. *Journal of Experimental Biology* 213, 901–911.

Woodroffe, C., McLean, R., Smithers, S., Lawson, E. 1999. Atoll reef-island formation and response to sea-level change: West Island, Cococ (Keeling) Islands. *Marine Geology* 160, 85–104.

# 14 Free-Ranging Sea Turtle Health

*Mark Flint*

## CONTENTS

## 14.1 INTRODUCTION

Assessing the health of free-ranging sea turtles has become increasingly important as evidence grows that environmental and animal health are intrinsically linked and sea turtle numbers continue to decline. Dr. George's Chapter 14 of *The Biology of Sea Turtles*, Volume I (George 1997) has an excellent review of the known nutritional anomalies, diseases, parasites, and environmental health problems affecting sea turtles at the time. Dr. Herbst's and Professor Jacobson's Chapter 15 of Volume II (Herbst and Jacobson 2003) also has an excellent comprehensive review of the systematic approaches and pitfalls to conducting a sea turtle disease investigation as at the turn of the new millennium. Since then, there have been advances in our understanding of the ecology and pathogenesis of certain diseases, such as fibropapillomatosis, as well as the development of more sophisticated and standardized tools to assess the health of wild and captive turtles. Further, in this field of veterinary conservation medicine, there has been a movement toward Dr. Schwabe's 1960s principle of *One Medicine* (Schwabe 1969), with the proposal that sea turtles are sentinel indicators of environmental health (Aguirre and Lutz 2004). Now known as *One World One Health One Medicine*, this concept, defined by the American Veterinary Medical Association as "the collaborative effort of multiple disciplines—working locally, nationally, and globally—to attain optimal health for people, animals and our environment" (http://www.avma.org/onehealth/responding.asp), appears to well suit the objectives and needs of a free-ranging sea turtle disease investigation.

While clinical management of sea turtles in captive situations will continue to play an important role, there is an increasing need, through epidemiological research, to understand causes of decline if many of these species are to recover successfully. As many diseases of sea turtles are complex, understanding their drivers and how to manage them will require that veterinarians collaborate with biologists, environmental scientists, and other professions. This management strategy will need to proceed in concert with efforts to develop innovative field and laboratory tools and robust baseline reference intervals to assess the health of wild turtles. Reptile medicine generally, including sea turtle medicine, has come a long way in recent years with comprehensive texts examining medicine, surgery, and pathology (Jacobson 2007a, Mader 2006); however, these advances in understanding clinical management of sea turtles have not significantly increased our understanding of wild turtle health. There is still a need to continue and refine the research currently being undertaken.

Using a veterinary approach to analyze sea turtle health within a collaborative team of investigators working in complimentary areas may increase our overall understanding of health and ill-health etiology and transmission in sea turtles, in addition to the health of the marine environment (Wilcox and Aguirre 2004). Findings may help mitigate further challenges to both sea turtle populations and the ecosystem. It is through this philosophy that recent sea turtle disease investigations have been undertaken, leading to the proposal that sea turtles and other marine vertebrates may act as a (proxy) sentinel indicator of environmental health due to (1) association of occurrence of certain diseases such as fibropapillomatosis and poor environmental conditions (Aguirre and Lutz 2004, Aguirre and Tabor 2004) and (2) their long-lived site fidelity in some species, such as green turtles that can reside in a single foraging site for decades, only leaving for courtship and/or nesting.

This chapter outlines factors affecting sea turtle survivorship, design of free-ranging sea turtle disease investigations, the current diseases of concern, and the employment of sea turtle population modeling in predicting environmental health.

## 14.2  SEA TURTLE SURVIVORSHIP

Sea turtles are an integral component of our oceanic ecosystems, and many populations continue to decline (Chaloupka and Limpus 2001, Chan 2006, Jackson et al. 2001) with all seven extant species continuing to be of conservation concern (IUCN/SSC 2008). Loss of nesting habitat, nest depredation, and fisheries bycatch are well-documented causes of wild turtle mortality and probably play an important role in declines of some populations (Dobbs and Pierce 2005, Limpus 2008a,b, 2009), particularly those that come in contact with human development or activities, but other populations that are not directly influenced by humans also continue to decline. Other than some isolated or opportunistic postmortem surveys and case reports producing some baseline references (Flint et al. 2010e, Glazebrook and Campbell 1990a, Gordon 2005, Limpus et al. 2009, Oros et al. 2005, Raidal et al. 1998, Work and Balazs 2010, Work et al. 2004), veterinary-orientated population-based analyses or monitoring of free-ranging turtles has not been used to determine the influence disease has on turtle survivorship. More information on the effects of diseases, their prevalence, route of transmission, infectivity, and potentiating factors is needed on the basic physiology and health status of free-ranging turtles as well as improved diagnostic techniques to gain a greater understanding of the interplay between the environment and their health (Aguirre and Lutz 2004, Herbst and Jacobson 2003, Morton et al. 2009, Ward and Lafferty 2004, Whiting et al. 2007). Potential reasons for continued population decline could include effects of climate change that has been proposed to cause disease emergence, skewed sex ratios of emerging hatchlings, or declines in available nesting habitat (Habib et al. 2010, Wallace et al. 2011). These environmental factors may act on the local or global level. Regional variation places individuals under different stressors, suggesting that diseases may be unique to a specific area (Jackson et al. 2001), but the migratory capacity of individuals, and as such the capability to transmit disease, suggests disease investigation and subsequent population trends should also be addressed on a global scale.

## 14.2.1 FACTORS AFFECTING SEA TURTLE SURVIVORSHIP

Surveillance programs have concentrated on establishing the essential basic parameters of biology and ecology. These global population surveillance programs identified sea turtle numbers were experiencing long-term declines and subsequent intervention strategies were developed (Dobbs and Pierce 2005). Recovery plans, such as the Recovery Plan for Marine Turtles in Australia (RPMTA 2003, 2006) and the recovery plans for U.S. Pacific and Atlantic populations of the green and loggerhead turtles (NMFS and USFWS 2008, NOAA 2008a,b), have synthesized data and issued mandates to various environmental protection and management agencies to design protocols to decrease mortality and increase the survivorship of sea turtles. With a focus on human impacts, one of the significant deficits in the recovery plans is a lack of understanding of the effects of environmental impacts on morbidity and mortality, the role of disease, and how to verify whether a turtle population was clinically healthy (Herbst and Jacobson 2003).

From a management perspective, the majority causes of death in sea turtles can be broken down into human impacts (e.g., boat strike, harvest, disorientation, fishing-related effects, malicious mischief, and drowning), natural (e.g., parasites, disease, nesting [trauma and thermal stress] and predation), and unknown (no cause of death determined due to advanced decomposition or unclear findings).

### 14.2.1.1 Human Impacts

Potential causes of anthropogenic population declines in sea turtles include (1) bycatch of turtles in trawl nets or longline fisheries resulting in turtle drownings; (2) commercial (large-scale) harvesting of turtles for saleable products such as jewelry, oils, meat, and ornaments; (3) traditional harvest of breeding females or eggs by indigenous peoples for food; (4) marine debris such as discarded/lost/unmonitored fishing gear (crab nets, lobster or crab pots, and hook and fishing line) causing entrapment resulting in drowning or intestinal disorders such as torsion caused by linear foreign bodies and/or gastric ruptures; (5) boat strike by recreational watercraft in high-traffic zones; (6) predation by terrestrial predators including native high-density predators such as raccoon and varanids and non-native pigs (e.g., *Sus scrofa* in Australia) and foxes (e.g., *Vulpes vulpes* in Australia) raiding nests to consume eggs; (7) environmental contamination caused by agricultural pesticide and herbicide runoff such as DDTs and pollutants including sewage and oils entering shallow waters via commercial and public drains causing loss of feeding grounds such as sea grass beds, primary toxicosis in individuals, and, in the case of oils, the formation of tar balls (e.g., nonvolatile petroleum or sand conglomerated by oil) that subsequently causes gastrointestinal impaction; (8) environmental degradation caused by habitat removal and/or flooding events; and (9) human-induced climate change, causing damage to oceanic systems such as coral reefs, sea grass species and survivorship, changing water and ambient temperature that affects sea level, as well as altering the sex ratio of hatchlings, and potentially creating previously absent niches for the emergence of new diseases (Dobbs and Pierce 2005, Habib et al. 2010, Limpus 2008a,b, Limpus 2009).

Strategies that have been implemented to combat these threats include (1) protected habitats (national parks, marine protected areas, and green zones), (2) turtle exclusion devices (TEDs) and bycatch reduction devices (BRDs), (3) regulations to cease the legal commercial harvesting of sea turtles, (4) indigenous involvement in conservation, (5) substantial fines for littering marine reserves, (6) "go-slow" zones (speed-limits) for boats in areas known to have turtles, (7) feral animal control strategies, (8) environmental monitoring and mitigation, and (9) unsuccessful global attempts to limit impacts or production of climate-change-inducing carbon dioxide (Dobbs and Pierce 2005, Habib et al. 2010, Limpus 2008a,b, 2009). The response of sea turtle populations to these single or multiple strategies have been variable, with some populations continuing to decline.

### 14.2.1.2 Natural Factors

With population numbers continuing to decline and adverse environmental events, both natural and anthropogenic, managers in many regions have increased efforts to determine the cause of death.

It is becoming apparent that an increasing number of sea turtle deaths cannot be fitted into the human impacts identified earlier, nor is it sufficient with respect to making management decisions to group other causes of death under the umbrella of "natural." Collaboration between biologists, ecologists, toxicologists, and veterinarians is starting to tease out the undisclosed "natural" causes using a structured, holistic approach to investigating disease (Flint et al. 2010e).

## 14.3   SEA TURTLE DISEASE INVESTIGATION

Disease investigation is the characterization of disease processes, sources, and contributing factors that result in abnormal findings in individuals and populations. This contrasts health assessment programs that describe and establish normal and abnormal values within a population (Herbst and Jacobson 2003). In the case of sea turtles, the investigator often needs to establish baseline parameters while simultaneously characterizing disease processes using both live and dead turtles to gather the required information to gain a comprehensive understanding of the diseases affecting sea turtle populations.

The first foray into researching the diseases of sea turtles arose with commercial green turtle farming on the Grand Cayman Island, British West Indies, in 1968 and the *head start* programs in the 1970s (Jacobson 1993). Recently, attention has turned to the concept of disease emergence as a major cause of population decline (George 1997, Ward and Lafferty 2004). Attempts are under way to understand etiologies and pathogenesis of specific diseases within turtle populations and to develop and refine diagnostic methods to identify and characterize these processes (Flint et al. 2009b, 2010c,d, Herbst and Jacobson 2003, Valente et al. 2006, 2007a,b).

Specific primary diagnostic tools such as reference intervals for hematology and blood biochemistry are an effective way of assessing the health of live free-ranging sea turtles. However, these reference intervals often do not exist for the cohort or population of turtle being examined and need to be derived. Absent or inadequate tools may be due to reference intervals derived from limited replicates or from captive sea turtles, which are known to vary from free-ranging sea turtles (Herbst and Jacobson 2003). Further, variations are known to be due to differences in the geographical location, gender, breeding status, age, and diet (Hamann et al. 2002, 2003, Jessop et al. 2004, Spotila 2004, Whiting et al. 2007).

To advance the field of sea turtle disease investigation, studies should be designed to incorporate components of health assessments and disease investigation. This requires examining sufficient numbers of clinically functional (sufficiently healthy live animals to not strand or die but may still carry subclinical disease) and nonfunctional (dead) sea turtles within a population to derive statistically sound comparisons. To achieve this, sea turtle disease investigations may require morphometric data, external examination, clinical examination, primary diagnostic screening tests, derivation of differential diagnoses, secondary diagnostic tests, and postmortem examinations before definitive diagnoses of the diseases processes within a population may be made (Herbst and Jacobson 2003).

### 14.3.1   Approach to a Disease Investigation

Health assessments and disease investigations have advantages and disadvantages. Assessing the health of live turtles lends itself to collecting blood samples for hematology and blood biochemistry, collecting fresh samples for microbial cultures, and performing radiographs, ultrasounds, and other advanced imaging. While these tools can indicate if the turtle is clinically healthy or not, they seldom give the exact cause of how any identified abnormalities were contracted or if this abnormality is going to affect the turtle's chances of surviving or breeding. By contrast, conducting disease investigations on dead turtles allows for gross and microscopic pathological examination and assessment of parasitism, which will more often provide a cause of death. However, this strategy biases sample collection toward the critically ill or dead subsets of the population and may miss subclinical diseases or nonfatal diseases of the functional population, reducing our understanding of the range of diseases within a population.

### 14.3.1.1 Live Sea Turtles

#### 14.3.1.1.1 Body Condition

Morphometric examination allows for the collection of fundamental biological information such as body condition that can indicate the health status of the individual. Determined as the ratio of straight or curved carapace length and weight (or girth), body condition can be expressed empirically providing a quantitative value (Bjorndal et al. 2000).

Other morphometric data collected should include species identification, presence of applied tags and/or microchips for individual identification, head dimensions, standard depth, plastron length, and tail length (Figure 14.1) to gain an understanding of the population structure, to allow

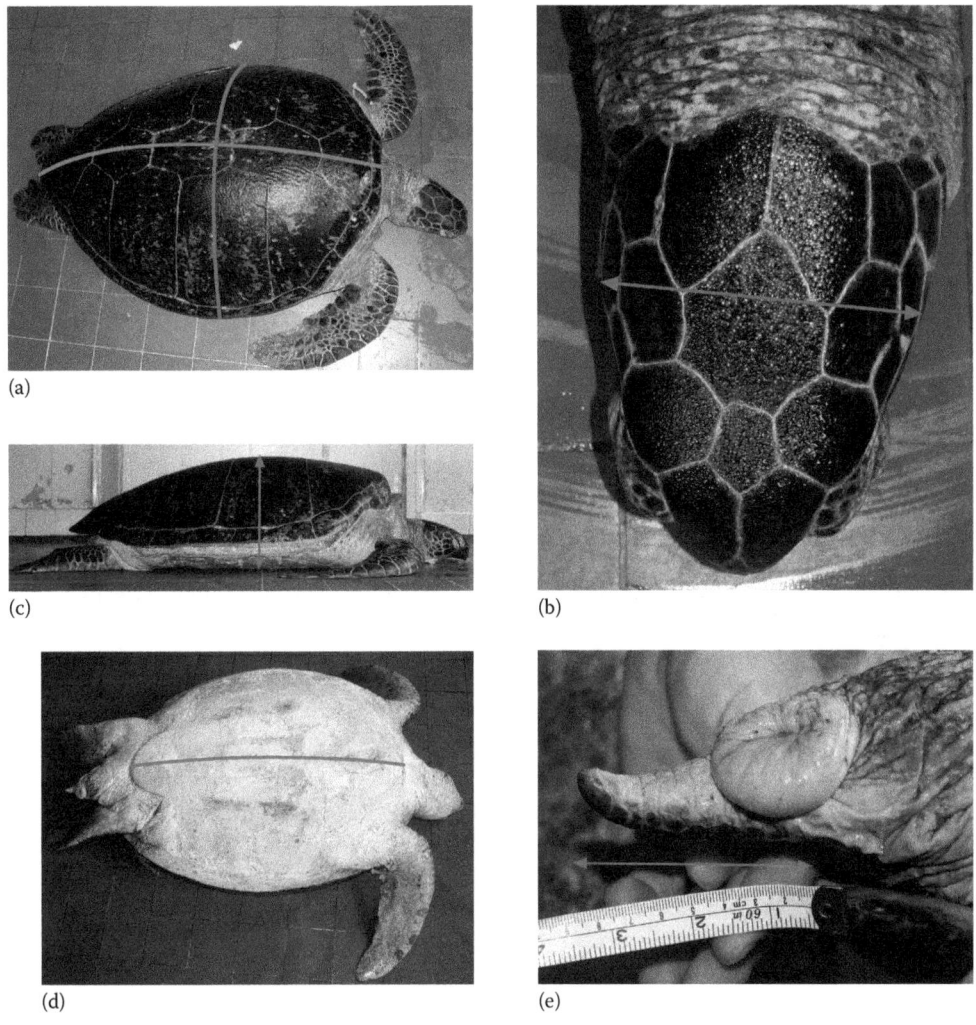

(a)

(c)

(b)

(d)

(e)

**FIGURE 14.1** Landmarks for the recommended morphometric measurements taken as part of the preliminary external examination: (a) curved carapace length, measured dorsally from the midline cranial suture of soft tissue and carapace and the caudal midline (blue line); and curved carapace width, measured dorsoventrally at the widest point of the carapace from the lateral edge of the marginal scutes (red line); (b) head width, measured dorsally at the widest part of the skull; (c) standard depth, measured dorsoventrally from the highest point of the carapace; (d) plastron length, measured ventrally midline from the cranial most to caudal most section of the plastron; and (e) tail length, measured from the caudal most point of the carapace to the tip of the tail.

managers the ability to confirm recorded data, and to provide cohort data that may be used to predict disease predilection for a specific age group, sex, or species (Flint et al. 2009b). In addition,
it is advantageous if the sex of the animal, its breeding status, and relative age can be determined
by laparoscopic examination (Limpus and Limpus 2003, Limpus and Reed 1985) based on gonad
development (Miller and Limpus 2003).

### 14.3.1.1.2  External Examination

External examination provides the initial subjective assessment of the health and well-being of
a sea turtle and may provide clues as to what factors (e.g., fresh boat strike wounds) impacted on
the animal's health. In addition to body condition, it should entail assessment for abnormalities,
asymmetry, epibiont load and growths, such as tumors (Bjorndal et al. 2000, Herbst and Jacobson
2003, Wolke and George 1981, Work 2000, Wyneken 2001). For example, a high epibiont load in a
loggerhead turtle (*Caretta caretta*) may be normal, but in a green turtle (*Chelonia mydas*), it could
indicate ill-health such as a buoyancy disorder causing the animal to spend extended periods of time
at the water's surface.

### 14.3.1.1.3  Clinical Examination

In addition to the external examination, clinical assessment should entail, but not be limited to,
cranial nerve examination, mentation, cloacal temperature (particularly in cases of suspected hypo-
thermia), respiratory rate, heart rate, symmetrical use of head and limbs, and assessment of the
accessible internal organs (Chrisman et al. 1997, Deem et al. 2006, 2009, Flint et al. 2010c, Herbst
and Jacobson 2003).

### 14.3.1.1.4  Screening Tests

Screening tests should be the same basic assessment techniques as those used in domestic
animal veterinary medicine. These may include hematology, blood biochemistry, microbiology,
parasitology, toxicology, and histology. Selection of which test may be used or what order tests
should be prioritized is based on a case-by-case basis in individual medicine. In population health
assessments, tests can be selected by those which will achieve the objectives (or test the hypotheses)
of the study. In most cases, assessments should prioritize hematology and blood biochemistry as
these tools can provide disease process information at the turtle's system level, which can provide
direction for the selection of follow-up tests.

#### 14.3.1.1.4.1  Hematology and Blood Biochemistry    Blood obtained from the external jugular
veins (Owens and Ruiz 1980) can be used to derive hematological and biochemical values (Arthur
et al. 2008, Bolten and Bjorndal 1992, Jacobson et al. 2007) as well as identify clinical ill-health
once normal reference intervals have been established using clinically healthy representatives
of a turtle population (Deem et al. 2006, 2009, Flint et al. 2009a, 2010c, Perrault et al. 2012,
Stamper et al. 2005, Whiting et al. 2007, Work and Balazs 1999).

With well-detailed descriptions of blood cytology and morphology (Casal 2007, Casal and Oros
2007, Samour et al. 1998, Work et al. 1998) and improved techniques for the establishing of reference intervals (Flint et al. 2010c, Horn et al. 1998, Pesce et al. 2005, Solberg 1987), risk factors such
as geographical location, habitat, genetics (Herbst and Jacobson 2003), maturity, sex (Hamann et al.
2006), breeding status (Deem et al. 2006), migratory status (Stamper et al. 2005), and diet (Whiting
et al. 2007) can be grouped to produce a robust and functional diagnostic tool that may be applicable
to multiple sea turtle cohorts (Flint et al. 2010c). With these improvements, hematology and blood
biochemistry reference intervals are becoming an increasingly useful and necessary tool in health
assessments for determining conservation management strategies (Hamann et al. 2006) and provide
a less invasive alternative to postmortem-based disease investigations.

*14.3.1.1.4.2   Microbiology, Parasitology, Toxicology, and Histology*   Microbiology, parasitology, and histology are important tools in refining the diagnoses of disease in animals within a population. Numerous species of bacteria and fungi have been identified in sea turtles (Work et al. 2003); however, the pathogenic nature of these microbes should be interpreted with caution as many may be present without causing significant pathology. Parasite infections have been proposed as a significant cause of clinical disease and mortality in sea turtles (Aguirre et al. 1998, Cribb and Gordon 1998, Dailey et al. 1991, Flint et al. 2010a,c, Glazebrook et al. 1989, Gordon et al. 1998, Jacobson et al. 2006, Raidal et al. 1998), but the association between parasitism and stranding/mortality is not straightforward, rather a part of a multifactorial disease process (Gordon et al. 1998, Jacobson et al. 2006). Researchers should view studies on parasites as an integral component of any disease investigation. Morphological identification (Platt 2001, Smith 1997) and characterization of the genetic sequences (Blair 2006, Nolan and Cribb 2005, Work et al. 2005) of the prevalent parasite species affecting sea turtles can be employed to identify which species are in selected populations or if body organ predilection (and secondary disease) is occurring. Collecting biopsies of tissues (soft tissue organs, blood, skeletal muscle, subcutaneous tissue, or bone) allows for the assessment of contaminants (toxins, heavy metals, PCBs, and DDTs) (Jacobson et al. 2006, Keller et al. 2004, 2006), for histological diagnoses (Flint et al. 2009b) and research investigations such as ageing (Klinger and Musick 1992). Currently under investigation is the development of bulk screening kits to cheaply screen multiple potential contaminants using relatively noninvasive whole blood or hematologic sampling (C. Gaus, personal communication).

*14.3.1.1.5   Secondary Diagnostic Tests*

The use of secondary tests in sea turtle diagnostics allows for the refinement of the diagnoses, and is often selected based on the clinical examination and results of the primary tests, such as hematology and blood biochemistry, in the same manner as for other species. Secondary tests include radiography (contrast or plain), ultrasonographic investigation, endoscopy, exploratory laparoscopy, exploratory surgery, computerized tomography scanning (CT scan), and magnetic resonance imaging (MRI) (Norton 2005, Nutter et al. 2000, Valente et al. 2006, 2007a,b, 2008. Although guidelines for the use of these tools are in the early stages of development, their application to individual patient care and the nonlethal approach is allowing easy collection of information that may be applicable at the population level if multiple sea turtles from within a population are examined to provide health extrapolation to the population as a whole.

### 14.3.1.2   Dead Sea Turtles

*14.3.1.2.1   Postmortem Examination*

Necropsies form the basis of disease investigations and are often the most useful tool for diagnosing causes of mortality required at the population level to indicate what appropriate actions, if any, should be taken to mitigate mortalities or morbidity (Flint et al. 2009b). Information derived from isolated single individuals compiled over time can contribute substantially to understanding what is causing mortality in wild turtles, thereby allowing managers to prioritize whether or not particular causes of mortality need to be mitigated or reduced. Like any other technique, necropsies have their limitations in that the population sample comprised only animals that are found dead and/or stranded and does not take into account animals with less severe or subclinical diseases. For each local area, any fresh dead turtles should be necropsied to determine the prevalent local causes of death and disease to contribute to the body of knowledge.

In the absence of medical history, one looks to environmental clues that could explain the mortality, for example, assessing the local food resources for depletion or change, determining if known foraging habitats or migratory paths have undergone anthropogenic environmental (dredging or redirecting) or usage (new shipping lanes or mining activity) changes, and factoring in occurrence of unusual events such as droughts or floods causing alteration of freshwater runoffs or concentrations of contaminants; all of which may contribute to interpretation of postmortem findings (Flint et al. 2009b, Herbst and Jacobson 2003). Similarly, one should also use the environmental clues to exclude potentially erroneous differential diagnoses. For example, multiple superficial abrasions on the carapace and soft tissue of a turtle carcass found at the low tide mark of a rocky beach may be due to wave action on a hard substrate and could have occurred postmortem.

Postmortem examination should consist of complete external and internal exams along with collections of tissues for microscopic examination (Flint et al. 2009b,c, Jacobson 1999, Wolke and George 1981, Work 2000, Wyneken 2001).

## 14.4  DISEASES OF CONCERN IN FREE-RANGING SEA TURTLES

### 14.4.1  Parasitism

Professor Greiner provides a comprehensive review in Chapter 16 of this volume concentrating on parasites found in sea turtles in Florida. As such, this section will present findings in sea turtles in Australia where two predominant parasite groups have been associated with diseases: Spirorchiidae (digenetic trematode parasites) and *Caryospora cheloniae* (coccidia). Under certain circumstances, high death rates in sea turtles have been associated with both parasites (Flint et al. 2010a, Gordon et al. 1993, 1998).

Spirorchiid trematode (cardiovascular or heart fluke) infection has been proposed as a significant cause of clinical disease and mortality in turtles in some regions (Flint et al. 2010e). Eggs and the lesions they cause are found in all organs, most significantly affecting the cardiovascular, gastrointestinal, lymphatic, and central nervous systems (Flint et al. 2010e, Gordon et al. 1998, Jacobson et al. 2006, Patterson-Kane et al. 2009, Stacy et al. 2008). Postmortem examinations show multiple granulomas and adult parasites throughout each organ with clinical pathology including complete occlusion of the primary vasculature, gastrointestinal stasis, lymphatic congestion, and space-occupying lesions of the brain (Figure 14.2; Flint et al. 2010e, Patterson-Kane et al. 2009, Stacy et al. 2010).

Globally, stranding investigations report a spirorchiid infection rate of up to 100% (Dailey et al. 1991, 1992, Gordon et al. 1998). Despite at least 14 species being reported in turtles, only 4 of the spirorchiid species have had their pathology of disease in sea turtles characterized (*Carettacola*, *Hapalotrema*, *Neospirorchis*, and *Learedius* spp). Similarly, the life cycles of these parasites have not been described, resulting in a poor understanding of how transmission occurs and the role of intermediate hosts in their life cycle (Flint et al. 2009b). A Hawaiian study demonstrated spirorchiid antibody could be detected by a crude ELISA test (Work et al. 2005), but it did not identify what species were infecting turtles nor did it give an indicator of parasite burden, thereby complicating our understanding of the effects of these parasites on sea turtles. To date, much of the parasite investigations have been conducted during postmortem examinations. Using molecular biological techniques on blood would provide a technique applicable to antemortem animals. Adjunct research into this area indicates the ITS region, which can be used to code for the Spirorchiid parasites of interest in terms of causing pathology in sea turtles (Platt 2001, Snyder 2004, Tkach et al. 2009), and mitochondrial DNA are showing promise as an accurate way to identify the known trematode parasite species found in sea turtles. When refined, these techniques may be employed to determine parasite status on initial examination of live sea turtles, response to treatment and prevalence of infection in a population. Similarly, as part of this study, a comparison is being made of the species of Spirorchiid parasites and their associated pathology in sea turtles of Florida, Hawaii, and Queensland to determine what role, if any, regional mortality and environmental factors are having on this complex disease process.

Coccidial infection by the species *Caryospora cheloniae* has been reported to cause epizootic mortality in Australian green turtles in Moreton Bay, Queensland (Gordon 2005, Gordon et al. 1993). Infection rates and severities, as for those of Spirorchiid parasites, may be related to environmental

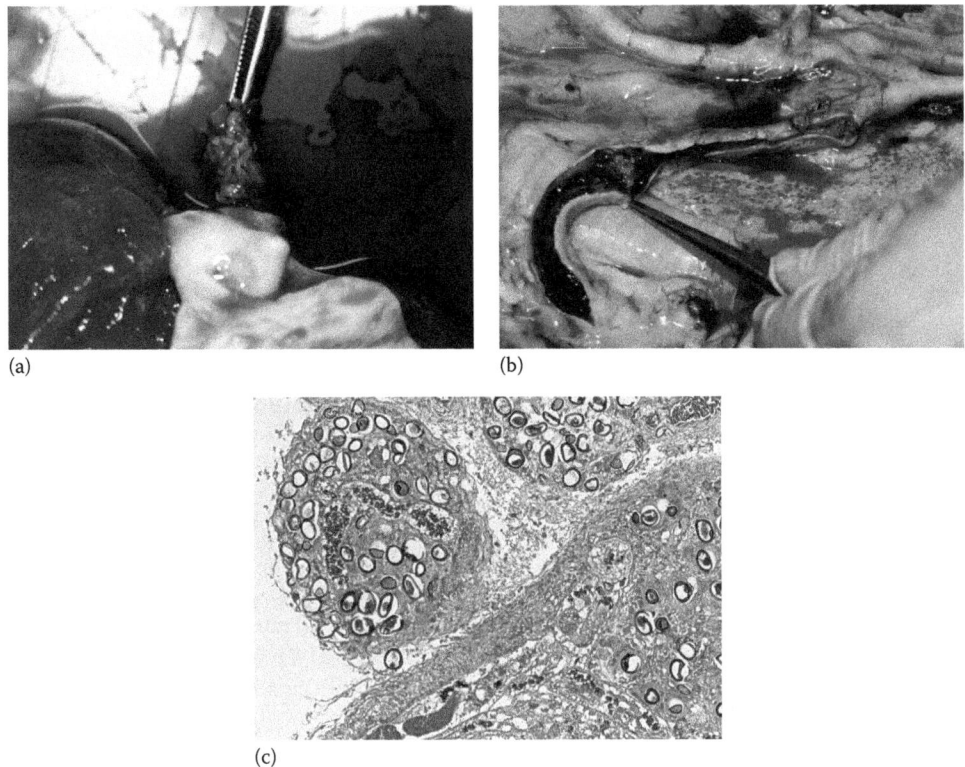

FIGURE 14.2 (a) Adult Spirorchiid parasites causing occlusion of the aorta; (b) Spirorchiid parasite granulomas, formed around the Spirorchiid parasite eggs, causing occlusion of the right descending aorta; and (c) histological micrograph (100×; H&E stain) adult Spirorchiid parasite and eggs within a granuloma in the brain of a large immature green sea turtle in Australia.

factors including temperature-related efficiency of turtle immune responses (Gordon et al. 1993). Sporadic fatalities due to *C. cheloniae* infection are still being observed in the Moreton Bay area but at a much reduced rate (Flint et al. 2010e). Reasons for these changes in these associated mortality rates and why Moreton Bay was a "hot spot" for such an outbreak are still unknown.

### 14.4.2 FIBROPAPILLOMATOSIS

Fibropapillomatosis (FP) is a common cutaneous, apparently infectious condition in all species of sea turtles, with gross and histological appearance similar to that in domestic species. Sea turtle FP was first reported in Florida in 1938 and has now caused numerous epidemics in green, loggerhead, and olive ridley (*Lepidochelys olivacea*) sea turtles worldwide (Greenblatt et al. 2005, Herbst 1994, Limpus and Miller 1990, Smith and Coates 1938, Stacy et al. 2008, Work et al. 2004). Juvenile turtles are most often affected, but not during or immediately after the pelagic life phase. Indeed, molecular evidence suggests that sea turtles acquire infection with the virus after recruitment to neritic habitat (Ene et al. 2005). Low levels of disease are observed in adult sea turtles (George 1997, Limpus and Miller 1990). Papillary or smooth, flat or nodular masses, ranging in size from <1 to >30 cm diameter, occur on soft tissues including the oral, ocular, orbital adnexa, neck, limbs, and tail, as well as the sutures of the plastron and carapace (Figure 14.3; Flint et al. 2009b). They may also involve the internal organs (Jacobson 2007b, Milton and Lutz 2003, Wyneken et al. 2006) depending on the biogeographic region, where in Florida and Hawaii it is seen more commonly than in places like Australia. Similarly, they may invade the corneal surface of the eye as has been

**FIGURE 14.3** Cutaneous fibropapillomatosis of the left axilla in a small immature green sea turtle in Australia. Tumors range in presentation from smooth to cauliflower, white to dark gray and from less than 1 cm in diameter to greater than 8 cm.

recorded in several parts of the world at different times (Flint et al. 2010b, Jacobson et al. 1989, Norton et al. 1990). Spontaneous regression of tumors has been observed in turtles in Hawaii, Florida, and Queensland, although the frequency of regression and the duration of FP in sea turtles appear to vary at each location. Further, significance of body part affected by the FP also influences potential survivorship. Animals with internal, glottal, or corneal FP are thought to have a poor prognosis, whereas in Hawaii approximately 30% of animals with cutaneous FP, and near 100% in Australia, are believed to make a full recovery without the need for medical or surgical intervention (Chaloupka and Balazs 2005).

Fibropapilloma masses are benign, although they often cause severe debilitation due to space-occupying effects or interference with systemic function (Aguirre and Lutz 2004). Advanced FP has been linked to lymphopenia, chronic inflammation, immunosuppression, and systemic Gram-negative bacterial infections (Norton et al. 1990, Work and Balazs 1999, Work et al. 2000, 2003). These factors may contribute to the overrepresentation of turtles with moderate to severe FP in strandings when compared with the FP frequency of the functional population.

The etiology and pathogenesis of FP in sea turtles are only partly understood. An alpha-herpes-virus (chelonid FP-associated herpesvirus; CFPHV) has been isolated from naturally occurring tumor masses (Greenblatt et al. 2004), but to date attempts to isolate and sequence the virus have not been successful (Herbst et al. 1995, Work et al. 2009). In the natural environment, the marine leech *Ozobranchus margoi* and cleaner fish saddleback wrasse *Thalassoma duperrey* are suspected mechanical vectors (Greenblatt et al. 2004, Lu et al. 2000). Progression to tumor development is likely to be multifactorial and involve environmental cofactors. For example, seasonal elevation of water temperature, high levels of anthropogenic activity in the immediate environment, and exposure to potential tumor-promoting compounds including okadaic acid produced by the benthic dinoflagellate *Prorocentrum* species and products of *Lyngbya majuscula* (toxic cyanobacterium) algal blooms are thought to be implicated (Aguirre et al. 1994, Arthur et al. 2006, Greenblatt et al. 2004, Herbst 1994, Work et al. 2005).

### 14.4.3 INFECTIOUS DISEASES

Reported bacterial infections in sea turtles have included *Vibrio, Aeromonas, Salmonella, Pseudomonas, Bacteroides, Fusobacterium, Flavobacterium, Clostridium,* and *Mycobacterium* spp. (Figure 14.4; Clary and Leong 1984, Glazebrook and Campbell 1990a,b, Greer et al. 2003, McArthur 2004, Obendorf et al. 1987, Sinderman 1977, Stewart 1990, Wiles and Rand 1987). A few infectious diseases with zoonotic potential have been identified in sea turtles. These include *Vibrio, Campylobacter, Salmonella,* and atypical *Mycobacterium* spp. (Glazebrook and Campbell 1990a).

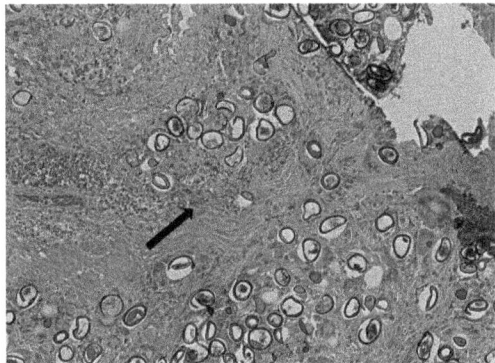

**FIGURE 14.4** Secondary infiltration of a Spirorchiid parasite granuloma with bacteria in a large immature green sea turtle in Australia.

These isolates are of particular importance to indigenous communities who hunt these turtle species and to veterinarians, who are likely to come in close contact with fluids and discharges during examination and treatment. Vigilance for zoonotic diseases should be maintained, and any occurrence reported to monitoring authorities to ensure the information may be circulated.

Fungal infections in free-living sea turtles are considered rare but may manifest as systemic mycoses, gastrointestinal disorders, or pulmonary airway diseases under stressful conditions such as captive environments or cold shock. Fungal infections including *Colletotrichum acutatum*, *Candida*, *Penicillium lilacinum*, *Cladosporium*, *Sporotrichum*, *Paecilomyces*, *Aspergillus*, and *Fusarium* spp. have been identified, predominantly in captive turtle populations (Flint et al. 2009b, Glazebrook and Campbell 1990a, Oros et al. 2004).

### 14.4.4 TOXINS

The encroachment of urbanization, with associated runoffs of herbicides, pesticides, and storm-water debris, pathogens, toxins, and wastes (Chilvers et al. 2005), can affect sea turtles either directly or indirectly by affecting food sources and breeding cycles. Covered comprehensively by Dr Keller in an earlier chapter of this volume, environmental pollutants including heavy metals and organochlorine pesticides have been identified in blood and tissues of various sea turtle populations, and are associated with subtle negative effects on immune function and other health parameters. However, until recently, most resulting impacts on the effects of contaminants on sea turtles were subject to conjecture (Day et al. 2005, 2007, Hermanussen et al. 2006, Hochscheid et al. 2004, Keller et al. 2004, 2006, Perrault et al. 2011). It is assumed the full suite of contamination effects acting on sea turtle populations have not yet been comprehensively measured due to the cost-inhibitory nature of mass toxin screening and consequently our understanding of their exact role in disease contribution and manifestation is limited.

### 14.4.5 GASTROINTESTINAL ANOMALIES

In wild populations of sea turtles, gastrointestinal anomalies may include linear foreign bodies and perforations, obstructions, obstipation, or gastrointestinal ileus (Figure 14.5). Although there is substantial literature highlighting techniques to treat impactions and devitalized gastrointestine, the causes of these nonmechanical anomalies are poorly understood. In cases such as linear foreign bodies caused by fishing lines and hooks, which may result in plication of the gut or coelomitis with intestinal perforation, there is a clear etiology. Other types of foreign body material that may cause obstruction include fiber, metal, plastic, rubber, and atypical food types. These are a common cause of pathology in some cohorts of stranded or sick sea turtles (Boyle and Limpus 2008), but are not

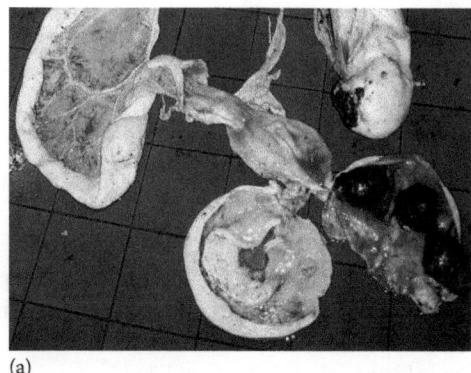
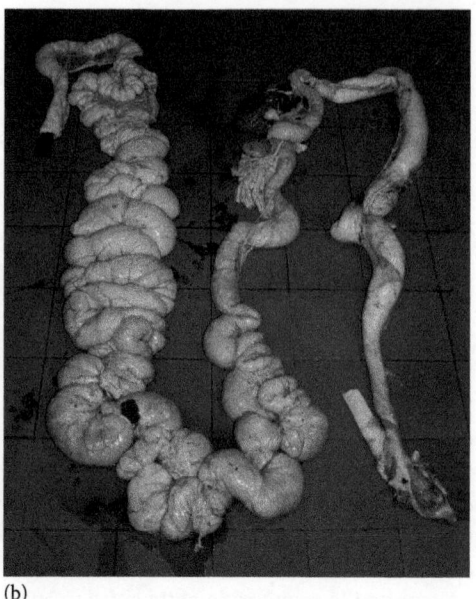

(a)                                                                                (b)

FIGURE 14.5    (a) Gastrointestinal torsion with associated natural-fiber fecal balls and secondary inflammation of the intestinal mucosa in a mature green sea turtle; and (b) a common gross presentation of fishing line
ingestion (linear foreign body) causing intestinal plication in a small green sea turtle.

noted in others (Chaloupka et al. 2008b, Flint et al. 2010e, Oros et al. 2005). Also with a straightforward etiology, obstructions can be due to torsions as seen in other species, or by digesta that occlude
or reduce the patency of the lumen of the intestine. By contrast, it is the etiology of obstipation that
are less obvious. It is often associated with fecal balls comprised solely of natural food products
(e.g., crushed shells in loggerheads and compressed sea grasses in green turtles), but whether the
impaction is secondary to a primary pathology such as ileus is unknown. A theory, which is yet to
be supported or rejected, is denervation of the vagal and associated nerves, which innervate and
regulate/control the peristaltic movement of the gut that enables ingesta and digesta to move. Viral,
bacterial, or parasitic agents may be implicated, rendering the gut functionally static.

## 14.5   POPULATION MODELING

Population modeling is becoming an important tool in wildlife management (Ward and Lafferty
2004). Since the first deterministic models were used to predict survivorship of sea turtles (Crouse
et al. 1987), modeling has improved to include population response to a wide range of parameters
(Bjorndal et al. 2000, Chaloupka 2002, Chaloupka and Balazs 2005, 2007, Wallace et al. 2011) that
take into account predicted mortality and recruitment within the population, but do not necessarily account for the effects of disease. While these models allow for overall changes in mortality
or survivorship to be investigated, they do so irrespective of the origin of the change. Modeling
using meta-analyses has concentrated on using published studies to elicit trends in populations
providing retrospective feedback to an environment's management plan (Dethmers et al. 2006,
Jackson 1997, Jackson et al. 2001). This resulted in a delayed feedback to implemented changes
(i.e., they were not predictive of the effect on environment). Population models, especially those
implemented in Marine Area Protection management, are now capable of dynamic predictive monitoring (Chaloupka and Limpus 2005, Chaloupka et al. 2008a). This allows small adjustments to
be incorporated into both the model and the management plan in "real time," creating an adaptive
management plan (Gerber et al. 2005).

Given the versatility of this analysis tool, disease modeling could be directly applied to sea turtles with minor modifications of principles used in existing models. Models could be improved and modified as a better understanding of the risk factors and diseases affecting sea turtles in each locale are gained. Such an approach would provide near-instant survivorship feedback under a range of disease and environmental conditions and allow the real-time adjustments to management plans required to mitigate negative effects on these threatened animals.

## 14.6 CONCLUSIONS

Disease in sea turtles has shown to be a multifactorial, understudied area of epidemiology and veterinary medicine. Efforts to address this have increased greatly in recent years. If the impact of disease on wild populations of sea turtles is to be mitigated, then disease investigations incorporating turtle biology and ecology, clinical assessment, postmortem examination to definitively diagnose disease (at a population level through collaborative efforts), environmental assessment, and disease modeling need to be conducted. This will allow prediction of the effects of environment on disease occurrence and prevalence in a timely manner that may allow for active intervention strategies to be employed.

## REFERENCES

Aguirre AA, G Balazs, B Zimmerman, and FD Galey. Organic contaminants and trace metals in the tissues of green turtles (*Chelonia mydas*) afflicted with fibropapillomas in the Hawaiian islands. *Marine Pollution Bulletin* 1994; 28: 109–114.

Aguirre AA and P Lutz. Sea turtles as sentinels of marine ecosystem health: Is fibropapillomatosis an indicator? *EcoHealth* 2004; 1: 275–283.

Aguirre AA, TR Spraker, GH Balazs, and B Zimmerman. Spirorchidiasis and fibropapillomatosis in green turtles from the Hawaiian Islands. *Journal of Wildlife Diseases* 1998; 34: 91–98.

Aguirre AA and GM Tabor. Introduction: Marine vertebrates as sentinels of marine ecosystem health. *EcoHealth* 2004; 1: 236–238.

Arthur KE, CJ Limpus, and JM Whittier. Baseline blood biochemistry of Australian green turtles (*Chelonia mydas*) and effects of exposure to the toxic cyanobacterium *Lyngbya majuscula*. *Australian Journal of Zoology* 2008; 56: 23–32.

Arthur K, G Shaw, CJ Limpus, and JW Udy. A review of the potential role of tumour-promoting compounds produced by *Lyngbya majuscula* in marine turtle fibropapillomatosis. *African Journal of Marine Science* 2006; 28: 441–446.

Bjorndal KA, AB Bolten, and M Chaloupka. Green turtle somatic growth model: Evidence fordensity dependence. *Ecological Applications* 2000; 10: 269–282.

Blair D. Ribosomal DNA variation in parasitic flatworms. In: Maule A, ed. *Parasitic Flatworms: Molecular Biology, Biochemistry, Immunology and Control*. Nosworthy Way, Wallingford, Oxfordshire, U.K.: CABI, 2006, pp. 96–123.

Bolten AB and KA Bjorndal. Blood profiles for a wild population of green turtles (*Chelonia mydas*) in the southern Bahamas: Size-specific and sex-specific relationships. *Journal of Wildlife Diseases* 1992; 28: 407–413.

Boyle MC and CJ Limpus. The stomach contents of post-hatchling green and loggerhead sea turtles in the southwest Pacific: An insight into habitat association. *Marine Biology* 2008; 155: 233–241.

Casal AB, F Freire, G Bautista-Harris, A Arencibia, and J Oros. Ultrastructural characteristics of blood cells of juvenile loggerhead sea turtles (*Caretta caretta*). *Anatomia, Histologia, Embryologia* 2007; 36: 332–335.

Casal AB and J Oros. Morphologic and cytochemical characteristics of blood cells of juvenile loggerhead sea turtles (*Caretta caretta*). *Research in Veterinary Science* 2007; 82: 158–165.

Chaloupka M. Stochastic simulation modelling of southern Great Barrier Reef green turtle population dynamics. *Ecological Modelling* 2002; 148: 79–109.

Chaloupka M and G Balazs. Modelling the effect of fibropapilloma disease on the somatic growth dynamics of Hawaiian green sea turtles. *Marine Biology* 2005; 147: 1251–1260.

Chaloupka M and G Balazs. Using Bayesian state-space modelling to assess the recovery and harvest potential of the Hawaiian green sea turtle stock. *Ecological Modelling* 2007; 205: 93–109.

Chaloupka M, N Kamezaki, and CJ Limpus. Is climate change affecting the population dynamics of the endangered Pacific loggerhead sea turtle? *Journal of Experimental Marine Biology and Ecology* 2008a; 356: 136–143.

Chaloupka M and CJ Limpus. Trends in the abundance of sea turtles resident in southern Great Barrier Reef waters. *Biological Conservation* 2001; 102: 235–249.

Chaloupka M and CJ Limpus. Estimates of sex- and age-class-specific survival probabilities for a southern Great Barrier Reef green sea turtle population. *Marine Biology* 2005; 146: 1251–1261.

Chaloupka M, T Work, G Balazs, S Murakawa, and R Morris. Cause-specific temporal and spatial trends in green sea turtle strandings in the Hawaiian Archipelago (1982–2003). *Marine Biology* 2008b; 154: 887–898.

Chan EH. Marine turtles in Malaysia: On the verge of extinction? *Aquatic Ecosystem Health and Management* 2006; 9: 175–184.

Chilvers BL, IR Lawler, F Macknight, H Marsh, M Noad, and R Paterson. Moreton Bay, Queensland, Australia: An example of the co-existence of significant marine mammal populations and large-scale coastal development. *Biological Conservation* 2005; 122: 559–571.

Chrisman CL, M Walsh, JC Meeks, H Zurawka, R LaRock, and LH Herbst. Neurological examination of sea turtles. *Journal of American Veterinary Medical Association* 1997; 211: 1043–1047.

Clary JC and JK Leong. Disease studies aid Kemp's ridley sea turtle headstart research. *Herpetological Review* 1984; 15: 69–70.

Cribb TH and AN Gordon. Hapalotrema (Digenea: Spirorchidae) in the green turtle (*Chelonia mydas*) in Australia. *Journal of Parasitology* 1998; 84: 375–378.

Crouse DT, L Crowder, and H Caswell. A stage-based population model for loggerhead sea turtles and implications for conservation. *Ecology* 1987; 68: 1412–1423.

Dailey MD, ML Fast, and GH Balazs. *Carettacola hawaiiensis* n. sp. (Trematoda: Spirorchidae) from the green turtle, *Chelonia mydas*, in Hawaii. *Journal of Parasitology* 1991; 77: 906–909.

Dailey MD, ML Fast, and G Balazs. A survey of the Trematoda (Platyhehninthes: Digenea) parasitic in green turtles, *Chelonia mydas* (L.) from Hawaii. *Bulletin of the Southern Californian Academy of Science* 1992; 91: 84–91.

Day RD, SJ Christopher, PR Becker, and DW Whitaker. Monitoring mercury in the loggerhead sea turtle, *Caretta caretta*. *Environmental Science and Technology* 2005; 39: 437–446.

Day RD, AL Segars, MD Arendt, AM Lee, and MM Peden-Adams. Relationship of blood mercury levels to health parameters in the loggerhead sea turtle (*Caretta caretta*). *Environmental Health Perspectives* 2007; 115: 1421–1428.

Deem SL, ES Dierenfeld, GP Sounguet, AR Alleman, C Cray, RH Poppenga, TM Norton et al. Blood values in free-ranging nesting leatherback sea turtles (*Dermochelys coriacea*) on the coast of the Republic of Gabon. *Journal of Zoo and Wildlife Medicine* 2006; 37: 464–471.

Deem SL, TM Norton, MA Mitchell, AL Segars, AR Alleman, C Cray, RH Poppenga et al. Comparison of blood values in foraging, nesting, and stranded loggerhead turtles (*Caretta caretta*) along the coast of Georgia, USA. *Journal of Wildlife Diseases* 2009; 45: 41–56.

Dethmers KE, D Broderick, C Moritz, NN Fitzsimmons, CJ Limpus, S Lavery, S Whiting et al. The genetic structure of Australasian green turtles (*Chelonia mydas*): Exploring the geographical scale of genetic exchange. *Journal of Molecular Ecology* 2006; 15: 3931–3946.

Dobbs K and S Pierce. Marine reptiles. In: Chin A, ed. *The State of the Great Barrier Reef On-line*. Townsville, Queensland, Australia: Great Barrier Reef Marine Park Authority, 2005. http://www.gbrmpa.gov.au/corp_site/info_services/publications/sotr/marine_reptiles/index.html (accessed March 23, 2012).

Ene A, M Su, S Lemaire, C Rose, S Schaff, R Moretti, J Lenz et al. Distribution of chelonid fibropaillomatosis-associated Herpesvirus variants in Florida: Molecular genetic evidence for infection of turtles following recruitment to neritic developmental habitats. *Journal of Wildlife Diseases* 2005; 41: 489–497.

Flint M, D Blair, JC Patterson-Kane, M Kway-Tanner, and PC Mills. Blood flukes (Spirorchiidae) as a major cause of marine turtle mortality in Queensland. XII *International Congress of Parasitology* 2010a; 57–61.

Flint M, CJ Limpus, JC Patterson-Kane, PJ Murray, and PC Mills. Corneal fibropapillomatosis in green sea turtles (*Chelonia mydas*) in Australia. *Journal of Comparative Pathology* 2010b; 142: 341–346.

Flint M, JM Morton, CJ Limpus, JC Patterson-Kane, PJ Murray, and PC Mills. Development and application of biochemical and haematological reference intervals to identify unhealthy green sea turtles (*Chelonia mydas*). *The Veterinary Journal* 2010c; 185: 299–304.

Flint M, JM Morton, JC Patterson-Kane, CJ Limpus, and PC Mills. Reference intervals for plasma biochemical and hematological measures in loggerhead sea turtles (*Caretta caretta*) from Moreton Bay, Australia. *Journal of Wildlife Diseases* 2010d; 46: 731–741.

Flint M, JM Morton, JC Patterson-Kane, CJ Limpus, PJ Murray, and PC Mills. Using plasma biochemistry and haematological blood reference ranges as a tool in diagnosing disease for green turtles (*Chelonia mydas*) in Queensland Australia. *Proceedings of the 29th International Sea Turtle Symposium*, February 17–19, 2009. Brisbane, Queensland, Australia, 2009a.

Flint M, JC Patterson-Kane, CJ Limpus, and PC Mills. Health surveillance of stranded green turtles in southern Queensland, Australia (2006–2009): An epidemiological analysis of causes of disease and mortality. *EcoHealth* 2010e; 7: 135–145.

Flint M, JC Patterson-Kane, CJ Limpus, TM Work, D Blair, and PC Mills. Post mortem diagnostic investigation of disease in free-ranging marine turtle populations: A review of common pathological findings and protocols. *Journal of Veterinary Diagnostic Investigation* 2009b; 21: 733–759.

Flint M, JC Patterson-Kane, PC Mills, and CJ Limpus. A veterinarian's guide to sea turtle post mortem examination and histological investigation, 2009c. http://www.uq.edu.au/vetschool/index.html?page=102248 (accessed March 23, 2012).

George RH. Health problems and diseases of sea turtles. In: Lutz PL and JA Musick, eds. *The Biology of Sea Turtles*. Boca Raton, FL: CRC Press, Inc., 1997.

Gerber LR, M Beger, MA McCarthy, and HP Possingham. A theory for optimal monitoring of marine reserves. *Ecology Letters* 2005; 8: 829–837.

Glazebrook JS and RSF Campbell. A survey of the diseases of marine turtles in northern Australia I. Farmed turtles. *Disease of Aquatic Organisms* 1990a; 9: 83–95.

Glazebrook JS and RSF Campbell. A survey of the diseases of marine turtles in northern Australia II. Oceanarium-reared and wild turtles. *Disease of Aquatic Organisms* 1990b; 9: 97–104.

Glazebrook JS, RS Campbell, and D Blair. Studies on cardiovascular fluke (Digenea: Spirorchiidae) infections in sea turtles from the Great Barrier Reef, Queensland, Australia. *Journal of Comparative Pathology* 1989; 101: 231–250.

Gordon AN. A necropsy-based study of green turtles (*Chelonia mydas*) in south-east Queensland. PhD thesis, School of Veterinary Science, The University of Queensland, Brisbane, Queensland, Australia, 2005, p. 234.

Gordon AN, WR Kelly, and TH Cribb. Lesions caused by cardiovascular flukes (Digenea: Spirorchidae) in stranded green turtles (*Chelonia mydas*). *Journal of Veterinary Pathology* 1998; 35: 21–30.

Gordon AN, WR Kelly, and RJ Lester. Epizootic mortality of free-living green turtles, *Chelonia mydas*, due to coccidiosis. *Journal of Wildlife Diseases* 1993; 29: 490–494.

Greenblatt RJ, TM Work, GH Balazs, CA Sutton, RN Casey, and JW Casey. The Ozobranchus leech is a candidate mechanical vector for the fibropapilloma-associated turtle herpesvirus found latently infecting skin tumors on Hawaiian green turtles (*Chelonia mydas*). *Virology* 2004; 321: 101–110.

Greenblatt RJ, TM Work, P Dutton, CA Sutton, TR Spraker, RN Casey, CE Diez et al. Geographic variation in marine turtle fibropapillomatosis. *Journal of Zoo and Wildlife Medicine* 2005; 36: 527–530.

Greer LL, JD Strandberg, and BR Whitaker. Mycobacterium chelonae osteoarthritis in a Kemp's ridley sea turtle (*Lepidochelys kempii*). *Journal of Wildlife Diseases* 2003; 39: 736–741.

Habib RR, K El Zein, and J Ghanawi. Climate change and health research in the eastern Mediterranean region. *EcoHealth* 2010; 7: 156–175.

Hamann M, CJ Limpus, and JM Whittier. Patterns of lipid storage and mobilisation in the female green sea turtle (*Chelonia mydas*). *Journal of Comparative Physiology* 2002; 172: 485–493.

Hamann M, CJ Limpus, and JM Whittier. Seasonal variation in plasma catecholamines and adipose tissue lipolysis in adult female green sea turtles (*Chelonia mydas*). *General and Comparative Endocrinology* 2003; 130: 308–316.

Hamann M, CS Schäuble, T Simon, and S Evans. Demographic and health parameters of green sea turtles *Chelonia mydas* foraging in the Gulf of Carpentaria, Australia. *Endangered Species Research* 2006; 2: 81–88.

Herbst L. Fibropapillomatosis of marine turtles. *Annual Review of Fish Diseases* 1994; 4: 389–425.

Herbst LH and ER Jacobson. Practical approaches for studying sea turtle health and disease. In: Lutz P, JA Musick, and J Wyneken, eds. *The Biology of Sea Turtles*, Vol. II. New York: CRC Press, 2003, pp. 385–410.

Herbst LH, ER Jacobson, R Moretti, T Brown, J Sundberg, and PA Klein. Experimental transmission of green turtle fibropapillomatosis using cell-free tumor extracts. *Disease of Aquatic Organisms* 1995; 22: 1–12.

Hermanussen S, CJ Limpus, O Papke, DW Connell, and C Gaus. Foraging habitat contamination influences green turtle PCDD/F exposure. *Organohalogen Compounds* 2006; 68: 592–595.

Hochscheid S, F Bentivegna, and JR Speakman. Long-term cold acclimation leads to high Q10 effects on oxygen consumption of loggerhead sea turtles *Caretta caretta*. *Physiological and Biochemical Zoology* 2004; 77: 209–222.

Horn PS, AJ Pesce, and BE Copeland. A robust approach to reference interval estimation and evaluation. *Clinical Chemistry* 1998; 44: 622–631.

IUCN/SSC. The IUCN Red List of Threatened Species. *IUCN Red List*. Cambridge, U.K. 2008. www.redlist.org

Jackson JB. Reefs since Columbus. *Coral Reefs* 1997; 16: S23–S32.

Jackson JB, MX Kirby, WH Berger, KA Bjorndal, LW Botsford, BJ Bourque, RH Bradbury et al. Historical overfishing and the recent collapse of coastal ecosystems. *Science* 2001; 293: 629–638.

Jacobson ER. Implications of infectious diseases for captive propogation and introduction programs of threatened/endangered reptiles. *Journal of Zoo and Wildlife Medicine* 1993; 24: 245–255.

Jacobson ER. Sea turtle biopsy and necropsy techniques. Gainesville, FL: University of Florida, 1999. http://www.vetmed.ufl.edu/college/departments/sacs/research/SeaTurtleBiopsyandNecropsyTechniques.html

Jacobson ER. *Infectious Diseases and Pathology of Reptiles: Color Atlas and Text*. Boca Raton, FL: CRC Press, 2007a.

Jacobson ER. Viruses and viral diseases of reptiles. In: Jacobson ER, ed. *Infectious Diseases and Pathology of Reptiles: Color Atlas and Text*. Boca Raton, FL: CRC Press, 2007b, pp. 395–460.

Jacobson ER, K Bjorndal, A Bolten, R Herren, G Harman, and L Wood. Establishing plasma biochemical and hematocrit reference intervals for sea turtles in Florida. 2007. http://accstr.ufl.edu/blood_chem.htm (accessed on October 23, 2012).

Jacobson ER, BL Homer, BA Stacy, EC Greiner, NJ Szabo, CL Chrisman, F Origgi et al. Neurological disease in wild loggerhead sea turtles *Caretta caretta*. *Disease of Aquatic Organisms* 2006; 70: 139–154.

Jacobson ER, JL Mansell, JP Sundberg, L Hajjar, ME Reichmann, LM Ehrhart, M Walsh et al. Cutaneous fibropapillomas of green turtles (*Chelonia mydas*). *Journal of Comparative Pathology* 1989; 101: 39–52.

Jessop TS, JM Sumner, CJ Limpus, and JM Whittier. Interplay between plasma hormone profiles, sex and body condition in immature hawksbill turtles (*Eretmochelys imbricata*) subjected to a capture stress protocol. *Journal of Comparative Biochemistry and Physiology* 2004; 137: 197–204.

Keller JM, JR Kucklick, MA Stamper, CA Harms, and PD McClellan-Green. Associations between organochlorine contaminant concentrations and clinical health parameters in loggerhead sea turtles from North Carolina, USA. *Environmental Health Perspectives* 2004; 112: 1074–1079.

Keller JM, PD McClellan-Green, JR Kucklick, DE Keil, and MM Peden-Adams. Effects of organochlorine contaminants on loggerhead sea turtle immunity: Comparison of a correlative field study and in vitro exposure experiments. *Environmental Health Perspectives* 2006; 114: 70–76.

Klinger RC and JA Musick. Annular growth layers in juvenile loggerhead turtles (*Caretta caretta*). *Bulletin of Marine Science* 1992; 51: 224–230.

Limpus CJ. *A Biological Review of Australian Marine Turtles. 1. Loggerhead Turtle, Caretta caretta (Linnaeus)*. Brisbane, Queensland, Australia: Queensland Environmental Protection Agency, 2008a.

Limpus CJ. *A Biological Review of Australian Marine Turtles. 2. Green Turtle, Chelonia mydas (Linnaeus)*. Brisbane, Queensland, Australia: Queensland Environmental Protection Agency, 2008b.

Limpus CJ. *A Biological Review of Australian Marine Turtles. 6. Leatherback Turtle, Dermochelys coriacea (Vandelli)*. Brisbane, Queensland, Australia: Queensland Environmental Protection Agency, 2009.

Limpus CJ and DJ Limpus. The biology of the loggerhead turtle, *Caretta caretta*, in western South Pacific Ocean foraging areas. In: Witherington B and A Bolten, eds. *Biology and Conservation of Loggerhead Turtles*. Washington, DC: Smithsonian Institution Press, 2003, pp. 93–113.

Limpus C and J Miller. The occurrence of cutaneous fibropapillomas in marine turtles in Queensland. *Proceedings of Australian Marine Turtle Conservation Workshop*, Canberra, ACT, Australia, 1990, pp 186–188.

Limpus CJ, JD Miller, and DJ Limpus. The occurrence of ectopic cloaca deformity in the green turtle in eastern Australia. *Chelonian Conservation and Biology* 2009; 8: 100–101.

Limpus CJ and P Reed. The green turtle, *Chelonia mydas*, in Queensland: A preliminary description of the population structure in a coral reef feeding ground. In: Grigg G, R Shine, and H Ehmann, eds. *Biology of Australasian Frogs and Reptiles*. Sydney, Australia: Royal Zoological Society of New South Wales, 1985, pp. 47–52.

Lu Y, Y Wang, Q Yu, AA Aguirre, GH Balazs, VR Nerurkar, and R Yanagihara. Detection of herpesviral sequences in tissues of green turtles with fibropapilloma by polymerase chain reaction. *Archives of Virology* 2000; 145: 1885–1893.

Mader DR. *Reptile Medicine and Surgery*, 2nd edn. St. Louis, MO: Saunders Elsevier, 2006.

McArthur S. Infectious agents. In: McArthur S, R Wilkinson, and J Meyer, eds. *Medicine and Surgery of Tortoises and Turtles*. Ames, IA: Blackwell Publishing, 2004, pp. 31–34.

Miller JD and CJ Limpus. Ontogeny of marine turtle gonads. In: Lutz P, JA Musick, and J Wyneken, eds. *The Biology of Sea Turtles*, Vol. II. Boca Raton, FL: CRC Press, 2003, pp. 199–224.

Milton SL and P Lutz. Physiological and genetic response to environmental stress. In: Lutz P, JA Musick, and J Wyneken, eds. *The Biology of Sea Turtles*, Vol. II. Boca Raton, FL: CRC Press, 2003, pp. 163–198.

Morton SR, O Hoegh-Guldberg, DB Lindenmayer, M Harriss Olson, L Hughes, MT McCulloch, S McIntyre et al. The big ecological questions inhibiting effective environmental management in Australia. *Austral Ecology* 2009; 34: 1–9.

NMFS and USFWS. *Recovery Plan for the Northwest Atlantic Population of the Loggerhead Sea Turtle (Caretta caretta)*, Second Revision. Silver Spring, MD: National Marine Fisheries Service, 2008.

NOAA. *Green Turtle (Chelonia mydas)*. NOAA Fisheries Office of Protected Resources, 2008a.

NOAA. *Loggerhead turtle (Caretta caretta)*. NOAA Fisheries Office of Protected Resources, 2008b.

Nolan MJ and T Cribb. The use and implications of ribosomal DNA sequencing for the discrimination of digenean species. *Advances in Parasitology* 2005; 60: 101–163.

Norton TM. Chelonian emergency and critical care. *Seminars in Avian and Exotic Pet Medicine* 2005; 14: 106–130.

Norton TM, ER Jacobson, and J Sundberg. Cutaneous and renal fibropapilloma in a green turtle, *Chelonia mydas. Journal of Wildlife Diseases* 1990; 26: 212–216.

Nutter FB, DD Lee, and MA Stamper. Hemiovariosalpingectomy in a loggerhead sea turtle (*Caretta caretta*). *Veterinary Record* 2000; 146: 78–80.

Obendorf DL, J Carson, and TJ McManus. Vibrio damsela infection in a stranded leatherback turtle (*Dermochelys coriacea*). *Journal of Wildlife Diseases* 1987; 23: 666–668.

Oros J, A Arencibia, L Fernandez, and HE Jensen. Intestinal candidiasis in a loggerhead sea turtle (*Caretta caretta*): An immunohistochemical study. *Veterinary Journal* 2004; 167: 202–207.

Oros J, A Torrent, P Calabuig, and S Deniz. Diseases and causes of mortality among sea turtles stranded in the Canary Islands, Spain (1998–2001). *Diseases of Aquatic Organisms* 2005; 63: 13–24.

Owens DW, and GJ Ruiz. New methods of obtaining blood and cerebrospinal fluid from marine turtles. *Herpetologica* 1980; 36: 17–20.

Patterson-Kane JC, M Flint, PC Mills, CJ Limpus, and D Blyde. A retrospective study of histological lesions in stranded sea turtles in the Gold Coast region, Queensland. *Proceedings of the 29th International Sea Turtle Symposium*, February 17–19, 2009. Brisbane, Queensland, Australia, 2009.

Perrault JR, D Miller, E Eads, C Johnson, A Merrill, LJ Thompson, and J Wyneken. Maternal health status correlates with nest success of leatherback sea turtles (*Dermochelys coriacea*) from Florida. *PLoS One* 2012; 7: e31841. doi:31810.31371/journal.pone.0031841.

Perrault JR, J Wyneken, LJ Thompson, C Johnson, and D Miller. Why are hatching and emergence success low? Mercury and selenium concentrations in nesting leatherback sea turtles (*Dermochelys coriacea*) and their young in Florida. *Marine Pollution Bulletin* 2011; 62: 1671–1682.

Pesce AJ, PS. Horn, and D Lewis. 2005. Reference interval software. Draft Version, Copyright 2005, University of Cincinnati, Cincinnati, Ohio.

Platt TR. Family Spirorchiidae Stunkard, 1921. In: Gibson DI, A Jones, and RA Bray, eds. *Keys to the Trematoda*, Vol. 1. Wallingford, Oxfordshire, U.K.: CABI Publishing, 2001, pp. 453–467.

Raidal SR, M Ohara, RP Hobbs, and RI Prince. Gram-negative bacterial infections and cardiovascular parasitism in green sea turtles (*Chelonia mydas*). *Australian Veterinary Journal* 1998; 76: 415–417.

RPMTA, 2003. *Recovery Plan for Marine Turtles in Australia*. Canberra, ACT, Australia: Department of Environment and Heritage. Commonwealth of Australia, Canberra.

RPMTA, 2006. *Recovery Plan for Marine Turtles in Australia*. In: MSS, ed: Canberra, ACT, Australia: Commonwealth of Australia, Canberra.

Samour JH, JC Howlett, C Silvanose, CR Hasbun, and SM Al-Ghais. Normal haematology of free-living green sea turtles (*Chelonia mydas*) from the United Arab Emirates. *Comparative Haematology International* 1998; 8: 102–107.

Schwabe CW. *Veterinary Medicine and Human Health*, 1st edn. Baltimore, MD: The Williams & Wilkins Company, 1969.

Sinderman CJ. Aeromonas disease in loggerhead turtles. In: Sinderman CJ, ed. *Disease Diagnosis and Control in North American Marine Aquaculture, Developments in Aquaculture and Fisheries Sciences*, Vol. 6. New York: Elsevier North-Holland, 1977, pp. 292–293.

Smith JW. The blood flukes (Digenea: Sanguinicolidae and Spirorchidae) of cold-blooded vertebrates: Part 2. Appendix 1: Comprehensive parasite-host list; Appendix 2: Comprehensive host- parasite list. *Helminthological Abstracts* 1997; 66: 329–344.

Smith GM and CW Coates. Fibro-epithelial growths of the skin in large marine turtles, *Chelonia mydas* (Linnaeus). *Zoologica (NY)* 1938; 23: 93–98.

Snyder SD. Phylogeny and paraphyly among tetrapod blood flukes (Digenea: Schistosomatidae and Spirorchiidae). *International Journal for Parasitology* 2004; 34: 1385–1392.

Solberg HE. International Federation of Clinical Chemistry (IFCC). Scientific Committee, Clinical Section. Expert Panel on Theory of Reference Values (EPTRV) and Internation Committee for Standardization in Haematology (ICSH) Standing Committee on Reference Values. Approved recommendation (1987) on the theory of reference values. Part 5. Statistical treatment of collected reference values. Determination of reference limits. *Clinica Chimica Acta* 1987; 170: S13–S32.

Spotila JR. *Sea Turtles: A Complete Guide to their Biology, Behavior, and Conservation.* Baltimore, MD: The Johns Hopkins University Press and Oakwood Arts, 2004.

Stacy BA, AM Foley, EC Greiner, LH Herbst, AB Bolten, PA Klein, CA Manire et al. Spirorchiidiasis in stranded loggerhead *Caretta caretta* and green turtles *Chelonia mydas* in Florida (USA): Host pathology and significance. *Disease of Aquatic Organisms* 2010; 89: 237–259.

Stacy BA, JF Wellehan, AM Foley, SS Coberley, LH Herbst, CA Manire, MM Garner et al. Two herpesviruses associated with disease in wild Atlantic loggerhead sea turtles (*Caretta caretta*). *Veterinary Microbiology* 2008; 126: 63–73.

Stamper MA, C Harms, SP Epperly, J Braun-McNeill, and MK Stoskopf. Relationship between barnacle epibiotic load and hematologic parameters in loggerhead sea turtles (*Caretta caretta*), a comparison between migratory and residential animals in Pamlico Sound, North Carolina. *Journal of Zoo and Wildlife Medicine* 2005; 36: 635–641.

Stewart JS. Anaerobic bacterial infections in reptiles. *Journal of Zoo and Wildlife Medicine* 1990; 21: 180–184.

Tkach VV, SD Snyder, and JA Vaughan. A new species of blood fluke (Digenea: Spirorchiidae) from the Malayan box turtle, *Cuora amboinensis* (Cryptodira: Geomydidae) in Thailand. *Journal of Parasitology* 2009; 95: 743–746.

Valente ALS, R Cuenca, ML Parga, S Lavin, J Franch, and I Marco. Cervical and coelomic radiology of loggerhead sea turtle (*Caretta caretta*). *Canadian Journal of Veterinary Research* 2006; 70: 285–290.

Valente ALS, R Cuenca, M Zamora, ML Parga, S Lavin, F Alegre, and I Marco. Computed tomography of the vertebral column and coelomic structures in the normal loggerhead sea turtle (*Caretta caretta*). *The Veterinary Journal* 2007a; 174: 362–370.

Valente ALS, ML Parga, Y Espada, S Lavin, F Alegre, I Marco, and R Cuenca. Ultrasonographic imaging of loggerhead sea turtles (*Caretta caretta*). *Veterinary Record* 2007b; 161: 226–232.

Valente ALS, ML Parga, Y Espada, S Lavan, F Alegre, I Marco, and R Cuenca. Evaluation of Doppler ultrasonography for the measurement of blood flow in young loggerhead sea turtles (*Caretta caretta*). *Veterinary Journal* 2008; 176: 385–392.

Wallace BP, AD DiMatteo, AB Bolten, M Chaloupka, BJ Hutchinson, FA Abreu-Grobois, JA Mortimer et al. Global conservation priorities for marine turtles. *PLoS One* 2011; 6: e24510. doi:24510.21371/journal.pone.0024510.

Ward JR and KD Lafferty. The elusive baseline of marine disease: Are diseases in ocean ecosystems increasing? *PLoS Biology* 2004; 2: 542–547.

Whiting SD, ML Guinea, CJ Limpus, and K Fomiatti. Blood chemistry reference values for two ecologically distinct population of foraging green turtles, eastern Indian Ocean. *Comparative Clinical Pathology* 2007; 16: 109–118.

Wilcox B and AA Aguirre. One ocean, one health. *EcoHealth* 2004; 1: 211–212.

Wiles M and TG Rand. Integumental ulcerative disease in a loggerhead turtle, *Caretta caretta*, at the Bermuda Aquarium: Microbiology and histopathology. *Disease of Aquatic Organisms* 1987; 3: 85–90.

Wolke RE and A George. *Sea Turtle Necropsy Manual.* Panama City, FL: National Oceanic and Atmospheric Administration, 1981.

Work TM. *Sea Turtle Necropsy Manual for Biologists in Remote Refuges.* Honolulu, Hawaii: USGS, 2000.

Work TM and GH Balazs. Relating tumor score to hematology in green turtles with fibropapillomatosis in Hawaii. *Journal of Wildlife Diseases* 1999; 35: 804–807.

Work T and G Balazs. Pathology and distribution of sea turtles landed as bycatch in the Hawaii-based North Pacific pelagic longline fishery. *Journal of Wildlife Diseases* 2010; 46: 422–432.

Work TM, GH Balazs, RA Rameyer, SP Chang, and J Berestecky. Assessing humoral and cell-mediated immune response in Hawaiian green turtles, *Chelonia mydas*. *Veterinary Immunology and Immunopathology* 2000; 74: 179–194.

Work TM, GH Balazs, RA Rameyer, and RA Morris. Retrospective pathology survey of green turtles *Chelonia mydas* with fibropapillomatosis in the Hawaiian Islands, 1993–2003. *Disease of Aquatic Organisms* 2004; 62: 163–176.

Work TM, GH Balazs, JL Schumacher, and M Amarisa. Epizootiology of spirorchiid infection in green turtles (*Chelonia mydas*) in Hawaii. *Journal of Parasitology* 2005; 91: 871–876.

Work TM, GH Balazs, M Wolcott, and R Morris. Bacteraemia in free-ranging Hawaiian green turtles *Chelonia mydas* with fibropapillomatosis. *Diseases of Aquatic Organisms* 2003; 53: 41–46.

Work TM, J Dagenais, G Balazs, JL Schumacher, TD Lewis, JC Leong, RN Casey et al. In vitro biology of fibropapilloma-associated turtle herpesvirus and host cells in Hawaiian green turtles (*Chelonia mydas*). *Journal of General Virology* 2009; 90: 1943–1950.

Work TM, RE Raskin, GH Balazs, and SD Whittaker. Morphologic and cytochemical characteristics of blood cells from Hawaiian green turtles. *American Journal of Veterinary Research* 1998; 59: 1252–1257.

Wyneken J. *The Anatomy of Sea Turtles*. Miami, FL: National Oceanic and Atmospheric Administration, 2001, p. 172.

Wyneken J, DR Mader, ES Weber, and C Merigo. Medical care of sea turtles. In: Mader DR, ed. *Reptile Medicine and Surgery*, Second edition. St Louis, MO: Saunders Elsevier, 2006, pp. 972–1007.

# 15 Sea Turtle Epibiosis

*Michael G. Frick and Joseph B. Pfaller*

## CONTENTS

## 15.1 INTRODUCTION

In the marine environment, any exposed, undefended surface will eventually be colonized by marine propagules (Wahl, 1989). Colonization of inanimate structures (e.g., dock pilings and boat hulls) is called *fouling*, while colonization of other marine organisms is called *epibiosis*. Epibiosis results in spatially close associations between two or more living organisms (Harder, 2009), in which a single host (or *basibiont*) supports one or more typically opportunistic colonizers (or *epibionts*) (Wahl and Mark, 1999). Epibiosis is the most common form of symbiosis in the marine environment and may be classified into several types of associations (e.g., mutualism, commensalism, parasitism) depending on the interactions between a host and its epibionts (Leung and Poulin, 2008).

Sea turtles often act as hosts to a wide variety of epibionts, most of which are unspecialized organisms normally found associated with inanimate structures in the surrounding marine environment (i.e., "free living"). These types of epibiotic associations are known as *facultative commensalisms* (Wahl and Mark, 1999). That is, the host receives no direct benefit from the epibiont and the epibiont demonstrates little to no substrate specificity. For these associations to occur, the various settlement cues that facultative commensal epibionts utilize when selecting substrata must also be present on sea turtles (Zardus and Hadfield, 2004). Alternatively, there are several epibionts that are found almost exclusively on sea turtles (Frick et al., 2011a). These associations are known as *obligate commensalisms*, whereby the epibiont is dependent on the host turtle for survival, but the welfare of the host turtle is not dependent on the presence or behavior of the epibiont. While some obligate commensal epibionts are known to perform activities that might be considered beneficial to the host turtle, there are no examples of *obligate mutualisms*, in which both the host turtle and the epibiont depend on each other for survival. Future studies, however, may identify such obligate mutualisms. Most obligate (and facultative) commensal epibionts do not derive nutrients from the tissue of the host turtle and are not parasitic; instead the host turtle simply provides a foraging platform (Frick et al., 2002a). On the contrary, several sea turtle epibionts are known to derive nutrients from the tissue of the host turtle and, therefore, represent associations known as *parasitism* (Leung and Poulin, 2008). Parasitic epibionts of sea turtles are rare, but these associations may have important consequences for the health of host turtles (Greenblatt et al., 2004).

Following a rich history of anecdotal reports dating back to Darwin (1851, 1854), the study of epibiosis in sea turtles has received considerable attention in recent years. The vast majority of studies describe the diversity of epibiota, and speculate on the possible causes and effects of these associations. From these descriptive studies, we have learned a great deal with respect to the wonderful diversity of epibiotic forms found associated with sea turtles (Appendix A). Fewer studies, however, approach sea turtle epibiosis from the community perspective. These studies not only describe diversity of epibiota but also consider the structuring of epibiotic communities and the complex suite of interactions occurring on the turtle across space and time. Finally, even fewer studies attempt to quantify and understand the ecological interactions between turtles and their epibiota. These studies have allowed researchers to better understand the ecological and evolutionary implications of epibiosis, and to decipher the valuable information that can be gleaned from studying sea turtle epibionts.

Despite the antiquity of some sea turtle epibiont observations, the study of sea turtle epibiosis remains in a prolonged state of infancy when compared to the breadth of information that has recently and quickly accrued on sea turtle migrations and home ranges (largely through the deployment of satellite tags). Likewise, our understanding of sea turtle genetics and molecular phylogeny exceeds that of basic facets of sea turtle ecology—including diet, foraging behavior, and epibiotic associations. Given the documented declines of turtle populations in some areas, it has become imperative for scientists to understand how sea turtles interact with the constituents of the habitats they occupy, be it while foraging or through epibiosis. Such information allows scientists to view sea turtles within the context of a complex and ecologically rich marine environment, and it aids in modeling the potential impacts that certain natural and anthropogenic-driven events may have upon sea turtles and the habitats they utilize.

In this chapter, we begin by introducing many of the common forms of epibionts known to be associated with sea turtles. Second, we describe several common epibiotic community types, and discuss the spatial and temporal factors by which epibiotic communities are structured. Third, we propose a number of costs and benefits that may affect sea turtle–epibiont interactions and discuss the ecological inferences and implications of sea turtle epibiosis. Lastly, we outline a conceptual model of epibiosis with which researchers may apply to better understand the factors that affect their particular epibiotic systems and more easily decipher the important biological information that can be gleaned from studying epibiotic interactions.

## 15.2   COMMON FORMS

The diversity of epibionts known from sea turtles is exceptional. For example, loggerhead (*Caretta caretta*) and hawksbill turtles (*Eretmochelys imbricata*) are known to host 200+ and 150+ epibiont taxa, respectively. For this reason, we have not included an itemized list of epibionts from each turtle species. Instead, we have included a list of references that include records of epibionts from sea turtles separated by geographic region (Appendix A) and encourage investigators to examine the studies cited in this chapter.

### 15.2.1   Sessile Forms

Sessile forms attach directly to a substrate and do not move around freely. These forms are the most common and conspicuous epibionts of sea turtles. Most sessile forms have motile, planktonic larvae that recruit to suitable substrata, where they attach and transform into adults. For these organisms, the carapace and skin of sea turtles must possess certain settlement cues that larvae recognize, including water flow characteristics, chemical signals, and surface rugosity. Of the sessile forms documented from sea turtles, the most noticeable are barnacles (Cirripedia). Barnacles attached to the carapace of sea turtles are considered "pioneer" species that facilitate the colonization of subsequent epibiota (see Section 15.3.5; Frick et al., 2002b). Some coronuloid barnacles embed themselves in the skin and soft tissues of sea turtles (e.g., *Chelolepas cheloniae*). Through chemical mediation, these barnacles become encased in connective tissue, which aids in strengthening the shell of the barnacle while protecting the host tissue from further injury (Frick et al., 2011a). Other sessile forms include algae, foraminiferans, poriferans, cnidarians (Hydrozoa and Anthozoa), mollusks (Bivalvia), bryozoans, and tunicates. Many of these sessile forms are colonial and can reproduce asexually. As a result, some colonies are known to grow quite large and overtake much of the carapace of the host turtle. In such situations, aggregations of sessile forms provide additional surface area for the recruitment of other sessile epibiota, and create numerous crevices and spaces for the colonization of various motile epibionts (see later).

### 15.2.2   Sedentary Forms

Sedentary forms live a semi-sessile existence, in which motile individuals construct refugia or tubes attached to a substrate. Sea turtles host a variety of sedentary forms, including polychaete worms, amphipods, and tanaids (Frick et al., 1998, 2004b). Some sedentary forms create only small (1–2 mm long) tubes to dwell in, while others, particularly sabellariid worms and *Corophium* amphipods, will aggregate into dense communities—creating reef-like structures consisting of hundreds of individual tubes bonded together. These "worm reefs" can become quite large (up to 10 cm high) and cover the entire carapace of the host turtle (Frick et al., 2004b). These complex structures also provide suitable habitat for the colonization of small motile epibionts.

### 15.2.3   Motile Forms

Motile forms do not directly attach to a substrate and are capable of free movement throughout their lives. These organisms may colonize sea turtles directly from the plankton (similar to sessile forms) or secondarily colonize turtles after initially recruiting to their primary habitat. In the latter case, colonization may occur when resting turtles contact pelagic or benthic substrata. Motile forms reported as sea turtle epibionts include protozoans, sipunculid worms, platyhelminth worms, annelid worms (hirudineans and polychaetes), mollusks (Polyplacophora and Gastropoda), dipterans (flightless marine midges), decapods (Brachyura, Anomura, Caridea), copepods, ostracods, peracarids (amphipods, isopods, and tanaids), echinoderms (Ophiuroidea and Echinoidea), and fish

(Genera *Echeneis* and *Remora*; "shark suckers"). Most motile forms are small and cryptic, and live within the gaps and sinuses provided by aggregations of sessile and sedentary epibionts. Moreover, the deposition of sediment between sessile aggregations provides habitat for small infaunal animals that live in the trapped mud layer (e.g., polychaete worms, amphipods, and clams). For these reasons, the presence of most motile forms is often dependent on the preceding colonization of other sessile and sedentary epibiota. Two exceptions are *Caprella* amphipods, which cling tightly to the host carapace via limbs with hooked dactyls, and *Planes* crabs, which hide in the inguinal notch between the carapace and tail (Chace, 1951). Not surprisingly, these are two of the more common motile epibionts of sea turtles around the world.

## 15.3   COMMUNITIES AND COMMUNITY DYNAMICS

### 15.3.1   Pelagic/Oceanic Communities

All extant sea turtles, except the flatback turtle (*Natator depressus*), utilize pelagic and oceanic habitats during juvenile life stages (Bolten, 2003) and some continue to use these habitats throughout adulthood (e.g., *Dermochelys coriacea* and eastern Pacific *Lepidochelys olivacea*). Adult and subadult loggerhead turtles (*C. caretta*) are considered mostly neritic, but some individuals make occasional forays into the pelagic/oceanic environment (Frick et al., 2009; Reich et al., 2010). During pelagic/oceanic life stages, sea turtles may host communities of pelagic organisms that are typically found associated with drifting flotsam (e.g., *Sargassum*) and jetsam. These organisms primarily include pedunculate barnacles of the genera *Lepas* and *Conchoderma*, and grapsid crabs of the genus *Planes*. *Lepas* spp. and *Conchoderma* spp. are ubiquitous throughout the world's oceanic environment and are known to colonize a variety of other nektonic hosts (e.g., Reisinger and Bester, 2010; Pfaller et al., 2012). Studies on *Planes* crabs from oceanic-stage sea turtles represent the most detailed information on sea turtle–epibiont symbiosis to date (Davenport, 1994; Dellinger et al., 1997; Frick et al., 2000a, 2003b, 2004a, 2006, 2011b; Pons et al., 2011). Other less frequent epibionts of the pelagic/oceanic community may include pelagic sea slugs (*Fiona pinnata*), sea spiders (*Endeis spinosa*), pelagic tunicates (*Diplosoma gelatinosum*), and crabs of the genera *Portunus* and *Plagusia* (Frick et al., 2003a, 2011b; Loza and López-Juardo, 2004). The presence of pelagic/oceanic epibionts on sea turtles outside these areas strongly suggests that these turtles have recently migrated from the pelagic/oceanic environment, providing valuable insights into cryptic migratory behaviors and habitat preferences of sea turtles.

### 15.3.2   Benthic/Neritic Communities

After early life stages in pelagic/oceanic areas, most cheloniid sea turtles transition to more coastal and benthic habitats—presumably in search of food, and later for mates (Bjorndal, 1997). In benthic/neritic habitats, sea turtles become exposed to intense colonization pressure by marine propagules (larvae and spores) seeking to colonize submerged substrata and begin their benthic existence. The skin and especially the carapace of sea turtles provide suitable substrata for a variety of benthic/neritic organisms (Frick et al., 1998, 2000a; Schärer, 2001). As previously mentioned, the recruitment of sessile and sedentary forms (e.g., barnacles, tubicolous worms, and tunicates) facilitates the colonization of smaller motile forms (e.g., crabs, amphipods, mollusks, etc.), which inhabit the gaps and crevices between sessile aggregations. After prolonged exposure to settlement by local plants and animals in a given area, the epibiotic communities of sea turtles begin to resemble the adjacent benthic environment. For this reason, the species composition of benthic/neritic communities is largely dependent on the geographic region or habitat in which the host turtle occupies (Frick et al., 1998; Schärer, 2001). Complex benthic/neritic communities are most evident on nesting female turtles, which tend to remain relatively sedentary and localized during the nesting period (Frick et al., 2000b).

### 15.3.3 Obligate Communities

Obligate communities are composed almost entirely of organisms that are known exclusively as epibionts of sea turtles and other motile marine organisms. That is, these communities are largely independent of the habitat in which the turtle occupies (i.e., pelagic/oceanic vs. benthic/neritic). The predominant epibiont of obligate communities is the coronuloid barnacle *Chelonibia testudinaria*. This ubiquitous species is the most frequently reported epibiont of sea turtles and is also known to colonize crabs, sirenians, and crocodilians (Newman and Ross, 1976; Zardus and Hadfield, 2004; Cupul-Magaña et al., 2011; Nifong and Frick, 2011). *Chelonibia testudinaria* occurs in great numbers on some turtles and appears to function as a "pioneer" for the development of more extensive and diverse epibiotic communities (Frick et al., 2002b; Rawson et al., 2003). Aggregations of *C. testudinaria* provide refugia for other obligate epibionts, such as the ruby-eyed amphipod (*Podocerus chelonophilus*) and the robust tanaid (*Hexapleomera robusta*). However, both species will also cling directly to the skin and carapace of host turtles, and *P. chelonophilus* will also aggregate around epidermal lesions and eat necrotic tissue from the wounds of host turtles (Moore, 1995). Other obligate epibionts of sea turtles include marine red alga (*Polysiphonia carettia*), which is known only from cheloniid sea turtles (Senties et al., 1999), and several other species of coronuloid barnacles that are wholly chelonophilic (Ross and Frick, 2011). While some individual turtles are known to host strictly obligate communities (Frick et al., 2010a), most communities composed primarily of obligate epibionts also contain some facultative forms.

### 15.3.4 Community Distribution

The spatial distribution of epibiont communities on host turtles may be influenced by a complex suite of factors, including recruitment dynamics, water flow patterns, differential disturbance among body regions, and inter- and intraspecific interactions (Pfaller et al., 2006). In general, studies that examine or anecdotally report on the distribution of sea turtle epibionts have found that epibiont communities tend to aggregate on the carapace, as opposed to the skin or plastron (Gramentz, 1988; Fuller et al., 2010). Extra-carapacial epibionts mostly include barnacles, parasitic leeches, and *Planes* crabs (Chace, 1951; Gramentz, 1988; Frick et al., 1998; Hayashi and Tsuji, 2008). Some barnacles occur only along the plastral sutures (e.g., *Stomatolepas transversa*) (Young, 1991), while others mostly occur along the leading edges of the front flippers (e.g., *Stephanolepas muricata*) (Frick et al., 2011a). Limb movements, unfavorable water flow patterns, and the sloughing of skin by the host turtle probably restrict the recruitment and development of extra-carapacial epibionts. Nevertheless, information on the distributions of extra-carapacial epibionts is still lacking (Frick et al., 2011a).

Most studies that examine the spatial distribution of epibiont communities on sea turtles have focused on the carapace, where the densest and most diverse communities are found (Frick et al., 1998). These studies indicate that epibiotic communities tend to be distributed in nonrandom patterns. Most studies report a tendency for epibiont communities to cluster along the vertebral scutes and across the posterior third of the carapace (Caine, 1986; Matsuura and Nakamura, 1993; Frick et al., 1998; Pfaller et al., 2006). Such nonrandom distributions are thought to reflect the preference of filter-feeding epibionts (e.g., barnacles) for elevated flow rates along the vertebral scutes and the favorable settlement conditions for other epibiota along the posterior of the carapace where flow rates are reduced (Pfaller et al., 2006). Recruitment of "pioneer" species in these areas (e.g., *Chelonibia* barnacles and *Polysiphonia* alga) will then facilitate the accumulation of more diverse epibiotic communities (Gramentz, 1988; Frick et al., 2000b; Fuller et al., 2010). Additionally, the colonization and persistence of epibionts on the anterior costal scutes may be reduced by contact from the front flippers (Caine, 1986; Dodd, 1988) and/or removal during "self-grooming" (Schofield et al., 2006; Frick and McFall, 2007). Other studies show mostly random distributions among barnacle species with some spatial structuring among different size classes of barnacles (Fuller et al., 2010).

404 The Biology of Sea Turtles, Volume III

Recently, Moriarty et al. (2008) confirmed that the obligate commensal barnacle, *Chelonibia testudinaria*, is capable of substantial (but slow) post-settlement locomotion. Individual *C. testudinaria* were shown to move across multiple scutes from areas of low water flow to areas with better filter-feeding conditions. Such movements may be triggered by differential flow rates over the carapace or/and the presence of conspecifics that disrupt flow patterns. As previously mentioned, *Chelonibia* spp. are important "pioneer" species for epibiotic communities (Frick et al., 2000) and post-settlement locomotion will certainly affect the spatial distribution of epibiotic communities. However, as the density of *C. testudinaria* and other epibiota increases, post-settlement locomotion and survival will be reduced, and the overall distribution may become more reflective of differences in recruitment patterns (Pfaller et al., 2006).

Debilitated turtles will host epibionts, especially barnacles, over their entire external surface area—including portions of the mouth regularly exposed to the outside environment. These "barnacle bill" turtles will often suffer severe deformations as a result of barnacle colonization. Current information indicates that such turtles are immunosuppressed or lethargic prior to barnacle colonization and that limited mobility by the host likely facilitates rapid and prolific colonization of barnacles (Deem et al., 2009). Nevertheless, because healthy turtles may also support massive aggregations of epibionts over much of their bodies, it is difficult to judge the health of a turtle simply by examining epibiont loads and percentage coverage (*see* Deem et al., 2009).

## 15.3.5 COMMUNITY SUCCESSION

Prior to the colonization of macroorganisms, all structures exposed to seawater initially undergo a similar sequence of events (Wahl, 1989): (1) biochemical conditioning, whereby surfaces absorb dissolved macromolecules; (2) bacterial colonization; and (3) unicellular eukaryote (e.g., yeasts, protozoa, and diatoms) colonization. To our knowledge, these critical stages in the process of epibiosis in sea turtles have never been explored.

The temporal succession of "macro"-epibiont communities on host turtles remains poorly understood, as well. To date, there is one study that examines temporal succession of epibiont communities from individual turtles over an extended period of time (Frick et al., 2002b). Using, flipper-tagging data, photography, and in situ assessments, epibiont data were collected from the carapaces of nesting loggerhead turtles (*C. caretta*) in Georgia, United States, over the course of 3 months. General observations of community succession were similar to those reported for neritic, epibenthic communities (Dean, 1981). Community succession is typically initiated when hard, sessile forms like barnacles (*C. testudinaria* in Frick et al., 2002b) colonize a relatively bare carapace. These "pioneers" facilitate the subsequent colonization of other epibiota by increasing the surface area for colonization and changing water flow patterns (Pfaller et al., 2006). Secondary colonizers include other sessile forms (e.g., hydrozoans and bryozoans) and sedentary forms, which take refuge within the interstices of the barnacles (e.g., tanaids). The accumulation of sediments among primary and secondary sessile forms then facilitates the colonization of sessile tunicates and many small, motile forms. Tunicates and other secondary sessile forms tend to overgrow and kill the barnacles beneath them. Tunicates (*Molgula manhattensis*) appear to be the climax species of the carapace epibiont community on nesting loggerheads in Georgia, United States. Aggregations of *M. manhattensis* occasionally cover the entire carapace at the end of the season, providing innumerable gaps and crevices for a diverse array of motile epibionts.

At or before reaching terminal succession, epibiont communities may be partially or catastrophically disturbed by various biotic and abiotic factors. Turtles that accumulate benthic/neritic communities may immigrate to different, less favorable habitats, causing the less tolerant epibionts to die and slough off. In some cases, this may completely clear the carapace of epibiota. Moreover, community succession may be disrupted when host turtles "groom" themselves by actively rubbing against submerged structures to remove epibiota (Heithaus et al., 2002;

Schofield et al., 2006; Frick and McFall, 2007). Evidence of such behaviors is often present in the form of longitudinal scratch marks on the carapace (Caine, 1986; Frick and McFall, 2007). Lastly, predatory epibionts (e.g., *Planes* crabs and several gastropods) and fish may systematically clean/ remove certain epibionts (Davenport, 1994; Losey et al., 1994; Frick et al., 2000a, 2011b; Pfaller et al., 2008; Sazima et al., 2010). These factors may lead to partial or complete turnover of the epibiotic communities of sea turtles.

## 15.4 ECOLOGICAL INTERACTIONS

### 15.4.1 EFFECTS ON EPIBIONTS

Epibionts may benefit from epibiosis through reduced competition and predation. These are major factors affecting the ability of marine propagules to successfully colonize a substratum (Enderlein and Wahl, 2004). Thus, when risk of predation is high or when settlement area is limited—whether by high population densities (e.g., on benthic structures) or by low substrata availability (e.g., on pelagic flotsam)—epibiosis of sea turtles may be beneficial for the survival of marine propagules (Wahl, 1989; Pfaller et al., 2012). Some "burrowing" barnacles may avoid predation by encasing themselves within the tissue of host turtles via chemical mediation (Frick et al., 2011b). Epibionts may also benefit from improved energetic positioning. Filter-feeding epibionts, such as barnacles, may benefit from favorable feeding currents on host turtles (Pfaller et al., 2006), while photosynthetic epibionts, such as algae, may benefit from increased oxygen and light availability (Shine et al., 2010). Furthermore, epibionts may benefit though range expansion and increased genetic mixing by hitchhiking on migratory turtles (termed *phoresis*). Researchers have hypothesized that sea turtles may act as long-distance dispersal vectors for benthic marine invertebrates (Schärer and Epler, 2007; Harding et al., 2011).

Epibiosis may be costly to epibionts when turtle behaviors cause physical disturbance and unfavorable fluctuations in physiological conditions (Wahl, 1989). Contact between turtles during mating, or between turtles and submerged structures (e.g., rock or coral ledges), may physically damage epibionts, especially those with fragile, erect body forms (e.g., leafy bryozoans and soft corals). As previously mentioned, sea turtles are also known to actively remove epibionts by scraping against submerged structures (Heithaus et al., 2002; Schofield et al., 2006; Frick and McFall, 2007). Moreover, epibionts that are sensitive to desiccation may die when turtles emerge to nest or bask at the surface (Caine, 1986; Bjorndal, 2003). Similarly, epibionts that are sensitive to fluctuations in temperature, salinity, or pressure may not survive when turtles migrate and/or dive. Another cost for certain epibionts might be reduced access to food resources and mates, which would ultimately cause reduced longevity and reproductive capacity. These costs might favor epibionts capable of asexual reproduction and dietary versatility.

### 15.4.2 EFFECTS ON HOST TURTLES

Epibiosis may be costly to host turtles when epibionts cause increased weight and drag. In extreme cases, epibiotic loads have been reported that effectively double the mass and volume of juvenile sea turtles (Bolten unpubl. data *in* Bjorndal, 2003). Epibionts attached to the carapace may increase drag by disrupting the laminar flow over the carapace (Logan and Morreale, 1994) and those embedded in the leading edge of the front flippers may increase drag while swimming (Wyneken, 1997; Frick et al., 2011a). The energetic costs of hosting epibionts are likely greatest when turtles undertake long-distance migrations and least when turtles remain relatively sedentary (e.g., females during internesting periods). Because otherwise healthy turtles will often support massive epibiont aggregations (Deem et al., 2009), turtles are apparently capable of overcoming the costs associated with "epibiotic drag" and should not be judged as healthy or unhealthy simply by examining epibiotic loads (see Deem et al., 2009). Furthermore, the aforementioned "barnacle bill" turtles tend to accumulate their prolific barnacle loads after (not before) becoming lethargic at the surface.

Epibiosis may also be costly to host turtles when certain epibionts detrimentally affect the health of host turtles. A number of common epibionts of sea turtles (e.g., platyhelminth worms, annelid worms and barnacles) are thought to be the cause of or related to infections of sea turtles (George, 1997; Alfaro, 2008). Tissue damage caused by burrowing epibionts may increase the vulnerability of host turtles to pathogens (George, 1997). Some coronuloid barnacles (e.g., *C. cheloniae, S. muricata*, and *Cylindrolepas darwiniana*) become embedded within hard and soft tissues of host turtles causing deep-tissue wounds that can sometimes leave impressions on the underlying bone (Hendrickson, 1958; Green, 1998; Frick and Zardus, 2010; Frick et al., 2010a). *Platylepas decorata* have also been found imbedded in the beaks of host turtles causing severe beak deformation, which may lead to reduced foraging capacity and death of the host turtle (see Green, 1998; Frick and Zardus, 2010). Other non-barnacle forms may act as disease vectors of pathogens. Parasitic marine turtle leeches (*Ozobranchus* sp.) not only consume host tissue but also are believed to act as disease vectors for the dispersal of the fibropapilloma-associated herpes virus found in latent tumors that often cover, deform, and debilitate host turtles (Greenblatt et al., 2004). Commensal gastropods of sea turtles may act as intermediate hosts for spirorchiid blood flukes (Frazier et al., 1985), which can have devastating effects on host turtles (George, 1997).

Host turtles may benefit from epibiosis through improved optical, chemical, or electrical camouflage. Predators may not recognize hosts as potential prey items if epibiotic communities visually or chemically resemble the surrounding benthic communities (Rathbun, 1925; Fishlyn and Phillips, 1980; Feifarek, 1987; Frazier et al., 1991). Moreover, dense epibiotic communities may disrupt electric fields produced by hosts, allowing hosts to avoid predation by predators that utilize electrolocation when searching for prey (e.g., sharks) (Ruxton, 2009). Hosts may also benefit from epibiosis through associational defense and cleaning. Epibionts with chemical or structural defenses (e.g., toxins, sharp projections, or hard outer coverings) may deter predation on host turtles (Wahl and Mark, 1999; Bjorndal, 2003). Predatory epibionts may provide a cleaning benefit by consuming other epibionts—some of which may be harmful—from the surface of host turtles (Davenport, 1994; Sazima et al., 2010).

### 15.4.3  Ecological Inferences

Studies of epibiosis have helped elucidate cryptic life history attributes of sea turtles and informed the implementation of conservation measures. While such studies will not and should not supplant the use of tag-return data, satellite telemetry, stable-isotope analyses, or population genetics, studying epibiosis can provide a time- and cost-effective alternative to elucidate the geographic ranges, habitat preferences, and migratory corridors of sea turtles. Using primarily examples from the well-studied epibiont community of loggerhead turtles in the northwestern Atlantic Ocean, we illustrate the types of ecological inferences that can be gained by studying the epibionts of sea turtles.

Epibiont data have been used to elucidate the foraging locations of loggerhead turtles nesting along the Atlantic coast of Florida, United States. These turtles occasionally host epibionts that are geographically restricted to far southern Florida, the Bahamas, and the Caribbean (Caine, 1986; Pfaller et al., 2008). Such associations suggest that these nesting turtles had recently migrated from more southerly areas where their range overlapped with free-living populations of the epibionts. Data from flipper-tag returns, satellite telemetry, and stable-isotope analyses have confirmed that turtles nesting in Florida frequently utilize these more southerly, tropical waters during nonbreeding seasons (Meylan, 1983; Foley et al., 2008; Pajuelo et al., 2012). Caine (1986) further extrapolated these epibiont data to suggest the presence of two discrete nesting assemblages along the southeastern United States, one to the north and one to the south of Daytona Beach, Florida (approximately 29° N latitude). Several years later this hypothesis was rather precisely confirmed by molecular data (Bowen et al., 1993; Encalada et al., 1998) and now these two nesting assemblages receive markedly different conservation status (Turtle Expert Working Group, 2009).

In another example from nesting loggerhead turtles in Florida, United States, Reich et al. (2010) supplemented stable-isotope data with epibiont community data to suggest a bimodal foraging strategy by female loggerheads prior to their arrival at breeding grounds. Because isotopic signatures (depleted vs. enriched $\delta^{13}C$) can vary along multiple environmental continua, the incorporation of epibiont data in this study provided additional support for an oceanic versus neritic dichotomy, as opposed to dietary or latitudinal gradients. These results have important implications for role of adult loggerhead turtles in the oceanic environment and the management policies that serve to protect them.

Epibiont data have also been used to assess the foraging migrations of juvenile and subadult loggerhead turtles. Killingley and Lutcavage (1983) used duel isotopic profiles ($\delta^{18}O$ and $\delta^{13}C$) from the shells of *C. testudinaria* to reconstruct the movements of subadult loggerheads between oceanic habitats in the northwest Atlantic and estuarine habitats in the Chesapeake Bay (Maryland and Virginia). Moreover, Limpus and Limpus (2003) used the presence of particular epibionts (*Planes* sp. and *S. muricata*) and morphological features to identify which juvenile turtles caught in neritic habitats in the southwest Pacific Ocean had recently recruited from the open ocean. In both studies, epibiont data provided valuable insights in to cryptic host movements that otherwise would have been very difficult to obtain.

Lastly, in another interesting application of epibiont data, Eckert and Eckert (1988) measured the size distribution of epibiotic barnacles (Conchoderma virgatum) on nesting leatherback turtles to extrapolate the time of arrival to the tropical nesting region. Because reproduction in these barnacles is typically restricted to tropical regions, their colonization of turtles is limited to the period when turtles also occupy tropical waters. Based on reproductive periodicity and established growth rates of barnacles (Eckert and Eckert, 1987), the authors determined that turtles do not arrive from temperate latitudes until just prior to nesting and orient directly toward their preferred nesting beach (Eckert and Eckert, 1988). These data have provided important information on the cryptic migratory behavior of leatherback turtles and have better informed the implementation of conservation measures.

## 15.4.4 ECOLOGICAL IMPLICATIONS

The ecological implications of sea turtle epibiosis remain one of the most poorly understood aspects of this nascent field. Aside from many of the direct effects of epibiosis on host turtles and epibionts discussed earlier (Sections 15.4.1 and 15.4.2), sea turtle epibiosis may have other less obvious, indirect effects on the marine communities and habitats that sea turtles inhabit.

Several authors have discussed the potential role of sea turtles as dispersal vectors for a diverse array of marine invertebrates over broad geographic regions (Bjorndal and Jackson, 2003; Schärer and Epler, 2007; Harding et al., 2011; Lezama et al., 2012). Hitchhiking on highly mobile hosts may facilitate genetic mixing and/or range expansion for epibionts capable of reproducing on turtles or after arriving in distant locations (Rawson et al., 2003). These factors may be particularly important for invertebrate taxa with limited dispersal capacities (Schärer and Epler, 2007). Turtle-mediated genetic mixing may aid in maintaining the genetic diversity and homogeneity of marine invertebrate populations (Rawson et al., 2003), but may also inhibit biological diversification by impeding local adaptation or random divergence. Moreover, turtle-mediated range expansion may promote biological diversification if newly established populations subsequently remain isolated from their source populations, or disrupt ecosystem functioning when invaders compete with or consume resident species.

A recent study has drawn attention to the potential for turtle-mediated introductions of nonindigenous and potentially invasive species. Harding et al. (2011) report the first records of the nonindigenous veined rapa whelk (*Rapana venosa*) as an epibiont of loggerhead turtles in Virginia and Georgia. *R. venosa* is a generalist shellfish predator native to Asia that has recently been introduced in to the Chesapeake Bay (Harding and Mann, 1999). However, the size and stage of the epibiotic individuals on turtles in Georgia indicate the presence of an extra-Chesapeake breeding population of this invasive species. The authors suggest that turtle-mediated dispersal is currently the only compelling explanation for the occurrence of *R. venosa* on turtles in Georgia. These findings have important implications for the future management of invasive marine invertebrates.

Sea turtles are known to modify the physical structure of their habitat in a number of ways (Bjorndal and Jackson, 2003). Thus, another unexplored ecological implication of sea turtle epibiosis might be the extent to which turtles modify hard-bottom habitats when actively removing epibiota. This behavior involves turtles pushing their carapace against the underside of rock ledges and vigorously scrapping against the rock to remove epibiota, particularly barnacles (Frick and McFall, 2007). The rock ledges often erode during such behaviors, leaving behind scours or arched ledges, which turtles may return to for subsequent "self-grooming." The extent to which these habitat modifications affect the surrounding reef or hard-bottom communities remains unknown.

## 15.5 CONCEPTUAL MODEL OF EPIBIOSIS

As we accumulate studies of epibiotic diversity in sea turtles, we have begun to formulate a conceptual framework to better understand and learn from these epibiotic interactions. While there have been several broad reviews on epibiosis (see Wahl, 1989; Wahl and Mark, 1999; Harder, 2009; Wahl, 2009), there has been no attempt to construct a conceptual framework to explain such associations. The conceptual model of epibiosis depicted in Figure 15.1 outlines three hierarchical factors inherent to epibiotic interactions: (1) geographic overlap (Figure 15.1A), (2) ecological overlap (Figure 15.1B), and (3) the balance of costs and benefits to hosts and epibionts that dictate the likelihood of epibiosis once in close proximity (Figure 15.1C). Because the factors that affect epibiotic interactions—as displayed in this conceptual model—are inherent to the biology of the species involved, we can learn about the ecology and evolution of these species by studying epibiosis. Such a conceptual framework will hopefully allow researchers to better understand the factors that affect their particular epibiotic systems and more easily decipher the important biological information that can be gleaned from studying epibiosis in sea turtles.

### 15.5.1 GEOGRAPHIC OVERLAP

A necessary prerequisite for epibiosis is geographic overlap between the range of the host turtle and the range of the epibiont (Figure 15.1A). Logically, without geographic overlap, epibiosis between a host turtle and any potential epibiont would never occur. This is an obvious criterion for epibiosis. However, because the host turtles are highly mobile, the occurrence of particular epibiont taxa with more limited distributions can reveal information about cryptic host movements. Studies of sea turtle epibionts have provided important information on the migratory behavior of loggerhead and leatherback turtles (Caine, 1986; Eckert and Eckert, 1988), and subsequently informed the implementation of conservation measures.

### 15.5.2 ECOLOGICAL OVERLAP

Where geographic ranges overlap, epibiosis will then depend on the spatial and temporal overlap in ecology of the host turtles and potential epibionts (Figure 15.1B). Local geographic areas are typically heterogeneous mosaics of different habitats, each characterized by different ecological communities of plants and animals (e.g., saltmarshes, coral reefs, pelagic areas). The species composition of local communities may also vary through time, especially for seasonal differences

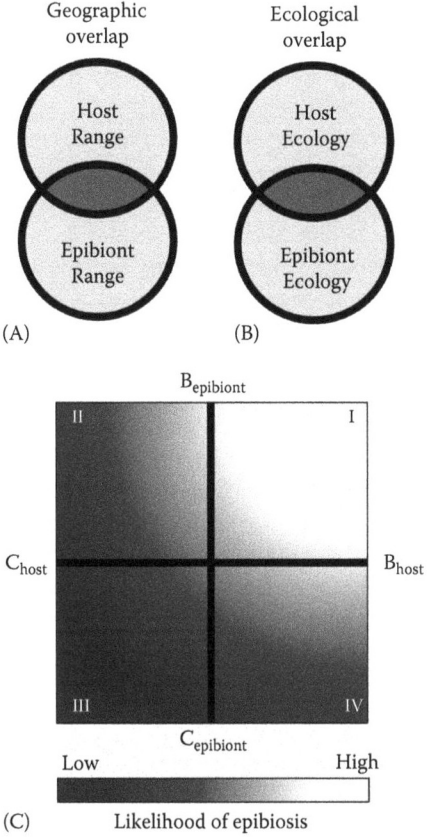

FIGURE 15.1  Conceptual model of epibiosis. (A,B) Venn diagrams showing the geographic and ecological overlap between hosts and epibionts, respectively. (C) Graph showing the likelihood of epibiosis based on the balance of cost and benefits to hosts and epibionts ($B_{epibiont}$, benefit to the epibiont; $B_{host}$, benefit to the host; $C_{host}$, cost to the host; $C_{epibiont}$, cost to the epibiont).

in recruitment of larval propagules. Host turtles may utilize many different habitats or may show preferences for certain habitats during different behaviors (e.g., foraging, resting, and mating) or life stages, or at different times of the year. In order for epibiosis to occur, the host turtles must occupy the same habitat at the same time as free-living populations of potential epibionts. Thus, the epibionts associated with a given host turtle should reflect the assemblage of plants and animals that occupy the habitats where the hosts spend time. For example, sea turtles that tend to inhabit benthic/neritic habitats tend to host different epibionts than turtles that tend to inhabit pelagic/oceanic habitats (Limpus and Limpus, 2003; Reich et al., 2010). Such information can be used to assess interspecific and intraspecific differences in habitat use, which is critical for the implementation of effective conservation strategies.

### 15.5.3  BALANCE OF COSTS AND BENEFITS

Once in close proximity, there is a complex balance of costs and benefits for host turtles and potential epibionts that ultimately determine the likelihood of epibiosis. Figure 15.1 displays a 2D likelihood surface in which each axis represents a continuum from high benefit to high cost. The various positions of different hosts and epibionts along these cost-benefit axes depend on the net cost or benefit experienced during epibiosis. Because the relative costs and benefits are different for different turtle–epibiont pairs, some associations are more likely and therefore more frequent than others. Epibiotic interactions in which both species experience a net benefit would have a high likelihood of occurring and therefore would be more frequent (quadrant I). Such mutually beneficial associations would favor mechanisms for active attraction and may develop into obligate associations over evolutionary time. On the other end of the continua, interactions in which both species suffer high costs would have a low likelihood and would effectively never occur (quadrant III). Interactions in which one species incurs a high cost while the other receives minimal benefit would also have a low likelihood (bottom left of quadrants II and IV), as the former species would actively avoid such interactions and the latter would gain very little by exploiting the former. Conversely, if one species receives a high benefit at a high cost to the other species (top left of quadrant II and bottom right of quadrant IV), then such associations might exhibit patterns similar to that of parasitic interactions (top left of quadrant II only). Lastly, interactions in which one species receives a high benefit while the other incurs little or no cost would have a higher likelihood and would be relatively frequent (top left of quadrants II and IV). This last scenario characterizes many of the interactions between sea turtles and their epibiota, and is typically referred to as commensalism (Leung and Poulin, 2008).

As previously mentioned, epibionts may benefit from epibiosis through reduced spatial competition and predation (Wahl, 1989; Enderlein and Wahl, 2004; Pfaller et al., 2012), improved energetic positioning (Pfaller et al., 2006; Shine et al., 2010), and range expansion (Schärer and Epler, 2007; Harding et al., 2011), while coping with costs associated with physical disturbance (Wahl, 1989; Schofield et al., 2006; Frick and McFall, 2007), transport to unfavorable physiological environments (Caine, 1986; Bjorndal, 2003), and reduced access to food resources and mates. Host turtles may benefit from epibionts through optical, chemical, or electrical camouflage (Rathbun, 1925; Fishlyn and Phillips, 1980; Feifarek, 1987; Frazier et al., 1991; Ruxton, 2009) and associational defense and cleaning (Davenport, 1994; Wahl and Mark, 1999; Bjorndal, 2003), while coping with costs associated with increased weight and drag (Logan and Morreale, 1994; Bjorndal, 2003), and tissue damage and associated susceptibility to pathogens (George, 1997; Greenblatt et al., 2004). The balance of costs and benefits to host turtles and epibionts will ultimately determine the likelihood—and therefore the frequency—of epibiosis for most turtle–epibiont associations.

## 15.6 CONSIDERATIONS

Studies that seek to elucidate the relationships that exist between sea turtles and other marine organisms require investigators to adopt an interdisciplinary approach to data collections and analyses. Knowledge of the standard measurements and preservation methods employed by taxon specialists is important to properly report and archive marine algae and invertebrate specimens (Lazo-Wasem et al., 2011). A familiarity with the life histories and general biology of the marine organisms that utilize the habitats occupied by sea turtles is essential for identifying situations that bring sea turtles into contact with the marine organisms they consume and those that attach to them. An understanding of the major systematic characters that define the major family-groups of local marine flora and fauna is helpful for identification, and to adequately ascertain evolutionary relationships between sea turtles and other marine organisms.

## ACKNOWLEDGMENTS

We sincerely thank Rebecca Pfaller for valuable editorial and technical assistance. We thank Kristina L. Williams of the Caretta Research Project, Karen Bjorndal, Alan Bolten, Peter Eliazar, and our other colleagues at the Archie Carr Center for Sea Turtle Research for their encouragement, advice, and support, and we greatly appreciate the helpful comments of anonymous reviewers that improved an earlier draft of the present chapter.

## 15.A APPENDIX A: ANNOTATED BIBLIOGRAPHY OF SELECTED SEA TURTLE EPIBIONT STUDIES AND REPORTS LISTED BY GEOGRAPHIC REGION

### 15.A.1 CARIBBEAN–WESTERN ATLANTIC

Bacon, 1976 (Trinidad); Bugoni et al., 2001 (Rio Grande do Sul: Brazil); Cardenas-Palomo and Maldonado-Gasca, 2005 (Yucatan: Mexico); Caine, 1986 (South Carolina, Florida); Farrapeira-Assunção, 1991 (Brazil); Frazier et al., 1985 (Georgia, Florida); Frazier et al., 1991 (Georgia); Frazier et al., 1992 (Georgia; Rio Grande do Sul: Brazil); Frick et al., 1998 (Georgia); Frick and Slay, 2000 (Georgia); Frick and Zardus, 2010 (Panama, Georgia, and Florida); Frick et al., 2000a, 2000b (Georgia); Frick et al., 2002a,b (Georgia); Frick et al., 2003a (Jumby Bay: Antigua); Frick et al., 2004b (Georgia); Frick et al., 2006 (Florida); Frick et al., 2010a (Nova Scotia, Georgia); Frick et al., 2010b (Georgia and Florida); Gruvel, 1905 (Antilles Sea); Henry, 1954 (Florida, Texas); Hunt, 1995 (Florida); Ives, 1891 (Yucatan: Mexico); Killingley and Lutcavage, 1983 (Virginia); Lutcavage and Musick, 1985 (Virginia); Nilsson-Cantell, 1921 (Florida); Nilsson-Cantell, 1939 (Bay of Chacopata: Venezuela); Pereira et al, 2006 (Almofala: Brazil); Pilsbry, 1916 (Cape Frio: Brazil; Delaware, Florida, New Jersey; Point Patuca: Honduras; West Indies); Plotkin, 1996 (Texas); Richards, 1930 (New Jersey); Rudloe et al., 1991 (Florida); Schwartz, 1960 (Maryland); Walker, 1978 (North Carolina); Wass, 1963 (Virginia); Wells, 1966 (Florida); Weltner, 1897 (Florida; Cuba; Bahia: Brazil); Young, 1991 (Brazil); Zavodnik, 1997 (Rovinj: Croatia); Zullo and Bleakney, 1966 (Massachusetts; Nova Scotia: Canada); Zullo and Lang, 1978 (South Carolina).

### 15.A.2 MEDITERRANEAN–EASTERN ATLANTIC

Badillo-Amador, 2007 (Mediterranean Sea); Barnard, 1924 (Table Bay: South Africa); Broch, 1924 (Baie du Levrier: Mauretania; Gambia); Broch, 1927 (Rabat: Morocco); Carriol and Vader, 2002 (Finmark: Norway); Caziot, 1921 (Nice: France); Chevereaux and de Guerne, 1893 (between Algeria and Balaeres); Darwin, 1854 (Africa; Mediterranean Sea); Davenport, 1994 (Madeira); Frazier et al., 1985 (Peloponnesus, Zakynthos Island: Greece); Frick et al., 2010a (Gabon: Africa); Gauld, 1957 (Accra: Ghana); Geldiay et al., 1995 (Koycegiz-Dalyankoy: Turkey); Gramentz, 1988 (Malta, Zacharo, Zakynthos: Greece; Lampedusa: Italy); Gruvel, 1903 (Palermo: Italy; Alexandria: Egypt); Gruvel, 1931 (Gulf of Alexandrette); Haelters and Kerckhof, 1999 (DeHaan: Belgium); Haelters and Kerckhof, 2001 (Oostende: Belgium); Holothuis, 1952 (Ouddorp: the Netherlands); Holothuis, 1969 (Ameland Island: the Netherlands); Kitsos et al., 2005 (Aegean Sea); Kolosvary, 1939 (Rovigno, d'Istria: Croatia); Kolosvary, 1943 (Alexandria: Egypt; Palermo, Sicily: Italy); Kolosvary, 1951 (Mediterranean Sea); Koukouras and Matsa, 1998 (Aegean Sea; Levantine Basin); Lanfranco, 1979 (St. Julian's: Malta); Lucas, 1968 (Mediterranean Sea); Margaritoulis, 1985 (Zakynthos: Greece); Nilsson-Cantell, 1921 (Bibundi: Cameroon); Nilsson-Cantell, 1931 (Mediterranean Sea); O'Riordan, 1979 (Dingle: Ireland); O'Riordan and Holmes, 1978 (Ventry Harbor: Ireland); Pilsbry, 1916 (Taranto: Italy; Cape of Good Hope: South Africa); Quigley and Flannery, 1993 (Dingle Bay: Ireland); Relini, 1968 (Gulf of Trieste: Italy); Relini, 1969 (Adriatic Sea); Relini, 1980 (Adriatic Sea); Sezgin et al., 2009 (Turkey); Smaldon and Lyster, 1976 (Skarvoy: Norway; Crail, Kirkcudbrightshire: Scotland; Cornwall: England); Stubbings, 1965 (Hann, Saloum River: Senegal); Stubbings, 1967 (Goree, Hann: Senegal); Utinomi, 1959 (Banyuls-sur-Mer: France); Zakhama-Sraieb et al., 2010 (Gulf of Gabès: Mediterranean Sea).

### 15.A.3 INDO–WEST PACIFIC

Annandale, 1906 (Rameswaram Island: India; Gulf of Manaar); Balazs, 1978 (Hawaii); Balazs, 1980 (Hawaii); Balazs et al., 1987 (Hawaii); Borradaile, 1903 (Minikoi Island: India); Broch, 1916 (Broome: Australia); Broch, 1931 (Gulf of Thailand; Nagasaki: Japan); Broch, 1947 (Ream: Cambodia; Indochina); Bustard, 1976 (Great Barrier Reef: Australia); Daniel, 1956 (Tuticorin, Drusadai Islands, Royapuram Coast, Madras Coast: India); Daniel, 1962 (Little Andaman

Island: India); Darwin, 1854 (Low Archipelago: French Polynesia; Australia); Dawydoff, 1952 (Pulo Condore: Vietnam; Ream: Cambodia); Deraniyagala, 1939 (Bentota: Ceylon); Dobbs and Landry, 2004 (Great Barrier Reef: Australia); Fernando, 1978 (Porto Novo: India); Glazebrook and Campbell, 1990 (Torres Strait: Australia); Fischer, 1886 (Pulo Condor: Vietnam); Foster, 1978 (North Island: New Zealand); Frazier, 1971 (Aldabra Atoll); Frazier et al., 1985 (Orissa: India; Tanzania: Africa); Frazier, 1989 (Dwarka Island: India); Frazier et al., 1992 (Orissa, Gujarat: India; Karachi, Pakistan); Gordon, 1970 (Hawaii); Gruvel, 1903 (Seychelles; Mallicolo: Vanuatu; Djibouti; Sandwich Island; Cochinchina: Vietnam); Gruvel, 1907 (Andaman Islands: India); Gruvel, 1912 (Tuamotu Archipelago: French Polynesia); Hayashi and Tsuji, 2008 (Okinawa: Japan); Hendrickson, 1958 (Johor, Sarawak: Malaysia); Hiro, 1936 (Wakayama Prefecture, Aichi Prefecture: Japan); Hiro, 1937a (Baberudaobu Island: Palau); Hiro, 1939 (Toyama Bay: Japan); Jones, 1990 (Australia); Jones et al., 1990 (Tasmania; Australia); Jones et al., 2000 (summary of distribution); Kruger, 1911b (Sagami Bay: Japan); Kruger, 1912 (Timor Sea); Lanchester, 1902 (Kota Bharu: Malaysia); Limpus et al., 1983a (Campbell Island: Australia); Limpus et al., 1983b (Crab Island: Australia); Limpus et al., 2005 (Raine Island: Australia); Loop et al., 1995 (Milman Island: Australia); Losey et al., 1994 (Hawaii); Matsuura and Nakamura, 1993 (Kagoshima Prefecture: Japan); McCann, 1969 (North Island: New Zealand); Monroe and Limpus, 1979 (Queensland: Australia); Mustaquim and Javed, 1993 (Sandspit Beach: Pakistan); Newman et al., 1969 (Hawaii); Newman and Abbott, 1980 (California); Nilsson-Cantell, 1921 (Western Australia: Australia); Nilsson-Cantell, 1930a (Enoe Island: Malaysia); Nilsson-Cantell, 1932 (Bentota: Sri Lanka); Nilsson-Cantell, 1937 (Singapore); Nilsson-Cantell, 1938 (Maldives; Kilakarai, Andaman Islands, River Hooghly, mouth of Ganges: India); Pillai, 1958 (Quilon: India); Pilsbry, 1916 (Hawaii; Caroline Islands; Ana: Japan; Saigon: Vietnam); Pilsbry, 1927 (Hawaii); Ren, 1980 (Xisha Islands); Ren, 1987 (China); Ross, 1981 (Oman); Smaldon and Lyster, 1976 (Kuala Lumpur: Malaysia); Tachikawa, 1995 (Japan); Utinomi, 1949 (Hakata Bay: Japan); Utinomi, 1958 (Sagami Bay: Japan); Utinomi, 1966 (Amakusa: Japan); Utinomi, 1969 (Kharg: Iran); Utinomi, 1950 (Tanabe Bay: Japan); Utinomi, 1970 (Hakui, Cape Kyoga-misaki, Kamo, Nezugaseki, Sado Island: Japan); Wagh and Bal, 1974 (Bombay: India); Weltner, 1897 (Massaua: New Guinea; Torres Strait); Weltner, 1910 (Ile Europa); Zann and Harker, 1978 (Queensland: Australia); Zardus and Balazs, 2007 (Hawaii).

### 15.A.4   EASTERN PACIFIC

Angulo-Lozano et al., 2007 (Sinaloa: Mexico); Beaumont et al., 2007 (Galapagos Islands: Ecuador); Brown and Brown, 1995 (Peru); Darwin, 1854 (Mexico; Galapagos Islands: Ecuador); Frazier et al., 1985 (Galapagos Islands: Ecuador); Frazier et al., 1992 (Santa Rosa: Ecuador); Frick et al., 2011a,b (Baja California: Mexico; Eastern Tropical Pacific; Galapagos: Ecuador); Green, 1998 (Galapagos Islands: Ecuador); Henry, 1941 (La Paz: Mexico); Henry, 1960 (Gulf of California, Guaymas: Mexico); Hernandez-Vasquez and Valadez-Gonzalez, 1998 (Jalisco: Mexico); Hubbs, 1977 (California); Kolosvary, 1943 (San Jose: Guatemala); Lazo-Wasem et al., 2011 (Jalisco: Mexico); MacDonald, 1929 (Cocos Island: Costa Rica); Newman et al., 1969 (Baja California: Mexico; Eastern Pacific); Pilsbry, 1916 (Baja California: Mexico; Galapagos Islands: Ecuador); Ross and Newman, 1967 (Baja California: Mexico); Stinson, 1984 (California); Vivaldo et al., 2006 (Michoacan, Oaxaca: Mexico); Weltner, 1897 (western Mexico; California; Valparaiso: Chile); Young and Ross, 2000 (Sonora: Mexico); Zullo, 1986 (Galapagos Islands: Ecuador); Zullo, 1991 (Galapagos Islands: Ecuador).

### REFERENCES

Alfaro, A. 2008. Synopsis of infections in sea turtles caused by virus, bacteria and parasites: An ecological review. In A.F. Rees, M.G. Frick, A. Panagoloulou and K. Williams, compilers. *Proceedings of the 27th Annual Symposium on Sea Turtle Biology and Conservation.* NOAA Technical Memorandum NMFS-SEFC-569, Miami, FL, p. 5.
Annandale, N. 1906. Report on the Cirripedia collected by Professor Herdman, at Ceylon, in 1902. *Ceylon Pearl Oyster Fisheries, Supplemental Report* 31: 137–150.

Bacon, P. 1976. Cirripedia of Trinidad. *Studies on the Fauna of Curacao and Other Caribbean Islands* 50: 3–55.

Badillo-Amador, F.J. 2007. Epizoítos y parásitos de la tortuga boba (*Caretta caretta*) en el Mediterráneo Occidental. PhD thesis, Facultat de Ciencies Biologiques, Univesitat de Valencia, Valencia, Spain, 262pp.

Balazs, G.H. 1978. A hawksbill turtle in Kaneohe Bay, Oahu. *Elepaio* 38: 128–129.

Balazs, G.H. 1980. Synopsis of the biological data on the green turtle in the Hawaiian Islands. NOAA Technical Memorandum NMFS-7, pp. 1–141.

Balazs, G.H., R.G. Forsyth, and A.K.H. Kam. 1987. Preliminary assessment of habitat utilization by Hawaiian green turtles in their resident foraging pastures. NOAA Technical Memorandum NMFSC-SWFC-71, pp. 1–107.

Barnard, K.H. 1924. Contributions to the crustacean fauna of South Africa. *Annotates of the South African Museum* 20: 1–103.

Beaumont, E.S., P. Zárate, J.D. Zardus, P.H. Dutton, and J.H. Seminoff. 2007. Epibiont occurrence in Galapagos green turtles (*Chelonia mydas*) at nesting and feeding grounds. In A.F. Rees, M.G. Frick, A. Panagoloulou, and K. Williams, compilers. *Proceedings of the 27th Annual Symposium on Sea Turtle Biology and Conservation.* NOAA Technical Memorandum NMFS-SEFC-569, Miami, FL, p. 8.

Bjorndal, K.A. 1997. Foraging ecology and nutrition of sea turtles. In P.L. Lutz and J.A. Musick, eds. *The Biology of Sea Turtles.* CRC Press, Boca Raton, FL, pp. 199–231.

Bjorndal, K.A. 2003. Roles of loggerheads in marine ecosystems. In A.B. Bolten and B.E. Witherington, eds. *Biology and Conservation of the Loggerhead Sea Turtle.* Smithsonian Institution Press, Washington, DC, pp. 235–254.

Bjorndal, K.A. and J.B.C. Jackson. 2003. Roles of sea turtles in marine ecosystems: Reconstructing the past. In P.L. Lutz, J.A. Musick, and J. Wyneken, eds. *The Biology of the Sea Turtles*, Vol. II. CRC Press, Boca Raton, FL, pp. 259–274.

Bolten, A.B. 2003. Variation in sea turtle life history patterns: Neritic vs. oceanic developmental stages. In P.L. Lutz, J.A. Musick, and J. Wyneken, eds. *The Biology of the Sea Turtles*, Vol. II. CRC Press, Boca Raton, FL, pp. 243–257.

Borradaile, L.A. 1903. VII. The Barnacles (Cirripedia). In S. Gardner, ed. *The Fauna and Geography of the Maldive and Laccadive Archipelagoes, Being the Account of the Work Carried on and of the Collections Made by and Expedition during the Years 1899 and 1900*, Vol. I. University Press, Cambridge, MA, pp. 440–443.

Bowen, B.W., J.C. Avise, J.I. Richardson, A.B. Meylan, D. Margaritoulis, and S.R. Hopkins-Murphy. 1993. Population structure of loggerhead turtles (*Caretta caretta*) in the northwestern Atlantic Ocean and Mediterranean Sea. *Conservation Biology* 7: 834–844.

Broch, H. 1916. Cirripedien. Results of Dr. E. Moberg's Swedish scientific expeditions to Australia 1910–13. *Kungliga Svenska Vetenskaps Akademiens Handlingar* 52: 1–16.

Broch, H. 1924. Cirripedia. Parasitologia Mauritanica, Materiaux pour la Faune Parasitologique en Mauritanie. Arthropoda (2e Partie). *Bulletin Comite d'Etudes Hist. Sci. l'Afrique Occidentale Francaise* (October–December, 1924): 1–21.

Broch, H. 1927. Studies on Moroccan cirripeds (Atlantic Coast). *Bulletin de la Societe des Sciences Naturelles, Maroc* 6–7: 11–38.

Broch, H. 1931. Indomalayan Cirripedia. Papers from Dr. Th. Mortensen's Pacific Expedition 1914–1916. *Videnskabelige Meddelelser Dansk Naturhistorisk Forening* 91: 1–146.

Broch, H. 1947. Cirripedes from Indochinese shallow-waters. *Avhandlinger Norske Videnskaps-akademi Oslo I* 1: 1–32.

Brown, C.A. and W.M. Brown. 1995. Status of sea turtles in the southeastern Pacific: Emphasis on Peru. In K.A. Bjorndal, ed. *Biology and Conservation of Sea Turtles*, Revised edition. Smithsonian Institution Press, Washington, DC, pp. 235–240.

Bugoni, L., L. Krause, A.O. de Almeida, and A.A. De Pádua Bueno. 2001. Commensal barnacles of sea turtles in Brazil. *Marine Turtle Newsletter* 94: 7–9.

Bustard, H.R. 1976. Turtles of coral reefs and coral islands. In O.A. Jones and R. Endean, eds. *Biology and Geology of Coral Reefs*, Vol. III (Biology 2). Academy Press, NY, pp. 343–368.

Caine, E.A. 1986. Carapace epibionts of nesting loggerhead turtles: Atlantic coast of U.S.A. *Journal of Experimental Marine Biology and Ecology* 95: 15–26.

Cardenas-Palomo, N. and A. Maldonado-Gasca. 2005. Epibiontes de tortugas de carey juveniles *Eretmochelys imbricata* en el Santuario de Tortugas Marinas de Rio Lagartos, Yucatan, Mexico. *CICIMAR Oceanides* 20: 29–35.

Carriol, R.P. and W. Vader. 2002. Occurrence of *Stomatolepas elegans* (Cirripedia: Balanomorpha) on a leatherback turtle from Finnmark, northern Norway. *Journal of the Marine Biological Association of the United Kingdom* 82: 1033–1034.

Caziot, E. 1921. Les Cirripedes de la mer de Nice. *Bulletin Society Zoology, France* 46: 51–54.

Chace, F.A. 1951. The oceanic crabs of the genera *Planes* and *Pachygrapsus*. *Proceedings of the United States National Museum* 101: 65–103.

Chevereaux, E. and J. de Guerne. 1893. Crustaces et Cirripeds commensaux des Tortues marines de la Mediterranee. *Comptes Rendus des Seances de l'Academie des Sciences* 116: 443–445.

Cupul-Magaña, F.G., A. Rubio-Delgado, A.H. Escobedo-Galván, and C. Reyes-Nuñez. 2011. First report of marine barnacles *Lepas anatifera* and *Chelonibia testudinaria* as epibionts on American crocodile (*Crocodylus acutus*). *Herpetology Notes* 4: 213–214.

Daniel, A. 1956. The Cirripedia of the Madras Coast. *Bulletin of the Madras Government Museum* 6: 1–40.

Daniel, A. 1962. A new species of platylepadid barnacle (Cirripedia: Crustacea) from the green turtle from Little Andaman Island. *Annals and Magazine of Natural History (Series 13)* 5: 641–645.

Darwin, C. 1851. *A Monograph on the Subclass Cirripedia, with Figures of All Species. The Lepadidae, or, Pedunculated Cirripedes*. Ray Society, London, U.K. 400pp.

Darwin, C. 1854. *A Monograph on the Subclass Cirripedia, with Figures of All the Species. The Balanidae, the Verrucidae, etc*. Ray Society, London, U.K. 684pp.

Davenport, J. 1994. A cleaning association between the oceanic crab *Planes minutus* and the loggerhead sea turtle *Caretta caretta*. *Journal of the Marine Biological Society of the United Kingdom* 74: 735–737.

Dawydoff, C. 1952. Contribution a l'etude des invertebrates de la faune marine benthique de l'Indochine. *Bulletin biologique de la France et de la Belgique* 37(Suppl): 127–131.

Dean, T.A. 1981. Structural aspects of sessile invertebrates as organizing forces in an Estuarine fouling community. *Journal of Experimental Marine Biology and Ecology* 53: 163–180.

Deem, S.L., T.M. Norton, M. Mitchell, A. Segars, A.R. Alleman, C. Cray, R.H. Poppenga, M. Dodd, and W.B. Karesh. 2009. Comparison of blood values in foraging, nesting, and stranded loggerhead turtles (*Caretta caretta*) along the coast of Georgia, USA. *Journal of Wildlife Diseases* 45: 41–56.

Dellinger, T., J. Davenport, and P. Witz. 1997. Comparisons of social structure of Columbus crabs living on loggerhead sea turtles and inanimate flotsam. *Journal of the Marine Biological Association of the United Kingdom* 77: 185–194.

Deraniyagala, P.E.P. 1939. The tetrapod reptiles of Ceylon. *Ceylon Journal of Science, Colombo Museum, Natural History Series* 1: 1–412.

Dobbs, K.A. and A.M. Landry, Jr. 2004. Commensals on nesting hawksbill turtles (*Eretmochelys imbricata*), Milman Island, northern Great Barrier Reef, Australia. *Memoirs of the Queensland Museum* 49: 674.

Dodd, C.K. Jr. 1988. Synopsis of the biological data on the loggerhead sea turtle *Caretta caretta* (Linnaeus 1758). *U.S. Fish and Wildlife Service Biological Report* 88: 1–110.

Eckert, K.L. and S.A. Eckert. 1988. Pre-reproductive movements of leatherback sea turtles (*Dermochelys coriacea*) nesting in the Caribbean. *Copeia* 1988: 400–406.

Encalada, S.E., K.A. Bjorndal, A.B. Bolten, J.C. Zurita, B. Schoeder, E. Possardt, C.J. Sears, and B.W. Bown. 1998. Population structure of loggerhead turtle (*Caretta caretta*) nesting colonies in the Atlantic and Mediterranean as inferred from mitochondrial DNA control region sequences. *Marine Biology* 130: 567–575.

Enderlein, P. and M. Wahl. 2004. Dominance of blue mussels versus consumer-mediated enhancement of benthic diversity. *Journal of Sea Research* 51: 145–55.

Farrapeira-Assunção, C.M. 1991. Revisão do gênero *Chelonibia* Leach, 1817 na costa Brasileira (Crustacea, Cirripedia). *Abstracts of XVIII Congresso Brasileiro de Zoologia*, Salvador, Brazil, 133pp.

Feifarek, B.P. 1987. Spines and epibionts as antipredator defenses in the thorny oyster *Spondylus americanus* Hermann. *Journal of Experimental Marine Biology and Ecology* 105: 39–56.

Fernando, S.A. 1978. Studies on the biology of barnacles (Crustacea: Cirripedia) of Porto Novo region, South India. PhD dissertation, Center of Advanced Study in Marine Biology, Annamalai University, Tamil Nadu, India, 213pp.

Fischer, P. 1886. Description d'un nouveau genre de Cirripedes (*Stephanolepas*) parisite des tortues marines. *Actes de la Société Linnéenne de Bordeaux* 40: 193–196.

Fishlyn, D.B. and D.W. Phillips. 1980. Chemical camouflaging and behavioural defenses against predatory seastar by three species of gastropods from the surfgrass Phyllospadix community. *Biological Bulletin* 158: 34–48.

Foley, A.M., B.A. Schroeder, and S.L. MacPherson. 2008. Post-nesting migrations and resident areas of Florida loggerheads. In H. Kalb, A. Rohde, K. Gayheart, and K. Shanker, compilers. *Proceedings of the 25th Annual Symposium on Sea Turtle Biology and Conservation*. NOAA Technical Memorandum NMFS-SEFSC-582, Miami, FL, pp. 75–76.

Foster, B.A. 1978. The marine fauna of New Zealand: Barnacles (Cirripedia: Thoracica). *New Zealand Oceanographic Institute Memorandum* 69: 1–160.

Frazier, J. 1971. Observations on sea turtles at Aldabra Atoll. *Philosophical Transactions of the Royal Society of London, Series B* 260: 373–410.

Frazier, J. 1989. Observations on stranded green turtles, *Chelonia mydas*, in the Gulf of Kutch. *Journal of the Bombay Natural History Society* 86: 250–252.

Frazier, J.G., I. Goodbody, and C.A. Ruckdeschel. 1991. Epizoan communities on marine turtles. II. Tunicates. *Bulletin of Marine Science* 48: 763–765.

Frazier, J.G., D. Margaritoulis, K. Muldoon, C.W. Potter, J. Rosewater, C. Ruckdeschel, and S. Salas. 1985. Epizoan communities on marine turtles. I. Bivalve and gastropod mollusks. *Marine Ecology* 6: 127–140.

Frazier, J.G., J.E. Winston, and C.A. Ruckdeschel. 1992. Epizoan communities on marine turtles. III. Bryozoa. *Bulletin of Marine Science* 51: 1–8.

Frick, M.G., K. Kopitsky, A.B. Bolten, K.A. Bjorndal, and H.R. Martins. 2011a. Sympatry in grapsoid crabs (genera *Planes* and *Plagusia*) from olive ridley sea turtles (*Lepidochelys olivacea*), with descriptions of crab diets and masticatory structures. *Marine Biology* 158: 1699–1708.

Frick, M.G., P.A. Mason, K.L. Williams, K. Andrews, and H. Gerstung. 2003a Epibionts of hawksbill turtles in a Caribbean nesting ground: A potentially unique association with snapping shrimp (Crustacea: Alpheidae). *Marine Turtle Newsletter* 99: 8–11.

Frick, M.G. and G. McFall. 2007. Self-grooming by loggerhead turtles in Georgia, USA. *Marine Turtle Newsletter* 118: 15.

Frick, M.G., A. Ross, K.L. Williams, A.B. Bolten, K.A. Bjorndal, and H.R. Martins. 2003b. Epibiotic associates of oceanic-stage loggerhead turtles from the southeastern North Atlantic. *Marine Turtle Newsletter* 101: 18–20.

Frick, M.G. and C.K. Slay. 2000. *Caretta caretta* (loggerhead sea turtle) epizoans. *Herpetological Review* 31: 102–103.

Frick, M.G., K.L. Williams, A.B. Bolten, K.A. Bjorndal, and H.R. Martins. 2004a. Diet and fecundity of Columbus crabs, *Planes minutus*, associated with oceanic-stage loggerhead sea turtles, *Caretta caretta*, and inanimate flotsam. *Journal of Crustacean Biology* 24: 350–355.

Frick, M.G., K.L. Williams, A.B. Bolten, K.A. Bjorndal, and H.R. Martins. 2009. Foraging ecology of oceanic-stage loggerhead turtles *Caretta caretta*. *Endangered Species Research* 9: 91–97.

Frick, M.G., K.L. Williams, M. Bresette, D.A. Singewald, and R.M. Herren. 2006. On the occurrence of Columbus crabs (*Planes minutus*) from loggerhead turtles in Florida, USA. *Marine Turtle Newsletter* 114: 12–14.

Frick, M.G., K.L. Williams, E.J. Markestyn, J.B. Pfaller, and R.E. Frick. 2004b. New records and observations of epibionts from loggerhead sea turtles. *Southeastern Naturalist* 3: 613–620.

Frick, M.G., K.L. Williams, and M. Robinson. 1998. Epibionts associated with nesting loggerhead sea turtles (*Caretta caretta*) in Georgia. *Herpetological Review* 29: 211–214.

Frick, M.G., K.L. Williams, and D. Veljacic. 2000a. Additional evidence supporting a cleaning association between epibiotic crabs and sea turtles: How will the harvest of *Sargassum* weed impact this relationship? *Marine Turtle Newsletter* 90: 11–13.

Frick, M.G., K.L. Williams, and D.C. Veljacic. 2002a. New records of epibionts from loggerhead sea turtles *Caretta caretta* (L.) *Bulletin of Marine Science* 70: 953–956.

Frick, M.G., K.L. Williams, D. Veljacic, J.A. Jackson, and S.E. Knight. 2002b. Epibiont community succession on nesting loggerhead sea turtles, *Caretta caretta*, from Georgia, USA. In A. Mosier, A. Foley and B. Brost, compilers. *Proceedings of the 20th Annual Symposium on Sea Turtle Biology and Conservation*. NOAA Technical Memorandum, NMFS-SEFSC-447, Miami, FL, pp. 280–282.

Frick, M.G., K.L. Williams, D. Veljacic, L. Pierrard, J.A. Jackson, and S.E. Knight. 2000b. Newly documented epibiont species from nesting loggerhead sea turtles (*Caretta caretta*) in Georgia. *Marine Turtle Newsletter* 88: 103–108.

Frick, M.G. and J.D. Zardus. 2010. First authentic report of the turtle barnacle *Cylindrolepas darwiniana* since its description in 1916. *Journal of Crustacean Biology* 30: 292–295.

Frick, M.G., J.D. Zardus, and E.A. Lazo-Wasem. 2010a. A new *Stomatolepas* barnacle species (Cirripedia: Balanomorpha: Coronuloidea) from leatherback sea turtles. *Bulletin of the Peabody Museum of Natural History* 51: 123–136.

Frick, M.G., J.D. Zardus, and E.A. Lazo-Wasem. 2010b. A new coronuloid barnacle subfamily, genus and species from cheloniid sea turtles. *Bulletin of the Peabody Museum of Natural History* 51: 169–177.

Frick, M.G., J.D. Zardus, A. Ross, J. Senko, D. Montano-Valdez, M. Bucio-Pacheco, and I. Sosa-Cornejo. 2011b. Novel records of the barnacle *Stephanolepas muricata* (Cirripedia: Balanomorpha: Coronuloidea); including a case for chemical mediation in turtle and whale barnacles. *Journal of Natural History* 45: 629–640.

Fuller, W.J., A.C. Broderick, R. Enever, P. Thorne, and B.J. Godley. 2010. Motile homes: A comparison of the spatial distribution of epibiont communities on Mediterranean sea turtles. *Journal of Natural History* 44:1743–1753.

Gauld, D.T. 1957. An annotated check-list of the Crustacea of the Gold Coast. I. Cirripedia. *Journal of the West African Science Association* 3: 10–11.

Geldiay, R., T. Koray, and S. Balik. 1995. Status of sea turtle populations (*Caretta caretta caretta* and *Chelonia mydas*) in the northern Mediterranean Sea, Turkey. In K.A. Bjorndal, ed. *Biology and Conservation of Sea Turtles*, Revised edition. Smithsonian Institution Press, Washington, DC, pp. 425–437.

George, R.H. 1997. Health problems and disease of sea turtles. In P.L. Lutz and J.A. Musick, eds. *The Biology of the Sea Turtles*. CRC Press, Boca Raton, FL, pp. 363–385.

Glazebrook, R.S. and R.S.F. Campbell. 1990. A survey of the diseases of marine turtles in northern Australia. 2. Oceanarium-reared and wild turtles. *Diseases of Aquatic Organisms* 9: 97–104.

Gordon, J.A. 1970. An annotated checklist of Hawaiian barnacles (Class Crustacea: Subclass Cirripedia) with notes on their nomenclature, habitats and Hawaiian localities. *Hawaii Institute of Marine Biology Technical Report* 19: 1–130.

Gramentz, D. 1988. Prevalent epibiont sites on *Caretta caretta* in the Mediterranean Sea. *Naturalista Siciliano* 12: 33–46.

Green, D. 1998. Epizoites of Galapagos green turtles. In R. Byles and Y. Fernandez, compilers. *Proceedings of the 16th Annual Symposium on Sea Turtle Biology and Conservation*. NOAA-Technical Memorandum NMFS-SEFSC-412, Miami, FL, p. 63.

Greenblatt, R.J., T.M. Work, G.H. Balazs, C.A. Sutton, R.N. Casey, and J.W. Casey. 2004. The Ozobranchus leech is a candidate mechanical vector for the fibropapilloma-associated turtle herpesvirus found latently infecting skin tumors on Hawaiian green turtles (*Chelonia mydas*). *Virology* 1: 101–110.

Gruvel, A. 1903. Revision des Cirrhipedes appartenant a la collection du Museum d'Histoire Naturelle. *Nouvelles Archives du Museum D'Histoire Naturelle de Paris (Series 4)*, 5: 95–170.

Gruvel, A. 1905. *Monographie des Cirrhipedes ou Theocostraces*. Masson et Cie, Editeurs, Paris, 472pp.

Gruvel, A. 1907. Cirrhipedes opercules de le l'Indian Museum de Calcutta. *Memories of the Asiatic Society of Bengal* 2: 1–10.

Gruvel, A. 1912. Mission Gruvel sur la cote occidentale d'Afriques (1909–1910) et collection du Museum d'Histoire Naturelle, Les Cirrhipedes. *Bulletin du Museum D'Histoire Naturelle, Series* 1: 344–350.

Gruvel, A. 1931. *Crustaces de Syrie*. Les etat de Syrie, Paris, France, pp. 397–435.

Haelters, J. and F. Kerckhof. 1999. Een waarneming van de lederschildpad *Dermochelys coriacea* (Linnaeus, 1758), en de eerste waarneming van *Stomatolepas dermochelys* Monroe and Limpus, 1979 aan de Belgische kust. *De Strandvlo* 19: 30–39.

Haelters, J. and F. Kerckhof. 2001. Opnieuw een klapmuts *Cystophora cristata* Erxleben, 1777 en een lederschildpad *Dermochelys coriacea* (Linnaeus 1758) aan onze kust. *De Strandvlo* 21: 81–83.

Harder T. 2009. Marine epibiosis: Concepts, ecological consequences and host defence. *Marine and Industrial Biofouling* 4: 219–31.

Harding, J.M. and R. Mann. 1999. Observations on the biology of the veined rapa whelk, *Rapana venosa* (Valenciennes, 1846) in the Chesapeake Bay, USA. *Journal of Shellfish Research* 24: 9–18.

Harding, J.M., W.J. Walton, C.M. Trapani, M.G. Frick, and R. Mann. 2011. Sea turtles as potential dispersal vectors for non-indigenous species: The veined rapa whelk as an epibiont of loggerhead sea turtles. *Southeastern Naturalist* 10: 233–244.

Hayashi, R. and K. Tsuji. 2008. Spatial distribution of turtle barnacles on the green sea turtle, *Chelonia mydas*. *Ecological Research* 2007 (*On-line Journal of the Ecological Society of Japan*), 5pp.

Heithaus, M.R., J.J. McLash, A. Frid, L.M. Dill, and G.J. Marshall. 2002. Novel insights into green turtle behaviour using animal-borne video cameras. *Journal of the Marine Biological Association of the United Kingdom* 82: 1049–1050.

Hendrickson, J.R. 1958. The green sea turtle, *Chelonia mydas* (Linn.), in Malaya and Sarawak. *Proceedings of the Zoological Society of London* 130: 455–535.

Henry, D.P. 1941. Notes on some sessile barnacles from Lower California and the west coast of Mexico. *Proceedings of the New England Zoological Club* 18: 99–107.

Henry, D.P. 1954. The barnacles of the Gulf of Mexico. Gulf of Mexico, its origin, waters and marine life. U.S. Fish and Wildlife Service, 55, *Fishery Bulletin* 89: 443–446.

Henry, D.P. 1960. Thoracic Cirripedia of the Gulf of California. *University of Washington Publication Oceanography* 4: 135–158.

Hernandez-Vasquez, S. and C. Valadez-Gonzalez. 1998. Observations on the epizoa found on the turtle *Lepidochelys olivacea* at La Gloria, Jalisco, Mexico. *Ciencias Marina* 24: 119–125.

Hiro, F. 1936. Occurrence of the cirriped *Stomatolepas elegans* on a loggerhead turtle found at Seto. *Annotationes Zoologicae Japonenses* 15: 312–320.

Hiro, F. 1937a. Cirripeds of the Palao Islands. *Palao Tropical Biological Station Studies* 1: 37–72.

Hiro, F. 1939. Studies on the cirripedien fauna of Japan. V. Cirripeds of the northern part of Honshyu. *Science Reports of the Tohoku Imperial University (Series 4)* 14: 201–218.

Holothuis, L.B. 1952. Enige interessante, met drijvende voorwerpen op de Nedelandse kust aangespoelde zeepissebedden en zeepokken. *De Levende Natuur* 55: 72–77.

Holothuis, L.B. 1969. Enkele interessante Nederlandse Crustacea. *Zoologische Bijdragen* 2: 34–48.

Hubbs, C.L. 1977. First record of mating ridley turtles in California, with notes on commensals, characters, and systematics. *California Fish and Game* 63: 263–267.

Hunt, T.L. 1995. Preliminary survey of commensals associated with *Caretta caretta*. In J.I. Richardson and T.H. Richardson, compilers. *Proceedings of the 12th Annual Workshop on Sea Turtle Biology and Conservation*. NOAA-Technical Memorandum NMFS-SEFSC-361, Miami, FL, pp. 204–207.

Ives, J.E. 1891. Crustacea from the northern coast of Yucatan, the harbor of Vera Cruz, the west coast of Florida and the Bermuda Islands. *Proceedings of the Academy of Natural Sciences of Philadelphia* 43: 176–207.

Jones, D.S. 1990. The shallow-water barnacles (Cirripedia: Lepadomorpha, Balanomorpha) of southern Western Australia. In F.E. Wells, D.I. Walker, H. Kirkman and R. Lethbridge, compilers. *Proceedings of the Third International Marine Biological Workshop: The Marine Flora and Fauna of Albany, Western Australia*. Western Australia Museum, Perth, Australia, pp. 333–437.

Jones, D.S., J.T. Anderson and D.T. Anderson. 1990. Checklist of the Australian Cirripedia. *Technical Report of the Australian Museum* 3: 1–38.

Jones, D.S., M.A. Hewitt, and A. Sampey. 2000. A checklist of the cirripedia of the South China Sea. *Raffles Bulletin of Zoology* 8(Suppl): 233–307.

Killingley, J.S. and M. Lutcavage. 1983. Loggerhead turtle movements reconstructed from 18O and 13C profiles from commensal barnacle shells. *Estuarine Coastal Shelf Science* 16: 345–349.

Kitsos, M., M. Christodoulou, C. Arvanitidis, M. Mavidis, I. Kirmitzoglou, and A. Koukouras. 2005. Composition of the organismic assemblage associated with *Caretta caretta*. *Journal of the Marine Biological Association of the United Kingdom* 2005: 257–261.

Kolosvary, G. 1939. Ueber Fundortsangaben adriatischer Balanen. *Bollettino dei Musei di Zoologia ed Anatomia Comparata della R. Università di Torino (Series III)* 47: 37–41.

Kolosvary, G. 1943. Cirripedi thoracica in der Sammlung des ungarischen National-Museums. *Annales Historico-Naturales Musei Nationalis Hungarici* 36: 67–120.

Kolosvary, G. 1951. Les balanids de la Mediterranee. *Acta Biologica* 2: 411–413.

Koukouras, A. and A. Matsa. 1998. The thoracican cirriped fauna of the Aegean Sea: New information, checklist of the Mediterranean species, faunal comparisons. *Senckenbergiana Maritima* 28: 133–142.

Kruger, P. 1911b. Zur Cirripedien fauna Ostasiens. *Zoologischer Anzeiger, Leipzig* 38: 459–464.

Kruger, P. 1912. Uber ostrasiatische Rhizocephalen. Anhang: Uber eninige intersante Vertreter der Cirripedia thoracica. *Abhandlungen der Mathematisch-Physikalische Klasse der Königlich Bayerischen Akademie der Wissenschaften, II Supplement Band* 8: 1–16.

Lanchester, W.F. 1902. On the Crustacea collected during the "Skeat Expedition" to the Malay Peninsula. *Proceedings of the Zoological Society of London* 2: 363–381.

Lazo-Wasem, E.A., T. Pinou, A. Peña de Niz, and A. Feuerstein. 2011. Epibionts associated with the nesting marine turtles *Lepidochelys olivacea* and *Chelonia mydas* in Jalisco, Mexico: A review and field guide. *Bulletin of the Peabody Museum of Natural History* 52(2): 221–240.

Lanfranco, G. 1979. *Stomatolepas elegans* Costa (Crustacea, Cirripedia) on *Dermochelys coriacea* Linn., taken in Maltese waters. *Central Mediterranean Naturalist* 1: 24.

Leung T.L.F. and Poulin R. 2008. Parasitism, commensalism, and mutualism: Exploring the many shades of symbioses. *Vie Milieu* 58: 107–115.

Lezama, C., A. Carranza, A. Fallabrino, A. Estrades, and M. López-Mendilaharsu. 2012. Unintended backpackers: Bio-fouling of the invasive gastropod *Rapana venosa* on the green turtle *Chelonia mydas* in the Río de la Plata Estuary, Uruguay. *Biological Invasions* doi:10.10007/s10530-012-0307-9.

Limpus, C.J. and D.J. Limpus. 2003. Biology of the loggerhead turtle in western south Pacific Ocean foraging areas, In A.B. Bolten and B.E. Witherington, eds. *Biology and Conservation of the Loggerhead Sea Turtle*. Smithsonian Institution Press, Washington, DC, pp. 93–113.

Limpus, C.J., D.J. Limpus, M. Munchow, and P. Barnes. 2005. Queensland turtle conservation project: Raine Island turtle study, 2004–2005. Queensland Government Conservation Technical and Data Report, pp. 1–37.

Limpus, C.J., J.D. Miller, V. Baker, and E. Mclachlan. 1983a. The hawksbill turtle, *Eretmochelys imbricata* (L.), in north-eastern Australia: The Campbell Island rookery. *Australian Wildlife Research* 10: 185–197.

Limpus, C.J., C.J. Parmenter, V. Baker, and A. Fleay. 1983b. The Crab Island sea turtle rookery in the north-eastern Gulf of Carpentaria. *Australian Wildlife Research* 10: 173–184.

Linnaeus, C. 1758. *Systema Naturae*. Holmiae, Editio Decima, Reformata, Vol. 1, 824pp.

Logan, P and S.J. Morreale. 1994. Hydrodynamic drag characteristics of juvenile *L. kempi*, *C. mydas*, and *C. caretta*. In B.A. Schroeder and B.E. Witherington, compilers. *Proceedings of the 13th Annual Symposium on Sea Turtle Biology and Conservation*. NOAA-Technical Memorandum NMFS-SEFSC-341, Miami, FL, pp. 248–252.

Loop, K.A., J.D. Miller, and C.J. Limpus. 1995. Nesting by the hawksbill turtle (*Eretmochelys imbricata*) on Milman Island, Great Barrier Reef, Australia. *Wildlife Research* 22: 241–252.

Losey, G., G.H. Balazs, and L.A. Privitera. 1994. Cleaning symbiosis between the wrasse, *Thalassoma duperry*, and the green turtle, *Chelonia mydas*. *Copeia* 1994: 684–690.

Loza, A.L. and L.F. Lopez-Juardo. 2004. Comparative study of the epibionts on the pelagic and mature female loggerhead turtles on the Canary and Cape Verde Islands. In R.B. Mast, B.J. Hutchinson, and A.H. Hutchinson, compliers. *Proceedings of the 24th Annual Symposium on Sea Turtle Biology and Conservation*. NOAA Technical Memorandum NMFS-SEFSC-567, Miami, FL, p. 100.

Lucas, M. 1968. Les cirrhipedes l'Europe. *Les Naturalistes Belges* 49: 105–160.

MacDonald, R. 1929. A report of some cirripeds collected by the S.S. "Albatross" in the eastern Pacific during 1891 and 1904. *Bulletin of the Museum of Comparative Zoology* 69: 527–538.

Margaritoulis, D. 1985. Preliminary observations on the breeding behaviour and ecology of *Caretta caretta* in Zakynthos, Greece. *2e Congrès international sur la zoogéographie et l'écologie de la Grèce et des régions adjacentes*, Athens, Greece, September 1981, 10, pp. 323–332.

Matsuura, I. and K. Nakamura. 1993. Attachment pattern of the turtle barnacle *Chelonibia testudinaria* on the carapace of nesting loggerhead turtles *Caretta caretta*. *Bulletin of the Japanese Society for the Science of Fish (Nippon Suisan Gakkaishi)* 59: 1803.

McCann, C. 1969. First southern hemisphere record of the platylepadine barnacle *Stomatolepas elegans* (Costa) and notes on the host *Dermochelys coriacea* (Linne). *New Zealand Journal of Marine and Freshwater Research* 3: 152–158.

Meylan, A.B. 1983. Marine turtles of the Leeward Islands, Lesser Antilles. *Smithsonian Institution Atoll Research Bulletin* 278: 1–43.

Moore, P.G. 1995. *Podocerus chelonophilus* (Amphipoda: Podoceridae) associated with epidermal lesions of the loggerhead turtle, *Caretta caretta* (Chelonia). *Journal of the Marine Biological Association of the United Kingdom* 75: 253–255.

Monroe, R. and C.J. Limpus. 1979. Barnacles on turtles in Queensland waters with descriptions of three new species. *Memoirs of the Queensland Museum* 19: 197–223.

Moriarty, J.E., J.A. Sachs, and K. Jones. 2008. Directional locomotion in a turtle barnacle, *Chelonibia testudinaria*, on green turtles, *Chelonia mydas*. *Marine Turtle Newsletter* 119: 1–4.

Mustaquim, J. and M. Javed. 1993. Occurrence of *Chelonibia testudinaria* (Linnaeus) (Crustacea: Cirripedia) in coastal waters of Pakistan. *Pakistan Journal of Marine Science* 2: 73–75.

Newman, W.A. and D.P. Abbott. 1980. Cirripedia: The barnacles. In R.H Morris, D.P. Abbott and E.C. Haderlie, eds. *Intertidal Invertebrates of California*. Stanford University Press, Stanford, CA, pp. 504–535.

Newman, W.A., V.A. Zullo, and T.H. Withers. 1969. Cirripedia. *Treatise on Invertebrate Paleontology, Part R, Arthropoda* 4: R206–R295.

Nifong, J.C. and M.G. Frick. 2011. First record of the American alligator (*Alligator mississippiensis*) as a host to the sea turtle barnacle (*Chelonibia testudinaria*). *Southeastern Naturalist* 10: 557–560.

Nilsson-Cantell, C.A. 1921. Cirripedian-Studien. Zur kenntnis der Biologie, Anatomie und Systematik dieser Gruppe. *Zoologiska Bidrag från Uppsala* 7: 75–395.

Nilsson-Cantell, C.A. 1930a. Diagnoses of some new cirripeds from the Netherlands Indies collected by the expedition of His Royal Highness the Prince Leopold of Belgium in 1929. *Bulletin de Musee Royal d'Histoire Naturelle de Belgique* 6: 1–2.

Nilsson-Cantell, C.A. 1931. Revision der Sammulung recenter Cirripedien des Naturhistorichen Museums in Bael. *Verhandlungen der Naturforschenden Gesellschaft in Basel* 42: 103–137.

Nilsson-Cantell, C.A. 1932. The barnacles *Stephanolepas* and *Chelonibia* from the turtle *Eretmochelys imbricata*. *Ceylon Journal of Science, Section B (Spolia Zeylanica)* 16: 257–264.

Nilsson-Cantell, C.A. 1937. On a second collection of Indo-Malayan cirripeds from the Raffles Museum. *Bulletin of the Raffles Museum* 13: 93–96.

Nilsson-Cantell, C.A. 1938. Cirripeds from the Indian Ocean in the collection of the Indian Museum, Calcutta. *Memoirs of the Indian Museum* 13: 1–81.

Nilsson-Cantell, C.A. 1939. Recent and fossil balanids from the north coast of South America. *Capita Zoologica* 8: 1–7.

O'Riordan, C.E. 1979. Marine fauna notes from the National Museum of Ireland 6. *Irish Naturalists Journal* 19: 356–358.

O'Riordan, C.E. and J.M.C. Holmes. 1978. Marine fauna notes from the National Museum of Ireland. 5. Passengers on the North Atlantic currents. *Irish Naturalists Journal* 19: 152–153.

Pajuelo, M., K.A. Bjorndal, K.J. Reich, M.D. Arendt, and A.B. Bolten. 2012. Distribution of foraging habitats of male loggerhead turtles (*Caretta caretta*) as revealed by stable isotopes and satellite telemetry. *Marine Biology* 159: 1255–1267.

Pereira, S., E.H.S.M. Lima, L. Ernesto, H. Matthews, and A. Ventura. 2006. Epibionts associated with *Chelonia mydas* from Northern Brazil. *Marine Turtle Newsletter* 111: 17–18.

Pfaller, J.B., K.A. Bjorndal, K.J. Reich, K.L. Williams, and M.G. Frick. 2006. Distribution patterns of epibionts on the carapace of loggerhead turtles, *Caretta caretta*. *Journal of the Marine Biological Association of the United Kingdom Marine Biodiversity Records* 1: e36.

Pfaller, J.B., M.G. Frick, F. Brischoux, C.M. Sheey III, and H.B. Lillywhite. 2012. Marine snake epibiosis: A review and first report of decapods associated with *Pelamis platurus*. *Integrative and Comparative Biology* 52: 296–310.

Pfaller, J.B., M.G. Frick, K.J. Reich, K.L. Williams, and K.A. Bjorndal. 2008. Carapace epibionts of loggerhead turtles (*Caretta caretta*) nesting at Canaveral National Seashore, Florida. *Journal of Natural History*. 42: 1095–1102.

Pillai, N.K. 1958. Development of *Balanus amphitrite*, with a note on the early larvae of *Chelonibia testudinaria*. *Bulletin of the Central Research Institute of Kerala, University of Kerala, Series C* 6: 117–130.

Pilsbry, H.A. 1916. The sessile barnacles (Cirripedia) contained in the collections of the U.S. National Museum; including a monograph of the American species. *U.S. National Museum Bulletin* 93: 1–366.

Pilsbry, H.A. 1927. Littoral barnacles of the Hawaiian Islands and Japan. *Proceedings of the Academy of Natural Sciences of Philadelphia* 79: 305–317.

Plotkin, P.T. 1996. Occurrence and diet of juvenile loggerhead sea turtles, *Caretta caretta*, in the northwestern Gulf of Mexico. *Chelonian Conservation and Biology* 2: 78–80.

Pons, M., A. Verdi, and A. Domingo. 2011. The pelagic crab *Planes cyaneus* (Dana, 1851) (Decapoda, Brachyura, Grapsidae) in the southwestern Atlantic Ocean in association with loggerhead sea turtles buoys. *Crustaceana* 84: 425–434.

Quigley, D.T. and K. Flannery. 1993. Southern marine fauna and flora from S.W. Ireland. *Porcupine Newsletter* 5: 152–155.

Rainbow, P.S. and G. Walker. 1977. The functional morphology of the alimentary tract of barnacles (Cirripedia: Thoracica). *Journal of Experimental Marine Biology and Ecology*. 28: 183–206.

Ranzani, C. 1817–1818. Osservazioni su i Balanidi. Bologna, opuscoli Scientifici I (1817): 195–202; II (1817): 269–276; III (1818): 63–93.

Rathbun, M.J. 1925. The spider crabs of America. *Bulletin— United States National Museum* 129: 1–598.

Rawson, P.D., R. Macnamee, M.G. Frick, and K.L. Williams. 2003. Phylogeography of the coronulid barnacle, *Chelonibia testudinaria*, from loggerhead sea turtles, *Caretta caretta*. *Molecular Ecology* 12: 2697–2706.

Reich, K.J., K.A. Bjorndal, M.G. Frick, B.E. Witherington, C. Johnson, and A.B. Bolten. 2010. Polymodal foraging in adult female loggerheads (*Caretta caretta*). *Marine Biology* 157: 651–663.

Reisinger, R.R. and M.N. Bester. 2010. Goose barnalces on seals and a penguin at Gough Island. *African Zoology* 45: 129–132.

Relini, G. 1968. Segnalazione di du cirripedi nuovi per l'Adriatico. *Bolletin de Societie du Adriatica Sciencia Trieste* 56: 218–225.

Relini, G. 1969. La distribuzione dei Cirripedi Toracice nei mari Italiani. *Archaeologie Botanica Biogreoria Italia 4*, 45: 169–186.

Relini, G. 1980. Cirripedi toracici. *Guide per il Riconoscimento delle Specie Animali delle Acque Lagunari e Costiere Italiane* 2: 1–122.

Ren, X. 1980. Turtle barnacles of the Xisha Islands, Guangdong Province, China. *Studia Marina Sinica* 17: 187–197.

Ren, X. 1987. Studies on Chinese Cirripedia (Crustacea) VIII. Supplementary Report. *Studia Marina Sinica* 28: 175–187.

Richards, H.G. 1930. Notes on the barnacles from Cape May County, New Jersey. *Proceedings of the Academy of Natural Sciences of the Philadelphia* 83: 143–144.

Ross, J. 1981. Hawksbill turtle Eretmochelys imbricata in the Sultanate of Oman. *Biological Conservation* 19: 99–106.

Ross, A. and M.G. Frick. 2011. Nomenclatural emendations of the family-group names Cylindrolepadinae, Stomatolepadinae, Chelolepadinae, Cryptolepadinae, and Tubicinellinae of Ross & Frick, 2007— Including current definitions of family-groups within the Coronuloidea (Cirripedia: Balanomorpha). *Zootaxa* 3106: 60–66.

Ross, A. and W.A. Newman. 1967. Eocene Balanidae of Florida, including a new genus and species with a unique plan of "turtle-barnacle" organization. *American Museum Novitates* 2288: 1–21.

Rudloe, J., A. Rudloe and L. Ogren. 1991. Occurrence of immature Kemp's ridley turtles, *Lepidochelys kempi*, in coastal waters of northwest Florida. *Northwest Gulf Science* 12: 49–53.

Ruxton, G.D. 2009. Non-visual crypsis: A review of the empirical evidence for camouflage to sense other than vision. *Philosophical Transactions of the Royal Society B* 364: 540–557.

Sazima, C., A. Grossman, and I. Sazima. 2010. Turtle cleaners: Reef fishes foraging on epibionts of sea turtles in the tropical southwestern Atlantic, with a summary of this association type. *Neotropical Ichthyology* 8: 187–192.

Schärer, M.T. 2001. A survey of the epibiota of *Eretmochelys imbricata* (Testudines: Cheloniidae) of Mona Island, Puerto Rico. *Revista de Biologia Tropical* 51: 87–89.

Schärer, M.T. and J.H. Epler. 2007. Long-range dispersal possibilities via sea turtle—A case study for *Clunio* and *Pontomyia* (Diptera: Chironomidae) in Puerto Rico. *Entomological News* 118: 273–277.

Schofield, G., K.A. Katselidis, P. Dimopoulos, J.D. Pantis, and G.C. Hays. 2006. Behaviour analysis of the loggerhead sea turtle *Caretta caretta* from direct in-water observation. *Endangered Species Research* 2: 71–79.

Schwartz, F.J. 1960. The barnacle, *Platylepas hexastylos*, encrusting a green turtle, *Chelonia mydas mydas*, from Chincoteague Bay, Maryland. *Chesapeake Science* 1: 116–117.

Senties, G.A., J. Espinoza-Avalos, and J.C. Zurita. 1999. Epizoic algae of nesting sea turtles *Caretta caretta* (L.) and *Chelonia mydas* (L.) from the Mexican Caribbean. *Bulletin of Marine Science* 64: 185–188.

Sezgin, M., A.S. Ateş, T. Katağan, K. Bakir, and Ş. Yalçin Özkilek. 2009. Notes on amphipods *Caprella andreae* Mayer, 1890 and *Podocerus chelonophilus* (Chevreux and Guerne, 1888) collected from the loggerhead sea turtle, *Caretta caretta*, off the Mediterranean and the Aegean coasts of Turkey. *Turkish Journal of Zoology* 33: 433–437.

Shine, R., F. Brischoux, and A.J. Pile. 2010. A seasnake's colour affects its susceptibility to algal fouling. *Proceedings of the Royal Society B* 277: 2459–64.

Smaldon, G. and I.H.J. Lyster. 1976. *Stomatolepas elegans* (Costa, 1840) (Cirripedia): New records and notes. *Crustaceana* 30: 317–318.

Stinson, M.L. 1984. Biology of sea turtles in San Diego Bay, California, and in the northeastern Pacific. Master thesis, San Diego State University, San Diego, CA, 578pp.

Stubbings, H.G. 1965. West African Cirripedia in the collections of the Institut Francais d'Afrique Noire, Dakar, Senegal. *Bulletin de l'Institut Français d'Afrique Noire, Series A* 27: 876–907.

Stubbings, H.G. 1967. The cirriped fauna of tropical West Africa. *Bulletin of the British Museum (Natural History) Zoology* 15: 229–319.

Tachikawa, H. 1995. Notes on three species of stalked barnacles found from a turtle barnacle on the carapace of a green turtle, *Chelonia mydas*. *Nanki Seibutu* 37: 67–68.

Turtle Expert Working Group. 2009. An assessment of the loggerhead turtle population in the western northern Atlantic Ocean. NOAA Technical Memorandum NMFS-SEFSC-575.

Utinomi, H. 1949. Studies on the cirripedian fauna of Japan. VI. Cirripeds from Kyusyu and Ryukuyu Islands. *Publications of the Seto Marine Biological Laboratory* 1: 19–37.

Utinomi, H. 1950. Cirripeds commonly taken by dredging near Tanabe Bay (Record of collections dredged from off Minabe Prov. Kii, IV). *Nanki Seibutu* 2: 60–65.

Utinomi, H. 1958. Studies on the cirripedian fauna of Japan. VII. Cirripeds from Sagami Bay. *Publications of the Seto Marine Biological Laboratory* 4: 281–311.

Utinomi, H. 1959. Thoracic cirripeds from the environs of Banyuls. *Vie et Milieu* 10: 379–399.

Utinomi, H. 1966. Fauna and flora of the sea around the Amakusa Marine Biological Laboratory. Part VI., Cirriped Crustacea. *Amakusa Marine Biological Laboratory*, July 1966, pp. 1–11.

Utinomi, H. 1969. Cirripedia of the Iranian Gulf. *Videnskabelige Meddelelser Dansk Naturhistorisk Forening* 132: 79–94.

Utinomi, H. 1970. Studies on the cirripedian fauna of Japan. IX. Distributional survey of thoracic cirripeds in the southeastern part of the Japan Sea. *Publications of the Seto Marine Biological Laboratory* 17: 339–372.

Vivaldo, S.G., D.O. Sarabia, C.P. Salazar, A.G. Hernandez, and J.R. Lezama. 2006. Identification of parasites and epibionts in the olive ridley turtle (*Lepidochelys olivacea*) that arrived to the beaches of Michoacan and Oaxaca, Mexico. *Veterinaria Mexico* 37: 431–440.

Wagh, A.B. and D.V. Bal. 1974. Observations on the systematics of sessile barnacles from the west coast of India-1. *Journal of Bombay Natural History Society* 71: 109–123.

Wahl, M. 1989. Marine epibiosis. I. Fouling and antifouling: Some basic aspects. *Marine Ecology Progress Series* 58: 175–89.

Wahl, M. 2009. Epibiosis: Ecology, effects and defences. In M. Wahl, ed. *Marine Hardbottom Communities. Ecological Studies Series*, Vol. 206. Springer, Berlin, Germany, pp. 61–72.

Wahl, M., O. Mark. 1999. The predominately facultative nature of epibiosis: Experimental and observational evidence. *Marine Ecology Progress Series* 187: 59–66.

Walker, G. 1978. A cytological study of the cement apparatus of the barnacle, *Chelonibia testudinaria* Linnaeus, an epizoite on turtles. *Bulletin of Marine Science* 28: 205–209.

Wass, M.L. 1963. Check list of the marine invertebrates of Virginia. Virginia Institute of Marine Science Gloucester Point, Virginia, Spec. Sci. Rept. No. 24 (revised): 1–56.

Wells, H.W. 1966. Barnacles of the northeastern Gulf of Mexico. *Quarterly Journal of the Florida Academy of Science* 29: 81–95.

Weltner, W. 1897. Verzeichnis der bisher beschriebenen recenten Cirripedienarten. Mit Angabe der im berliner Museum vorhandenen Species und ihrer Fundorte. *Archiv für Naturgeschicthe* 1: 227–280.

Weltner, W. 1910. Cirripedien von Ostafrika. In: *Reise in Ostafrika*. V.A. Voeltzkow, Stuttgart, Germany, 2: 525–528.

Wyneken, J. 1997. Sea turtle locomotion: Mechanisms, behavior, and energetics. In P.L. Lutz and J.A. Musick, editors. *The Biology of Sea Turtles*. CRC Press, Boca Raton, FL, pp. 165–198.

Young, P.S. 1999. Subclasse Cirripedia. In: L. Buckup and G. Bond-Buckup, eds. *Os Crustáceos do Rio Grande do Sul*. Porto Alegre, Ed. Universidade/UFRGS, pp. 24–53.

Young, P.S. and A. Ross. 2000. Cirripedia. In: J.E.L. Bosquets, E.G. Soriano and N. Papavero, eds. *Bioversidad, Taxonomia y Biogeographia de Arthropodos de Mexico: Hacia una Sintesis de su Conocimiento*. Vol. II. Universidad Nacional Autonomia de Mexico, Mexico, pp. 213–237.

Zakhama-Sraieb, R., S. Karaa, M.N. Bradai, I. Jribi, and F. Charfi-Cheikhrouha. 2010. Amphipod epibionts of the sea turtles *Caretta caretta* and *Chelonia mydas* from the Gulf of Gabès (central Mediterranean). *Journal of the Marine Biological Association of the United Kingdom Marine Biodiversity Records* 3: e38.

Zann, L.P. and B.M. Harker. 1978. Egg production of the barnacles *Platylepas ophiophilus* Lanchester, *Platylepas hexastylos* (O. Fabricius), *Octolasmis warwickii* Gray and *Lepas anatifera* Linnaeus. *Crustaceana* 35: 206–214.

Zardus, J.D. and G.H. Balazs. 2007. Two previously unreported barnacles commensal with the green sea turtle, *Chelonia mydas* (Linnaeus, 1758), in Hawaii and a comparison of their attachment modes. *Crustaceana* 80: 1303–1315.

Zardus, J.D. and M.G. Hadfield. 2004. Larval development and complemental males in *Chelonibia testudinaria*, a barnacle commensal with sea turtles. *Journal of Crustacean Biology* 24: 409–421.

Zavodnik, D. 1997. *Chthamalus montagui* and *Platylepas hexastylos* two cirriped crustaceans new to the eastern Adriatic Sea. *Natura Croatica* (Croatian Nat. Hist. Mus.) 6: 113–118.

Zullo, V.A. 1986. Quaternary barnacles from the Galapagos Islands. *Proceedings of the California Academy of Sciences* 44: 55–66.

Zullo, V.A. 1991. Zoogeography of the shallow water cirriped fauna of the Galapagos Islands and adjacent regions in the tropical Eastern Pacific. In M.J. Jones, ed. *Galapagos Marine Invertebrates. Taxonomy, Biogeography and Evolution in Darwin's Islands*. Plenum Publishing Company, New York.

Zullo, V.A. and J.S. Bleakney. 1966. The cirriped *Stomatolepas elegans* (Costa) on leatherback turtles from Nova Scotian waters. *Canadian Field-Naturalist* 80: 163–165.

Zullo, V.A. and W.H. Lang. 1978. Order Cirripedia. In: R.G. Zingmark, ed. *Annotated Checklist of the Biota of the Coastal Zone of South Carolina*. University of South Carolina, Columbia, SC, pp. 158–160.

## BIBLIOGRAPHY

Allee, W.C., A.E. Emerson, O. Park, T. Park, and K.P. Schmidt. 1949. *Principles of Animal Ecology.* W.B. Saunders Co., Pennsylvania, PA.

Allen, E.R. and W.T. Neill. 1952. Know your reptiles: The diamondback terrapin. *Florida Wildlife* 6: 42.

Angulo-Lozano, L., P.E. Nava-Duran, and M.G. Frick. 2007. Epibionts of olive ridley turtles (*Lepidochelys olivacea*) nesting at Playa Ceuta, Sinaloa, Mexico. *Marine Turtle Newsletter* 118: 13–14.

Aradas, A. 1869. Desrizione di una nuova specie del genere *Coronula. Atti della Accademia Gioenia di Scienze Naturali in Catania* 43: 215–224.

Arndt, R.G. 1975. The occurrence of barnacles and algae on the red-bellied turtle, *Chrysemys r. rubiventris* (LeConte). *Journal of Herpetology* 9: 357–359.

Aurivillius, C.W.S. 1894. Studien uber cirripedien. *Kongliga Svenska Vetenskaps-Akademien* 26: 1–107.

Ayling, A.M. 1976. The strategy of orientation in the barnacle *Balanus trigonus. Marine Biology* 36: 335–342.

Aznar, F.J., J.A. Balbuena, and J.A. Raga. 1994. Are epizoites indicators of a western Mediterranean striped dolphin die-off? *Diseases of Aquatic Organisms* 18: 159–163.

Baer, J.G. 1951. *Ecology of Animal Parasites.* University of Illinois Press, Urbana, IL.

Barnes, M. 1989. Egg production in cirripedes. *Oceanography and Marine Biology Annotated Review* 27: 91–166.

Bleakney, J.S. 1967. Food items in two loggerhead sea turtles, *Caretta caretta caretta* (L.) from Nova Scotia. *Canadian Field-Naturalist* 81: 169–272.

Booth, J. and J.A. Peters. 1972. Behavioral studies on the green turtle (*Chelonia mydas*) in the sea. *Animal Behaviour* 20: 808–812.

Bourget, E. 1977. Shell structure in sessile barnacles. *Naturaliste Canadien* 104: 281–323.

Briggs, K.T. and G.V. Morejohn. 1972. Barnacle orientation and water flow characteristics in California grey whales. *Journal of Zoology* 167: 287–292.

Cake, E.W., Jr. 1983. Symbiotic associations involving the southern oyster drill *Thais haemastoma floridana* (Conrad) and macrocrustaceans in Mississippi waters. *Journal of Shellfish Research* 3: 117–128.

Carr, A. 1952. *Handbook of Turtles.* Cornell University Press, New York.

Carr, A. 1964. Transoceanic migrations of the green turtle. *Bioscience* 14: 49–52.

Carr, A. 1965. The navigation of the green turtle. *Scientific American.* 12: 79–86.

Chen, Y. 1989. Cirripedia. In C. Wei and Y. Chen, eds. *Fauna of Zhejiang: Crustacea.* Zhejiang Science and Technology Publishing House, Hangzhou, Zhejiang Province, China, pp. 38–73.

Crisp, D.J. 1960. Mobility of barnacles. *Nature* 188: 1208–1209.

Crisp, D.J. 1974. Factors influencing the settlement of marine invertebrate larvae. In P.T. Grant and A.M. Mackie, eds. *Chemoreception in Marine Organisms.* Academic Press, Inc., NY, pp. 177–265.

Crisp, D.J. 1983. *Chelonibia patula* (Ranzani), a pointer to the evolution of the complemental male. *Marine Biology Letters* 4: 281–294.

Crisp, D.J. and J.D. Costlow, Jr. 1963. The tolerance of developing cirripede embryos to salinity and temperature. *Oikos* 14: 22–34.

Crisp, D.J. and H.G. Stubbings. 1957. The orientation of barnacles to water currents. *Journal of Animal Ecology* 26: 179–196.

Dall, W.H. 1872. On the parasites of the cetaceans of the N.W. coast of America, with descriptions of new forms. *Proceedings of the California Academy of Sciences* 4: 299–301.

deAlessandri, G. 1895. Contribuzione allo studio dei Cirripedi fossili d'Italia. *Bollettino della Società Geologica Italian* 13: 234–314.

deAlessandri, G. 1906. Studi monografici sui cirripedi fossili d'Italia. *Palaeon Italica* 12: 207–324.

Eckert, K.L. and S.A. Eckert. 1987. Growth rate and reproductive condition of the barnacle *Conchoderma virgatum* on gravid leatherback sea turtles in Caribbean waters. *Journal of Crustacean Biology* 7: 682–690.

Fabricius, O. 1790. Beschribung zweiter neuer Gattung Meereichein (Lepades) nebst der Islandischen Kammuschel (Ostrea islandica) mit abbildungen. *Schriften der Berlinischen Gesellschaft Naturforschender Freunde* 1: 101–111.

Fabricius, O. 1798. Tillaeg-til Conchylie-Slaegterne *Lepas, Pholas, Mya, Solen. Skrivter av Naturhiftorie Selskabet Kiobenhavn* 4: 34–51.

Felix, F., B. Bearson, and J. Falconi. 2006. Epizoic barnacles removed from the skin of a humpback whale after a period of intense surface activity. *Marine Mammal Science* 22: 979–984.

Fischer, P. 1884. Cirrhipedes de l'archipel de la Nouvelle-Caledonie. *Bulletin de la Société Zoologique de France* 9: 355–360.

Frazier, J. 1986. Epizoic barnacles on pleurodiran turtles: Is the relationship rare? *Proceedings of the Biological Society Washington* 99: 472–477.

Frazier, J.G. and D. Margaritoulis. 1990. The occurrence of the barnacle, *Chelonibia patula* (Ranzani, 1818) on an inanimate substratum (Cirripedia, Thoracica). *Crustaceana* 59: 213–218.

Frick, M.G. and A. Ross. 2002. Happenstance or design: An unusual association between a turtle, and octocoral and a barnacle. *Marine Turtle Newsletter* 97: 10–11.

Gamez Vivaldo, S., D. Osorio Sarabia, C. Peneflores Salazar, A. Garcia Hernandez, and J. Ramirez Lezama. 2006. Identificacion de parasitos y epibiontes de la tortuga golfina (*Lepidochelys olivacea*) que arribo a playas de Michoacan y Oaxaca, Mexico. *Veterinaria Mexico* 37: 431–440.

Geraci, J.R. and D.J. St. Aubin. 1987. Effects of parasites on marine mammals. *International Journal of Parasitology* 17: 407–414.

Gittings, S.R., G.D. Denis, and H.W. Harry. 1986. Annotated guide to the barnacles of the northern Gulf of Mexico. *Texas A&M University Sea Grant College Program* 86–402: 1–36.

Grant, C. 1956. Aberrant lamination in two hawksbill turtles. *Herpetologica* 12: 302.

Gray, J.E. 1825. A synopsis on the genera of Cirripedes arranged in natural families, with a description of some new species. *Annals of Philosophy* 10: 97–107.

Greef, S.R. 1885. Ueber die Fauna der Guinea-Inseln S. Thome und Rolas. *Sitzungsberichte der Gesellschaft zur Beforderung der gesammten Naturwissenschaften zu Marburg* 41–80.

Guess, R.C. 1982. Occurrence of a Pacific loggerhead turtle, *Caretta caretta gigas* Deraniyagala, in the waters off Santa Cruz Island, California. *California Fish and Game* 68: 122–123.

Gutmann, W.F. 1960. Funktionelle morphologie van Balanus balanoides. *Abhandlungen der Senckenbergischen Naturforschenden Gesellschaft* 500: 1–43.

Healy, J.M. and D.T. Anderson. 1990. Sperm ultrastructure in the Cirripedia and its phylogenetic significance. *Records of the Australian Museum* 42: 1–26.

Hiro, F. 1937b. Studies on the cirripedien fauna of Japan. II. Cirripeds found in the vicinity of the Seto Marine Biological Laboratory. *Memoirs of the College of Science, University of Kyoto (Series B)* 12: 385–478.

Jackson, C.G., Jr. and A. Ross. 1971. The occurrence of barnacles on the alligator snapping turtle, *Macroclemys temmincki* (Troost). *Journal of Herpetology* 5: 188–189.

Jackson, C.G., Jr. and A. Ross. 1972. Balanomorph barnacles on *Chrysemys alabamensis*. *Quarterly Journal of the Florida Academy of Science* 35: 173–176.

Jackson, C.G., Jr. and A. Ross. 1975. Epizoic occurrence of a bryozoan, *Electra crustulenta*, on the turtle *Chrysemys alabamensis*. *Transactions of the American Microscopy Society* 94: 135–136.

Jackson, C.G., Jr., A. Ross, and G.L. Kennedy. 1973. Epifaunal invertebrates of the ornate diamondback terrapin, *Malaclemys terrapin macrospilota*. *American Midland Naturalist* 89: 495–497.

Johnson, T.W., Jr. and R.R. Bonner. 1960. *Lagenidium callinectes* Couch in barnacle ova. *Journal of the Elisha Mitchell Society* 76: 147–149.

Kadovich, J. 1961. Relationship of some marine organisms of the northeast Pacific to water temperatures, particularly during 1957 through 1959. *California Fish and Game, Fisheries Bulletin* 112: 1–62.

Karuppiah, S., A. Subramanian, and J.P. Obbard. 2004. The barnacle, *Xenobalanus globicipitis* (Cirripedia, Coronulidae), attached to the bottle-nosed dolphin, *Tursiops truncatus* (Mammalia, Cetacea) on the southeastern coast of India. *Crustaceana* 77: 879–882.

Kasuya, T. and D.W. Rice. 1970. Notes on the baleen plates and arrangement of parasitic barnacles of the gray whale. *Scientific Reports of the Whales Research Institute* 22: 39–43.

Kato, M., K. Hayasaka, and T. Matsuda. 1960. Ecological studies on the morphological variation of a sessile barnacle, *Chthamalus challengeri*. III. Variation of the shell shape and of the inner anatomical features introduced by the population density. *Bulletin of the Marine Biological Station of Asamushi, Tohoku University* 10: 19–25.

Key, M.M., Jr., J.W. Volpe, W.B. Jeffries, and H.K. Voris. 1997. Barnacle fouling of the blue crab *Callinectes sapidus* at Beaufort, North Carolina. *Journal of Crustacean Biology* 17: 424–439.

Kim, I.H. and H.S. Kim. 1980. Systematic studies on the cirripeds (Crustacea) from Korea. 1. Balanomorph barnacles (Cirripedia, Thoracica, Balanomorpha). *Korean Journal of Zoology* 23: 161–194.

Kitsos, M., M. Christodoulou, S. Kalpakis, M. Noidou, and A. Koukouras. 2003. Cirripedia Thoracica associated with *Caretta caretta* (Linnaeus, 1758) in the northern Aegean Sea. *Crustaceana* 76: 403–409.

Klepal, W. 1987. A review of the comparative anatomy of the males in cirripedes. *Oceanography and Marine Biology: Annual Review* 25: 285–351.

Kolosvary, G. 1942. Studien an Cirripedien. *Zoologischer Anzeige* 137, pp. 138–150.

Kolosvary, G. 1947. Der balaniden der Adria. *Annales Historico-Naturales Musei Nationalis Hungarici* 39: 1–88.

Kruger, P. 1911a. Beitrage zur Cirripedien fauna Ostasiens. *Abhandlungen der Mathematisch-Physikalische Klasse der Königlich Bayerischen Akademie der Wissenschaften,* II Supplement-Band 6: 1–72.

Lamarck, J.B.A. de M. de. 1802. Mémoire sur la Tubicinelle. *Annales du Museum National d'Histoire Naturelle* 1: 461–464.

Lang, W.H. 1979. Larval development of shallow water barnacles of the Carolinas (Cirripedia: Thoracica) with keys to naupliar stages. NOAA-Technical Report-NMFS Circular-421, pp. 1–39.

Leach, W.E. 1817. Distribution systematique de la class Cirripèdes. *Journal de Physique de Chimie et d'Histoire Naturelle* 85: 67–69.

Leach, W.E. 1818. Cirripedes. In: Supplement to the fourth and fifth editions of the *Encyclopedia Britannica* 3: 168–171.

Lezama, C., M. Lopez-Mendilaharsu, F. Scarabino, A. Estrades, and A. Fallabrino. 2006. Interaction between the green sea turtle (*Chelonia mydas*) and an alien gastropod (*Rapana venosa*) in Uruguay. In M.G. Frick, A. Panagopoulou, A.F. Rees and K. Williams, compilers. *Proceedings of the 26th Annual Symposium on Sea Turtle Biology and Conservation,* Miami, FL, pp. 64–65. ISBN: 960-87926-1-4.

Linnaeus, C. 1767. Systema naturae per regna tria naturae—Edito duodecima, reformata. *Holmiae* 1: 533–1327.

Lutcavage, M.E. and J.A. Musick. 1985. Aspects of the biology of sea turtles in Virginia. *Copeia* 1985: 449–456.

Meischner, D. 2001. Seepocken auf einer Meeres-Schildkröte, ein ökologisches Idyll. *Nature und Museum* 131: 1–7.

Mignucci-Giannoni, A.A., C.A. Beck, R.A. Montoya-Ospina and E.H. Williams Jr. 1999. Parasites and commensals of the West Indian manatee from Puerto Rico. *Journal of the Helminthological Society of Washington* 66: 67–69.

Miranda, L. and R.A. Moreno. 2002. Epibionts from *Lepidochelys olivacea* (Eschscholtz, 1829) (Reptilia: Testudinata: Cheloniidae) in the central south region of Chile. *Revista de Biologia Marina y Oceanografia* 37: 145–146.

Mokaday, O., Y. Loya, Y. Achituv, E. Geffen, D. Graur, S. Rozenblatt and I. Bricker. 1999. Speciation versus phenotypic plasticity in coral inhabiting barnacles: Darwin's observation in an ecological context. *Journal of Molecular Evolution* 49: 367–375.

Monroe, R. 1981. Studies on the Coronulidae (Cirripedia): Shell morphology, growth, and function, and their bearing on subfamily classification. *Memoirs of the Queensland Museum* 20: 237–251.

Mörch, O.A.L. 1852. Cephalophora Catalogus Conchyliorum (Cirripedia) 1: 65–68.

Newman, W.A. 1996. Sous-classe des Cirripèdes (Cirripedia Burmeister, 1834), Super-ordres des Thoraciques et des Acrothoraciques (Thoracica Darwin, 1854—Acrothoracica Gruvel, 1905). *Traité de Zoologie* 7: 453–540.

Newman, W.A. and A. Ross. 1971. Antarctic Cirripedia. *American Geophysical Union, Antarctic Research Series* 14: 1–257.

Newman, W.A. and A. Ross. 1976. Revision of the balanomorph barnacles; including a catalog of the species. *San Diego Society of Natural History* 9: 1–108.

Newman, W.A. and A. Ross. 1977. A living *Tesseropora* (Cirripedia: Balanomorpha) from Bermuda and the Azores: First records from the Atlantic since the Oligocene. *Transactions of the San Diego Society of Natural History* 18: 207–216.

Newman, W.A. and A. Ross. 2001. Prospectus on larval cirriped setation formulae, revisited. *Journal of Crustacean Biology* 21: 56–77.

Nilsson-Cantell, C.A. 1930b. Cirripedes. In: Resultates Scientifiques du Voyage aux Indes Orientales Neerlandaises de LL. AA. RR. le Prince et la Princesse Leopold de Belgique. *Mémoires du Musee Royal d'Histoire Naturelle de Belgique* 3: 1–24.

Nilsson-Cantell, C.A. 1957. Thoracic cirripeds from Chile. *Lunds Universitets Arsskrift N.F. (Series 2),* 53: 1–25.

Nogata, Y. and K. Matsumura. 2006. Larval development and settlement of a whale barnacle. *Biology Letters* 2: 92–93.

Orams, M.B. and C. Schuetze. 1998. Seasonal and age/size-related occurrence of a barnacle (*Xenobalanus globicipitis*) on bottlenose dolphins (*Tursiops truncatus*). *Marine Mammal Science* 14: 186–189.

Ortiz, M., R. Lalana, and C. Varela. 2004. Caso extremo de epibiosis de escaramujos (Cirripedia: Balanomorpha), sobre una esquilla (Hoplocarida: Stomatopoda), en Cuba. *Revista de Investigaciones Marinas* 25: 75–76.

Pasternak, Z, A. Abelson, and Y. Achituv. 2002. Orientation of *Chelonibia patula* (Crustacea: Cirripedia) on the carapace of its crab host as determined by the feeding mechanism of the adult barnacles. *Journal of the Marine Biological Association of the United Kingdom* 82: 583–588.

Peterson, M.N.A. 1966. Calcite: Rates of dissolution in a vertical profile in the central Pacific. *Science* 154: 1542–1544.

Pilsbry, H.A. 1910. *Stomatolepas*, a commensal barnacle in the throat of the loggerhead turtle. *American Naturalist* 44: 304–306.

Pitombo, F.B. 2004. Phylogenetic analysis of the Balanidae (Cirripedia: Balanomorpha). *Zoologica Scripta* 33: 261–276.

Rees, E.I.S. and G. Walker. 1977. A record of the turtle barnacle *Chelonobia* [sic] *testudinaria* (L.) in the Irish Sea. *Porcupine Newsletter* 5: 189.

Rice, D.W. and A.A. Wolman. 1971. The life history and ecology of the gray whale (*Eschrichtius robustus*). *American Society of Mammalogists, Special Publication* 3: 1–142.

Ridgway, S.H., E. Linder, K.A. Mahoney, and W.A. Newman. 1997. Grey whale barnacles *Cryptolepas rhachinecti* infest white whales, *Delphinapterus leucas*, housed in San Diego Bay. *Bulletin of Marine Science* 61: 377–385.

Riedl, R. 1963. Fauna und Flora der Adria. Cirripedia only, pp. 10–15, pls. 1–2, pp. 18–19, 252–258, Figs. 83–84, 2 maps. Paul Parey Verlag. Hamburg.

Ross, A. 1963a. A new Pleistocene *Platylepas* from Florida. *Quarterly Journal of the Florida Academy of Science* 26: 150–158.

Ross, A. 1963b. *Chelonibia* in the Neogene of Florida. *Quarterly Journal of the Florida Academy of Science* 26: 221–233.

Ross, A. 1964. Type locality of *Platylepas wilsoni* Ross. *Quarterly Journal of the Florida Academy of Science* 27: 278.

Ross, A. and W.K. Emerson. 1974. *Wonders of Barnacles*. Dodd, Mead and Co., New York.

Ross, A. and M.G. Frick. 2007. From Hendrickson (1958) to Monroe and Limpus (1979) and beyond: An evaluation of the turtle barnacle *Tubicinella cheloniae*. *Marine Turtle Newsletter* 118: 2–5.

Ross, A. and W.A. Newman. 1995. A coral-eating barnacle, revisited (Cirripedia, Pyrgomatidae). *Contributions to Zoology* 65: 129–175.

Ross, A. and W.A. Newman. 2000. *Pyrgoma kuri* Hoek, 1913: a case study in morphology and systematics of a symbiotic coral barnacle (Cirripedia: Balanamorpha). *Contributions to Zoology* 68: 245–260.

Ryder, J.A. 1879. Strange habitat of a barnacle on a gar pike. *American Naturalist* 8: 453.

Samaras, W.F. and F.E. Durham. 1985. Feeding relationship of two species of epizoic amphipods and the gray whale, *Eschrichtius robustus*. *Bulletin of the Southern California Academy of Sciences* 84: 113–126.

Scaravelli, D. 1998. Segnalazioni faunistiche. 29. *Stomatolepas elegans* (O.G. Costa 1838) (Crustacea Thoracica Balanidae). *Quaderno di Studi e Notizie di Storia Naturale Della Romagna* 10: 78.

Scarff, J.E. 1986. Occurrence of the barnacles, *Coronula diadema*, *C. reginae* and *Cetopirus complanatus* (Cirripedia) on right whales. *Scientific Reports of the Whales Research Institute* 37: 129–153.

Schmitt, W.L. 1965. *Crustaceans*. University of Michigan Press. Ann Arbor, MI.

Seigel, R.A. 1983. Occurrence and effects of barnacle infestation on diamondback terrapins (*Malaclemys terrapin*). *American Midland Naturalist* 109: 34–39.

Shields, J.D. 1992. Parasites and symbionts of the crab *Portunus pelagicus* from Moreton Bay, eastern Australia. *Journal of Crustacean Biology* 12: 94–100.

Southward, A.J. 1986. Class Cirripedia (barnacles). In W. Sterrer, ed. *Marine Fauna and Flora of Bermuda. A Systematic Guide to the Identification of Marine organisms*. John Wiley & Sons, New York, pp. 299–305.

Southward, A.J., ed. 1987. *Barnacle Biology*. A.A. Balkema, Rotterdam, the Netherlands, i–xxii + 443pp.

Southward, A.J. 1998. New observations on barnacles (Crustacea: Cirripedia) of the Azores region. Arquipelago. *Life and Marine Sciences* 16: 11–27.

Spears, T.L., L.G. Abele, and M.A. Applegate. 1994. A phylogenetic study of cirripeds and their relatives (Crustacea: Thecostraca). *Journal of Crustacean Biology* 14: 641–656.

Steenstrup, J.J.S. 1851. Videnskabelige Meddelelser fra den Naturhist. Forening i Kjöbenhavn, for Aaret, 1851. Table 3, Fig. 11–15.

Stunkard, H.W. 1955. Freedom, bondage and the welfare state. *Science* 121: 811–816.

Walker, L.W. 1949. Nursery of the gray whales. *Natural History* 58: 248–256.

Weltner, W. 1895. Die Cirripedien von Patagonien, Chile and Juan Fernandez. *Archiv für Naturgeschicthe* 61: 288–292.

Williams, K.L. and M.G. Frick. 2008. Tag returns from loggerhead turtles from Wassaw Island, GA. *Southeastern Naturalist* 7: 165–172.

Williams, A.B. and H.J. Porter. 1964. An unusually large turtle barnacle (*Chelonibia patula*) on a blue crab from Delaware Bay. *Chesapeake Science* 5: 150–153.

Withers, T.H. 1928. The cirriped *Chelonibia caretta* Spengler, in the Miocene of Zanzibar Protectorate. *Annals and Magazine of Natural History, Series* 10, 2: 390–392.

Withers, T.H. 1929. The cirriped *Chelonibia* in the Miocene of Gironde, France and Vienna, Austria. *Annals and Magazine of Natural History, Series* 10, 4: 566–572.

Withers, T.H. 1953. *Catalogue of Fossil Cirripedia in the Department of Geology*, Vol. 3, Tertiary. British Museum (Natural History), 396 pp.

Yasuyuki, N. and K. Matsumura. 2006. Larval development and settlement of a whale barnacle. *Biology Letters* 2: 92–93.

Young, P.S. 1991. The superfamily Coronuloidea Leach (Cirripedia, Balanomorpha) from the Brazilian coast, with redescription of *Stomatolepas* species. *Crustaceana* 61: 190–212.

Zangerl, R. 1948. The vertebrate fauna of the Selma Formation of Alabama. Part I. *Fieldiana: Geology Memoirs* 3: 116.

Zann, L.P. 1975. *Biology of a Barnacle (Platylepas ophiophilus Lanchester) Symbiotic with Sea Snakes.* University Park Press, Baltimore, MA, pp. 267–286.

Zullo, V.A. 1963. A preliminary report on systematic and distribution of barnacles (Cirripedia) of the Cape Cod region. Marine Biology Laboratory, Woods Hole, MA. Systematics Ecology Program, 33pp.

Zullo, V.A. 1969. Thoracic Cirripedia of the San Diego formation, San Diego County, California. *Contributions in Science Las Angeles County Museum* 159: 1–25.

Zullo, V.A. 1979. Marine flora and fauna of the northeastern United States, Arthropoda: Cirripedia. *NOAA Technical Report NMFS Circular* 425: 1–29.

Zullo, V.A. 1982. A new species of the turtle barnacle *Chelonibia* Leach, 1817, (Cirripedia, Thoracica) from the Oligocene Mint Spring and Byram Formations of Mississippi. *Mississippi Geology* 2: 1–6.

# 16 Parasites of Marine Turtles

*Ellis C. Greiner*

## CONTENTS

# 16.1  INTRODUCTION

The parasite fauna of marine turtles is varied and diverse. It includes helminths, protozoa, arthropods, and annelids. The amazing thing is that some of these marine turtles migrate great distances through the oceans and yet they must reside in waters that allow them to be infected by microscopic infective stages at some point in their journeys. These parasites are evidently well adapted to their reptilian hosts as some are very common parasites and others are more rarely detected. The parasites have been associated with these turtles for a very long time as many organs of the turtles harbor parasites. Marine turtles are usually the final or definitive host of the parasites, but sometimes they serve as a source of infection to other vertebrates that prey upon them. Due to the intended shortness of this chapter, focus will be on the parasites in the marine turtles in Florida waters as work on these turtles has been done in my former laboratory over the past 25 years. There is a wealth of information on parasites of marine turtles, but much of it is based on examinations of one or few turtles. My research reflects taking advantage of turtles that have died in Florida. Carcasses were submitted to my laboratory through cooperation with turtle rehabilitation facilities and state and federal turtle biologists. This work was done through federal permits held by Dr. Elliott R. Jacobson.

## 16.1.1  Methods of Helminth Recovery and Parasite Diagnosis

The plastron and legs were removed from the carcass and the viscera were removed into a container. The entire gastrointestinal system was separated from the remainder. The esophagus and stomach were placed into separate containers and the intestine was divided into equal thirds and each section labeled and isolated. Each section was slit open and the mucosa was washed to free most parasites into the container and then the mucosa was visually inspected. One end of a section of intestine was then placed between two fingers of one hand and the entire length was pulled through the fingers to strip remaining worms and ingesta into the container. Solid organs like the liver and kidney were sliced and pressed in a container with water to allow worms in the blood vessels and bile ducts to be expressed into the water. The urinary bladder was removed and opened into its own container. The heart and associated major vessels were slit open and washed into a container. The contents of each container were then placed into a standard #50 sieve separately and the contents were washed with gentle water spray. This process cleaned and concentrated the worms. The sieve was back flushed into a container and contents were then viewed in a light box. The worms were removed, separated by species, counted, fixed, and stored in 70% ethanol in vials with host ID and organ and species name. It becomes obvious that helminths have a primary normal site of infection within the turtle, but you must realize that post mortem these worms move into adjacent organs. So, depending upon the time span between the turtle death and the necropsy, where the parasites might be found might change.

The primary means of diagnosing mature helminths is by fecal examination to detect eggs. Two basic methods are used. The most common is the fecal flotation in which a salt or sugar solution of an appropriate specific density will be used to cleanse and concentrate many helminth eggs, protozoan cysts, and coccidian oocysts. I prefer sodium nitrate, available commercially as Fecasol, as I find it floats out a diversity of parasite stages. Whereas some parasitologists feel one must use centrifugation to find these stages, I disagree and use a standing flotation. The problem with marine turtle parasite diagnostics is that the majority of helminths in these hosts are trematodes. Their eggs do not float and thus we use a fecal sedimentation method in which the cleansing and concentrating are done by allowing the eggs to sediment to the bottom of a tube, rather than float to the top. A very simple system is to place a drop or two of liquid dish soap in a 500 mL squeeze bottle, fill it with water, and shake to mix. Feces (1–2 g if possible) is placed into a small container such as a urine sample cup and 30–40 mL of the soapy water is added. The feces must be broken apart as well as possible. Place a single layer of gauze over the opening and pour the contents into a 50 mL centrifuge tube. You may need to rewash the sediment back into the container and strain the contents into the tube a second time. Do not allow the tube to overflow. Place the tube in a rack so it remains vertical and allow it to stand for 5 min. Decant or aspirate the fluid down close to the sediment and then refill with the soapy water. You will repeat this twice and then switch from the soapy water to plain water as this will reduce the soap bubble problem. Continue the process until the diluent remains clear and then let it settle once more. Aspirate off the diluent down to the sediment. All of the aspirated diluent may be discarded. Place some of the sediment onto a microscope slide and add a coverslip. If you add too much sediment, you may miss the eggs. Scan the coverslip systematically until it is all viewed. You may wish to make and scan more than one slide preparation. You should also know that all diagnostic stages including those that float, such as nematode eggs and oocysts, will sediment. Some eggs will be numerous and others will be in low numbers. Compare what you see with either procedure to the photos of eggs of some of the helminths of turtles (Figure 16.1, **23** through **50**). If you are able to examine very fresh feces, you may also use a direct smear procedure to look for motile protozoa. Place a few drops of normal saline on a microscope slide and with an applicator stick, add some feces and mix it thoroughly in the saline and then add a coverslip. Then scan the entire coverslip as described earlier. You will possibly find eggs on these slides, but not in the quantity you will after concentrating them. You need to have an ocular micrometer on your microscope that is calibrated so you can measure the eggs as that is a crucial element in the identification of these eggs along with shape, and the presence of absence of an operculum.

### 16.1.2 TREMATODES

Flukes are the most diverse and numerous parasites in marine turtles. Most of these in marine turtles are digenetic trematodes (flukes), which have complex life cycles sometimes requiring one to three other species of animal to complete their life history. They are flat worms, no body cavity is present, and both sets of reproductive systems are present in each adult. They may be found in most body organs in these turtles, but are usually restricted to a primary site for development. Because different species of marine turtles eat different things as a normal dietary regimen, this might influence the diversity of flukes present in turtles, as those that have a more select or restricted diet such as hawksbills and leatherbacks seem to have a less diverse fauna than omnivores as loggerheads. Flukes exist in all the major marine turtle species and sometimes numbered in the thousands per infected host. A single species of aspidograstrid trematode, a distinct and separate type of fluke, occurs only in the loggerhead.

### 16.1.3 NEMATODES

Nematodes are another component of the helminthic community and these are not nearly as diversified as the trematodes in marine turtles. They are referred to as round worms as they are tubular in morphology, they have a body cavity, and have a single sex in each adult. Their life cycles may be complex, but the most of those in marine turtles usually only infect the turtles and have direct life cycles. Most nematode parasites of marine turtles reside in the gastrointestinal tract, but one taxon lives in the lungs. Infections of one species of nematode (*Anisakis*) resulted from captive greens being fed sardines and larvae did not develop to maturity and were found embedded in the stomach walls at necropsy causing gastric ulcerations (Burke and Rogers, 1962), and this parasite has been reported in free-ranging turtles in the Mediterranean region and will be discussed later.

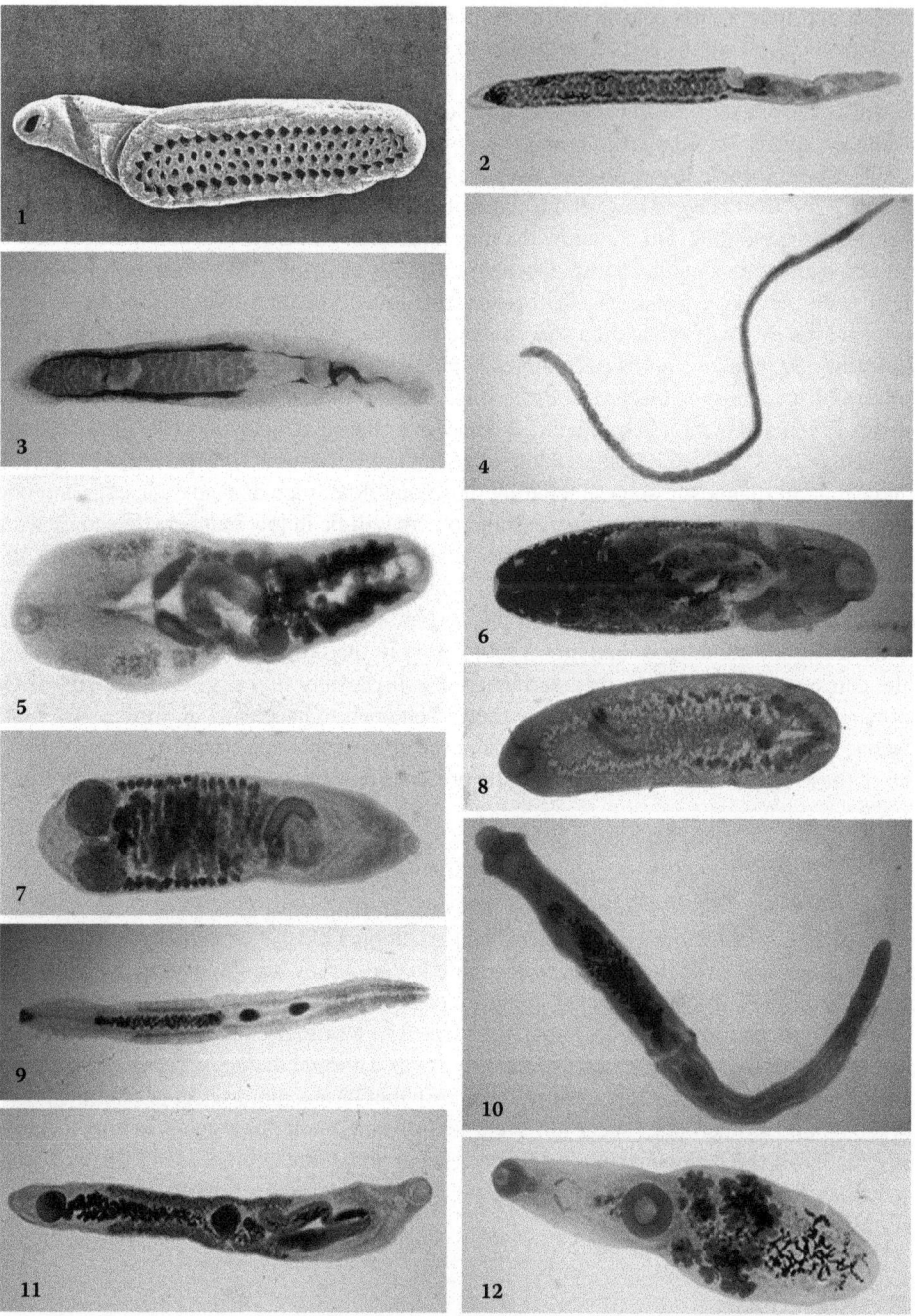

**FIGURE 16.1** Identifying adult trematodes requires measurements of size, examination of the shape, realizing the relative position and number of the testes, position of the ovary, extent and position of the vitellaria, one versus two suckers and their location on the body, length of the intestinal caeca, and characteristics of the termination region of the male. Some of these flukes may not be in proper proportion with respect to others in order to assemble the Plate. Figures 1 through 12 are adult flukes from loggerheads and 13 through 22 are from green turtles. **1**. *Lophotaspis vallei*, 7 mm, **2**. *Carettacola bipora*, 8 mm, **3**. *Hapalotrema mistroides*, 5 mm, **4**. *Neospirorchis pricei*, 10 mm, **5**. *Cymatocarpus undulatus*, 6 mm, **6**. *Pachypsolus irroratus*, 4 mm, **7**. *Pleurogonius trigonocephalus*, 4 mm, **8**. *Diaschistorchis pandus*, 5 mm, **9**. *Rhytidodes gelatinosus*, 8 mm, **10**. *Calycodes anthos*, 12 mm, **11**. *Orchidasma amphiorchis*, 9 mm, **12**. *Plesiochorus cymbiformis*, 6 mm,

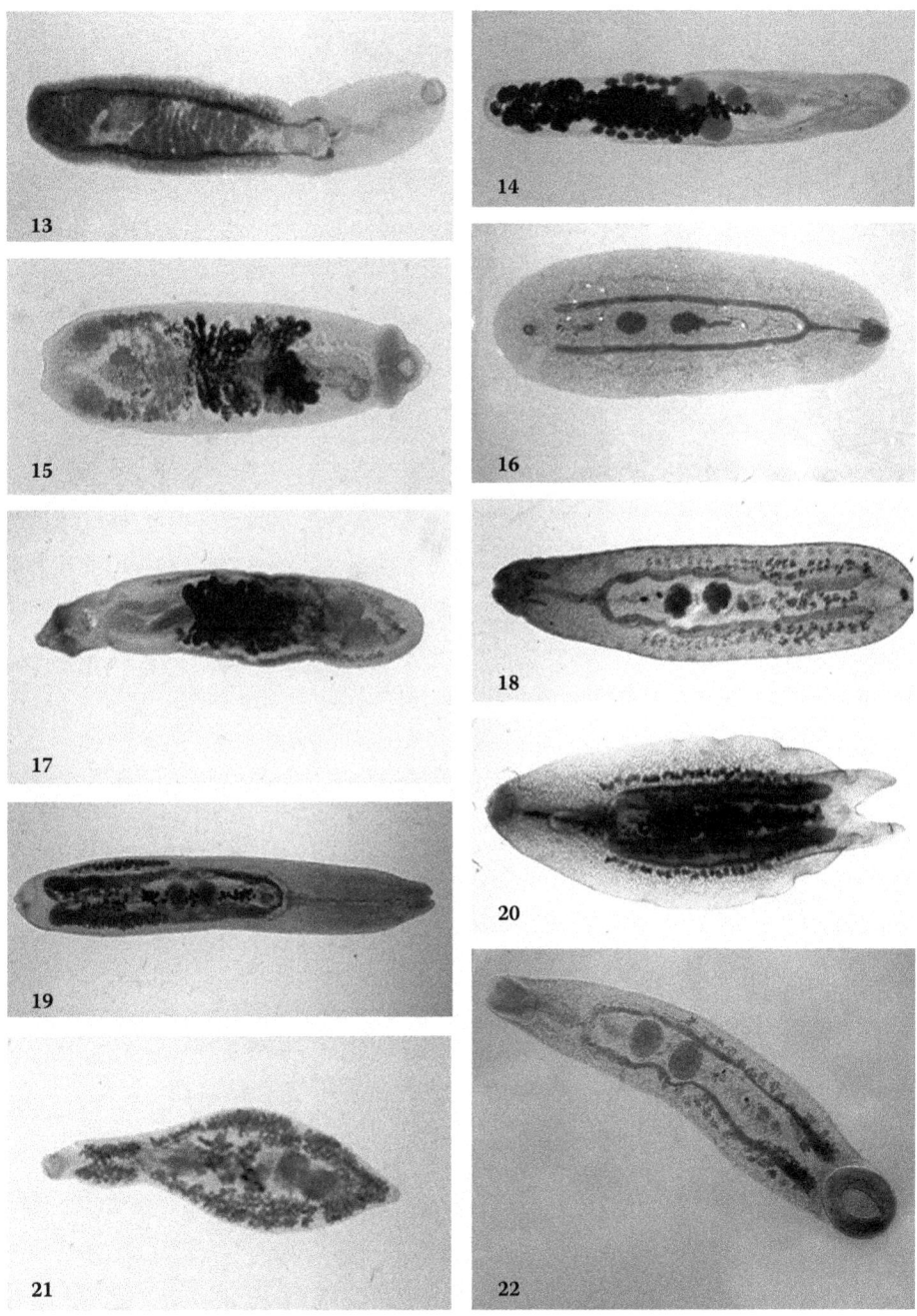

FIGURE 16.1 (continued)   Identifying adult trematodes requires measurements of size, examination of
the shape, realizing the relative position and number of the testes, position of the ovary, extent and position
of the vitellaria, one versus two suckers and their location on the body, length of the intestinal caeca, and
characteristics of the termination region of the male. Some of these flukes may not be in proper proportion
with respect to others in order to assemble the Plate. Figures 1 through 12 are adult flukes from loggerheads
and 13 through 22 are from green turtles. **13**. *Learedius learedi*, 5 mm, **14**. *Enodiotrema megachondrus*,
9 mm, **15**. *Cricocephalus albus*, 3.5 mm, **16**. *Deuterobarus proteus*, 8 mm, **17**. *Pronocephalus obliquus*,
4.5 mm, **18**. *Microscaphidium reticulare*, 5 mm, **19**. *Angiodyctium parallelum*, 8 mm, **20**. *Octangium
sagitta*, 4 mm, **21**. *Rhytidoides similis*, 5 mm, **22**. *Schizamphistomoides spinulosus*, 7 mm,

*(continued)*

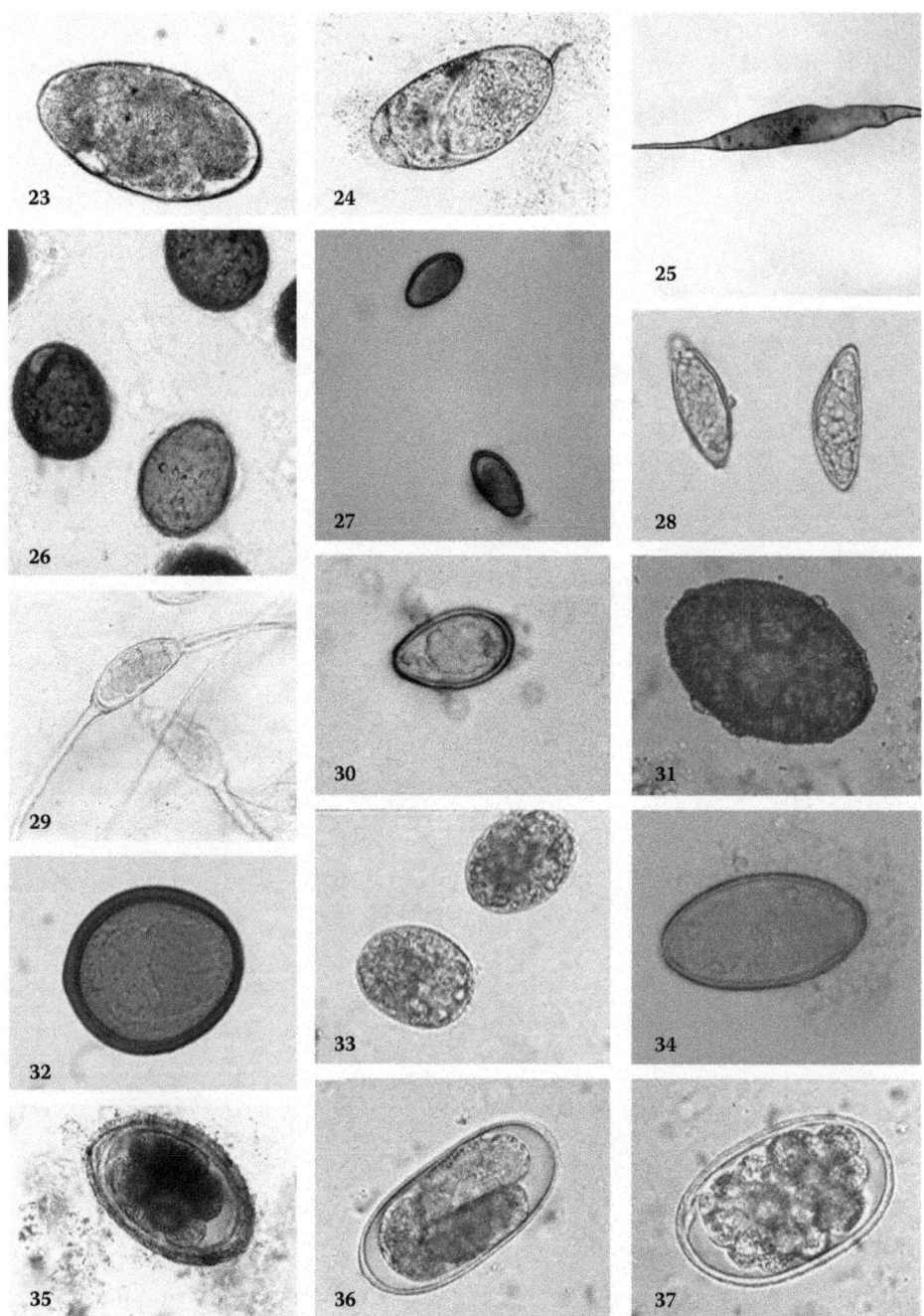

FIGURE 16.1 (continued)  Fluke eggs are based on size and morphology. Most of them have an operculum (trap door for larval escape, see Figure 45 which has its operculum open), what has developed inside the egg, shape of the egg, and if there are polar spines or filaments and how many and their length. **23.** *Lophotaspis vallei* egg 169×68 µm, **24.** *Carettacola bipora* egg 88×38 µm, **25.** *Hapalotrema mistroides* egg 414×36 µm, **26.** *Neospirorchis* sp. egg 45×35 µm. **27.** *Cymatocarpus undulatus* egg 37×16 µm, **28.** *Pachypsolus irroratus* egg 51×18 µm, **29.** *Pleurogonius sindhi* egg capsule 52×22 µm, **30.** *Diaschistorchis pandus* egg 40×22 µm, **31.** *Rhytidodes gelatinosus* egg 75×53 µm, **32.** *Orchidasma amphiorchis* egg 60×50 µm, **33.** *Plesiochorus cymbiformis* egg 41×28 µm. **34.** *Styphlotrema solitaria* egg 32×20 µm, **35.** *Sulcascaris sulcata* egg 91×57 µm. **36.** *Tonaudia/Kathlania* egg 104×55 µm, **37.** *Cuculanus carretae* egg 91×57 µm,

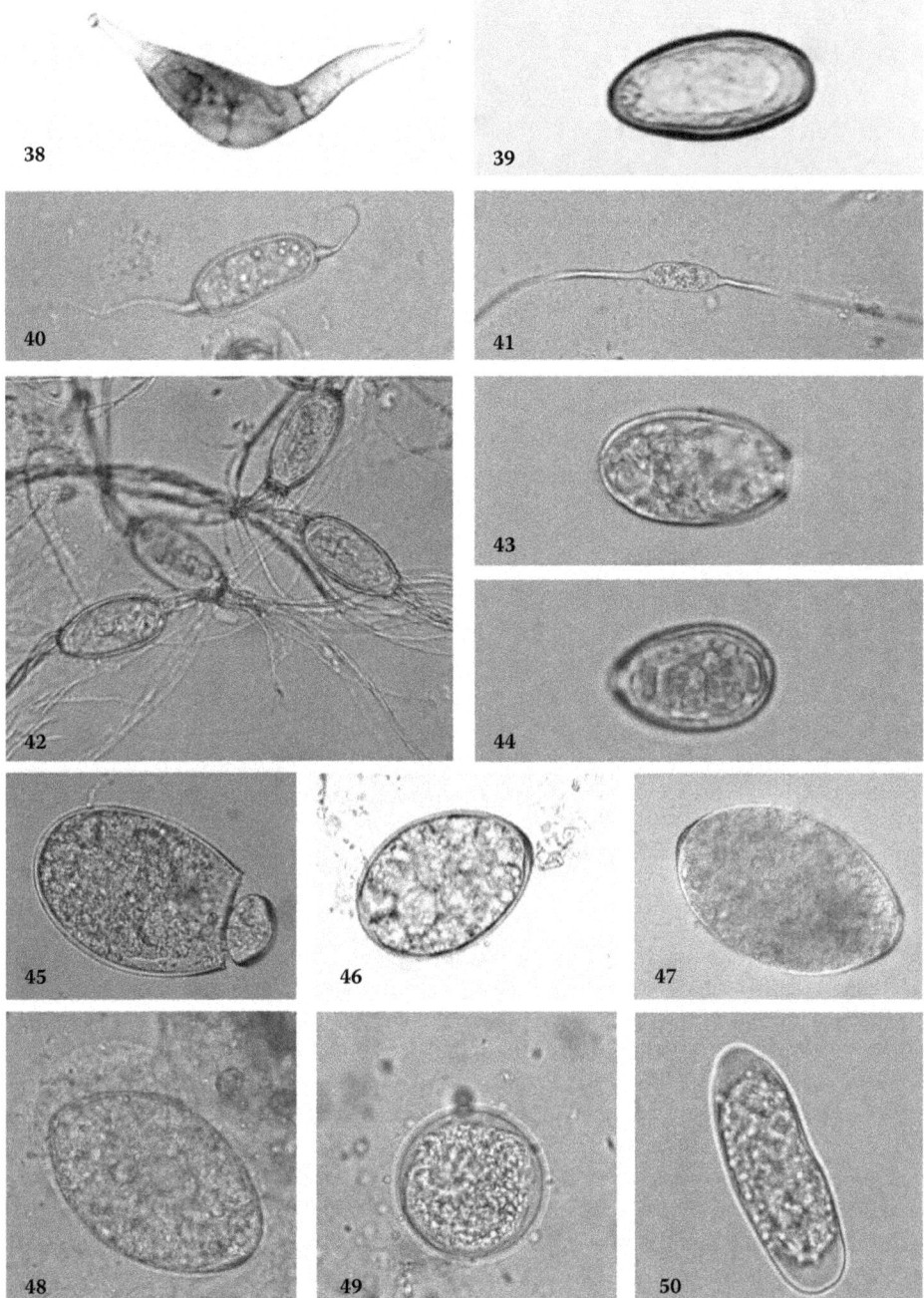

FIGURE 16.1 **(continued)** Fluke eggs are based on size and morphology. Most of them have an opercu-lum (trap door for larval escape, see Figure 45 which has its operculum open), what has developed inside the egg, shape of the egg, and if there are polar spines or filaments and how many and their length **38**. *Learedius learedi* egg 297×42 μm, **39**. *Enodiotrema megachondrus* egg capsule 42×21 μm, **40**. *Cricocephalus albus* egg capsule 33×14 μm, **41**. *Pronocephalus obliquus* egg capsule 33×14 μm, **42**. *Desmogonius desmogonius* egg capsule 32×13 μm, **43**. *Charaxicephalus robustus* egg 26×16 μm 44. *Angiodyctium parallelum* egg 25×15 μm, **45**. *Deuterobarus proteus* egg 94×34 μm, **46**. *Rhytidodes gelatinosus* egg 68×37 μm, **47**. *Schizamphistomum sceloporum* egg, 98×60 μm, **48**. *Octangium sagitta* egg 89×57 μm, **49**. *Balantidum bacteriophorus* cyst 58 μm, and **50**. *Caryospora cheloniae* sporocyst 41×17 μm.

### 16.1.4 CESTODES

Adult cestodes are segmented and are flat worms with solid bodies and both sexes are in the same individual worm. Tapeworms need an intermediate host to complete their life cycles. Marine turtles function as intermediate hosts for tapeworms (trypanorhynchs) that mature in sharks. The shark becomes infected by feeding on turtles containing these larvae. The ornate larvae have four long spiny tentacles, which anchor the worm into the intestinal mucosa of the final host. These larvae may be distributed throughout the peritoneal cavity, often associated with the mesenteries or attached to the peritoneal surface of the gut.

### 16.1.5 PROTOZOA

While there are many types of these single-celled eukaryotic parasites, only a few have been reported from marine turtles. Two species of coccidian have been named from marine turtles along with one species of ciliate. Amoebae and flagellates also have been reported. All of these appear to have a direct life cycle; and while some develop within cells, others remain in the lumen of the intestines.

### 16.1.6 ANNELIDS AND ARTHROPODS

Some parasites live inside the host and others live on the outside and the latter are ectoparasites. Most arthropods would be ectoparasites, but one mite is internal and that is *Chelonacarus elongatus*, which lives in the wall of the cloaca of green turtles (Pence and Wright, 1998). Another arthropod, which is associated with marine turtles occasionally, is a biting midge. The concept of female turtles being fed upon by hematophagous arthropods while they are on shore laying eggs has been demonstrated. A biting midge (*Culicoides phlebotomus*) has been recovered feeding on nesting leatherbacks in Costa Rico (Borkent, 1995). The epibiota will be dealt with in Chapter 10 by Frick and Pfaller and will not be covered here.

### 16.1.7 PARASITOLOGICAL PRINCIPLES

The intrigue of parasites is amazing. Many questions remain unsolved, such as how host-specific these helminths are. We know that many parasites demonstrate loose host specificity, but some are only able to infect a single definitive host species. Others fit on a continuum between these extremes. Examples of species-specific host specificity in marine turtles would include *Lophotaspis vallei*, *Learedius learedi*, and *Styphlotrema solitaria* (see Tables 16.1, 16.2, 16.3 and 16.4). Species with a wider host range would include *Plesiochorus cymbiformis*, *Rhytidodes gelatinosus*, *Enodiotrema carettae*, and *Pleurogonius trigonocephalus* (see Tables 16.1, 16.2, 16.3 and 16.4). These conclusions are based on findings from turtles examined in Florida and other regions of the world as will be realized in the discussions on individual turtles. How much harm these parasites cause to their final hosts is poorly understood. We know that the Spirorchiidae (blood flukes) cause a great deal of damage (Stacy et al., 2010a). There might be a threshold in numbers of a single species when the host will suffer consequences and when one adds up insults caused by the presence of several parasite species. The extent of damage could vary from case to case depending on the species involved. The site where most of the individuals of a species were recovered is indicated in the Tables, but this does not show the range of habitats utilized by each species. The beauty of this variation is that the helminthic community has evolved to use a wide range in habitats and niche selection. Moreover, not all parasites infect all hosts. Variation in the intensity of infection may be influenced by the complexity of the life cycle, the availability of intermediate hosts where they are needed, interspecific effects of one parasite species on another, immune response by the host, and turtle population density. Obviously if too many of a pathogenic species of helminth are present,

TABLE 16.1

**Summary of Adult Worms from 44 Loggerheads 1991–2006 from Florida**

| Species | #Pos. | Prevalence (%) | Intensity Mean | Intensity Range | Total Worms | Normal Habitat[a] |
|---|---|---|---|---|---|---|
| Lophotaspis vallei | 10 | 22.7 | 30.7 | 2–119 | 307 | St |
| Carettacola bipora | 19 | 43.1 | 4.9 | 1–13 | 94 | H, Ar |
| Hapalotrema mistroides | 11 | 25.0 | 2.6 | 1–9 | 29 | L v |
| Neospirorchis pricei | 3 | 6.8 | 11.0 | 10–12 | 33 | H, Ar |
| Cymatocarpus undulatus | 25 | 56.8 | 288.8 | 2–1,350 | 7,224 | UI |
| Pachypsolus irroratus | 14 | 31.8 | 54.6 | 3–202 | 764 | Es, St |
| Enodiotrema megachondrus | 13 | 29.5 | 57.9 | 2–574 | 753 | St |
| Enodiotrema carettae | 16 | 36.3 | 7.7 | 1–42 | 123 | UI |
| Pleurogonius trigonocephala | 24 | 54.5 | 51.7 | 1–289 | 1,280 | MI |
| Diaschistorchis pandus | 11 | 25.0 | 53.1 | 2–138 | 584 | Es, St |
| Pyelosomum renicapite | 6 | 13.6 | 16.8 | 1–54 | 101 | MI, LI |
| Pyelosomum chelonei | 4 | 9.1 | 79.5 | 2–286 | 318 | LI |
| Rhytidodes gelatinosus | 23 | 52.6 | 161 | 1–882 | 3,703 | UI |
| Styphlotrema solitaria | 11 | 25.0 | 70.1 | 1–395 | 771 | UI |
| Orchidasma amphiorchis | 25 | 56.8 | 825.8 | 6–3,918 | 20,644 | UI |
| Calycodes anthos | 12 | 27.3 | 61.6 | 1–164 | 739 | L, GB |
| Plesiochorus cymbiformis | 30 | 68.2 | 68.4 | 2–190 | 2,051 | UB |
| Trypanorhynch cysts | 2 | 4.5 | — | — | — | Me |
| Cuculanus carretae | 20 | 45.4 | 180.4 | 1–465 | 3,609 | UI |
| Kathlania leptura | 19 | 43.2 | 713.6 | 2–10,212 | 13,558 | MI |
| Tonaudia tonaudia | 22 | 50.0 | 529.8 | 1–3,714 | 11,655 | LI |
| Sulcascaris sulcata | 23 | 52.7 | 230.6 | 1–2,416 | 5,072 | St |

No loggerheads without worms present. Number of worm species/infected host mean = 8.7 (2–17).

[a] Ar, aorta; H, heart; Lv, liver blood vessels; L, liver; GB, gall bladder; Es, esophagus; St, stomach; UI, upper intestine; MI, middle intestine; LI, lower intestine; UB, urinary bladder; Me, mesenteries.

there could be a negative impact on the parasite as well because if the host dies, there will be no more progeny from it. So through evolutionary change in long-standing relationships, parasites usually become adapted to co-exist with their hosts.

Terms used in this chapter that need definition include the following (see Bush et al., 1997):

Prevalence is the number of hosts infected by a parasite species divided by the number of hosts examined and then multiplied by 100 and is expressed as a percentage.

Intensity is the number of parasites of a species divided by the number of hosts infected with that species.

Mean intensity is the total number of individual parasites of a species divided by the number of infected hosts.

## 16.2 LOGGERHEAD (*CARETTA CARETTA*) PARASITES

Seventeen trematodes species, four species of nematodes, and larval trypanorhynch tapeworms were recovered from 44 loggerheads from Florida at necropsy (Table 16.1). The number of species per host ranged from 2 to 17 with an average of 8.7 species per turtle. The intensity of infection ranged from 1 to 10,212 for individual species. All loggerheads were infected with helminths. While this

TABLE 16.2

## Summary of Adult Worms from 74 Green Turtles 1991–2006 from Florida

| Species | #Pos. | Prevalence (%) | Intensity Mean | Intensity Range | Total Worms | Normal Habitat[a] |
|---|---|---|---|---|---|---|
| *Learedius learedi* | 10 | 13.1 | | 1–16 | 53 | H, MV |
| *Neospirorchis* sp. | 2 | 2.6 | 3.5 | 1–7 | 7 | H, MV |
| *Cymatocarpus undulatus* | 1 | 1.3 | 26 | 26 | 26 | UI, MI |
| *Enodiotrema megachondrus* | 6 | 8.1 | 13.8 | 1–79 | 85 | LI |
| *Enodiotrema carettae* | 2 | 2.6 | 40.5 | 2–79 | 81 | L, GB |
| *Pleurogonius trigonocephalus* | 3 | 3.9 | 71 | 1–216 | 219 | MI |
| *Pleurogonius sindhi* | 15 | 19.7 | 41.1 | 1–221 | 617 | MI |
| *Pleurogonius longisculus* | 7 | 9.4 | 8.8 | 2–17 | 62 | MI |
| *Pleurogonius lobatus* | 3 | 3.9 | 8.0 | 2–18 | 24 | LI |
| *Pleurogonius linearis* | 2 | 2.6 | 1 | 1 | 2 | St |
| *Diaschistorchis pandus* | 1 | 1.3 | 10 | 10 | 10 | Es, St |
| *Cricocephalus albus* | 47 | 63.5 | 57.5 | 1–524 | 2,702 | St |
| *Pronocephalus obliquus* | 30 | 40.5 | 74.4 | 1–416 | 2,232 | MI |
| *Metacetabulum invaginatum* | 27 | 36.5 | 15.6 | 1–172 | 421 | UI |
| *Desmogonius desmogonius* | 4 | 5.4 | 6.0 | 1–11 | 24 | Es, St |
| *Charaxicephalus robustus* | 8 | 10.8 | 9.5 | 2–31 | 76 | St |
| *Rhytidodes gelatinosus* | 10 | 13.5 | 10.1 | 1–49 | 101 | UI |
| *Rhytidoides similis* | 16 | 21.6 | 19.0 | 1–98 | 304 | L, GB |
| *Orchidasma amphiorchis* | 4 | 5.4 | 56.2 | 10–114 | 225 | UI |
| *Plesiochorus cymbiformis* | 4 | 5.4 | 2.5 | 1–6 | 9 | UB |
| *Deuterobarus proteus* | 30 | 40.5 | 125 | 1–1,006 | 3,750 | MI, LI |
| *Angiodictyum paralellum* | 17 | 22.9 | 112.8 | 1–787 | 1,918 | MI |
| *Microscaphidium reticulare* | 20 | 27.0 | 244.3 | 1–2,449 | 4,887 | UI |
| *Neoctangium travassosi* | 25 | 33.8 | 67.2 | 1–117 | 1,680 | MI |
| *Octangium sagitta* | 41 | 55.4 | 378.6 | 1–4833 | 15,524 | MI |
| *Polyangium linguatula* | 18 | 24.3 | 38.3 | 1–128 | 689 | MI |
| *Schizamphistomoides spinulosus* | 35 | 47.3 | 14.3 | 1–63 | 501 | MI |
| *Schizamphistomum sceloporum* | 1 | 1.3 | 4 | 4 | 4 | LI |
| Trypanorhynch larvae | 14 | 16.9 | — | — | — | ME |

Six greens contained no helminths. Number of worm species/infected host mean and range=6.3 (1–16).

[a] H, heart; MA, major arteries; L, liver; GB, gall bladder; Es, esophagus; St, stomach; UI, upper intestine; MI, middle intestine; LI, lower intestine; UB, urinary bladder; Me, mesenteries.

chapter will deal primarily with parasites recovered in the parasitology laboratory at the College of Veterinary Medicine, University of Florida, there are other worms that have been reported from loggerheads not found in this survey.

## 16.2.1 ASPIDOGASTRID FLUKES

A single species of this order of trematodes infects loggerheads and no other marine turtle species. *L. vallei* (Figure 16.1, **1**) has a subdivided sucker (= holdfast) covering most of its ventral surface, which readily distinguishes it from all other flukes in these hosts. It is believed that the aspidogastrids in vertebrates use a single gastropod as an intermediate host (Wharton, 1939). It was found in 22.7% of the loggerheads with an average of 30.7 individuals per infected host.

TABLE 16.3

**Summary for Adult Worms from Four Hawksbills from Florida 1991–2006**

| Species | #Pos./#Exam. | Intensity | | Total Worms | Normal Habitat[a] |
|---|---|---|---|---|---|
| | | Mean | Range | | |
| *Neospirorchis* sp. | 1/4 | 9 | 9 | 9 | B |
| *Rhytidodes gelatinosus* | 3/4 | 16 | 9–45 | 54 | UI |
| *Enodiotrema carettae* | 1/4 | 1 | 1 | 1 | L |
| *Cricocephalus albus* | 2/4 | 3 | 2–4 | 6 | St, UI |
| *Diaschistorchis pandus* | 2/4 | 12 | 12 | 24 | St |
| *Plesiochorus cymbiformis* | 1/4 | 3 | 3 | 3 | UB |
| Trypanorhynch cysts | 1/4 | — | — | — | — |

All hawksbills contained parasites. Number of worm species/infected host mean = 3 (2–4).

[a] B. brain; L, liver; St, stomach; UI, upper intestine; UB, urinary bladder.

TABLE 16.4

**Summary of Adult Worms from Four Kemp's Ridley Turtles from Florida 1991–2006**

| Species | #Pos./#Exam. | Intensity | | Total Worms | Normal Habitat[a] |
|---|---|---|---|---|---|
| | | Mean | Range | | |
| *Carettacola bipora* | 1/4 | 1 | 1 | 1 | L vessels |
| *Cymatocarpus undulatus* | 2/4 | 5,904 | 222–11,809 | 11,809 | MI |
| *Cricocephalus albus* | 2/4 | 4 | 1–7 | 8 | St |
| *Pleurogonius trigonocephalus* | 1/4 | 5 | 5 | 5 | MI |
| *Rhytidodes gelatinosus* | 1/4 | 2 | 2 | 2 | UI |
| *Enodiotrema megachondrus* | 1/4 | 22 | 22 | 22 | UI |
| *Enodiotrema carettae* | 1/4 | 5 | 5 | 5 | UI |
| *Calycodes anthus* | 1/4 | 4 | 4 | 4 | UI |
| *Tonaudia tonaudia* | 2/4 | 932 | 639–1,226 | 1,865 | MI |
| *Sulcascaris sulcata* | 1/4 | 1 | 1 | 1 | St |
| Trypanorhynch cysts | 2/4 | | | | St, UI walls |

All Kemp's Ridleys were infected. Number of worm species/infected host mean = 6 (2–10).

[a] L, liver; St, stomach; UI, upper intestine; MI, middle intestine.

### 16.2.2 Blood Flukes (Digenea: Spirorchiidae)

Adults of three species of this family were recovered from blood vessels. They were *Carettacola bipora* (Figure 16.1, **2**), *Hapalotrema mistroides* (Figure 16.1, **3**), and *Neospirorchis pricei* (Figure 16.1, **4**). These are the most pathogenic of the parasites in these turtles. A recent mortality of loggerheads in Florida suggested that these flukes may have contributed to this die off (Jacobson et al., 2006). The massive infections in the brain are undoubtedly causing neurological damage, but the extent of the damage caused by less intense infections or infections in other sites is unknown (Stacy, 2012, personal communication). The adults sometime are numerous enough that they inhibit blood flow. Maybe more importantly, if these flukes are to pass their progeny onto new hosts, the eggs have to exit the host in some manner. So the eggs have to exit the vessels, often those in association with the gut, then pass through the intestinal wall to gain access to the lumen of the gut. They may cause extensive damage as they pass through normal tissue. Often eggs are found in most organs when the host has a number of blood flukes present. Some spirorchiids distribute their eggs close to the site of the adult and others embolize their eggs throughout the host. If they gain entrance into the gut lumen, they may pass along with the feces and exit the host. Many eggs are trapped in the tissues by host response. Finding adults of this family is difficult as they are not always washed out of the vessels, especially when they occur in small vessels. Therefore, the prevalence for these parasites here is underestimated. Stacy et al. (2010a) reported 96% of the loggerheads were infected with *Neospirorchis* spp., 78% with *Hapalotrema* spp., and 22% for *Carettacola* spp. From the data in Table 16.1, *C. bipora* is the most common blood fluke in loggerheads in Florida, but from related data species of *Neospirorchis* are more common than *C. bipora*. These blood flukes are usually found in low numbers, and whether this is a true reflection of their intensities or failure to locate the adults at necropsy is uncertain. Stacy et al. (2010a) demonstrated the normal sites for adult *H. mistroides* are heart, aorta, and mesenteric arteries and *H. pambanensis* (not found in my survey) was the heart and aorta, whereas *C. bipora* is in hepatic vessels and *N. pricei* also uses the heart and associated great vessels.

Stacy et al. (2010a) have recorded infections of *Neospirorchis* in the thymus, endocrine organs, intestine, and brain. We are unsure of how flukes in vessels feeding the brain are capable of exiting the turtle. The data in this reference (resulting from the PhD dissertation of Dr. Brian Stacy) were collected in an incredibly meticulous manner producing these fantastic results and giving us an in-depth understanding of these parasites. Because zoological species are based on morphology and because the adults could not be isolated from these other organs, these taxa have not been named. Stacy et al. (2010a) also discusses the pathology associated with these spirorchiids in great depth, and the reader is referred to this paper as the detail is beyond what will be discussed in the present chapter. For extensive bibliographies on spirorchiids (see Smith, 1973, 1997).

### 16.2.3 Gastrointestional Flukes (Digenea: Brachycoelidae, Pachypsolidae, Plagiorchiidae, Pronocephalidae, Styphlotrematidae, and Telorchiidae)

Most digenetic flukes live in the gastrointestinal tract. Their distribution begins in the esophagus with *Pachypsolus irroratus* (Figure 16.1, **6**) and *Diaschistorchis pandus* (Figure 16.1, **8**) and range through to the lower intestine where species of *Pyelosomum* reside (Table 16.1). The stomach is home to *Enodiotrema megachondrus* and the two species found in the esophagus. The intestine is divided by distribution of fluke habitats as the upper intestine houses *Cymatocarpus undulatus* (Figure 16.1, **5**), *E. carettae*, *R. gelatinosus* (Figure 16.1, **9**), *Orchidasma amphiorchis* (Figure 16.1, **10**), and *S. solitaria*, the middle region has *Pleurogonius trigonocephala* (Figure 16.1, **7**) and *Pyelosoma renicapite*, and the lower gut also has *P. renicapite* and *P. chelonei* (Table 16.1). The highest numbers and diversity of fluke are present in the upper intestine and this may reflect the quality and/or quantity of nutrients there before the turtle removes them to nourish itself. *C. undulatus*, *O. amphiorchis*, and *P. trigonocephala* were the more numerous and more common flukes found in loggerheads (Table 16.1).

### 16.2.4 Liver Flukes (Digenea: Calycoididae)

One species resides in the liver and gall bladder and that is *Calycoides anthus* (Figure 16.1, **10**).

### 16.2.5 Urinary Bladder Flukes (Digenea: Gorgoderidae)

Only one species, *P. cymbiformis* (Figure 16.1, **12**), lives in the lumen of the urinary bladder.

## 16.2.6 LARVAL TAPEWORMS (CESTODA: TRYPANORHYNCHA)

The only tapeworms detected were larval forms of the Trypanorhyncha. We have not identified these to even genus, so we are unsure of how many species might be present in these turtles.

## 16.2.7 NEMATODES (NEMATODA: CUCULIDAE, KATHLANIDAE, ANISAKIDAE)

The nematodes, like the flukes, have evolved to use different habitats. *Sulcascaris sulcata* resides in the stomach, *C. carettae* in the upper intestine, *Kathlania leptura* in the middle gut, and *Tonaudia tonaudia* in the lower intestine. *S. sulcata* is the largest of these and is attached to the mucosa and may cause ulcers, but most specimens will be found unattached in necropsies as the worms will detach after the host is dead. This species has been found in post-hatchling loggerheads and may have a greater potential to cause damage due to the large size of the worm in a small host (Stacy, 2012, personal communication). This species uses a variety of bivalves and snails as intermediate hosts (Lester et al., 1980; Lichtenfels et al., 1980; Berry and Cannon, 1981). The other intestinal inhabiting nematodes have direct cycles. Although I did not find lungworms in our hosts (maybe due to their fragility, tiny size, and condition of the carcasses), *Angiostoma carettae* was detected in living loggerheads from Florida (Bursey and Manire, 2006; Manire et al., 2008). I did not find larval *Anisakis* in the stomach walls of our turtles, but they have been reported and were recently identified by molecular methods as *A. pegreffi* (Santoro et al., 2010b).

## 16.2.8 COMPARISON WITH OTHER HELMINTH SURVEYS

More studies on loggerheads have been done in the Mediterranean Sea than elsewhere. Five studies from that region examined a total of over 300 turtles and found a relatively depauperate helminth community. These references are for loggerheads in Italy, Spain, and Portugal (Manfredi et al., 1998; Aznar et al., 1998; Piccolo and Manfredi, 2002; Valente et al., 2009; Santoro et al., 2010a). While many studies list parasites found in a single turtle or name a new parasite, these have reported on the prevalence and intensity or abundance of helminths in loggerheads. Collectively they recovered 10 species of digenetic flukes, 4 species of nematode, 2 larval acanthocephalans, and 1 post larval tapeworm. *E. megachondrus* was the most common and most widely distributed fluke and it had the highest intensities. It was followed by *R. gelatinosus* and *C. anthos*. The remaining flukes were more restricted in geographical distribution and had lower prevalence and intensities and these were *O. amphiorchis*, *P. trigonocephalus*, *P. cymbiformis*, *P. renicapite*, *P. irroratus*, *S. solitaria*, and *Adenogaster serialis*. Nematodes included *S. sulcata*, *C. carettae*, *K. leptura*, and *Anisakis* sp. Sey (1977) examined loggerheads from Egypt and added only *D. pandus* to the known helminths from Mediterranean marine turtles. The prevalence of *R. gelatinosus* was the only species that was comparable with those in Florida turtles and the prevalence of the remaining species were usually less than half of that determined in Florida. All of these species were recovered from Florida loggerheads, with the exception of *A. serialis* and *Anisakis*. In all cases the intensity of infections was much lower in the Mediterranean as compared to Florida. Twelve juvenile loggerheads were examined in Brazil (Werneck et al., 2008c). They recovered 5 species including *O. amphiorchis* in 4 loggerheads with a range of 1–335 adults per turtle, 12 specimens of *P. renicapite* in 1 host, a single *C. anthos* in 1, *S. sulcata* in 2 with 15–18 adults and *K. leptura* in 2 hosts and 51–406 worms present. Chen et al. (2012) examined two loggerheads in Taiwan for spirorchiids and found none present.

## 16.2.9 PROTOZOA (EUCOCCIDIORIDA: EIMERIDAE)

A single coccidian species, *Eimeria carettae*, has been described from loggerheads. This is a typical appearing oocyst of *Eimeria* containing four sporocysts each with two sporozoites, except that there are thin filaments originating from the Stieda bodies on the sporocysts (Upton et al., 1990). Nothing more has been published on this parasite since it was described.

## 16.3  GREEN TURTLE (*CHELONIA MYDAS*) PARASITES

Twenty-eight digenetic trematodes and larval cestodes were removed from 74 green turtles (Table 16.2). No helminths were recovered from 6 green turtles. An average of 6.2 and a range of 1–16 fluke species were recovered per infected turtle. The intensity of infections ranged from 2 to 15,524 for individual species. No adult nematodes were found in green turtles, but larval forms have been seen in histological sections from green turtles from Florida (Stacy, 2012, personal communication). Protozoa include two coccodia, *Caryospora cheloniae* and *Cryptosporidium* sp., unidentified species of *Octomitus* (flagellate) and *Entamoeba* (ameba) have been isolated as has a species of *Balantidium*. More papers have been published on green turtles parasites than any of the other marine turtles, and it is beyond the scope of this chapter to deal with that wealth of information.

### 16.3.1  BLOOD FLUKES (DIGENEA: SPIRORCHIIDAE)

Two species of blood-inhabiting flukes were found in green turtles and these were *L. learedi* (Figure 16.1, **13**) and a species of *Neospirorchis* both from the heart and major associated vessels. This family was discussed in the section on loggerheads and most of the same information applies here. *L. learedi* is the only spirorchiid for which we have an indication of the intermediate host based on molecular methods and it is the knobby keyhole limpit (*Fissurella nodosa*) (Stacy et al., 2010b). The pathological changes observed in greens associated with spirorchiids were also reported by (Stacy et al., 2010a). Again, there is a specialization in *Neospirorchis* species by organ selection as there was in loggerheads, and there are new species in vessels of the brain, thyroid gland, and gastrointestinal tract (Stacy et al., 2010a).

### 16.3.2  GASTROINTESTINAL FLUKES (DIGENEA: BRACHYCOELIDAE, PLAGIORCHIIDAE, PRONOCEPHALIDAE, RHYTIDODIDAE, TELORCHIIDAE, MICROSCAPHIDIIDAE, AND CLADORCHIIDAE)

Greens have a more diverse fluke fauna than found in loggerheads. The gastrointestinal tract had the most diverse parasite fauna as was the case with the loggerheads. Their distribution (Table 16.2) within the host varied from the esophagus (*D. pandus* and *Desmogonius desmogonius*) to the lower intestine *E. megachondrus* (Figure 16.1, **14**), *Pleurogonius lobatus*, and *Schizamphistomum scelo-porum*. Flukes from the stomach included *D. pandus*, *D. desmogonius*, *Charaxicephalus robustus*, and *Cricocephalus albus* (Figure 16.1, **15**). Upper intestine flukes were *C. undulatus*, *Metacetabulum invaginatum*, *R. gelatinosus*, and *Microscaphidium reticulare* (Figure 16.1, **18**). The middle intestine community comprised *P. trigonocephalus*, *P. signhi*, *P. longisculus*, *Deuterobarus proteus* (Figure 16.1, **16**), *Pronocephalus obliquus* (Figure 16.1, **17**), *Angiodictyum parallelum* (Figure 16.1, **19**), *Neoctangium tranvassoi*, *Octangium sagitta* (Figure 16.1, **20**), *Polyangium linguatula*, and *Schizamphistomoides spinulosus* (Figure 16.1, **22**). The lower gut inhabitants were *E. megachon-drus*, *P. lobatus*, *D. proteus*, and *S. sceloporum*. The middle intestine had the highest diversity of fluke species. *O. sagitta* was the most numerous species followed by *M. reticulare* and *D. proteus*, all of which are in the Microscaphidiidae.

### 16.3.3  LIVER FLUKES (DIGENEA: RHYTIDODIDAE AND PLAGIORCHIIDAE)

*E. carettae* and *Rhytidoides similis* (Figure 16.1, **21**) lived in the liver (bile ducts) and gall bladder.

### 16.3.4  URINARY BLADDER FLUKES (DIGENEA: GORGODERIDAE)

*P. cymbiformis* was the sole helminth in the urinary bladder lumen.

## 16.3.5 COCCIDIANS (EUCOCCIDIORIDA: EIMERIDAE, CRYPTOSPORIDAE)

A species of *Caryospora* resides in greens. It has eight sporozoites, which were originally thought to be in a single sporocyst within an oocyst which defines *Caryospora*, but it may be in a sporocyst without an oocyst. *C. cheloniae* was named from mariculture-reared green turtle hatchlings at the Turtle Farm in Grand Cayman (Leibovitz et al., 1978). They indicated it was pathogenic and included some descriptions of the damage done. More recently, an epizootic caused by this coccidian in free-ranging adult green turtles off the coast of Australia was published (Gordon et al., 1993). They necropsied 24 of the 70 that died within a 6 week time span. This coccidian was present in the intestinal as well as non-intestinal sites and it caused encephalitis well as enteritis. Much more is known from the dissertation of Dr. Anita Gordon (1995) in which she has demonstrated developing stages in small intestine, stomach, large intestine, urinary bladder, brain, thyroid gland, kidney, and adrenal gland. Her evidence suggests this is a serious pathogen of greens in the wild and we have detected oocysts in a few green turtles from Florida, but know nothing about its pathogenicity here.

*Cryptosporidium* oocysts have been reported from greens in Hawaii, but numerous loggerhead and green turtles histologic sections from Florida did not reveal evidence of this parasite (Grazyk et al., 1997; Stacy, 2012, personal communication).

## 16.3.6 CILIATES (VESTIBULOFERIDA; BALANTIDIIDAE)

The ciliate is *Balantidium bacteriophorus* and it was found in the stomach of a green turtle collected off the shore of Nicaragua (Fenchel, 1980). Nothing further has been published on this ciliate. It is presumably present in greens and loggerheads in Florida, based upon the cyst-like structures found in feces in the current study.

## 16.3.7 FLAGELLATES (DIPLOMONADIDA: HEXAMITIDAE, TRICHOMONADIDA: TRICHOMONADIDAE)

*Octomitus* sp. and a trichomonad whose identification was not taken to genus exist in greens (Fenchel, 1980). A brief description and a line drawing of the *Octomitus* are the extent of our knowledge of these flagellates.

## 16.3.8 AMOEBAE (ENTAMOEBIDAE)

*Entamoeba* sp. was recorded from green turtles. This parasite is poorly characterized and the one citation suggests it might kill turtles (Frank et al., 1976). It has been seen in histological sections from both greens and loggerheads in Florida (Stacy, 2012 personal communication).

## 16.3.9 COMPARISON WITH OTHER HELMINTH SURVEYS

Green turtle populations have been examined for parasites in Egypt (Sey, 1977), mariculture-reared greens in Grand Cayman (Greiner et al., 1980), Hawaii (Dailey et al., 1992), and nesting greens in Costa Rica (Santoro et al., 2006). The Mediterranean green turtles were infected with 8 species of digenetic flukes and the highest prevalence was 42.8 (3/7) which was detected for *M. reticulare*, *O. sagitta*, and *P. linguatula*. The remainders were in 28% or fewer, namely *C. robustus*, *D. proteus*, *A. parallelum*, *Microscaphidium aberrans*, and *Cricocephalus resectus* and the intensity of infections were much lower than those found in Florida. That study only examined the gastrointestinal tract, gall bladder, lungs, and urinary bladder. The same organs plus the heart of greens were examined in Hawaii. Seven digenetic flukes were recovered and those were *L. learedi*, *Carettacola hawaiiensis*, two species of *Haplalotrema*, *P. linguatula*, *Angiodyctium longum*, and *Pyelosoma cochlear*. The spirorchiid fauna was more diverse than in Florida in the sample used in this chapter, but other data of Stacy et al. (2010a) erase that difference. The intensity of infections in Hawaii was less than those in Florida as well. Only two digenes were collected from the mariculture-reared greens on Grand Cayman and they were *L. learedi* in 40% of the greens with an intensity of 1–49 with a mean of 14 per infected turtle. A single specimen of *Pleurogonius mehrai* was removed from the upper intestine. The most recent study was in Costa Rica and that study had a sample of 40 nesting females and is a nice example of how such studies should be done. Twenty-nine species of digenetic flukes were recovered including 4 spirorchiids (*L. learedi, Hapalotrema postorchis, Monticellius indicum,* and *Amphiorchis solus**), 5 angiodictyids (*M. reticulare, P. linguatula, Deuterobarus intestinalis, Octangium hyphalum,** and *Microscaphidium warui**), 1 clinostomid (*Clinostomum complanatum**), 2 cladorchiids (*S. sceloporum and Schistosomoides erratum*), 2 rhytidodids (*R. similis and Rhytidoides intestinalis*), and 15 pronocephalids (*P. cochlear, C. resectus, C. magastomus, C. albus, P. obliquus, D. desmogonius, Pleurogonius linearis,** *P. longisculus, P. singhi, P. lobatus,** *P. solidus,** *Pleurogonius* sp.,** *Charaxicephaloides* sp., *C. robustus,* and *Rameshwarotrema uterocrescens*). Eight species* of these were found in only one or two turtles with only one or two specimens in each, and only two of these minor flukes were identified from Florida greens and they were *P. lobatus* and *P. linearis*, which also were detected in very low prevalence. Costa Rican data indicated that species with the highest intensities were similar to those in Florida as *M. recticulare* was their highest followed by *D. intestinalis* (see Table 16.2 for comparison). It is possible that the two studies identified their flukes differently in some instances and that will have to be sorted out with future research. Typically greens from Florida had higher intensities when the same taxon was in both locations. A major difference between the two studies was that the Costa Rica turtles were all nesting females and the Florida specimens ranged in size from 23 cm straight carapace length to mature adults (unfortunately size data for a number of Florida turtles were lost). Four other recent papers examined blood flukes in green turtles in different locations. Three studies in Brazil found a single *A. solus* (Werneck et al., 2011), *M. indicum* was found in two juvenile greens (Werneck et al., 2008a), and in the final study 11 juveniles were examined and 6 were infected with *L. learedi* (Werneck et al., 2006). A study in Taiwan recovered *L. learedi* in 11/13 turtles, *H. mehrai* in 7/13, *H. postorchis* in 6/13, and *C. hawaiiensis* in 2/13 (Chen et al., 2012).

## 16.4  HAWKSBILL (*ERETMOCHELYS IMBRICATA*) PARASITES

Relatively few studies have been conducted on the parasites of the hawksbills. Only four hawksbills were examined in my laboratory culminating in a fauna of six species of digenetic trematodes and trypanorhynch larvae. An average of three species (range of two to four) per infected turtle was detected. The mean intensity ranged from 3 to 16 flukes (Table 16.3). This species primarily eats sponges, which would limit the potential for ingesting molluskan hosts serving as intermediate hosts to helminths. Thus the low number of parasites and low diversity of parasites might reflect this behavior. No nematodes or protozoa have been reported from this species.

### 16.4.1  Blood Flukes (Digenea: Spirorchiidae)

One hawksbill was infected with a species of *Neospirorchis* in the cerebral vasculature (Table 16.3). Nine adults were observed, but it is impossible to extract them from these vessels due to the fact that the diameter of the adult fluke approximates the lumen diameter of the infected vessels and thus species identification was not possible. Whether this is the same undescribed species in the cranial vessels of the green turtle is unknown.

### 16.4.2  Gastrointestinal Flukes (Digenea: Plagiorchiidae, Pronocephalidae, and Rhytidodidae)

A single specimen of *E. carettae* was recovered from one individual. Two hawksbills were infected with *C. albus* and two to four flukes were in each and one *D. pandus* was in one turtle. *R. gelatinosus* was in three individuals and from 9 to 45 were present in each (Table 16.3).

### 16.4.3 URINARY BLADDER FLUKES (DIGENEA: GORGODERIDAE)

A single species, *P. cymbiformis*, was present in the urinary bladder of one hawksbill and four adults were recovered.

### 16.4.4 COMPARISON WITH OTHER SURVEYS AND A LISTING OF SPECIES KNOWN FROM HAWKSBILLS

Because of the few hawksbills examined, the following is a list of helminths reported from this species. Fischthal and Acholonu (1976) have the most complete listing from hawksbills from their necropsies in Puerto Rico and which included seven new species, namely *Amphiorchis caborojoensis*, *Epibathra stenobursata*, *Glyphicephalus latus*, *Pleurogonius laterouterus*, *P. puertoricensis*, *Pachypsolus puertoricensis*, and *Calycodes caborojoensis* and 21 previously reported species which were *L. orientalis*, *Hapalotrema spirorchis*, *Amphiorchis amphiorchis*, *M. reticulare*, *O. sagitta*, *O. travassosi*, *S. sceloporum*, *P. cymbiformis*, *C. albus*, *C. magastomus*, *P. linearis*, *P. trigonocephalus*, *P. lobatus*, *Pyelosoma posteriorchis*, *D. pandus*, *M. invaginatum*, *Pachypsolus ovalis*, *Enodiotrema reductum*, *S. solitaria*, *O. amphiorchis*, and *R. gelatinosus*. Most of their intensities were less than five worms per infected turtle and *D. pandus* was the most prevalent species followed by *O. sagitta* and *O. tranvassosi,* which also had the highest intensities. More recent reports from hawksbills from Puerto Rico provided *Angiodictyum mooreae*, *A. anteroporum*, and *A. parallelum* (Dyer et al., 1995a), and an extensive checklist of hawksbill helminths was provided by Dyer et al. (1995c). Chattopadhyaya (1972) added *Octangium microrchis* and *Pyelsomum solum*, both from the lower intestine. Two more species from Australia were *Pleurogonius truncatus* and *Pyelosomum parvum*, both new species (Prudhoe, 1944). Two other new species from Bermuda were *Diaschistorchis megas* and *Pachypsolus brachus* (Barker, 1922). The only report of *Neospirorchis* sp. and *E. carettae* is from the Florida sample. Five *Amphiorchis cabarojoensis* and two *Carettacola stunkardi* were recovered from a single Hawksbills from Brazil (Werneck et al., 2008b).

## 16.5 LEATHERBACK (*DERMOCHELYS CORIACEA*) PARASITES

Two Florida leatherback turtles were examined and one of those was still partially frozen. The only parasites from it were three *Pyelosomum renicapite* from the intestine. These specimens were about 2.5–3 cm in length, the largest flukes I have encountered in marine turtles. A fresh leatherback was killed by a trawler and it provided a better look at the parasites present. This resulted in *P. renicapite*, *C. undulates*, *Enodiotrema instar*, *C. anthos*, and larval tapeworms.

### 16.5.1 COMPARISON WITH OTHER SURVEYS AND A LISTING OF SPECIES KNOWN FROM LEATHERBACKS

A north Atlantic leatherback examination occurred in Newfoundland where two leatherbacks were killed. Three digenetic flukes were recovered and they were *P. renicapite*, *C. anthos*, and *Cymatocarpus* sp. (Threlfall, 1979). An examination of one leatherback in Puerto Rico yielded 25 *P. renicapite* from the intestine (Dyer et al., 1995b) and two of the same species were removed from a leatherback in France (Almore et al., 1989). A Mediterranean leatherback yielded 44 *P. renicapite*, and 60 *E. instar* from the stomach and 71 *E. carettae* from the liver (Manfredi et al., 1996); and a second leatherback collected a few years later had the same three flukes (Piccolo and Manfredi, 1999). Four of eight leatherbacks examined from Brazil and Uruguay were infected with *P. renicapite,* and the one from Uruguay contained a single *C. anthos* (Werneck et al., 2012).

## 16.6  KEMP'S RIDLEY (*LEPTOCHELYS KEMPII*) PARASITES

Four Kemp's Ridleys were examined in our study. They yielded eight species of digenetic trematodes and two species of nematodes and trypanorhynch larvae (Table 16.4). Each was infected with a mean of 6 species with a range of 2–10. Mean intensities ranged from 1–5904 and no more than 2 of the 4 were infected with the same species of parasite (Table 16.4). *C. undulatus* had the highest intensity, with one turtle having nearly 12,000 in its intestine. This turtle species shared its entire parasite fauna with loggerheads, but the fauna was more restricted than in the loggerheads (see Tables 16.1 and 16.4). One study examined cold-stunned Kemp's Ridleys in Massachusetts; and while they examined tissues histologically, they only identified a trypanorhynch post-larva as presumptively identified as *Tentacularia coryphaenae*, but did not attempt other identifications (Innis et al., 2009). No other reports were found on the parasites of this turtle species.

## 16.7  OLIVE RIDLEY (*LEPIDOCHELYS OLIVACEA*) PARASITES

We did not examine any olive ridleys as they rarely occur in Florida. However, two published reports present an indication of their parasite fauna. One study in Mexico examined the gastrointestinal tracts of 32 olive ridleys and recovered 8 digenetic flukes and 8 turtles were not infected. These were *A. serialis*, *P. lobatus*, *P. renicapite*, *P. irroratus*, *E. megachondrus*, *O. amphiorchis*, *C. anthos*, and *Prosorchis psenopis*. All of these except the last one are shared with other marine turtles. *E. megachondrus* had the highest abundance and had the second highest prevalence, following *A. serialis* (Perez Ponce de Leon, 1996). A study of 3 olive ridleys in Costa Rica recovered 3 *E. megachondrus* in the upper intestine from 1 turtle, 1 *P. irroratus* from the stomach and 31 *P. cymbiformis* from the urinary bladder of another host (Santoro and Morales, 2007). Nothing more is known on the parasites of this marine turtle.

## ACKNOWLEDGMENTS

I would like to thank a number of people who worked with me and helped amass these data. First I would like to thank Dr. Elliott Jacobson who got me started on these fascinating parasites. I appreciate Dr. Donald Forrester who did the first survey of green turtles with me soon after I arrived at the University of Florida. I also extend my gratitude to Dr. Brian Stacy for showing me what I had missed in finding blood flukes and for constructive suggestions on this chapter. Garry Foster, Toni McIntosh, Diane Heaton-Jones, and Xiwan Zeng and student assistants Leslie White Gillette, Kristen Munsterman Sotos, and Andy Killian all spent time in my laboratory recovering, counting, and helping to identify the parasites. I thank the turtle biologists and veterinarians for teaching me about marine turtles.

## REFERENCES

Almor, P., J.A. Raga, E. Abril et al. 1989. Parasitisme de la tortue luth, *Dermochelys coriacea* (Linnaeus, 1766) dans les eaux Europeennes par *Pyelosomum renicapite* (Leidy, 1856). *Vie Milieu* 39: 57–59.

Aznar, F.J., F.J. Badillo, and J.A. Raga. 1998. Gastrointestinal helminths of loggerhead turtles (*Caretta caretta*) from the Western Mediterranean: Constraints on community structure. *J. Parasitol.* 84: 474–479.

Barker, F.D. 1922. The parasitic worms of the animals of Bermuda. I. Trematodes. *Proc. Am. Acad. Arts Sci.* 57: 215–236.

Berry, G.N. and L.R.G. Cannon. 1981. The life history of *Sulcascaris sulcata* (Nematoda: Ascaroidea), a parasite of marine molluscs and turtles. *Int. J. Parasitol.* 11: 43–54.

Borkent, A. 1995. Biting midges (Ceratopogonidae: Diptera) feeding on leatherback turtles in Costa Rica. *Brenesia* 43–44: 25–30.

Burke J.B. and L.J.Rogers. 1962. Gastric ulcerations associated with larval nematodes (*Anisakis sp.* Type 1) in pen reared green turtles (*Chelonia mydas*) from Torres Strait. *J. Wildl. Dis.* 18: 41–46.

Bursey, C.R. and C.A. Manire. 2006. *Angiostoma carettae* n. sp. (Nematoda: Angiostomatidae) from the loggerhead turtle *Caretta carreta* (Testudines: Cheloniidae), Florida, U.S.A. *Comp. Parasitol.* 73: 253–256.

Bush, A.O., K.D. Lafferty, J.M. Lotz et al., 1997. Parasitology meets ecology on its own terms: Margolis et al., revisited. *J. Parasitol.* 83: 575–583.

Chattopadhyaya, D.R. 1972. Studies on the trematode parasite of reptiles found in India. Contribution to our knowledge of the family Pronocephalidae Looss, 1902. *Riv. Parassitol.* 33: 99–124.

Chen, H., R.J. Kuo, T.C. Chang et al. 2012. Fluke (Spirorchiidae) infections in sea turtles stranded on Taiwan: Prevalence and pathology. *J. Parasitol.* 98: 437–459.

Dailey, M.D., M.L. Fast, and G.H. Balazs. 1992. A survey of the Trematoda (Platyhelminthes: Digenea) parasitic in green turtles, *Chelonia mydas*, from Hawaii. *Bull. South. Calif. Acad. Sci.* 91: 84–91.

Dyer, W.G., E.H. Williams, Jr., and L. Bunkley-Williams.1995a. *Angiodictyum mooreae* n. sp. (Digenea: Microscaphidiidae) and other digeneans from an Atlantic hawksbill turtle, *Eretmochelys imbricata imbricata,* from Puerto Rico. *J. Aquat. Anim. Health* 7: 38–41.

Dyer, W.G., E.H. Williams, Jr., and L. Bunkley-Williams. 1995b. Digenea of the green turtle (*Chelonia mydas*) and the leatherback turtle (*Dermochelys coriacea*) from Puerto Rico. *Caribb. J. Sci.* 31: 269–273.

Dyer, W.G., E.H. Williams, Jr., and L. Bunkley-Williams. 1995c. Some digeneans (Trematoda) of the Atlantic hawksbill turtle, *Eretmochelys imbricata imbricata* (Testudines: Cheloniidae) from Puerto Rico. *J. Helm. Soc. Wash.* 62: 13–17.

Fenchel, T.M. 1980. The protozoan fauna from the gut of the green turtle, *Chelonia mydas* L. with a description of *Balantidium bacteriophorus* sp. nov. *Arch. Protistenk.* 123: 22–26.

Fischthal, J.H. and A.D. Acholonu. 1976. Some digenetic trematodes from the Atlantic hawksbill turtle, *Eretmochelys imbricata imbricata* (L.). *Proc. Helm. Soc. Wash.* 43: 174–185.

Frank, W., U. Bachmann, and R. Braun. 1976. Unusual incidents of death caused by amoebiasis in an arch-turtle (*Spehnodon punctatus*), young green turtles (*Chelonia mydas*) and a pseudo-loggerhead turtle (*Caretta caretta*). 1. Amoebiasis in *Sphenodon punctatus* (in German). *Salamandra* 12: 94–102.

Gordon, A.N. 1995. A necropsy based study of green turtles (*Chelonia mydas*) in south-East Queensland. PhD dissertation. School of Veterinary Science, University of Queensland, Queensland, Australia. 234 pp.

Gordon, A.N., W.R. Kelly, and R.J.G. Lester. 1993. Epizootic mortality of free-living green turtles, *Chelonia mydas*, due to coccidiosis. *J. Wild. Dis.* 29: 490–94.

Grazyk, T.K., G.H. Balazs, T. Work et al. 1997. *Cryptosporidium* sp. infections in green turtles, *Chelonia mydas*, as a potential source of marine waterborne oocysts in the Hawaiian Islands. *Appl. Environ. Microbiol.* 63: 2925–2927.

Greiner, E.C., D.J. Forrester, and E.R. Jacobson. 1980. Helminths of mariculture-reared green turtles (*Chelonia mydas*) from Grand Cayman, British West Indies. *Proc. Helm. Soc. Wash.* 47: 142–144.

Innis, C., A.C. Nyaoke, C.R. Williams III et al. 2009. Pathologic and parasitologic findings in cold-stunned Kemp's ridley sea turtles (*Lepidochelys kempi*) stranded in Cape Cod, Massachusetts, 2001–2006. *J. Wild. Dis.* 45: 594–610.

Jacobson, E.R., B.L. Homer, B.A. Stacy et al. 2006. Neurological diseases in wild loggerhead sea turtles *Caretta caretta*. *Dis. Aquat. Organ.* 70: 139–154.

Leibovitz, L., G. Rebell, and G.C. Boucher. 1978. *Caryospora cheloniae* sp. n.: A coccidial pathogen of mariculture-reared green sea turtles (*Chelonia mydas mydas*). *J. Wild. Dis.* 14: 269–275.

Lester, R.J.G., D. Blair, and D. Heald. 1980. Nematodes from scallops and turtles from Shark Bay, Western Australia. *Aust. J. Mar. Freshwater Res.* 31: 713–717.

Lichtenfels, J.E., T.K. Sawyer, and G.C. Miller. 1980. New hosts for larval *Sulcascaris* sp. (Nematoda, Anisakidae) and prevalence in the calico scallop (*Argopecten gibbus*). *Trans. Am. Microsc. Soc.* 99: 448–451.

Manfredi, M.T., G. Piccolo, and C. Meotti. 1998. Parasites of Italian turtles. II. Loggerhead turtles (*Caretta caretta* [Linnaeus, 1758]). *Parassitologia* 40: 305–308.

Manfredi, M.T., G. Piccolo, F. Prato et al. 1996. Parasites in Italian sea turtles. I. The leatherback turtle *Dermochelys coriacea* (Linnaeus, 1766). *Parassitologia* 38: 581–583.

Manire, C.A., M.J. Kinsel, E.T. Anderson et al. 2008. Lungworm infection in three loggerhead sea turtles *Caretta caretta*. *J. Zoo Wildl. Med.* 39: 92–98.

Pence, D.B. and S.D. Wright. 1998. *Chelonacarus elongatus* n. gen., n. sp. (Acari: Cloacaridae) from the cloaca of the green turtle *Chelonia mydas* (Cheloniidae). *J. Parasitol.* 84: 835–839.

Perez-Ponce de Leon, G., L. Garcia-Prieto, and V. Leon-Regagnon. 1996. Gastrointestinal digenetic trematodes of olive ridley's turtle (*Lepidochelys olivacea*) from Oaxaca, Mexico. *Proc. Helm. Soc. Wash.* 63: 76–82.

Piccolo, G. and M.T. Manfredi. 1999. Reperti parasitologici in tartarughe marine del Mediter. *Rev. Idrobiol.* 38: 77–82.

Piccolo, G. and M.T. Manfredi. 2002. Nematode infection in loggerhead turtles from the coasts of Italy. *Parassitologia* 44 (Suppl 1): 136.

Prudhoe, S. 1944. Two new pronocephalid trematodes from Australia. *Ann. Mag. Nat. Hist.* 11: 481–486.

Santoro, M., F.J. Badillo, S. Mattiucci et al. 2010a. Helminth communities of loggerhead turtles (*Caretta caretta*) from central and western Mediterranean Sea: The importance or host's ontogeny. *Parasitol. Int.* 59: 367–375.

Santoro, M., E.C. Greiner, J.A. Morales et al. 2006. Digenetic trematode community in nesting green sea turtles (*Chelonia mydas*) from Tortuguero National Park, Costa Rica. *J. Parasitol.* 92: 1202–1206.

Santoro, M., S. Mattiucci, M. Paroletti et al. 2010b. Molecular identification and pathology of *Anisakis pegreffi* (Nematoda: Anisakidae) infection in the Mediterranean loggerhead see turtle (*Caretta caretta*). *Vet. Parasitol.* 174: 65–71.

Santoro, M. and J.A. Morales. 2007. Some digenetic trematodes of the olive ridley sea turtle, *Lepidochelys olivacea* (Testudines, Cheloniidae) in Costa Rica. *Helminthologia* 44: 25–28.

Sey, O. 1977. Examination of helminth parasites of marine turtles caught along the Egyptian coast. *Acta Zool. Acad. Sci. Hung.* 23: 387–394.

Smith, J.W. 1973. The blood flukes (Digenea: Sanguincolidae and Spirorchiidae) of cold blooded vertebrates and some comparisons with schistosomes. *Helm. Abs.* 41: 161–204.

Smith, J.W. 1997. The blood flukes (Digenea: Sanguicolidae and Spirorchiidae) of cold blooded vertebrates Part1. A review of the literature published since 1971 and bibliography. *Helm. Abs.* 66: 255–294.

Stacy, B.A. 2012. Personal communication.

Stacy, B.A., A.M. Foley, E.C. Greiner et al. 2010a. Spirorchiidiasis in stranded loggerhead *Caretta caretta* and green turtles *Chelonia mydas* in Florida (USA): Host pathology and significance. *Dis. Aquat. Organ.* 89: 237–259.

Stacy, B.A., T. Frankovich, E.C. Greiner et al. 2010b. Detection of spirorchiid trematodes in gastropod tissues by polymerase chain reaction: Preliminary identification of an intermediate host of *Learedius learedi*. *J. Parasitol.* 96: 752–757.

Threlfall, W. 1979. Three species of Digenea from the Atlantic leatherback turtle (*Dermochelys coriacea*). *Can. J. Zool.* 57: 1825–1829.

Upton, S.J., D.K. Odell, and M.T. Walsh. 1990. *Eimeria caretta* sp. nov. (Apicomplexa: Eimeriidae) from the loggerhead turtle, *Caretta caretta* (Testudines). *Can. J. Zool.* 68: 1268–1269.

Valente, A.L., C. Delgado, C. Moreira et al. 2009. Helminth component community of the loggerhead sea turtle, *Caretta caretta*, from Madeira Archipelago, Portugal. *J. Parasitol.* 95: 249–252.

Werneck, M.R., J.H. Becker, B.G. Gallo, and R.J. Silva. 2006. *Learedius learedi* Price, 19334 (Digenea, Spirorchiidae) in *Chelonia mydas* Linnaeus 1758 (Testudines, Chelonidae) in Brazil: Case report. *Arq. Bras. Med. Vet. Zootec.* 58: 550–558.

Werneck, M.R., B.M.G. Gallo, and R.J. da Silva. 2008a. First report of *Monticellius indicum* Mehra, 1939 (Digenea: Spirorchiidae) infecting *Chelonia mydas* Linnaeus, 1758 (Testudines, Chelonidae) from Brazil. *Brazil. J. Biol.* 68: 455–456.

Werneck, M.R., B.M.G. Gallo, and R.J. Silva. 2008b. Spirorchiids (Digenea: Spirorchiidae) infected a hawksbill turtle *Eretmochelys imbricata* (Linnaeus 1758) from Brazil. *Arq. Bras. Med. Vet. Zootec.* 60: 663–666.

Werneck, M.R., E.H.S.M. Lima, B.M.G. Gallo et al. 2011. Occurrence of *Amphiorchis solus* (Simha & Chattopadhyaya, 1970) (Digenea: Spirorchiidae) infecting the green turtle *Chelonia mydas* Linnaeus, 1758 (Testudines; Chelonidae) in Brazil. *Comp. Parasitol.* 78: 200–203.

Werneck, M.R., T.M. Thomazini, E.S. Mori et al. 2008c. Gastrointestinal helminth parasites of loggerhead turtle *Caretta caretta* Linnaeus, 1758 (Testudines, Cheloniidae) in Brazil. *Pan Am. J. Aquat. Sci.* 3: 351–354.

Werneck, M.R., L. Verissimo, P. Baldassinet et al. 2012. Digenetic trematodes of *Dermochelys coriacea* from the southwestern Atlantic Ocean. *Mar. Turtle Newslett.* 132: 13–14.

Wharton, G.W. 1939. Studies on *Lophotaspis vallei* (Stossich, 1899) (Trematoda: Aspidogastridae). *J. Parasitol.* 25: 83–86.

# Index

For Product Safety Concerns and Information please contact our EU
representative GPSR@taylorandfrancis.com
Taylor & Francis Verlag GmbH, Kaufingerstraße 24, 80331 München, Germany

www.ingramcontent.com/pod-product-compliance
Ingram Content Group UK Ltd.
Pitfield, Milton Keynes, MK11 3LW, UK
UKHW050926180425
457613UK00003B/33